T0303913

Statistics and Data Analysis for Microarrays Using R and Bioconductor

Second Edition

CHAPMAN & HALL/CRC
Mathematical and Computational Biology Series

Aims and scope:

This series aims to capture new developments and summarize what is known over the entire spectrum of mathematical and computational biology and medicine. It seeks to encourage the integration of mathematical, statistical, and computational methods into biology by publishing a broad range of textbooks, reference works, and handbooks. The titles included in the series are meant to appeal to students, researchers, and professionals in the mathematical, statistical and computational sciences, fundamental biology and bioengineering, as well as interdisciplinary researchers involved in the field. The inclusion of concrete examples and applications, and programming techniques and examples, is highly encouraged.

Series Editors

N. F. Britton
Department of Mathematical Sciences
University of Bath

Xihong Lin
Department of Biostatistics
Harvard University

Hershel M. Safer
School of Computer Science
Tel Aviv University

Maria Victoria Schneider
European Bioinformatics Institute

Mona Singh
Department of Computer Science
Princeton University

Anna Tramontano
Department of Biochemical Sciences
University of Rome La Sapienza

Proposals for the series should be submitted to one of the series editors above or directly to:
CRC Press, Taylor & Francis Group
4th, Floor, Albert House
1-4 Singer Street
London EC2A 4BQ
UK

Published Titles

Algorithms in Bioinformatics: A Practical Introduction
Wing-Kin Sung

Bioinformatics: A Practical Approach
Shui Qing Ye

Biological Computation
Ehud Lamm and Ron Unger

Biological Sequence Analysis Using the SeqAn C++ Library
Andreas Gogol-Döring and Knut Reinert

Cancer Modelling and Simulation
Luigi Preziosi

Cancer Systems Biology
Edwin Wang

Cell Mechanics: From Single Scale-Based Models to Multiscale Modeling
Arnaud Chauvière, Luigi Preziosi, and Claude Verdier

Clustering in Bioinformatics and Drug Discovery
John D. MacCuish and Norah E. MacCuish

Combinatorial Pattern Matching Algorithms in Computational Biology Using Perl and R
Gabriel Valiente

Computational Biology: A Statistical Mechanics Perspective
Ralf Blossey

Computational Hydrodynamics of Capsules and Biological Cells
C. Pozrikidis

Computational Neuroscience: A Comprehensive Approach
Jianfeng Feng

Data Analysis Tools for DNA Microarrays
Sorin Draghici

Differential Equations and Mathematical Biology, Second Edition
D.S. Jones, M.J. Plank, and B.D. Sleeman

Dynamics of Biological Systems
Michael Small

Engineering Genetic Circuits
Chris J. Myers

Exactly Solvable Models of Biological Invasion
Sergei V. Petrovskii and Bai-Lian Li

Gene Expression Studies Using Affymetrix Microarrays
Hinrich Göhlmann and Willem Talloen

Glycome Informatics: Methods and Applications
Kiyoko F. Aoki-Kinoshita

Handbook of Hidden Markov Models in Bioinformatics
Martin Gollery

Introduction to Bioinformatics
Anna Tramontano

Introduction to Bio-Ontologies
Peter N. Robinson and Sebastian Bauer

Introduction to Computational Proteomics
Golan Yona

Introduction to Proteins: Structure, Function, and Motion
Amit Kessel and Nir Ben-Tal

An Introduction to Systems Biology: Design Principles of Biological Circuits
Uri Alon

Kinetic Modelling in Systems Biology
Oleg Demin and Igor Goryanin

Knowledge Discovery in Proteomics
Igor Jurisica and Dennis Wigle

Meta-analysis and Combining Information in Genetics and Genomics
Rudy Guerra and Darlene R. Goldstein

Methods in Medical Informatics: Fundamentals of Healthcare Programming in Perl, Python, and Ruby
Jules J. Berman

Modeling and Simulation of Capsules and Biological Cells
C. Pozrikidis

Niche Modeling: Predictions from Statistical Distributions
David Stockwell

Published Titles (continued)

Normal Mode Analysis: Theory and Applications to Biological and Chemical Systems
Qiang Cui and Ivet Bahar

Optimal Control Applied to Biological Models
Suzanne Lenhart and John T. Workman

Pattern Discovery in Bioinformatics: Theory & Algorithms
Laxmi Parida

Python for Bioinformatics
Sebastian Bassi

Spatial Ecology
Stephen Cantrell, Chris Cosner, and Shigui Ruan

Spatiotemporal Patterns in Ecology and Epidemiology: Theory, Models, and Simulation
Horst Malchow, Sergei V. Petrovskii, and Ezio Venturino

Statistics and Data Analysis for Microarrays Using R and Bioconductor, Second Edition
Sorin Drăghici

Stochastic Modelling for Systems Biology
Darren J. Wilkinson

Structural Bioinformatics: An Algorithmic Approach
Forbes J. Burkowski

The Ten Most Wanted Solutions in Protein Bioinformatics
Anna Tramontano

Chapman & Hall/CRC Mathematical and Computational Biology Series

Statistics and Data Analysis for Microarrays Using R and Bioconductor

Second Edition

Sorin Drăghici

CRC Press
Taylor & Francis Group
Boca Raton London New York

CRC Press is an imprint of the
Taylor & Francis Group, an **informa** business
A CHAPMAN & HALL BOOK

First published 2012 by Chapman and Hall

Published 2019 by CRC Press
Taylor & Francis Group
6000 Broken Sound Parkway NW, Suite 300
Boca Raton, FL 33487-2742

© 2012 by Taylor & Francis Group, LLC
CRC Press is an imprint of Taylor & Francis Group, an Informa business

No claim to original U.S. Government works

ISBN 13: 978-1-4398-0975-4 (hbk)

Visit the Taylor & Francis Web site at
http://www.taylorandfrancis.com

and the CRC Press Web site at
http://www.crcpress.com

To Jeannette, my better half,
to Tavi, who brightens every day of my life,
and to Althea, whom I miss every day we are not together

Contents

List of Figures

List of Tables

Art is science made clear.

—Jean Cocteau

Any good poet, in our age at least, must begin with the scientific view of the world; and any scientist worth listening to must be something of a poet, must possess the ability to communicate to the rest of us his sense of love and wonder at what his work discovers.

—Edward Abbey, The Journey Home

The most erroneous stories are those we think we know best - and therefore never scrutinize or question.

—Stephen Jay Gould

My definition of an expert in any field is a person who knows enough about what's really going on to be scared.

—P.J. Plauger

Preface

Although the industry once suffered from a lack of qualified targets and candidate drugs, lead scientists must now decide where to start amidst the overload of biological data. In our opinion, this phenomenon has shifted the bottleneck in drug discovery from data collection to data analysis, interpretation and integration.

—Life Science Informatics, UBS Warburg Market Report, 2001

One of the most promising tools available today to researchers in life sciences is the microarray technology. Typically, one DNA array will provide hundreds or thousands of gene expression values. However, the immense potential of this technology can only be realized if many such experiments are done. In order to understand the biological phenomena, expression levels need to be compared between species or between healthy and ill individuals or at different time points for the same individual or population of individuals. This approach is currently generating an immense quantity of data. Buried under this humongous pile of numbers lays invaluable biological information. The keys to understanding phenomena from fetal development to cancer may be found in these numbers. Clearly, powerful analysis techniques and algorithms are essential tools in mining these data. However, the computer scientist or statistician that does have the expertise to use advanced analysis techniques usually lacks the biological knowledge necessary to understand even the simplest biological phenomena. At the same time, the scientist having the right background to formulate and test biological hypotheses may feel a little uncomfortable when it comes to analyzing the data thus generated. This is because the data analysis task often requires a good understanding of a number of different algorithms and techniques and most people usually associate such an understanding with a background in mathematics, computer science, or statistics.

Because of the huge amount of interest around the microarray technology, there are quite a few books available on this topic. Many of the few available texts concentrate more on the wet lab techniques than on the data analysis aspects. There are several books that review the topic in a somewhat superficial manner, covering everything there is to know about data analysis of microarrays in a couple of hundred pages or less. Other available books focus on excruciating details that only developers of analysis packages find useful. Others are simple proceedings of conferences, gathering together unrelated papers that focus on very specific aspects and topics. Overall, I felt there was a need for a good, middle-of-the-road book that would cover topics in sufficient details to make it possible for readers to really understand what is going on, but without overwhelming details or intimidating heavy formalisms or notations.

At the same time, the R environment has started to dominate heavily everything that is done in terms of data analysis in this area. Again, while many good books on R as a programming and analysis language are available, I felt that a book that would allow the reader to become competent in the analysis of microarray data by providing: i) everything needed to learn the basics of R, ii) the basics of the microarray technology, as well as iii) the understanding necessary in order to apply the right tools to the right problems would be beneficial.

Audience and prerequisites

The goal of this book is to fulfill this need by presenting the main computational techniques available in a way that is useful to both life scientists and analytical scientists. The book tries to demolish the imaginary concrete wall that separates biology and medicine from computer science and statistics and allow the biologist to be a refined user of the available techniques, as well as be able to communicate effectively with computer scientists and statisticians designing new analysis techniques. The intended audience includes as a central figure the researcher or practitioner with a background in the life sciences that needs to use computational tools in order to analyze data. At the same time, the book is intended for the computer scientists or statisticians who would like to use their background in order to solve problems from biology and medicine. The book explains the nature of the specific challenges that such problems pose as well as various adaptations that classical algorithms need to undergo in order to provide good results in this particular field.

Finally, it is anticipated that there will be a shift from the classical compartmented education to a highly interdisciplinary approach that will form people with skills across a range of disciplines crossing the borders between traditionally unrelated fields, such as medicine or biology and statistics or computer science. This book can be used as a textbook for a senior undergraduate or graduate course in such an interdisciplinary curriculum. The book is suitable for a data analysis and data mining course for students with a background in biology, molecular biology, chemistry, genetics, computer science, statistics, mathematics, etc.

Useful prerequisites for a biologist include elementary calculus and algebra. However, the material is designed to be useful even for readers with a shaky mathematical foundation since those elements that are crucial for the topic are fully discussed. Useful prerequisites for a computer scientist or mathematician include some elements of genetics and molecular biology. Once again, such knowledge is useful but not required since the essential aspects of the technology are covered in the book.

Aims and contents

The first and foremost aim of this book is **to provide a clear and rigorous description of the algorithms without overwhelming the reader with the usual cryptic notation or with too much mathematical detail**. The presentation level is appropriate for a scientist with a background in life sciences. Little or no mathematical training is needed in order to understand the material presented here. Those few mathematical and statistical facts that are really needed in order to understand the techniques are completely explained in the book at a level that is fully accessible to the non-mathematically minded reader. The goal here was to keep the level as accessible as possible. The mathematical apparatus was voluntarily limited to the very basics. The most complicated mathematical symbol throughout the book is the sum of n terms: $\sum_{i=1}^{n} x_i$. In order to do this, certain compromises had to be made. The definitions of many statistical concepts are not as comprehensive as they could be. In certain places, giving the user a powerful intuition and a good understanding of the concept took precedence over the exact, but more difficult to understand, formalism. This was also done for the molecular biology aspects. Certain cellular phenomena have been presented in a simplified version, leaving out many complex phenomena that we considered not to be absolutely necessary in order to understand the big picture.

A second specific aim of the book is **to allow a reader to learn the R environment and programming language using a hands-on and example-rich approach**. From this perspective, the book should be equally useful to a large variety of readers with very different backgrounds. No previous programming experience is required or expected from the reader. The book includes chapters that describe everything from the basic R commands and syntax to rather sophisticated procedures for quality control, normalization, data analysis and machine learning. Everything from the simplest commands to the most complex procedures is illustrated with R code. All analysis results shown in the book are actual results produced by the code shown in the text and therefore, all code shown is free from spelling or syntax errors.

A third specific aim of the book is **to allow a microarray user to be in a position to make an informed choice as to what data analysis technique to use in a given situation, even if using other analysis packages**. The existing software packages usually include a very large number of techniques, which in turn use an even larger number of parameters. Thus, the biologist trying to analyze DNA microarray data is confronted with an overwhelming number of possibilities. Such flexibility is absolutely crucial because each data set is different and has specific particularities that must be taken into account when selecting algorithms. For example, data sets obtained in different laboratories have different characteristics, so the choice of normalization procedures is very important. However, such wealth of choices can be overwhelming for the life scientist who, in most cases, is not very familiar with all intricacies of data analysis and ends up by always using the default choices. This book is designed to help such a scientist by emphasizing at a high level of abstraction the characteristics of various techniques in a biological context.

As a text designed to bridge the gap between several disciplines, the book includes chapters that would give all the necessary information to readers with a variety of backgrounds. The book is divided into two parts. The first part is designed to offer an overview of microarrays and to create a solid foundation by presenting the elements from statistics that constitute the building blocks of any data analysis. The second part introduces the reader to the details of the techniques most commonly used in the analysis of microarray data.

Chapter 2 presents a short primer on the central dogma of molecular biology and why microarrays are useful. This chapter is aimed mostly at analytical scientists with no background in life sciences. **Chapter 3** briefly presents the microarray technology. For the computer scientist or statistician, this constitutes a microarray primer. For the microarray user, this will offer a bird's-eye view perspective on several techniques emphasizing common as well as technology-specific issues related to data analysis. This is useful since many times the users of a specific technology are so engulfed in the minute details of that technology that they might not see the forest for the trees.

Chapter 4 discusses a number of important issues related to the reliability and reproducibility of microarray data. This discussion is important both for the life scientist who needs to understand the limitations of the technology used, as well as for the computer scientist or statistician who needs to understand the intrinsic level of noise present in this type of data.

Chapter 5 constitutes a short primer on digital imaging and image processing. This chapter is mostly aimed at the life scientists or statisticians who are not familiar with digital image processing.

Chapter 6 is an introduction to the R programming language. This chapter discusses basic concepts from the installation of the R environment to the basic syntax and concepts of R.

Chapter 7 presents the Bioconductor project and briefly illustrates its capabilities with some simple examples. This chapter was contributed by one of the founders of the Bioconductor project, Vincent Carey, currently an Associate Professor of Medicine (Biostatistics) at Harvard Medical School, and an Associate Biostatistician in the Department of Medicine at the Brigham and Women's Hospital in Boston.

Chapters 8, **9**, and **11** focus on some elementary statistics notions. These chapters will provide the biologist with a general perspective on issues very intimately related to microarrays. The purpose here is to give only as much information as needed in order to be able to make an informed choice during the subsequent data analysis. The aim of the discussion here is to put things in the perspective of somebody who analyzes microarray data rather than offer a full treatment of the respective statistical notions and techniques. **Chapter 9**

discusses several important distributions, **Chapter 11** discusses the classical hypothesis testing approach, and **Chapter 12** applies it to microarray data analysis. **Chapter 10** uses R to illustrated the basic statistical tools available in R for descriptive statistics and basic built-in distributions.

Chapter 13 presents the family of ANalysis Of VAriance methods intensively used by many researchers to analyze microarray data. **Chapter 14** discusses the more general linear models and illustrates them in R. **Chapter 15** uses some of the ANOVA and linear model approaches in the discussion of various techniques for experiment design.

Chapter 16 discusses several issues related to the fact that microarrays interrogate a very large number of genes simultaneously and its consequences regarding data analysis.

Chapters 17 and **18** present the most widely used tools for microarray data analysis. In most cases, the techniques are presented using real data. Chapter 17 includes several techniques used in exploratory analysis, when there is no known information about the problem and the task is to identify relevant phenomena as well as the parameters (genes) that control them. The main techniques discussed here include box plots, histograms, scatter plots, volcano plots, time series, principal component analysis (PCA), and independent component analysis (ICA). The clustering techniques described in Chapter 18 include K-means, hierarchical clustering, biclustering, partitioning-around-medoids, and self-organizing feature maps. Again, the purpose here is to explain the techniques in an unsophisticated yet rigorous manner. The all-important issue of when to use a specific technique is discussed on various examples emphasizing the strengths and weaknesses of each individual technique.

Chapter 19 discusses specific quality control issues characteristic to Affymetrix and Illumina data. These are illustrated using R functions and packages applied on real data sets. Tools such as intensity distributions, box plots, RNA degradation curves, and quality control metrics are used to illustrate problems ranging from array saturation, to RNA degradation, and annotation issues. Plots illustrating various problems are shown side-by-side with plots showing clean data such that the reader can understand and learn what to look for in such plots.

Chapter 20 concentrates on data preparation issues. Although such issues are crucial for the final results of the data mining process, they are often ignored. Issues such as color swapping, color normalization, background correction, thresholding, mean normalization, etc., are discussed in detail. This chapter will be extremely useful both to the biologist, who will become aware of the different numerical aspects of the various preprocessing techniques, and to the computer scientist, who will gain a deeper understanding of various biological aspects, motivations, and meanings behind such preprocessing. Again, all normalization issues are illustrated using R functions and packages applied on real data sets.

Chapter 21 presents several methods used to select differentially regulated genes in comparative experiments.

Chapter 22 discusses the Gene Ontology, including its goal, structure, annotations, and some statistics about the data currently available in it. **Chapter 23** shows how GO can be used to translate lists of differentially expressed genes into a better understanding of the underlying biological phenomena. Just when you think this is easy, **Chapter 24** comes to tell you about the many mistakes and issues that could appear in this GO profiling. **Chapter 25** reviews more than a dozen tools that are currently available for this type of functional analysis.

Chapter 26 somehow reverses the direction considering the problem of how to select the microarrays that are best suited for investigating a given biological hypothesis.

Chapter 27 discusses some of the problems that can be caused by the fact that the same biological entity may have different IDs in different public databases. Some tools that allow a mapping from one type of ID to another are discussed and compared.

Chapter 28 takes the analysis to the next level, using a systems biology approach that aims to take into consideration the way genes are known to interact with each other. This type of knowledge is captured in collections of signaling pathways available from various sources. This chapter discusses various approaches currently available for the analysis of signaling pathways.

Chapter 29 is a brief review of several machine learning techniques that are widely used with microarray data. Since unsupervised methods are discussed in **Chapter 18**, this chapter focuses on supervised methods including linear discriminants, feed-forward neural networks, and support vector machines.

Finally, the last chapter of the book presents some conclusions as well as a brief presentation of some novel techniques expected to have a great impact on this field in the near future.

Road map

This book can be used in several ways, depending on the background of the reader and the goals pursued. The chapters can be combined in various ways, allowing an instructor to tailor a course to the specific background and expectations of a given audience.

Some of the courses (with or without a laboratory component) that can be easily taught using this book include:

1. Introduction to statistics: Chapters 8, 9, 11, 12, 13, 14, 15, 16

2. Introduction to R and Bioconductor: Chapters 6, 7, 10, 14, 17, 18, 29

3. Microarray data analysis for life scientists: Chapters 3 – 30

4. Microarray data analysis for computer scientists (including R and Bioconductor): Chapter 2 – 30

5. Quality control and normalization techniques: Chapters 19, 20

6. Interpretation of high-throughput data: GO profiling and pathway analysis: Chapters 22 – 28

This book focuses on R and Bioconductor. The reader is advised to install the software and actually use it to perform the analysis steps discussed in the book. The accompanying CD includes all code used throughout the book.

Acknowledgments

The author of this book has been supported by the Biological Databases Program of the National Science Foundation under grants number NSF-0965741 and NSF-0234806, the Bioinformatics Cell, MRMC US Army – DAMD17-03-2-0035, National Institutes of Health – grant numbers 1R01DK089167-01, R01-NS045207-01 and R21-EB000990-01, and Michigan Life Sciences Corridor grant number MLSC-27. Many thanks to Sylvia Spengler, Director of the Biological Databases program, National Science Foundation, Salvatore Sechi, Program Manager NIDDKD, Peter McCartney, Program Director, Division of Biological Infrastructure, National Science Foundation, Peter Lyster, Program Director in the Center for Bioinformatics and Computational Biology, NIGMS without whose support our research and this book would not have been possible.

I would like to express my gratitude to Dr. Robert Romero, Chief of the Perinatology Research Branch of the National Institute of Child Health and Human Development, who has been a tremendous source of inspiration and a role model in my research. I am also very grateful for his strong support which allowed me to dedicate more time to my research and to other scholarly endeavors such as writing this book.

My deepest gratitude goes to Robert J. Sokol, M.D., The John M. Malone, Jr., M.D., Endowed Chair, and Director of the C.S. Mott Center for Human Growth and Development, Distinguished Professor of Obstetrics and Gynecology in the Wayne State University School of Medicine. His efforts and support were the main reasons that kept me at Wayne State when I had great opportunities elsewhere. Also, his generosity in creating the endowed chair in systems biology that bears his name – and that I am currently holding – made possible the creation of the strong research group of which I am a part.

My colleague Adi Laurentiu Tarca had a substantial contribution to this book. This consisted of a first draft of the linear models chapter, as well as numerous code snippets used throughout several other chapters to illustrate various concepts and techniques. He also contributed with a very thorough critical reading of many other chapters followed by very useful comments and suggestions.

My thanks also go to Vincent Carey, Associate Professor of Medicine (Biostatistics) at Harvard Medical School, and an Associate Biostatistician in the Department of Medicine at the Brigham and Women's Hospital in Boston. Vincent is one of the founders of the Bioconductor project and contributed Chapter 7 to this book. He was also very helpful in answering several questions regarding Bioconductor and various packages used throughout this book.

My Ph.D. students Calin Voichita, Michele Donato, Cristina Mitrea, Dorina Twigg, Munir Islam, Rebecca Tagett, and my visiting postdoctoral researcher Josep Maria Mercader (currently back at his home institution – Barcelona Supercomputer Center) contributed enormously to this book, first by accepting to be my guinea pigs for the examples and explanations used in the book, and then by proof-reading all 1,081 pages of this book – several times. In spite of their tremendous efforts, typos and small errors probably continue to exist here and there. If this is the case, the fault is entirely mine. I am sure that for each such surviving typo, somewhere in my inbox there is an email from one of them, bringing it to my attention.

My research associate, Zhonghui Xu, had a substantial contribution to this book by writing various R routines and snippets of code, in particular most of those in the chapters on normalization and quality control.

Gary Chase kindly reviewed the chapters on statistics (for the first incarnation of this book) and did so on incredibly short deadlines that allowed the book to be published on time. His comments were absolutely invaluable and made the book stronger than I initially conceived it.

I owe a lot to my colleague, friend and mentor, Michael A. Tainsky. Michael taught me everything I know about molecular biology and I continue to be amazed by his profound grasp of some difficult data analysis concepts. I have learned a lot during our collaboration over the past 10 years. Michael also introduced me to Judy Abrams who provided a lot of encouragement and has been a wonderful source of fresh and exciting ideas, many of which ended up in grant proposals. My colleague and friend Steve Krawetz was the first one to identify the need for something like Onto-Express (Chapter 23). During my leave of absence, he continued to work with my students and participated actively in the creation of our first Onto-Express prototype. Our collaboration over the past few years produced a number of great ideas and a few papers. Thanks also go to Jeff Loeb for our productive interaction as well as for our numerous discussions on corrections for multiple experiments.

Many thanks to Otto Muzik for sticking to the belief that our collaboration is worthwhile through the first, less than productive, year and for being such a wonderful colleague and friend. In spite of some personal differences, Otto did not hesitate to offer his help when my father was diagnosed with cancer.

I am very grateful to him for his offer to help, even though my father passed away before he could take advantage of Otto's kind offer.

Jim Granneman and Bob Mackenzie knocked one day on my door looking for some help analyzing their microarray data. One year later, we were awarded a grant of over 3.5 million from the Michigan Life Science Corridor. This allowed me to shift some of my effort to this book. Currently, Jim and Bob are my coPIs on a 1.5 million NIH grant which is funding some very exciting research. Thank you all for involving me in your work.

I am grateful to Soheil Shams, Bruce Hoff, Anton Petrov and everybody else with whom I interacted during my one year sabbatical at BioDiscovery. Bruce, Anton, and I worked together at the development of the ANOVA based noise sampling method described in Chapter 21 based on some initial experiments by Xiaoman Li. Figures 18.6, 18.12, and 18.15 were initially drawn by Bruce.

During my stay in Los Angeles, I met some top scientists, including Michael Waterman and Wing Wong. I learned a lot from our discussions and I am very grateful for that. Mike Waterman provided very useful comments and suggestions on the noise sampling method in Chapter 21 and later, on Onto-Express described in Chapter 23. Wing Wong and his students at UCLA and Harvard have done a great job with their software dChip, which is a great first stop every time one has to normalize Affymetrix data.

Gary Churchill provided useful comments on the work related to the ANOVA based noise sampling technique presented in Chapter 21. The discussions with him and Kathy Kerr helped us a lot. Also, their ANOVA software for microarray data analysis represent a great tool for the research community. We used some of their software to generate the images illustrating the LOWESS transform in Chapter 20.

John Quackenbush kindly provided a wonderful sample data set that we used in the PCA examples in Chapter 17. He and his colleagues at TIGR designed, implemented and made available to the community a series of very nice tools for microarray data analysis including the Multiple Experiment Viewer (MEV) used to generate some images for Chapter 17.

About the author

Sorin Drăghici has obtained his B.Sc. and M.Sc. degrees in Computer Engineering from "Politehnica" University in Bucharest, Romania followed by a Ph.D. degree in Computer Science from University of St. Andrews, United Kingdom (third oldest university in UK after Oxford and Cambridge). Besides this book, he has published two other books, several book chapters, and over 100 peer-reviewed journal and conference papers. He is inventor or co-inventor on several patent applications related to biotechnology, high-throughput methods and systems biology. He is currently an editor of IEEE/ACM Transactions on Computational Biology and Bioinformatics, the Journal of Biomedicine and Biotechnology, and the International Journal of Functional Informatics and Personalized Medicine. He is also active as a journal reviewer for 15 international technical journals related to bioinformatics and computational biology, as well as a regular NSF and NIH panelist on biotechnology and bioinformatics topics.

Currently, Sorin Drăghici is holding the Robert J. Sokol, MD Endowed Chair in Systems Biology in the Department of Obstetrics and Gynecology in the School of Medicine, as well as joint appointments as Professor in the Department of Clinical and Translational Science, School of Medicine, and the Department of Computer Science, College of Engineering, Wayne State University. He is also the Chief of the Bioinformatics and Data Analysis Section, Perinatology Research Branch, National Institute for Child Health and Development, NIH, as well as the head of the Intelligent Systems and Bioinformatics Laboratory (ISBL, http://vortex.cs.wayne.edu).

Disclaimer

Any opinions, findings, and conclusions or recommendations expressed in this material are those of the author and do not necessarily reflect the views of the National Science Foundation, National Institutes of Health, any other funding agency, nor those of any of the people mentioned above.

Chapter 1

Introduction

If we begin with certainties, we shall end in doubts; but if we begin with doubts, and are patient in them, we shall end in certainties.

—*Francis Bacon*

1.1 Bioinformatics – an emerging discipline

Life sciences are currently at the center of an informational revolution. Dramatic changes are being registered as a consequence of the development of techniques and tools that allow the collection of biological information at an unprecedented level of detail and in extremely large quantities. The human genome project is a compelling example. Initially, the plan to sequence the human genome was considered extremely ambitious, on the border of feasibility. The first serious effort was planned over 15 years at a cost of $3 billion. Soon after, the schedule was revised to last only 5 years. Eventually, the genome was sequenced in less than 3 years, at a cost much lower than initially expected [361]. The nature and amount of information now available open directions

1

of research that were once in the realm of science fiction. Pharmacogenomics [359], molecular diagnostics [86, 125, 260, 360, 376, 455] and drug target identification [302] are just a few of the many areas [163] that have the potential to use this information to change dramatically the scientific landscape in the life sciences.

During this informational revolution, the data-gathering capabilities have greatly surpassed the data analysis techniques. If we were to imagine the Holy Grail of life sciences, we might envision a technology that would allow us to fully understand the data at the speed at which it is collected. Sequencing, localization of new genes, functional assignment, pathway elucidation, and understanding the regulatory mechanisms of the cell and organism should be seamless. Ideally, we would like knowledge manipulation to become tomorrow the way goods manufacturing is today: high automatization producing more goods, of higher quality, and in a more cost-effective manner than manual production. In a sense, knowledge manipulation is now reaching its pre industrial age. Our farms of sequencing machines and legions of robotic arrayers can now produce massive amounts of *data* but using them to manufacture highly processed pieces of *knowledge* still requires skilled masters painstakingly forging through small pieces of raw data one at a time. The ultimate goal in life science research is to automate this knowledge discovery process.

Bioinformatics is the field of research that presents both the opportunity and the challenge to bring us closer to this goal. Bioinformatics is an emerging discipline situated at the interface between analytical sciences such as statistics, mathematics, and computer science on one side, and biological sciences such as molecular biology, genomics, and proteomics on the other side. Initially, the term bioinformatics was used to denote very specific tasks such as the activities related to the storage of data of biological nature in databases. As the field evolved, the term has started to also encompass algorithms and techniques used in the context of biological problems. Although there is no universally accepted definition of bioinformatics, currently the term denotes a field concerned with the application of information technology techniques, algorithms, and tools to solve problems in biological sciences. The techniques currently used have their origins in a number of areas, such as computer science, statistics, mathematics, etc. Essentially, **bioinformatics** is the science of refining biological information into biological knowledge using computers. Sequence analysis, protein structure prediction, and dynamic modeling of complex biosystems are just a few examples of problems that fall under the general umbrella of bioinformatics. However, new types of data have started to emerge. Examples include protein–protein interactions, protein–DNA interactions, signaling and biochemical pathways, population-scale sequence data, large-scale gene expression data, and ecological and environmental data [191].

A subfield of particular interest today is genomics. The field of **genomics** encompasses investigations into the structure and function of very large number of genes undertaken in a simultaneous fashion. The term genomics comes from **genome**, which is the entire set of genes in a given organism. **Structural**

genomics includes the genetic mapping, physical mapping, and sequencing of genes for entire organisms (genomes). **Comparative genomics** deals with extending the information gained from the study of some organisms to other organisms. **Functional genomics** is concerned with the role that individual genes or subsets of genes play in the development and life of organisms.

Just as genomics refers to the large-scale studies involving the properties and functions of many genes, **proteomics** is concerned with the large-scale study of proteins. The **proteome** of an organism is the set of all proteins in the given organism. Following the same lines of linguistic synthesis, the "-omics" suffix has been appended to a number of other terms generating a large variety of novel terms. Among those, some have been well accepted by the community, such as transcripts → **transcriptome** → **transcriptomics**, metabolite → **metabolome** → **metabolomics**, etc. In all cases, the "X-ome," represents the set of all entities of type X in a given organism, while the "X-omics" represents the field of research studying them using high-throughput approaches.

Currently, our understanding of the role played by various genes and their interactions seems to be lagging far behind the knowledge of their sequence information. Table 1.1 presents some data that reflect the relationship between sequencing an organism and understanding the role of its various genes [311]. The yeast is an illustrative example. Although the 6,200 genes of its genome have been known since 1997, only approximately 94% of them have inferred functions. The situation is similar for *E. coli, C. elegans, Drosophila*, and *Arabidopsis*.[1]

Most researchers agree that the challenge of the near future is to analyze, interpret, and understand all data that are being produced [56, 148, 290, 440]. In essence, the challenge faced by the biological scientists is to use the large-scale data that are being gathered to discover and understand fundamental biological phenomena. At the same time, the challenge faced by computer scientists is to develop new algorithms and techniques to support such discoveries [191].

The explosive growth of computational biology and bioinformatics has just started. Biotechnology and pharmaceutical companies are channeling many resources towards bioinformatics by starting informatics groups. The tendency has been noted by the academic world, and there are a number of universities that have declared bioinformatics a major research priority. The need for bioinformatics-savvy people as well as experts in bioinformatics is enormous and will continue to accentuate in the near future. Two important phenomena are at play here. On the one hand, modern life science research has irreversibly adopted the use of very large throughput technologies such as DNA microarrays, mass-spectrometry, high-throughput sequencing, etc. These techniques generate terabytes of data on a daily basis. On the other hand, the academic

---

[1]Even genomes that are considered substantially complete, in reality may still have small gaps [6]. Until we understand better the function of various genes, we cannot discount the functional relevance of the genetic material in those gaps.

Organism	Number of genes	Genes with inferred function	Genome Completion date
S. cerevisiae	6,201	94%	1996 [181]
E. coli	4,467	62%	1997 [55]
C. elegans	21,185	63%	1998 [448]
D. melanogaster	18,462	82%	1999 [6]
A. thaliana	33,264	76%	2000 [370]
Homo sapiens	39,920	46%	2001 [437]
B. anthracis	5,415	80%	2002 [351]
Rattus norvegicus	37,533	37%	2004 [173]

TABLE 1.1:   The unbalance between obtaining the data and understanding it. Although the complete genomes of several simpler organisms are available, understanding the role of various genes lags far behind. *Saccharomyces cerevisiae* is the baker's yeast; *Escherichia coli* is a bacterium that lives in the gut of worm-blooded animals (and some of its varieties sometimes infect the human food chain); *C. elegans* is a nematode (worm); *Drosophila melanogaster* is the fruit fly; *Arabidopsis thaliana* is a plant; *Bacillus anthracis* is the anthrax pathogen; *Rattus norvegicus* is the Norwegian brown rat, and *Homo sapiens* is the human. The yeast, fruit fly, *C. elegans*, arabidopsis, and the rat are often used as model organisms.

system is not yet able to produce enough people with the truly multidisciplinary background and knowledge requested in areas such as bioinformatics. Hence, there is a large gap between the need to effectively analyze the mountains of data being generated continuously and the number of people able to perform such analyses in the best possible ways. Bioinformatics experts able to analyze data are and will continue to be very valuable to many employers. It is hope that this book will help you get closer to becoming such an expert.

The avalanche of data resulting from the progress in the field of molecular biology was started by understanding the nature of molecular information in living organisms and the development of new and precise high throughput screening methods. Molecular biology deals primarily with the information that macromolecules, such as the **deoxyribonucleic acid (DNA)** and the **ribonucleic acid (RNA)** carry, their interrelationship, and role in cells. Therefore, a brief description of the cell and its very basic mechanisms follows in the next chapter. Although these concepts are not absolutely necessary in order to understand the data analysis methods and techniques presented in the rest of the book, it is always better if one has a basic understanding of where the numbers analyzed come from and what they actually mean.

Chapter 2

The cell and its basic mechanisms

> . . . I could exceedingly plainly perceive it to be all perforated and porous, much like a Honey-comb, but that the pores of it were not regular. . . . these pores, or **cells**, . . . were indeed the first microscopical pores I ever saw, and perhaps, that were ever seen, for I had not met with any Writer or Person, that had made any mention of them before this. . .

> —Robert Hooke, Micrographia, 1665

2.1 The cell

The cell is the building block of all organisms. Fig. 2.1 shows a typical eukaryotic cell as well as a typical prokaryotic cell. A **eukaryotic** cell (top panel in Fig. 2.1) has a nucleus and is found in more evolved organisms. A **prokaryotic** cell (bottom panel in Fig. 2.1) does not have a nucleus and is always single-cellular, e.g., bacteria. As shown in these figures, each cell is a very complex system that includes a number of parts and structures.

The main parts of a eukaryotic cell include the **membrane**, the **cytoplasm**, the **mitochondria**, the **microtubules**, the **lysosomes**, the **ribo-**

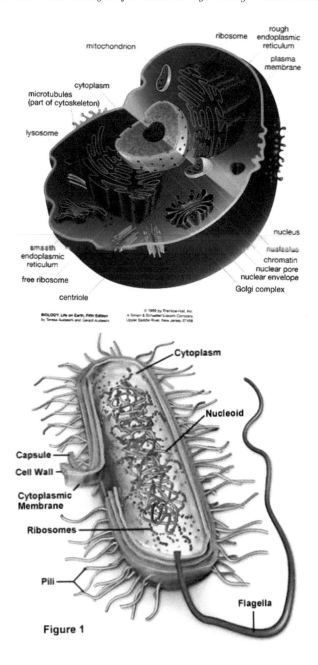

FIGURE 2.1: Two cells. A eukaryotic cell (top panel) has a nucleus and is found in more evolved organisms. A prokaryotic cell (bottom panel) does not have a nucleus and is mostly found in bacteria. The eukaryotic cell figure is from *Life on Earth*, Audesirk et al., Prentice Hall. Printed with permission. The prokaryotic cell is copyrighted Michael W. Davidson. Printed with permission.

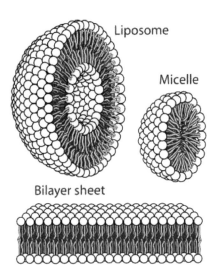

FIGURE 2.2: Cross section of the different structures that phospholipids can take in an aqueous solution. The circles are the hydrophilic heads and the wavy lines are the fatty acid side chains. The bilayer appears in the cell membrane, nuclear membrane, vesicles, etc. Author: Mariana Ruitz, released in the public domain.

somes, the **smooth** and **rough endoplasmic reticula**, etc. In a eukaryotic cell, there is also a **nucleus** that hosts a **nucleolus** and the **chromatin** within a **nuclear envelope** that features some **pores**. In the following, we will briefly discuss these.

The **cellular membrane** generally consists of two layers of phospholipid molecules. Each such molecule has a polar hydrophilic head and two non-polar (hydrophobic) tails. Since both the cytoplasm inside the cell as well as the extracellular environment contain a lot of water, the membrane molecules are aligned in a double layer, each layer presenting the hydrophilic head on the surface of the membrane (both inside and outside the cell), while all the hydrophobic tails point towards each other within the membrane. Fig. 2.2 shows the basic structure of a membrane, as well as those of two other structures that can be formed by the phospholipid bilayer. However, the cellular membrane has a structure that is far more complex than a simple phospholipid bilayer. Among other molecules, it also includes some proteins called integral membrane proteins. These can appear on the inside surface of the membrane,

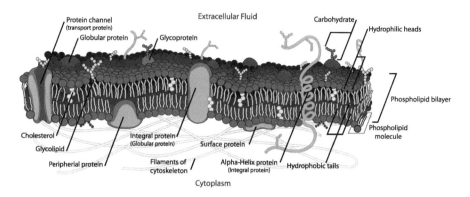

FIGURE 2.3: The cell membrane generally consists of two layers of phospholipid molecules (the basic bilayer shown in Fig. 2.2) but also has a number of other features, including surface proteins present either on the inside or on the outside surface of the membrane, integral proteins of various types crossing both layers of the membrane, channel-forming proteins acting like gateways for certain molecules, transmembrane proteins that have receptors on the outside surface and are able to trigger specific intracellular responses when their target ligand is present in the extracellular space, etc. Author: Mariana Ruitz, released in the public domain.

on its outside surface or crossing the membrane (in which case they are called transmembrane proteins). The outside part of such a protein is called a receptor. Its role is to bind to a given molecule, called ligand, when this molecule is present outside the cell. Generally, when this happens, the transmembrane protein will initiate an intracellular response. Other transmembrane proteins act as gateways, allowing certain molecules from outside the cell to enter the cell through a channel formed by the protein. The Fig. 2.3 shows a crosscut through the cellular membrane illustrating some of these additional features of the cellular membrane. The membrane is involved in several very important processes such as cell adhesion, cell signaling, and ion channel conductance.

The **cytoplasm** includes everything that is in the cell (organelles, water, other chemical molecules, etc.), except the nucleus. The **nucleoplasm** includes everything that is in the nucleus. Together, the cytoplasm and the nucleoplasm form the **protoplasm**.

Organelles are specialized sub-cellular structures of the cytoplasm. In some sense, the organelles do for the cell what the organs do for complex organisms: each organelle has a very specific function in the complex mechanisms that keep the cell alive. Some authors define the organelles as being membrane-bound structures that have some specific function in the cell. Other authors use a less restrictive definition considering that any structure that carries out a particular and specialized function is an organelle, whether it is membrane-bound or not. For instance, the ribosome (see details below) is an organelle according to the latter definition, but not according to the former.

The **chromatin** is a structure made of proteins and highly packed DNA. The DNA contains the **genes** that code for all proteins, as well as other functional and control elements. A prokaryotic cell does not have a nucleus, and the DNA material is found directly in the cytoplasm.

The **mitochondria** (singular **mitocondrion**) are the power plants of the cell (see the eukaryotic cell in Fig. 2.1). Their main role is to produce energy for the cell. There are several fascinating facts about mitochondria. First, the mitochondria have their own DNA (called **mitochondrial DNA** or **mtDNA**), which is circular, very much like the DNA of a bacteria. Based on this as well as other data, it has been suggested that the mitochondria are in fact what is left from a small prokaryote cell that was swallowed millions of years ago by an eukaryote, or perhaps a larger prokaryote [299]. Rather than digesting the smaller prokaryote, the bigger cell found out that a symbiotic partnership would be much better for both of them. The smaller cell benefits from the free basic fuel, as well as the safe environment provided by the larger cell. In turn, the larger cell benefits from the energy produced by the smaller cell. This is an example of mutualism, a type of symbiosis in which both organisms benefit and neither is harmed. This partnership is so strong now that eukaryotic cells cannot survive without mitochondria, and the **endosymbionts** (the smaller cells which were incorporated in the eukaryotic cell) also cannot survive on their own. The same mutualism is found in plants and algae whose cells contain chloroplasts, organelles able to transform sunlight into energy during the process of photosynthesis.

The second fascinating fact about mitochondria is that in most multicellular organisms (including human), the mitochondrial DNA is inherited from mother to child. This is unlike the DNA in the nucleus, which is formed in the offspring by combining the nucleic DNA from both mother and father. In fact, this very unusual property of the mitochondrial DNA is at the center of a book by Bryan Sykes, titled *The Seven Daughters of Eve*. In this book, the author describes how the entire population of Europe can be traced back to only seven women (hence the title) using this property of the mtDNA. In fact, in the same book, Bryan Sykes uses the same argument to refute Thor Heyerdahl's hypothesis that the population of Polynesia originated in South America and reached the Polynesian islands by crossing the Pacific on primitive bamboo rafts.[1]

---

[1]This hypothesis may seem rather far-fetched, but actually in 1947, Thor Heyerdahl built such a bamboo raft and sailed it more than 4,300 miles (8,000 Kms) from South America to the Tuamotu Islands. The raft, Kon-Tiki, was built using exclusively materials and technologies available at a time to those populations. For instance, since those materials and technologies did not include either iron or nails, the logs of the raft were held together with manually weaved hemp ropes [207]. Even though his theory regarding the origins of the Polynesian population was ultimately proved incorrect, Thor Heyerdahl will remain in history for the courage and determination to put his life on the line in order to prove the feasibility of his scientific hypothesis.

Since the mitochondria have their own DNA, it follows that they can reproduce independently of the host cell. Thus, the number of mitochondria can vary in time within a given cell or cell type. Various treatments such as some antifungal treatments can actually kill many mitochondria. After the treatment is stopped, the mitochondria will start multiplying again and their population will eventually recover to normal levels. Recently, it has been discovered that mitochondria also play an essential role in mechanisms other than energy generation. For instance, mitochondria can be the target of the immune system response [301]. Thus, it has recently been shown that the killer T cells of the immune system can trigger the programmed cell death (or apoptosis) of virus-infected or cancer cells by releasing two serine proteases called granzymes in the target cells. One of these, Granzyme A, targets a certain protein NDUFS3 which is degraded. In turn, this causes mitochondria to produce damaging reactive oxygen which eventually causes cell death. The other granzyme, Granzyme B, causes the breakdown of the outer mitochondrial membrane which releases a number of death-promoting proteins activating a chain-reaction known as the caspase protease cascade resulting in massive DNA damage and cell death.

The **microtubules** are cylindrical hollow structures with a diameter of approximately 25 nm and length varying from 200 nanometers to 25 micrometers, which are found in the cytoplasm of eukaryotic cells (see Fig. 2.4). They are part of the cytoskeleton of the cell, providing structural support (e.g., giving the cell its shape), and assisting in cellular locomotion, transport, and cell division.

The **lysosomes** are vesicles that contain digestive enzymes called acid hydrolases. The term lysosome comes from *lysis* (dissolution or destruction in Greek) and *soma* (body in Greek). The lysosome is involved in the breakdown of various materials such as food particles, viruses, or bacteria that managed to penetrate into the cell. Because the lysosome membrane isolates the contents of the lysosome from the rest of the cell, the inside of the lysosome can be maintained at a pH that is much more acidic (around 4.5–4.8) than the neutral pH (7.0) of the intracellular fluid, or **cytosol**. This acidic pH is necessary for the functioning of the hydrolases. This plays the role of a safety mechanism since if the digestive enzymes are released in the cell by accident they will not harm it, as long as the pH in the cytosol is normal. However, if the cell is dead, dying, or injured, the acid hydrolases released from lysosomes can self-digest the cell in a process of **autolysis**. This is why sometimes lysosomes are also called suicide bags.

The **ribosomes** are complexes of RNA and proteins whose main role is to translate messenger RNA (mRNA) into chains of polypeptides using amino acids delivered by the transfer RNA (tRNA) molecules. The term ribosome comes from *ribo*nucleic acid (RNA) and *soma* (body in Greek). The role of ribosomes will be discussed in more detail in Section 2.3.

The **endoplasmic reticulum** (plural endoplasmic reticula) (ER) is an interconnected structure composed of **tubules**, **vesicles**, and **cisternae**. A

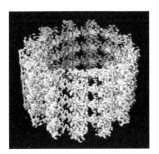

FIGURE 2.4: A microtubule is a hollow cylindrical structure made out of the α and β tubulin proteins. The microtubules have many roles, including providing structural support, assisting in cellular locomotion, and cell division. Image in the public domain.

tubule is a small tube-like structure. A vesicle is a small sac surrounded by a membrane similar to the cellular membrane. Vesicles store, transport, or digest various cellular products or waste. A cisterna (plural cisternae) is a flattened disc surrounded by a membrane. They also carry proteins. The ERs are involved in the translation of certain specialized proteins, the transport of proteins to be used in the cell membrane or to be secreted from the cell, sequestration of calcium, and production and storage of certain macromolecules. The ERs can be subdivided into **rough endoplasmic reticula**, **smooth endoplasmic reticula**, and **sarcoplasmic reticula**, each having some slightly different characteristics and roles.

Fig. 2.5 shows the endomembrane system (from *endo* meaning internal in Greek, and membrane) in a eukaryotic cell, including the rough and smooth endoplasmic reticula, secretory vesicles, lysosomes, the Golgi apparatus, etc.

The **nucleus** (plural **nuclei**) is the largest organelle in the cell. In mammalian cells, the nucleus measures about 11–22 micrometers in diameter and occupies about 10% of its volume. The nucleus contains most of the cell's nuclear genetic material. This is in the form of very long linear DNA molecules organized most of the time into a DNA-protein complex structure called chromatin, which is essentially very tightly packed double-stranded DNA. During cell division, the chromatin forms well-defined structures called chromosomes. Each chromosome contains many genes as well as long sequences of intergenic DNA. The main roles of the nucleus are to protect the nucleic DNA, to control

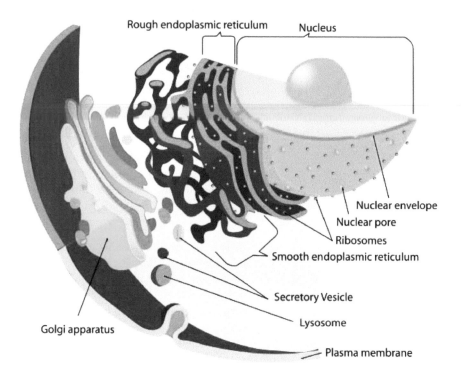

FIGURE 2.5: The endomembrane system is a system of intracellular membranes that divide the eukaryotic cells into various organelles. The endomembrane system includes the cell membrane itself, the nuclear envelope that separates the nucleus from the cytoplasm, the smooth and rough endoplasmic reticula, the Golgi apparatus, the lysosomes, vesicles, etc.

the gene expression process, and to mediate the replication of DNA during the cell cycle.

The nucleus is surrounded by a nuclear membrane that separates it from the cytoplasm. Since this membrane is impenetrable to most molecules, it has a number of small orifices, or **pores**, which allow certain small, water-soluble molecules to penetrate the nuclear membrane in very specific conditions. Larger molecules such as proteins must be transported in a very carefully controlled way by specialized transporter proteins. The surface of the nucleus is also studded with ribosomes much like the surface of the rough endoplasmic reticulum which continues it. Although the interior of the nucleus is not separated by other membranes, its content is not uniform. Fig. 2.6 shows that the nucleus has a central part called nucleolus, which is mainly involved in the assembly of ribosomes, and two types of chromatin: **heterochromatin** and **euchromatin**. The euchromatin is the less dense of the two and contains those genes that are expressed often by the cell. The structure of the euchromatic resembles that of a set of beads on a string (see Fig. 2.7 and Fig. 2.8). The heterochromatin is the more compact form and contains genes that are transcribed only infrequently, as well as chromosome constitutive elements such as **telomeres** (repetitive DNA that appears at the end of the chromosomes protecting them from destruction) and **centromeres** (the central region of a chromosome, where the arms of the chromosome are joined together).

2.2   The building blocks of genomic information

2.2.1   The deoxyribonucleic acid (DNA)

DNA is most commonly recognized as two paired chains of chemical bases, spiraled into what is commonly known as the double helix. DNA is a large polymer with a linear backbone of alternating sugar and phosphate residues. The sugar in DNA molecules is a 5 carbon sugar (deoxyribose); successive sugar residues are linked by strong (covalent) phosphodiester bonds. A nitrogenous base is covalently attached to carbon atom number $1'$ (one prime) of each sugar residue. There are four different kinds of bases in DNA, and this why it simple to understand its basic function and structure. The order in which the bases occur determines the information stored in the region of DNA being looked at.

The four types of bases in DNA are adenine (A), cytosine (C), guanine (G), and thymine (T) each consisting of heterocyclic rings of carbon and nitrogenous atoms. The bases are divided into two classes: purines (A and G) and pyrimidines (C and T). When a base is attached to a sugar, we speak of a nucleoside. If a phosphate group is attached to this nucleoside, then it becomes a nucleotide. The nucleotide is the basic repeat unit of a DNA strand.

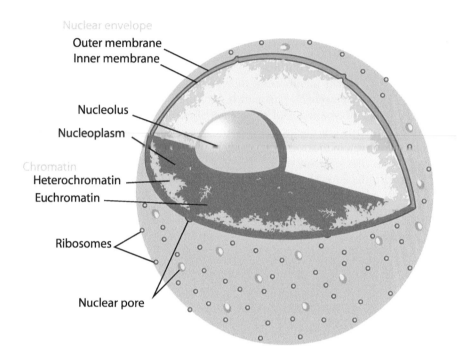

FIGURE 2.6: The nucleus is the largest organelle in a cell. The nucleus has a central part called nucleolus, which is mainly involved in the assembly or ribosomes, and two types of chromatin: heterochromatin and euchromatin. The nucleus is surrounded by a nuclear membrane that separates it from the rest of the cytoplasm. This membrane is studded with pores and ribosomes. The pores allow certain small, water-soluble molecules to penetrate the nuclear membrane in very specific conditions. Author: Mariana Ruiz, released in the public domain.

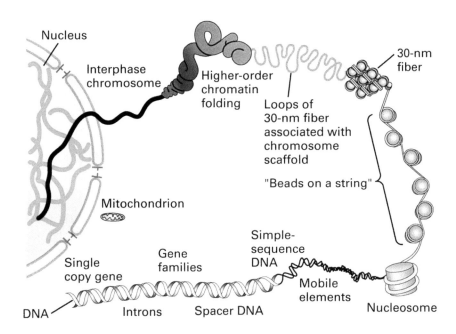

FIGURE 2.7: The DNA material in the nucleus is tightly packed in a complex way. The very long double-stranded DNA that contains the genes is sometimes compared with a string. From place to place along this string, there are cylindrical structures called **histones**. The scale is such that if the DNA is compared with a string, the histones could be compared with some beads on this string, only that instead of the string going through each bead, the string is wrapped around each bead. The double-stranded DNA wrapped around a histone forms a **nucleosome**. The "beads on a string" structure can be further folded and packed even tighter in loops of DNA fiber that are further folded and compacted to form the chromatin. In order to be transcribed, the DNA encoded for a gene needs to be accessible so the location of a gene in relationship with the histones and other structures may be important for the gene expression process. Figure from *Molecular Cell Biology*, 5th Edition (2004), Lodish, H., et al., printed with permission.

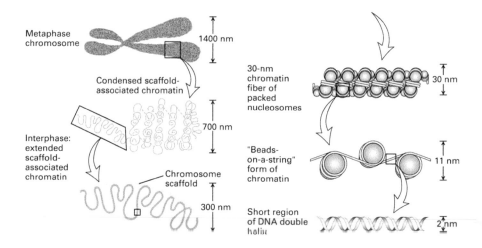

FIGURE 2.8: The scales of various structures used in chromatin packing. Figure from *Molecular Cell Biology*, 5th Edition (2004), Lodish, H., et al., printed with permission.

The formation of the double helix is due to the hydrogen bonding that occurs between laterally opposed bases. Two bases form a **base pair** (bp). The chemical structure of the bases is such that adenine (A) specifically binds to thymine (T) and cytosine (C) specifically binds to guanine (G). These are the so called Watson-Crick rules. Since no other interactions are possible between any other combination of base pairs, it is said that A is complementary to T and C is complementary to G.[2] Two strands are called complementary if, for any base on one strand, the other strand contains this base's complement. Two complementary single-stranded DNA chains that come into close proximity react to form a stable double helix (see Fig. 2.9) in a process known as **hybridization** or **annealing**. Conversely, a double-stranded DNA can be split into two complementary, single-stranded chains in a process called **denaturation** or **melting**. Hybridization and denaturation play an extremely important role both in the natural processes that happen in the living cells and in the laboratory techniques used in genomics. Because of the base complementarity, the base composition of a double-stranded DNA is not random. The amount of A equals the amount of T, and the amount of C is the same as the amount of G.

Let us look at the backbone of the DNA strand again. Phosphodiester

---

[2]In fact, other interactions are possible but the A-T and C-G are the ones that occur normally in the hybridization of two strands of DNA. This is because the A-T and C-G pairings are the ones that introduce a minimal distortion to the geometrical orientation of the backbones.

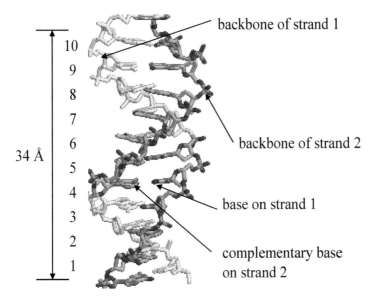

FIGURE 2.9: A short fragment (10 base pairs) of double-stranded DNA. Image obtained with Protein Explorer.

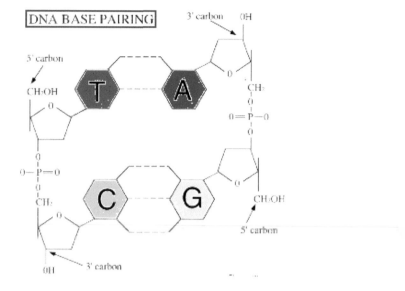

FIGURE 2.10: Each end of a single strand of DNA is identified by the carbon atom that terminates the strand (5' or 3'). Two strands of DNA always associate or anneal in such a way that the 5' → 3' direction of one DNA strand is the opposite to that of its partner.

bonds link carbon atoms number 3' and 5' of successive sugar residues. This means that in the terminal sugar the 5' is not linked to a neighboring sugar residue. The other end is termed 3' end, and it is characterized by the lack of a phosphodiester bond on that particular carbon atom. This gives a unique direction to any DNA strand. By convention, the DNA strand is said to run from the 5' end to the 3' end. The two strands of a DNA duplex are considered to be antiparallel. They always associate or anneal in such a way that the 5' → 3' direction of one DNA strand is the opposite to that of its partner (see Fig. 2.10).

The two uprights of the DNA ladder are a structural backbone, supporting the rungs of the ladder. These are also the information-carrying parts of the DNA molecule. Each rung of the ladder is made up of two bases that are paired together. This is what makes the steps of the spiral staircase. The two-paired bases are called a base pair as described earlier. The length of any DNA fragment is measured in base pairs (bp), similarly to how we measure length in inches. However, since the DNA is formed with base pairs, the length of a DNA fragment can only be a discrete number of such pairs, unlike a length which can include fractions of an inch.

Each nucleotide has a discrete identity. The sequence of the nucleotides in a DNA can be read by the "machinery" inside the cell. Genes, which represent large sequences of DNA, can be looked at as instructions telling the cell how much protein to make, when it should be made, and the sequence that can be used to make it. The information in the DNA is like a library. In the library, you will find books, and they can be read and reread many times, but they are never used up or given away. They are retained for further use. Similarly, the information in each gene is read (see below), perhaps millions of times in the life of an organism, but the DNA itself is never used up.

2.2.2 The DNA as a language

Each base can be thought of as a specific letter of a 4-letter alphabet, combining to form words and sentences. In essence, a gene is a recipe for making a protein. Let us consider for example the following very simple recipe for making an omelette:

Take three eggs; scramble; add oil in a pan; heat it up; add the eggs; get the omelette.

From a syntactic perspective, this is a simple string of characters from a set that includes the 52 alphabetic characters in the English language (lower and upper case) plus some special characters (space, semicolon, period, etc.). Usual grammatical conventions tells us to start a phrase using a capital letter and end it using a period. If we were to be minimalistic and restrict ourselves to the 26 lower-case characters, we can give up the spaces and be explicit about the initial capital letter and the final period. In this case, we could write the recipe as:

capitaltakethreeeggsscraambleaddoilinapanheatitupaddtheeggsgettheomeletteper iod

In the coding above, the "capital" and "period" markers are there just to indicate to us when a recipe starts and ends, in case we want to put together a collection including many such recipes. Similarly, since one chromosome contains many genes in a unique, very long DNA sequence, the beginning and end of a gene are indicated by special markers called **start** and **stop codons**.

Furthermore, in a digital computer system, the recipe above would be stored in a binary format, usually using 8 bits (or one byte) for every alphanumeric and special character. The recipe above would now look something like this:

01100011	01100001	01110000	01101001	01110100	01100001	...
c	a	p	i	t	a	...

From this, it follows that any time such a recipe is accessed in the memory of a computer, a translation has to take place from the binary alphabet {0,1} used by the computer to the English alphabet used by humans. Similarly, since a protein is a sequence of amino acids, and its recipe is stored as a sequence of

DNA bases, every time a protein is produced a translation has to take place that would map the recipe written in the 4-letter alphabet of the DNA, to the necessary protein sequence that uses amino acids from a 20 letter alphabet.

The DNA bases are grouped in triplets, or **codons**, for the same reason bits are grouped in octets, or bytes: if each symbol is limited to only two values (in the binary alphabet), or four values (in the DNA alphabet), groups of several symbols are needed in order to represent symbols from larger alphabets such as the set of 20 amino acids (for proteins) or the set of 26 letters of the English alphabet (for text). In fact, soon after the discovery of the DNA, the existence of a three-letter code to map from the DNA alphabet to the amino acid alphabet was postulated by George Gamov based on the fact that n=3 is the smallest value of n that satisfies $4^n > 20$. In other words, n=3 is the smallest size of a tuple for which there are more tuples than the 20 amino acids that needed to be coded for.

Each triplet of DNA nucleotides, or each codon, corresponds to a certain amino acid. Fig. 2.11 shows the correspondence between all possible codons and their respective amino acids, as well as some structural information, chemical properties and post-translational modifications of the various amino acids. The mapping from codons to amino acids is known as the genetic code. There is a start codon that indicates where the translation should start and several end codons that indicate the end of a coding sequence. Note that since there are $4^3 = 64$ different codons and only 20 different amino acids, it follows that either several different codons have to code for the same amino acid or many codons have to code for no amino acids at all. In fact, the former is true, with most amino acids being coded for by more than one codon. This may be more easily visible in Fig. 2.12. For instance, if a codon has the first two nucleotides C and T^3, the codon will be translated into leucine independently of the third nucleotide. Similarly, the CC* codon will be translated into proline, etc.

It turns out that there is another level of complexity about the genetic code. At the same time with the protein coding information, the genome has to also carry other types of signals such as regulatory signals telling the cell when to start and stop protein production for each protein, signals for splicing, etc. It has been shown recently that when these additional requirements are taken into consideration, the universal genetic code that we know is nearly optimal with respect to all other possible codes [15]. The optimality here was defined as the ability to minimize the effects of the most disastrous type of errors, the frame-shifts, as well as the property that close codons (codons that differ by only one letter) are mapped to either the same amino-acid or to chemically related ones. This means that, if a translation process misreads a single letter, the error introduced will have no or little consequences. In the same context, it has also been noted that amino acids with a simple chemical structure tend to have more codons assigned to them [177].

---

[3] For reasons that will be explained soon, the CT tuple appears as CU in Fig. 2.11.

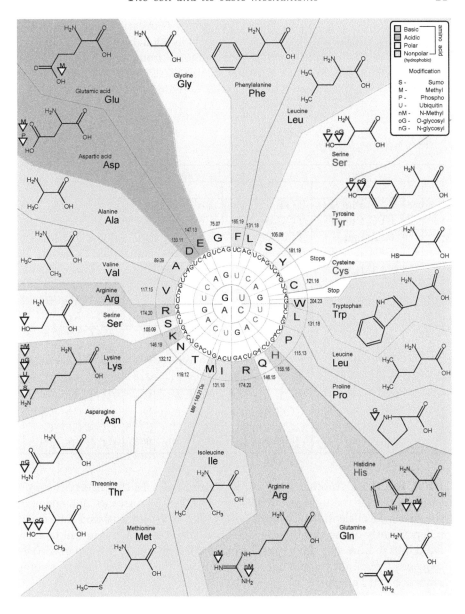

FIGURE 2.11: The genetic code is the mapping between the 4 letter-alphabet of the DNA/RNA nucleotides that appear in the genes, and the 20 letter-alphabet of the amino acids that form the proteins coded for by the genes. In this image, one starts in the middle with the first nucleotide of a codon and goes outwards following the remaining nucleotides. The figure also shows some structural information, chemical properties, as well as various possible post-translational modifications. Original image in public domain by Kosi Gramatikoff courtesy of Abgent; modified by Seth Miller and the author.

Second Position					
		T	C	A	G	
First Position	T	TTT Phe [F]	TCT Ser [S]	TAT Tyr [Y]	TGT Cys [C]	T
		TTC Phe [F]	TCC Ser [S]	TAC Tyr [Y]	TGC Cys [C]	C
		TTA Leu [L]	TCA Ser [S]	TAA *Ter* [end]	TGA *Ter* [end]	A
		TTG Leu [L]	TCG Ser [S]	TAG *Ter* [end]	TGG Trp [W]	G
	C	CTT Leu [L]	CCT Pro [P]	CAT His [H]	CGT Arg [R]	T
		CTC Leu [L]	CCC Pro [P]	CAC His [H]	CGC Arg [R]	C
		CTA Leu [L]	CCA Pro [P]	CAA Gln [Q]	CGA Arg [R]	A
		CTG Leu [L]	CCG Pro [P]	CAG Gln [Q]	CGG Arg [R]	G
	A	ATT Ile [I]	ACT Thr [T]	AAT Asn [M]	AGT Ser [S]	T
		ATC Ile [I]	ACC Thr [T]	AAC Asn [N]	AGC Ser [S]	C
		ATA Ile [I]	ACA Thr [T]	AAA Lys [K]	AGA Arg [R]	A
		ATG Met [M]	ACG Thr [T]	AAG Lys [K]	AGG Arg [R]	G
	G	GTT Val [V]	GCT Ala [A]	GAT Asp [D]	GGT Gly [G]	T
		GTC Val [V]	GCC Ala [A]	GAC Asp [D]	GGC Gly [G]	C
		GTA Val [V]	GCA Ala [A]	GAA Glu [E]	GGA Gly [G]	A
		GTG Val [V]	GCG Alo [A]	GAG Glu [E]	GGG Gly [G]	G

FIGURE 2.12: Another view of the genetic code. In this figure, it may be easier to see the redundancy intrinsic to the code. For instance, if a codon has the first two nucleotides C and T, the codon will be translated into leucine independently of the third nucleotide. Similarly, the CC* codon will be translated into proline, etc. There are 3 stop codons that mark the end of a coding sequence ("*end*" above) and one start codon marking the begining of a coding sequence ("*M*" above).

2.2.3 Errors in the DNA language

The DNA sequence of a chromosome can also contain errors. There are several types of errors as follows:

1. **Mutations**. A mutation is an situation in which one nucleotide is substituted by another one. For instance, in our original recipe:

   ...takeeggsscrambleaddoil...

   a mutation could substitute the second "g" with an "o" to yield:

   ...takeegosscrambleaddoil...

   leading to a rather different result:

   ...take egos scramble add oil...

   Mutations as above are also known as single point mutations or single nucleotide mutations. Since the genetic code is such that different codons can correspond to the same amino acid, a single point mutation in the DNA sequence of a gene may not change at all the structure of the protein coded for by the given protein. However, it is also possible that a single point mutation is extremely disruptive or even lethal for a given organism.

2. **Deletions**. A deletion is a situation in which one of the nucleotides is missing. For instance, in our original recipe:

   ...takeeggsscrambleaddoilinapanheatitupaddegsandgetomellette ...

   a deletion could remove the highlighted "h" to yield:

   ...takeeggsscrambleaddoilinapaneatitupaddegsandgetomellette...

   leading again to a departure from the intended outcome, in this case, an early consumption of the rather raw dish:

   ...take eggs, scramble, add oil in a pan, eat it up...

3. **Insertions**. An insertion is a situation in which an additional nucleotide is inserted in the middle of an existing sequence. For instance, in our original recipe:

   ...takeeggsscrambleaddoil...

   an insertion could add the highlighted "s" to yield:

   ...takeeggsscrambleaddsoil...

   leading to the consumption of something rather different from the intended omelette:

   ...take eggs scramble add soil...

4. **Frame shifts**. A frame shift is a situation in which the DNA sequence is shifted. Such a shift can be caused for instance by an insertion or

deletion. Because of this shift, the identity of all subsequent codons will be changed. For instance, in our original binary-coded recipe above:

01100011	01100001	01110000	01101001	01110100	01100001	...
c	a	p	i	t	a	...

a frame shift can cause the reading to start at the second digit instead of the first one, which will mean that the same bits will be grouped, and hence interpreted, very differently:

...0	11000110	11000010	11100000	11010010	11101000	11000010
	ä	-	Ó	Ê	ц	—

In this example, a frame shift mutation that moved the start point by just one position to the right, caused changes so substantial in the subsequent bytes that all the characters became special characters and the text fragment itself became completely unintelligible. Similarly, most frame shift mutations in a DNA sequence will cause very substantial changes in the corresponding amino acid sequence. In fact, this is how the organization of the protein-coding DNA in triplets was originally proven. In a 1961 experiment, Crick, Brenner et. al. [105] showed that either inserting or deleting only one or two nucleotides prevents the production of a functional protein while inserting or deleting 3 nucleotides at a time still allows the functional protein to be produced.

Together, DNA works with protein and RNA (ribonucleic acid) in a manner that is similar to the way a computer works with programs and data. DNA has a code that is continuously active, having instructions and commands such as "if-then," "go to," and "stop" statements. It is involved with regulating protein levels, maintaining quality control, and providing a database for making proteins. The best analogy to use when describing DNA and its function in cells is to look at a cell as a little factory. In a cell, much like in a factory, specific items are produced in specific places. There is a certain flow of materials through the cell, and there are various communication and feedback mechanisms regulating the speed with which various processes happen in various parts of the cell. Such communication and feedback mechanisms also allow the cell to adapt the speed of its internal processes to the demands of the environment much like a factory can adjust its production in response to the changing needs of the market.

2.2.4 Other useful concepts

A **gene** is a segment or region of DNA that encodes specific instructions, that allow a cell to produce a specific product. This product is typically a protein, such as an enzyme. There are many different types of proteins. Proteins are used to support the cell structure, to break down chemicals, to build new chemicals, to transport items, and to regulate production. Every human being has about 22,000 putative genes[4] that produce proteins. Many of these genes are totally identical from person to person, but others show variation in different people. The genes determine hair color, eye color, sex, personality, and many other traits that in combination make everyone a unique entity. Some genes are involved in our growth and development. These genes act as tiny switches that direct the specific sequence of events that are necessary to create a human being. They affect every part of our physical and biochemical systems, acting in a cascade of events, turning on and off the expression, or production, of key proteins that are involved in the different steps of development.

The key term in growth and development is **differentiation**. Differentiation involves the act of a cell changing from one type of cell, when dividing through mitosis, into two different types of cells. The most common cells used to study differentiation are the stem cells. These are considered to be the "mother cells," and are thought to be capable of differentiating into any type of cell. It is important to understand the value of differentiation when learning about genetics and organism development. Everyone starts life as one single cell, which divides several times before any differentiation takes place. Then, in a specific stage, differentiation begins to take place, and internal organ cells, skin cells, muscle cells, blood cells, etc. are created.

Another very important process is the DNA **replication**. The DNA replication is the process of copying a double-stranded DNA molecule to create two identical double-stranded molecules. The original DNA strand is called the template DNA. Each strand of the double-stranded template DNA becomes half of a new DNA double helix. Because of this, the DNA replication is said to be **semi-conservative**. This process, illustrated in Fig. 2.13, must take place before a cell division can occur.

The replication process starts at some fixed locations on the DNA strand called **replication origins**. These replication origins have certain features such as repeats of a short sequence that create a pattern that is recognized by an **initiation protein**. Once this protein binds to the double strand, the template DNA is unwound and separated into the two component single strands (see the topoisomerase and the helicase in Fig. 2.13). This region is called the **replication fork** and travels along the DNA sequence as the replication takes place. The single-stranded regions are trapped by **single-stranded binding proteins** making them accessible to the DNA polymerase. The DNA polymerase is an enzyme that can add single nucleotides to the $3'$ end of an existing single strand by matching the nucleotides in the template. However, the DNA polymerase needs an existing sequence with a free $3'$ end before it can do its job. The short nucleotide sequence that provides the starting point for the

---

[4]Initially, the number of human genes was estimated at 100,000 to 140,000. Subsequently, the estimate was revised down to about 20,000 to 30,000. At the moment of this writing, there still exists a controversy over this number [340].

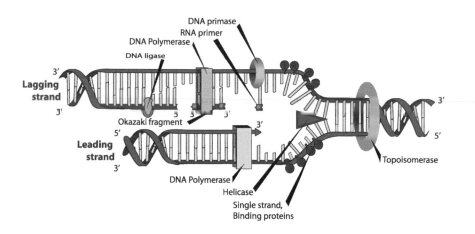

FIGURE 2.13: The main processes involved in the DNA replication. The topoisomerase is an enzyme that facilitates the unwinding of the double-stranded DNA, allowing the helicase enzyme to separate the two DNA strands. These single strands are kept apart by certain single-strand binding proteins allowing the DNA polymerase and the other enzymes to have access to the single strands. The construction of the new strands starts from primers built by a primase enzyme. On the lower strand, called the **leading strand**, the newly constructed strand is continuously elongated by the DNA polymerase that adds new nucleotides to the free 3′ end of the new strand (left to right in this figure), replicating the template. On the upper strand, called the **lagging strand**, the new strand is elongated intermittently by adding new Okazaki fragments which are continuously constructed by The gaps are repaired by a DNA ligase enzyme. Image in public domain by Mariana Ruiz, modified by the author.

DNA polymerase is called a **primer**. During the cell replication, the primer is constructed *ex novo* by an enzyme called **primase**. The primase actually synthesizes an RNA primer that will be later removed by an enzyme called **RNase H**, allowing the polymerase to add the correct DNA nucleotides instead.

As stated above, the DNA polymerase can only add nucleotides to the $3'$ end of a strand. Hence, it follows that a DNA strand can only be constructed in one direction, from the $3'$ to the $5'$ end. However, an examination of Fig. 2.13 shows that apparently both new strands grow at the same time. The new strand being constructed in the lower part of the figure grows from the $5'$ end to the $3'$ end with new nucleotides being added, as expected to the free $3'$ end of the newly formed strand. However, the strand being constructed in the upper part of the figure appears to grow from the $3'$ end by extending its $5'$ end which we know the polymerase cannot do. In fact, this strand does grow by extending its $5'$ end, but the job is not done by the polymerase. The polymerase only constructs short fragments of DNA, still going right to left, by extending the free $3'$ end. These short fragments are called **Okazaki fragments** after the Japanese scientists who discovered them in 1968. These short fragments attach themselves to the template strand as dictated by the complementarity rules. Finally, the **DNA ligase**, an enzyme that can suture or ligate, gaps in double-stranded DNA comes and adds the missing link in the newly constructed strand. In essence, the lower strand in Fig. 2.13 is being elongated continuously, seamlessly by the DNA polymerase, while the upper strand is elongated by concatenating short Okazaki fragments.

Another important concept related to genes is that of a **single nucleotide polymorphism** or SNP (pronounced "snip"). A SNP is a single nucleotide difference between different individuals of the same species, or between paired chromosomes of an individual. For instance, if the DNA sequence of most individuals of a species at a certain location reads ACGTTACG while another specific individual has the sequence ACGT**A**ACG at the same location, it is said that this individual has a SNP at that particular location. In essence, a SNP is a mutation that is more widespread in the population (and perhaps not very harmful). The two alternatives associated to a SNP location are called **alleles**. Sometimes a combination of alleles at multiple loci are transmitted together from one generation to another. Such a combination of alleles is called a **haplotype**, a term obtained from contracting words "haploid genotye." The set of all alleles of an individual, in other words the entire set of differences characteristic to this individual, is called the **genotype** of that individual. The term genotype is also used to refer to a subset of genetic traits, sometimes a single such trait. In contrast, the **phenotype** is one (or more) observable characteristic(s) of an organism. In general, the phenotype is determined by the genotype as well as environmental and developmental conditions. However, there are certain phenotypes that are determined directly by the genotype, sometime by a single gene.

Organisms such as *Homo sapiens*, which are **diploid**, that is their genomes contain two copies of each chromosome (with the exception of the sex chromosomes), can also have SNPs between the two copies of a given chromosome. These are also referred to as alleles. Most SNPs have only two alleles. The allele that is less frequent is called the **minor allele**.

If a phenotype is determined by a single gene and this gene has, let us say, two alleles, sometime the presence of one of these in any one of the two copies of the chromosome determines the phenotype independently of the allele present on the other copy of the chromosome. In this case, this allele is called the **dominant** allele. By convention, dominant alleles are written in uppercase letters, and recessive alleles in lowercase letters. For instance, A is dominant to a (and a is **recessive** to A), which means that two individuals with the AA and Aa genotypes have the same phenotype, while a third individual with the aa genotype has a different phenotype. There are also other, more complex mechanisms of inheritance.

2.3 Expression of genetic information

Genes make up only a subset of the entire amount of DNA in a cell. The human genome, for instance, contains approximatively 3.1 billion base pairs. However, less that 2% of the genome codes for proteins. Active research in the past 20 years has identified signals (specific sequences of bases) that delimit the beginning and ending of genes. These signaling areas can be thought of as regulatory elements. They are used to control the production of protein, and work as a biofeedback system. They are usually located near the beginning of a sequence that is used to code for a protein. Through protein-DNA interactions they are used to turn on and off the production of proteins. Noncoding DNA conveys no known protein-coding or regulatory information and makes up the bulk of the DNA in our cells. These intergenic regions can include highly repetitive sequences. Although this DNA is sometimes called "junk DNA," several functions have been proposed for it, including playing a role in reshaping and rearranging the genes in the genome, acting as a buffer to decrease the damage introduced by random mutations, or acting as a spacer such that genes that are transcribed often are accessible to the DNA polymerase and the other enzymes involved in transcription after all the twists, turns, and folding of the double-stranded DNA into the chromatin structure.

The flow of genetic information is from DNA to RNA to proteins. This one-way process is the expression of genetic information in all cells and has been described as the **central dogma** of molecular biology.

To make products from a gene, the information in the DNA is first copied, base for base, into a similar kind of information carrier, called a **transcript**, or **messenger RNA** (mRNA). The RNA copy of the gene sequence acts as a messenger, taking information from the nucleus (where the DNA is found in its chromosomal form) and transporting it into the cytoplasm of the cell (where the machinery for making gene products is found). Once in the cytoplasm, the messenger RNA is translated into the product of the gene, a protein. The sequence of the protein is defined by the original sequence of the DNA bases found in the gene.

DNA is the hereditary material in all present-day cells. However, in early evolution, it is likely that RNA served this role. As a testimony for this, there still exist organisms, such as RNA viruses, that use RNA instead of DNA to carry the hereditary information from one generation to another. Retroviruses such as the Human Immunodefficiency Virus (HIV) are a subclass of RNA viruses, in which the RNA replicates via a DNA intermediate, using the enzyme **reverse transcriptase** (RT). This enzyme is an RNA-dependent DNA polymerase that, simply said, makes DNA from RNA. This enzyme plays a very important role in microarray technology as will be discussed later.

Let us re-examine the several processes that are fundamental for life and important to understand the need for microarray technology and its emergence. Every cell of an individual organism contains the same DNA, carrying the same information.[5] In fact, this is the very basis of the use of DNA as evidence in criminal cases: a single droplet of bodily fluid, a single hair, or a few cells can uniquely identify an individual. However, in spite of carrying the very same DNA, a liver cell is obviously different from a muscle cell, for example. These differences occur because not all genes are expressed in the same way in all cells. The differentiation between cells is given by different patterns of gene activations, which in turn, control the production of proteins. Much as studying the different levels of expression of various genes in different tissues can help us understand why the tissues are different, studying the different levels of expression between the same tissue in different conditions can help us understand the differences between those conditions.

Proteins are long, linear molecules that have a crucial role in all life processes. Proteins are chains of amino acid molecules. As previously discussed, there are 20 amino acid molecules that can be combined to build proteins. The number of all possible sequences of amino acids is staggering. For instance, a sequence with length 10 can contain $20^{10} = 10,240$ billion different combinations of amino acids, which is a very large number indeed. Although proteins are linear molecules, they are folded in complex ways. The protein-folding process is a crucial step since a protein has to be folded into a very specific

---

[5]Like many other things in life sciences, this is true *most of the time*, rather than always. There could be, in fact, individuals of various species, including humans, that can be composed of two or more populations of genetically distinct cells, that originated from different zygotes. Such an individual is called chimera, after a mythological creature that had a body composed of parts of different animals. Chimeras are extremely rare.

way for it to function properly.[6] Enzymes are specialized proteins[7] that act as catalysts and control the internal chemistry of the cells. This is usually done by binding to specific molecules in a very precise way such as certain atoms in the molecules can form bonds. Sometimes, the molecules are distorted in order to make them react more easily. Some specialized enzymes process DNA by cutting long chains into pieces or by assembling pieces into longer chains. A gene is active, or expressed, if the cell makes the gene product (e.g. protein) encoded by the gene. If a lot of gene product is produced, the gene is said to be highly expressed. If no gene product is produced, the gene is not expressed (unexpressed).

2.3.1   Transcription

The process of using the information encoded into a gene to produce a protein involves reading the DNA sequence of the gene. The first part of this process is called **transcription** and is performed by a specialized enzyme called **RNA polymerase**. Essentially, the transcription process converts the information coded into the DNA sequence of the gene into an RNA sequence. This "expression" of the gene will be determined by various internal or external factors. The objective of researchers is to detect and quantify gene expression levels under particular circumstances.

The RNA molecule is a long polynucleotide very similar to DNA, but with several important differences. First, the backbone structure of RNA is not the same. Second, RNA uses the base uracil (U) instead of thymine (T).[8] And finally, RNA molecules in cells exist as single stranded entities, in contrast to the double helix structure of DNA.

The transcription process is somewhat similar to the DNA replication. Much like in the DNA replication, the part of the DNA sequence that contains the gene to be transcribed has to be unfolded and accessible, the two DNA strands are temporarily separated and the RNA polymerase moves along one strand, reading its succession of bases and constructing an RNA sequence containing the same information. The enzyme RNA polymerase attaches itself to a specific DNA nucleotide sequence situated just before the beginning of a gene. This special sequence is called a **promoter** and it works by setting up the RNA polymerase on the correct DNA strand and pointing in the right direction. The two DNA strands are separated locally such that the RNA polymerase can do its work and transcribe DNA into RNA. The RNA molecule

---

[6]The so-called "mad cow disease" is apparently caused by a brain protein folded in an unusual way. When a wrongly folded protein comes into contact with a normal protein, it induces the normal protein to fold itself abnormally. This abnormal folding prevents the protein from performing its usual role in the brain, which leads to a deterioration of brain functions and, eventually, death.

[7]Most but not all enzymes are proteins. Some RNA molecules called ribozymes act like enzymes.

[8]This is why the CT tuple appears as CU in Fig. 2.11 which shows the mapping from RNA to codons, rather than DNA to codons.

is synthesized as a single strand, with the direction of transcription being $5' \rightarrow 3'$. The RNA polymerase starts constructing RNA using ribonucleotides freely available in the cell. The RNA sequence constructed will be complementary to the DNA sequence read by the RNA polymerase. When the DNA sequence contains a G for instance, the polymerase will match it with a C in the newly synthesized RNA molecule; likewise, an A will be matched with a U in the chain under construction (because U substitutes for T in the RNA molecule), etc. The process will continue with the RNA polymerase moving into the gene, reading its sequence, and constructing a complementary RNA chain until the end of the gene is reached. The end of the gene is marked by a special sequence that signals the polymerase to stop. When this sequence is encountered, the polymerase ends the synthesis of the RNA chain and detaches itself from the DNA sequence.

The RNA sequence thus constructed contains the same information as the gene. This information will be used to construct the protein coded for by the gene. However, the structure of the protein is not yet completely determined by the RNA sequence synthesized directly from the DNA sequence of the gene. The RNA chain synthesized by the RNA polymerase is called a primary transcript or pre-mRNA and is only the initial transcription product. In fact, the sequence of nucleotides in a gene may not be used in its entirety to code for the gene product. Thus, for more complex organisms, a great part of the initial RNA sequence is disposed of during a **splicing** process to yield a smaller RNA molecule called **messenger RNA** or **mRNA**. Its main role is to carry this information to some cellular structures outside the nucleus called ribosomes where proteins will be synthesized. The non-coding stretches of sequence that are eliminated from the primary transcript to form a mature mRNA are called **introns**. Conversely, the regions that will be used to build the gene product are called coding regions, or **exons**. Thus, RNA molecules transcribed from genes containing introns are longer than the mRNA that will carry the code for the construction of the protein.

The mechanism that cuts the transcribed RNA into pieces, eliminates the introns, and reassembles the exons together into mRNA is called **RNA splicing**. The RNA splicing takes place in certain places determined by a specific DNA sequence that characterizes the intron/exon boundaries. Fig. 2.14 shows the consensus sequence for the intron/exon boundaries. In general, the splicing is carried out by small, nuclear RNA particles (**snRNPs**, pronounced snurps) that get together with some proteins to form a complex called **spliceosome**. During the splicing, at each splicing site, the spliceosome bends the intron to be eliminated in the shape of a loop called lariat, bringing together the two exon ends to be connected. In a subsequent step, the two exon ends are connected, the lariat is cut off, and the spliceosome detaches from the mRNA.

Depending on the circumstances, the pre-mRNA can be cut into different pieces, and these pieces can be assembled in different ways to created different proteins. This mechanism that allows the construction of different mRNAs from the same DNA sequence is called **alternative splicing**. The mechanism

5′		3′
— AG GURAGU ————	YNYYRAY ————	$(Y)_{12}$NYAG R
splice donor	lariat site	splice acceptor

FIGURE 2.14: The consensus sequence of a splicing site. The symbol R denotes any puRine (A or G); the symbol Y denotes any pYrimidine (C or U); the symbol N stands for aNy of A, C, G, or U. The lines stand for arbitrary sequences. The blue color represents the ends of the exons. As shown in the figure, there is a lot of variability in these sites with the exception of the bases in red which are required in order for the splicing to occur. During the splicing, the spliceosome bends the intron to be eliminated in the shape of a loop called *lariat*, bringing together the two exon ends to be connected. In a subsequent step, the two exon ends are connected, the lariat is cut off and the spliceosome detaches from the mRNA.

of alternative splicing greatly increases the protein coding abilities of genes by allowing a gene to code for more than one protein. Fig. 2.15 shows a number of alternative **splice variants** encoded by a single gene. The pre-mRNA shown at the top of Fig. 2.15 includes a number of introns and exons. An mRNA is obtained in each case by eliminating some introns and concatenating the remaining exons. Which particular mRNA is constructed at any one time depends on the circumstances and is controlled through a number of mechanisms. Depending on their effect, these mechanisms are divided into enhancers and silencers, and subdivided by their target into exon-splicing enhancers and silencers, or intron-splicing enhancers and silencers.

Another important reaction that occurs at this stage is called **polyadenylation**. This reaction produces a long sequence of A nucleotides concatenated onto the 3′ end of a mature mRNA. This reaction is of interest because some protocols in microarray technology use the final product of polyadenylation. Transcription of the RNA is known to stop after the enzymes (and some specialized small nuclear RNAs) responsible for the transcription process recognize a specific termination site. Cleavage of the RNA molecule occurs at a site with sequence AAUAAA and then about 200 adenylate (i.e., AMP) residues are sequentially added in mammalian cells by the enzyme poly(A) polymerase to form a poly(A) tail. This tail is used as a target in the process of reverse transcription.

2.3.2 Translation

After the post-transcriptional processing, the mRNA transcribed from the genes in the nuclear DNA leaves the nucleus and moves into the cytoplasm.

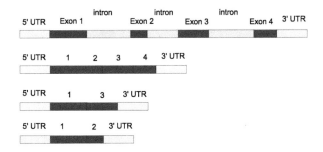

FIGURE 2.15: Alternative splice variants encoded by a single gene. The pre-mRNA shown at the top of the figure includes a number of introns and exons. The mRNA is obtained in each case by eliminating some introns and concatenating the remaining exons. In general, the splicing is carried out by small, nuclear RNA particles (snRNPs, pronounced snurps) that get together with some proteins to form a complex called spliceosome. UTRs represent untranslated regions. The mechanism of alternative splicing greatly increases the protein-coding abilities of genes by allowing a gene to code for more than one protein.

The mRNA containing the sequence coding for the protein attaches to ribosomes (see Section 2.1). Here, the information contained in the mRNA is mapped from a sequence of RNA nucleotides into a sequence of amino acids forming the protein. This process is called **translation**. As a mnemonic help, this process translates the information necessary in order to construct a protein from the 4 base alphabet of the DNA/RNA to the 20 letter alphabet of the amino acids.

The ribosome attaches to the messenger RNA near a specific start codon that signals the beginning of the coding sequence. The various amino acids that form the protein are brought to the ribosome by molecules of RNA that are specific to each type of amino acid (see Fig. 2.16). This RNA is called **transfer RNA (tRNA)**. The tRNA molecules recognize complementary-specific codons on the mRNA and attach to the ribosome. The first tRNA to be used will have a sequence complementary to the sequence of the first codon of the mRNA. In turn, this first tRNA molecule will bring to the ribosome the first amino acid of the protein to be synthesized. Subsequently, a second tRNA molecule with a sequence complementary to the second codon on the mRNA will attach to the existing ribosome-mRNA-tRNA complex. The shape of the complex is such that the amino acids are brought into proximity and they bind to each other. Then, the first tRNA molecule is released and the first two amino acids linked to the second tRNA molecule are shifted on the ribosome bringing the third codon into position. The tRNA bringing the third amino acid can now attach to the third codon because of its complementary sequence and the process is repeated until the whole protein molecule is synthesized.

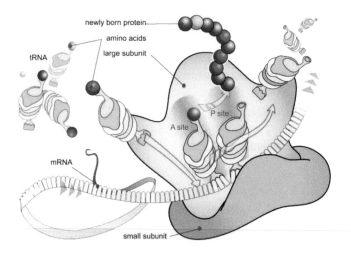

FIGURE 2.16: The protein translation process. The tRNA molecules recognize complementary-specific codons on the mRNA and attach to the ribosome. Each such tRNA molecule brings the amino acid corresponding to its respective codon, which is attached to the newly-formed polypeptide chain that will become the protein. Once the amino acid has been attached, its corresponding tRNA molecule is detached, and the process is repeated for the next amino acid, until the entire protein is assembled.

The process stops when a special stop codon is encountered, which signals the mRNA to fall off the ribosome together with the newly constructed protein.[9] After it is released, the protein may suffer a set of final changes called post-translational modifications. Such modifications might include cleavage, folding, phosphorylation, methylation, etc. Once these are done, the protein starts performing the cellular function for which it was designed. At this stage, it is said that the protein is active.

It must be mentioned that the process described above is greatly simplified. For instance, the complex between tRNA molecules and their corresponding amino acids is in turn controlled by another enzyme called aminoacyl-tRNA synthetase. There is at least one type of synthetase for each type of amino-acid. Since other complex steps, such as tRNA–amino acid reaction, are involved in the protein synthesis, it is clear that the amount of protein produced in the cell is also dependent on the successful completion of all these intermediate steps. Furthermore, as explained above, post-translational modifications can be crucial in making a protein active. Having abundant amounts of inactive protein will not help the cell perform the necessary functions. However, in

---

[9]See [121] for an excellent introduction to DNA and gene cloning for the nonspecialist and [406] for a more complete treatment of the subject.

general, there is a quantitative correspondence between the amount of mRNA produced by the enzyme reading the gene and the amount of protein produced. Therefore, the amount of mRNA produced from various genes is usually directly proportional to the amount of protein produced from that mRNA, i.e. to the expression level of that gene. This is the main assumption at the basis of most experiments that try to characterize the gene's expression levels using DNA microarrays. Nevertheless, one should keep in mind that the measured levels of mRNA do not always map to proportional levels of protein, and even if they did, not all those proteins may be in an active form, etc. A number of other techniques are available to obtain information at other levels of this complex process. For instance, proteomics techniques can provide information about the amounts of proteins available, phosphorylation assays can provide information about the amount of phosphorylated protein available, etc. A complete understanding of the cellular processes will inevitably require the integration of many heterogeneous types of data.

2.3.3 Gene regulation

The regulation of gene expression is the process that living cells use to control the amount of a gene product that is produced in the cell at any one time. As discussed above, most gene products are proteins. However, there are genes that produce RNA that is never translated into protein and yet they play some role in the cell. This process of controlling the amount of gene product is also called **gene modulation**. Gene regulation is continuously active for many genes in many cells. Because of gene regulation, a cell or organism can modify its response depending on environmental factors, signals from other cells or organisms, and even time of the day. For instance, *E. coli* is able to use various types of nutrients such as lactose and glucose. However, glucose is much more energy efficient so *E. coli* prefers to consume it if it's available. If glucose is not available but lactose is, *E. coli* regulates several of its genes to produce an enzyme, β-*galactosidase*, which is able to digest lactose. Similarly, the yeast can switch between a metabolism that uses oxygen to one that does not (beer versus bread).

Gene regulation can happen during any stage of the process that leads from a gene to a functional protein. During transcription, for instance, regulation can happen through one or more of the following mechanisms:

1. **Transcription factors** – these are proteins that bind DNA and control the production of RNA from DNA. These can be subdivided into:

    (a) **Repressors** – bind to the DNA strand nearby or overlapping the promoter region, preventing the RNA polymerase from transcribing the gene

    (b) **Activators** – bind to the DNA strand nearby of overlapping the promoter region facilitating the interaction between the RNA polymerase and a particular promoter, increasing the expression of the given gene

2. **Regulatory elements** – these are sites on the DNA helix that are involved in transcription control and regulation. These can be:

    (a) **Promoters** – - are the sites on the DNA helix that the activators bind to. Promoters are usually found close to the beginning of the gene

    (b) **Enhancers** – are also sites on the DNA helix that are bound by activators; an enhancer may be located upstream or downstream of the gene that it regulates. Furthermore, an enhancer does not need to be located near to the transcription initiation site to affect the transcription of a gene,

Gene regulation can also happen during RNA processing as well as after translation, through **post-translational modifications**. These post-translational modifications are chemical modifications that happen to the protein after it is translated. Such modifications usually involve the addition or removal of a functional group such as phosphate (phosphorylation), acetate (acetylation), lipids, carbohydrates, etc., or by making structural changes. Many proteins have an **active** form, in which they can perform their role in the cell, and an **inactive** form, in which they cannot perform their usual activity.

A more recently discovered mechanism of gene regulation involves **RNA interference** performed either through **micro-RNAs** (miRNA) [365] or **small interfering RNAs** (siRNA) [149]. The miRNAs are short (21–23 nucleotides), single-stranded, RNA molecules that are partially complementary to one or more mRNA sequences corresponding to other genes. Since the miRNA molecules will bind to their target mRNA molecules, fewer such molecules will be available for subsequent translation so the effect of the miRNAs will be to down-regulate their target genes. The siRNAs are short (20–25 base pairs), double-stranded RNA molecules that can interfere with the process of gene transcription-translation of other genes, either by degradation of the targeted RNA, or by histone and DNA methylation.

2.4 The need for high-throughput methods

Why bother measuring the expression of all genes? A simple answer involves the fact that the genomes of many model organisms have been sequenced, and we would like to simply have the luxury of looking at the whole genome expression profile under the influence of a particular factor. Several methods have long been available to measure expression levels but, alas, only for a few genes at a time. Large-scale screenings of gene expression signatures were not possible the way they are routinely performed nowadays with microarrays. Therefore, a need for a quick snapshot of all or a large set of genes was pressing. Another important reason for the emergence of microarrays is the necessity to understand the networks of biomolecular interactions at a global scale. Each particular type of cell (e.g., tissue) will be characterized by a different pattern of gene expression levels, i.e. each type of cell will produce a different set of proteins in very specific quantities. A typical method in genetics was to use some method to render a gene inactive (knock it out) and then study the effects of this knockout in other genes and processes in a given organism. This approach, which was for a long time the only approach available, is terribly slow, expensive, and inefficient for a large-scale screening of many genes. Microarrays allow the interrogation of thousands of genes at the same time. Being able to take a snapshot of a whole gene expression pattern in a given tissue opens innumerable possibilities. One can compare various tissues with each other, or a tumor with the healthy tissue surrounding it. One can also study the effects of drugs or stressors by monitoring the gene expression levels. Gene expression can be used to understand the phenomena related to aging or fetal development. Screening tests for various conditions can be designed if those conditions are characterized by specific gene expression patterns. Drug development, diagnosis, comparative genomics, functional genomics, and many other fields may benefit enormously from a tool that allows accurate and relatively inexpensive collection of gene expression information for thousands of genes at a time.[10]

2.5 Summary

Deciphering the genomes of several organisms, including that of humans, led to an avalanche of data that needed to be analyzed and translated into biological meaning. This sparked the emergence of a new scientific field called bioinformatics. This term is generally used to denote computer methods, statistical and data mining techniques, and mathematical algorithms that are used to solve biological problems. The field of bioinformatics combines the efforts of experts from various disciplines who need to communicate with each other and understand the basic terms in their corresponding disciplines. This chapter was written for computer engineers, statisticians, and mathematicians to help them refresh their biological background knowledge. The chapter de-

---

[10]This is not to be interpreted that microarrays will substitute gene knockouts. Knocking out a gene allows the study of the more complex effects of the gene, well beyond the mRNA abundance level. Microarrays are invaluable as screening tools able to simultaneously interrogate thousands of genes. However, once interesting genes have been located, gene knockouts are still invaluable tools for a focused research.

scribed the basic components of a cell, the structure of DNA and RNA, and the process of gene expression. There are 4 types of DNA building blocks called nucleotide bases: A, C, G, and T. These 4 bases form the genetic alphabet. The genetic information is encoded in strings of variable length formed with letters from this alphabet. Genetic information generally flows from DNA to RNA to proteins; this is known as the central dogma of molecular biology. Genetic information is stored in various very long strings of DNA. Various substrings of such a DNA molecule constitute functional units called genes and contain information necessary to construct proteins. The process of constructing proteins from the information encoded into genes is called gene expression. First, the information is mapped from DNA to RNA. RNA is another type of molecule used to carry genetic information. Similarly to DNA, there are 4 types of RNA building blocks and a one-to-one mapping from the 4 types of DNA bases to the 4 types of RNA bases. The process of converting the genetic information contained in a gene from the DNA alphabet to the RNA alphabet is known as transcription. The result of the transcription is an RNA molecule that has the informational content of a specific gene. In higher organisms, the transcription process takes place in the cell's nucleus, where DNA resides. The RNA molecules are subsequently exported out of the cell nucleus into the cytoplasm where the information is used to construct proteins. This process, known as translation, converts the message from the 4-letter RNA alphabet to the 20-letter alphabet of the amino acids used to build proteins. The amounts of protein generated from each gene determine both the morphology and the function of a given cell. Small changes in expression levels can determine major changes at the organism level and trigger illnesses such as cancer. Therefore, comparing the expression levels of various genes between different conditions is of extreme interest to life scientists. This need stimulated the development of high throughput techniques for monitoring gene expression such as microarrays.

Chapter 3

Microarrays

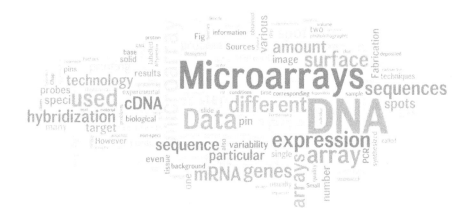

If at first you don't succeed, you are running about average.

—M. H. Alderson

3.1 Microarrays – tools for gene expression analysis

In its most general form, a DNA array is usually a substrate (nylon membrane, glass or plastic) on which one deposits single-stranded DNAs (ssDNA)with various sequences. Usually, the ssDNA is printed in localized features that are arranged in a regular grid-like pattern. In this book, we will conform with the nomenclature proposed by Duggan et al. [139], and we will refer to the ssDNA printed on the solid substrate as a **probe**.

What exactly is deposited depends on the technology used and on the purpose of the array. If the purpose is to understand the way a particular set of genes function, the surface will contain a number of regions dedicated to those individual genes. However, arbitrary strands of DNA may be attached to the surface for more general queries or DNA computation. The array thus fabricated is then used to answer a specific question regarding the DNA on its surface. Usually, this interrogation is done by washing the array with a solution containing ssDNA, called a **target**, that is generated from a particular

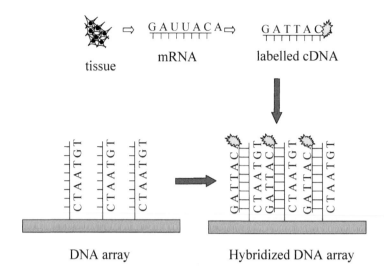

FIGURE 3.1: A general overview of the DNA array used in gene expression studies. The mRNA extracted from tissue is transformed into complementary DNA (cDNA), which is hybridized with the DNA previously spotted on the array.

biological sample under study as described below. The idea is that the DNA in the solution that contains sequences complementary to the sequences of the DNA deposited on the surface of the array will hybridize to those complementary sequences. The key to the interpretation of the microarray experiment is in the DNA material that is used to hybridize on the array. Since the target is labeled with a fluorescent dye, a radioactive element, or another method, the hybridization spot can be detected and quantified easily.

When used in gene expression studies, the DNA target used to hybridize the array is obtained by reverse transcription of the mRNA extracted from a tissue sample to a double stranded complementary DNA (cDNA) (see Fig. 3.1). This DNA is fluorescently labeled with a dye, and a subsequent illumination with an appropriate source of light will provide an image of the array of features (sets of probes on GeneChips, spots on cDNA arrays, or beads on Illumina arrays). The intensity of each spot or the average difference between matches and mismatches can be related to the amount of mRNA present in the tissue and, in turn, with the amount of protein produced by the gene corresponding to the given feature.

This step can also be accomplished in many different ways. For instance, the labeling can be done with a radioactive substance and the image obtained by using a photosensitive device. Or several targets can be labeled with different dyes and used at the same time in a competitive hybridization process in a multichannel experiment. A typical case is a two-channel experiment using cy3 and cy5 as dyes, but other dyes can also be used. After an image-processing step is completed, the result is a large number of expression values. Typically, one DNA array will provide expression values for hundreds or thousands of genes.

3.2 Fabrication of microarrays

Two main approaches are used for microarray fabrication: deposition of DNA fragments and *in situ* synthesis. The first type of fabrication involves two methods: deposition of PCR-amplified cDNA clones, and printing of already synthesized oligonucleotides. *In situ* manufacturing can be divided into photolithography, ink-jet printing, and electrochemical synthesis.

3.2.1 Deposition

In deposition-based fabrication, the DNA is prepared away from the chip. Robots dip thin pins into the solutions containing the desired DNA material and then touch the pins onto the surface of the arrays. Small quantities of DNA are deposited on the array in the form of spots. Unlike *in situ* manufacturing in which the length of the DNA sequence is limited, spotted arrays can use small sequences, whole genes or even arbitrary PCR products.

As discussed in Chapter 2, the living organism can be divided into two large categories: eukaryotes and prokaryotes. The group of eukaryotes includes the organisms whose cells have a nucleus. Prokaryotes are organisms whose cells do not have a nucleus, such as bacteria. In general, eukaryotes have a much more complex intracellular organization than prokaryotes. Gene expression in most eukaryotes is studied by utilizing complementary DNA (cDNA) clones, which allow the amplification of sufficient quantities of DNA for deposition. Mature mRNA is reverse transcribed into short cDNAs and introduced into bacterial hosts, which are grown, isolated, then selected out if they carry foreign DNA. As discussed in Chapter 2, bacteria are prokaryotes, which, unlike eukaryotes, do not have a nucleus and do not have introns in their DNA. Therefore the prokaryotic gene expression machinery is different, and it is less complicated to amplify their genes.

The cloning strategy leverages bacterial properties in order obtain large quantities of eukaryotic DNA. Single-pass, inexpensive sequencing of entire clone libraries results in sets of **expressed sequence tags** (ESTs), which are partial sequences of the clone inserts that are long enough to uniquely identify the gene fragments. The **polymerase chain reaction** (PCR) is used to amplify clones containing desired fragments, using primers flanking the inserts, or oligonucleotide primers designed specifically for selective amplification. Once the cDNA cloned inserts are amplified by PCR, they are purified, and the final PCR products are then spotted on a solid support.

Another method of microarray fabrication is the attachment of short, synthesized oligonucleotides to the solid support. One advantage of this method is that oligonucleotide probes can be designed to detect multiple variant regions of a transcript or the so-called splice variants (see Fig. 2.15). These oligonucleotides are short enough to be able to target specific exons. Measuring the abundance of specific splice variants is not possible with spotted cDNA arrays because cDNA arrays contain probes of long and variable length, to which more than one different splice variant might hybridize.

3.2.1.1 The Illumina technology

A more recent technology for the fabrication of microarrays is the BeadArray technology, developed by Illumina (see Fig. 3.2). This technology uses Bead-Chips, which are microarrays composed of very small (3 μm) silica beads that are placed in small wells etched out in one of two substrates: fiber-optic bundles or planar silica slides. These beads are randomly self-assembled on the substrate in a uniform pattern that places the beads approximately 5.7 microns apart. Each bead is covered with hundreds of thousands of copies of a given nucleotide sequence forming the probe specific to the given assay. Each such oligonucleotide sequence is approximately 50 bp long and is concatenated with another custom-made sequence that can be used as an address. This address sequence is used to locate the position of each bead on the array and to uniquely associate each bead with a specific target site.

After the self-assembly is complete, individual bead types are decoded and identified. Figure 3.3 illustrates how the decoding process works in a simplified example using 16 different bead types. The randomly assembled array is sequentially hybridized to 16 "decoder oligonucleotides," each of which is a perfect match for one of the assay oligos bound to a particular bead type. In this example, the first four decoder oligos are labeled with the same blue fluorescent dye, and the second set of four decoder oligos are labeled with a green dye, and so on. The array is hybridized to the first set of 16 decoder oligos, labeled as described above, then imaged and stripped. The second hybridization includes the same 16 decoder oligos, labeled in a different order with fluorescent dyes. Following the second round of hybridization and imaging, it is simple to precisely identify the exact bead type in each position on the array. For example, a location that is blue in the first round and then yellow in the second round is bead type number 3, while a location that is yellow in the first round and then green in the second round is bead type number 10.

Since the space needed for each bead is so small, a high density can be achieved on the array, allowing thousands, or even millions, of target sites to

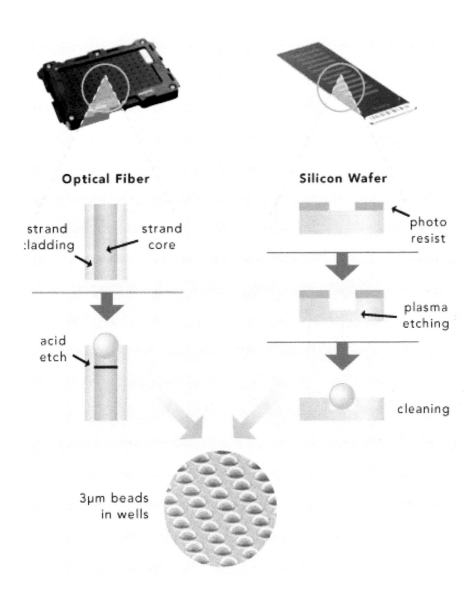

FIGURE 3.2: The Illumina BeadArray Technology. Very small (approximately 3 microns) silica beads are placed in small wells etched out of either optical fibers or a silicon wafer. The beads are held in place by Van der Waals forces as well as hydrostatic interactions with the walls of the well. The surface of each bead is covered with multiple (hundreds of thousands) copies of the sequence chosen to represent a gene. Courtesy of Illumina, Inc.

The Decoding Process

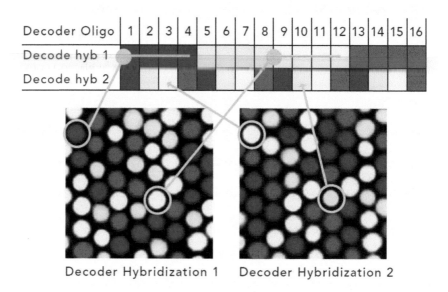

Decoder Hybridization 1   Decoder Hybridization 2

FIGURE 3.3: Decoding the Illumina bead types. The array is hybridized twice with the colors shown on the top. The combinations of colors that a given bead has in the two hybridization allows the unique identification of their type. For instance, the bead near the top left corner was blue in the first hybridization and yellow in the second one. The only type that matches this is 3. The second bead shown was yellow in the first hybridization and green in the second one yielding type 10. Courtesy of Illumina, Inc.

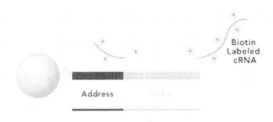

FIGURE 3.4: An Illumina Direct Hybridization probe is more similar to the probes used in other manufacturing technologies with the difference that the probe is attached to a bead, rather than to a flat surface. The beads are either fixed in optical fibers or distributed across a silicon wafer as shown in Fig. 3.2. The labeled target cDNA reverse-transcribed from the sample RNA hybridizes to the probe as usual. Courtesy of Illumina, Inc.

be analyzed simultaneously. Furthermore, the arrays can be formatted to test several samples in parallel.

Illumina produces whole-genome arrays, as well as more focused arrays that include probes for a subset of genes related to the specific condition. There are either one or two probes per gene, depending on the type of array (focused set or whole genome). For gene expression analysis, the Illumina arrays come in two flavors, using slightly different approaches. The Direct Hybridization Assay, illustrated in Fig. 3.4, uses a single DNA sequence per bead much like in all other technologies. This single-stranded sequence is meant to hybridize with the labeled target sequence present in the sample. The amount of fluorescence produced will provide a measure of the amount of target present in the sample.

The other approach to expression level measurements is called DASL, which stands for c**D**NA-mediated **A**nnealing, **S**election, **E**xtension and **L**igation [153]. This approach is described in Fig. 3.5 and Fig. 3.6. In the Whole-Genome DASL HT (WG DASL) Assay,[1] a pair of oligonucleotides is annealed to each target site, and more than 29,000 oligonucleotide pairs can be multiplexed together in a single reaction. A high specificity is obtained by requiring that both members of an oligonucleotide pair must hybridize in close proximity for the assay to generate a strong signal. The main advantage of the DASL approach with respect to other commercially available assays is related to the quality of the mRNA that can be evaluated. Since the WG DASL Assay uses two short sequences that in the gene are separated by a gap, there is a lot of flexibility in choosing the sequences. Furthermore, since these probes span only about 50 bases, partially degraded RNA, such as that from formalin-fixed paraffin-embedded (FFPE) samples, can be used in the

---

[1]More information about the WG DASL can be found at: http://www.illumina.com/documents/products/datasheets/datasheet_whole_genome_dasl_ht.pdf

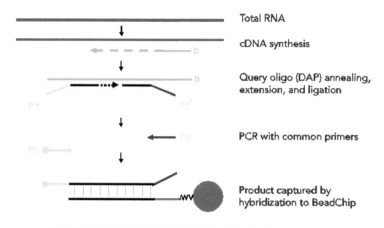

Total RNA

cDNA synthesis

Query oligo (DAP) annealing, extension, and ligation

PCR with common primers

Product captured by hybridization to BeadChip

FIGURE 3.5: The Illumina WT DASL Technology. A pair of oligonucleotides is annealed to each target site. A high specificity is obtained by requiring that both members of an oligonucleotide pair must hybridize in close proximity for the assay to generate a strong signal. Since these probes span only about 50 bases, partially degraded RNA, such as that from formalin-fixed paraffin-embedded (FFPE) samples, can be used in the assay. Courtesy of Illumina, Inc.

assay. There are estimated to be more than 400 million of these FFPE samples archived in North America for cancer alone. Many of these samples represent clinical outcomes with the potential to provide critical insight into expression profiles associated with complex disease development. Unfortunately, FFPE archival methods often lead to partial RNA degradation, often limiting the amount of information that can be derived from such samples. In contrast to the WG DASL arrays, cDNA arrays using very long sequences require good quality RNA that can only be obtained from fresh tissue, or tissue frozen very soon after collection.

Genotyping with Illumina Arrays. In addition to gene expression analysis, the Illumina BeadArray platform can also be used for genotyping applications as well. The genotyping arrays span a much larger multiplex range than the expression arrays (up to 5 million markers per sample). Illumina offers two genotyping assays with the BeadArray platform: the GoldenGate Assay for custom, low-multiplex studies, and the Infinium HD Assay for high-multiplex studies.

The GoldenGate Assay. The Illumina GoldenGate Genotyping Assay is a flexible, pre-optimized assay that uses a discriminatory DNA polymerase and ligase to interrogate up to 3,072 SNP loci simultaneously (see Section 2.2.4 for more details about SNPs). This assay is illustrated in Fig. 3.7. The genomic

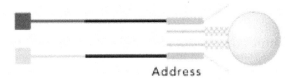

Address

FIGURE 3.6: An Illumina DASL probe. The address is a short sequence that is specific to each targeted site. The address is used to attach the probes to the beads. In the DASL assay, each targeted site is represented by two sequences, one from upstream of the targeted site and one from downstream of it. Because each targeted site is represented by two short sequences that can be separated by an arbitrary gap, this assay is better able to work with partially degraded RNA, such as the one coming from older formalin-fixed paraffin-embedded samples. In contrast, cDNA arrays using long sequences usually require good-quality mRNA. Courtesy of Illumina, Inc.

DNA (gDNA) sample used in this assay is first fragmented and then bound to paramagnetic particles in preparation for hybridization with the assay oligonucleotides. Three oligonucleotides are designed for each SNP locus. Two oligos are specific to each allele of the SNP site, called the Allele-Specific Oligos (ASOs). A third oligo that hybridizes several bases downstream from the SNP site is the Locus-Specific Oligo (LSO). All three oligonucleotide sequences contain regions of genomic complementarity and universal PCR primer sites; the LSO also contains a unique address sequence that targets a particular bead type on the array. Up to 3,072 SNPs may be interrogated simultaneously in this manner using GoldenGate technology. During the hybridization process, the assay oligonucleotides hybridize to the genomic DNA sample bound to paramagnetic particles. Because hybridization occurs prior to any amplification steps, no amplification bias can be introduced into the assay. Following hybridization, extension of the appropriate ASO (the one containing the complementary SNP) and ligation of the extended product to the LSO joins information about the genotype present at the SNP site to the address sequence on the LSO. These joined, full-length products provide a template for PCR using universal PCR primers P1, P2, and P3.Universal PCR primers P1 and P2 are Cy3- and Cy5-labeled. After downstream-processing, the single-stranded, dye-labeled DNAs are hybridized to their complement bead type through their unique address sequences. Hybridization of the GoldenGate Assay products onto the BeadChip allows for the separation of the assay products in solution, onto a solid surface for individual SNP genotype readout. After hybridization, a high-precision scanner is used to analyze fluorescence signal on the BeadChip, which is in turn analyzed using software for automated genotype clustering and calling. The GoldenGate assay is designed for low-plex, custom studies. It should be mentioned that the DASL assay described above actually uses the GoldenGate extension/ligation chemistry.

The Infinium HD Assay. The Infinium HD assay is designed for high-multiplex studies, with the ability to assay up to 5 million markers simultaneously. This assay is illustrated in Fig. 3.8.

Genomic markers are interrogated though a two-step detection process. Carefully designed 50-mer probes selectively hybridize to the loci of interest, stopping one base before the interrogated marker. Marker specificity is conferred by enzymatic single-base extension to incorporate a labeled nucleotide. Subsequent dual-color florescent staining allows the labeled nucleotide to be detected by Illumina's imaging scanners, which identify both color and signal intensity. For genotyping assays, the red and green color signals specify each allele, where homozygotes are indicated by red/red or green/green signals, and heterozyotes are indicated by red/green (yellow) signals. Signal intensity information can be used to detect structural aberrations, such as copy number variants, inversions, or translocations.

3.2.2 *In situ* synthesis

During array fabrication based on *in situ* synthesis, the probes are photochemically synthesized on the chip. There is no cloning, no spotting, and no PCR carried out, which is advantageous since these steps introduce a lot of noise in the cDNA system.

Probe selection is performed based on sequence information alone. This means that every probe synthesized on the array is known in contrast to cDNA arrays, which deal with expressed sequence tags, and, in many cases, the function of the sequence corresponding to a spot is unknown. Additionally, this technology can distinguish and quantitatively monitor closely related genes just because it can avoid identical sequence among gene family members.

There are currently three approaches to *in situ* probe synthesis. The first method is photolithographic (Affymetrix, Santa Clara, CA) and is similar to the technology used to build very large scale integrated (VLSI) circuits used in modern computers. This fabrication process uses photolithographic masks for each base. If a probe should have a given base, the corresponding mask will have a hole allowing the base to be deposited at that location. Subsequent masks will construct the sequences base by base. This technology allows the fabrication of very high density arrays but the length of the DNA sequence constructed is limited. This is because the probability of introducing an error at each step, while very small, is different from zero. In order to limit the overall probability of an error, one needs to limit the length of the sequences. To compensate for this, a gene is represented by several such short sequences. The particular sequences must be chosen carefully to avoid cross-hybridization between genes.

The second approach is the ink-jet technology (Agilent, Protogene, etc.), which employs the technology used in ink-jet color printers. Four cartridges

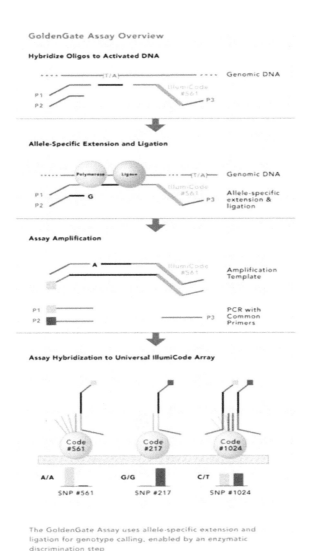

FIGURE 3.7: The Illumina Golden Gate assay. The GoldenGate assay is based on the BeadArray technology: assay oligonucleotides, containing an address sequence, hybridize to gDNA to identify the allele at a given loci. After processing and amplification, the amplified product binds to a bead on the array that contains a complementary address sequence. Dual-color fluorescence dyes, specific to each ASO, indicate the genotype of the SNP from the gDNA fragment. Courtesy of Illumina, Inc.

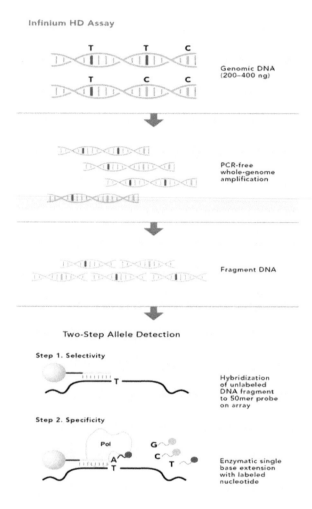

FIGURE 3.8: The Illumina Infinium assay. Genomic markers are interrogated though a two-step detection process. 50-mer probes selectively hybridize to the loci of interest, stopping one base before the interrogated marker. Subsequent dual-color fluorescent staining allows the labeled nucleotide to be detected by Illumina's imaging scanners, which identify both color and signal intensity. Courtesy of Illumina, Inc.

are loaded with different nucleotides (A, C, G, and T). As the print head moves across the array substrate, specific nucleotides are deposited where they are needed.

Finally, the electrochemical synthesis approach (CombiMatrix, Bothel, WA) uses small electrodes embedded into the substrate to manage individual reaction sites. Solutions containing specific bases are washed over the surface and the electrodes are activated in the necessary positions in a predetermined sequence that allows the sequences to be constructed base by base.

The Affymetrix technology includes the steps outlined in Figs. 3.9, 3.10, and 3.11. Synthetic linkers modified with photochemical removable protecting groups are attached to a glass surface. Light is shed through a photolithographic mask to a specific area on the surface to produce a localized photodeprotection (Fig. 3.9). The first of a series of hydroxyl-protected deoxynucleosides is incubated on the surface. In this example, it is the protected deoxynucleoside T. In the next step, the mask is directed to another region of the substrate by a new mask, and the chemical cycle is repeated (Fig. 3.10). Thus, one nucleotide after another is added until the desired chain is synthesized. Recall that the sequence of this nucleotide corresponds to a part of a gene in the organism under scientific investigation. The synthesized oligonucleotides are called probes. The material that is hybridized to the array (the reverse transcribed mRNA) is called the target, or the sample.

The **gene expression arrays** have a match/mismatch probe strategy. This is illustrated in Fig. 3.11. Probes that match the target sequence exactly are referred to as reference probes. For each reference probe, there is a probe containing a nucleotide change at the central base position – such a probe is called a mismatch. These two probes – reference and mismatch – are always synthesized adjacent to each other to control for spatial differences in hybridization. Additionally, the presence of several such pairs per gene (each pair corresponding to various parts – or exons – of the gene) helps to enhance the confidence in detection of the specific signal from background in case of weak signals.

More recently, technological advances allowed the fabrication of oligonucleotide arrays with extremely large numbers of features. At the same time, the sequencing of the entire genome has been completed for several organisms of interest, including *Homo sapiens*. Given both facts above, one could envisage arrays that cover the entire genome of a given organism. In fact, Affymetrix currently manufactures and sells such arrays, called **tiling arrays**. Affymetrix tiling arrays use short sequences, currently 25-mer oligonucleotides, that are equally spaced across the entire genome (see Fig. 3.12). The gap between two such sequences in the genome is referred to as the **resolution** of the tiling array. At the moment of this writing, Affymetrix offers tiling arrays with a resolution of 35 base pairs.

Unlike the expression arrays where probes are designed considering the direction of the strand each gene is on, tiling arrays are designed based on the direction of the genome, rather than that of a particular transcript. Tiling

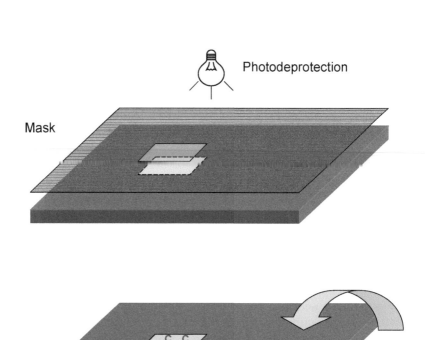

FIGURE 3.9: Photolithographic fabrication of microarrays. Synthetic linkers modified with photochemical removable protecting groups are attached to a glass surface. Light is shed through a photolithographic mask to a specific area on the surface to produce a localized photodeprotection. The first of a series of hydroxyl-protected deoxynucleosides is incubated on the surface. In this example, it is the protected deoxynucleoside C. The surface of the array is protected again, and the array is ready for the next mask.

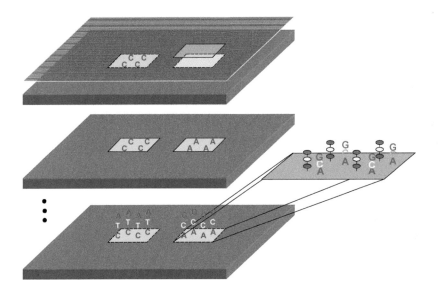

FIGURE 3.10: Photolithographic fabrications of microarrays. The second mask is applied and light is used to deprotect the areas that are designed to receive the next nucleoside (A). The fabrication process would generally require 4 masking steps for each element of the probes. Several steps later, each area has its own sequence as designed.

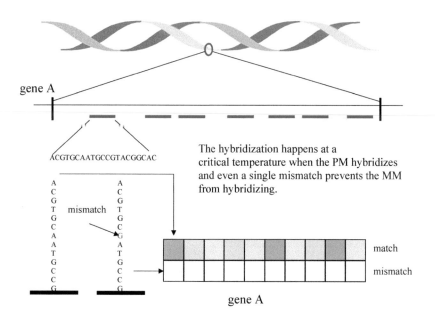

FIGURE 3.11: The principles of the Affymetrix technology. The probes correspond to short oligonucleotide sequences thought to be representative for the given gene. Each oligonucleotide sequence is represented by two probes: one with the exact sequence of the chosen fragment of the gene (perfect match or PM) and one with a mismatch nucleotide in the middle of the fragment (mismatch or MM). For each gene, the value that is usually taken as representative for the expression level of the gene is the average difference between PM and MM. Reprinted from S. Draghici, "Statistical intelligence: effective analysis of high-density microarray data" published in *Drug Discovery Today*, Vol. 7, No. 11, p. S55–S63, 2002, with permission from Elsevier.

cDNA arrays	Oligonucleotide arrays
Long sequences	Short sequences due to the limitations of the synthesis technology
Spot unknown sequences	Spot known sequences
More variability in the system	More reliable data
Easier to analyze with appropriate experimental design	More difficult to analyze

TABLE 3.1: A comparison between cDNA and oligonucleotides arrays.

arrays labeled with "F" are complementary to the forward direction $(+)$, while tiling arrays labeled with "R" are complementary to the reverse $(-)$ strand of the given genome. At the time of this writing, there are several tiling arrays available. The GeneChip Human Tiling 1.0R Array Set is a set of 14 arrays that include both a perfect match as well as a mismatch probe for each target location on the genome. These arrays can be used for both transcript mapping as well as in **chromatin immunoprecipitation** (ChIP) experiments. The GeneChip Human Tiling 2.0R Array Set is a set of 7 arrays that include only the perfect match probes for each location. Both sets cover the genomic sequence left after the repetitive elements were removed by RepeatMasker. Each array within the sets above contain more than 6.5 million probes. Another tiling array, the GeneChip Human Promoter 1.0R, uses the same tiling technique but focuses only on the known human promoter regions. This array includes approximatively 4.6 million probes covering 22,500 known promoter regions.

3.2.3 A brief comparison of cDNA and oligonucleotide technologies

It is difficult to make a judgment as to the superiority of a given technology. At this point in time, the cDNA technology seems to be more flexible, allowing spotting of almost any PCR product whereas the Affymetrix technology seems more reliable and easier to use. This field is so dynamic that this situation might change rapidly in the near future. Table 3.1 summarizes the advantages and disadvantages of cDNA and high-density oligonucleotide arrays. Table 3.2 shows the current performance of the Affymetrix oligonucleotides arrays [36, 286].

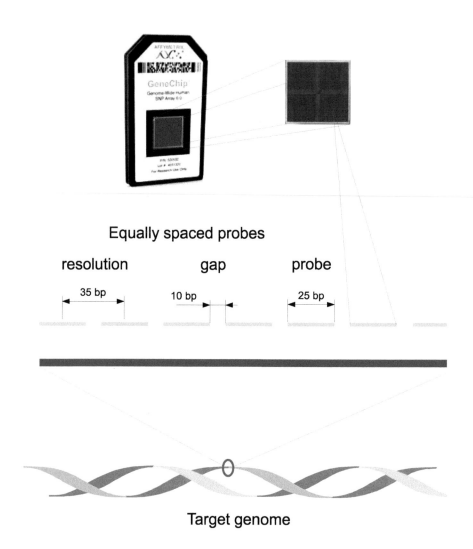

FIGURE 3.12: Tiling arrays cover the entire length of the genome after the repetitive elements have been removed. Probes of 25 oligonucleotides are tiled at an average resolution of 35 bps, with an average gap of 10 bps. Some tiling array sets contain both perfect match and mismatch sequences. Others contain only perfect matches.

	Current limit	Practical use
Density (genes/array)	40,000	20,000
Gene representation (probe pairs/gene)	4	2
Linear dynamic range	4 logs	3 logs
Fold change detection	10%	100%
Similar sequences separation	93% identical	70–80% identical
Starting material	$2ng$ total RNA	$5\mu g$ total RNA
Detection specificity	$1 : 10^6$	$1 : 10^5$

TABLE 3.2: The performance of the Affymetrix technology.

3.3 Applications of microarrays

Microarrays have been used successfully in a range of applications, including sequencing [373], SNP detection [152, 443], genotyping, disease association [386, 463], genetic linkage, genomic loss and amplification (copy number variation (CNV)) [388, 442, 449], detection of chromosomal rearrangements, etc. However, this book will focus on the (arguably) mainstream application for microarrays, which is the investigation of the genetic mechanisms in the living cells through expression analysis [147, 289, 374, 373, 384, 183, 413]. A few typical examples would include comparing healthy and malignant tissue [14, 183, 16, 54, 339], studying cell phenomena over time [113, 400] as well as study the effect of various factors such as interferons [112], cytomegalovirus infection [488], and oncogene transfection [260] on the overall pattern of expression. Perhaps even more important than the success in any individual application, the large number of papers reporting results obtained with microarrays (and subsequently validated) have increased the overall confidence that microarrays are tools that can be used to generate accurate, precise, and reliable gene expression data [84, 479, 373, 392, 391].

Microarrays can also be used for purely computational purposes such as in the field of DNA computing [249]. In these cases, the microarray can contain sequences of DNA encoding various possible solutions of the problem to be solved. Several successive steps are performed in order to solve the problem. Each such step consists of three sub-steps: a hybridization, the destruction of the single-stranded DNA not hybridized, and a denaturation that will prepare the chip for the next computational step. The role of the DNA used in each step is to prune the large number of potential solutions coded on the surface of the array. Specific sequences added in a specific step hybridize to the single-stranded DNA attached to the surface. This marks the partial solutions by binding them in double strands. Subsequently, the chip is washed with a solution that destroys the single-stranded DNA. A denaturation step will break the double-stranded DNA and bring the chip to a state in which it is ready for the next computational step.

In this book, we will concentrate on the use of microarrays in gene expression studies, focusing on specific challenges that are related to this particular application. Although the microarray data will be our main motivation and source of examples, the concepts discussed in this book, as well all analysis methods presented, are general and can be applied to a very large class of data.

3.4 Challenges in using microarrays in gene expression studies

Compared to other molecular biology techniques, microarrays are relatively new. As such, their users are challenged by a number of issues as follows:

1. **Noise.**

   Because of their nature, microarrays tend to be very noisy. Even if an experiment is performed twice with exactly the same materials and preparations in exactly the same conditions, it is likely that after the scanning and image processing steps, many genes will probably be characterized by different quantification values. In reality, noise is introduced at each step of various procedures[2] [377]: mRNA preparation (tissues, kits, and procedures vary), transcription (inherent variation in the reaction, enzymes), labeling (type and age of label), amplification, pin type (quill, ring, ink-jet), surface chemistry, humidity, target volume (fluctuates even for the same pin), slide inhomogeneities (slide production), target fixation, hybridization parameters (time, temperature, buffering, etc.), unspecific hybridization (labeled cDNA hybridized on areas that do not contain perfectly complementary sequences), nonspecific background hybridization (e.g., bleeding with radioactive materials), artifacts (dust), scanning (gain settings, dynamic range limitations, inter-channel alignment), segmentation (feature/background separation), quantification (mean, median, percentile of the pixels in one spot), etc.

   The challenge appears when comparing different tissues or different experiments. Is the variation of a particular gene due to the noise or is it a genuine difference between the different conditions tested? Furthermore, when looking at a specific gene, how much of the measured variance is due to the gene regulation and how much to noise? The noise is an inescapable phenomenon and the only weapon that the researcher seems to have against it is replication (Chapters 13 and 21).

---

[2]Not all steps apply to all types of arrays.

2. **Normalization.**

The aim of the normalization is to account for systematic differences across different data sets (e.g. overall intensity) and eliminate artifacts (e.g., nonlinear dye effects). The normalization is crucial if results of different experimental techniques are to be combined. While everybody agrees on the goal of normalization, the consensus seems to disappear regarding how exactly the normalization should be done. Normalization can be necessary for different reasons such as different quantities of mRNA (leading to different mean intensities), dye nonlinearity and saturation towards the extremities of the range, etc. Normalization issues and procedures are discussed in detail in Chapter 20.

3. **Experimental design.**

The experimental design is a crucial but often neglected phase in microarray experiments. A designed experiment is a test or several tests in which a researcher makes purposeful changes to the input variables of a process or a system in order to observe and identify the reasons for changes in the output response. Experiment design issues are discussed in details in Chapter 15.

4. **Large number of genes.**

The fact that microarrays can interrogate thousands of genes in parallel is one of the features that led to the wide adoption of this technology. However, this characteristic is also a challenge. The classical metaphor of the needle in the haystack can easily become an accurate description of the task at hand when tens of thousands of genes are investigated. Furthermore, the sheer number of genes can change the quality of the phenomenon and the methods that need to be used. The classical example is that of the p-values in a multiple testing situation (Chapter 16).

5. **Significance.**

If microarrays are used to characterize specific conditions (e.g., [14, 183]), a crucial question is whether the expression profiles differ in a significant way between the groups considered. The classical statistical techniques that were designed to answer such questions (e.g., chi-square tests) cannot be applied directly because in microarray experiments the number of variables (usually thousands of genes) is much greater than the number of experiments (usually tens of experiments). Novel techniques need to be developed in order to address such problems.

6. **Biological factors.**

In spite of their many advantages, microarrays are not necessarily able to completely substitute other tools in the arsenal of the molecular biologist. For instance, knocking out genes is slow and expensive but offers an unparalleled way of studying the effects of a gene well beyond its mRNA expression levels. In normal cells, the RNA polymerase transcribes the DNA into mRNA, which carries the information to ribosomes, where the protein is assembled by tRNA in the translation process. Most microarrays measure the amount of mRNA specific to particular genes and the expression level of the gene is associated directly with the amount of mRNA. However, the real expression of the gene is the amount of protein produced, not the amount of mRNA. Although in most cases the amount of mRNA reflects accurately the amount of protein, there are situations in which this may not be true. If nothing else, this is a fundamental reason for which microarrays cannot be trusted blindly. However, there are also other reasons. Even if the amount of protein were always directly proportional to the amount of mRNA, the proteins may require a number of post-translational modifications in order to become active and fulfill their role. Any technology that works exclusively at the mRNA level, such as microarrays, will be blind with respect to these changes. In conclusion, microarrays are great tools for screening many genes and focusing hypotheses. However, conclusions obtained with microarrays must be validated with independent assays, using different techniques that investigate the phenomenon from various perspectives.

Furthermore, even if we assume that all the genes found to be differentially regulated in a microarray experiment are indeed so, a nontrivial issue is to translate this information into biological knowledge. The way various biological processes are affected and the degree to which genes interact within various regulatory pathways is much more important than which particular genes are regulated. The issue of translating lists of differentially regulated genes into biological knowledge is discussed in Chapters 23 and 28.

7. **Array quality assessment.**

It is useful if data analysis is not seen as the last step in a linear process of microarray exploration but rather as a step that completes a loop and provides the feedback necessary to fine-tune the laboratory procedures that produced the microarray. Thus, array quality assessment is an aspect that should be included among the goals of the data analysis. It would be very useful if the analysis could provide besides the expression values some quality assessment of the arrays used. Such quality measures will allow discarding the data coming from below standard arrays as well as the identification of possible causes of failure in the microarray process. The issue of asessing the quality of a given array or data set is discussed thoroughly in Chapter 19.

One of the advantages of DNA arrays is that they can be used to query many genes at the same time. A single array can contain anywhere from a few hundreds to a few tens of thousands of spots. However, this advantage is also a challenge. After hybridization and image processing, each spot will generate

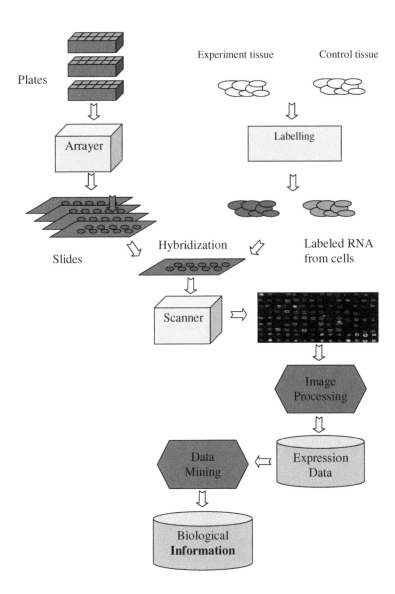

FIGURE 3.13: An overview of the cDNA array processing. The cylinders represent data that needs to be stored. The parallelepipeds represent pieces of hardware, while the hexagons represent information-processing steps in which the computers play an essential role.

a quantified number. Therefore, each array can generate up to a few tens of thousands of values. Furthermore, the immense potential of the DNA arrays can be realized only if many such experiments are done for various biological conditions and for many different individuals. Therefore, the amount of data generated by this type of experiments is staggering. Techniques classically developed in computer science (such as data mining and machine learning) can prove to be extremely useful for the interpretation of such data.

Another aspect of microarray experiments that relates to their use in medicine and drug discovery is their reproducibility. A DNA array experiment tends to be a complicated process as shown schematically in Fig. 3.13. Furthermore, each step in this process is usually associated to a very complicated laboratory protocol that needs to describe in great detail the substances, conditions, and procedures used [372]. There are an amazing number of factors that can influence the results in a dramatic way. In order to be able to validate the results obtained using such DNA arrays (e.g. a new drug), a very large amount of data needs to be stored and made accessible in a convenient way [68]. In an effort to alleviate data sharing and storing problems, **standards** are being developed. There are current international standardization efforts undertaken jointly by various agencies in the US and Europe. The Microarray Gene Expression Data (MGED) society has ongoing efforts in the following directions [68]:

- MIAME – The formulation of the minimum information about a microarray experiment required to interpret and verify the results.

- MAGE – The establishment of a data exchange format (MAGE-ML) and object model (MAGE-OM) for microarray experiments.

- Ontologies – The development of ontologies for microarray experiment description and biological material (biomaterial) annotation in particular.

- Normalization – The development of recommendations regarding experimental controls and data normalization methods.

Clearly, novel techniques in databases (e.g., multi-dimensional indexing) can make life a lot easier at this stage.

According to Murphy's law, if anything can go wrong, it will. In DNA arrays, the number of things that can go wrong is very large [377, 461]. Debugging the global process involving DNA arrays requires simultaneous access to different types of data usually stored in different databases. Being able to do transversal queries across multiple databases might help locate problems. For instance, combining queries in a database containing known expression values with queries in a database containing functional information and with queries in a database containing the laboratory protocols might point out that a particular subset of experimental results are questionable. Combining further queries across the various databases might help pinpoint the problem (e.g., the clones prepared by a certain lab technician provided questionable results or the preparations involving a solution from a particular fabrication batch need to be discarded). Techniques developed for data warehouses for business applications may be used successfully here.

3.5 Sources of variability

It is interesting and important to follow the microarray process from the point of view of the amount of variability introduced by each step. Our goal is to compare gene expression in two phenotypes, such as tumor tissue versus healthy tissue. The ultimate goal of such an experiment is to see the differences in gene expression between these two states. If the same experiment is performed several times, each run will provide slightly different expression values. When the experiments are done with the two RNA samples that are to be compared, the exact expression values obtained for all the genes will be different between the two samples. Unfortunately, it is difficult to distinguish between the variation introduced by the different expression levels and the variation inherent to the laboratory process itself. From this point of view, any variation that appears even if the same sample is processed twice can be considered noise. The main sources of such variations corresponding to each individual step in the process are shown in Table 3.3 [377, 461]. Much of the variability introduced by the microarray process itself can be determined using technical replicates, where the same RNA sample is run on different chips, resulting in slightly different expression values. Chapter 15 will discuss some of the techniques that can be used to separate the noise from the interesting differences between the phenotypes studied.

One of the important factors is the preparation of mRNA. Even if the kits used for RNA isolation are from the same company and the same batch, two preparations may yield different results. Another important contributor to overall experimental variability is the target preparation. This particular step of the microarray protocol comprises the enzyme-mediated reverse transcription of mRNA and concomitant incorporation of fluorescently labeled nucleotides. A careful analysis of all components of this step has been shown to maximize the reaction and increase the signal-to-noise ratio [461].

The sources of variability can be divided into several categories as follows. The sources related to the sample preparation include the mRNA preparation, transcription (RT-PCR), and labeling. The slide preparation stages contribute with variability caused by pin type variation, surface chemistry, humidity, target volume, slide inhomogeneities, and target fixation issues. Hybridization-related variability is determined by hybridization parameters, nonspecific spot hybridization and nonspecific background hybridization.

One of the sources of fluctuations in microarray data is the pin geometry.

Factor	Comments
mRNA preparation	Tissues, kits, and procedures vary
transcription	Inherent variation in the reactions, type of enzymes used
Labeling	Depends on the type of labeling and procedures as well as age of labels
Amplification (PCR protocol)	PCR is difficult to quantify
Pin geometry variations	Different surfaces and properties due to production random errors
Target volume	Fluctuates stochastically even for the same pin
Target fixation	The fraction of target cDNA that is chemically linked to the slide surface from the droplet is unknown
Hybridization parameters	Influenced by many factors, such as temperature of the laboratory, time, buffering conditions, and others
Slide inhomogeneities	Slide production parameters, batch-to-batch variations
Nonspecific hybridization	cDNA hybridizes to background or to sequences that are not their exact complement
Gain setting (PMT)	Shifts the distribution of the pixel intensities
Dynamic range limitations	Variability at low end or saturation at the high end
Image alignment	Images of the same array at various wavelengths corresponding to different channels are not aligned; different pixels are considered for the same spot
Grid placement	Center of the spot is not located properly
Nonspecific background	Erroneous elevation of the average intensity of the background
Spot shape	Irregular spots are hard to segment from background
Segmentation	Bright contaminants can seem like signal (e.g., dust)
Spot quantification	Pixel mean, median, area, etc.

TABLE 3.3: Sources of fluctuations in a typical cDNA microarray experiment.

As already outlined, microarrays can be obtained by printing the spots on solid surfaces such as glass slides or membranes. This printing involves deposition of small volumes of DNA on a solid surface. The volumes deposited by microarray printing pins are within the range of nanoliters (10^{-9} L) or picoliters (10^{-12} L). Because these volumes are much below the range of normal liquid dispensing systems new technologies for printing of microarrays emerged [109, 359].

The technologies for printing can be divided into contact and non-contact printing. The first category involves direct contact between the printing device and the solid support. Contact printing includes solid pins, capillary tubes, tweezers, split pins, and microspotting pins. Non-contact printing uses a dispenser borrowed from the ink-jet printing industry.

The most commonly used pins are the split pins. Such a split or a quill pin has a fine slot machined into its end. When the split pin is dipped into the well of a PCR plate, a small amount of volume (0.2 to 1.0 μL) is loaded into the slot. The pin then touches the surface of the slide, the sample solution is brought into contact with the solid surface, the attractive force of the substrate on the liquid withdraws a small amount of volume from the channel of the pin, and a spot is formed. However, the amount of transported target fluctuates stochastically even for the same pin. The geometry of the pin and its channel introduce variability into the system between the genes spotted with different pins. The surface chemistry and the humidity also play important roles in the spot formation and, in consequence, introduce variability in the process.

All the above sources of variation need to be addressed at various stages of data acquisition and analysis. Some sources of variance may not produce results observable on the scanned image of the microarray. However, several sources of variance do produce results that are very clearly visible. Several examples are shown in Fig. 3.14 and include incorrect alignment, non-circular spots, spots of variable size, spots with a blank center, etc. The microarray jargon uses colorful terms that are very descriptive if somehow less technical. For instance, blooming is an extension of a particular spot in its immediate neighborhood, which is very frequent on filter arrays when the labeling is done with a radioactive substance (see panel A in Fig. 3.14). If the amount of radioactive mRNA is very abundant in a particular spot, the radioactivity will affect the film in a larger area around the spot sometime covering several of the neighboring spots. Doughnuts are spots with a center having a much lower intensity than their circumference. A possible cause may be the pin touching the surface too hard, damaging it and thus preventing a good hybridization in the center of the spot. Other factors may involve the superficial tension of the liquid printed. Comets are long, tail-like marks coming out of the spots, while splashes resemble the pattern created by a drop of liquid splashing on a hard surface (notice a splash on the right of panel D in Fig. 3.14). Hybridization, washing, and drying can also create problems. Typically, such problems involve larger areas and many spots (lower right corner of panel D in Fig. 3.14). Dust is a problem for printing and a drastic example is illustrated in the same panel D in Fig. 3.14. This is why most arrayers are enclosed in a glass or plexiglass

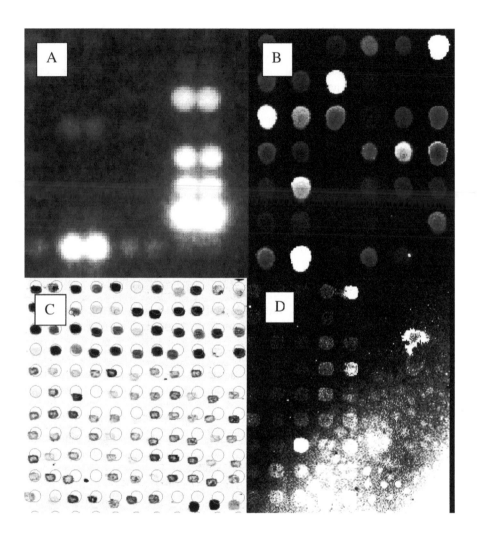

FIGURE 3.14: Examples of microarray defects. A. Radioactively labeled spots expanding over neighboring spots (blooming). B. Non-circular and weak spots as well as some weak-center spots (doughnuts). C. Imperfect alignment. D. Hybridization and contamination causing a nonspecific, high background with various gradients.

container to prevent dust deposition during printing. Some facilities adhere to the "clean environment policy" according to which researchers attending the arrayers and performing manipulations in the printing facility wear special uniforms and caps, the room is equipped with air filters, and other technical precautions are taken.

The fact that many of these problems are so conspicuous in the image emphasizes that image processing and quantification are crucial steps in the sequence of analyzing the microarray results. These issues will be discussed in detail in Chapter 5.

3.6   Summary

Microarrays are solid substrates with localized features each hosting hundreds of single stranded DNAs, representing a specific sequence. These molecules, called probes, will hybridize with single-stranded DNA molecules, named targets, that have been labeled during a reverse transcription procedure. The targets reflect the amount of mRNA isolated from a sample obtained under a particular influence factor. Thus, the amount of fluorescence emitted by each spot will be proportional with the amount of mRNA produced from the gene having the corresponding DNA sequence. The microarray is scanned and the resulting image is analyzed such that the signal from each feature or probe can be quantified into some numerical values. Such values will represent the expression level of the gene in the given condition. Microarrays can be fabricated by depositing cDNAs or previously synthesized oligonucleotides; this approach is usually referred to as printed microarrays. In contrast, *in situ* manufacturing encompasses technologies that synthesize the probes directly on the solid support. Each technology has its advantages and disadvantages and serves a particular research goal. The microarray technology has a very high throughput, interrogating thousands of genes at the same time. However, the process includes numerous sources of variability. Several tools such as statistical experimental design and data normalization can help to obtain high quality results from microarray experiments.

Chapter 4

Reliability and reproducibility issues in DNA microarray measurements

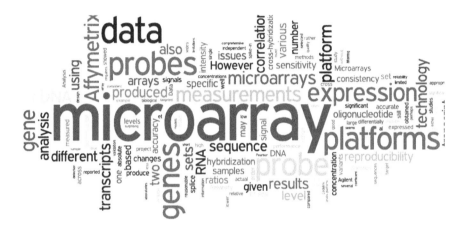

If reproducibility may be a problem, conduct the test only once.

—*Velilind's Laws of Experimentation*

4.1 Introduction

Since its introduction approximatively 15 years ago [374], the DNA microarray technology has evolved rapidly.[1] However, microarray results are still marred by several technical issues that are often neglected. This chapter will survey and in some cases expose the most relevant issues.

Although DNA microarrays are all based on the same fundamental principle of the complementary hybridization of nucleic acid strands, the available

---

[1] Most of the material in this chapter appeared in the article "Reliability and reproducibility issues in DNA microarray measurements" by Sorin Draghici, Purvesh Khatri, Aron C. Eklund and Zoltan Szallasi, published in Trends in Genetics Volume 22, Issue 2, February 2006, pages 101-109 and reproduced here with permission.

69

technical choices differ widely. An important distinction is the length of the probes used. Microarrays can be roughly categorized into cDNA arrays, usually using probes constructed with polymerase chain reaction (PCR) products of up to a few thousands base pairs, and oligonucleotide arrays, using either short (25–30mer) or long oligonucleotide (60–70mer) probes. In addition, the probes can be either contact-spotted, ink-jet deposited, or directly synthesized on the substrate. Each of these approaches has its own requirements in terms of the amount of RNA needed, data acquisition, transformation and normalization techniques, etc. [35]. These, as well as other differences in types and composition of probes, deposition technologies, labeling and hybridization protocols, etc., ultimately manifest as differences in accuracy, specificity, sensitivity, and robustness of different microarray platforms, which in turn have led to claims of reduced reproducibility of results from one platform to another. The goal of this chapter is to discuss such factors and the extent to which they can affect the outcome of a microarray experiment.

4.2   What is expected from microarrays?

Searching for key determinants of a phenotype at the gene expression level requires suitable coverage of the genome coupled with reasonable reproducibility, accuracy, and sensitivity. These limitations matter less if microarrays are used for screening purposes, because changes in individual genes are expected to be verified by independent means. However, the stakes were raised by the proposed application of microarray technology as a diagnostic tool in molecular disease classification [428, 447] because regulatory agencies, such as the Food and Drug Administration, require solid, empirically supported data about the accuracy, sensitivity, specificity, reproducibility, and reliability of microarrays (http://www.fda.gov/cdrh/oivd/guidance/1210.pdf). As it will be apparent later on, the first decade of microarray technology produced rather limited data pertinent to these issues.

4.3   Basic considerations of microarray measurements

There have been few methodologies in molecular biology where expectations were raised so high with so little actual evidence for the capabilities of the technology. This is not surprising considering that the academic community, perhaps with the exception of a few laboratories with above average funding and influence, had essentially no control over the development of the microarray technology. For example, for the first six years in which microarrays were available commercially, no probe sequence information was available whatsoever from most mainstream manufacturers. End users had to trust the manufacturer that a given probe actually quantified a specific transcript. It is surprising that researchers accepted the technology without this type of information. After the first few proof-of-concept papers, and under the influence of the commercial marketing hype, researchers adopted the technology in a scientific equivalent of keeping up with the Jones, without waiting for solid reproducibility data. Many cDNA microarrays had a substantial number of incorrect probes deposited [450, 236, 193, 420] and a surprisingly large portion of Affymetrix microarray probes (up to 30–40% depending on the actual chip) cannot be verified by high-quality sequence databases such as Refseq [306, 195]. Given this, it comes as no surprise that removing these probes from further analysis significantly improves the accuracy and reliability of the measurements [305]. In order to be fair to microarray manufacturers, it should be recognized that the design of highly specific microarray probes is a complex and difficult problem, and a certain percentage of failed probe design was probably unavoidable.

To its credit, Affymetrix subsequently released its probe sequences, but few competitors followed. This situation seems to be changing, in no small part due to a community-wide microarray validation project initiated and led by FDA researchers, the MicroArray Quality Control Project [336, 392, 88]. Making the microarray probe sequences accessible was not, however, a battle easily won. Only a few years ago, Affymetrix threatened to withdraw information that made it possible for open-source software projects, such as Bioconductor, to identify individual probes on their microarray chips, leaving the end users in the dark again [460].

The discrepancies between an intended probe sequence and the actual probe sequence synthesized or deposited on the microarray chip also deserve some attention. The synthesis of nucleotide chains, especially those performed on solid surfaces, such as the technology used by Affymetrix, Agilent, Combimatrix, and NimbleGen, is not 100% accurate. This means that microarray probes directly synthesized on substrates will contain a significant number of nucleotide chains that are shorter and consequently have a different probe sequence from the design sequence, due to base skipping [161]. Microarray platforms using probes that are HPLC purified and then deposited on the solid surface, such as the CodeLink Arrays [348], have the advantage of containing a nearly homogeneous population of probes. HPLC purification significantly increases the specificity of 30-mer, spotted microarray probes [348].

There are other important concerns in microarray analysis in addition to the problem of incorrect probes. Perhaps the most important issue is that microarray analysis is based on the assumption that most microarray probes produce specific signals under a single, rather permissive hybridization condition. As we will see further on, this is probably not true as testified by widespread cross hybridization of transcripts on microarrays [484].

Even if the probe design issues were solved, users would still be interested in a critical assessment of the sensitivity, accuracy, and reproducibility of microarray results. These issues are discussed below.

4.4 Sensitivity

Determining the sensitivity threshold of microarray measurements is essential in order to define the concentration range in which accurate measurements can be made. In order to do this, one needs to know the approximate concentration of a reasonable number of gene transcripts in the RNA sample that can be detected and quantified on a given microarray platform.

In an attempt to assess the dynamic range of microarrays Holland measured the range of transcript abundance for 275 genes in yeast using kinetically monitored reverse transcriptase-initiated PCR (kRT-PCR) [216]. These data were then compared with data from cDNA and oligonucleotide arrays. The yeast data obtained by cDNA and Affymetrix oligonucleotide arrays were in reasonable agreement down to the level of 2 copies per cell. However, below that threshold microarrays failed to produce meaningful measurements.

It was expected that microarray probes of varying length would provide various levels of trade-off between sensitivity, signal strength, and specificity, hence the wide variety of probe lengths tested for microarray application. Signal strength increases with probe length in a certain range. For example, on average, 30-mers provide twice the intensity of the signal produced by a 25-mer probe [352]. However, a further increase in the length of probes produced limited enhancement of signal intensity while the specificity of probes, as quantified by the relative intensity of the perfect match probes versus single base pair mismatch probes, actually decreased [352]. The sensitivity of 30-mer probes was also calibrated by known concentrations of target transcripts and found to be on the order of $1:10^6$ mass ratio [348]. A side-by-side comparison of RNA aliquots on the CodeLink (30 nt) and Affymetrix platforms (25 nt) also suggested an up to 10-fold higher sensitivity of the former platform [394].

The relative merit of microarray probes of various lengths inspired further studies. Kane et al. compared the sensitivity of probes using PCR products versus probes using spotted 50-mer oligonucleotides [245]. These results showed that for rat liver RNA microarrays had a minimum reproducible detection limit of approximately 10 mRNA copies/cell. Czechowski et al. also obtained similar results when comparing RT-PCR profiling of more than 1400 Arabidopsis transcription factors with the 22K Arabidopsis Affymetrix array [107]. While 83% of those genes could be reliably quantified by RT-PCR, Affymetrix gene chips could detect less than 55% percent of the 1400 transcription factors, which are usually expressed at the lower end of the dynamic range of the transcriptome (for most of them less than 100 copies/cell).

In summary, the detection limit of current microarray technology appears to be between 1 and 10 copies of mRNA per cell. This sensitivity threshold is probably lower for cell types with a more limited concentration range of transcripts such as yeast [216]. While this sensitivity is impressive, it might still be insufficient to detect relevant changes in low abundance genes, such as transcription factors [216, 107]. It remains to be seen whether novel technological developments such as labeling with quantum dots [282] may be able to further increase the sensitivity of microarray platforms.

4.5 Accuracy

Microarrays can be used to measure either: a) absolute transcript concentrations or b) relative transcript concentrations, i.e. expression ratios. In principle, accurate absolute concentration measurements will also provide accurate measurements of expression ratios, but the reverse does not necessarily hold true. Estimating ratios requires a less detailed understanding of how the signal intensity of a given microarray probe is related to the concentration of the measured transcript. As long as a probe binds its target specifically and the produced signal intensity is directly proportional to the amount of transcripts bound, the expression ratios will reflect reality to a significant extent. On the other hand, estimating absolute concentrations, especially at the current insufficient level of understanding DNA and RNA hybridization, requires careful calibration with known concentrations of the transcripts. As suggested by several types of evidence, the slope of the curves correlating transcript concentration with signal intensity seems to be rather uniform across a given microarray platform [107, 480, 96].

Traditionally, two-channel cDNA array data (e.g. cy3/cy5) is usually used to measure ratios[2] while single channel oligonucleotide array data (e.g., Affymetrix) have been used as representative of absolute expression values. However, some issues become apparent when examining the signal intensities produced by Affymetrix probes in two different probes of the same probe set that are targeted against closely placed or overlapping sequences on a given transcript. These are likely to hybridize to the same labeled RNA fragment and still may produce signals varying by orders of magnitude (see Fig. 4.1). This suggests that the same transcript concentration can produce rather different probe signal intensities depending on the probe, which in turn means that interpreting the measured signals as proportional to the absolute concentrations is not necessarily advisable. Ratios can be measured with a higher accuracy [107], which is reflected by the fact that probes in a given Affymetrix

---

[2]Usually but not always, see for instance the loop designs proposed by Churchill and co-workers [98, 259, 256, 124].

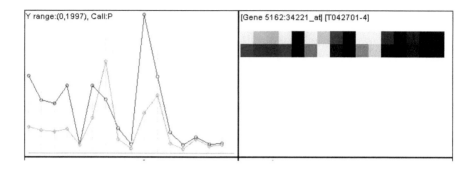

FIGURE 4.1: Different probes corresponding to the same gene can yield widely different signals. For the gene shown, probe 10 (from left to right) is close to the saturation level, while probes 5, 9, 13, 15, and 16 are close to the background level. This gene is called present and a reasonable high expression level calculated as average difference. This illustrates how sensitive the actual numbers are to the choice of the specific probes.

probe set (i.e., probes designed to recognize the same gene transcript) while producing significantly different intensity may still produce consistent ratio values across the very same probe set when two RNA samples are compared (see Fig. 4.2).

Assessing the accuracy of microarray measurements requires that true concentrations, or true concentration ratios, be available for a large number of transcripts. True concentrations can be obtained by a) spike-in or dilution experiments [231], or b) measuring transcript levels by independent means such as quantitative RT-PCR or Northern blots. A few spike-in or dilution data sets were provided by Affymetrix (Affymetrix-Latin square data) and its industrial partners. On the Affycomp website a wide variety of Affymetrix DNA-chip analysis methods were evaluated based on these data sets. (http://affycomp.biostat.jhsph.edu). However, the variance of average chip intensity among the spike-in data sets is much lower than those measured in most real-life data sets, casting doubts on the general applicability of these data for developing analytical tools for highly diverse clinical gene expression profiles. Furthermore, the limited number of spike in genes, 42 at most, also makes it difficult to use this data set for the comprehensive evaluation of both the technology and the data analysis methods. Even assuming that the genes were selected in a completely unbiased manner (i.e., assuming that genes known to produce good results were not favored and problematic transcripts were not ignored), this is a very small number of genes. In light of the strong dependency of the measurements on the specific target genes and the specific probe sequence selected, extrapolating the accuracy measured on 42 genes to another 10,000–30,000 genes appears more like an act of faith than

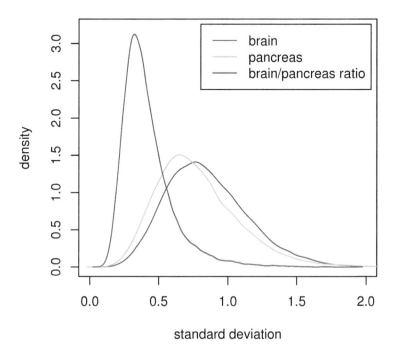

FIGURE 4.2: Probe intensity ratios are generally more consistent than absolute probe intensities. Probe-level intensities from two Affymetrix HG-U133A arrays were log2 transformed, and the standard deviation was calculated for the 11 probes in each probe set. The smoothed distributions of these standard deviations are plotted for brain (blue), pancreas (green), and the probe-level ratios of brain to pancreas (red). Data are from the GNF expression atlas [407]
.

a scientific inference. Some spike-in data sets were produced by some academic laboratories, as well. Choe and coworkers produced Affymetrix gene chip data for Drosophila RNA samples with about 1,300 spiked-in genes against a fairly well-defined background of roughly 2,500 genes [94]. This data set produced reassuring results when combined with the most appropriate of their data analysis methods. The agreement between observed and actual fold changes was good ($R^2 = 0.86$) when the probe sets in the lowest quartile of signal intensity were filtered out. This latter filtering step seems to be the single most important preparatory step to ensure accurate and consistent microarray measurements. Their results also suggested that the detection of approximately 70% of true positives can be achieved before reaching a 10% false-discovery rate. However, this seems to add further support to the idea that microarray measurements may not be reliable for genes expressed at low levels. A similar, large-scale spike-in data set for human genes would also be most welcome for calibration purposes.

The alternative to spike-in based validation requires the independent quantification of transcripts by RT-PCR or Northern blot analysis. Because of the cost associated with independent verifications, most such studies measure expression levels by independent means only for a limited number of transcripts, typically on the order of a few tens of genes [182, 272]. The transcripts selected for verification usually belong to the class of widely studied genes, with well agreed upon sequences, and for which reliable, often commercially available RT-PCR primers exist [272]. For these genes, considering the high quality of the associated annotations, one would expect a reasonable level of accuracy. Indeed, widely used microarray platforms, such as the Affymetrix GeneChip, produce verifiable differential expression calls in about 85–90% of the cases [182]. However, this applies only to the range of expression levels, which is above the sensitivity threshold of the given microarray platform effectively eliminating from analysis perhaps up to 40-50% of the transcripts present in the RNA samples. Furthermore, the failure to detect the correct fold changes for a highly relevant gene such as the Epidermal Growth Factor Receptor (EGFR) [182], a gene often implicated in cancer diagnostics, should also encourage a pause for reflection for those interested in the diagnostic applications of microarrays.

A few studies produced independent measurements by RT-PCR for a more comprehensive set of transcripts ranging in number from about 50 to 1400 [216, 107, 480]. The MicroArray Quality Control Project produced the first such comprehensive data set for human, covering about one thousand genes [392]. Based on these studies the picture emerging for the most widely used microarray platforms such as cDNA microarrays and Affymetrix GeneChips seems to warrant the following conclusions:

- In their appropriate dynamic range, microarray measurements accurately reflect the existence and direction of expression changes in approximately 70–90% of the genes. However, the magnitude of such expression changes as reflected by microarray data tends to be different from the magnitude of the changes measured with other technologies such as RT-PCR.

- Microarrays (both single and dual channel) tend to measure ratios more accurately than absolute expression levels. For example, in the most comprehensive such data set, quantifying 1,400 genes by RT-PCR Czechowski et al. found a poor correlation between normalized raw data produced by RT-PCR and normalized raw data produced by Affymetrix arrays in the same RNA sample. However, when two different samples, RNA from shoots and roots of Arabidopsis, were compared, a more promising result emerged. The ratios between these two RNA samples extracted from RT-PCR and array measurements showed a Pearson correlation of 0.73 for the most highly expressed set of 50 genes. Similar results were obtained in other studies as well. However, one should note that a correlation of around 0.7 is not impressive for two platforms that are expected to measure the abundance of given transcripts in the same samples.

- The relatively good correlation between microarray-based and RT-PCR-based gene expression ratios does not necessarily mean that microarray technology directly produces accurate estimates on gene expression ratios. In fact, it has been well known for a long time that microarray based expression ratios are compressed [480], i.e., the ratios of mRNA expression levels are consistently underestimated. It seems that the ratio compression is a significantly more consistent phenomenon for cDNA microarrays than for short oligonucleotide chips [480].

In conclusion, a handful of independent validation studies and spike-in data sets have allowed an empirical assessment of accuracy of microarray technology. While accurate measurement of absolute transcript levels by microarrays is probably beyond the current capabilities of the technology, ratios can be estimated reasonably well, especially when the significant level of ratio compression is taken into consideration and corrected for. However, this favorable assessment applies only to the measurement of transcripts that are expressed well above the sensitivity level of microarrays, rendering perhaps half of the transcriptome beyond the reach of microarrays.

4.6 Reproducibility

Reproducibility is the most readily assessable characteristic of any microarray platform. Unfortunately, a specific platform can have an excellent reproducibility without necessarily producing measurements that are accurate or consistent with measurements from other platforms. A good reproducibility simply requires that a given probe bind the same number of labeled transcripts in repeated measurements of the same sample. Badly designed probes that perhaps cross-hybridize with a number of other transcripts besides the targeted one can easily provide highly reproducible and yet useless data. Therefore, reproducibility is a necessary but completely insufficient requirement.

Indeed, in their appropriate sensitivity range, most microarray platforms produce highly reproducible measurements. Oligonucleotide arrays (Affymetrix, Agilent, and Codelink) [30, 32] seem to have an advantage providing correlation coefficients of above 0.9. For other platforms, such as cDNA microarrays or the Mergen platform, the reproducibility may still be an issue. On such platforms, the reported Pearson correlation coefficient between technical replicates can range between the disappointing level of 0.5 and the reassuring level of 0.95 [236, 32, 237]. It is, therefore, not surprising, as discussed below, that these latter platforms show poor correlation with commercial oligonucleotide-based platforms [32, 474].

4.7 Cross-platform consistency

If microarray data were highly reproducible across various platforms and if they provided information about the absolute transcript levels, one could use appropriately normalized gene expression data without regard to the platform on which the data was obtained. This in turn would reduce the need to replicate experiments and allow researchers to build universal gene expression databases that would compile many different data sets from a variety of experimental conditions. This consideration is particularly relevant to microarray analysis of clinical samples with limited amounts of mRNA.

Because of the relative scarcity of comprehensive, large-scale, spike-in or independently measured gene expression data sets, cross-platform consistency has been used as a surrogate measure of microarray reliability. In this approach, aliquots from the same RNA sample, or RNA isolated from the same biological source, are profiled on different microarray platforms. The consistency of these results is considered an indication of the reliability of all platforms compared. Lack of consistency can be caused by the inferior performance of at least one of platforms, without clear indication of their relative merit. Interpreting the cross-platform consistency as a proof of accuracy and reliability is tempting but not scientifically sound because highly similar results across platforms could be simply caused by consistent cross-hybridization patterns without either platform measuring the true level of expression. Nevertheless, a high level of cross-platform consistency is desirable, because, if both platforms performed accurate measurements, then cross-platform consistency would automatically follow. In other words, cross-platform consistency is a necessary but not sufficient requirement in order to validate the technology. Despite its obvious limitations, cross-platform consistency studies produced several useful lessons for microarray users.

Cross-platform comparison of various microarray platforms depends on the availability of data sets based on the same RNA aliquots profiled on different microarray platforms. Until recently there was only one widely available such data set, the NCI60 cell line panel profiled using cDNA microarray [376] and the Affymetrix platform [401]. These two data sets were reanalyzed several times for cross-platform consistency with gradually improving results, highlighting the importance of probe sequence verification. One of the difficulties in the cross-platform comparison of microarray data is to ascertain that probes on the various platforms aimed at the same gene do in fact quantify the same mRNA transcript. The various strategies to match probes between different platforms can be constrained by the amount of information provided by the manufacturers of the given microarray. Before actual probe sequence information was released probe matching could be based only on gene identifiers such as the Unigene ID [271]. This strategy is known to produce a significant number of incorrect pairings [274, 305]. Therefore, it is not surprising that in an early study, while comparing the two NCI60 data sets using this microarray probe matching strategy, Kuo et al. found an alarming level of inconsistencies (Pearson correlation being less than 0.34). In order to measure the extent of within-array cross-hybridization, Kuo et al. observed that the genes represented by cDNA probes with a higher number of cross-matches to other genes (defined as sequence similarity using BLAST) have lower correlation with the oligonucleotide data. This suggested that cross-hybridization may be a possible cause for the poor cross-platform consistency. They also found very low correlations for genes with low intensity values on cDNA arrays (Pearson coefficient 0.03 and Spearman coefficient 0.02) and low average difference in the Affymetrix arrays, indicating that the low-abundance transcripts were not measured reliably on either platform.

As partial or complete probe sequence data have become available, more accurate strategies could be implemented. Probes could be matched across the various platforms based on whether they can be sequence-mapped to the same transcript. When microarray probe pairs across the platforms that obviously mapped to different transcripts, while still sharing, erroneously, the same UniGene ID, were filtered out, the mean correlation of gene expression between the two previously described NCI60 data sets increased to 0.6 [274]. Similar results were obtained in a study when RNA aliquots were profiled and compared across several Affymetrix platforms and the Agilent Human 1 cDNA microarray platform [305]. UniGene matched probes that failed the direct sequence mapping test showed significantly lower expression correlation across the two microarray platforms [305].

Finally, probe sequences can be used to ascertain that microarray probes on different platforms are targeted against the same region of a given transcript. This ensures that the two platforms are quantifying the same splice variants but also increases the chance of similar undesired cross-hybridization patterns. For the two NCI60 data sets probes targeting the same region of the transcripts showed the highest correlation (mean Pearson coefficient of around 0.7) [81]. This relatively high correlation, however, also required filtering out genes producing low intensity signals. The rigorous sequence mapping strategy employed by Carther et al. also ensured that only those Affymetrix probes that could be verified by high quality sequence databases were used in the final analysis. By the application of an appropriate, "in silico" produced common reference for Affymetrix gene chips, the two data sets (Affymetrix and cDNA microarray) could be pooled, and hierarchical cluster analysis produced meaningful results on the thus combined gene expression profiles [81]. The results of this study seem to answer positively the important question regarding whether microarray data coming from different labs can be assembled into a coherent unique database. However, this study also shows that if such a "universal database" is to ever be constructed, this should not be done by merely storing in a common database expression data reported by the various platforms but rather but revisiting the biochemical foundations of the technology and using such knowledge in order to interpret and filter the numerical data generated by the arrays.

In a much-cited report, Tan et al. produced gene expression profiles on both technical and biological replicates of RNA samples on three different platforms: the oligonucleotide-based CodeLink arrays from GE Healthcare, Affymetrix GeneChips, and cDNA arrays from Agilent [414]. The intraplatform consistency was high for both types of replicates for all platforms (>0.9), as consistently reported by both manufacturers and other laboratories. However, the Pearson correlation coefficient was more moderate across the various platforms: the correlation of matched gene measurements between oligonucleotide arrays (Affymetrix and Codelink) was the highest (0.59), while the correlation between cDNA and oligonucleotide arrays was lower (0.48–0.50). However, because of the lack of comprehensive probe sequence information, it was not possible to apply rigorous sequence mapping criteria, and the above listed correlation coefficients were calculated without filtering out genes with low expression levels. Without the application of these noise-reducing strategies, it is not surprising that the three platforms showed a rather disappointing level of concordance in their ability to predict gene expression changes: close to 200 genes were predicted to be differentially expressed by at least one platform, but only four genes were detected as differentially expressed by all three platforms. This type of result should again invite caution for any set of results in which probe-level sequence matching and low-level filtering has not been performed. While disappointing in terms of concordance between the specific genes reported as differentially regulated, this work also showed that meaningful biological conclusions can still be obtained by a higher-level analysis, in which the sets of differentially regulated genes are mapped on their associated biological processes, cellular locations, etc., using GO annotations. While the specific differentially regulated genes were very different between the platforms, all platforms agreed very well in terms of the biological processes involved. Fortunately, several tools and techniques now exist that can automate this type of analysis for large number of genes, thus helping researchers circumvent some of the limitations of the technology [264, 129, 261, 265, 263]. The above-described discrepancies reported in the literature prompted several well-controlled studies aimed at determining the overall level of cross-platform consistency across a wide variety of platforms. Jarvinen et al. compared Affymetrix, Agilent, cDNA, and custom cDNA microarrays using human breast cancer cell lines [236]. At first glance, the results looked promising with correlation coefficients ranging between 0.6 and 0.86 for the various combinations of platforms and data analysis methods. However, a closer look reveals that the excellent correlation is produced by fewer than 20% of the genes contained on these microarray platforms. Matching microarray probes by Unigene IDs and using various other filtering steps, such as removing genes with absent calls on Affymetrix arrays or cDNA probes that did not pass quality control criteria, eliminated the majority of the well over 10,000 genes contained on these platforms, leaving only one to two thousand genes. This and other studies also provided experimental evidence that in up to 50% of the cases contradictory results between cDNA microarrays and short oligonucleotide-based platforms can be explained by the placement of incorrect clones on the cDNA microarray [193, 420]. This further emphasizes the importance of verifying the sequence of the microarray probes. Similar results were produced by a more recent study using mouse RNA samples using the corresponding Affymetrix and cDNA microarrays [272]. The differential gene expression calls produced by the platforms yielded a good correlation for about 90% of the so-called "good" genes. This was a subset of about six thousand genes selected by various quality filtering procedures from the 30,000–40,000 transcripts represented on the two platforms.

The performance of less frequently used platforms were also assessed recently. In their analysis of mouse microarray platforms, Yauk et al. added the Mergen and Agilent oligonucleotide platforms to the Affymetrix, Codelink and cDNA microarray platforms [474]. After intensity filtering for "present" calls, ratios measured by the Affymetrix, Codelink, and Agilent oligo arrays showed a more satisfactory correlation among themselves (above 0.7), whereas the custom cDNA microarray and the Mergen platform showed a significantly lower correlation (correlation coefficient of 0.5 or below) with the other platforms. The Agilent cDNA microarray platform placed in between these two groups. The Toxicogenomics Research Consortium has run an even wider comparison of the various mouse microarrays [32]. In addition to the commercial oligo arrays from Affymetrix, Agilent, and GE Healthcare (Codelink), spotted oligo arrays from Compugen and spotted cDNA microarrays from two different sources (TIGR and NIA) were also analyzed. In addition to cross-platform comparability, this project also examined the bias introduced when the same platform was used by different laboratories analyzing aliquots from the same RNA sample. For the five hundred genes represented on all platforms, the cross-platform consistency varied between 0.11 (Codelink versus spotted cDNA) and 0.76 (two different version of spotted cDNA microarrays). When the same platform was used by two different laboratories, the Affymetrix platform produced by far the highest cross-laboratory correlation (0.91).

Considering the often conflicting interests of government regulatory agencies, academia, and industry, the continuing interest of both the scientific and public media in microarrays as the "spearheading technology" of the postgenomic biology, it is not surprising that the above summarized body of literature failed to produce a generally accepted assessment of the technology. It is the typical "half-empty/half-full" situation, depending on the direction from which one is approaching the question. A significant portion, probably between 30% and 50%, of the transcriptome can be reliably analyzed by microarray technology in terms of the direction of expression changes. After appropriate corrections, gene expression ratios can also be estimated with a reasonable accuracy for these genes. The rest of the transcriptome, including the majority of transcription factors, may be below the sensitivity level of the technology or cannot be reliably quantified by microarrays yet. Since relatively simple approaches, such as removing incorrect probes from further analysis [305], can significantly improve the accuracy of microarray measurements it is worth discussing the causes of these inaccuracies and inconsistencies. Understanding these can lead to further improvement of the technology either by improving probe design or by removing obviously misleading microarray probes.

4.8 Sources of inaccuracy and inconsistencies in microarray measurements

As a reasonable approximation, signals produced by any given microarray probe can be considered as the composite of three signals: 1) specific signal produced by the originally targeted labeled transcript; 2) cross-hybridization signal produced by transcripts that have a non-perfect but still significant sequence similarity with the probe; 3) nonspecific background signal, which is present in the absence of any significant sequence similarity.

On an ideal, high-specificity microarray platform the second and third component would be negligible relative to the targeted specific signal. However, even under such ideal conditions microarray technology in its current state would face significant limitations for a number of reasons as discussed below.

First, the relationship between probe sequences, target concentration, and probe intensity is rather poorly understood. A given microarray probe is designed as a perfect complementary strand to a given region of the transcript. Based on the Watson-Crick pairing, the probe will capture a certain number of the transcripts. This number is proportional to the concentration of the transcript, but the actual relationship between transcript concentration and the number of molecules bound to the probe, and thus the signal produced, also depends on the affinity of the probe, or free energy change values, under the given hybridization conditions. This affinity is determined to a large extent by the actual nucleotide sequence stretch participating in the binding. This sequence-affinity relationship is rather poorly understood. While the sequence dependence of DNA/DNA hybridization in solutions has been studied in detail [367], DNA/RNA hybridization received significantly less attention. Remarkably, the results of Sugimoto et al. [410] suggest that the sequence dependence of DNA/RNA hybridization may still hold surprises. For example, for certain sequences, the binding energy of a DNA/RNA duplex can be stronger for a single mismatch than for the corresponding perfectly complementary strands [410, 319]. The kinetics of hybridization is further complicated by the incorporation of modified nucleotides into the target transcripts during the most widely used labeling protocols. Furthermore, the results obtained in solutions cannot be directly applied to the hybridization of microarray probes attached to surfaces [341]. Various authors have tried to investigate the dependence of affinities on the microarray probe sequence, but no convincing model has emerged yet. In some cases the lack of appropriate data constrained the analysis [485, 200]. In others, the general strategy of model generation was published without releasing the accompanying data or the actual model [307].

Second, splice variants constitute another dimension that can introduce difficulties in the microarray analysis. It is estimated that at least half of the human genes are alternatively spliced, and a single gene may have a large number of potential splice variants [314]. A given short oligonucleotide probe is targeted at either a constitutive exon (present in all splice variants) or at an exon specific for certain splice variants. In the first case, the probe intensity will reflect the concentration of all splice variants present in the sample, therefore obscuring expression changes occurring in certain splice variants. In the latter case, the specific splice variant will be measured, but other splice variants of the same gene will be ignored. Covering the various types of exons on short oligonucleotide-based arrays is necessary to dissect the splice variant associated composite signals. cDNA microarrays usually have a unique long probe with which they have to measure the abundance of all splice variants. This may be one other phenomenon explaining the discrepancies often observed between cDNA and short oligonucleotide microarray.

Third, folding of the target transcripts [313] and cross-hybridization [484] can also contribute to the variation between different probes targeting the same region of the transcript. It has been shown previously that a large fraction of the microarray probes produce significant cross-hybridization signals [484, 464] for both oligonucleotide-based and cDNA microarrays. Even a limited stretch of sequence complementarity may be sufficient to allow binding between unrelated sequences. It is not an easy task, however, to evaluate the overall impact of cross-hybridization on the accuracy of microarray measurements. In the case of Affymetrix arrays, for example, the effect of a single cross-hybridizing probe can be down-weighted by the rest of the probe set (currently 10 other probes). Furthermore, the impact of cross-hybridization strongly depends on the relative concentration and the relative affinities of the correct target and the cross-hybridizing target(s). The latter must be present at sufficient quantities in order to interfere with specific signals. Cross-hybridization, in conjunction with splice variants, is probably a prime candidate to explain the discrepancies in differential gene expression calls between various microarray platforms, although no systematic study has been undertaken along those lines yet.

The greatest impact of cross-hybridization is perhaps on the tiling array measurements. In this, no attempt is made to select highly specific probes, which are selected in an automatic fashion in order to cover a certain segment of the genome at regular, closely placed intervals. The significant level of discrepancies between the various "tiling array"–based data sets [241] is certainly caused, at least in part, by the widespread cross-hybridization due to non-optimized probe design. Removing and/or redesigning the microarray probes prone to cross-hybridization is a reasonable strategy to increase the hybridization specificity and hence, the accuracy of the microarray measurements. However, this requires a good understanding of cross-hybridization towards which only limited progress has been made due to the lack of appropriate experimental data.

In light of the above-described complexity of microarray signals, issues such as the compression of expression ratios can be reasonably explained. The presence of cross-hybridization signals on a given probe, for example, may prevent the detection of large changes in gene expression levels since a probe will always produce a certain level of "false" signal, even if the true signal is much lower or perhaps undetectable.

As our understanding of splice variants, specific and nonspecific nucleic acid hybridization and other relevant issues deepens, we will be able to design probes that will quantify transcripts in an increasingly optimal fashion. The quest for increasing microarray performance by regularly eliminating and redesigning probes can be, for example, easily tracked in the case of the Affymetrix technology, when only a fraction of probes are retained on successive generations of microarray chips even for probe sets targeting the same gene [324]. Although noble in purpose, this constant probe redesign has the very undesirable side effect that data sets obtained on different generations of arrays cannot be combined in any easy way.

4.9   The MicroArray Quality Control (MAQC) project

The MicroArray Quality Control (MAQC) project is a community effort to assess the quality and reliability of microarray technologies. This project was divided into 3 phases. The goals of phase I (MAQC-I) were as follows:[3]

- Provide quality control (QC) tools to the microarray community to avoid procedural failures

- Develop guidelines for microarray data analysis by providing the public with large reference data sets along with readily accessible reference RNA samples

- Establish QC metrics and thresholds for objectively assessing the performance achievable by various microarray platforms and

- Evaluate the advantages and disadvantages of various data analysis methods

According to the FDA source cited above: "MAQC-I involved six FDA Centers, major providers of microarray platforms and RNA samples, EPA, NIST, academic laboratories, and other stake-holders. Two human reference RNA samples have been selected, and differential gene expression levels between the two samples have been calibrated with microarrays and other technologies (e.g., QRT-PCR). The resulting microarray datasets have been used for assessing the precision and cross-platform/laboratory comparability of microarrays, and the QRT-PCR datasets enabled evaluation of the nature and magnitude of any systematic biases that may exist between microarrays and QRT-PCR."

The results of this effort were published in the September 2006 issue of *Nature Biotechnology* [392]. The authors concluded that the experiments showed "intraplatform consistency across test sites as well as a high level of interplatform concordance in terms of genes identified as differentially expressed." However, as shown in Fig. 4.3, the overlap of the top 50 genes found to be differentially expressed at 2 different sites in exactly the same biological samples is only about 80% even when the selection is done with the most permissible methods. This is probably acceptable for most experiments but also shows that there is room for improvement.

The goals of the second phase of the MAQC project (MAQC-II) were:[4]

- Assess the capabilities and limitations of various data analysis methods in developing and validating microarray-based predictive models

---

[3]http://www.fda.gov/ScienceResearch/BioinformaticsTools/
MicroarrayQualityControlProject/default.htm#MAQC-I

[4]http://www.fda.gov/ScienceResearch/BioinformaticsTools/
MicroarrayQualityControlProject/default.htm#MAQC-II

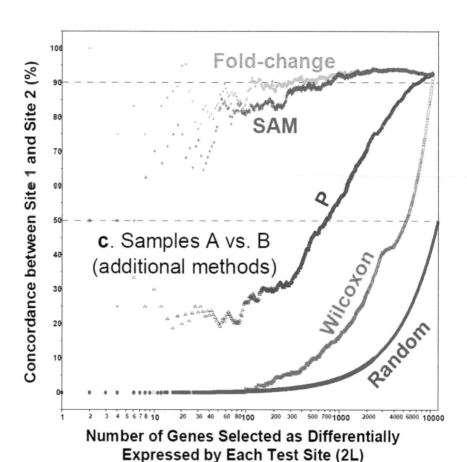

FIGURE 4.3: Some results from the MAQC-I project comparing two samples: A= Universal Human Reference RNA from Stratagene and B= Human Brain Reference RNA from Ambion. Note that if only 50 genes are to be selected as differentially expressed (DE), the overlap between the list of DE genes obtained at two sites ranges from zero (Wilcoxon) to 20% (various p-values, to 70% (SAM), and to about 80% (fold change). Note that these are experiments performed on exactly the same samples.

- Reach consensus on the "best practices" for development and validation of predictive models based on microarray gene expression and genotyping data for personalized medicine

Thirty-six teams developed classifiers for 13 endpoints from six relatively large training data sets. These analyses collectively produced more than 30,000 models that were assessed by independent and blinded validation sets generated for this project. The results of this study were reported in the August 2010 issue of Nature Biotechnology. According to the FDA source cited above: "The cross-validated performance estimates for models developed under good practices are predictive of the blinded validation performance. The achievable prediction performance is largely determined by the intrinsic predictability of the endpoint, and simple data analysis methods often perform as well as more complicated approaches. Multiple models of comparable performance can be developed for a given endpoint and the stability of gene lists correlates with endpoint predictability. Importantly, similar conclusions were reached when more than 12,000 new models were generated by swapping the original training and validation sets." The authors also found that "model performance depended largely on the endpoint and team proficiency," which perhaps was one of the reasons that determined the journal to summarize this paper as follows: "The MAQC consortium's latest study suggests that human error in handling DNA microarray data analysis software could delay the technology's wider adoption in the clinic."

4.10 Summary

Microarrays are a popular research and screening tool for differentially expressed genes. Their ability to simultaneously monitor the expression of thousands of genes is unsurpassed. However, certain limitations of the current technology exist and became more apparent during the past couple of years. In its appropriate sensitivity range, the existence and direction of gene expression changes can be reliably detected for the majority of genes. However, accurate measurements of absolute expression levels and the reliable detection of low abundance genes are currently beyond the reach of microarray technology. Because of this, the ability of detecting changes in the expression of specific individual genes may be affected. Various efforts to assess the reliability, accuracy, and reproducibility of microarrays, including the MAQC project, reported mixed, but generally optimistic results. The analysis of the microarray results at a higher level (e.g., biological processes or pathways) may be more meaningful than the analysis of fold changes of specific genes as reported by any given platform. Basic research in the areas of nucleic acid hybridization, and technological advances in detection methods and hybridization conditions will certainly increase the measurement capabilities of microarray technology.

Chapter 5

Image processing

Not everything that can be counted counts, and not everything that counts can be counted.

—Albert Einstein

5.1 Introduction

The main goal of array image processing is to measure the intensity of the spots and quantify the gene expression values based on these intensities. A very important and often neglected goal is also assessing the reliability of the data, and generating warnings signaling possible problems during the array production and/or hybridization phases. This chapter is divided into two parts. Section 5.2 provides a very short description of the basic notions involved in digital imaging. Section 5.3 and following focus on image processing issues specific to microarrays.

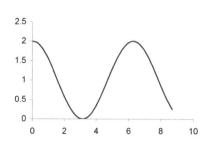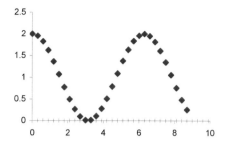

FIGURE 5.1: Sampling is the process of taking samples (or reading values) at regular intervals from a continuous function.

5.2 Basic elements of digital imaging

In computer processing, the analog image must undergo an analog to digital (A/D) conversion. This procedure consists of sampling and quantification. **Sampling** is the process of taking samples or reading values at regular intervals from a continuous function (see Fig. 5.1). In image processing, the continuous function is a continuous voltage waveform that is provided by an analog sensor. The sampling process transforms the original continuous function into an array of discrete values. This is the sampled function. There is a complex theory that established how many samples need to be taken in order to capture the information present in the continuous function and what happens when fewer samples are captured. However, these issues are beyond the scope of the present text.

Analog images can be seen as two-dimensional continuous functions. If the image is scanned along a particular direction, the variation of the intensity along that direction forms a one-dimensional intensity function. For instance, if we follow a horizontal line in an image, we can record how the intensity increases and decreases according to the informational content of the image. When the image is sampled, the continuous intensity variation along the two directions of the image is captured in a rectangular array of discrete values (see Fig. 5.2). Such rectangular array of sampled intensity values forms a **digital image**. Each of these picture elements is called a **pixel**.

Typically, the image array is a rectangle with dimensions that are a power of 2: $N = 2^M$. This is done in order to simplify the processing of such images on numerical computers that use base 2. Table 5.1 shows the dimensions in pixels of a square image of size $N = 2^M$ for increasing values of M. It is important to understand that the number of pixels is completely independent of the physical dimensions of the image. The same image can be scanned at 128×128 pixels

columns

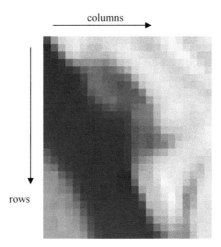

rows

FIGURE 5.2: A digital image. When the image is sampled, the continuous intensity variation along the two directions of the image is captured in a rectangular array of discrete values.

M	Resolution	Total number of pixels
7	128×128	$2^{14} = 16,384$
8	256×256	$2^{16} = 65,536$
9	512×512	$2^{18} = 262,144$
10	1024×1024	$2^{20} = 1,048,576$

TABLE 5.1: Total number of pixels for various resolutions.

or 1024×1024. This is simply an issue of how often do we read values from any given line and how many different lines we will consider. The number of pixels used for scanning is called resolution and is usually specified as a combination of two numbers: number of lines read and number of values read for each line. For instance, an image taken at 1600×1200 will contain 1600 lines and each line will have 1200 values for a total of $1600x \times 1200 = 1,920,000$ pixels. This is the figure used to describe the capabilities of computer monitors or digital cameras. It is essential that the image be sampled sufficiently densely or the image quality will be severely degraded.

For continuous functions that change in time, this requirement is expressed mathematically by the **sampling theorem**, which relates the number of samples to the highest frequency contained in the signal. Intuitively, the sampling theorem says that at least two samples must be taken for each period of the highest frequency signal. If the sampling is performed at this minimum rate or better, the original signal can be reconstructed perfectly from the sampled signal. If the sampling is performed at a rate below this minimum, distortions called **aliasing effects** occur.

An analogue image can be seen as a continuous function that relates intensity to spatial coordinates. For images, performing a sufficiently dense sampling reduces to having a sufficient resolution for the digital image that will be used to capture the information in the analogue image. The importance of the resolution is illustrated in Fig. 5.3. In a low-resolution image, a pixel will represent a larger area. Since a pixel can only have a unique value, when such a digital image is displayed, one can notice large rectangular areas of uniform intensity corresponding to the individual pixels. Robust quantification can be obtained only if the image is scanned with a good resolution. A rule of thumb is to have the pixel size approximatively $1/10$ of the spot diameter. If the average spot diameter is at least 10–12 pixels, each spot will have a hundred pixels or more in the signal area ($A = \pi r^2$, where A is the area and r is the radius), which is usually sufficient for a proper statistical analysis that will help the segmentation.

Another important notion in digital image processing is the color depth. This term refers to how many different values can be stores in a single pixel, i.e. how many different colors or shades of gray that pixel can take. The color depth is directly dependent on the amount of memory available for each pixel. This is another typical example of digitization. **Digitization** is the process of converting analogue data to digital data. If an analogue quantity or measurement such as the intensity level in a gray-level image needs to be stored in a digital computer, this analogue quantity has to be converted to digital form first. The first step is to decide the amount of memory available to store one such intensity value. Let us assume that one **byte** is available for this purpose. One byte equals eight bits. Eight bits can store $2^8 = 256$ different values: $0, 1, \ldots, 254, 255$. Once this value is known, the range of the analogue value to be represented is divided into 256 equal intervals. Let us say the intensity can take values from 0 to 1000. The intervals will be $[0, 1/256], [1/256, 2/256], \ldots, [254/256, 255/256]$. This mapping between intervals of the analogue value and the fixed number of discrete values $0, 1, \ldots, 254, 255$ available to represent it is the digitization. Every time an analogue value is measured, the interval that contains it is found and the corresponding digital value is stored. For instance, if the analogue reading is in between $1/256$ and $2/256$, which is the second interval available, the value will be stored as 1 (the first value is always 0).

The **color depth** of an image is the number of bits used to store one pixel of that image. An image with a color depth of 8 will be able to use 256 different colors. The computer makes no distinction between storing colors or gray levels so 256 colors may actually be 256 different levels of gray. Table 5.2 presents a few common color depths together with the number of colors available in each case. The first row of the table corresponds to a binary image in which each pixel is represented on one bit only and can only be black or white. The last row corresponds to a color depth of 22 bits, which allows for 4 million colors. A color depth of 24 bits or more is sometimes referred to as "true color"

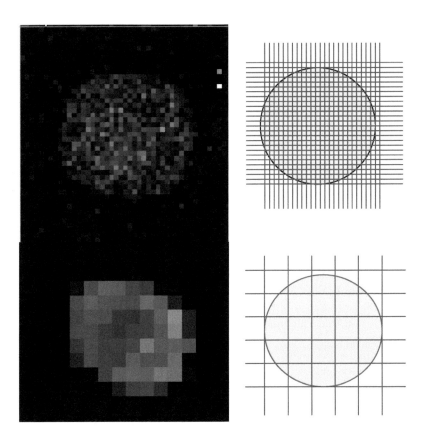

FIGURE 5.3: The effect of the resolution. Two spots scanned at high resolution (top) and low resolution (bottom). A low resolution means that a pixel, which is a unique value, will represent a larger area of the original image. Usually the value of the pixel will be some type of average of the intensity in the area represented by the given pixel.

Color depth	Number of colors available	Colors available
1	2	0, 1
2	4	0, 1, 2, 3
8	256	0, 1, . . . , 255
16	65,536	0, 1, . . . , 65, 535
22	4,194,304	0, 1, . . . , 4, 194, 303
24	16,777,216	0, 1, . . . , 16, 777, 215

TABLE 5.2: Color depths and number of colors available.

since an image using this many colors will be able to reproduce accurately the natural colors that surround us (i.e., in such a way that our eyes would not be able to distinguish the difference).

The color depth issue hides another possible danger. Fig. 5.4 shows what happens when the color depth is not sufficient. The image to the left uses 8 bits/pixel and is of reasonable quality. The middle image uses 4 bits/pixels. Artifacts can already be noted in particular on the sky behind the photographer. What was a smooth gradual transition in the original 8 bits/pixel image is now substituted by three clearly delimited areas. The original image used a range of many values to represent that particular area of the sky. When the color depth was reduced, the same interval was mapped to only three different gray levels. Thus, the gradual smooth transition was replaced by three areas each displayed with one of the three available shades of gray. The examination of the third image displayed at 2 bits/pixel is even more interesting. Now, only 4 different shades of gray are available. A superficial look may lead to the conclusion that this image is actually better than the 4 bits/pixels image in the middle or even the original at 8 bits/pixels. This is because the contrast was increased by bringing the coat of the photographer to a darker level (actually black) while bringing the sky to a lighter level. However, a more thorough examination shows that whole areas of the image have lost their information content (see for instance the tower behind the photographer which disappeared completely in this image). This is again because different shades of gray (the ones used for the tower and the sky immediately behind it) are now mapped to the same value, making those pixels indistinguishable. Thus, pixels which were part of different objects now have exactly the same color and the objects simply disappear.

It is important to note that the color depth and resolution are orthogonal. In other words, what is lost in terms of resolution cannot be recovered by using a higher color depth or vice versa. In practical terms, this means that we need to make sure that *both* color depth and resolution are suitable for our purposes when using digital images. For microarray applications, the usual color depth is 16 bits/pixels, which allows for 65,535 shades of gray. As mentioned elsewhere, the scanning resolution should be such that the diameter of a spot is 10 pixels or more.

FIGURE 5.4: The effects of an insufficient color depth. Left to right the images use the same resolution and a color depth of 8, 4, and 2 bits/pixel, respectively. Note what happens to the sky in the middle image (4 bits/pixel) and to the tower in the background in the image to the right (2 bits/pixel). (Reprinted with permission from Al Bovik Ed., *Handbook of Image and Video Processing*, Academic Press, 2000.)

5.3 Microarray image processing

A typical two-channel or two-color microarray experiment involves two samples such as a tumor sample and a healthy tissue sample. RNA is isolated from both samples. Reverse transcription is carried out in order to obtain cDNA and the products are labeled with fluorescent dyes. One sample (for instance, the tumor) is labeled with the red fluorescent dye and the other (the healthy tissue) with the green dye. The labeled cDNAs are hybridized to the probes on the glass slides and the slides are scanned to produce digital images.

For each array, the scanning is done in two phases. First, the array is illuminated with a laser light that excites the fluorescent dye corresponding to one channel, for instance, the red channel corresponding to the tumor sample. An image is captured for this wavelength. In this image, the intensity of each spot is theoretically proportional to the amount of mRNA from the tumor with the sequence matching the given spot. Subsequently, the array is illuminated with a laser light having a frequency that excites the fluorescent dye used on the green channel corresponding to the healthy tissue. Another image is captured. The intensity of each spot in this second image will be proportional to the amount of matching mRNA present in the healthy tissue. Both images are black and white and usually stored as high resolution Tag Image File Format (.tiff) files. For visualization purposes, most of the software available create a composite image by overlapping the two images corresponding to the individual channels. In order to allow a visual assessment of the relationship between the quantities of mRNA corresponding to a given gene in the two channels, the software usually uses a different artificial color for each of the two channels. Typically, the colors used are red and green to make the logical connection with the wavelength of the labeling dye. If these colors are used, overlapping the images will produce a composite image in which spots will have colors from green through yellow to red as in Fig. 5.5. Let us consider a certain gene that is expressed abundantly in the tumor tissue and scarcely in the healthy (e.g., spot 2 in Fig. 5.5). The spot corresponding to this gene will yield an intense spot on the red channel due to the abundant mRNA labeled with red coming from the tumor sample (upper right in Fig. 5.5). The same spot will be dark on the green channel since there is little mRNA from this gene in the healthy tissue (upper left in Fig. 5.5). Superposing the two images will produce a red spot (lower panel in Fig. 5.5). A gene expressed in the healthy tissue and not expressed in the tumor will produce a green spot (e.g., spot 3 in Fig. 5.5); a gene expressed in both tissues will provide equal amounts of red and green and the spot will appear as yellow (spot 4), and a gene not expressed in either tissue will provide a black spot (spot 1).

The main steps of data handling in a microarray process are described in Fig. 5.6. The microarray process is initiated as two threads: the production of the microarray itself and the collection and treatment of the sample(s) which will provide the mRNA that will be queried. The two threads are joined in the hybridization step in which the labeled mRNA is hybridized on the array. Once this is done, the hybridized array is scanned to produce high-resolution tiff files. These files need to be processed by some specialized software that quantifies the intensity values of the spots and their local background on each channel. Ideally, this step would provide expression data as a large matrix traditionally visualized with genes as rows and conditions as columns. A final process of data analysis/data mining should extract the biologically relevant knowledge out of these quantitative data.

5.4   Image processing of cDNA microarrays

cDNA microarray images consist of spots arranged in a regular grid-like pattern. Often, it is useful to think of the array as being organized in **sub-grids**. Such sub-grids are usually separated by small spaces from neighboring sub-grids and form a **meta-array**. Each sub-grid is created by one pin of the printing-head. A spot can be localized on the array by specifying its location in terms of **meta-row**, **meta-column**, **row**, and **column**).

The image processing of a microarray can be divided into:

1. Array localization – spot finding.

2. Image segmentation – separating the pixels into signal, background, and other.

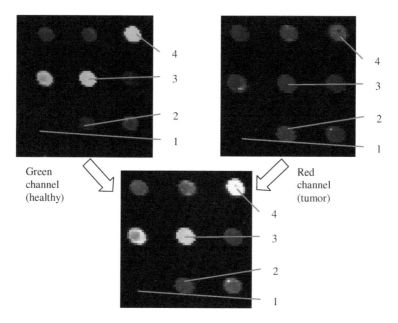

Green channel (healthy)

Red channel (tumor)

FIGURE 5.5: A synthetic image is obtained by overlapping the two channels. A gene that is expressed abundantly in the tumor tissue and scarcely in the healthy will appear as a red spot (e.g., spot 2), a gene expressed in the healthy tissue and not expressed in the tumor will appear as green spot (e.g., spot 3), a gene expressed in both tissues will provide equal amounts of red and green and the spot will appear as yellow (spot 4) and a gene not expressed in either tissue will provide a black spot (spot 1).

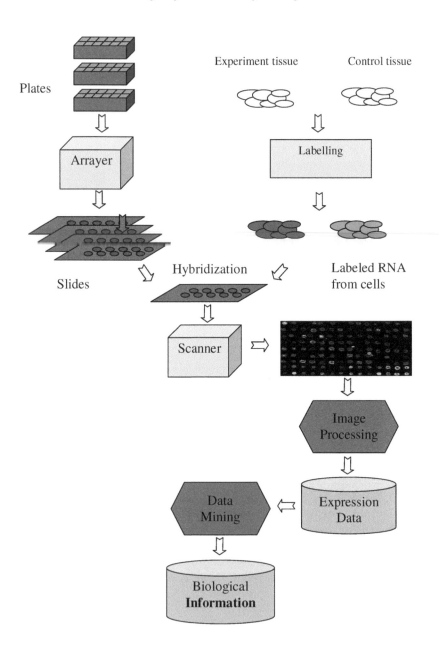

FIGURE 5.6: An overview of the DNA array processing. The cylinders represent data that needs to be stored. The parallelepipeds represent pieces of hardware while the hexagons represent information processing steps in which the computers play an essential role.

3. Quantification[1] – computation of values representative for the signal and background levels of each spot.

4. Spot quality assessment – computations of quality measures.

5.4.1 Spot finding

Since it is known in advance how many spots there are, the pattern according to which they were printed, as well as their size, a simple computer program could apparently accomplish the image processing task by superimposing an array of circles with the defined dimensions and spacing on the given image. The pixels falling inside these circles would be considered signal and those outside would be background. Unfortunately, this is not possible. In the real world, the exact location of each grid may vary from slide to slide even if the grid itself were perfect. Furthermore, the relative position of the sub-grids may again vary even if the sub-grids themselves were perfect. Finally, within a single sub-grid, individual spots can be severely misaligned. There are a number of sources contributing to the problem of an imperfect grid, mainly related to mechanical constraints in the spotting process, hybridization inconsistencies, and the necessity to print dense arrays in order to increase the throughput of the approach. Because of these reasons, the first step of the image processing has to deal with finding the exact position of the spots which sometimes can be rather far from their expected location according to the grid.

The spot finding operation aims to locate the signal spots in images and estimate the size of each spot. There are three different levels of sophistication in the algorithms for spot finding, corresponding to the degree of human intervention in the process. These are described below in the order given by the most to the least amount of manual intervention.

Manual spot finding. This method is essentially a computer-aided image processing approach. The computer does not have any abilities to "see" the spots. It merely provides tools to allow the users to tell the software where each of the signal spots are in the image. Historically, this was the first method used only in the very early days of microarray technology. This method is prohibitively time-consuming and labor intensive for images that have thousands of spots. Users had to spend a day or so to adjust the circles over the spots such that an acceptable level of accuracy was achieved. Furthermore, considerable inaccuracies may be introduced at this time due to human errors, particularly with arrays having irregular spacing between the spots and large variation in spot sizes.

Semiautomatic spot finding. The semiautomatic method requires some level of user interaction. This approach typically uses algorithms for automatically adjusting the location of the grid lines, or individual grid points after the

---

[1]The term of *quantitation* is sometimes used in the microarray jargon to describe the same process.

user has specified the approximate location of the grid. What the user needs to do is to tell the program where the outline of the grid is in the image. For example, the user may need to put down a grid by specifying its dimensions (in number of rows and columns) and by clicking on the corner spots. Subsequently, the spot-finding algorithm adjusts the location of the grid lines, or grid points, to locate the arrayed spots in the image. User interface tools are usually provided by the software in order to allow for manual adjustment of the grid points if the automatic spot-finding method fails to correctly identify each spot. This approach offers great timesaving over the manual spot finding method since the user needs only to identify a few points in the image and make minor adjustments to a few spot locations if required. Such capabilities are offered by most software vendors.

Automatic spot finding. The ultimate goal of array image processing is to build an automatic system, which utilizes advanced computer vision algorithms, to find the spots reliably without the need for any human intervention. This method would greatly reduce the human effort, minimize the potential for human error, and offer a great deal of consistency in the quality of data. Such a processing system would require the user to specify the expected configuration of the array (e.g., number of rows and columns of spots) and would automatically search the image for the grid position. Having found the approximate grid position, which specify the centers of each spot, the neighborhood can be examined to detect the exact location of the spot. Knowledge about the image characteristics should be incorporated to account for variability in microarray images. The spot location, size, and shape should be adjusted to accommodate for noise, contamination, and uneven printing.

5.4.2 Image segmentation

Image segmentation is the process of partitioning an image into a set of non overlapping regions whose union is the entire image. The purpose of segmentation is to decompose the image into parts that are meaningful with respect to a particular application, in this case spots separated from background. Once the spots have been found, an image segmentation step is necessary in order to decide which pixels form the spot and should be considered for the calculation of the signal, which pixels form the background, and which pixels are just noise or artifacts and should be eliminated.

In the following discussion of various approaches used for segmentation, we will assume that the image contains high-intensity pixels (white) on a low-intensity background (black) as in Fig. 5.7 A, B, and D. This is usually the case and all the computation is done on such data. However, images are often displayed in negative (i.e., with low-intensity pixels on high-intensity background – black pixels on white background as in Fig. 5.7 C) or in false color where the intensity of the spots is displayed as a color (e.g., green or red). False colors allow composite displays where two or more images each corresponding to a different mRNA and each encoded with a different false color are overlapped. In such composite images, the relative intensity between the various channels is displayed as a range of colors obtained from the superposition of various amounts of the individual colors, as in Fig. 5.5.

Pure spatial-based signal segmentation. The simplest method is to place a circle over a spot and consider that all the pixels in this circle belong to the signal. In fact, it is safer to use two circles, one slightly smaller than the other (see Fig. 5.8). The pixels within the inner circle are used to calculate the signal value while the pixels outside the outer circle are used to calculate the background. The pixels between the two circles correspond to the transition area between the spot and its background and are discarded in order to improve the quality of the data. All the pixels outside the circle within the boundary of a square determined by the software are considered as background. This approach is very simple but poses several big problems as illustrated in the figure. First, several spots are smaller than the inner circle they are in. Counting the white pixels as part of the spots just because they fall in the circle will artifactually lower the signal values for those spots. Second, there are dust particles, contaminants and sometimes fragments of spots that are outside the outer circle and will be counted as background. Again, this would artificially increase the background and distort the real relationship between the spot and its local background. For these reasons, this method is considered not to be very reliable.

Intensity-based segmentation. Methods in this category use exclusively intensity information to segment out signal pixels from the background. They assume that the signal pixels are brighter on average than the background pixels. As an example, suppose that the target region around the spot taken from the image consists of 40×40 pixels. The spot is about 20 pixels in diameter. Thus, from the total of 1600 (40×40) pixels in the region, about 314 ($\pi \cdot 10^2$) pixels, or 20%, are signal pixels, and they are expected to have their intensity values higher than that of the background pixels. To identify these signal pixels, all the pixels from the target region are ordered in a one-dimensional array from the lowest intensity pixel to the highest one, $\{p_1, p_2, p_3, ..., p_{2500}\}$, in which p_i is the intensity value of the pixel of the i-th lowest intensity among all the pixels. If there is no contamination in the target region, the top 20% pixels in the intensity rank may be classified as the signal pixels. The advantage of this method is its simplicity and speed; it is good for obtaining results using computers of moderate computing power. The method works well when the spots are of high intensities as compared to the background. However, the method has disadvantages when dealing with spots of low intensities, or noisy images. In particular, if the array contains contaminants such as dust or other artifacts, the method will perform poorly since any pixel falling in the top 20% of the intensity range will be classified as signal even if it is situated way out of any spot.

Mann-Whitney segmentation. This approach combines the use of spatial information with some intensity-based analysis. Based on the result of the spot-finding operation, a circle is placed in the target region to include the

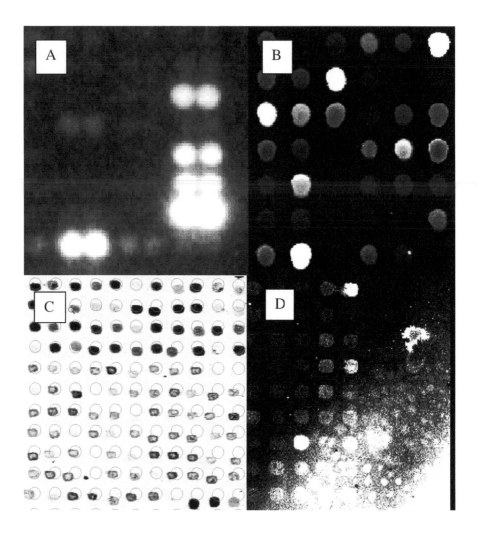

FIGURE 5.7: Examples of image processing challenges. A. Radioactively labeled spots expanding over neighboring spots (blooming). B. Non-circular and weak spots as well as some weak-center spots (doughnuts). C. Imperfect alignment. D. Hybridization and contamination causing a nonspecific, high background with various gradients.

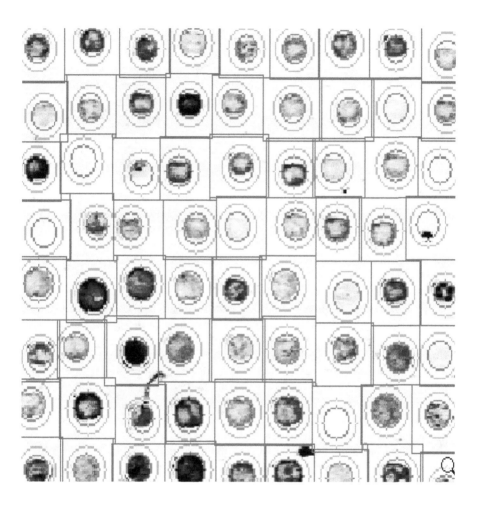

FIGURE 5.8: Spatial segmentation. A circle is placed over each spot. The pixels within the inner circle are used to calculate the signal value; the pixels outside the outer circle are used to calculate the background. The pixels between the two circles correspond to the transition area between the spot and its background and are discarded in order to improve the quality of the data.

region in which the spot is expected to be found. Since the pixels outside of the circle are assumed to be the background, the statistical properties of these background pixels can be used to determine which pixels inside the circle are signal pixels. A Mann-Whitney test is used to obtain a threshold intensity level that will separate the signal pixels from other pixels (e.g., background) even if they are inside the expected area of the spot. Pixels inside of the circle having a higher intensity than the threshold intensity are treated as signal. This method works very well when the spot location is found correctly and there is no contamination in the image. However, when contamination pixels exist inside of the circle, they can be treated as signal pixels. This is because some contamination can have an intensity higher than the background. Furthermore, if there are contamination pixels outside of the circle, or the spot location is incorrect such that some of the signal pixels are outside of the circle, the distribution of the background pixels can be incorrectly calculated, resulting in a threshold level higher than appropriate. Consequently, signal pixels with intensity lower than the threshold will be misclassified as background. This method also has limitations when dealing with weak signals and noisy images. When the intensity distribution functions of the signal and background are largely overlapping with each other, classifying pixels based on an intensity threshold is prone to classification errors, resulting in measurement biases. This is similar to what has been discussed in the pure intensity-based segmentation method.

Combined intensity-spatial segmentation (or the trimmed measurements approach). This approach combines both spatial and intensity information in segmenting the signal pixels from the background in a manner similar to the Mann-Whitney approach. The logic of this method proceeds as follow. Once the spot is localized and a target circle is placed in the target region, most of the pixels inside of the circle will be signal pixels and most of the pixels outside of the circle will be background. However, due to the shape irregularity, some signal pixels may leak out of the circle and some background pixels may get into the circle. Background pixels within the circle may be considered as outliers in the intensity distribution of the signal pixels. Similarly, signal pixels that fall outside the circle will also appear as outliers with respect to the intensity distribution of the background pixels. Contamination pixels anywhere will appear as outliers in the intensity domain for either signal and background. These outliers would severely change the measurement of the mean and total signal intensity if they are not eliminated. To remove the effect of outliers on these measurements, one may simply "trim-off" a fixed percentage of pixels from the intensity distribution of the pixels for both signal and background regions.

The Mann-Whitney approach described above performs a statistical analysis on the pixels outside the presumed spot area and then uses the threshold calculated there to segment the pixels inside the target area. The trimmed measurements approach performs a statistical analysis of both distribution (outside as well as inside the presumed spot) and eliminates the outliers from each such distribution without making the leap of faith that the characteristics of the distribution outside will also reflect the properties of the distribution inside. Eliminating approximately 5–10% of each distributions allows this approach to cope very elegantly with artifacts such as doughnuts, dust particles, or other impurities even if they lay both inside and outside the spot.

Although this method performs extremely well in general, a potential drawback is related to its statistical approach if the spots are very small (3–4 pixels in diameter) since in this case the distribution will have relatively few pixels.

Two circles can be used in order to further improve the accuracy of this method. The inside of the inner circle will be the area of the prospective spot while the outside of the outer circle will be the area considered as background. The small region between the two circles is considered a buffer zone between the spot and its background. In this area, the intensities may vary randomly due to an imperfect spot shape and are considered unreliable. The accuracy of the analysis is improved if the pixels in this area are discarded from the analysis.

A comparison between spatial segmentation and trimmed measurement segmentation is shown in Fig. 5.9. The figure shows the spatial segmentation on top and the trimmed measurement segmentation at the bottom. The first row of each image represents the cy3 image (Sample A), while the second row represents the cy5 image (Sample B). This spot has a stronger signal on the cy3 channel and a weaker signal on the cy5 channel. However, the cy3 channel also has a large artifact in the background to the left of the spot. This artifact also covers the spot partially. This is a good example of a spot on which many methods will perform poorly. Each row shows from left to right: i) the raw image, ii) the segmented image, iii) the overlap between the raw image and the segmented image, and iv) the histograms corresponding to the background and signal. Signal pixels are red, background pixels are green, and ignored pixels are black. The pixels that are black in the segmented image are eliminated from further analysis. The spatial segmentation will only consider the spatial information and will consider all pixels marked red in the top two images as signal. These pixels also include a lot of pixels belonging to the artifact as can be seen from the overlap of the segmented image and the original image (image 3 in the top row). The same phenomenon happens for the background. Note that the spatial segmentation will consider all green pixels in the top two images as background. These pixels include many high-intensity pixels due to the artifact and the average of the background will be inappropriately increased. The trimmed measurement segmentation does an excellent job of eliminating the artifact from both background and signal. Note in row 3, image 2 (second from the left), the discontinuity in the green area and note in the overlapped image (same row 3, image 3) that this discontinuity corresponds exactly to the artifact. Similarly, the red circle showing the segmentation of the signal pixels has a black area corresponding to the same artifact overlapping the spot itself. In effect, this method has carved a good approximation of the shape of the artifact and has prevented its pixels from affecting the computation of the hybridization signal.

The two methods discussed above use minimal amount of spatial information, i.e. the target circle obtained from spot localization is not used to improve the detection of signal pixels. Their design priority is to make the measurements of the intensity of the spots with minimal computation. These methods are useful in semi-automatic image processing because the speed is paramount (the user is waiting in front of the computer) and the user can visually inspect the quality of data.

In a fully automated image processing system, the accuracy of the signal pixel classification becomes a central concern. Not only that the correct segmentation of signal pixels must offer accurate measurement of the signal intensity, but it must also provide multiple quality measurements based on the geometric properties of the spots. These quality measures can be used to draw the attention of a human inspector to spots having questionable quality values after the completion of an automated analysis. Fig. 5.10 shows the result of the trimmed measurement segmentation on two adjacent spots partially covered by an artifact. The spot to the right is the same as the one in Fig. 5.9. The pixels corresponding to the artifact are removed, a reliable measurement can be extracted from it and the spot is salvaged. The spot to the left has too few pixels left after the removal of the artifact and is marked as a bad spot. Its quantification value will be associated to a flag providing information about the type of problem encountered.

5.4.3 Quantification

The final goal of the image processing is to compute a unique value that hopefully is directly proportional with the quantity of mRNA present in the solution that hybridized the chip. One such value needs to be obtained for each gene on the chip. The purpose of the **spot quantification** is to combine pixel intensity values into a unique quantitative measure that can be used to represent the expression level of a gene deposited in a given spot. Such a unique value could be obtained in several ways. Typically, spots are quantified by taking the mean, median, or mode of the intensities of all signal pixels. Note that a simple sum of the pixel intensities would be dependent on the size of the spot through the number of pixels in a spot. Therefore, values obtained from microarrays printed at different spot densities could not be compared directly.

The key information that needs to be recorded from microarrays is the expression strength of each target. In gene expression studies, one is typically interested in the difference in expression levels between the test and reference mRNA populations. On two-channel microarrays, each channel of the control (reference) and experiment (test) is labeled with a different fluorochrome and the chip is scanned twice, once for each channel wavelength. The difference in expression levels between the two conditions under study now translates to

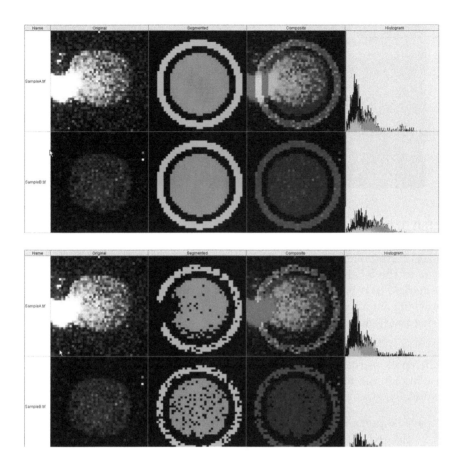

FIGURE 5.9: Spatial segmentation (top) compared to trimmed measurement segmentation (bottom). First row of each image represents the cy3 while the second row represents the cy5 image. Each row shows from left to right: i) the raw image, ii) the segmented image, iii) the overlap between the raw image and the segmented image, and iv) the histograms corresponding to the background and signal. Signal pixels are red, background pixels are green, ignored pixels are black. The pixels that are black in the segmented image are eliminated from further analysis. Note that the spatial segmentation includes the artifact pixels both in the computation of the background and in the computation of the signal. The trimmed measurement segmentation does an excellent job of eliminating the artifact from both background and signal. Image obtained using ImaGene, BioDiscovery, Inc.

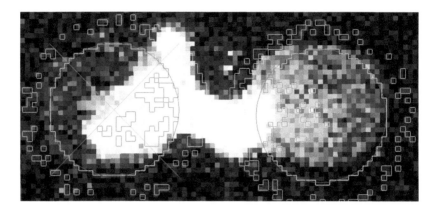

FIGURE 5.10: Artifact removal and bad spot detection in fully automatic image processing. The spot to the right is the same as the one in Fig. 5.9. The pixels corresponding to the artifact are removed and the spot is salvaged. The spot to the left has too few pixels left after the removal of the artifact and is marked as a bad spot. Its quantification value will be associated to a flag providing information about the type of problem encountered. Image obtained using ImaGene, BioDiscovery, Inc.

differences in the function of intensities on the two images. Under idealized conditions, the total fluorescent intensity from a spot is proportional to the expression strength. These idealized conditions are:

- The preparation of the target cDNA (through reverse transcription of the extracted mRNA) solution is done appropriately, such that the cDNA concentration in the solution is proportional to that in the tissue.

- The hybridization experiment is done appropriately, such that the amount of cDNA binding on the spots is proportional to the target cDNA concentration in the solution.

- The amount of cDNA deposited on each spot during the chip fabrication is constant.

- There is no contamination on the spots.

- The signal pixels are correctly identified by image analysis.

In the following discussion, we assume that the first two conditions are satisfied. Whether these two conditions are truly satisfied should be controlled through the design of the experiments. For the measurements obtained based on image analysis algorithms, we are mainly concerned about the remaining three conditions. In most cases, the last three conditions are all violated. The amount of DNA deposited during the spotting procedure may vary from time to time and spot to spot. Higher amounts may result in larger spot sizes so the size of the spots cannot be considered constant, not even for spots on the same array. When a spot is contaminated, the signal intensity covered by the contaminated region is not measurable. The image processing may not correctly identify all signal pixels; thus, the quantification methods should not assume an absolute accuracy in the segmentation stage.

The values commonly computed for individual spots are: total signal intensity, mean signal intensity, median signal intensity, mode signal intensity, volume, intensity ratio, and the correlation ratio across two channels. The underlying principle for judging which one is the best method is based on how well each of these measurements correlates to the amount of the DNA target present at each spot location.

The **total signal intensity** is the sum of the intensity values of all pixels in the signal region. As it has been indicated above, this total intensity is sensitive to the variation of the amount of DNA deposited on the spot, the existence of contamination and the anomalies in the image processing operation. Because these problems occur frequently, the total signal intensity is not an accurate measurement and is rarely used.

The **mean signal intensity** is the average intensity of the signal pixels. This method has certain advantages over the total. Very often the spot size correlate to the DNA concentration in the wells during the spotting processing. Using the mean will reduce the error caused by the variation of the amount of DNA deposited on the spot by eliminating the differences introduced by the size of the spot. With advanced image processing allowing for accurate segmentation of contamination pixels from the signal pixels, the mean should be a very good measurement method.

The **median of the signal intensity** is the intensity value that splits the distribution of the signal pixels in halves. The number of pixels above the median intensity is the same as the number of pixels below. Thus, this value is a landmark in the intensity distribution profile. An advantage of choosing this landmark as the measurement is its resistance to outliers. As it has been discussed in the previous section, contamination and problems in the image processing operation introduce outliers in the sample of identified signal pixels. The mean measurement is very vulnerable to these outliers. A unique erroneous value much higher or much lower than the others can dramatically affect the mean. However, such a unique outlier will not affect the median.[2]

When the distribution profile is unimodal, the median intensity value is very stable and it is close to the mean. In fact, if the distribution is symmetric (in both high- and low-intensity sides), the median is equal to mean. Thus, if the image processing techniques used are not sophisticated enough to ensure the correct identification of signal, background, and contamination pixels, the median might be a better choice than the mean. An alternative to the median

---

[2]See also the detailed discussion of the mean, median and mode in Chapter 8.

measurement is to use a trimmed mean. The trimmed mean estimation is done after certain percentage of pixels have been trimmed from the tails of the intensity distribution.

The **mode of the signal intensity** is the "most-often-found" intensity value and can be measured as the intensity level corresponding to the peak of the intensity histogram. The mode is another landmark in the intensity distribution enjoying the same robustness against outliers offered by the median. The trade-off is that the mode becomes a biased estimate when the distribution is multi-modal, i.e. when the intensity histogram has more than one peak. This is because the mode will be equal to one of the peaks in the distribution, more specifically to the highest. When the distribution is uni-modal and symmetric, mean, median, and mode measurements are equal. Often the difference between mode and median values can be used as an indicator of the degree to which a distribution is skewed (elongated on one side) or multi-modal (have several peaks).

The **volume of signal intensity** is the sum of the signal intensity above the background intensity. It may be computed as (mean of signal – mean of background)× area of the signal. This method adopts the argument that the measured signal intensity has an additive component due to the nonspecific binding and this component is the same as that from the background. This argument may not be valid if the nonspecific binding in the background is different from that in the spot. In this case, a better way is to use blank spots for measuring the strength of nonspecific binding inside of spots. It has been shown that the intensity on the spots may be smaller than it is on the background, indicating that the nature of nonspecific binding is different between what is on the background and inside of the spots. Furthermore, by incorporating the area of the signal, the volume becomes sensitive to the spot size. Thus, if an intelligent segmentation algorithm detects a defect on a spot and removes it, the remaining signal area will be artificially smaller. In consequence, the volume will be artificially decreased even if the expression level of the gene is high.

If the hybridization experiments are done in two channels, then the **intensity ratio between the channels** might be a quantified value of interest. This value will be insensitive to variations in the exact amount of DNA spotted since the ratio between the two channels is being measured. This ratio can be obtained from the mean, median or mode of the intensity measurement for each channel.

Another way of computing the intensity ratio is to perform **correlation analysis** across the corresponding pixels in two channels of the same slide. This method computes the ratio between the pixels in two channels by fitting a straight line through a scatter plot of intensities of individual pixels. This line must pass through the origin and the slope of it is the intensity ratio between the two channels. This is also known as **regression ratio**. This method may be effective when the signal intensity is much higher than the background intensity. Furthermore, the assumption is that the array was scanned in such a way that the pixels in the two channels can be mapped exactly to each other. This may not always be possible, depending on the type of scanner used. The motivation behind using this method is to bypass the signal pixel identification process. However, for spots of moderate to low intensities, the background pixels may severely bias the ratio estimation of the signal towards the ratio of the background intensity. Then the advantage of applying this method becomes unclear and the procedure suffers the same complications encountered in the signal pixel identification methods discussed above. Thus, its theoretical advantage over intensity ratio method may not be present. One remedy to this problem is to identify the signal pixels first before performing correlation analysis. Alternatively, one could identify the pixels that deviate within a specified amount from the mean of the intensity population.

5.4.4 Spot quality assessment

In a fully automated image processing system, the accuracy of the quantification becomes a central concern. Not only that the correct segmentation of signal pixels must offer accurate measurement of the signal intensity, but it must also permit multiple quality measurements based on the geometric properties of the spots. These quality measures can be used to draw the attention of a human inspector to spots having questionable quality values. The following quality measures are of interest in microarray image analysis.

Spot signal area to spot area ratio. Spot area is the spot signal area plus the area occupied by ignored regions caused by contamination or other factors, which are directly connected to the signal region ("touching" the signal area as in Fig. 5.10). This measure provides information about the size of the ignored area located nearby the signal. The smaller the ratio the larger this ignored area is and therefore the lower the quality of the spot. The signal area to spot area ratio is a measure of local contamination. A researcher may generate a scatter plot of data from two sub-grids (pins) and evaluate the quality of the spot printing. This is a very convenient way to identify defective pins that might fail to be revealed by a visual inspection.

Shape regularity. This measure considers all the ignored and background pixels that fall within the circle proposed by the spot-finding method. The shape regularity index can be calculated as the ratio of number of those pixels to the area of the circle. This technique measures how deformed the actual signal region is with respect to the expected circular shape. Clearly, round, circular spots are to be trusted more than badly deformed ones so the lower the shape regularity ratio, the better the spot is.

Spot area to perimeter ratio. This ratio will be maximum for a perfectly circular spot and will diminish as the perimeter of the spot becomes more irregular. This measure is somewhat overlapping with the shape regularity above in the sense that both will allow the detection of the spots with a highly irregular shape.

Displacement. This quality measure is computed using the distance from the expected center of the spot to its actual location: $1 -$ (offset to grid). The expected position in the grid is computed by the grid placement algorithm. This measure can be normalized to values from 0 to 1, by dividing it by half of the snip width (grid distance). A spot closer to its expected position will be more trustworthy than a spot far away from it.

Spot uniformity. This measure can be computed as: $1 -$ (signal variance to mean ratio). This measure uses the ratio of the signal standard deviation and the signal mean. The ratio is subtracted from 1 such that a perfect spot (zero variance) will yield a uniformity indicator of 1. A large variation of the signal intensity will produce lower values for this quality measure and will indicate a less trustworthy spot. Dividing by the mean is necessary because spots with higher mean signal intensity also have stronger signal variation.

All of the above measures have been or can be normalized to vary from 0 to 1. Furthermore, they can also be adjusted such that a value of 1 corresponds to an ideal spot and a lower value shows spots degradation. Once this is done, a weighted sum of these measures can provide a global, unifying quality measure, ranging from 0 (bad spot) to 1 (good spot). This approach also allows users to vary the relative weights of the various individual quality indicators to reflect their own preferences.

Automatic flagging of spots that are not distinguishable from the background (empty spots) or have questionable quality (poor spots) is a necessity in high-throughput gene expression analysis. Flagged spots can be excluded in the data mining procedure as part of the data preparation step. In spite of the great importance of assessing the quality of the spots and carrying this information forward throughout the data analysis pipeline, relatively few software packages provide such measurements. It is hoped that, as the users become more aware of the issues involved, they will require software companies to provide such quality measures in a larger variety of products.

As a final observation, note that typically a gene is printed in several spots on the same array. Such spots are called replicates and their use is absolutely necessary for quality control purposes. Quantification values obtained for several individual spots corresponding to a given gene still need to be combined in a unique value representative for the gene. Such a representative value can in turn be computed as a mean, median or mode of the replicate values. Alternatively, individual replicate values can be analyzed as a part of a more complex model such as ANalysis Of Variance (ANOVA) (see Chapter 13). More details about the image processing of cDNA arrays are available in the literature [244].

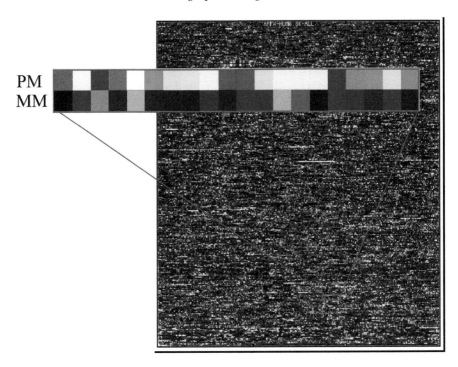

FIGURE 5.11: The image of an Affymetrix microarray. A gene is represented by a set of 20 probes. Each probe consists of 25 nucleotides. The top row contains the perfect match (PM) probes while the bottom row contains the mismatch (MM) probes. The MM probes are different from the PM probes by a single nucleotide. If the mRNA corresponding to a gene was present during the hybridization, the PM probes have a higher intensity than the MM probes. The average difference between the PM and MM probes is considered proportional to the expression level of the gene.

5.5   Image processing of Affymetrix arrays

Because the Affymetrix technology is proprietary, virtually all image processing of the Affymetrix arrays is done using the Affymetrix software. The issues are slightly different due to several important differences between technologies.

A first important difference between cDNA and oligonucleotide arrays (oligo arrays) is the fact that cDNA arrays can use long DNA sequences while oligonucleotide arrays can ensure the required precision only for short sequences. In order to compensate for this, oligo arrays represent a gene using several such short sequences. A first challenge is to combine these values to obtain a meaningful value that is proportional to the level of expression of the gene.

A second important difference is that in oligo arrays, there is no background. The entire surface of the chip is covered by probes and a background value cannot be used as an indication of the level of intensity in the lack of hybridization. However, as described in Chapter 3, the Affymetrix arrays represent a gene using a set of match/mismatch probes. In general, these arrays contain 20 different probes for each gene or EST. Each probe consists of 25 nucleotides (thus called a 25-mer). Let the reference probe be called a perfect match and denoted as PM, and the partner probe containing a single different nucleotide be called a mismatch (MM) (see Fig. 5.11). RNAs are considered present on the target mixture if the signals of the PM probes are significant above the background after the signal intensities from the MM probes have been subtracted.

Thus, a first analysis of the image can provide two types of information for each gene represented by a set of PM/MM probes. The number of probes for which the PM value is considerably higher than the MM value can be used to extract a qualitative information about the gene. The Affymetrix software captures this information in a set of "calls": if many PM values are higher than their MM values, the gene is considered "present" (P); if only a few (or no) PM values are higher than their MM values, the gene is called "absent" (A). If the numbers are approximately equal the gene is called "marginal" (M). The exact formulae for determining the calls are provided by Affymetrix and can change in time.

The ultimate goal of the microarray technology is to be quantitative, i.e. to provide a numerical value directly proportional to the expression of the gene. A commonly used value provided by the Affymetrix software is the average difference between PM and MM:

$$AvgDiff = \frac{\sum_i^N (PM_i - MM_i)}{N} \tag{5.1}$$

where PM_i is the PM value of the i-th probe, the MM_i is the corresponding MM value and N is the number of probes. In general, this value is high for expressed genes and low for genes that are not expressed.

The fact that there are two measures of which one is qualitative (the call) and the other one is quantitative (the average difference) can lead to situations apparently contradictory. Let us consider the example given in Fig. 5.12. We have two cases. For the gene to the left, calculating the pixel intensities from the array features yields an average difference of 1,270. The call of the software is "Absent" for this RNA. The gene on the right has an average difference of 1,250 but this time the software indicates a "present" RNA. This is an apparent counterexample to the general rule that expressed genes have higher average differences than not expressed genes. How can we explain this? What is happening here?

Let us consider two genes, A and B. Gene B is not expressed, but has a

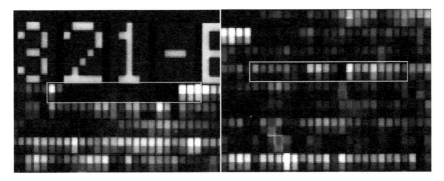

FIGURE 5.12: Two areas on an Affymetrix array corresponding to two different genes. The gene represented to the left is reported absent while the one to the right is reported as present. However, the average difference calculated for the gene on the left is higher than the average difference calculated for the gene on the right. Image obtained with dChip [277].

very short sequence, which is identical to a subsequence of gene A which is highly expressed. In this case, the probes of B identical to the substrings of A will give a strong signal artificially increasing the average difference for B. At the same time, if a different gene C is expressed at a moderate level, all its probes might have moderate intensities. The average difference calculated for C might be comparable or even lower than the average difference calculated for B. Such an example is illustrated in Fig. 5.13.

5.6 Summary

Digital images are rectangular arrays of intensity values characterized by several resolution and color depth. Each intensity value corresponds to a point in the image and is called a pixel. The resolution is the number of pixels in the image and is usually expressed as a product between the number of rows and the number of columns (e.g. 1024×768). The color depth is the number of bits used to store the intensity value of a single pixel. The digital image of a microarray has to satisfy minimum requirements of resolution and color depth. The resolution used for cDNA arrays should be such that the diameter of a spot measured in pixels is at least 10 pixels (but more is preferable). The usual color depth used for microarrays is 16 bits, which allows each pixel to represent 65,536 different intensity levels. The image processing of microarray digital images aims at obtaining numerical values proportional to the level of mRNA present in the tested sample(s) for each of the genes interrogated. For

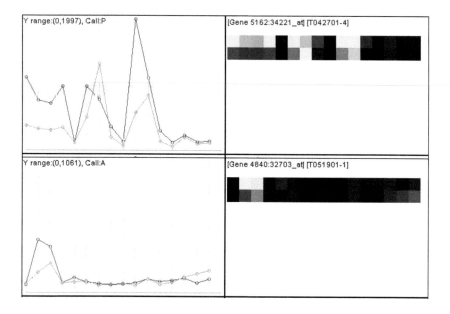

FIGURE 5.13: An apparent contradiction between the calls of two genes and their average differences. The figure shows the hybridization of the probes corresponding to each gene to the right and a graphical representation of the intensities corresponding to each probe to the left. The gene on top is reported as present, the gene on the bottom is reported as absent. However, the average difference of the gene on top is lower (1250) than the average difference of the gene at the bottom (1270). Image obtained with dChip [277].

cDNA arrays, this is done in several steps: array localization, image segmentation, quantification, and spot quality assessment. The array localization is the process that localizes the spots (spot finding). The spot finding can be done manually (obsolete), semi-automatically or automatically. The image segmentation is the process that decides which pixels belong to the spots and which pixels belong to the background. The segmentation can be done based on spatial information, intensity information, using a Mann-Whitney analysis of the distribution of the pixels or using a combined intensity-spatial approach. The quantification combines the values of various pixels in order to obtain a unique numerical value characterizing the expression level of the gene. Quantification must be done selectively such that the pixels corresponding to various possible contaminations or defects are not taken into consideration in the computation of the value representative for the expression level of the gene. This representative value can be calculated as the total, mean, median, or mode of the signal intensity, the volume of the signal intensity of the intensity ratio between the channels. The spot quality can be assessed using the ratio between the signal area and the total spot area, the shape regularity, ratio between spot area to perimeter, displacement from the expected position in the grid and spot uniformity. The image processing of Affymetrix arrays produces several types of information including calls and average differences. Calls are meant to provide qualitative information about genes. A gene can be present, marginally present or absent. The average difference is calculated as the average of the differences between the perfect match (PM) and mismatch (MM) of all probes corresponding to a given gene. The average difference is commonly used as the quantitative measure of the expression level of the gene.

Chapter 6

Introduction to R

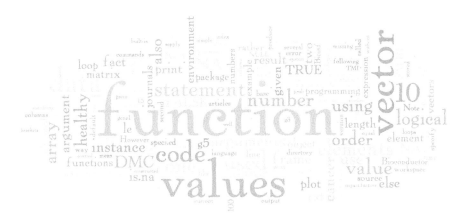

R is an open source implementation of the well-known S language.

—*Peter Daalgard, Introductory Statistics with R*

6.1   Introduction to R

6.1.1   About R and Bioconductor

R is a name used to refer to a language, a development environment, as well as an integrated suite of software routines that allow efficient data manipulations, calculations, and graphical display. The R language can be regarded as an open source implementation of the S language developed at Bell Laboratories by Rick Becker, John Chambers, and Allan Wilks. S subsequently evolved into S-PLUS, which was for a long time a preferred choice for commercial data analysis software. Since both S-PLUS and R evolved from S, R, and S-PLUS are closely related. Although R is best known as an environment in which many statistical routines and packages have been implemented, there is

119

nothing special about statistics and R, and many other application domains could use the R environment equally well.

Bioconductor is an open source and open development software project for the analysis and comprehension of genomics data [169]. Bioconductor is based primarily on the R programming language although contributions are accepted in any programming language, as long as the package containing the added functionality is a standard R package, and can be used from within R. There are two releases of Bioconductor every year corresponding to the two annual releases of the R environment. At any given time, there is a **release version** of Bioconductor, which corresponds to the released version of R, and a **development version** of Bioconductor, which corresponds to the development version of R. Generally, the release version will be more stable and bug-free but may not contain the latest routines and may not include the most recently added features. The release version is strongly recommended for the audience of this book. When experimenting with a new language and environment, there are always moments when one cannot quite understand what is happening and wonders whether there is some problem with the system. Using a release version rather than a development one, would pretty much ensure that such doubts regarding the system can be effectively eliminated. With a release version, if unexpected results are encountered, one can be reasonably sure that the problem is with the user: either there is a syntax error or a conceptual misunderstanding.

A more detailed description of Bioconductor, as well as some examples of analyzes performed with packages from Bioconductor are included in Chapter 7.

6.1.2 Repositories for R and Bioconductor

The latest version of R can be downloaded from `http://www.r-project.org/` or from any one of the sites belonging to the Comprehensive R Archive Network (CRAN) at `http://cran.r-project.org/mirrors.html`. Since R is a dynamic environment, and packages and/or the specific syntax may change over time, the version of R current at the time this book is printed is included on the accompanying CD. This would guarantee that the routines and snippets of code discussed throughout the book will actually run as described.

The latest version of Bioconductor can be found at `www.bioconductor.org/`. For the same reasons as above, as well as in order to ensure the compatibility between the Bioconductor and R versions, the version of Bioconductor current at the time of writing this book is also included here. The code included with this book tends to use fairly standard functions and syntax which is unlikely to change in the future. However, since anything can happen in the future, the possibility of incompatibilities between this code and future releases of R and/or Bioconductor cannot be completely eliminated. One should note that if a new version of Bioconductor is downloaded from the web, in most cases the underlying R platform should also be updated to the corresponding version required by the latest Bioconductor.

6.1.3 The working setup for R

Today, most C/C++/Java programmers work in an integrated development environment (IDE). IDEs normally consist of a source code editor, a compiler and/or interpreter, build-automation tools, and (usually) a debugger. Capabilities include source code editing (usually with syntax highlighting and auto-completion), the ability to execute the code, browse through the data structures, trace and debug errors, etc. These features are not quite as advanced in the R programming environments available today as they are in the environments designed for older programming languages such as C++ or Java. Most R environments are either GUI interfaces for the basic R engine, or additional packages for specific text editors that provide these editors with basic R-oriented features such as syntax highlighting and easy execution. Examples of R GUIs could include the Java-based JGR, pronounced "Jaguar" (`http://rosuda.org/JGR/index.shtml`, the GTK+2 based pmg (Poor Man's GUI) (`http://wiener.math.csi.cuny.edu/pmg`) and Rattle (the R Analytical Tool To Learn Easily) (`http://rattle.togaware.com/`), rcmdr (The R Commander) (`http://socserv.mcmaster.ca/jfox/Misc/Rcmdr/`), SciViews R GUI (`http://www.sciviews.org`), the Statistical Lab (`www.statistiklabor.de/en/`), etc.

Most notable editors that have R-dedicated extension packages include Emacs (ESS or Emacs Speaks Statistics at `http://ess.r-project.org/`), Vim (Vi IMproved), Eclipse, and WinEdt. Emacs and/or Vim would probably be very popular with Unix/Linux user, Eclipse with C/C++/Java programmers, and WinEdit with Latex/Windows users.

Nevertheless, the simplest way to work with R, probably familiar to most R users is to open two windows: one for the native R GUI, the other for one's preferred text editor. It is essential that the chosen text editor do not try to provide any "editing help" beyond the text that is typed by the user. In particular, word processors such as Microsoft Word or WordPerfect would not be a good choice due to their formatting, as well as numerous attempts to correct spelling mistakes. In the Windows environment, Notepad or WinEdt would be much better choices. Another version of an R-friendly text-editor would be TINN (`http://www.sciviews.org/Tinn-R/`). In case you wonder, TINN stands for "**T**his **I**s **N**ot **N**otepad" and is also provided by SciViews.

The typical working session would include typing the R commands in the text editor in fragments as appropriated for the given task, then selecting, copying, and pasting into the R console, where the commands are also executed. If errors or warning are produced, one goes back to the text editor, corrects the problems, and repeats the cycle until the desired results are obtained. The edited text file opened in the text editor provides a way to store the code for future use. This file is much better than what could be extracted from the R console because it will not contain the R prompts, outputs, and error messages so it can be easily run again later on. This file is also better than what can be obtained with `history()` command since many times a development or debugging session will contain many repeated calls to same or similar code, which would be both useless and misleading if included in a permanent record of the code developed.

6.1.4 Getting help in R

The R environment has a number of built-in entities, such as classes and functions (methods). Information about any of these can be obtained by typing a question mark "?" followed by the name of the entity we are requesting help about. For instance, we can find out the choice of parameters for the plot function by invoking "`?plot`" which will open the window shown in Fig. 6.1. Exactly the same result can be obtained with the command "`help(plot)`" where `plot` can be substituted with the name of the built-in function or class we are inquiring about.

The ? help can only be used if we know exactly the name of the class or function we need information for. If the exact name is not known, one can get some help using the help search function available in R. The syntax of this command is "`help.search("things to be searched for")`." For instance, if we wanted to find out what is the exact name and syntax of the R command that calculates the standard deviation, we could type "`help.search("standard deviation")`." This would produce a window showing there are 4 built-in functions that relate to standard deviation: i) row standard deviation of a numeric array, ii) pooled standard deviation, iii) standard deviation and iv) a plot function that draws the row standard deviations as a function of row means. Further detailed help about any of them can be now obtained with the "?" command.

6.2 The basic concepts

6.2.1 Elementary computations

The R environment has a **console** in which the user can type commands. There are also menus that allow things such as package installation, package updating, etc. In the following, we will focus on the use of R through its console.

The R console can be used as a hand-held calculator on which the user simply types the operands separated by the operator and then presses return in order to obtain the result.

```
> 1 + 1 <CR>
```

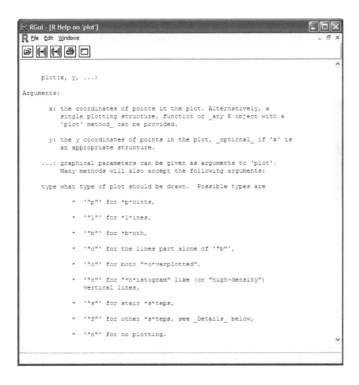

FIGURE 6.1: The help window in the R environment. This can be obtained by typing a "?" followed by the name of the entity whose description we want. The window above, which explains the various parameters of the plot function, was obtained with "?plot".

[1] 2

In the snippet above, the ">" character is the R prompt, showing that R is ready to accept an input from the user, "1+1" is typed by the user and is followed by the carriage return ("<CR>") key.[1] For simplicity, the <CR> character will not be shown henceforth even though the user is always expected to press return at the end of a command. Besides the expected result "2", R also produces a "[1]". This is R's way of telling us that the result is a scalar, that is a single value. In general, if more output values are shown on the same line, the number in brackets is the index of the first element printed on that line.

More complex computations can also be performed in the same way. For instance, the value of the expression:

$$3 * e^2 + \frac{\ln(10)}{\sqrt{2}}$$

can be calculated with the command:

```
> 3*exp(2) + log(10)/sqrt(2)
```

```
[1] 23.79534
```

This command is pretty easy to understand: `exp(2)` is the exponential function in base e that calculates e^2, log is the logarithm in base e, and `sqrt` calculate the square root of its argument. This type of expressions allow us to use R as we would any scientific calculator.

It may be worth noting that the number of decimals displayed is often only a truncation of the number of decimals used internally. The number of digits displayed can be set with the function `options` for instance:

```
> options(digits=12)
> 3*exp(2)+log(10)/sqrt(2)
```

```
[1] 23.7953418303
```

Note however that the value used as the argument of the function `options` does not represent the number of digits displayed after the decimal point but rather the total number of digits displayed:

```
> 10*(3*exp(2)+log(10)/sqrt(2))
```

```
[1] 237.953418303
```

```
> 100*(3*exp(2)+log(10)/sqrt(2))
```

---

[1] This is the key normally used to mark the end of a line when typing in most applications. On some keyboards this is "Return" or "Enter."

```
[1] 2379.53418303
```

```
> 1000*(3*exp(2)+log(10)/sqrt(2))
```

```
[1] 23795.3418303
```

Also, this value is interpreted by the R environment as a suggestion only rather than as a fixed width. For instance, if the result would be affected by the requested limitation on the number of digits, R will ignore the request and display the correct result, even though it uses more than 12 digits:

```
> 10^12*(3*exp(2)+log(10)/sqrt(2))
```

```
[1] 23795341830307
```

6.2.2 Variables and assignments

The R language allows the use of **symbolic variables**. A **variable** is a memory location that can store a value or a result of a computation. Variables have **names** that allow convenient manipulations of data. The values stored in a variable, and hence the variables, can be of several types including scalars, vectors, and matrices.

Values are assigned to variables using the **assignment operator**, "$< -$". This operator consists of two symbols: '$<$' followed by '$-$'. These symbols are meant to be interpreted together as an arrow showing that the value to the right of it will be stored in the variable to the left of it:

```
> x <- 7
```

Although nothing is shown, the value has been stored in the given variable as requested. From now on, until this variable is discarded or overwritten, using the name of the variable will recall the value stored in the variable:

```
> x
```

```
[1] 7
```

This will also happen if the name of the variable is used in computations:

```
> x^2
```

```
[1] 49
```

or

```
> y = x+2
> y
```

```
[1] 9
```

Note that in the first line immediately above, the value $x + 2$ was stored in the variable y and no result was shown. In order to see the value stored in y, we had to explicitly ask the system what the value of y was in the second line above. Also note the user of another operator for assignment, the character '='. This has the same effect as the $< -$ with the advantage that is only one character to type, and the disadvantage that it can be confused with the logical operator for equality (as in asking the question "is X = 4?").

Names of variables are almost arbitrary strings constructed with letters, digits, and the dot symbol. However, a variable name cannot start with a digit or other special character such as '_'. Unfortunately, there are some built-in objects used by the system, that have unassuming names such as: c, q, t, C, T, etc. In some situations, confusions may occur if one of the system names is used by the user, but such confusions are not a major problem. Any doubts about a potential name conflict can be easily eliminated by asking for help. For instance, "help(c)" shows quickly that "c" is a built in function that combines its arguments to form a vector. However, R takes into consideration the way a name is used when trying to link it to an object. For instance, we can use c as the name of a variable even though there is a system function with the same name, as long as every time we use the name "c" is clear whether a variable or a function is appropriate in that place:

```
> c=c(1,2,3)
> c
```

```
[1] 1 2 3
```

In this example, the name c to the left of the assignment operator was linked to a variable, whereas the name c(...) to the right of the assignment was linked to the function that concatenates its arguments to construct a vector. The third use of c, without brackets and without any arguments was again correctly linked to the variable c and made R return its content.

For the readers familiar with other programming languages and their implementation, this is the expected behavior. When a variable name is encountered, the system would go through the stack of function calls and link the variable name to the most recent object with that name that is appropriate for the given context.

6.2.3 Expressions and objects

The basic interaction mode in R is that of an interpreter. In this mode, the user enters an **expression** and the system immediately evaluates it and prints the results as in:

```
> 1+2
```

```
[1] 3
```

Operators	Meaning
[[[	indexing
:: :::	access variables in a name space
$ @	component / slot extraction
^	exponentiation (right to left)
− +	unary minus and plus
:	sequence operator
%any%	special operators
* /	multiply, divide
+ −	(binary) add, subtract
< > <= >= == !=	ordering and comparison
!	negation
& &&	and
| ||	or
~	as in formulae
−> −>>	rightward assignment
=	assignment (right to left)
<− <<−	assignment (right to left)
?	help (unary and binary)

TABLE 6.1: Operators and their meaning in R.

All expressions return a value but sometimes these **return values** are not printed. This happens with assignments such as:

```
> y=1+3
```

but also with expressions or functions that are evaluated for their **side effects** rather than their return values. Examples can include functions that open or close a file, plot a graph, etc. Expressions can involve **variables, objects, operators**, and **functions**.

A **variable** is a memory location that can store a value or a result of a computation.

As in any object-oriented programming language, an R **class** is an abstract data type that describes a data structure together with some **methods** can be used to manipulate objects of that type. Methods are very similar to the functions used in a non-object-oriented language with the difference that a method is associated with a class rather than being an independent entity. An **object** is an instance of a class. Multiple objects can be instantiated from a given class, etc. Examples would include integers, vectors, lists, etc. An **operator** is a descriptor for an operation that can be performed between **operands**. Examples can include '+', '−','/', etc. From an object-oriented perspective operators are methods defined for their corresponding abstract data type. Table 6.1 shows the operators available in R and their meaning.

6.3 Data structures and functions

6.3.1 Vectors and vector operations

One of the features that make R a very powerful language is its ability to efficiently perform computation involving **vectors** and **matrices**. A vector is an array of values. Vectors can be created in several ways. One of simplest ways of creating a vector from an arbitrary list of numbers uses the "c(...)" function mentioned above, which combines its arguments into a vector. The elements of a vector are not restricted to numerical values. For instance, we can construct a **vector of character strings** to store the titles of three journals concerned with interdisciplinary applications of computer science: *Journal of the American Medical Informatics Association* (JAMIA), *IEEE Transactions on Biomedical Imaging* (IEEE TMI), and *Bioinformatics*. Character strings are specified by surrounding them with quotes. In this case, we will use double quotes although single quotes are also allowable:

```
> journals <- c("JAMIA", "IEEE TMI", "Bioinf")
> journals

[1] "JAMIA"    "IEEE TMI" "Bioinf"
```

Now, we would like to store the number of articles published by these journals during 2003 and 2004, according to the Institute for Scientific Information (ISI). These values can be stored in a **vector of numerical values**, which can be constructed as:

```
> articles <- c(124, 278, 1003)
> articles

[1]   124  278 1003
```

However, the number of article published by a given journal has to do more with the efficiency of the editorial process and willingness to print more pages than it has to do with the quality of the journal, or its impact in the scientific community. A more important indicator for a quality of a journal is how often articles published in the given journal are cited in other scientific publications. These data are also tracked by ISI as the number of citations to papers in a given journal. For our three journals above, the numbers of citations are:

```
> cites <- c(538, 1095, 6037)
> cites

[1]   538 1095 6037
```

Also, the raw number of citations is clearly dependent on the number of papers published. It is clear that if a journal publishes more papers, the likelihood of subsequent citations is higher. Hence, we need to normalize the number of citations to the number of papers published in each journal. In R, this can be achieve very easily as follows:

```
> impact.factors = cites/articles
> impact.factors

[1] 4.33870967742 3.93884892086 6.01894317049
```

In the above expression, the dot character, ".", is used a part of the array name and does not have any other special meaning. Note that even though the syntax used the names of the arrays containing the data, the result is an array of values obtained by dividing each element of the first array by the corresponding element of the second array. The impact factors thus obtained can be interpreted as the average number of citations that a paper published in the given journals was expected to have during the given period.

In the example above, the lengths of the two vectors involved in the division were the same. However, it is possible to perform operations on vectors of different length. In this case, the elements of the shorter vector are **recycled**. The most common such situation occurs when one vector is divided for instance by a scalar. For instance, the average number of papers published annually during this 2-year period can be obtained as:

```
> annual.average <- articles/2
> annual.average

[1]   62.0 139.0 501.5
```

If the lengths of the two arrays are not the same, R will proceed to calculate the requested operation taking elements from each array, one by one. Whenever one of the arrays runs out of elements, R will restart using its elements from the beginning. In this case, the second array had only one element, 2, which was used over and over again until all elements of the first array were divided by it. If the length of the longer vector is not a multiple of the length of the shorter vector, R will produce a warning, letting the user know that something awkward, and most likely unintended, happened.

Besides the numerical and character vectors introduced above, R allows for **vectors of logical values**. The logical values are TRUE and FALSE, which can be abbreviated to T and F. A vector of logical values can be constructed much like any other vector, using the c(...) function:

```
> logical <- c(T, F, T, F)
> logical

[1]   TRUE FALSE   TRUE FALSE
```

A powerful feature of R allows it to evaluate **relational expressions** on vectors, producing logical vectors. For instance, if we wanted to test which of the journals above have an impact factor larger than 4, we could do:

```
> better.journals <- impact.factors > 4
> better.journals
```

```
[1]   TRUE FALSE   TRUE
```

R will evaluate the given relation for each pair of corresponding elements of the vectors involved in the relational expression and construct a vector of logical values. The resulting logical vector can be subsequently used to control how processing is done on the elements of the original vector. For instance, we can now find out which are the better journals by using:

```
> journals[better.journals]
```

```
[1] "JAMIA"   "Bioinf"
```

As we see here, an array can be **indexed by a logical array**. The result is an array constructed with only the elements that correspond to the TRUE values in the logical array used as the index.

A reasonable question is what happens if a vector is constructed using the c(...) function with elements of different types, for instance, logical and integer at the same time. The answer is that all arguments are forced, or **coerced**, to a common type, which is the most accommodating type, the simplest type that can represent all values specified. This is similar to implicit type casting in programming languages such as C or C++. Mixing logical and integer values will produce a vector of integers:

```
> c( T, F, 1, 1)
```

```
[1] 1 0 1 1
```

Mixing integers and real values will yield real numbers:

```
> c( 1.5, 2, 0, 1)
```

```
[1] 1.5 2.0 0.0 1.0
```

And finally, mixing logical, integer, and real values will also yield real values:

```
> c( T, F, 1.5, 1)
```

```
[1] 1.0 0.0 1.5 1.0
```

6.3.2 Referencing vector elements

There are several ways in which we can refer to a specific element of a vector. The simplest and most intuitive is referencing **using an index**, much as in the usual mathematical notation. For instance, if we want to find out what the second element of the vector *journals* is, we can use:

```
> journals[2]
```

```
[1] "IEEE TMI"
```

Note that in R, the indexes run from 1 to the length of the vector. For a vector x with n elements, $x[1]$ will be the first element and $x[n]$ will be the last.

The second possibility is to reference an element **using its name**. Names can be assigned with the function "**names**". For instance, if we wanted to associate the number of articles published with the titles of the journals, we could do that with:

```
> names(articles)=c("JAMIA", "IEEE TMI", "Bioinf")
> articles
```

```
  JAMIA IEEE TMI    Bioinf
    124      278      1003
```

Once a vector has names for its elements, these names will be printed when the content of the vector is displayed. Given that we already had the names of the journals in the vector *journals*, the same results could have been achieved with the simple assignment:

```
> names(articles)=journals
> articles
```

```
  JAMIA IEEE TMI    Bioinf
    124      278      1003
```

This can also be done with the other vectors containing journal citation data:

```
> names(cites)=journals
> cites
```

```
  JAMIA IEEE TMI    Bioinf
    538     1095      6037
```

```
> names(impact.factors)=journals
> impact.factors
```

```
        JAMIA       IEEE TMI         Bioinf
 4.33870967742 3.93884892086 6.01894317049
```

Once names are associated with a vector, specific elements can also be referenced using these names. Now we can ask directly for the impact factor of Bioinformatics:

```
> impact.factors["Bioinf"]

      Bioinf
6.01894317049
```

Note how the array was indexed using the name, "Bioinf", specified as a character string. A common mistake would be to use round rather than square brackets:

```
> impact.factors("Bioinf")
Error: couldn't find function "impact.factors"
```

which fails because round brackets are used for function calls and there is no function called "impact.factors". Another common mistake is to forget that character strings are case sensitive:

```
> impact.factors["bioinf"]

<NA>
  NA
```

which produces NA (not available) because "bioinf" is different from "Bioinf", and there is no element in the vector "impact.factors" with the name "bioinf".

Elements of a vector or matrix can also be referenced using **a logical array**. Earlier, we constructed a logical array by testing which journals had an impact factor greater than 4:

```
> better.journals <- impact.factors > 4
> better.journals

  JAMIA IEEE TMI    Bioinf
   TRUE    FALSE     TRUE
```

and then we used it to select from the journals array, only those journals that satisfy this requirement:

```
> journals[better.journals]

[1] "JAMIA"  "Bioinf"
```

6.3.3 Functions

Functions are named pieces of code that take a number of **arguments** (or **parameters**) and produce some results. Functions need to be first defined by specifying the name of the function, the type and number of arguments, and what exactly the function will do. Most of the functions used in this book will be already written by somebody else. The code that describes what a function does can be seen by typing the function name. For instance, typing "log" will show the code of the function that calculates the logarithm of a number in a given base:

```
> log
function (x, base = exp(1))
if (missing(base)) .Internal(log(x))
else .Internal(log(x, base))
<environment: namespace:base>
>
```

In most cases, the code describing the function would be rather difficult to follow for R beginners. Luckily, this will hardly ever be necessary. A more useful and more often used capability is that of obtaining a detailed description of the arguments of a function. This can be can be obtained with the function args:

```
> args(log)

function (x, base = exp(1))
NULL
```

This tells us the the function log takes two arguments, called x and base. The arguments used in the definition of the function are **formal arguments**. The formal arguments serve as placeholders in order to specify the number of objects that can be passed to the function, as well as how these objects will be used during the execution of the function. Note like unlike in other programming languages, such as C/C++/Java, the type of these variables is not specified. In the example above, there are two formal arguments: x, the number whose log with be calculated, and *base*, the base of the logarithmic function to be used. Some formal arguments have **default values**, i.e. values that will be passed to the function at run-time if no other values are specified by the user for those arguments. In the example above, we can see that the second formal argument, *base*, has a default value equal to $\exp(1)$. Thus, $\exp(1) = e = 2.71\ldots$ will be used as the base if the log function is called without specifying any base.

Once a function has been defined, it can be used through a **function call**. A function is called by specifying its name together with a set of **actual arguments**. The actual arguments specify the values that will be used during that specific function call. The actual arguments need to be matched to, or bind to, their corresponding formal arguments. In R this process of binding actual arguments to formal arguments happens at run-time. The binding can be done in two ways. If the function has very few formal arguments, one can use the actual arguments in the exact order of the formal arguments. For instance, the function `log` can be used with only one argument, the number x whose log should be calculated as in:

```
> log(10)
```

```
[1] 2.30258509299
```

or with two arguments, the number x and the base of the log, as in:

```
> log(10,exp(1))
```

```
[1] 2.30258509299
```

Since the base is optional, the log function called with only one actual argument will bind that argument to the formal argument x. The log function called with two actual arguments will bind the first argument to x and the second one to *base*. This is an example of **positional matching**. This means that we do not need to specify the names of the formal arguments, as long as we respect their position. This is efficient and convenient because the parameters can be passed to the function just by specifying them in the correct order, with no additional hassle. However, if a given function has many arguments, one would need to specify values for all the arguments in order to maintain the proper order. This can be cumbersome for functions which can take a large number of arguments. A good example of such function is `plot`, which can take more than 20 arguments:

```
> args(plot.default)
function (x, y = NULL, type = "p", xlim = NULL,
    ylim = NULL, log = "", main = NULL, sub = NULL,
    xlab = NULL, ylab = NULL, ann = par("ann"),
    axes = TRUE, frame.plot = axes, panel.first = NULL,
    panel.last = NULL, col = par("col"), bg = NA,
    pch = par("pch"), cex = 1, lty = par("lty"),
    lab = par("lab"), lwd = par("lwd"),
    asp = NA, ...)
NULL
>
```

For this function, if we wanted to specify the character to be used when plotting, *pch*, and if we were to use the positional matching, we would need to specify all arguments from the very first one until at least *pch*. This would dramatically reduce anybody's enthusiasm to ever use anything but the default plotting character. Fortunately, R also allows **keyword matching** of arguments. In order to use keyword matching, we only need to use the name of the formal argument together with the value of the actual argument as in:

```
> plot(articles, cites, pch="x")
```

This would generate the desired plot using **x**-es instead of **o**-s to plot the points. In fact, the two types of argument specification – by position and by name – can be mixed in the same function call. For instance, in the example above, *articles* and *cites* use positional matching while *pch = x* is a named actual argument. In fact, parameters can also be specified by using a less than their full name. As long the string of characters used to specify the named argument can be uniquely matched to a formal argument, R will not complain.

The dots at the end of the list of arguments indicate that this function can accept even more arguments, not yet specified. This mechanism is used to pass parameters to other functions used within the definition of the calling function. This is possible because R performs a **lazy evaluation** of its parameters. This means that the actual value of a parameter is not calculated until it is needed. This mechanism allows the default values of some arguments to depend on other arguments, among other things.

6.3.4 Creating vectors

Let us assume we want to calculate the value of a mathematical function across a certain interval. In order to do this, a common approach is to create a vector of equidistant points from across the interval desired. Let us assume that the interval of interest is $[1, 10]$ and we would like to calculate the value of our function x^2 at regular intervals of 1. In order to do this, we need a vector that contains all values: $1, 2, \ldots, 10$. There are several functions that can be used to create such vectors. The function **concatenate** (c), that we have already used, combines its arguments to build a vector. Hence, we can simply specify all values for which we would like to calculate the value of our function:

```
> x <- c(1,2,3,4,5,7,8,9,10)
> x
```

```
[1]  1  2  3  4  5  7  8  9 10
```

This was okay in this situation, in which the number of values we wanted was relatively small. However, if we wanted to plot a graph of the function, for instance, the quality of the graph would be much better if we could increase the resolution and calculate the values of the function in 100 points rather than 10 points across the same $[1, 10]$ interval. Clearly, we could use a function that would take only the end points of the interval and use a step to automatically generate all the intermediate values. Such a function exists and is called **seq**, or **sequence**. For instance, a vector with the numbers from 1 to 100 can be generated with:

```
> x <- seq(1,100)
```

In this case, **seq** used a default step of 1 to generate all numbers between 1 and 100 and create a vector with them. Clearly, this is much more convenient than typing all values involved, as required if we were to use the c construct. The function seq can also be used in other versions, too. In seq(from, to, by=), we can specify the length of the step to be used. For instance, the vector we would need in order to sample the interval [1,10] in 100 points can be obtained as follows:

```
> x=seq(1,10,by=0.1)
> x
```

```
 [1]  1.0  1.1  1.2  1.3  1.4  1.5  1.6  1.7  1.8  1.9  2.0
[12]  2.1  2.2  2.3  2.4  2.5  2.6  2.7  2.8  2.9  3.0  3.1
[23]  3.2  3.3  3.4  3.5  3.6  3.7  3.8  3.9  4.0  4.1  4.2
[34]  4.3  4.4  4.5  4.6  4.7  4.8  4.9  5.0  5.1  5.2  5.3
[45]  5.4  5.5  5.6  5.7  5.8  5.9  6.0  6.1  6.2  6.3  6.4
[56]  6.5  6.6  6.7  6.8  6.9  7.0  7.1  7.2  7.3  7.4  7.5
[67]  7.6  7.7  7.8  7.9  8.0  8.1  8.2  8.3  8.4  8.5  8.6
[78]  8.7  8.8  8.9  9.0  9.1  9.2  9.3  9.4  9.5  9.6  9.7
[89]  9.8  9.9 10.0
```

In the version seq(from, to, length.out=), we can specify how many values we want, rather than the length of the step. Another very useful variation of seq actually omits the function name and only specifies the from and to values:

```
> 1:10
```

```
 [1]  1  2  3  4  5  6  7  8  9 10
```

Another function that can be used to create vectors is rep, which stands for **replicate**. As the name says, this function will replicate its argument. The length of the result can be specified as the number of times the repetition of the entire argument should be performed:

```
> rep(1:4, times=2)
```

```
[1] 1 2 3 4 1 2 3 4
```

or explicitly as the length of the output, length.out,:

```
> rep(1:4, length.out=6)
```

```
[1] 1 2 3 4 1 2
```

or the number of times **each** individual element should be repeated.

```
> rep(1:4, each=3)
```

```
 [1] 1 1 1 2 2 2 3 3 3 4 4 4
```

With the 3 functions above, we can generate pretty much any array we might ever need. For instance, let us say that we want an array with one 1, two 2s, three 3s, two 4s and one 5. This can be constructed with:

```
> rep(1:5, c(1,2,3,2,1))
```

```
[1] 1 2 2 3 3 3 4 4 5
```

In this example, the `1:5` generated an array with the values 1,2,...,5; the `c(...)` generated another array with the desired number of repetitions for each element and `rep` put the two together to generate the desired output.

6.3.5 Matrices

A **matrix** is a two-dimensional array of values. Matrices can be constructed with the "`matrix`" function. For instance, we can put together all our journal citation data above in a single matrix:

```
> jdata <- matrix(c(cites,articles,impact.factors),
+ nrow=3,ncol=3)
> jdata
```

```
      [,1] [,2]          [,3]
[1,]   538  124 4.33870967742
[2,]  1095  278 3.93884892086
[3,]  6037 1003 6.01894317049
```

This is a 3×3 matrix in which the columns are the number of citations, articles and impact factors of the 3 journals considered above. Note the '+' symbol at the beginning of the second line. This symbol it was not typed by the user but rather it is displayed by the system every time the user presses Return or Enter and the characters typed so far do not form a syntactically correct command. This symbol tells us that the command has not been fully parsed and R is still expecting some characters. This is a way to split long commands over several lines if necessary. Also note the use of the c construct in the definition of the `jdata` above. The `c()` function concatenates the 3 vectors into a single vector. This was necessary because the `matrix` function expects a vector as its first argument. Elements are read from this vector and organized in rows and columns according to the other two parameters, `nrow` and `ncol`. Implicitly, the matrix is filled in column order, i.e. elements from the input vector are used to fill the first column, then the second one, etc. A matrix can also be constructed in row order by using the argument `byrow=TRUE` in the call of the matrix function.

As it was constructed above, the first row of the jdata matrix will correspond to the first journal, JAMIA, and can be addressed as `jdata[1,]`:

```
> jdata[1,]
```

```
[1] 538.00000000000 124.00000000000   4.33870967742
```

The number of articles published by JAMIA will be the element situated on row 1, column 2 in this matrix and can be addressed as `jdata[1,2]`:

```
> jdata[1,2]
```

```
[1] 124
```

At the moment, all our journal citation data is stored in a nice matrix. However, we still have to remember what we stored in each row and each column. It would be nice if we could assign names to each row and column. This can be done with the functions `rownames` and `colnames`, respectively:

```
> rownames(jdata) <- journals
> colnames(jdata) <- c( "articles", "citations",
+ "impact factor")
> jdata
```

```
          articles citations impact factor
JAMIA          538       124  4.33870967742
IEEE TMI      1095       278  3.93884892086
Bioinf        6037      1003  6.01894317049
```

A common operation with matrices is **transposition**. This flips the matrix in such a way that the rows become columns, and vice versa. The function that performs this operation is `t`:

```
> t(jdata)
```

```
                    JAMIA             IEEE TMI
articles      538.00000000000 1095.00000000000
citations     124.00000000000  278.00000000000
impact factor   4.33870967742    3.93884892086
                    Bioinf
articles      6037.00000000000
citations     1003.00000000000
impact factor    6.01894317049
```

The computer memory is actually a linear array of memory locations. Hence, the elements of a matrix are also stored sequentially somewhere in the computer memory. The number of rows and columns of a matrix is only important in order to translate the usual notation `[i,j]`, which uses indexes for rows and columns, into an index in the linear array of stored values. This is very clear when using programming languages such as "C" or "C++" in which the elements of the matrix can also be addressed using a linear index that simply counts from the beginning of the array. The same phenomenon happens in R. We can first construct a vector as a one-dimensional array of numbers:

```
> x<-1:12
> x
```

```
[1]  1  2  3  4  5  6  7  8  9 10 11 12
```

then we can tell the system that it should treat this object as a matrix rather than a one-dimensional array. We can do that by specifying the **dimensions** of the matrix with the function `dim`:

```
> dim(x) <- c(3,4)
> x
```

```
     [,1] [,2] [,3] [,4]
[1,]    1    4    7   10
[2,]    2    5    8   11
[3,]    3    6    9   12
```

From now on, the system will treat x as a two-dimensional array, a matrix with 3 rows and 4 columns. However, in reality, nothing has changed in x. Its values continue to be stored in the same memory locations as before and indeed, they can be accessed using either a single index, specifying the position of the desired element from the beginning of the array:

```
> x[6]
```

```
[1] 6
```

or using two indexes, the way one normally does within a matrix:

```
> x[3,2]
```

```
[1] 6
```

Note that the default matrix representation is by columns, i.e. the elements of the matrix are stored in memory in the order of the columns. If we create another object y, by copying in it the transpose of x, the new object will also be stored by columns:

```
> y=t(x)
> y
```

```
     [,1] [,2] [,3]
[1,]    1    2    3
[2,]    4    5    6
[3,]    7    8    9
[4,]   10   11   12
```

```
> x
```

```
     [,1] [,2] [,3] [,4]
[1,]    1    4    7   10
[2,]    2    5    8   11
[3,]    3    6    9   12

> x[6]

[1] 6

> y[6]

[1] 5
```

If we wanted to create a matrix by filling its values in by rows, rather than by columns, we can use the `byrow=T` argument of the function `matrix`:

```
> z <- matrix(1:12,nrow = 3, ncol = 4, byrow=T)
> z

     [,1] [,2] [,3] [,4]
[1,]    1    2    3    4
[2,]    5    6    7    8
[3,]    9   10   11   12
```

Finally, matrices can also be constructed by binding together columns while also specifying their names. This can be achieved with the function `cbind`:

```
> new.jdata = cbind( "articles"=c(538, 1095, 6037),
+   "citations"=c(124,278,1003),
+   "impact factor"=c(4.33,3.93,6.01))
> new.jdata

     articles citations impact factor
[1,]      538       124          4.33
[2,]     1095       278          3.93
[3,]     6037      1003          6.01
```

A similar operation can be performed with rows by using the function `rbind`.

```
> rbind(c(1:3),c(4:6),c(7:9))

     [,1] [,2] [,3]
[1,]    1    2    3
[2,]    4    5    6
[3,]    7    8    9
```

6.3.6 Lists

In some situations, it is convenient to create and manipulate more complex data structures. A collection of heterogeneous objects can be combined to construct a larger, more complex data structure with the `list` function. For instance, we can construct an list that stores both the titles of the journals and their impact factors as follows:

```
> list_of_journals <- list(titles=journals,
+    impact_factors=impact.factors)
> list_of_journals

$titles
[1] "JAMIA"    "IEEE TMI" "Bioinf"

$impact_factors
    JAMIA   IEEE TMI     Bioinf
4.3387097 3.9388489 6.0189432
```

Using the code above, we created a **list** that has two elements: an array of character strings, currently storing the titles of the journals, and an array of numbers, currently storing the impact factors of those journals. When creating the list, we specified two **names**, one for each compoment of the list: `titles` and `impact_factors`. When we requested to see the object we created, `list_of_journals`, R showed separately each of the two components, with their respective names. The main difference between a vector with named components and a list, is that all elements of a vector have to have the same type, whereas the compoments of a list can be of heterogeneous, arbitrary types. For instance, in this case, the first component of the list we created is an array of character strings while the second component is an array of real values. In this example, the two arrays had the same dimensions, but in general, the different components of the list could have different dimensions.

The named components of a list can be referenced using the name of the list and the name of the list component, separated by a '$' character:

```
> list_of_journals$titles

[1] "JAMIA"    "IEEE TMI" "Bioinf"
```

6.3.7 Data frames

Data frames are data structures that are meant to store data that involves multiple variables. A data frame is essentially a list of vectors of the same length, in which the elements in a given position of each vector correspond to the same variable. For instance, if an experiment is to study the effect of a given drug, a common experiment design is to measure the expression level of a number of genes before and after administering the drug. Let us assume that the measurements of 10 genes are recorded before and after the treatment with the drug studied in two vectors *before* and *after*, respectively. These vectors can be read from their respective files (assumed to be in the current R directory) with:

```
> before = read.table("before.dat")
> before
```

```
         x
1    10.3058593
2    10.8856691
3    11.2277810
4    10.8644878
5     9.1144952
6     9.0166140
7    11.8154077
8     9.7103607
9    10.4707687
10    9.7584028
```

```
> after = read.table("after.dat")
> after
```

```
         x
1    12.2282625
2    11.4250025
3    11.3917311
4    11.9081940
5    12.6692205
6    12.4416190
7    10.4959695
8     9.3952821
9    10.4289047
10   11.4901844
```

We construct two vectors with the names and ages of the subjects:

```
> subjects = c("John","Mary", "Susan","Robert", "Mike",
+    "Peter", "David", "Anita", "Dan", "George")
> ages = c(15, 21, 23, 25, 26, 28, 51, 38, 42, 45)
```

All these various types of data about these subjects can be now put together in a data frame:

```
> d<- data.frame(subjects, ages, before, after)
> d
```

```
   subjects ages     x    x.1
1      John   15 10.306 12.228
2      Mary   21 10.886 11.425
3     Susan   23 11.228 11.392
4    Robert   25 10.864 11.908
5      Mike   26  9.114 12.669
6     Peter   28  9.017 12.442
7     David   51 11.815 10.496
8     Anita   38  9.710  9.395
9       Dan   42 10.471 10.429
10   George   45  9.758 11.490
```

In this structure, the data are *paired*, i.e., the data on any given row refers to the same person. Note that the names of the columns are different. The first two columns bear the names of the vectors used to create the data frame. However, the last two columns need some meaningful names. We can do this with the function **names**:

```
> names(d)=c("Subjects", "Ages", "before", "after")
> d
```

```
   Subjects Ages before  after
1      John   15 10.306 12.228
2      Mary   21 10.886 11.425
3     Susan   23 11.228 11.392
4    Robert   25 10.864 11.908
5      Mike   26  9.114 12.669
6     Peter   28  9.017 12.442
7     David   51 11.815 10.496
8     Anita   38  9.710  9.395
9       Dan   42 10.471 10.429
10   George   45  9.758 11.490
```

Specific data from the frame can be accessed using the '$' notation used for lists:

```
> d$before
```

```
 [1] 10.306 10.886 11.228 10.864  9.114  9.017 11.815  9.710
 [9] 10.471  9.758
```

```
> d$before[4]
```

```
[1] 10.86
```

The main advantage of a data frame with respect to a matrix is the fact that **a data frame may contain elements of different types** as shown above.

6.4 Other capabilities

6.4.1 More advanced indexing

We have seen that specific elements of a vector can be accessed using indexing such as d\$before[4], above. A more powerful way of accessing elements is by referring to several such elements at the same time. For instance, if we wanted to get the values corresponding to subjects 1, 3, and 5 in the data frame above, we could do so by constructing a vector with the indices of interest and then using this vector to index the data frame:

```
> d$before[c(1,3,5)]
```

```
[1] 10.306 11.228  9.114
```

Note that simply enumerating the indices without constructing a vector first would not have worked:

```
> d$before[1,3,5]
Error in d$before[1,3,5]:incorrect number of dimensions
>
```

In this case, the R environment interpreted the given statement as an attempt to index in a 3-dimensional structure (a 3D-matrix) and complain about the fact that the vector d\$before only had one dimension rather than the 3 dimensions implied by the given syntax.

Several vectors can be indexed in the same way, at the same time, by storing the index vector in a variable and using it as the index:

```
> ind = c(1,3,5,7,8)
> d$before[ind]
```

```
[1] 10.306 11.228  9.114 11.815  9.710
```

```
> d$after[ind]
```

```
[1] 12.228 11.392 12.669 10.496  9.395
```

A very convenient feature of R is the ability to use **negative indexing** as in "give me all values with the exception of the ones that I specify." In the example above, the remaining patients (2, 4, 6, 9, and 10) can be referenced as:

```
> d$after[-ind]
```

```
[1] 11.43 11.91 12.44 10.43 11.49
```

6.4.2 Missing values

In many situations, some measurements within a set of measurements are not available. For instance, a gene can be spotted in quadruplicate on a given DNA array. The intent is to measure this gene several times in order to get a better estimate about its real expression value. However, in many instances, one of these spots can be affected by any one of a number of problems (see Chapter 5). In such circumstances, many image processing software packages discard such spots and provide no intensity values for them. In R, this can be conveniently represented by a special value, NA, that stands for "not available." This value is carried through in computations. The result of an operation involving NA values will also be an NA value. For instance, if we try to multiple by 2 all elements of an array that contains some missing values, we will get what we one would expect, an array in which all values but the missing one are doubled:

```
> incomplete.array <- c(1, 2, NA, 4, 5)
> incomplete.array

[1]   1   2 NA   4   5

> double.incomplete.array <- incomplete.array * 2
> double.incomplete.array

[1]   2   4 NA   8 10
```

However, if we tried to calculate the mean of a set of values in which there are some missing values, we will get an NA that will signal us that at least one of the values involved in calculating the mean was not available:

```
> mean(incomplete.array)

[1] NA
```

Note that since any operation involving a missing value will produce a missing value, testing to see whether a value is NA cannot be done as a simple equality test. Thus, "incomplete.array[3]==NA" will produce NA, independently of what the value in incomplete.array[3] is, just because the right operand is NA. Given a vector, x, one can use the function "is.na(x)" to find out if there are any NA elements within the x. This function will produce a logical vector with TRUE in the position of any element which is NA.

```
> incomplete.array <- c(1, 2, NA, 4, NA)
> incomplete.array

[1]   1   2 NA   4 NA

> is.na(incomplete.array)

[1] FALSE FALSE  TRUE FALSE  TRUE
```

The same function can be used to set the NA values to specific values.

```
> incomplete.array[is.na(incomplete.array)] <- c(100,200)
> incomplete.array

[1]    1    2 100    4 200
```

The statement is.na(incomplete.array) is evaluated to a logical vector with 5 elements, the one in positions 3 and 5 being TRUE, the others FALSE. This array is used to index incomplete.array and store the new values, 100 and 200, in the appropriate positions. As we have seen when we looked for the better journals, an array indexed by a logical array, is evaluated to an array constructed just with elements that correspond to TRUE in the logical array.

It is often useful to find out how many NA values there are in a given vector, or whether there are any NA values at all. One way to achieve this is to use the fact that the logical values TRUE and FALSE are mapped to the integer values 1 and 0, respectively. Thus, if a vector were to have missing values at all, the function is.na() would return a number of TRUE values equal to the number of such missing values. Hence, a test for the existence of any missing values can be obtained by either testing whether the number of TRUE values returned by is.na() is zero, or by testing whether the number of FALSE values is equal to the length of the vector:

```
> v=c(1,2,3,NA,5,6,NA,7)
> v

[1]    1    2    3 NA    5    6 NA    7

> is.na(v)

[1] FALSE FALSE FALSE  TRUE FALSE FALSE  TRUE FALSE

> sum(is.na(v))

[1] 2

> sum(!is.na(v))

[1] 6

> length(v)

[1] 8

> sum(!is.na(v))==length(v)

[1] FALSE
```

In this case, the function is.na produces a vector of T/F values with 2 TRUE values, corresponding to the 2 missing values in v, and 6 FALSE values. The NOT operator, !, will reverse the logical values to 6 TRUE values, or ones and 2 FALSE values, or zeros. The function sum will then sum up the ones, corresponding to the elements of v which hold proper values and this sum is then compared with the length of v. A result of TRUE will mean v has no missing values, while a result of FALSE will mean that there is at least one missing value.

Any one of the expressions testing for missing values can be used in an if statement as in:

```
> if( sum(is.na(v)) == 0){
+   print("v does not have NAs")
+   }else{
+   print("v has NAs")
+   }

[1] "v has NAs"
```

As a common mistake, note that the expression sum(!is.na(v)==length(v)) which is very similar to the one used above, sum(!is.na(v))==length(v), will produce a rather different result:

```
> sum(!is.na(v)==length(v))

[1] 8
```

In this case, the function is.na produces a vector of 8 T/F elements, which is compared, element by element, with the vector of one element, 8, returned by the function length(v). Each individual test will produce a FALSE value, which yields a vector of 8 FALSE values, or 0-s. The not operator will transform that in a vector of 8 TRUE values, or 1-s, which will be summed up by the function sum to produce the result 8. Not only that this is not the TRUE/FALSE result expected, but also this result does not really depend on the presence or absence of N/A-s in the tested vector, v.

The approach above can be used to count the number of NA values in a vector, as shown. However, if the goal is to simply test whether a vector contains any NA values, this can be also achieve with the function any(), as follows:

```
> incomplete.array[3]=NA
> any(is.na(incomplete.array))

[1] TRUE
```

Furthermore, most functions handling numeric values have an argument called "na.rm" that can be set to TRUE in order to remove the NA values from the computation and thus obtain meaningful values even though the input vector(s) contain some NAs. For instance:

```
> mean(incomplete.array)
```

```
[1] NA
```

```
> mean(incomplete.array,na.rm=TRUE)
```

```
[1] 51.75
```

6.4.3 Reading and writing files

R provides a few powerful routines that allow us to **read** and **write** files. Let us assume that in a given microarray experiment a number of 5 genes, X1 through X5, are monitored across a group of 10 patients. Five of these patients have cancer while the other 5 are healthy. Let us assume that the log-transformed and centered data are available in a spreadsheet in the following format:

```
> ge
         g1       g2        g3       g4       g5   group
1    0.5531 -0.6477   0.23759  0.1150  0.20041  cancer
2   -0.0783  0.7444  -0.45264 -0.5911 -0.31485  cancer
3   -0.1981 -0.7442   1.88031  1.2154 -0.70994  cancer
4    0.4200 -0.1346  -0.24201 -0.2394  1.92980  cancer
5   -0.0275 -1.3090   1.09862  1.0839  0.73767  cancer
6    0.1621  1.8620   0.02371  0.3551 -0.69382 healthy
7    0.8024 -0.4652  -0.55831 -0.9892  0.14503 healthy
8   -2.3553 -0.6038   0.20400  0.6201  0.21308 healthy
9   -0.1469  0.5195   2.23747  1.2842 -0.40770 healthy
10   1.1306 -0.3171   0.63924  0.2249  0.03792 healthy
```

This format can be saved as a space- or tab-delimited text file, let us say "inputfile.txt" found in the current directory. Such a file can be read in R with a command as simple as:

```
> ge = read.table("./inputfile.txt")
> ge
         g1       g2        g3       g4       g5   group
1    0.5531 -0.6477   0.23759  0.1150  0.20041  cancer
2   -0.0783  0.7444  -0.45264 -0.5911 -0.31485  cancer
3   -0.1981 -0.7442   1.88031  1.2154 -0.70994  cancer
4    0.4200 -0.1346  -0.24201 -0.2394  1.92980  cancer
5   -0.0275 -1.3090   1.09862  1.0839  0.73767  cancer
6    0.1621  1.8620   0.02371  0.3551 -0.69382 healthy
7    0.8024 -0.4652  -0.55831 -0.9892  0.14503 healthy
8   -2.3553 -0.6038   0.20400  0.6201  0.21308 healthy
9   -0.1469  0.5195   2.23747  1.2842 -0.40770 healthy
10   1.1306 -0.3171   0.63924  0.2249  0.03792 healthy
```

Note that, if the entire path of the file is not specified, R will try to find it in the current directory which initially is something like "c:\R" (in Windows, the current directory can be changed from the File menu of the R window). From the command line, the directory can be changed with setwd:

```
> old=getwd()
> setwd("c:/Users/Sorin")
> getwd()

[1] "c:/Users/Sorin"

> setwd(old)
```

As another observation, the read.table function has a header=T/F option that tells the function whether the file contains a header or not. If the argument header=TRUE, the content of the first line will be used as variable names (column names). However, the function is smart enough to auto-detect whether the file has a header. In particular, if the first row contains one fewer field than the number of columns (as in our example above), the header option will be set to TRUE automatically and the function will read the file as expected.

In R, the type, or class, of an object can be obtained with the function class. An inquiry into the type of the object returned by read.table shows this object is a data frame:

```
> class(ge)

[1] "data.frame"
```

The function read.table read the given input file and created a data frame, that was returned and assigned to the variable ge. Let us now add another column to this data frame, indicating where the samples were collected: DMC will indicate Detroit Medical Center, and KCI will indicate Karmanos Cancer Institute. Also, we will name this new column "source" using the function names:

```
> ge[,7]=c("DMC","KCI","KCI","DMC","DMC",
+  "KCI","DMC","DMC","KCI","DMC")
> names(ge)[7]="source"
>  ge
```

```
          g1       g2       g3       g4       g5    group source
1     0.5531  -0.6477  0.23759   0.1150  0.20041   cancer    DMC
2    -0.0783   0.7444 -0.45264  -0.5911 -0.31485   cancer    KCI
3    -0.1981  -0.7442  1.88031   1.2154 -0.70994   cancer    KCI
4     0.4200  -0.1346 -0.24201  -0.2394  1.92980   cancer    DMC
5    -0.0275  -1.3090  1.09862   1.0839  0.73767   cancer    DMC
```

```
6    0.1621   1.8620   0.02371   0.3551 -0.69382 healthy   KCI
7    0.8024 -0.4652  -0.55831  -0.9892  0.14503 healthy   DMC
8   -2.3553 -0.6038   0.20400   0.6201  0.21308 healthy   DMC
9   -0.1469  0.5195   2.23747   1.2842 -0.40770 healthy   KCI
10   1.1306 -0.3171   0.63924   0.2249  0.03792 healthy   DMC
```

In fact, adding the data and the new column name can also be done simultaneously:

```
> ge$source=c("DMC","KCI","KCI","DMC","DMC",
+ "KCI","DMC","DMC","KCI","DMC")
```

If we now wanted to write these data into an output file, we could accomplish this with `write.table`:

```
> write.table(ge,"outputfile.txt")
```

which will dutifully produce the desired file as a space-delimited text file. This file can now be opened in Excel or any other program that can read text files.

An examination of the help pages for `write.table` shows that it is very easy to customize the way the file it is written by specifying for instance the character that will separate the elements of a row (by default `sep=' '`, the character to be used to separate consecutive lines (by default `eol ='\n'`), whether to include the row and column names (both `TRUE` by default), etc. R also provides two more convenient wrapper functions for writing comma separated values (CSV) files: `write.csv` and `write.csv2`. These functions are designed to produce a valid CSV file and hence they are deliberately inflexible with respect to those aspects that are required by the CSV format.

6.4.4   Conditional selection and indexing

We have seen in section 6.3.2 that elements of an array can be selected using a logical array, which in turn can be obtained from a logical operation. In fact, the same results can be achieved by indexing directly with a **logical expression**, without explicitly constructing a logical array. A logical expression is an expression involving operands that can be evaluated to logical values, and logical operators such as logical **and**, logical **or**, etc. The operands can be logical values or variables, and/or **relational expression** such as "`ind > 5`" or "`d$after <=10`". Table 6.2 lists the logical and relational operators in R. A reader familiar with the C programming language will notice that R uses the same relational operators as C. The logical operators are also similar to the bit-wise logical operators in C. The C logical operators "`&&`" and "`||`" also exist in R, but they are used as relational operators in flow control.

For instance, if we wanted to select from the gene expression data frame ge only those patients who have a positive expression value of gene5, we could do:

Operator	Meaning
&	logical and
|	logical or
!	logical not
<	less than
>	greater than
<=	less than or equal to
>=	greater than or equal to
==	equal to
!=	not equal to

TABLE 6.2:   Logical and relational operators in R.

```
> subset.ge = ge[ge$g5>0,]
> subset.ge
```

```
         g1       g2       g3       g4       g5    group source
1    0.5531 -0.6477  0.2376   0.1150 0.20041   cancer    DMC
4    0.4200 -0.1346 -0.2420  -0.2394 1.92980   cancer    DMC
5   -0.0275 -1.3090  1.0986   1.0839 0.73767   cancer    DMC
7    0.8024 -0.4652 -0.5583  -0.9892 0.14503  healthy    DMC
8   -2.3553 -0.6038  0.2040   0.6201 0.21308  healthy    DMC
10   1.1306 -0.3171  0.6392   0.2249 0.03792  healthy    DMC
```

6.4.5   Sorting

R has two functions related to sorting. The first one, `sort`, take as an argument a vector and returns another vector containing the sorted values. Let us extract the gene expression values corresponding to gene g5 in a vector **g5** and sort the values:

```
> g5=ge[,5]
> g5
```

```
 [1]  0.20041 -0.31485 -0.70994  1.92980  0.73767 -0.69382
 [7]  0.14503  0.21308 -0.40770  0.03792
```

```
> sort(g5)
```

```
 [1] -0.70994 -0.69382 -0.40770 -0.31485  0.03792  0.14503
 [7]  0.20041  0.21308  0.73767  1.92980
```

Parameters to `sort` allow us to specify whether the sorting is to be done in increasing or decreasing order (**decreasing=F** by default), whether the missing values should be placed at the end of the sorted vector (**na.last=T**), at the beginning (**na.last=F**), or removed (**na.last=NA**), etc.

A more interesting and useful case occurs when we need to sort *something else* in the order given by the sorted elements of a vector. For example, let us assume we want to sort the expression values of gene 3, in the order given by the expression values of gene 5, in such a way that the elements of each sorted vectors will still correspond to the same person.

This can be easily achieved with the function order. Given a vector as an argument, order will produce another vector containing the positions of the elements in the original vector that correspond to the desired ordering. In our case, order(g5) produces the following:

```
> g5
```

```
[1]   0.20041 -0.31485 -0.70994   1.92980   0.73767 -0.69382
[7]   0.14503   0.21308 -0.40770   0.03792
```

```
> order(gb)
```

```
[1]   3   6   9   2 10   7   1   8   5   4
```

This tells us that the smallest element of g5 is in position 3, the second smallest element is in position 6, etc. This results produced by order, can be used to index another object, for instance g3, in order to sort it in the order given by the elements of g5.

```
> g5
```

```
[1]   0.20041 -0.31485 -0.70994   1.92980   0.73767 -0.69382
[7]   0.14503   0.21308 -0.40770   0.03792
```

```
> order(g5)
```

```
[1]   3   6   9   2 10   7   1   8   5   4
```

```
> g3=ge[,3]
> g3
```

```
[1]   0.23759 -0.45264   1.88031 -0.24201   1.09862   0.02371
[7] -0.55831   0.20400   2.23747   0.63924
```

```
> sorted.g3=g3[order(g5)]
> sorted.g3
```

```
[1]   1.88031   0.02371   2.23747 -0.45264   0.63924 -0.55831
[7]   0.23759   0.20400   1.09862 -0.24201
```

Or course, order(g5) can also be used in order to sort g5 itself:

```
> sorted.g5 = g5[order(g5)]
> sorted.g5
```

```
[1] -0.70994 -0.69382 -0.40770 -0.31485  0.03792  0.14503
[7]  0.20041  0.21308  0.73767  1.92980
```

Note that the sorted vectors maintain the pairing between the values obtained from the same sample: -0.70994 corresponds to 1.88031, etc., as in the original data frame in Section 6.4.3.

Now we can sort the entire data frame by the values of one of the columns, such as g5:

```
> ge.sorted.by.g5 = ge[order(ge$g5),]
> ge.sorted.by.g5
```

```
        g1       g2       g3       g4       g5   group source
3  -0.1981 -0.7442  1.88031  1.2154 -0.70994  cancer    KCI
6   0.1621  1.8620  0.02371  0.3551 -0.69382 healthy    KCI
9  -0.1469  0.5195  2.23747  1.2842 -0.40770 healthy    KCI
2  -0.0783  0.7444 -0.45264 -0.5911 -0.31485  cancer    KCI
10  1.1306 -0.3171  0.63924  0.2249  0.03792 healthy    DMC
7   0.8024 -0.4652 -0.55831 -0.9892  0.14503 healthy    DMC
1   0.5531 -0.6477  0.23759  0.1150  0.20041  cancer    DMC
8  -2.3553 -0.6038  0.20400  0.6201  0.21308 healthy    DMC
5  -0.0275 -1.3090  1.09862  1.0839  0.73767  cancer    DMC
4   0.4200 -0.1346 -0.24201 -0.2394  1.92980  cancer    DMC
```

The function `order` can also be used to sort by several criteria. This is accomplished by having several arguments, for instance, `order(ge$source,ge$group)`:

```
> ge1 = ge[order(ge$source,ge$group),]
> ge1
```

```
        g1       g2       g3       g4       g5   group source
1   0.5531 -0.6477  0.23759  0.1150  0.20041  cancer    DMC
4   0.4200 -0.1346 -0.24201 -0.2394  1.92980  cancer    DMC
5  -0.0275 -1.3090  1.09862  1.0839  0.73767  cancer    DMC
7   0.8024 -0.4652 -0.55831 -0.9892  0.14503 healthy    DMC
8  -2.3553 -0.6038  0.20400  0.6201  0.21308 healthy    DMC
10  1.1306 -0.3171  0.63924  0.2249  0.03792 healthy    DMC
2  -0.0783  0.7444 -0.45264 -0.5911 -0.31485  cancer    KCI
3  -0.1981 -0.7442  1.88031  1.2154 -0.70994  cancer    KCI
6   0.1621  1.8620  0.02371  0.3551 -0.69382 healthy    KCI
9  -0.1469  0.5195  2.23747  1.2842 -0.40770 healthy    KCI
```

As it can be seen above, this sorts first by the content of the column "`source`" and then by the content of "`group`." This is similar to the lexicographical order commonly found in dictionaries and phone books where the names are ordered by the first letter of the last name, then by the second letter, etc.

6.4.6   Implicit loops

In many situations, we need to perform the same operation to all elements of a vector, matrix or data frame. In other words, we would like to loop over all elements of a certain object. An example could be the ge data frame in which we would like to convert all values into positive values, or perform a normalization by subtracting the overall mean from each individual expression value. In R, this can be achieved with the function sapply(X, FUN, ...). This function takes a vector X (either atomic or list) as the first argument, the name of a function, FUN, as the second argument, and applies the given function FUN to each element of X. The results returned by sapply will be a vector or matrix, if appropriate. The function lapply does the same thing but always returns a list. For instance, this is what sapply and lapply return when their argument is a vector:

```
> x=c(2,3,4)
> sapply(x,log)

[1] 0.6931 1.0986 1.3863

> lapply(x,log)

[[1]]
[1] 0.6931

[[2]]
[1] 1.099

[[3]]
[1] 1.386
```

And this is what sapply and lapply return when their argument is a list:

```
> y=list();y[[1]] = 2; y[[2]]=4
> sapply(y,log)

[1] 0.6931 1.3863

> lapply(y,log)

[[1]]
[1] 0.6931

[[2]]
[1] 1.386
```

In the first example, lapply returned a list even though x was a vector, while in the second example sapply figured it out that since all the elements of the resulting list are scalars, the entire result can be represented as a vector rather than as a list.

Going back to our example, the following code applies the abs function, which returns the absolute value of a real or complex number, to all numerical elements of ge:

```
> sapply(ge[1:10,1:5],abs)
```

```
          g1     g2      g3     g4      g5
 [1,]  0.5531 0.6477 0.23759 0.1150 0.20041
 [2,]  0.0783 0.7444 0.45264 0.5911 0.31485
 [3,]  0.1981 0.7442 1.88031 1.2154 0.70994
 [4,]  0.4200 0.1346 0.24201 0.2394 1.92980
 [5,]  0.0275 1.3090 1.09862 1.0839 0.73767
 [6,]  0.1621 1.8620 0.02371 0.3551 0.69382
 [7,]  0.8024 0.4652 0.55831 0.9892 0.14503
 [8,]  2.3553 0.6038 0.20400 0.6201 0.21308
 [9,]  0.1469 0.5195 2.23747 1.2842 0.40770
[10,]  1.1306 0.3171 0.63924 0.2249 0.03792
```

A more useful thing would be to normalize all values in ge, for instance, by subtracting the overall mean. This would correspond to a global normalization.[2] In this case, we need to specify not only the function to be applied, in this case a simple subtraction, but also the value of the overall mean. This is actually easy because sapply passes any extra parameters to the function to be applied:

```
> m = mean(as.matrix(ge[1:10,1:5]))
> m

[1] 0.169

> nge = sapply(ge[1:10,1:5],"-",m)
> nge

            g1        g2        g3       g4       g5
 [1,]  0.384088 -0.8167  0.06858 -0.05398  0.03140
 [2,] -0.247306  0.5754 -0.62165 -0.76011 -0.48386
 [3,] -0.367150 -0.9133  1.71130  1.04641 -0.87895
 [4,]  0.250954 -0.3036 -0.41102 -0.40845  1.76079
 [5,] -0.196511 -1.4780  0.92961  0.91488  0.56866
 [6,] -0.006954  1.6930 -0.14530  0.18604 -0.86283
 [7,]  0.633365 -0.6342 -0.72733 -1.15818 -0.02398
 [8,] -2.524328 -0.7729  0.03499  0.45110  0.04406
 [9,] -0.315930  0.3505  2.06846  1.11521 -0.57671
[10,]  0.961586 -0.4862  0.47022  0.05590 -0.13109
```

---

[2]See Chapter 20 for a detailed description of several specific ways to normalize microarray data.

```
> mean(nge)
```

```
[1] -7.751e-18
```

In the code above, the function passed to `sapply` was the subtraction "-"; the additional parameter was actually the mean value to be subtracted from each element. At the end, we calculated the mean of the normalized matrix and verified that it is in fact, practically zero (10^{-18}).

Another type of normalization is an array normalization in which the mean of each array is subtracted from all the values coming from that array in order to make the arrays comparable to each other. This can be done as follows:

```
> ge
```

	g1	g2	g3	g4	g5	group	source
1	0.5531	-0.6477	0.23759	0.1150	0.20041	cancer	DMC
2	-0.0783	0.7444	-0.45264	-0.5911	-0.31485	cancer	KCI
3	-0.1981	-0.7442	1.88031	1.2154	-0.70994	cancer	KCI
4	0.4200	-0.1346	-0.24201	-0.2394	1.92980	cancer	DMC
5	-0.0275	-1.3090	1.09862	1.0839	0.73767	cancer	DMC
6	0.1621	1.8620	0.02371	0.3551	-0.69382	healthy	KCI
7	0.8024	-0.4652	-0.55831	-0.9892	0.14503	healthy	DMC
8	-2.3553	-0.6038	0.20400	0.6201	0.21308	healthy	DMC
9	-0.1469	0.5195	2.23747	1.2842	-0.40770	healthy	KCI
10	1.1306	-0.3171	0.63924	0.2249	0.03792	healthy	DMC

```
> new_ge=as.data.frame(t(apply(ge[,1:5],1,
+ function(x){return(x-mean(x))})))
> new_ge$group=ge$group;new_ge$source=ge$source
> new_ge
```

	g1	g2	g3	g4	g5	group	source
1	0.46142	-0.7394	0.1459	0.02335	0.1087	cancer	DMC
2	0.06021	0.8829	-0.3141	-0.45260	-0.1763	cancer	KCI
3	-0.48682	-1.0329	1.5916	0.92674	-0.9986	cancer	KCI
4	0.07323	-0.4814	-0.5887	-0.58617	1.5831	cancer	DMC
5	-0.34424	-1.6257	0.7819	0.76715	0.4209	cancer	DMC
6	-0.17974	1.5202	-0.3181	0.01326	-1.0356	healthy	KCI
7	1.01543	-0.2521	-0.3453	-0.77612	0.3581	healthy	DMC
8	-1.97092	-0.2194	0.5884	1.00451	0.5975	healthy	DMC
9	-0.84424	-0.1778	1.5402	0.58690	-1.1050	healthy	KCI
10	0.78749	-0.6603	0.2961	-0.11819	-0.3052	healthy	DMC

Here, we used the function `apply` which applies the function given as an argument to the margins of an array (i.e., to its rows, columns or both). The `apply` function takes 3 arguments: the array, the margin and the function to be applied. A margin equal to 1 specifies that the function will be applied

Matches	Prize
10 of 22	$250,000
9 of 22	$2,500
8 of 22	$250
7 of 22	$25
6 of 22	$7
0 of 22	$1 ticket

TABLE 6.3: In the game of Keno, the player picks 10 numbers from 1 to 80. 22 winning numbers are drawn from the same pool of 1 to 80. The table shows the winning combinations and their respective prizes. See Chapter 10 for a computation of the odds of winning each of these prizes.

to the rows of the array; a margin equal to 2 specifies that the function will be applied to the columns, while a margin equal to c(1,2) specifies that the function will be applied to both rows and columns. In this case, we defined a function on the fly to subtract the mean of an array x from (each element of) x, and we passed this function as the third argument to `apply`.

We can now verify that the normalization was successful by calculating the means of each row:

```
> rowMeans(new_ge[,1:5])
```

```
          1           2           3           4           5
0.000e+00  -2.220e-17   4.441e-17  -1.110e-17   2.220e-17
          6           7           8           9          10
-1.110e-17   3.331e-17  -5.551e-17   0.000e+00  -5.551e-18
```

Let us consider the Keno lottery[3] as another example. In this game, the player can pick 10 numbers from 1 to 80. The lottery draws 22 numbers and a ticket will win if it has 0, 6, 7, 8, 9, or 10 out of the 22 winning numbers (see Table 6.3). Let us assume we have a function that calculates the probability of winning for a given number of matches:[4]

```
> keno_prob <- function(n,m,k,l){
+ # n numbers available, m numbers on a ticket,
+ # k numbers extracted, l have to match
+     prob = choose(k,l)*choose(n-k,m-l)/choose(n,m)
+     #odds = 1/prob
+     prob
+ }
```

---

[3] See http://www.michigan.gov/lottery/0,1607,7-110-46442_812_872---,00.html.

[4] The reader should not be concerned at this point with how to write such a function. This topic is extensively covered in Chapter 10, which explains this function in details.

We would like to use this function to calculate the probability of winning each specific prize plus the overall probability of winning any of the prizes. In order to do this, we can construct a vector containing all number of matches necessary to win the various prizes:

```
> allwins = c(10:6,0)
> allwins

[1] 10  9  8  7  6  0
```

We would like to apply the `keno_prob` function defined above a number of times equal to the number of elements in `allwins`, each time using the current element of the vector as the last argument of `keno_prob`. Here, the reader experienced in other procedural programming languages such as C/C++ or Java might be tempted to try to achieve this using a `for` loop. This is, of course, possible, but such an approach would be both inefficient and against the R philosophy. The power of R comes from the fact that it can perform vector and matrix operations very efficiently, so we should try to exploit this. One approach would be to use the function `sapply`. The apparent difficulty is related to the fact that the function to be applied takes not one but four arguments and the current element of the vector should always be the fourth argument of this function. We can work around this by creating another function that takes a single argument, the number of matches in a ticket, and calls `keno_prob` with the appropriate arguments in order to calculate each probability:

```
>   prob_wins <- function(l){
+       keno_prob(80,10,22,l)
+       }
>   winprob = sapply(allwins,prob_wins)
>   winprob

[1] 3.927e-07 1.752e-05 3.210e-04 3.196e-03 1.923e-02
[6] 3.169e-02
```

Fortunately, we do not need to write such a wrapper function in such cases. Rather than this, it is simply sufficient to call `sapply` explicitly specifying the other arguments to be passed to `keno_prob`:

```
> winprob = sapply(allwins,keno_prob,n=80,m=10,k=22)
> winprob

[1] 3.927e-07 1.752e-05 3.210e-04 3.196e-03 1.923e-02
[6] 3.169e-02
```

In fact, in R there is another solution to this problem, which is even more elegant. This solution relies on R's built-in mechanism of dealing with vectors. In fact, we can simply call our `keno_prob` with our vector of values *as an argument*:

```
> keno_prob(80,10,22,allwins)
```

```
[1] 3.927e-07 1.752e-05 3.210e-04 3.196e-03 1.923e-02
[6] 3.169e-02
```

This is probably as good as it gets in terms of simplicity and efficiency. Furthermore, note that the function `keno_prob` was defined without any special precautions about the eventuality of any of the arguments being a vector. This is a good example of R's philosophy. Flexibility and efficiency are held in greatest regard at the expense of the ability to catch errors. Indeed, the definition of the function did not require us to specify any specific type for any of its arguments. R will accept almost anything as an actual argument and pass it down to the function. As long as the type of the actual parameter is compatible with the operations attempted on it (or the argument can be coerced into something acceptable), processing will be done. This provides tremendous power and flexibility as illustrated in the example above. The prices paid for this is that sometimes, R's way of interpreting or coercing arguments might be different from the intention of the programmer. The outcome may be *some* result but this result may not be at all what was intended. Hence, in order to avoid unexpected results it is probably safer to pass vectors as arguments only to functions who have been designed to accept such arguments (e.g., their documentation specifies that they accept vectors as arguments). Most built-in R functions fall into this category.

6.5 The R environment

6.5.1 The search path: `attach` and `detach`

In the example above in which we wanted to sort using several columns from the same data frame, we had to specify the name of the data frame every time we wanted to refer to one of its columns: `order(ge$source,ge$group)`. Clearly, this can become very cumbersome, especially if the name of the data frame were longer and more columns were to be needed. In order to address this issue, R provides the means to specify the name of a data frame as the implicit environment in which R should be looking for objects when the user specify an object name. This is achieved with the command `attach`, which takes the name of a data frame as its argument as in: `attach(ge)`:

```
> attach(ge)
  The following object(s) are masked _by_ .GlobalEnv :
        g3 g5
  The following object(s) are masked from package:base :
        source
>
```

By using `attach`, we attempted to tell R that whenever an object name is specified, it should look into the data frame `ge` for a column with that name, i.e. to treat an object name such as `g5` as if it had been specified as `ge$g5`. In this case, however, R is telling us that there are some conflicts between objects that already exist and the column names in `ge`. More specifically, the first conflict is between the column `g3` and `g5` of the data frame `ge` and the vectors `g3` and `g5` defined in the Section 6.4.5. The error message tells us that the columns `g3` and `g5` of the data frame `ge` will be *masked by* the objects with the same names from the global environment. This means that if we give a command specifying `g3` for instance, R will use the object `g3` from the global environment rather than the column from the data frame. In this case, the two happen to contain the same data, but in general, any such name conflicts should be avoided. The second conflict involves an object called `source` from the `base` package. In this case, R is telling us that the column `source` in our data frame *will mask* the object with the same name in the package `base`.

In order to avoid such conflicts, we can use the `names` function to change the names of the columns in the ge data frame, after which the `attach` command does not complain anymore:

```
> names(ge)=c("gene1","gene2","gene3","gene4",
+    "gene5","group","origin")
> ge
```

	gene1	gene2	gene3	gene4	gene5	group	origin
1	0.5531	-0.6477	0.23759	0.1150	0.20041	cancer	DMC
2	-0.0783	0.7444	-0.45264	-0.5911	-0.31485	cancer	KCI
3	-0.1981	-0.7442	1.88031	1.2154	-0.70994	cancer	KCI
4	0.4200	-0.1346	-0.24201	-0.2394	1.92980	cancer	DMC
5	-0.0275	-1.3090	1.09862	1.0839	0.73767	cancer	DMC
6	0.1621	1.8620	0.02371	0.3551	-0.69382	healthy	KCI
7	0.8024	-0.4652	-0.55831	-0.9892	0.14503	healthy	DMC
8	-2.3553	-0.6038	0.20400	0.6201	0.21308	healthy	DMC
9	-0.1469	0.5195	2.23747	1.2842	-0.40770	healthy	KCI
10	1.1306	-0.3171	0.63924	0.2249	0.03792	healthy	DMC

```
> attach(ge)
```

An alternative and perhaps simpler solution would have been to just deleted the g3 and g5 objects from the work space.

Note that `attach` creates a virtual copy of the original data frame. Hence, any changes performed on the variables after the data frame is attached will not affect the original data frame. For the readers familiar with other programming languages, this is similar to passing parameters by value. In general, the argument of `attach` is a "database," which can be a data frame, a list, or an R file created with the function `save`.

A complete explanation of `attach` and its counterpart `detach` requires an explanation about how R manages its environment and libraries. R maintains a **search path** which is a list of places in which R will look in its attempt to link a name specified by the user to an object. The first item on this list is always a global environment .GlobalEnv, which will contain all objects explicitly created by the user during a session. The search path can then contain a list of packages and/or data frames that are loaded by the user. The next to last item on the search path is Autoloads. This will contain names of packages that are not loaded into the memory yet but would be loaded as soon as one of their functions is called. For readers with a computer science background, this is R's implementation of the dynamic library loading idea. This mechanism offers a good compromise between offering a larger collection of functions to the user while still maintaining a small memory footprint. Finally, the last item on the search path is the package base, which is the system library containing the basic functions that constitute the core of the R environment.

When a name is specified in a command or function call, R starts looking at the items in the order they occur on the search path until the name is matched to an object. Because the global environment which contains the objects explicitly created by the user is always first, the user can always override existing functions or objects, if (s)he so desires. In order to avoid accidental overrides, most system variables in R have special names that start with a dot: ".", such as ".RData" or ".GlobalEnv". Hence, it is best not to use such names in the day-to-day use. Any package or data frame that is loaded later will be placed by default on the second position, after the .GlobalEnv. Subsequent loading of further packages or data frames will produce a list such as:

```
> search()
```

```
 [1] ".GlobalEnv"        "ge"
 [3] "package:patchDVI"  "package:stats"
 [5] "package:graphics"  "package:grDevices"
 [7] "package:utils"     "package:datasets"
 [9] "package:methods"   "Autoloads"
[11] "package:base"
```

In the context of the search path, the function performed by attach is very simple to understand. Its effect is to insert the specified database in the search path in a given position that can be specified by the user. By default, attach(my.data.frame) will insert the data frame on the second position, after .GlobalEnv but before any other packages or data frames previously loaded. The opposite of attach is detach, which detaches the name of the specified database by removing it from the search path.

6.5.2 The workspace

The R **workspace** is an environment in which R maintains all objects created during a working session. This environment can be saved at the end of a session using the save.image function:

```
>  save.image("mywork.Rdata")
```

This will save the workspace in a file called "mywork.Rdata" in the current working directory. The same function called without any arguments will save the workspace in a file called ".RData" in the current directory (.RData is the default extension of such workspace files). In Windows, double-clicking this file's icon will launch R, which in turn will load the content of this workspace. In Windows, the workspace can also be saved and loaded using choices from the File menu of the R GUI environment. By default, when using the menu choice "Load workspace," the system will only show the files of type .Rdata but without their extensions so one should look for "mywork" if attempting to reload the same workspace.

The workspace can also be later reloaded from within R with the function load() (e.g. load("mywork.RData")).

In Windows, R starts by default with the directory "My Documents" as the current working directory. The function R.home() returns R's installation directory:

```
>  getwd()
```

```
[1] "C:/Users/Sorin/Documents/Books/Book revision"
```

```
>  R.home()
```

```
[1] "C:/PROGRA~1/R/R-213~1.0"
```

The current directory can be changed with the function setwd(dir) or, in Windows, from the "File→Change dir" menu choice. The function getwd() returns the current working directory:

```
> setwd("d:/Sorin/Book revision")
> getwd()
[1] "d:/Sorin/Book revision"
>
```

Note the use of the forward slashes on the path even in the Windows environment. Trying to use the backslashes will result in an error with a less than explicit error message:

```
> setwd("d:\Sorin\Book revision")
Error in setwd(dir) : cannot change working directory
>
```

Also note that the space in the name of the "Book revision" directory did not need to be protected by a backslash although it could have been:

```
> setwd("d:/Sorin/Book\ revision")
> getwd()
[1] "d:/Sorin/Book revision"
>
```

Function	Description
ls()	list the content of the workspace
objects()	list the content of the workspace
rm(name)	remove *name* from the workspace
save.image("mywork")	save the workspace in the file *mywork*
load("mywork")	load the workspace from the file *mywork*
setwd("c:/Work")	set the current working directory to "$c : \backslash Work$"
getwd()	get current working directory
R.home()	returns the directory in which R is installed
history()	displays the most recent commands (default max=25)
savehistory("myhist")	saves the most recent commands (by default 512) in the file *myhist*
loadhistory("myhist")	loads the command history from the file *myhist*

TABLE 6.4: Commands and functions related to the workspace management.

The content of the workspace can be listed at any time with the function `ls()`. An object can be deleted from the workspace using the function `rm(object)`. Note that the argument of `rm` can be a list of objects, possibly including all objects in the workspace as in `rm(list=ls())`. Readers familiar with the Unix system will note that these commands for listing and removing files in R are the same as in the Unix shell. An alternative way of listing the content of the workspace is `objects()`.

Much like a Unix shell, the R command environment maintains a list of recent commands given by the user. These commands are actually easily accessible using the vertical arrows on the keyboard which will scroll up and/or down through the history, bringing the respective commands on the current line, ready to be modified. A frequent use of this feature is correcting a spelling or syntax error in one of the previous commands.

A summary of the functions related to the workspace and its management is given in Table 6.4.

6.5.3 Packages

Packages are collections of functions and data related to a certain specific problem or range of problems. Packages can include data sets, R code, or dynamically loaded libraries (DLLs). DLLs can be written in other languages such as C or Fortran. C is usually used for its complete control and efficient code generation while Fortran is known for its extensive and efficient numerical libraries. The idea is that a package will contain data and/or functions that are not needed all the time. A package can be loaded into the memory when needed and discarded either explicitly, or automatically at the end of the session.

The packages are stored locally in a subdirectory `library` of the R home directory. Usually, each package will appear as a subdirectory in the local `R.home/library` file such as `R.home/library/utils`. In turn, each package subdirectory will usually contain a number of subdirectories with R code, data, demos, etc.

The function `library()` can be used to list all packages available locally. A specific package is loaded into memory with the same function, by providing the name of the desired package as an argument:

```
> search()
```

```
 [1] ".GlobalEnv"         "ge"
 [3] "package:patchDVI"   "package:stats"
 [5] "package:graphics"   "package:grDevices"
 [7] "package:utils"      "package:datasets"
 [9] "package:methods"    "Autoloads"
[11] "package:base"
```

```
> library(class)
> search()
```

```
 [1] ".GlobalEnv"         "package:class"
 [3] "ge"                 "package:patchDVI"
 [5] "package:stats"      "package:graphics"
 [7] "package:grDevices"  "package:utils"
 [9] "package:datasets"   "package:methods"
[11] "Autoloads"          "package:base"
```

In the example above, the first call to `search()` lists the content of the search path, showing the currently loaded packages: `stats`, `graphics`, `grDevices`, `utils`, `datasets`, and `methods`. The `library` function is used to load the package `class` containing functions for classification. The second call to `search()` shows the requested package was loaded in the second position on the search path.

A package can be unloaded with the function `detach()`:

```
> detach("package:class")
```

This function is seldom used because packages are not considered part of the user environment and hence they are discarded every time the R session is terminated. This has the advantage that the amount of memory used by R in every session is kept under control. The disadvantage is that the required packages must be loaded every session.

6.5.4 Built-in data

The function `data` can be used to load into the memory a data set. In most cases, the data will be loaded into a data frame with the name specified as the argument. When called, the function `data` goes through each package in the search path and for each such package, looks for a "data" subdirectory that can contain the data in different formats such as .RData, text, or R code (.R extension). Each type of file will be treated appropriately. For instance, data files with a `.tab` extension will be read with read.table, code files with an `.R` extension will be executed, etc.

6.6 Installing Bioconductor

Bioconductor is available as a set of packages that can be downloaded and installed locally. As already mentioned, the latest version of Bioconductor can always be found at: `http://www.bioconductor.org/`. In spite of the large number of packages in Bioconductor and in spite of the somewhat complex dependencies between various packages, the installation is made extremely easy by a couple of scripts. In practice, in order to install Bioconductor, it is sufficient to execute a few lines of R code as follows:

```
> source("http://www.bioconductor.org/biocLite.R")
> biocLite()
```

The function `source` tells R to read and execute commands from the file or url provided as an argument. The first command above reads the file that contains the definition of the function `biocLite()`. The second line above actually calls it. The function `biocLite` is the installation script that will proceed to download, unpack, and install the missing packages. For each package, `biocLite` will download the .zip file containing the package, unzip it and install it in the local `R.home/library` file. More specifically, the script tries to download all files first, then individually unpacks them, checks for errors and installs them.

The `biocLite()` only installs a minimal, "default," group of the Bioconductor packages. The entire set of bioconductor packages can be installed with:

```
> biocLite(groupName="all")
```

but this is a large installation and it is not recommended for all users. The most practical approach is to install the default group of Bioconductor packages and subsequently install additional packages as needed, using `biocLite("mypackage")`.

Sometimes the installation script can report errors such as:

```
...
package 'affy' successfully unpacked and MD5 sums checked
Error in gzfile(file, "r") : unable to open connection
In addition: Warning messages:
1: downloaded length 4645936 != reported length 8138950
2: error 1 in extracting from zip file
3: cannot open compressed file 'matchprobes/DESCRIPTION'
>
```

In such cases, the best strategy is to re-launch the installation script biocLite(). Assuming that the above installation process eventually finishes without errors, individual packages from Bioconductor can now be loaded individually with the library command:

```
> library(marray)
```

In this case, the requested package marray required several other packages, which were automatically loaded as well.

A list of all packages installed can be obtained with the function sessionInfo() as illustrated below:

```
> sessionInfo()

R version 2.13.0 (2011-04-13)
Platform: i386-pc-mingw32/i386 (32-bit)

locale:
[1] LC_COLLATE=English_United States.1252
[2] LC_CTYPE=English_United States.1252
[3] LC_MONETARY=English_United States.1252
[4] LC_NUMERIC=C
[5] LC_TIME=English_United States.1252

attached base packages:
[1] stats     graphics  grDevices utils     datasets
[6] methods   base

other attached packages:
[1] marray_1.30.0    limma_3.8.2       patchDVI_1.8.1580

loaded via a namespace (and not attached):
[1] class_7.3-3
```

6.7 Graphics

The R environment has a nice collection of built-in graphics capabilities. The power of this environment comes from the fact that simple things can be done with simple commands while it is still possible to customize almost any aspect of a plot. If two variables are stored in two vectors, x and y, and we would like to plot y as a function of x, such a plot can be obtained with a command as simple as `plot(x,y)`. For instance, if we were to plot expression level of gene 1 across the entire set of 10 samples in the `ge` data frame, we could do so with:

```
> plot(ge$gene1,col="darkgreen")
```

The result of this command is shown in Fig. 6.2. Note that R took care of all cumbersome details. Reasonable choices are made automatically regarding all aspects of the plot: the ranges of the axes, the tick marks, the labeling for both tick marks and axes, the type of plot, etc. In this figure, the graph was constructed using points plotted for each (x,y) pair. Different types of graph can be easily requested using a supplementary parameter, the **graph type**. The various possible choices for this parameter are:

1. "p" for points,

2. "l" for lines,

3. "b" for both,

4. "c" for the lines part alone of "b",

5. "o" for both overplotted,

6. "h" for histogram like (or "high-density") vertical lines,

7. "s" for stair steps,

8. "S" for other steps,

9. "n" for no plotting.

Fig. 6.3 shows two such alternatives. In order to get nicer labels, we attach the data frame before plotting the graphs:

```
> plot(gene1,type="l",col="blue")

> plot(gene1,type="b",col="red")
```

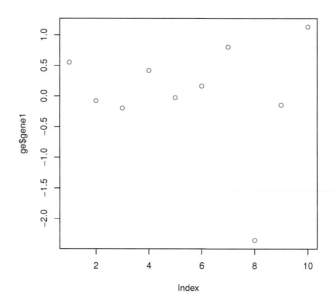

FIGURE 6.2: A simple plot in R showing the expression of **gene1** across the 10 samples. Reasonable choices are made automatically regarding all aspects of the plot: the ranges of the axes, the tick marks, the labeling for both tick marks and axes, the type of plot, etc.

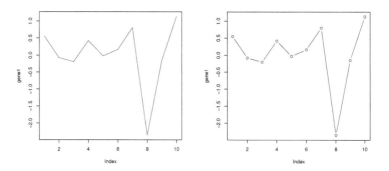

FIGURE 6.3: Different types of graphs can be easily obtained with various parameter values. The command `plot(gene1,type="l",col="blue")` generates a blue line graph, while `plot(gene1,type="b",col="red")` generates a red graph with both symbols and lines.

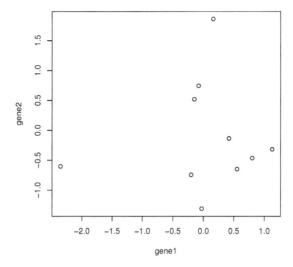

FIGURE 6.4: The command `plot(gene1,gene2)` generates a graph showing gene2 plotted as a function of gene1.

The command `type="l"` option generates a plot in which the (x,y) pairs are united with line segments, generating the graph to the left. The `type="b"` option generates a graph with both symbols and lines, shown in the right panel. Details regarding all other parameters and choices for a graph can be obtained with `help(plot)`.

A graph showing gene2 plotted against gene1 can be easily obtained with:

```
> plot(gene1,gene2)
```

In this plot, gene1 is plotted on the x axis and gene2 is plotted on the y axis. This plot is shown in Fig. 6.4.

6.8 Control structures in R

Like any other programming language, R offers the usual control structures: conditional statements and loops. These allow the programmer to control the flow of the execution. The syntax of these control structures is described in the following. In describing the syntax of these structures, we will use bold for the keywords and the syntactical elements that are intrinsic to the structure and italics for those parts that will be different for each utilization of the given control structure. For instance, "**if (** *logical-expression* **)** *statement*" shows that the word "if" and the round brackets are intrinsic parts of the structure, whereas the condition to be tested, *logical-expression*, as well as the statement to be executed if the condition is true, *statement* will be different every time this control structure is actually used. In computer science terminology, the terminal symbols are denoted by bold whereas the nonterminals are in italics.

6.8.1   Conditional statements

The conditional statement in R looks very similar to that used in C/C++ or Java:

> **if (** *logical-expression* **)** *statement*

In the above syntax, *logical-expression* has to be evaluated to a logical value that is not NA. As in most other cases, a vector of values (logical in this case) is also accepted in place of this logical expression. However, in this case, R will only use the first value in such a vector, rather than iterate through all its elements. A warning will also be issued in such a case. The statement can be a simple statement, such as an assignment or a function call, or a compound statement such as one formed with curly brackets around a set of simple statements:

> **{** *statement1*; ⋯ *statementN* **}**

The **if** conditional statement also has another variant:

> **if (** *condition* **)** *statement1* **else** *statement2*

In this case, *statement2* will only be executed if the logical expression tested is not true. As an example, the following piece of code prints the absolute value of a random number x, drawn from a normal distribution with zero mean and standard deviation equal to 1.

```
> x=rnorm(1)
> x

[1] -0.7084

> if( x > 0) {
+        print(x)
+  } else {
+        print(-x)
+  }

[1] 0.7084
```

R also provides a convenient function, **ifelse**, that can often be used

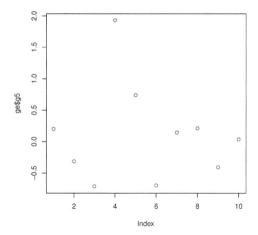

FIGURE 6.5: The command `colors =ifelse(ge$g5>0,"red","black")` returns an array with values selected from the two alternatives based on the truth value of the condition. The resulting array can be used to plot the negative and positive values with different colors, as above.

instead of a loop including an if statement. This function takes as an argument an object that contains logical values (or that can be coerced to such) and returns an object of the same type that is filled with elements selected from two alternatives, based on the given logical values. For instance, if we wanted to plot the values of the g5 variable in the ge dataframe using two colors: red for positive values and black for negative values, we could do this as follows:

```
> ge = read.table("./inputfile.txt")
> colors=ifelse(ge$g5>0,"red","blue")
> plot(ge$g5,col=colors)
```

The resulting plot is shown in Fig. 6.5.

6.8.2   Pre-test loops

In a pre-test loop the logical expression is tested before the statement in the body of the loop is executed in each iteration. The syntax is:

while(*logical-expression* **)** *statement*

where again the *statement* may be a compound statement formed with several simple statements enclosed within curly brackets, as above. As an example, the following piece of code calculates the factorial of a given number,

n:

$$n! = n \cdot (n-1) \cdot (n-2) \cdots 3 \cdot 2 \cdot 1 \qquad (6.1)$$

using a `while` loop:

```
>   n=5
>   fact=1
>   while( n > 1) {
+       fact = fact*n
+       n=n-1
+       }
>   print(fact)

[1] 120
```

6.8.3 Counting loops

A counting loop is meant to be used when the body of the loop needs to be executed a certain number of times, that can be specified in advance. The syntax is:

for(*variable* **in** *sequence* **)** *statement*

In the above, *variable* is a variable that takes values from the specified *sequence*. The *statement* can be simple or compounded. As an example, one could calculate the factorial of a number n using a `for` loop:

```
>   n=5
>   fact=1
>   values = 2:n
>   for( i in values ) {
+       fact = fact*i
+ }
>   print(fact)

[1] 120
```

In this example, the brackets in the body of the loop are not necessary since there is only one statement to be executed there. However, in most cases, the body of the loop will contain more than one statement and the brackets will be essential in order to obtain the correct results.

It is important to remember that the *sequence* on which the iteration will be performed is evaluated only once, at the beginning of the loop. Hence, the execution of the loop cannot be influenced by any changes brought to this sequence in the body of the loop. Similarly, for the duration of the loop execution, the loop index *variable* is read-only. Trying to control the number of iterations by making direct assignments to this variable, like one could do in C/C++, for instance, will not work in R.

6.8.4 Breaking out of loops

The R language also includes statements that allow breaking out of a loop. These are **break** and **next**. Like in C/C++, when a **break** is encountered, the control is transferred to the first statement after the loop. The **next** statement will interrupt the current loop iteration and advance the loop index to its next value. If several loops are nested, both **break** and **next** will only apply to the innermost loop.

6.8.5 Post-test loops

In most programming languages, the post-test loops actually test a condition at the end of the loop (e.g., **do** *statement* **while** (*condition*) in the C programming language, or **repeat** *statement* **until** (*condition*) in Pascal). In R, however, the syntax does not specify such a test explicitly. The syntax for this loop is simply:

repeat *statement*

Like for the other loops, the statement will usually be a compound statement surrounded by curly brackets. The lack of an explicit test effectively means that the **repeat** loop will behave like an infinite loop unless a condition is tested explicitly in a conditional statement having a **break** on one of the branches:

if(*condition*) **then break**

As an example, here is the factorial calculated with a **repeat** loop terminated with a **break** statement as discussed above:

```
> n=5
> fact=1
> repeat {
+       fact = fact*n
+       n=n-1
+       if ( n==1 ) break
+ }
> fact

[1] 120
```

6.9 Programming in R versus C/C++/Java

6.9.1 R is "forgiving" – which can be bad

As an example, let us consider we need to test whether a given vector v contains NA values. We will use the built-in function is.na(v), which returns TRUE if the value tested is NA and FALSE otherwise. Our test can be easily implemented with the built-in function any():

```
>   v=c(1,2,NA,3,4,NA,5)
>   any(is.na(v))
```

```
[1] TRUE
```

However, our goal here is to illustrate some particulars of the R programming environment so we will write our own function to test whether any of the elements of the vector v is NA. A first attempt at writing this code would apply the function is.na() to the vector v and use its return value in a conditional statement to provide the appropriate output. The code could look something like this:

```
>     v=c(1,2,3,NA,5,6,NA,7)
>   if( is.na(v) ){
+     print("v has NAs")
+     }else{
+     print("v does not has NAs")
+   }
```

Even though the syntax "looks" right, and that no errors or warnings of any kind are produced, this fragment of code is incorrect and will not provide the desired result but rather:

```
[1] "v does not has NAs"
```

There are several problems with this code snippet. First, is.na() is called on a vector rather than a scalar value. As we have already discussed, R has no problem with this and will apply the function on each element of the vector. The result will be a vector of logical elements:

```
>   is.na(v)
```

```
[1] FALSE FALSE FALSE   TRUE FALSE FALSE   TRUE FALSE
```

However, the conditional statement does expect to perform its test on a single logical value rather than on a vector of logical values. In this case, it does not make any sense to apply the if to a vector of logical values. Even though this makes no sense, R is faithful to its philosophy of being as forgiving as possible, and applies the test only to the first element of the vector. In this case, this element is FALSE which produces the incorrect statement "v does not have NAs" and a warning message. A correct implementation of the test desired can look like this:

```
>  v=c(1,2,3,NA,5,6,NA,7)
>  if( sum(is.na(v)) == 0){
+    print("v does not have NAs")
+    }else{
+    print("v has NAs")
+    }
```

```
[1] "v has NAs"
```

Here, one uses the fact that the logical values TRUE/FALSE can be converted to the integer values 1/0 and summed up. The sum will be zero if there are no TRUE elements in the result of is.na(), which happens only if the vector v does not have any NA elements.

6.9.2 Weird syntax errors

Paradoxically, R can be a source of some frustration for an unexpected category of readers – those that have a substantial programming experience in other languages. Let us reconsider briefly the piece of code that tests whether a vector has NA elements, and does something appropriate in each of the cases:

```
> if( sum(is.na(v)) == 0){
+   print("v does not have NAs")
+   # and do something appropriate...
+   } else {
+   print("v has NAs")
+   # and do something else...
+   }
```

This piece of code runs without errors or warnings and produces the desired behavior. However, when given the task to implement such a test, and given the syntax described in Section 6.8.1, most readers with programming experience in C/C++ could write code such as:

```
1   if( sum(is.na(v)) == 0)
2   {
3   print("v does not have NAs")
4   # and do something appropriate...
5   }
6   else
7   {
```

```
8   print("v has NAs")
9     # and do something else...
10  }
```

To the endless frustration of the experienced C/C++ programmer, this code –
apparently identical to the one above – does not work, producing an apparently
oxymoronic syntax error:

```
> if( sum(is.na(v)) == 0)
+ {
+ print("v does not have NAs")
+ # and do something appropriate...
+ }
> else
Error: syntax error, unexpected ELSE in " else"
> {
+ print("v has NAs")
+   # and do something else...
+ }
```

Like many other computer-generated messages, the explanation "unexpected
ELSE in " else"" is cryptic at best. The programmer's frustration can only
increase if the syntax described in Section 6.8.1 (or in R's help pages for
that matter) is re-examined. The code above appears to respect that syntax:
the condition to be tested is enclosed within round brackets, the compound
statement is enclosed within paired curly brackets, etc. Furthermore, this code
would execute correctly in many other environments using a similar syntax.
If the cause of such syntactical errors is not properly understood, the reader
might develop a religious (and obviously unnecessary) fear of altering, even
in the slightest, snippets of code taken from examples or existing packages.
Clearly, this would greatly impair the reader's abilities to use R effectively so
it is very useful to try to understand exactly the source of such problems.

In this case, the problem is caused by the fact that R does not parse
correctly an `if` statement that spans several lines. In particular, it would
not accept a new line character between the statement in the first part of
the `if` and the subsequent `else`. This has to do with R's intrinsic execution
mechanism which uses an *interpreter* rather than a *compiler*. A compiler reads
all statements of a program, analyzes them, and *then* generates executable
code. This code is subsequently executed to produce the output. In contrast,
an interpreter reads statements one by one, translates them into executable
code on-the-fly, and executes them immediately. This means that a compiler
has an opportunity to optimize the code written by the programmer while an
interpreter is not able to do any optimizations. In the previous example, the R
interpreter reads the code line by line, until it obtains a syntactically correct
statement that can be executed. In this case, this will happen at the end of
line 5, just before the `else`:

```
1  if( sum(is.na(v)) == 0)
2  {
3    print("v does not have NAs")
4    # and do something appropriate...
5  }
```

At the end of the last line show above, R has parsed a syntactically correct statement that can be executed, so the R interpreter will do so immediately. Once this is done, the interpreter is ready and eager to accept more commands, so it starts reading the subsequent else considering it to be a new, different statement, rather than the continuation of the previous if statement. At the end of this line, the interpreter would be confused and will try to report back to us that there was a syntax error caused by the presence of the token ELSE in the line " else":

```
> else
Error: syntax error, unexpected ELSE in " else"
> {
+   print("v has NAs")
+ }
[1] "v has NAs"
>
```

Essentially, what the R interpreter is telling is us that there is no instruction in its set that starts with "else." Note that immediately after the error is reported, R is continuing to read the next statement, which is a simple '{'. This does not form a syntactically complete statement so the R interpreter continues with the '+' prompt indicating it is in the middle of parsing a statement. This statement is completed two lines further down, when the closing curly bracket is encountered. At this time, the statement is executed and the phrase "v has NAs" is printed. Of course, at this point, the message was generated by the statement print("v has NAs"), and unfortunately, this assertion has nothing to do with the content of the vector v.

This is a good illustration of another piece of advice for good practice in R: if the execution of any portion of R code has generated any warnings or errors, the final output of the program should be treated with the outmost skepticism, while the error message should receive your full and undivided attention. In this case, for instance, the final output looked as expected, and even happened to be a true statement. However, as explained above, this output was due to happenstance rather than a real test on the elements of the v vector.

This problem is actually caused by the combination between the choice to interpret the code rather than compile it, and the syntax chosen. The choice of interpreting the code means that a statement should be executed as soon as it is completely parsed. The syntax is a factor because in R, the if statement does not have a closing token to indicate where it ends. For instance, the syntax used in the Unix shell is:

```
if commands;   then
    commands;
else
    commands;
fi
```

Even though the difference is minimal, confusions such as above cannot occur. In this case, the syntax requires the closing `fi` token, and there is no ambiguity in parsing an `if` statement without the `else` clause such as this:

```
if commands;   then
    commands;
```

Given the syntax shown above, this is clearly an incomplete statement which cannot be executed yet. The following token will either be an `else`, telling the interpreter that the `if` statement has an alternative, or the closing token `fi`, telling the interpreter that there is no `else` clause and that the statement is complete and can be executed.

With these clarifications in mind, it is very easy to avoid most such errors in R. In essence, one should keep in mind that the R interpreter will analyze its input at the end of each line. If the input string read up to that point forms a valid statement, no matter how short, R will execute it. If the intention of the programmer is to continue that statement on a subsequent line, the statement should be broken in such a way that it is clear for the interpreter that the statement is incomplete. An easy way to achieve this is to break the lines in such a way that not all open brackets are closed:

```
> if( sum(is.na(v)) == 0) {
+    print("v does not have NAs")
+    } else {
+    print("v has NAs")
+    }

[1] "v has NAs"
```

Another way to avoid this problem is to encompass the fragment of code that should not be split with brackets as in:

```
>    v=c(1,2,3,NA,5,6,NA,7)
>    {
+    if( sum(is.na(v)) == 0) {
+    print("v does not have NAs")
+    }
+    else  {
+    print("v has NAs")
+    }
+    }
```

```
[1] "v has NAs"
```

In the fragment above, the first open bracket (on line 2 above) will tell the interpreter that a single statement is being parsed. Thus, the closed bracket immediately preceding the `else` cannot be interpreted as the end of an `if` statement without the `else` clause, because up to that point, there are two open brackets and only one closed bracket. Hence, the parsing will continue and will correctly interpret and execute the entire snippet of code.

6.9.3 Programming style

Given that the usual control structures from other procedural languages such as C or C++ are also available in R, the temptation to write procedural code in the C or C++ style looms large, especially for the reader coming from such a programming background and without any previous exposure to other vector-based environments, such as MATLAB. For instance, if a certain task is to be performed 10 times, the C/C++–like programmer would be tempted to write code such as:

```
for i=1 to 10
    task(i)
```

This is the style characteristic to many procedural programming languages. However, this style is not the best in the R environment. R has very efficient implementations of vectors and vector operations, so the most efficient as well as most compact code can be written by taking advantage of these implementations.

Let us consider the example of calculating the factorial of the first 6 natural numbers. The factorial of an integer n is defined as:

$$n! = n \cdot (n-1) \cdot (n-2) \cdots 3 \cdot 2 \cdot 1 \tag{6.2}$$

Hence, a piece of code that calculates the required factorials could look like this:

```
>   result = NULL
>   for( i in 1:6 ) {
+        fact = 1
+        for ( j in 1:i )
+            fact = fact * j
+        result[i] = fact
+   }
>   result
```

```
[1]    1   2   6  24 120 720
```

This piece of code would probably look very natural to any C/C++/Java programmer, with the exception of the first assignment `result = NULL`. This statement plays the approximative role of a C/C++ declaration, essentially telling R that we want to create a variable named `result`. Note that R does not require us to specify the type of this variable at this time.[5] Beyond this point, we can use the variable result to store almost anything. The type of the variable will be decided when the variable is used for the first time. The code above does work and does produce the desired results. However, this is a very inelegant and very inefficient way of doing things in R. The better way of accomplishing the same goal is to use the built-in R functions for vector manipulation. Essentially, the goal is to leave all or most of the iteration housekeeping to R, since its built-in implementation will be much more efficient than anything written at the user level.

The implementation above has two nested loops, one iterating over the numbers whose factorial needs to be calculated, the other iterating over the numbers that need to be multiplied in order to calculate each factorial. A first attempt at improving our code by using R's built-in iteration mechanisms can substitute the inner loop with the function `prod`, which iterates over the elements of a vector calculating their product. Thus, the factorial of *i* could be simply calculated as `prod(1:i)`

```
>   for( i in 1:6 ) {
+        result[i] = prod(1:i)
+   }
>   result
```

```
[1]   1   2   6   24 120 720
```

The outer loop can also be eliminated using `sapply`. However, sapply would require a function so we could define such a function naming it for instance, `fact`:

```
>   fact <- function( n ) {
+        prod(1:n)
+   }
```

Using this function, we can now rewrite the code that calculate the factorials of all numbers between 1 and 6 without using any explicit loops:

```
>   sapply(1:6,fact)
```

```
[1]   1   2   6   24 120 720
```

---

[5]In this particular case, we know both the type (numeric) and the size (6) of the array we need here, and we could have declared the variable result appropriately: result=numeric(6). However, we used result=NULL in order to illustrate R's weak typing. In computer science terms, a programming language that allows the use of a variable without specifying its type is known as a *weakly typed* language. In contrast, a programming language that requires the programmer to specify the type of each variable or object will be *strongly typed*.

This implementation would iterate more efficiently than our first, nested-loop implementation but does include a function call (**fact**) for each number that needs to be calculated. Function calls do involve some overhead so in general, such an implementation would only be justified if the computation to be performed in the function were to be more complex (which probably will happen for most real-world functions). However, in this case, the computation within the function is very simple and the overhead involved in a function call is not justified. Hence, we should try to write yet another piece of code performing the same computation but without any (user-defined) function calls. This can actually be done in just one line of code:

```
>   sapply(sapply(1:6,seq),prod)
```

```
[1]   1   2   6  24 120 720
```

The inner **sapply** repeatedly calls the function **seq** (sequence) for each element of the array (1,2,3,4,5,6) constructed with 1:6. When **seq** is called with only one argument n, it uses the default values of its other arguments from=1 and by=1, to construct the sequence from 1 to *n*, with the step 1. For instance, seq(3) produces:

```
>   seq(3)
```

```
[1] 1 2 3
```

Hence, **sapply(1:6,seq)** will produce a list of vectors, in which each vector contains one of the sequences successively generated by **seq**:

```
>   sapply((1:6),seq)
```

```
[[1]]
[1] 1

[[2]]
[1] 1 2

[[3]]
[1] 1 2 3

[[4]]
[1] 1 2 3 4

[[5]]
[1] 1 2 3 4 5

[[6]]
[1] 1 2 3 4 5 6
```

The outer `sapply` repeatedly applies the function `prod` to each of the elements in the list above. This function calculates the product of the elements of a given vector and hence produces the factorials desired in this case.

Finally, the most efficient implementation that calculates the factorials of the numbers from 1 to 6 would use the built-in function `cumprod`, or cumulative product. This function returns a vector whose elements are the cumulative products of the elements of the argument. Similar functions exists for the computation of cumulative sums (`cumsum`) and cumulative extremes (`cummin` and `cummax`). For our problem, the desired factorials can be obtained with a single function call:

```
> cumprod(1:6)

[1]   1   2   6  24 120 720
```

In summary, in this section we showed several alternative implementations for a computation involving a double iteration. These implementations included a least efficient approach that used two explicit loops, a more efficient implementation using a single explicit loop, an alternative using a user-defined function and no explicit loops and finally, two one-line implementations, that used no function calls and no explicit loops.

The main ideas illustrated by these examples are that: i) in general, it is best to use the built-in iteration mechanisms rather than have explicit loops and ii) the most efficient implementation will be obtained by using the most appropriate functions available in the environment.

6.10 Summary

This chapter starts with a brief introduction to the R programming language and environment in Section 6.2. Section 6.3 continues with a discussion of the basic concepts, including elementary computations, variable and assignments, and expression and objects. Subsequently, the chapter covers the main data structures in R: vectors, matrices, lists and data frames, as well as the main operations on these data structures. The use of R functions in general, as well as examples including the most commonly used predefined functions are covered in the same Section 6.3. A more in-depth discussion of R's capabilities is provided in Section 6.4. This section discusses more advanced indexing techniques, treatment of missing values, reading and writing files, conditional selection and indexing, sorting and implicit loops. The R environment, including workspace, search path, packages and built-in data, are covered in Section 6.5. A short separate section, 6.6, covers the installation of Bioconductor, the collection of R packages that deal with the analysis of genomic data. An introduction to the very basic graphics facilities of R is included in Section 6.7. This section covers the very minimum necessary in order to produce a graph in R. A full discussion of the control structures available in the R programming language such as conditional statements and loops is included in Section 6.8. Here as well as elsewhere, the attention of the reader is brought to common mistakes that are usually associated with the initial phase of learning R. Throughout the chapter, the presentation emphasizes similarities and differences between R and other programming languages and environments such as C/C++, Unix shell, etc. This is meant to facilitate the learning of R for those users familiar with other programming languages. A common tendency for users migrating from other programming languages to R is to write code that is far less efficient than it can be. Another rather surprising phenomenon is that sometimes programmers well experienced in other procedural languages experience some frustration when migrating to R. Section 6.9 addresses these issues by discussing in details topics related to R's syntax, the programming style in R, as well as including useful advice as to how to deal with errors.

6.11 Solved Exercises

1. Consider the following snippet of code :

```
v=c(1,2,3,NA,5,6,NA,7)
if( sum(is.na(v)) == 0) {
 print("v does not have NAs")
 }
 else {
 print("v has NAs")
 }
```

Would this be executed correctly or would it generate a syntax error? If the code will be executed correctly, what will the output be? If the code will generate an error message, what would the error be and what caused it? How about the following snippet:

```
v=c(1,2,3,NA,5,6,NA,7)
if( sum(is.na(v)) == 0)
 {
 print("v does not have NAs")
 } else
 {
 print("v has NAs")
 }
```

Answer:

The first snippet will generate an error message on line 5. R will complain about a syntax error caused by "an unexpected ELSE in ' else'". This is caused by the fact that at the end of line 4, R has a syntactically complete if statement that can and will be executed. The subsequent line starting with "else" will be considered a new statement and generate a syntax error message since no statement starts with else. In the second snippet of code, this does not happen because only the very last end of line corresponds to a syntactically complete statement.

2. Consider the following piece of code:

```
> sum = 0
> values = c(1,2,3,4,5)
> for( i in values ) {
+     sum=sum+i
+     print(values)
+     print(i)
+     values[i+1]=100
+ }
> sum
```

Write down what you think the output of this piece of code will be in R (without actually running it). Run this code and compare the output obtained from R with your answer. Explain any differences. How does the R for loop differ from a for loop in C/C++?

Answer: This exercise illustrates the point that in R, the sequence on which the iteration is performed does not change after the initial evaluation. Even though the vector values is explicitly changed in the loop, the iteration is performed on its initial content:

```
> sum = 0
> values = c(1,2,3,4,5)
> for( i in values ) {
+     sum=sum+i
+     print(values)
+     print(i)
+     values[i+1]=100
+ }

[1] 1 2 3 4 5
[1] 1
[1]   1 100   3   4   5
[1] 2
[1]   1 100 100   4   5
[1] 3
```

```
[1]    1 100 100 100    5
[1] 4
[1]    1 100 100 100 100
[1] 5

> sum

[1] 15
```

Furthermore, even though the last assignment in the last iteration of the loop (values[6]=100) assigns a value beyond the boundary of the array values, there are no errors and no warnings.

3. Consider the following piece of code:

```
> sum = 0
> values = c(1,2,3,4,5)
> for( i in values ) {
+    sum=sum+i;
+    print(i);
+    i=3;
+    cat(paste("i=",i,"\n"));
+    cat(paste("sum=",sum,"\n"));
+ }
```

Write down what you think the output of this piece of code will be in R (without actually running it). Run this code and compare the output obtained from R with your answer. Explain any differences. How does the R for loop differ from a for loop in C/C++?

Answer:

This piece of code illustrates the fact that, unlike in C/C++, the value of the loop counter cannot be changed by an explicit assignment in the body of the loop:

```
> sum = 0
> values = c(1,2,3,4,5)
> for( i in values ) {
+    sum=sum+i;
+    print(i);
+    i=3;
+    cat(paste("i=",i,"\n"));
+    cat(paste("sum=",sum,"\n"));
+ }

[1] 1
i= 3
```

```
sum= 1
[1] 2
i= 3
sum= 3
[1] 3
i= 3
sum= 6
[1] 4
i= 3
sum= 10
[1] 5
i= 3
sum= 15

> sum

[1] 15
```

Here, the `print(i)` statement prints the value of loop counter i, pre-fixed by the index [1]. This counter takes each value from 1 to 5, as initially specified in the `for` loop. At each iteration, the variable i is assigned the value 3, which is dutifully printed by the `cat(paste("i=", i,"\n"));` statement. This statement produces the "3" displayed at each iteration. However, at the end of each iteration, the variable i will take the next value from the list initially specified in the `for` loop. For the reader familiar with programming languages theory, this is similar with the mechanism of *parameter passing by value* where any changes to the values passed to the function called would not be accessible from the calling function.

4. Given the matrix d below, construct an array containing the first column of the given matrix, but with an additional element equal to 100 inserted between the 4th and the 5th elements:

```
> d=matrix(data=c(
+    10.305859, 12.228262,
+    10.885669, 11.425003,
+    11.227781, 11.391731,
+    10.864488, 11.908194,
+     9.114495, 12.669221,
+     9.016614, 12.441619,
+    11.815408, 10.495970,
+     9.710361,  9.395282,
+    10.470769, 10.428905,
+     9.758403, 11.490184),
+    byrow=T,ncol=2,nrow=10,
```

```
+     dimnames=list(NULL,c("before","after")))
> d
```

```
      before  after
 [1,] 10.306 12.228
 [2,] 10.886 11.425
 [3,] 11.228 11.392
 [4,] 10.864 11.908
 [5,]  9.114 12.669
 [6,]  9.017 12.442
 [7,] 11.815 10.496
 [8,]  9.710  9.395
 [9,] 10.471 10.429
[10,]  9.758 11.490
```

Answer:

```
> new=c(d[1:4,1],100,d[5:length(d[,1]),1])
> new
```

```
[1]   10.306   10.886   11.228  10.864 100.000   9.114   9.017
[8]   11.815    9.710   10.471   9.758
```

5. Write a function `insrt` that will take a vector x, a position p, and a new value v, and insert the new value in the vector x at position p, by shifting the remaining elements to the right (the length of the vector will be increased by 1). For instance, if the function `insrt` were to be called on the vector $x = c(1,2,3,4,5)$, with actual parameters $p = 3$ and $v = 100$, the result would be $(1,2,100,3,4,5)$. At this time, assume that the user will always provide a valid position (i.e., between 1 and the last position in the existing vector), and a valid value v.

Answer:

```
> insrt<-function(x,p,v){
+     y<-NULL;
+     y[1:(p-1)]<-x[1:(p-1)];
+     y[p]<-v;
+     y[(p+1):(length(x)+1)]<-x[p:length(x)];
+     y
+ }
>   x = c(1,2,3,4,5)
>   insrt(x,3,100)
```

```
[1]   1   2 100   3   4   5
```

6. Write a function that tests whether two numerical vectors x and y that could contain NA values are identical.

Answer:

```
> # this function tests if the two vectors have NA
> # in exactly the same positions
> # if so,  it then tests whether the elements that
> # are not NA are the same in the two vectors
> # if not so, then the vectors are different and
> #the function returns FALSE
> are_equal<- function(x,y) {
+ if ( sum( is.na(x) == is.na(y)) == length(x) ) {
+     #print("NAs in same positions,
+     #testing for equal values...")
+     result = ( sum( x [!is.na(x)]== y[!is.na(x)])
+                 == length(x [!is.na(x)]) )
+     # note the use of sum to test whether
+     # all elements are TRUE;
+     # if the sum of all elements is equal
+     # to the length, then all values were TRUE
+ }
+ else {
+     result = FALSE
+ }
+ result
+ }
> y= c(1,NA,3,4,5)
> z= c(1,2,3,4,5)
> are_equal(x,y)

[1] FALSE

> are_equal(x,z)

[1] TRUE
```

An alternative, more compact version would be:

```
> shorter_are_equal<- function(x,y){
+ (length(x)==length(y))& #have same length
+ (sum(is.na(x)==is.na(y))==length(x))&
+ #have NAs in same positions
+ (sum(x==y,na.rm=TRUE)==sum(!is.na(x)))
+ #have all values equal
+ }
> shorter_are_equal(x,y)

[1] FALSE

> shorter_are_equal(x,z)
```

```
[1] TRUE
```

or

```
> shortest_are_equal<- function(x,y){
+ (length(x)==length(y))& #have same length
+ (sum(is.na(x)==is.na(y))==length(x))&
+ #have NAs in same positions
+ (all(na.omit(x)==na.omit(y)))
+ #have all values equal
+ }
> shortest_are_equal(x,y)
```

```
[1] FALSE
```

```
> shortest_are_equal(x,z)
```

```
[1] TRUE
```

Here the **na.omit** function returns a vector from which the NA values have been removed. Subsequently, the function **all** tests to see if all the values of its vector argument are true.

7. Construct a matrix to store 5 random draws of 10 numbers from a random normal distribution. Store each set of 10 numbers as a column.

   Answer: Five random draws of 10 numbers are essentially 50 random numbers. Hence:

```
>  num=matrix(rnorm(50),nrow=10,ncol=5)
>  num
```

```
          [,1]      [,2]      [,3]      [,4]      [,5]
 [1,] -2.1636 -1.29564  0.41504  1.4710  0.76975
 [2,]  0.6815 -0.02802 -0.43805  2.2661 -1.13413
 [3,]  1.0058  1.12447 -1.89758 -0.4374 -1.78739
 [4,] -0.6513 -0.09601 -0.69510  1.2273  0.42676
 [5,] -0.6360  0.82864 -0.74193 -0.1507 -1.01697
 [6,]  1.9117 -0.17552  0.04952 -0.8585 -0.26065
 [7,]  0.1402 -1.14238  0.69242 -1.6997  0.23875
 [8,]  2.0939 -0.51739  0.58970  0.9989  0.04139
 [9,]  0.5711  0.74460  0.80752 -0.2760 -0.09750
[10,]  0.1027  0.32073 -1.37552 -1.8395  0.60154
```

8. Construct a data frame to test a new classification algorithm when the null hypothesis is true. The data frame should contain "microarray data"

from 5 genes monitored across 10 patients. Since this is the null hypothesis, the numbers should be all random. Construct a data frame containing 50 numbers from a random normal distribution. Label the first 5 samples with "cancer" and the remaining 5 samples with "healthy."

Answer:

```
> num=data.frame(array(rnorm(50),dim=c(10,5)),
+ group=rep(c("healthy","cancer"),each=5))
> num
```

```
        X1       X2       X3       X4       X5    group
1    0.43818 -0.6740 -1.42272  0.7485 -0.9295 healthy
2   -1.97693  0.6024  1.29259  0.4738 -0.6281 healthy
3   -0.62475  0.1306 -0.07879  1.4145  1.0239 healthy
4    0.58851  2.1573  0.62945  0.8424 -0.8805 healthy
5   -0.89186  1.0044 -0.79073 -0.2320 -1.0979 healthy
6    0.44142 -0.1097  0.64102  0.1087 -0.4702  cancer
7   -0.73958  0.6032  0.67840 -0.2995  0.2173  cancer
8   -0.03191  1.1549 -0.82647  1.0059  0.5926  cancer
9    0.84614  0.1884  0.14761  0.2643 -0.3483  cancer
10   0.22494 -0.7740  0.81366 -1.4672 -0.8547  cancer
```

9. Write the content of the above data frame into a text file "testfile.txt" in the current directory using the function "write.table" using the default values of the other arguments. Open this file with Excel.

Answer:

```
> write.table(num,"./testfile.txt", quote=F)
```

10. Read the content of the file you have just created using the function "read.table" and store the values in a new data frame. Compare the content of the original data frame with the one read from the file.

Answer:

```
> new_num = read.table("./testfile.txt")
> new_num
```

```
        X1       X2       X3       X4       X5    group
1    0.43818 -0.6740 -1.42272  0.7485 -0.9295 healthy
2   -1.97693  0.6024  1.29259  0.4738 -0.6281 healthy
3   -0.62475  0.1306 -0.07879  1.4145  1.0239 healthy
4   -0.58851  2.1573  0.62945  0.8424 -0.8805 healthy
5   -0.89186  1.0044 -0.79073 -0.2320 -1.0979 healthy
6    0.44142 -0.1097  0.64102  0.1087 -0.4702  cancer
7   -0.73958  0.6032  0.67840 -0.2995  0.2173  cancer
8   -0.03191  1.1549 -0.82647  1.0059  0.5926  cancer
```

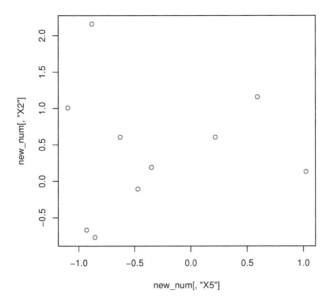

FIGURE 6.6: Gene X2 as a function of gene X5.

```
9    0.84614   0.1884   0.14761   0.2643 -0.3483   cancer
10   0.22494 -0.7740   0.81366 -1.4672 -0.8547   cancer
```

11. Plot the expression level of X2 as a function of X5 with the default options. Plot the same graph with the option `type="b"`. Explain the results obtained.

    Answer:

    ```
    > new_num = read.table("./testfile.txt")
    > plot(new_num[,"X5"],new_num[,"X2"],col="red")
    ```

6.12   Exercises

1. Consider the following piece of code:

    ```
    > sum = 0
    > values = c(1,2,3,4,5)
    ```

```
> for( i in values ) {
+    sum=sum+i
+ }
> sum
```

Is this code correct? If yes, write down what you think the output of this piece of code will be in R (without actually running it). Run this code and compare the output obtained from R with your answer. Explain any differences.

2. Modify the function `insrt` above such that if the position parameter specified by the user is less than 1, the new value will be inserted in the first position of the vector; if the position parameter is beyond the last position of the existing vector, the function will insert the new element in the last position.

3. Construct a data frame to test a new classification algorithm when the null hypothesis is true. The data frame should contain "microarray data" from 5 genes monitored across 15 patients. Since this is the null hypothesis, the numbers should be all random. Construct a data frame containing 50 numbers from a random normal distribution. Label the first 5 samples with "cancer," the following 5 with "treated," and the remaining 5 samples with "healthy." Calculate the mean of each gene in each group. Plot the distribution of the gene values using a different color for each group.

Chapter 7

Bioconductor: principles and illustrations

Chapter contributed by Vincent Carey
Associate Professor of Medicine (Biostatistics), Harvard Medical School
Associate Biostatistician, Department of Medicine,
Brigham and Women's Hospital

7.1 Overview

The Bioconductor project (*www.bioconductor.org*) is a software and documentation repository for components useful in workflows for the analysis of genome-scale data. All the software provided through Bioconductor is coded so that it can be used with the R language (*www.r-project.org*). Thus, the majority of source code for Bioconductor is actually written in R, but, thanks to the emphasis accorded to interoperability in the R project, algorithms coded in C, C++, Java, and other languages are also present. The basic requirements for software contributions in the Bioconductor project are

- code is delivered in the form of an *R package* – a formally specified set of folders with reserved names including checkable documentation and configuration files;

- a computable *literate programming* document called a *vignette* is included, that illustrates use of the software in an analysis process – this is distinct from the software *manual pages*, also required, that provide details on the calling sequences and behaviors of specific routines;

- the code must pass a formal checking process for the current version of R, which is revised approximately every six months;

- the code must be formally licensed using a license compatible with open source distribution – base Bioconductor packages use Artistic 2.0 at this time.

The Bioconductor project was created to support fully transparent deployment of algorithms for analysis of genome-scale data in biology. It was recognized *ca.* 2000 CE that researchers in statistical genomics were frequently duplicating effort in software development, and that the R programming language would be a sensible vehicle for uniting available data analytic software, new statistics research, and new developments in data structures and algorithms for high-throughput biology. Thus, a development core was formed, comprising half-dozen or so researchers at various universities, with a commitment to creating open and shareable workflow components for analyzing DNA microarray data. Contributions from the research community were invited, and many came forth, so that the project now includes hundreds of R packages addressing many aspects of high-throughput experimentation in biology, ranging from cDNA microarrays to the most current tools for assessment of copy-number variation and epigenomics. The remainder of this chapter provides selected details on operational aspects and contents of the project, with a focus on illustrating thorough use of Bioconductor components. It is not possible to provide a comprehensive outlook on the project's capabilities or performance; see the Springer monograph *Computational Biology and Bioinformatics Solutions with R and Bioconductor*, for a more thorough treatment [168].

7.2   The portal

Inspection of the `bioconductor.org` web site reveals opportunities to learn about the project through posted documentation, to learn about development processes, and to download software. Under the Documentation tab, links are provided to the project FAQ and to the mailing lists. The list at `bioconductor@r-project.org` is fairly high-traffic and includes considerable dialogue on a wide range of topics related to analysis with Bioconductor.

7.2.1 The main resource categories

Visiting the URL `bioconductor.org/packages/release/BiocViews.html` shows three prominent links: Software, AnnotationData, and ExperimentData. These are the three broad topics addressed by resources of the project.

The *ExperimentData* node collects packages that are primarily defined by structured experimental data, often harvested from public repositories, and documentation. Resources delivered here can be used to illustrate new methods or to learn about findings of current experiments. Of particular note is the fact that Bioconductor **container designs** are used to create *self-documenting* representations of complex experimental data. Thus, a complex experiment that employs multiple assay modalities, as found for example in the package *Neve2006*, is housed in a single R object, with components defining the provenance of the experiment (e.g., the MIAME schema), and, in this case, details on several dozen array CGH experiments, united with transcriptome-wide mRNA abundance measures obtained through DNA microarray hybridization. More details on container designs are provided below.

The *AnnotationData* node collects packages that manage mappings from identifiers of assay reporters to genomic and biological metadata. Up to 2007, the primary form of an annotation package was a collection of "hash tables," mapping from identifiers such as `1007_s_at` (a probe set identifier in Affymetrix(TM) expression platforms) to tokens such as HUGO gene symbol, chromosomal location and genomic coordinates, Gene Ontology term tags, KEGG pathway tags, OMIM concepts, PubMed reference tags, and so forth. In 2007, a new modality of annotation packaging has been introduced, responding to the burgeoning volume of metadata resources. The hash table system has been replaced by relational database tables deployed under SQLite (`sqlite.org`).

The *Software* node collects packages that help perform specific analyses, or that provide infrastructure for data management and analysis. Clicking on this link reveals a new collection of links called *Subviews*, along with a catalog of all software packages; subviews are provided to filter the catalog for simpler browsing. See Figure 7.1 for illustration.

In any catalog view, each package is represented by its own page. See Figure 7.2 for an example. The page provides direct links to the PDF version of the package vignette(s), along with links to downloadable compressed versions of package binaries or source codes.

7.2.2 Working with the software repository

As shown above, one can get comprehensive access to software, data, and annotation resources by browsing the Bioconductor web site. For R users, acquisition of resources is simplified by using special functions that will download and install software from any network-connected R session. To use these facilities, start R and issue the command

Bioconductor Task View: Visualization

Subview of

- Software

Packages in view

Package	Maintainer	Title
AffyExpress	Xuejun Arthur Li	Affymetrix Quality Assessment and Analysis Tool
apComplex	Denise Scholtens	Estimate protein complex membership using AP-MS protein data
arrayQuality	Agnes Paquet	Assessing array quality on spotted arrays
BioMVCClass	Elizabeth Whalen	Model-View-Controller (MVC) Classes That Use Biobase
cellHTS	Ligia Bras	Analysis of cell-based screens
cellHTS2	Ligia Bras	Analysis of cell-based screens - revised version of cellHTS
CGHcall	Sjoerd Vosse	Calling aberrations for array CGH tumor profiles.
ChromoViz	Jihoon Kim	Multimodal visualization of gene expression data
copa	James W. MacDonald	Functions to perform cancer outlier profile analysis.
ctc	Antoine Lucas	Cluster and Tree Conversion.

FIGURE 7.1: Excerpt from the Visualization subcatalog of Software.

CGHcall

Calling aberrations for array CGH tumor profiles.

Calls aberrations for array CGH data using a six state mixture model as well as several biological concepts that are ignored by existing algorithms. Visualization of profiles is also provided.

Author   Sjoerd Vosse & Mark van de Wiel
Maintainer Sjoerd Vosse

To install this package, start R and enter:

```
source("http://bioconductor.org/biocLite.R")
biocLite("CGHcall")
```

Vignettes (Documentation)

CGHcall.pdf

Package Downloads

Source	CGHcall_1.0.0.tar.gz
Windows binary	CGHcall_1.0.0.zip
OS X binary	CGHcall_1.0.0.tgz

FIGURE 7.2: Presentation page for the *cghCall* package.

```
> source("http://www.bioconductor.org/biocLite.R")
```

Now the command *biocLite()* can be issued, and a default selection of packages will be downloaded and installed in the working version of R's package library. These packages will be persistent, and this operation does not need to be repeated until it is desired to update versions. Focused acquisition of resources can be performed by supplying arguments to `biocLite`. For instance, a specific package can be installed with `biocLite("newpackage")` where *newpackage* is the name of the package to be loaded:

```
> biocLite("randomForest")
```

7.3   Some explorations and analyses

7.3.1   The representation of microarray data

T. Golub's 1999 paper [183] on leukemia classification is widely cited. We can use Bioconductor to explore the data in detail. For the code fragments given below to run, we assume that `biocLite("golubEsets")` has been successfully executed in a version of R that is at least 2.6. First, we attach the `golubEsets` package, bring an object called `Golub_Merge` into scope, and mention it to R.

```
> require(golubEsets)

> library(golubEsets)
> data(Golub_Merge)
> Golub_Merge

ExpressionSet (storageMode: lockedEnvironment)
assayData: 7129 features, 72 samples
  element names: exprs
protocolData: none
phenoData
  sampleNames: 39 40 ... 33 (72 total)
  varLabels: Samples ALL.AML ... Source (11 total)
  varMetadata: labelDescription
featureData: none
experimentData: use 'experimentData(object)'
  pubMedIds: 10521349
Annotation: hu6800
```

We can get experiment-level metadata:

```
> experimentData(Golub_Merge)
```

```
Experiment data
  Experimenter name: Golub TR et al.
  Laboratory: Whitehead
  Contact information:

  Title: ALL/AML discrimination
  URL: www-genome.wi.mit.edu/mpr/data_set_ALL_AML.html
  PMIDs: 10521349

  Abstract: A 133 word abstract is available. Use 'abstract' method.
```

The PubMed ID of the main paper is provided. The abstract of this paper is also included, but is hard to print as it is a single long string. We show the first 55 characters:

```
> substr(abstract(Golub_Merge), 1, 55)

[1] "Although cancer classification has improved over the pa"
```

The numerical/categorical data about the microarray assay and the samples to which it was applied are readily interrogated:

```
> dim(exprs(Golub_Merge))

[1] 7129    72

> dim(pData(Golub_Merge))

[1] 72 11
```

How are the assay reporters named?

```
> featureNames(Golub_Merge)[1001:1010]

 [1] "HG4390-HT4660_at" "HG4411-HT4681_at" "HG4433-HT4703_at"
 [4] "HG4458-HT4727_at" "HG4460-HT4729_at" "HG4462-HT4731_at"
 [7] "HG4480-HT4833_at" "HG4533-HT4938_at" "HG4542-HT4947_at"
[10] "HG4582-HT4987_at"
```

How are the samples named?

```
> sampleNames(Golub_Merge)[1:10]

 [1] "39" "40" "42" "47" "48" "49" "41" "43" "44" "45"
```

The primary outcome of the study is contained in a variable called ALL.AML. What is its distribution?

```
> table(Golub_Merge$ALL.AML)
```

```
ALL AML
 47  25
```

We see that there are simple function calls (e.g., `featureNames`) or oper- ator applications (e.g., `$`) that tell us about Golub's data. This is facilitated by the use of a special container structure called `ExpressionSet`:

```
> getClass("ExpressionSet")

Class "ExpressionSet" [package "Biobase"]

Slots:

Name:      experimentData     assayData        phenoData
Class:            MIAME     AssayData AnnotatedDataFrame

Name:         featureData     annotation      protocolData
Class: AnnotatedDataFrame      character AnnotatedDataFrame

Name:    .__classVersion__
Class:            Versions

Extends:
Class "eSet", directly
Class "VersionedBiobase", by class "eSet", distance 2
Class "Versioned", by class "eSet", distance 3
```

More details on the representation can be obtained by using the command `help("ExpressionSet-class")` when the *Biobase* package is in scope.

7.3.2 The annotation of a microarray platform

We can determine the type of microarray used with the `annotation` accessor. Continuing with Golub's data:

```
> annotation(Golub_Merge)

[1] "hu6800"
```

In this illustration, we use the more modern `hu6800.db` representation of annotation mappings.

```
> library(hu6800.db)
```

The constituents of the mapping package are:

```
> objects("package:hu6800.db")
```

```
 [1] "hu6800"              "hu6800_dbconn"
 [3] "hu6800_dbfile"       "hu6800_dbInfo"
 [5] "hu6800_dbschema"     "hu6800ACCNUM"
 [7] "hu6800ALIAS2PROBE"   "hu6800CHR"
 [9] "hu6800CHRLENGTHS"    "hu6800CHRLOC"
[11] "hu6800CHRLOCEND"     "hu6800ENSEMBL"
[13] "hu6800ENSEMBL2PROBE" "hu6800ENTREZID"
[15] "hu6800ENZYME"        "hu6800ENZYME2PROBE"
[17] "hu6800GENENAME"      "hu6800GO"
[19] "hu6800GO2ALLPROBES"  "hu6800GO2PROBE"
[21] "hu6800MAP"           "hu6800MAPCOUNTS"
[23] "hu6800MIM"           "hu6800ORGANISM"
[25] "hu6800ORGPKG"        "hu6800PATH"
[27] "hu6800PATH2PROBE"    "hu6800PFAM"
[29] "hu6800PMID"          "hu6800PMID2PROBE"
[31] "hu6800PROSITE"       "hu6800REFSEQ"
[33] "hu6800SYMBOL"        "hu6800UNIGENE"
[35] "hu6800UNIPROT"
```

We can see the "keys" of the HUGO symbol mapping as follows:

```
> ls(hu6800SYMBOL)[1:5]
```

```
[1] "A28102_at"    "AB000114_at"    "AB000115_at"
[4] "AB000220_at"  "AB000381_s_at"
```

The actual mappings are retrieved using mget:

```
> mget( ls(hu6800SYMBOL)[1:5], hu6800SYMBOL )
```

```
$A28102_at
[1] "GABRA3"

$AB000114_at
[1] "OMD"

$AB000115_at
[1] "IFI44L"

$AB000220_at
[1] "SEMA3C"

$AB000381_s_at
[1] "GML"
```

This approach is very general, but for past versions of Bioconductor, mapping exploration required detailed knowledge of the annotation environment components for each platform. The approach using a relational database will be more convenient for those schooled in SQL query resolution.

SQLite tables may be queried in R via the *DBI* package infrastructure. We can list the tables available for NCBI Entrez Gene metadata as follows:

```
> hh = org.Hs.eg_dbconn()
> dbListTables(hh)

 [1] "accessions"          "alias"
 [3] "chrlengths"          "chromosome_locations"
 [5] "chromosomes"         "cytogenetic_locations"
 [7] "ec"                  "ensembl"
 [9] "ensembl2ncbi"        "ensembl_prot"
[11] "ensembl_trans"       "gene_info"
[13] "genes"               "go_bp"
[15] "go_bp_all"           "go_cc"
[17] "go_cc_all"           "go_mf"
[19] "go_mf_all"           "kegg"
[21] "map_counts"          "map_metadata"
[23] "metadata"            "ncbi2ensembl"
[25] "omim"                "pfam"
[27] "prosite"             "pubmed"
[29] "refseq"              "sqlite_stat1"
[31] "ucsc"                "unigene"
[33] "uniprot"
```

Suppose we are interested in finding out the probe set identifiers for genes annotated as sarcoma oncogenes. We can form a partial matching query:

```
> SQL = paste( "select * from gene_info inner join genes
+ using(_id)", " where gene_name like '%sarcoma%oncogene%'")
> sops = dbGetQuery(hh, SQL)
> dim(sops)

[1] 29   4

> sops$symbol

 [1] "ARAF"         "BRAF"        "CRK"        "CRKL"
 [5] "FES"          "FGR"         "FOS"        "FOSB"
 [9] "HRAS"         "KIT"         "KRASP1"     "KRAS"
[13] "LYN"          "MAF"         "MAFG"       "MOS"
[17] "PDGFB"        "SKI"         "SRC"        "YES1"
[21] "YESP"         "MAFK"        "MAFB"       "MAFF"
[25] "ARAF3P"       "BRAFPS2"     "MAFA"       "ARAF2P"
[29] "LOC100421741"
```

We see that gene LYN is in the 13th row of the result, and so

```
> boxplot(exprs(Golub_Merge)["M16038_at",]~Golub_Merge$ALL.AML,col=
```

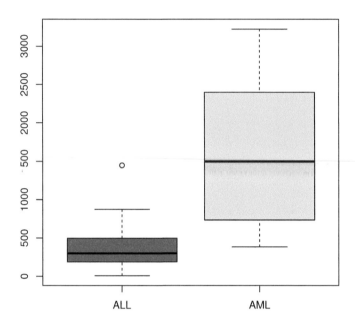

FIGURE 7.3: Code and display of distributions of expression of LYN by ALL vs. AML status.

```
> sops[13,]
```

```
      _id                                        gene_name
13 3389 v-yes-1 Yamaguchi sarcoma viral related oncogene homolog
      symbol gene_id
13      LYN     4067
```

gives us the Gene ID for LYN, and

```
> get("4067", revmap(hu6800ENTREZID))
```

```
[1] "M16038_at"
```

yields the probe set identifier.

Now we can display the data by disease group; see Figure 7.3.

7.3.3 Predictive modeling using microarray data

To conclude this overview of Bioconductor facilities for analysis of genome-scale data, we undertake a simple machine learning exercise. We will use Breiman's "random forest" procedure to create and evaluate a classifier, a mapping from the hu6800-based quantification of the transcriptome to the clinical diagnosis labels "ALL," "AML". This is easily carried out using the *MLInterfaces* package. We will filter the expression data sharply, to the annotated genes with variance across samples in the top 10% of the distribution of variances.

```
> library(MLInterfaces)
> GMF = nsFilter(Golub_Merge, var.cutoff=.9)[[1]]
> rf1 = MLearn(ALL.AML~., data=GMF,
+   randomForestI, xvalSpec("NOTEST"), importance=TRUE)
> rf1

MLInterfaces classification output container
The call was:
MLearn(formula = ALL.AML ~ ., data = GMF, .method = randomForestI,
    trainInd = xvalSpec("NOTEST"), importance = TRUE)
```

Calls to the `MLearn` method specify a formula (typically with a . on the right hand side, indicating that all genes available should be used), a learner schema (in this case `randomForestI`, and a training sample (or cross-validation) specification. In this case, because the random forests procedure employs internal bootstrapping, no cross-validation is really necessary, and we use the `NOTEST` setting, indicating that we will not reserve a test set for model appraisal. Any options accepted by the native learning function (`randomForest`) can also be passed to `MLearn`, and we ask for the variable importance measures to be computed.

The object returned by the `MLearn` method is a specialized container that holds some derived computations of general use for summarization and reporting. The actual object returned by `randomForest` is also held in this container, accessible with the `RObject` method:

```
> RObject(rf1)

Call:
 randomForest(formula = formula, data = trdata, importance = TRUE)
               Type of random forest: classification
                     Number of trees: 500
No. of variables tried at each split: 23

        OOB estimate of  error rate: 4.17%
Confusion matrix:
    ALL AML class.error
```

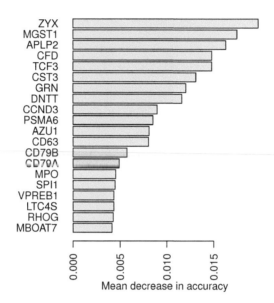

FIGURE 7.4: Variable importance for discriminating ALL and AML in Golub's full dataset, on the basis of the default random forest run illustrated in the text.

```
ALL  46   1   0.0212766
AML   2  23   0.0800000
```

We can obtain a provisional ordering of variables with respect to their contribution to the classification task as follows:

```
> savpar = par(no.readonly=TRUE)
> par(las=2, mar=c(6,9,6,6))
> plot(getVarImp(rf1,fixNames=FALSE), plat="hu6800",
+ toktype="SYMBOL",col="gold")
> par(savpar)
```

The figure produced by the code above is shown in Figure 7.4.

A more detailed explanation of various machine learning approaches is provided in Chapter 29.

7.4 Summary

We have reviewed the principles, portal, and some examples of use of resources of the Bioconductor project. Central aims of the project include a) development of a repository of software and data resources allowing statisticians to explore and create new procedures for investigation with genome-scale data, and b) establishment of standard practices of software packaging and testing that ensure transparency and portability of these procedures across widely used computing platforms. We examined a basic container structure for a collection of DNA microarrays, the `ExpressionSet`. This structure has been devised to respect, in broad terms, a basic idiom of data management in R: the data are accessible through reference to a single R variable, say `X`, and the notation `X[R, S]` expresses a selection of reporters on the basis of the expression `R`, and of samples on the basis of the expression `S`. The subscripting notation using `[` enforces closure of the set of `ExpressionSet` under selection operations, so that `class(X[R,S]) == class(X)`. This closure property simplifies method invocation under filtering in a highly efficient manner.

We conclude by mentioning other major components: the `graph` and `RBGL` packages for managing graphs and network structures, the `GO.db` and `KEGG.db` annotation packages for the contents of gene ontology and pathway catalogues, the `Biostrings` package for the representation of sequences (even to genome scale), and the `GGtools` package for joint representation and analysis of genotype (SNP-chip) and expression array data. The Bioconductor monograph [168] is a useful resource for greater acquaintance with details and scope of the project.

Chapter 8

Elements of statistics

"Organic chemist!" said Tilley expressively. "Probably knows no statistics whatever."

—Nigel Balchin, The Small Back Room

Definition of a Statistician: A man who believes figures don't lie, but admits that under analysis some of them won't stand up either.

—Evan Esar

8.1 Introduction

The basics of microarray data analysis historically relied on the ratio of two signals from a spot. One signal came from a red dye and the other from a green dye. Initially, it was postulated that the relevant information from two-dye microarray experiments is captured in the ratio of these two signals from each spot. In addition, tools such as clustering techniques were applied to data analysis without too much attention to classical statistical analysis. Gary Churchill and his postdoctoral researcher Kathleen Kerr were among the first to observe that the needs and challenges of the data coming from microarray experiments are about as old as statistics itself. It is very worthwhile to spend some time revisiting some basic notions of statistics.

The dictionary definition of statistics is: "a branch of mathematics dealing with the collection, analysis, interpretation, and presentation of masses of numerical data" [308]. This definition places statistics within the realm of mathematics, thus emphasizing its rigor and precision. However, statistics has also been defined as "*the art and science* of collecting, analyzing, presenting, and interpreting data" [72]. This is perhaps a better definition since it also emphasizes the fact that statistical analysis is far from being a mechanical process, and that creativity and imagination are often required in order to extract most information from the data.

Nowadays, statistics has become an integral part of microarray data analysis, and it is of paramount importance for anybody dealing with such data to understand the basic statistical terms and techniques. The objective of this chapter is to review some fundamental elements of statistics as related to microarray experiments. Our primary goal here is not to construct a rigorous mathematical edifice but to give the reader a few working definitions and useful intuitions. This chapter is intended as a statistical primer for the life scientist. As such, we preferred the use of simpler mathematical tools even if we had to sacrifice at times the generality of some definitions and proofs.

8.2   Some basic concepts

Define your terms, you will permit me again to say, or we shall never understand one another.

— *Voltaire*

8.2.1   Populations versus samples

The term **population** denotes the ensemble of entities considered or the set of all measurements of interest to the sample collector. If one were to study the opinion of the American people regarding a new presidential election in the United States, then one would need to study the opinion of the whole population of the country about their votes for their future president.[1] In the same way, if one wanted to see what genes are over-expressed in people suffering from obesity, one would need to consider the whole population of clinically overweight people registered in the United States.

In the above example it is obvious that, in most cases, one cannot study directly whole populations. Therefore, a subset of the population, called **a**

---

[1]This assumes that everybody votes.

sample, is considered instead. A sample is any subset of measurements selected from the population. The objective of statistics is to make an inference about a population based on information contained in the sample. However, choosing this sample is a science by itself since **bias** can affect the results. A classic example is the sampling that occurred in the United States in 1936. A magazine contacted its readers and subscribers who numbered 10 million people and asked them whom they were going to vote for. Based on the 2,300,000 replies, a prediction was made that the Republican candidate Landon would be elected. However, it turned out that the Democratic candidate Franklin D. Roosevelt won the elections by a very large margin. The error occurred because the more than 2 million who responded did not represent correctly the American population at that time. In some situations, the bias might be very apparent. For instance, let us consider a potential bill banning possession of firearms. If a sample chosen for an opinion poll were to be taken exclusively from the members of the National Rifle Association, a majority rejecting such a bill would not be surprising. However, should the sample be taken from the members of a support group for firearms victims, the exactly opposed conclusion might be expected. Of course, both such polls would provide little information about the results of a potential referendum involving the entire United States population.

8.2.2 Parameters versus statistics

A variable whose values are affected by chance is called a **random variable**. To convey an idea about the general appearance of objects or phenomena, people need numerical descriptive measures. Numerical descriptive measures of random variables for a population are called **parameters**. For example, the mean weight of people in the United States is a parameter of the United States population. A numerical descriptive measure of a sample is called a **statistic**. In most cases, this numerical value is obtained by applying a mathematical function (e.g., the average) to the values in the sample. For example, the mean weight of people in a classroom is a statistic calculated for the sample of the United States population given by their presence in the classroom. These people represent a subset of the whole population living in the United States. The very same numerical descriptor of a group of measurements can be either a parameter or a statistic, depending on the given context. For instance, the mean weight of the people in the United States is a parameter of the United States population, but only a statistic in the larger context of the world population.

In general, the parameters of a population are unknown. In many cases, the goal of the statistical analysis is to calculate statistics based on various random samples, and to use these statistics to estimate the parameters of the population.

The data can be of different types as follows:

1. Quantitative

- Continuous: these data take continuous values in an interval. For example, spot intensity levels, gene expression levels.

- Discrete: these data take discrete values such as number of spots, number of genes, number of patients, etc.

2. Ordinal (ranked). In this type of data, there is ordering but no numerical value. For example, the top 10 rock hits on the music charts, top 10 companies performing on the stock market, etc.

3. Categorical data: genotype, phenotype, healthy/disease groups and others.

Subtle differences exist between discrete data such as the number of times a student has attempted a given exam before passing (taking the discrete values 1,2,3,4,...) and ordinal data such as stocks ordered by their performance over a given period. At a superficial view, stocks can be ordered with the best performer as 1, second best as 2, etc. Students can also be ordered in the order given by their number of attempts. The crucial difference between the stocks (ordinal data in this context) and students (discrete data) is that discrete data provide more information than categorical ones. Thus, stock ordering tells us that the second best is worse than the best, but it does not give us any information about *how much worse it is*. The difference may be very small or may be huge. We just cannot tell. However, knowing the number of times various students attempted an exam before passing will allow us to get an idea about the difference between consecutive students. If the best student passed at the first attempt and the second best student passed only after 6 attempts, we will know that the second best student is actually much worse than the best one. In essence, discrete, quantitative data can be ranked but ranked data cannot become quantitative.

Measurements can be characterized in terms of **accuracy** and **precision**. Accuracy refers to how close the measurement value is to its true value. Precision refers to how close repeated measurements are to each other. A biased but sensitive instrument will give precise but inaccurate values. An insensitive instrument might provide an accurate value on a given occasion, but the value would be imprecise (a repeated measurement would probably provide a different value). There is a trade off related to the precision. Too much precision can be cumbersome. For instance, when asked about their weight, nobody provides a number with three decimal places. This begs the question: what is a "good" precision then? In general, the measurements should be such that an error in the last digit would introduce an error of less than 5%, preferably less than 1%. In microarray experiments, there are also other issues. As shown in Chapter 3, microarrays involve many steps and a rather complicated data processing pipeline (Fig. 3.13). The focus has to be on improving the accuracy and precision of the overall results as opposed to improving the accuracy of a single step. Many times, a small increase in accuracy in all steps involved can be translated in an improvement of the overall performance while a large increase in accuracy or precision in a single specific step may or may not improve the global performance. As a specific example, increasing the performances of the scanning without improving the printing, mRNA quality, hybridization procedure, etc., might actually degrade the overall performance by adding more noise into the subsequent data analysis.

When comparing samples, it is important to evaluate their **central tendency** and **variability**. Central tendency measures provide information regarding the behavior of the group of measurements (sample) as a whole, while variability measures provide information regarding the degree of variation between members of the same sample.

8.3   Elementary statistics

8.3.1   Measures of central tendency: mean, mode, and median

At this time, we will consider that the number of values n is finite. This is done in order to keep the mathematical apparatus simple and make the material as accessible as possible. All terms below also have more general definitions that are applicable to infinite sets, as well.

8.3.1.1   Mean

The most widely used measure of central tendency is the arithmetic **mean** or average. This is defined as the sum of all values divided by their number:

$$\frac{\sum_{i=1}^{n} x_i}{n} \tag{8.1}$$

In statistics, in spite of the fact that there is only one formula for the arithmetic mean, it is customary to differentiate between the mean of a population, which is the mean of all values in the population considered:

$$\mu = \frac{\sum_{i=1}^{N} X_i}{N} \tag{8.2}$$

and the mean of a sample, which is the mean of the subset of the population members included in the sample under consideration:

$$\overline{X} = \frac{\sum_{i=1}^{n} X_i}{n} \tag{8.3}$$

The sample mean is usually reported to one more decimal place than the data. Also, the sample mean is measured in the same measurement units as the data.

There are a few interesting phenomena that are relevant to the mean as a measure of central tendency. Let us consider the following set of measurements for a given population: 55.20, 18.06, 28.16, 44.14, 61.61, 4.88, 180.29, 399.11, 97.47, 56.89, 271.95, 365.29, 807.80, 9.98, 82.73.

The population mean can be computed as:

$$\mu = \frac{\sum_{i=1}^{15} X_i}{15} = \frac{55.20 + 18.06 + 28.16 + \cdots + 82.73}{15} = 165.570 \qquad (8.4)$$

Let us now consider two samples from this population. The first sample is constructed by picking the first four measurements in the population: 55.20, 18.06, 28.16, 44.14 and has the mean 36.640. The second sample is obtained by picking measurement 8 through 12: 399.11, 97.47, 56.89, 271.95, 365.29, 807.80. This sample has the mean 333.085. An immediate observation is that the mean of any particular sample $\overline{X}$ can be very different from the true population mean μ. This can happen because of chance or because the sampling process is biased. The larger the sample, the closer its mean will be to the population mean. At the limit, a sample including all members of the population will provide exactly the population mean. This phenomenon is related to the very important notion of expected value. The **expected value** of a random variable X, denoted $E[X]$, is the average of X in a very long run of experiments. This notion can be used to define a number of other statistical terms, as it will be discussed in the following sections.

If a sample is constructed by picking a value and then eliminating that value from the population in such a way that it cannot be picked again, it is said that the sampling is done **without replacement**. If a value used in a sample is not removed from the population such that the same value can potentially be picked again, it is said that the sampling is done **with replacement**. In the example above, the sampling was probably done without replacement because in such a small population it is likely that sampling with replacement would have picked a value more than once. However, there is no way to be sure. Conversely, if a sample includes a value more than once and we know a priori that the population has only distinct values, we can conclude that the sampling was done with replacement.

8.3.1.2 Mode

The **mode** is the value that occurs most often in a data set. For instance, let us consider the following data set: 962, 1005, 1033, 768, 980, 965, 1030, 1005, 975, 989, 955, 783, 1005, 987, 975, 970, 1042, 1005, 998, 999. The mode of this sample is 1005 since this value occurs four times. The second most frequent value would be 975 since this value occurs twice.

According to the definition above, any data set can only have one mode since there will be only one value that occurs most often. However, many times certain distributions are described as being "bimodal." This formulation describes a situation in which a data set has two peaks. Sometimes, this is because the sample includes values from two distinct populations, each characterized by its own mode. Also, there are data sets that have no mode, for instance if all values occur just once.

An easy way of finding the mode is to construct a histogram. A histogram is a graph in which the different measurements (or intervals of such measurements) are represented on the horizontal axis and the number of occurrences is represented on the vertical axis.

8.3.1.3   Median, percentiles, and quantiles

Given a sample, the **median** is the value situated in the middle of the ordered list of measurements. Let us consider the example of the following sample:

$$96, 78, 90, 62, 73, 89, 92, 84, 76, 86$$

In order to obtain the median, we first order the sample as follows:

$$62, 73, 76, 78, \mathbf{84}, \mathbf{86}, 89, 90, 92, 95$$

Once the sample is ordered, the median is the value in the middle of this ordered sequence. In this case, the number of values is even (10) so the median is calculated as the midpoint between the two central values 84 and 86:

$$median = \frac{84 + 86}{2} = 85 \tag{8.5}$$

The median can also be described as the value that is lower than 50% of the data and higher than the other 50% of it. A related descriptive statistic is the percentile. The **p-th percentile** is the value that has p% of the measurements below it and 100-p% above it. Therefore, the median can also be described as an estimate of the 50th percentile.

The **quantiles** are generalizations of percentiles. Percentiles are those values that divide the data into 100 equally sized subsets. Quantiles are values that divide the data into an arbitrary number of equal q subsets. The 4-quantiles (q=4) are also called **quartiles**, the 5-quantiles are called **quintiles**, the 10-quantiles are called **deciles**, and of course, the 100-quantiles are called percentiles. The quantiles can be defined and calculated in several ways distinguished by subtle statistical properties. However, for the purposes of this book, the k-th q-quantile of the population parameter X, can be defined as the value x such that:

$$P(X \leq x) \geq p \tag{8.6}$$

and

$$P(X \geq x) \geq 1 - p \tag{8.7}$$

where $p = k/q$. The quantiles can also be defined when p is a real number between 0 and 1. In these circumstances, the p-quantile of the random variable X is the value x such that:

$$P(X \leq x) \geq p \tag{8.8}$$

and

$$P(X \geq x) \geq 1 - p \qquad (8.9)$$

Regardless of the exact mathematical definition used, the most useful intuition about quantiles visualizes them as the boundaries between equal-sized bins of data when the data are sorted. For instance, the first quartile would be the data value that is higher than a quarter of the data points, etc.

8.3.1.4 Characteristics of the mean, mode, and median

It is useful to compare the properties of the mean, mode, and median. The mean is the arithmetic mean of the values. The mean has a series of useful properties from a statistical point of view, and because of this, it is the most used measure of central tendency. However, if even a single value changes, the mean of the sample will change because the exact values of the measurements are taken into account in the computation of the mean. In contrast, the median relies more on the ordering of the measurements and thus it is less susceptible to the influence of noise or outliers. The mode also takes into consideration some properties of the data set as a whole as opposed to the individual values so it should also be more reliable. However, the mode is not always available. Let us consider the sample in the example above:

$$96, 78, 90, 62, 73, 89, 92, 84, 76, 86$$

The mean of this sample is 82.6, the median is 85, and the mode does not really exist since all values occur just once. Now, let us assume that, due to a measurement error, the lowest value of the sample was recorded as 30 instead of 62. The "noisy" sample is:

$$96, 78, 90, 30, 73, 89, 92, 84, 76, 86$$

which can be ordered as follows:

$$30, 73, 76, 78, \mathbf{84, 86}, 89, 90, 92, 96$$

The measures of the central tendency for this "noisy" data set are as follows: *mean* $= 79.4$ and *median* $= 85$. Note that the mean was changed by approximately 4% while the median remained the same. The fact that the median was not affected at all is somehow due to fortune. This happened because the noisy measurement substituted one of the lowest values in the set with a value which was even lower. Thus, the ordering of the data was not changed and the median was preserved. In many cases, the median will be perturbed by noisy or erroneous measurements but never as much as the mean. In general, the median is considered the most reliable measure of central tendency. For larger samples that have a mode, the mode tends to behave similarly to the median inasmuch that it changes relatively little if only a few measurements are affected by noise. These properties of the mean, median, and mode are particularly important for microarray data, which tend to be characterized by a large amount of noise.

A symmetric distribution will have the mean equal to the median. If the distribution is unimodal (i.e., there is only one mode), the mode will also coincide with the mean and the median. When a distribution is skewed (i.e., has a longer tail on one side), the mode will continue to coincide with the peak of the distribution, but the mean and median will start to shift towards the longer tail.

Another issue worth discussing is the informational content of the measurements of central tendencies. A well-known joke tells the story of two soldiers on the battlefield, waiting to ambush enemy tanks. A tank appears and one of the soldiers aims his artillery gun and shoots at the tank. Unfortunately, he misses the target and his missile impacts half a mile to the left of the tank. His comrade scorns him and says: "Now let *me* show you how this is done." He aims and misses too, but this time the missile hits half a mile to the right. The officer looks at them and says, "Well done! On average, the tank is destroyed!" Another classical joke, relying on the same idea, tells the story of the statistician who drowned while trying to cross a river with an average depth of 6 inches. The moral of both stories is that the data are not completely described by measures of central tendency. A very important piece of information that is not captured at all by such measures is the degree of variation, or dispersion, of the samples.

8.3.2 Measures of variability

I abhor averages. I like the individual case. A man may have six meals one day and none the next, making an average of three meals per day, but that is not a good way to live.

— *Louis D. Brandeis, quoted in Alpheus T. Mason's Brandeis: A Free Man's Life*

The arithmetic mean is a single number, which reflects a central tendency of a set of numbers. For example, the mean of the intensities of several replicates of a gene (microarray spots corresponding to the same gene whether on a single array or from several arrays) will give us an idea about the strength of signal or expression of this gene. However, the mean is incomplete as a descriptive measure, because it does not disclose anything about the scatter or dispersion of the values in the set of numbers from which it is derived. In some cases, these values will be clustered closely to the arithmetic mean, whereas in others, they will be widely scattered.

8.3.2.1 Range

The simplest measurement of variability is the **range**:

$$X_{max} - X_{min} \qquad (8.10)$$

The range is simply the interval between the smallest and the largest measurement in a group. A wider range will indicate a larger variability than a narrower range. This indicator can be quite informative. For instance, the statistician above might not have attempted to cross the river if she knew the range of the depth measurements was, for instance, 1 inch to 15 feet. However, it is still possible to have very different data sets that have exactly the same measures of central tendency *and* the same range. Therefore, more measures of variability are needed.

8.3.2.2 Variance

Measurements in a sample differ from their mean. For instance, microarray spot replicates differ from their computed mean. The differences between each individual value and the mean is called the **deviate**:

$$X_i - \overline{X} \tag{0.11}$$

where X_i is an individual measurement and $\overline{X}$ is the mean of all X_i measurements.

The deviates will be positive for those values above the mean and negative for those values below it. Many different measures of variability can be constructed by using the deviates. Let us consider, for example, that a set of microarray spots have their mean intensities as follows:

$$435.02, \ 678.14, \ 235.35, \ 956.12, \ldots, 1127.82, \ 456.43$$

The mean of these values is 515.13 and their deviates are as follows:

$$
\begin{aligned}
435.02 - 515.13 &= -80.11 \\
678.14 - 515.13 &= 163.01 \\
235.35 - 515.13 &= -279.78 \\
956.12 - 515.13 &= 440.99 \\
&\vdots
\end{aligned}
$$

We would like to have a measure for the deviations able to tell us about their magnitude. A first idea would be to calculate the mean of these deviations. However, a moment of thought reveals that such a mean would be zero since the arithmetic mean is exactly what the name suggests: a mean with respect to addition. Thus, the sum of all positive deviations will equal the sum of all negative deviations and the mean deviation will be zero. This can also be shown by trying to calculate such a mean of deviates:

$$
\frac{\sum_{i=1}^{n}(X_i - \overline{X})}{n} = \frac{(X_1 - \overline{X}) + (X_2 - \overline{X}) + \cdots + (X_n - \overline{X})}{n} =
$$

$$
= \frac{(X_1 + X_2 + \cdots + X_n) - (\overline{X} + \overline{X} + \cdots + \overline{X})}{n} = \frac{\sum_{i=1}^{n} X_i - n\overline{X}}{n} =
$$

$$= \frac{\sum_{i=1}^{n} X_i}{n} - \frac{n\overline{X}}{n} = \overline{X} - \overline{X} = 0 \tag{8.12}$$

This difficulty can be overcome in two ways: by taking the average of the absolute values of the deviations or by using their squared values. It turns out that the most useful measure has the following expression:

$$\sigma^2 = \frac{\sum_{i=1}^{N} (X_i - \mu)^2}{N} \tag{8.13}$$

This is called **population variance** and is a measure that characterizes very well the amount of variability of a population. The variance is usually reported with two more decimal places than the data and has as measurement units the square of the measurement units of the data. This is actually the average square distance from the mean. Note that this is the *population* variance. The **variance of a sample** is calculated as:

$$s^2 = \frac{\sum_{i=1}^{n} (X_i - \overline{X})^2}{n - 1} \tag{8.14}$$

Note that now each individual measurement is corrected by subtracting the *sample mean* $\overline{X}$ instead of population mean μ, the variable used to refer to this measurement is s^2 instead of σ^2, the sum is over the n measurements in the sample instead of the N measurements of the population and the denominator is $n - 1$. This latter difference has to do with the bias of the sample, but this is beyond the scope of this overview.

The quantity:

$$\sum_{i=1}^{n} (X_i - \overline{X})^2 \tag{8.15}$$

is called the **corrected sum of squared** (CSS) because each observation is adjusted for its distance from the mean.

A widely used measure of variability is the **standard deviation**, which involves the square root of the variance. The standard deviation of a set of measurements is defined to be the positive square root of the variance. Thus, the quantity:

$$s = \sqrt{\frac{\sum_{i=1}^{n} (X_i - \overline{X})^2}{n - 1}} \tag{8.16}$$

is the **sample standard deviation** and:

$$\sigma = \sqrt{\frac{\sum_{i=1}^{N} (X_i - \mu)^2}{N}} \tag{8.17}$$

is the corresponding **population standard deviation**.

In most practical cases, the real population parameters such as population mean, μ, or variance σ^2 are unknown. The goal of many experiments is precisely to estimate such population parameters from the characteristics of one or several samples drawn from the given population.

8.3.3 Some interesting data manipulations

In certain situations, all measurements of a sample are affected in the same way by a constant quantity. As an example, microarrays are quantified by scanning them and then processing the intensity values corresponding to individual spots. However, the device used to scan these arrays has some parameters that can be chosen by the user. One of these settings is the gain of the photomultiplier tube (PMT). If this value is increased between two scans of the same array, all spots intensities will be higher. Such a transformation is sometimes called **data coding**. Similar transformations are often used to change systems of measurements such as converting from miles to kilometers or from degrees Fahrenheit to degrees Celsius. It is interesting to see how descriptive statistics such as mean and variance behave when such data manipulations are applied.

Let us consider first the case in which all measurements are changed by adding a constant value. We would like to express the mean of the changed values $\overline{X}_c$ as a function of the mean of the original values X:

$$\overline{X}_c = \frac{\sum_{i=1}^{n}(X_i + c)}{n} = \frac{\sum_{i=1}^{n} X_i + \sum_{i=1}^{n} c}{n} = \frac{\sum_{i=1}^{n} X_i}{n} + \frac{n \cdot c}{n} = \overline{X} + c \tag{8.18}$$

Therefore, if we know that all values have been changed by addition and we have the old mean, we can calculate the new mean just by adding the offset to the old mean.

The computation of the new variance can be performed in a similar way. However, we can now use the fact that $\overline{X}_c = \overline{X} + c$:

$$s_c^2 = \frac{\sum_{i=1}^{n} \left((X_i + c) - (\overline{X} + c) \right)^2}{n-1} = \frac{\sum_{i=1}^{n} \left(X_i + c - \overline{X} - c \right)^2}{n-1} =$$

$$= \frac{\sum_{i=1}^{n}(X_i - \overline{X})^2}{n-1} = s^2 \tag{8.19}$$

Therefore, the variance of a sample coded by addition remains unchanged.

The results are a bit different when the data are changed by multiplying it with a constant value:

$$\overline{X}_c = \frac{\sum_{i=1}^{n} c \cdot X_i}{n} = \frac{c \cdot \sum_{i=1}^{n} X_i}{n} = c\overline{X} \tag{8.20}$$

The corresponding computation for the variance yields:

$$s_c^2 = \frac{\sum_{i=1}^{n} \left(cX_i - c\overline{X} \right)^2}{n-1} = \frac{c^2 \cdot \sum_{i=1}^{n} \left(X_i - \overline{X} \right)^2}{n-1} = c^2 \cdot s^2 \tag{8.21}$$

In consequence, the standard deviation will be affected by the same multiplicative factor as the data:

$$s_c = c \cdot s \tag{8.22}$$

8.3.4 Covariance and correlation

The covariance is a measure that characterizes the degree to which two different variables are linked in a linear way. Let us assume that an experiment measures the height of people in a population sample x. In this experiment, another interesting variable might be the height of the knee. Let us consider this as another variable y. Such a variable is called a covariate or concomitant variable. Ultimately, we want to characterize the relationship between the variables x and y.

Let X and Y be two random variables with means μ_X and μ_Y, respectively. The quantity:

$$Cov(X,Y) = E\left[(X - \mu_X)(Y - \mu_Y)\right] = E\left[XY\right] - E\left[X\right]E\left[Y\right] \qquad (8.23)$$

is the covariance between X and Y where $E[X]$ is the expectation of variable X.

In practice, if the two variables X and Y are sampled n times, their covariance can be calculated as:

$$Cov_{xy} = \frac{\sum_{i=1}^{n}(x_i - \bar{x})(y_i - \bar{y})}{n-1} \qquad (8.24)$$

which is the sample **covariance**. This expression is very similar to the expression of the variance 8.14. Indeed, the covariance of a variable with itself is equal to its variance. Unlike the variance which uses the square of the deviates, the covariance multiplies the deviates of two different variables. The product of these deviates can be positive or negative. In consequence, the covariance can be positive or negative. A positive covariance indicates that the two variables vary in the same way: they are above and below their respective means at the same time, which suggests that they might increase and decrease at the same time. A negative covariance shows that one variable might increase when the other one decreases.

This measure is very informative about the relationship between the two variables but has the disadvantage that is unbounded. Thus, it can take any real value, depending on the measurements considered. Furthermore, if we were to calculate the covariance between the size of apples and their weight on the one hand and between the size of oranges and the thickness of their skin on the other hand, the two covariance values obtained could not be compared directly. The diameter of the apple is a length and would be measured in meters, whereas the weight would be a mass and would be measured in kilograms. Therefore, the first covariance would be measured in meters × kilograms. The second covariance would be measured in meters (orange diameter) × meters (skin thickness) = square meters. Comparing directly the numerical values obtained for the two covariances would be meaningless and worthy of being characterized as a comparison between "apples and oranges." In consequence, it would be impossible to assess whether the linear relationship between the size of the apples and their weight is stronger or weaker than the linear relationship between the size of the oranges and the thickness of their skin.

In order to be able to compare directly the degree of linear relationship between heterogenous pairs of variables, one would need a measure that is bounded in absolute value, for instance between 0 and 1. Such a measure can be obtained by dividing the covariance by the standard deviation of the variables involved.

Let X and Y be two random variables with means μ_X and μ_Y and variances σ_X^2 and σ_Y^2, respectively. The quantity:

$$\rho_{xy} = \frac{Cov(X,Y)}{\sqrt{(VarX)(VarY)}} \tag{8.25}$$

is called correlation coefficient (Pearson correlation coefficient, to be exact).

The Pearson correlation coefficient takes care of both measurement unit and range. The correlation coefficient does not have a measurement unit and takes absolute values between 0 and 1. In order to see this, let us compute the correlation coefficient as:

$$\rho_{xy} = \frac{Cov(XY)}{s_x \cdot s_y} = \frac{\frac{\sum_{i=1}^{n}(X_i-\overline{X})(Y_i-\overline{Y})}{n-1}}{\sqrt{\frac{\sum_{i=1}^{n}(X_i-\overline{X})^2}{n-1}} \cdot \sqrt{\frac{\sum_{i=1}^{n}(Y_i-\overline{Y})^2}{n-1}}} =$$

$$= \frac{\sum_{i=1}^{n}(X_i-\overline{X})(Y_i-\overline{Y})}{\sqrt{\sum_{i=1}^{n}(X_i-\overline{X})^2} \cdot \sqrt{\sum_{i=1}^{n}(Y_i-\overline{Y})^2}} \tag{8.26}$$

In this expression, it is clear that both the numerator and the denominator have the same measurements units and therefore their ratio is without dimension. Furthermore, it is clear that this expression will take a maximum value of 1 if the two variables are "perfectly correlated" and a minimum value of -1 if the two variables are exactly opposite. Values towards both extremes show the two variables have a strong linear dependence. Values near zero show the variables do not have a strong linear dependence.

The following examples will illustrate situations with low, medium, and high correlation as follows:

```
> x=rnorm(20)
> x=sort(x)
> y1=rnorm(20,mean=0,sd=1)
> y2=x+rnorm(20,mean=0,sd=1.5)
> y3=x+rnorm(20,mean=0,sd=0.2)
> y4=-x+rnorm(20,mean=0,sd=0.2)
> cor(x,y1)

[1] -0.03930589

> cor(x,y2)

[1] 0.319344
```

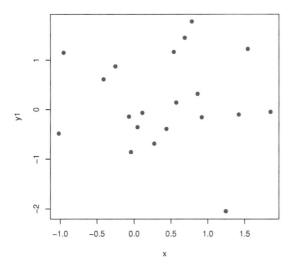

FIGURE 8.1: An example of two variables that have a low correlation.

```
> cor(x,y3)

[1] 0.9628555

> cor(x,y4)

[1] -0.9667046
```

Fig. 8.1 shows an example of two variables, x and y1, that have a low correlation obtained by drawing their values from two independent standard normal distributions. Fig. 8.2 shows two variables, x and y2, with a somewhat high correlation; y2 was obtained by adding to x some random noise with zero mean and a standard deviation equal to 1.5. Fig. 8.3 shows two variables, x and y3, with a high positive correlation obtained by adding noise with zero mean and a standard deviation of 0.2. Finally, Fig. 8.4 shows two variables, x and y4, with a high but negative correlation.

Given a set of variables $x_1, x_2, \ldots, x_k$, one could calculate the covariances and correlations between all possible pairs of such variables x_i and x_j. These values can be arranged in a matrix in which each row and column corresponds to a variable. Thus, the element σ_{ij} situated at the intersection of row i and column j will be the covariance of variables x_i and x_j while the elements

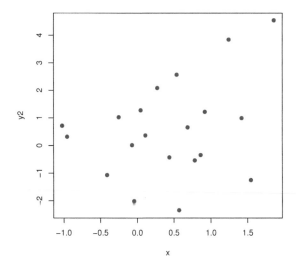

FIGURE 8.2: An example of two variables that have a medium positive correlation.

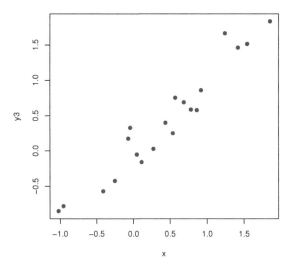

FIGURE 8.3: An example of two variables that have a high positive correlation.

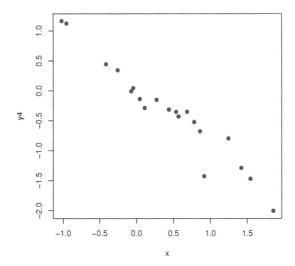

FIGURE 8.4: An example of two variables that have a high negative correlation.

situated on the diagonal will be the variances of their respective variables:

$$
\Sigma =
\begin{bmatrix}
\sigma_1 & \sigma_{12} & \sigma_{13} & \cdots & \sigma_{1k} \\
\sigma_{21} & \sigma_2 & \sigma_{23} & & \sigma_{2k} \\
\vdots & & & & \\
\sigma_{k1} & \sigma_{k1} & \sigma_{k3} & & \sigma_k
\end{bmatrix}
\tag{8.27}
$$

The correlation matrix is formed by taking the ijth element from Σ and dividing it by $\sqrt{\sigma_i^2 \sigma_j^2}$:

$$
\rho_{ij} = \sigma_{ij} / \sqrt{\sigma_i^2 \sigma_j^2}
\tag{8.28}
$$

A relation such as covariance is symmetric if $\sigma_{ij} = \sigma_{ji}$. Since correlations (and covariances) are symmetrical (check in Eq. 8.24 and Eq. 8.26 that nothing changes if X and Y are swapped), the correlation and covariance matrices will be symmetrical, i.e. the part above the first diagonal will be equal to the part below it.

8.3.5 Interpreting correlations

The correlation coefficient is a very useful measure and is very often used in practice. However, a few words of caution are in order here. First, the

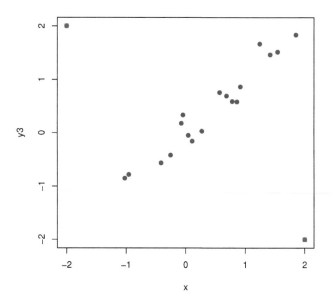

FIGURE 8.5: The effect of outliers on the correlation. Only two outliers (red squares) reduce the correlation of this data set from 0.999 to 0.365.

correlation is very sensitive to outliers. Two or three outliers can bring the correlation coefficient down quite considerably. As an example, let us take the two highly correlated variables from the previous example and modify two of the 20 points:

```
>  x[10]=-2
>  y3[10]=2
>  x[11]=2
>  y3[11]=-2
>  cor(x,y3)
```

```
[1] 0.1791453
```

The modified data are shown in Fig. 8.5. The original data set had a correlation of 0.989. The two outliers shown as red squares reduced the correlation of the X and Y variables to 0.365.

Another important observation is that *correlation is not equivalent to dependence*. Many times, the results of a correlation study have a negative conclusion, showing that certain variables x and y are not correlated. A superficial reading may interpret such a result as a conclusion that the variables X and

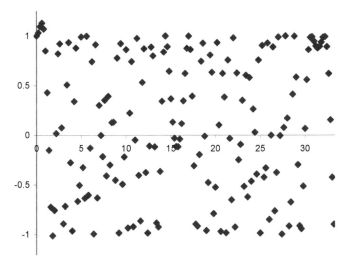

FIGURE 8.6: Two variables with a very low correlation (correlation coefficient $r_{xy} = 0.03$). Such results might be mistakenly extrapolated to the conclusion that x and y are independent variables. Fig. 8.7 shows that the two variables are highly dependent: all values of y are in fact precisely determined by the values of x.

Y are independent. This is not true. For instance, Fig. 8.6 shows a data set in which the variables x and y have an extremely low correlation (correlation coefficient $r_{xy} = 0.03$). Such results might be mistakenly extrapolated to the conclusion that x and y are independent variables. However, Fig. 8.7 shows that the data plotted in Fig. 8.6 are in fact 166 samples from the function $y(x) = sin(x^2) \cdot e^{-x} + cos(x^2)$. Not only that the two variables are dependent, but the value of x determines *exactly* the value of y through the given function. In this case, the low value of the *linear* correlation coefficient is explained by the high degree of non-linearity of the function.

In order to understand this phenomenon, it is important to keep in mind that the linear correlation coefficient defined above only tries to answer the question whether the values of the variables X and Y are compatible with a *linear* dependency. In other words, the question is whether there is a dependency of the form:[2] $Y = aX + b$. The answer to this question will be negative if either there is no dependency between X and Y, or if the dependency between X and Y exists but is approximated only very poorly by a linear function. In conclusion, if two variables are independent, they will probably have a low linear correlation coefficient. However, if two variables have a low correlation

---

[2]Note that if Y can be expressed as $aX + b$ then it is also possible to express X as $X = cY + d$ where c and d are appropriately determined.

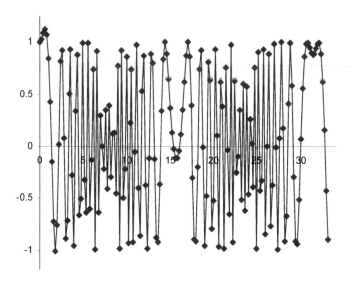

FIGURE 8.7: Limitations of the correlation analysis. The figure shows that the data plotted in Fig. 8.6 are in fact 166 samples from the function $y(x) = sin(x^2) \cdot e^{-x} + cos(x^2)$. The two variables are functionally dependent: the value of x determines *exactly* the value of y through the given function. The low value of the *linear* correlation coefficient is explained by the high degree of non-linearity of the function.

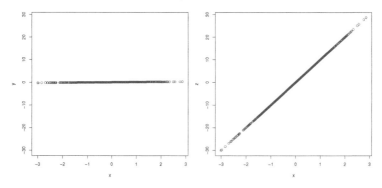

FIGURE 8.8: The correlation does not provide any information about the slope of the linear dependency (if such a linear dependency exists). In the left panel, $y = 0.01 \cdot x$. In the right panel, $z = 10 \cdot x$. The correlation between the variables x and y (left panel) is the same as the correlation between the variables x and z (right panel) $cor(x,y) = cor(x,z) = 1$.

coefficient, they may or may not be independent. In other words, *a high correlation coefficient is not a necessary condition for variable dependency.*

Another observation is that the correlation does not provide any information about the slope of the linear dependency if such a linear dependency exists. For instance, in Fig. 8.8, there is a perfect correlation (equal to 1) between the variables x and y in the left panel, as well as between the variables x and z in the right panel. However, the slopes of the data shown in the two panels are very different: $y = 0.01x$ versus $z = 10x$.

Probably the most important observation regarding the correlation is that *a high correlation does not imply or prove causality.* This is best illustrated by the story of the little girl who was one day watching her parents dress for a party. When she saw her dad donning his tuxedo, she warned, "Daddy, you shouldn't wear that suit." "And why not, darling?" "You know that it always gives you a headache the next morning."

In the scientific world, a recent paper caused a minor sensation when it showed that the consumption of beer is negatively correlated with the scientific output as measured by number of publications, total number of citations, and the number of citations per publications [187]. The author placed these results in the context of the well-known negative effects of alcohol consumption on cognitive performance to imply that the increased alcohol consumption leads to a decreased scientific productivity. In reality, the causality is not that clear. Yes, in principle one could easily visualize a scientist not being able to write a good paper after having had a few beers. However, one could just as easily visualize a scientist going to have a few beers after receiving yet another rejection from a journal. In other words, by simply observing this increased correlation, we cannot make the distinction between low scientific output as a results of high alcohol consumption and the exact opposite, a high alcohol consumption as a result of a low scientific output.

In another example of a similar nature, a 1999 *Nature* paper reported a high correlation between myopia and sleeping with room lighting before the age of 2 [346]. Citing previous work showing that the duration of the daily light period affects eye growths in chicks [404], the paper hypothesized an influence of ambient lighting during sleep on refractive development. This conclusion reached the mainstream media, which quickly amplified it. A subsequent CNN story explained that: "Even low levels of light can penetrate the eyelids during sleep, keeping the eyes working when they should be at rest. Taking precautions during infancy, when eyes are developing at a rapid pace, may ward off vision trouble later in life."[3] As of July 2011, this article is still available on CNN's web site in spite of a 2000 *Nature* paper that found no such effect [481], but rather a strong link between parents' myopia, their children's myopia, and the myopic parents' preference to leave some light on in their children's bedroom.

The reasoning that correlation proves causality is a common logical fallacy also known as *cum hoc ergo propter hoc* (Latin for "with this, therefore because of this"). This fallacy assumes that if two events A and B occur together, there is a cause and effect relationship between them such as A causes B. A related fallacy is *post hoc ergo propter hoc* (Latin for "after this, therefore because of this"), which introduces a temporal dimension by requiring the two events to be successive in time in order to claim an (equally false) causal relationship. In reality, a high correlation between A and B can be observed in at least four other situations besides the one in which A causes B, as follows:

1. B may be the cause of A (rather than A be the cause of B). For instance, it is often the case that traffic congestions are observed before major sporting events. However, this correlation by itself can hardly be used to suggest that on those days in which lots of people get frustrated driving at a snail's pace for hours, many of them decide to get out of their cars and congregate on the nearest football field.

2. There is some unknown factor X that causes both A and B. As an example, there is a high correlation between the size of the vocabulary (A), and the number of tooth cavities (B) in children. Rather than concluding that an expanding vocabulary causes cavities, one should consider the fact that both the number of words in a child's vocabulary and the number of cavities in their teeth tend to increase with age (X). An example for the *post hoc* variation would be to use the correlation between sleeping with one's shoes on and waking up with a headache, in order to infer that sleeping with one's shoes on causes headaches. A more plausible explanation is that both are caused by a third factor, such as alcohol intoxication.

---

[3] http://www.cnn.com/HEALTH/9905/12/children.lights/index.html

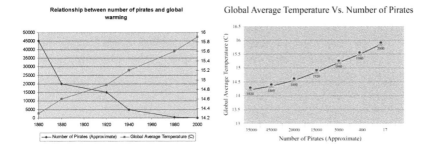

FIGURE 8.9: The relationship between the decreasing number of pirates worldwide and the global warming as illustrated by the increase in the average global temperature [201]. The left panel shows the changes in number of pirates and annual average temperatures during the period from 1860 to 2000. The correlation between these two quantities is 0.90, which was used to prove the "direct effect of the shrinking number of Pirates since the 1800s" by Bobby Henderson, the founder of the Pastafarianism (`http://www.venganza.org/`). The original data shown in the right panel has a correlation of 0.92, and includes one extra data point, which was eliminated from the left panel for didactic reasons. The graph in the right panel was created by Robert Henderson, used with permission.

3. The relationship between A and B is so complex that the correlation may be considered coincidental. Arguably, the most famous example in this category would be the strong negative correlation between the number of pirates and global warming (see Fig. 8.9) used to prove the "direct effect of the shrinking number of Pirates since the 1800s" by Bobby Henderson, the founder of the Church of the Flying Spaghetti Monster [201].

4. Finally, A and B could cause each other, being linked in a positive feedback loop. An example could be an autocatalytic reaction in which the product B is a catalyst for the reaction involving A.

The examples above included situations in which: i) two quantities were functionally related and yet their correlation was very low (as x and y in $y(x) = sin(x^2) \cdot e^{-x} + cos(x^2)$); ii) two quantities had a large correlation and likely had no causal relationship (the number of pirates and average global temperature); iii) two quantities had a large correlation but neither of them caused the other although a common causal factor did exist (the vocabulary and number of cavities); and iv) a situation in which a large correlation was due to a causal relationship in one direction only (sport events cause traffic jams but not the other way around). These examples show that *no conclusion between the either existence or the direction of a causal relationship can be inferred from correlation alone.*

However, one should not conclude that investigating the correlation between various variables is useless. In fact, correlation studies can be very important in practice and they are often used as a starting point to suggest specific hypotheses and design experiments that are able to validate or invalidate such hypotheses.

8.3.6 Measurements, errors, and residuals

A **statistical error** is the amount by which a measurement is different from its **expected value**, which is the population mean:

$$\varepsilon_i = X_i - \mu \tag{8.29}$$

Since the population mean is usually unknown and unobservable, it follows that the statistical error will also be *unobservable*. However, as discussed above, when a sample is available, the sample mean is used as an estimate of the population mean. In this case, one can calculate the amounts by which each measurement is different from the sample mean. These will be called **residuals** and will represent estimates of the statistical errors:

$$\widehat{\varepsilon}_i = X_i - \overline{X} \tag{8.30}$$

The hat symbol is a more general notation denoting an observable estimate of the variable below it which in general is unobservable.

Note that there is another qualitative difference between errors and residuals. The residuals over a given sample will always sum up to zero:

$$\sum_{i=1}^{n} \widehat{\varepsilon}_i = \sum_{i=1}^{n}(X_i - \overline{X}) = (X_1 - \overline{X}) + (X_2 - \overline{X}) + \cdots + (X_n - \overline{X}) =$$

$$= (X_1 + X_2 + \cdots + X_n) - (\overline{X} + \overline{X} + \cdots + \overline{X}) = \sum_{i=1}^{n} X_i - n\overline{X} =$$

$$= \sum_{i=1}^{n} X_i - n\frac{\sum_{i=1}^{n} X_i}{n} = \sum_{i=1}^{n} X_i - \sum_{i=1}^{n} X_i = 0 \tag{8.31}$$

In other words, *the residuals are not independent*: if we know only $n-1$ of them, we can always calculate the remaining one. This is because each of them is calculated as a deviate from a sample mean whose computation actually involves all measurements. Intuitively, the lack of independence is easy to see since changing any one measurement will change the sample mean and hence, all the residuals. This phenomenon is discussed in more detail in the next section on degrees of freedom. In contrast, the errors are all *independent*, with no constraints whatsoever. Knowing $n-1$ out of a set of n measured errors will provide little information about the n-th measurement.

8.4 Degrees of freedom

The notion of **degrees of freedom** is a perennial cause of confusion and misunderstandings, as well as a source of inspiration for many jokes. For instance, it has been said that "the number of degrees of freedom is usually considered self-evident – except for the analysis of data that have not appeared in a textbook" [142]. The number of degrees of freedom (df) can be defined in a number of different ways. Nevertheless, many definitions are confusing or cannot be easily applied in practice. For instance, as of February 2008, Wikipedia[4] defined the notion of degrees of freedom in the statistical context as "the number of categories or classes being tested minus 1" and then proceeded to further explain: "Degrees of freedom can be thought of as opportunities for change. For example, if five random samples are drawn from a given population, there are four opportunities for change, or four degrees of freedom." Such a definition can be utterly confusing. For instance, it is not clear why there would be only four "opportunities for change" if there are five measurements. Since the five measurements are independent, it might appear that there should be five "opportunities for change" rather than four.

Going beyond the web resources, most textbooks define the notion of degrees of freedom in a very cursory manner or neglect it altogether. Probably one of the worst definitions for this notion is stating that the number of degrees of freedom is simply an index to be used in the statistical tables associated with certain distributions such as the *t* or chi-square distributions. However, several very good treatments of this topic also exist in the literature. A classical interpretation of the number of degrees of freedom defines it as the difference between the dimensionality of the parameter spaces associated with the hypotheses tested [185]. This is a great approach but here we would like to define the notion before discussing hypothesis testing. Also, another classical interpretation that discusses the number of degrees of freedom from a geometrical perspective was originally proposed by Fisher [158] and detailed by Walker [441]. This is probably the most intuitive interpretation for people with a quantitative training and previous exposure to n-dimensional geometry. However, since this group of readers is only a subset of the intended audience for this text, we refrained from going into details solely in this direction.

Here, we will discuss this concept from a number of different perspectives, with the goal of giving the reader a solid understanding of this concept, as well as the practical ability to determine the number of degrees of freedom in most real-world applications.

---

[4]http://en.wikipedia.org/wiki/Degrees_of_freedom

8.4.1 Degrees of freedom as independent error estimates

Often, the goal of an experiment is to measure some quantity X, such as the weight of a object. In reality, we know that no individual measurement will provide the *exact* value, so what we are actually doing is take a number of measurements and **estimate** the quantity of interest to the best of our abilities. The simplest experiment that would provide any information would involve taking a single measurement of this object's weight. Let's denote this measurement by X_1. After performing this experiment, one could say that one now "knows" the object's weight as being X_1.

This is indeed what happens in most everyday life situations. If we find a rock and we want to know how much it weighs, we put it on a scale, read the weight, and that is pretty much it. Similarly, when buying fruits in a supermarket, the cashier will just put our bag of fruits on the scale and charge us according to the price per unit of weight and the weight indicated by the scale. In reality, every time we take one single measurement, we only estimate the amount to be equal to the value measured: $\widehat{X} = X_1$ (the hat notation used here is the general notation used to denote an estimate of a quantity).

This single-measurement estimate is sufficient in most everyday life situations. For instance, at the supermarket, the difference between the real weight of the bag of fruits and the weight indicated by the scale will probably correspond to a difference in price of less than one cent in most situations. Hence, it is not worth putting more effort into measuring the weight with any more accuracy. However, in many cases, a higher accuracy *is* needed. In those cases, what is also required is some estimate of this accuracy, which is often provided by estimating the **errors** involved in these measurements. For instance, if an object is to be sent in space, even small errors in the estimation of its weight can dramatically change the actual trajectory of the object.[5] Therefore, a legitimate issue is how well can we estimate the error in a given experiment such as the one above? This is exactly what the degrees of freedom try to capture. In this situation, we will be talking about the **degrees of freedom of the error**. In a first attempt, the concept of *degrees of freedom of the error can be defined as the number of independent values that can be used in order to estimate the error*. Note that the number of degrees of freedom is a structural aspect of the experiment, determined by the *number of measurements* rather than by their *values*.

---

[5] For instance, in the movie *Apollo 13*, there is a scene in which the return trajectory of the spaceship needed a correction involving an extremely risky manual firing of the jet engines because the actual weight of the spaceship was only a few kilograms less than the calculated weight. This in turn, was due to the fact that the mission could not actually land on the Moon where they were expected to collect some samples from the lunar surface.

8.4.2 Degrees of freedom as number of additional measurements

Clearly, if we have only one measurement, X_1, we will have to accept that number at face value. Our estimate of the quantity to be measured, $\widehat{X}$, will be the only value that can be, our only measurement, $\widehat{X} = X_1$, and we will have no measurements that could be used in order to estimate the error. We will say that in this data set composed of a single measurement $\{X_1\}$, there are zero degrees of freedom available to estimate the error. Let us consider again the measurement above X_1 and let us assume that we perform another measurement, X_2. Now, we have a set of two values $\{X_1, X_2\}$. We still choose the first value as our estimate of this quantity $\widehat{X} = X_1$. If we do this, we can now use the second measurement to compute an estimate of the error involved in the measurement process. This estimate of the error, or residual, can be calculated as:

$$\widehat{e_1} = X_2 - \widehat{X} = X_2 - X_1$$

This is the only value that we can possibly use from this set of measurements in order to estimate the measurement error. We will say that the number of degrees of freedom of the error in this case is one. Let us now consider that we have a set of n measurements of the same object, $\{X_1, X_2, ..., X_{n-1}, X_n\}$. For consistency's sake, we will still choose to take the first value as our estimate, $\widehat{X} = X_1$. In this case, we will be able to calculate $n - 1$ estimates of the error:

$$
\begin{aligned}
\widehat{e_1} &= X_2 - \widehat{X} \\
\widehat{e_2} &= X_3 - \widehat{X} \\
&\vdots \\
\widehat{e_{n-1}} &= X_n - \widehat{X}
\end{aligned}
\tag{8.32}
$$

In this case, we will say that the set of n measurements has $n - 1$ degrees of freedom available for the estimation of the error. The intuition gained here is that the number of degrees of freedom can be thought of as *the number of extra measurements beyond what is absolutely necessary to measure the quantity of interest.*

8.4.3 Degrees of freedom as observations minus restrictions

Now, we can choose to use an estimate for the quantity measured better than just taking the first measurement. One such better estimate might be the sample mean:

$$\widehat{X} = \overline{X} = \frac{\sum_{i=1}^{n} X_i}{n} \tag{8.33}$$

This, nevertheless, will not change the quality of the situation. The number of independent values that can be used to estimate the error will remain equal to $n - 1$. Now, we can calculate n estimates of the error:

$$
\begin{aligned}
\widehat{e}_1 &= X_1 - \overline{X} \\
\widehat{e}_2 &= X_1 - \overline{X} \\
&\vdots \\
\widehat{e}_n &= X_n - \overline{X}
\end{aligned}
\tag{8.34}
$$

but these estimate will be tied up by a further restriction, that their sum will always be zero:

$$
\widehat{e}_1 + \widehat{e}_2 + \cdots + \widehat{e}_n = 0
\tag{8.35}
$$

as shown by the equation 8.31 above. This illustrate another way of thinking about the number of degrees of freedom: *the degrees of freedom can be calculated as the number of observations n, minus the number of relations (or restrictions) that these observations must satisfy, r.* In this example, we can calculate n estimates of the error but we need to subtract $r = 1$ constraints, yielding the same number of degrees of freedom, $df = n - 1$, for a sample of n measurements.

8.4.4 Degrees of freedom as measurements minus model parameters

There is yet another way of thinking about degrees of freedom which is related to the one above. The number of degrees of freedom can be thought of as *the number of values that are available to estimate the errors minus the number of any other model parameters that are also involved in estimating those errors.* Again, equations 8.34 would provide n different estimates of the error, but in order to calculate these n values, we need to also calculate one intermediate value $\overline{X}$. This can be considered a parametric model for the errors, model which involves one parameter, its sample mean, $\overline{X}$. Hence, again, the number of degrees of freedom would be the number of error estimates, n minus the number of parameters $p = 1$, recovering the same $df = n - 1$.

8.4.5 Degrees of freedom as number of measurements we can change

This view that interprets the experiment as an attempt to estimate the errors using a parametric model with a single parameter $\widehat{X} = \overline{X}$ is also key to understanding certain discussions in which the number of degrees of freedom is referred to as the number of measurements can be changed without changing the mean. One's first reaction would be to question: why should the mean be fixed? If we were really to take another set of measurements, most likely the mean of that sample would be different, so where is this condition coming from? The answer is related to the fact that we are not trying to assess the variability of all possible samples, but rather trying to estimate the variability involved in *all samples that would have the same mean.* If we were to keep the overall mean fixed to the value obtained from this particular sample, the number of degrees of freedom would be the number of measurements we could change independently. Again, in this case, this number will be the total number of values in minus one, since the n-th value can always be calculated from the other $n-1$ chosen values and the fixed mean, $\overline{X}$.

8.4.6 Data split between estimating variability and model parameters

Finally, an interesting alternative interpretation of the degrees of freedom has been proposed by Gerard Dallal.[6] In this interpretation, a data set contains a number of n observations which constitute individual pieces of information. These pieces of information can be used in one of two ways: i) to estimate the parameters of the model used, or ii) to estimate the variability. In general, each model parameter being estimated will subtract one from the number of pieces of information that are available to estimate the variability. The number of degrees of freedom can then be defined as *the number of pieces of information left to estimate the variability.* This view is perfectly compatible with any of the other alternative perspectives above.

8.4.7 A geometrical perspective

Let us consider a sample of n measurements $\{X_1, X_2, ..., X_{n-1}, X_n\}$. This sample can be viewed as a point P in a space with n dimensions with the origin at the true population mean. If this true population mean is known, each measurement can be used to calculate an independent error from the true mean. Thus, the number of degrees of freedom for the error (from the true mean) is n since all X_i can take any values. However, in most cases, the true population mean is not known and is estimated by the mean of the sample $\overline{X}$. Let us assume that this mean has a specific value, M. If the true population mean is unknown, the only errors that can be calculated are those with respect to the mean of this sample. These would be residuals rather than true errors. The set of all points in space (corresponding to all samples of n measurements) that will have the same mean, M, are defined by the equation:

$$\frac{\sum_{i=1}^{n} X_i}{n} = M \tag{8.36}$$

which can be rewritten as:

$$X_1 + X_2 + \cdots + X_n - nM = 0 \tag{8.37}$$

This equation defines a hyperplane in the n-dimensional space that will contain all n-dimensional points corresponding to all samples of n measurements

---

[6]http://www.StatisticalPractice.com

having the mean M. This hyperplane will have $n-1$ dimensions corresponding to the $n-1$ degrees of freedom available to estimate the errors from the sample mean. This example illustrates how the number of degrees of freedom of the error can be seen as *the dimensionality of the vector space in which these errors can vary.*

8.4.8 Calculating the number of degrees of freedom

In this section, we will try to calculate the number of degrees of freedom in a number of situations that are often used in practice. In each case, we will try to use one or more of the perspective above in order to illustrate their equivalence and how each of them can be used in practice.

8.4.8.1 Estimating k quantities from n measurements

The first such case involves a situation in which k independent quantities are to be measured or, in statistical jargon, *estimated*. The bare minimum we can do here is to measure each quantity once, yielding k measurements. However, as above, we are also interested in estimating the errors with which these quantities are measured so we will go beyond the bare minimum and take several more measurements for each quantity. Let us assume that we take $n_1 > 1$ measurements for the first quantity, $n_2 > 1$ measurements for the second quantity, and $n_k > 1$ measurements for the k-th quantity. For each such quantity, we can pick one of the measurements as the estimate of the quantity studied which will lave $n_1 - 1$ estimates for the error for the first quantity, $n_2 - 1$ for the second one, etc. In total, there will be:

$$df = (n_1 - 1) + (n_2 - 1) + \cdots + (n_k - 1) = n - k \qquad (8.38)$$

degrees of freedom, where $n = \sum_{i=1}^{k} n_i$ is the total number of measurements taken for the k independent variables.

Alternatively, let us assume we use the sample mean as the estimate for each quantity. In this case, we can count the number of restrictions involved in the system above. For the each quantity (or variable) i, there will be one restriction, that the sum of its n_i residuals (or deviates from the mean) is zero:

$$\hat{e}_1 + \hat{e}_2 + \cdots + \hat{e}_{n_i} = 0 \qquad (8.39)$$

There are $n_1 + n_2 + \cdots + n_k = n$ independent measurements and k restrictions, one for each measured quantity, so we can calculate the number of degrees of freedom as $n - k$. The same results can be retrieved by thinking about the number of model parameters necessary in order to calculate the $n_1 + n_2 + \cdots + n_k = n$ estimates of the error. In this case, there is one parametric model for each variable m_i, with each model using one single parameter, the mean $\overline{X}_i$ of the measurements for the given quantity i. The number of degrees of freedom will be the total number of estimates of the error n, minus the number of parameters used in their computation, k, retrieving the same $n - k$.

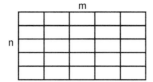

FIGURE 8.10: A table with n rows and m columns of measurements. The goal of the experiment is to estimate the variable measured (and implicitly the deviation from each measurement to this estimate). This estimate will be the overall mean. Here, the number of degrees of freedom of the error is the number of total (error) measurements $n \times m$ minus the number of parameters estimated in the process (one, which is the mean), or minus the number of restrictions imposed on these error estimates (one, which is that their sum should be zero).

8.4.9 Calculating the degrees of freedom for an $n \times m$ table

Let us now consider a table with n rows and m columns like the one shown in Fig. 8.10. For instance, such a table can be used to hold data that depends on two variables (e.g., sex and education). One variable depends on the row categories (e.g., male and female), the other variable defines the column categories (high school degree or less, B.Sc., M.Sc., Ph.D.). Each cell can contain for instance, the mean salary of people with the given degree and sex, working in a given organization, as shown in Fig. 8.11.

As previously stated, the number of degrees of freedom is a property of the experiment design and goal rather than a property of the data collected. Not only that the degrees of freedom depend on the number of measurements collected but it also depends on the aim of the experiment. In this example, let us assume the experimenter has no particular interest in the individual categories (neither rows nor columns), but is only interested in assessing the overall mean of the values in the table, i.e. the mean salary within the organization. If an overall estimate of the salary is desired, implicitly, we will also want to estimate the deviations (or errors) from each value in the table to the overall mean. In this case, there are $n \times m$ salary values, and implicitly $n \times m$ error estimates. These error estimates will all be calculated with respect to the estimated overall mean, which will be the only model parameter necessary in order to calculate these error estimates. Hence, the number of degrees of freedom will be:

$$df = n \times m - 1 \tag{8.40}$$

Alternatively, we can think that we are trying to estimate the variability involved in all samples that would have the same mean. If we were to keep the overall mean fixed to the value calculated in this particular sample, the

	HS	B.Sc.	Ph.D
M	42.3	53.7	87.2
F	41.2	54.1	86.4

FIGURE 8.11: A table showing the average salaries of the employees in a company divided by sex (Males and Females) and education level (HS=high school diploma, B.Sc., Ph.D.). If we are only interested in estimating the average salary within this company, the number of df is $3 \times 2 - 1 = 5$. In this case, this is equivalent to taken 6 arbitrary sample means. The information regarding the sex and the education level is not used in any way.

number of degrees of freedom would be the number of measurements we could change independently. Again, in this case, this number will be the total number of values in the table $n \times m$ minus one, which can always be calculated from the others and the mean.

Let us now consider that we are interested in estimating the mean salaries for every education level, i.e. in every column. In this case, we will estimate m column means which will act as model parameters. Hence, the number of df will be the total number of measurements, $n \times m$, minus the number of parameters estimated, m:

$$df = n \times m - m = (n - 1) \cdot m \qquad (8.41)$$

An alternative way of thinking is to imagine the column totals fixed[7] and count how many values we can vary freely without changing these totals. In Fig. 8.12, the fixed column sums (or means) are shown by the dots at the bottom of each column. In the right panel of the same figure, the circles show the values could be calculated from the remaining values and the column totals or means. The number of degrees of freedom is the number of empty cells, cells whose values can be allowed to vary arbitrarily and yet obtain samples with the same column totals (and implicitly the same column means).

Going back to the example shown in Fig. 8.11, let us now consider that we do not care to estimate the salaries for each education level, but we would like to estimate the salaries for each sex (each row). This means that we are now concerned with the degree of variability that can be found in samples within those categories that have the same row means. As before, having the same means with a constant number of measurements in each sample means that the totals are also fixed, this time in every row. This situation is illustrated in Fig. 8.13. The number of the df can be calculated following any of the ways of reasoning described above as:

$$df = n \times (m - 1) \qquad (8.42)$$

---

[7]This is equivalent to saying that the column means are fixed since the number of measurements in each column is also fixed.

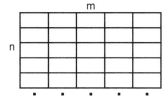 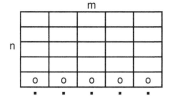

FIGURE 8.12: A table with n rows and m columns in which the column means are of interest (left panel). In such situation, we are interested to assess the variability of all samples of equal size that would have the same column means. In turn, this means that we will be looking at all possibilities that would keep the column totals fixed (since the number of measurements is also fixed). In the right panel the empty cells correspond to values that can be varied arbitrarily. There are $n-1$ such values in each column. The last value in each column (the cells with a circle) can be calculated from the others and the column totals or column means. In this situation, there are $(n-1)\cdot m$ degrees of freedom, obtained for instance by summing across the m columns, each with $n-1$ independent values, or by subtracting the m parameters/restrictions from the total number of $n \times m$ measurements: $n \times m - m = (n-1)\cdot m$.

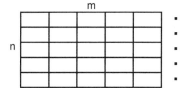 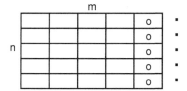

FIGURE 8.13: A table with n rows and m columns in which the row means are of interest (left panel). In such situation, we are interested to assess the variability of all samples of equal size that would have the same row means. In turn, this means that we will be looking at all possibilities that would keep the row totals fixed (since the number of measurements is also fixed). In the right panel the empty cells correspond to values that can be varied arbitrarily. There are $m-1$ such values in each row. The last value in each row (the cells with a circle) can be calculated from the others and the row totals or row means. In this situation, there are $(m-1)\cdot n$ degrees of freedom, obtained for instance by summing across the n rows, each with $m-1$ independent values, or by subtracting the n parameters/restrictions from the total number of $n \times m$ measurements: $n \times m - n = (m-1)\cdot n$.

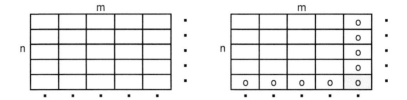

FIGURE 8.14: A table with n rows and m columns in which both column and row means are of interest (left panel). In the right panel, the empty cells correspond to values that can be varied arbitrarily. There are $m-1$ such values in each row. The last value in each row (the cells with a circle) can be calculated from the others and the row totals or row means. The number of degrees of freedom is equal to the number of empty cells: $(n-1)\cdot(m-1)$. The df can also be obtained by subtracting from the number of measurements, $n\times m$, the number of elements that can be determined from the totals (circles), $n+m-1$. The minus one comes from the yellow cell which would be counted twice otherwise. Thus, $n\cdot m-(n+m-1)=n\cdot m-n-m+1=(n-1)(m-1)$.

Finally, let us consider a situation in which one is interested in estimating the salaries in each row and column. In other words, one would like to know the mean salaries for men, women, high school graduates, B.Sc.s and Ph.D.s. This situation is illustrated in Fig. 8.14. In the right panel of this figure, the empty cells correspond to values that can be varied arbitrarily. There are $m-1$ such values in each row. The last value in each row (the cells with a circle) can be calculated from the others and the row totals or row means. The number of degrees of freedom is equal to the number of empty cells: $(n-1)\cdot(m-1)$. The df can also be obtained by subtracting from the number of measurements, $n\times m$, the number of elements that can be determined from the totals (circles), $n+m-1$. The minus one comes from the yellow cell which would be counted twice if we did not subtract it in the previous expression. Thus, the number of degrees of freedom is:

$$df = n\cdot m-(n+m-1)=n\cdot m-n-m+1=(n-1)(m-1) \qquad (8.43)$$

In summary, all examples above showed two important properties of the number of degrees of freedom that do not depend of how this number is calculated: i) the number of df does not depend on the data (note that the actual data in Fig. 8.11 was never used in any of the calculations above); and ii) it does depend on the goal of the experiment, more specifically on the quantity that needs to be estimated.

8.5   Probabilities

The **probability** of an event is the numerical measure of a likelihood (or degree of predictability) that the event will occur. From this point of view, an **experiment** is an activity with an observable outcome. A **random experiment** is an experiment whose outcome cannot be predicted with certainty. Examples of experiments can include:

- Flip a coin and record how many times heads or tails are up.

- Perform several microarray experiments and record whether a certain gene is up or down regulated.

- Perform several microarray experiments with the same array and see whether a particular spot has high or low intensities.

The set of all possible outcomes of an experiment is called the **sample space**. An **event** is a subset of the sample space. We say that an event occurred every time the outcome of the experiment is included in the subset of the sample space that defines the event. For instance, the sample space of rolling a die is the set of all possible outcomes: $\{1,2,3,4,5,6\}$. An event can be "obtaining a number less than 4." In this case, the event is the subset $\{1,2,3\}$. When we roll a die, we compare the outcome with the subset of the sample space corresponding to the event. For instance, if we rolled a "2," we check to see whether 2 is in the set $\{1,2,3\}$. In this case, the answer is affirmative so we say that the event "obtaining a number less than 4" occurred. Note that an event has been defined as a set. If this is the case, mutually exclusive events will be events whose subsets of the sample space do not have any common elements. For instance, the event of obtaining an odd number after rolling a die ($\{1,3,5\}$) is mutually exclusive with the event of obtaining an even number ($\{2,4,6\}$). However, the events "number less than or equal to 4" and "greater than or equal to 4" are not mutually exclusive since if 4 is rolled, both events occur at the same time.

Probabilities play an important role in understanding the data mining and the statistical approaches used in analyzing microarray data. Therefore, we will briefly introduce some basic terms, which will be useful for our further discussions.

One can define at least two different probabilities: empirical and classical. The **empirical probability** is the relative frequency of the occurrence of a given event. Let us assume that we are interested in an event A. The probability of A can be estimated by running a trial. If there are n cases in the sample and n_A cases in which event A occurred, then the empirical probability of A is its relative frequency of occurrence:

$$P(A) = \frac{n_A}{n} \tag{8.44}$$

Let us consider the following example. We are trying to estimate probability of occurrence of an albino pigeon in a given area. This probability can be estimated by catching a number of pigeons in the area. Most of the pigeons will be gray but a few will also be white (albino). If 100 birds are caught of which 25 birds are white, the empirical probability of an albino pigeon in the area is 0.25 or 25%.[8]

The **classical probability** of an event A is defined as:

$$P(A) = \frac{n(A)}{n(S)} \tag{8.45}$$

where S is the entire sample space. The definition simply says that the classical probability can be computed as the number of elements in event A divided by the number of elements in the whole sample space. In the example above, the classical probability will be the actual number of albino pigeons divided by the total number of pigeons in the area. In this example, as in many other cases, the exact probability cannot be computed directly and is either estimated or computed by some other means. If the entire sample space is completely known, the exact probability can be calculated directly and no trials are necessary.

The classical probability relies on certain axioms. These axioms are:

1. For any event, $0 \leq P(A) \leq 1$

2. $P(\emptyset) = 0$ and $P(S) = 1$

3. If $\{A_1, A_2, \ldots, A_n\}$ are mutually exclusive events, then

$$P(A_1 \cup A_2 \cup \ldots \cup A_n) = P(A_1) + P(A_2) + \ldots + P(A_n)$$

4. $P(A') = 1 - P(A)$

where A' is the complement of A.

These axioms might look intimidating, but in reality, they merely formalize our intuitions. Thus, the first axiom says that a probability is a real value between 0 and 1. The second axiom says that the probability of an event that is defined as the empty set (Φ) is zero. This is easy to understand. No matter what we do in a trial, we will observe an outcome. That value will not be included in the empty set because the empty set does not have any elements so the event corresponding to the empty set will never occur. In a similar way, the probability associated with the entire sample space S is 1 because no matter what the outcome of the trial is, it will be an element from the sample space. An event with a zero probability is an impossible event, whereas an event with a probability of 1 is certain.

---

[8]This is just a pedagogical example. Real-world estimates of such quantities are calculated using more sophisticated techniques.

The third axiom formalizes the idea that the probability of a union of disjoint events is the sum of the probabilities of the individual events. As an example, if A is the event of "rolling an odd number" and B is the event of "rolling an even number," then the probability of "A or B" is $P(A) + P(B) = 1/2 + 1/2 = 1$. In general, if the sample space is small and countable (e.g., 10 possible outcomes), it is very intuitive that if an event A has 2 favorable outcomes, an event B has, let us say, 3 favorable outcomes and we are interested in either A or B, then it is clear that the number of favorable outcomes is now $2 + 3 = 5$. Since the classical probability is defined as the ratio of the favorable outcomes to the total number of outcomes in the sample space, the probability of A or B is the sum of the two individual probabilities corresponding to A and B.

Finally, the fourth axiom states that the probability of the complement of an event A, denoted by A', can be computed as 1 minus the probability of the event A. This is because the complement A' of an event A is defined as the opposite event. For instance, the complement of the event of "rolling an even number" or ($\{2,4,6\}$) is "rolling an odd number" or ($\{1,3,5\}$). If the sample space is S and A is a subset, the subset corresponding to A' will be the complement of A with respect to S, i.e. the set of those elements in S that do not belong to A.

The connection between the two types of probabilities, empirical and classical, is made by the **law of large numbers** which states that as the number of trials of the experiment increases, the observed empirical probability will get closer and closer to the theoretical probability.

8.5.1 Computing with probabilities

8.5.1.1 Addition rule

Let us consider the following example.

EXAMPLE 8.1
We decide to declare that a microarray spot can be trusted (i.e., we have high confidence in it) if the spot is perfectly circular (the shape is a circle), perfectly uniform (the standard deviation of the signal intensities is less than a chosen threshold), or both. Let us assume that there are 10,000 spots printed on the array; 6,500 are circular, 7,000 spots are uniform, and 5,000 spots are both circular and uniform. What is the probability that a randomly chosen spot is trustworthy according to our chosen standard of high confidence?

SOLUTION The probability of a spot being circular is $6,500/10,000 = 0.65$. The probability of a spot being uniform is $7,000/10,000 = 0.70$. Note that these are classical probabilities, not estimates, because in this case the spots on the array constitute the entire sample space. We need to calculate the probability of the spot being either circular or uniform. However, it is clearly

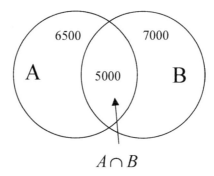

FIGURE 8.15: A Venn diagram. A is the set of 6,500 circular spots, B is the set of 7,000 uniform spots and $A \cap B$ is the set of 5,000 spots that are both circular and uniform. Just adding the probabilities of A and B would count the spots in $A \cap B$ twice.

stated in the problem that some spots are both circular and uniform. If we just add the probabilities of circular spots and uniform spots, the result would not be correct since spots having both properties will be counted twice, as shown in Fig. 8.15. The correct result can be obtained by adding the number of circular and uniform spots but subtracting the number of spots that are both circular and uniform. Thus, there will be 6,500 + 7,000 − 5,000 acceptable spots. In consequence, the probability that a spot is trustworthy on this array is:

$$P(trustworthy) = 0.65 + 0.70 - 0.50 = 0.85$$

In general, the **addition rule** states that if A_1 and A_2 are two events, the probability of either A_1 or A_2 equals the probability of A_1, plus the probability of A_2 minus the probability of both happening at the same time:

$$P(A_1 \cup A_2) = P(A_1) + P(A_2) - P(A_1 \cap A_2) \tag{8.46}$$

8.5.1.2 Conditional probabilities

Sometimes, two phenomena influence each other. Let us consider two events caused by these two phenomena, respectively. When one such event occurs, the probability of the second event is changed. The notion of conditional probability is needed in order to describe such dependencies. Let us consider two events, i.e. A and B with non-zero probabilities. The **conditional probabil-**

ity of event B given that event A has already happened is:

$$P(B|A) = \frac{P(A \cap B)}{P(A)} \tag{8.47}$$

Intuitively, the conditional probability of B given A, $P(B|A)$ stands for the probability of B and A occurring together as a percentage of the probability of A. The notion of conditional probability is useful in various situations as illustrated in the following example.

EXAMPLE 8.2

Let us consider the following experiment. During a reverse transcription reaction of a mRNA fragment to a cDNA, a dye is incorporated into the newly synthesized fragment, called a target. Dyes hybridize differentially depending on their chemical properties. Let the dye Y be incorporated with probability 0.9 into a given target. The probability that we read a signal from a spot is 0.8. What is the probability that a labeled target hybridizes on a spot?

SOLUTION   Let labeling be event A and hybridization event B. In order to detect a signal, it is necessary for the spot to be hybridized with a labeled target. Therefore, if we observed a signal, both events A and B occurred for that particular spot. In other words, 0.8 is the probability of A and B occurring together $P(A \cap B)$. The labeling occurs first and does not depend on anything else. Its probability is $P(A) = 0.9$. The question refers to a labeled target hybridizing on the array. In other words, the requested quantity is the probability of the hybridization occurring given that the labeling occurred already. This is $P(B|A)$. From Eq. 8.47:

$$P(B|A) = \frac{P(A \cap B)}{P(A)} = \frac{0.8}{0.9} = 0.88 \tag{8.48}$$

☐

Two events that do not influence each other are called **independent events**. Suppose that the probability of event B is not dependent on the occurrence of event A or, simply, that B and A are independent events. Two events A and B are independent if and only if (iff):

$$P(B|A) = P(B) \tag{8.49}$$

This is read as "the probability of B occurring given that A has already occurred is equal to the probability of B alone." In other words, B "does not care" whether A has happened or not. The "if and only if" formulation has the precise meaning of logical equivalence. This means that the definition can be used both ways. If we know the probabilities involved, $P(B|A)$ and $P(B)$ and they do equal each other, then we can conclude that the two events A and B are independent. Conversely, if we know a priori that A and B are independent and we know only one of $P(B|A)$ and $P(B)$, we can calculate the unknown one based on (8.49) above.

Let us use the definition of independent events (8.49) together with the definition of conditional probabilities (8.47):

$$P(B|A) = P(B)$$

$$P(B|A) = \frac{P(A \cap B)}{P(A)}$$

Essentially, these are two different expressions for the same quantity $P(B|A)$. Therefore, the two expressions must be equal in the given circumstances:

$$P(B) = \frac{P(A \cap B)}{P(A)} \tag{8.50}$$

From this, we can extract $P(A \cap B)$ as:

$$P(A \cap B) = P(A) \cdot P(B) \tag{8.51}$$

This is important. If two events A and B are independent, the probability of them happening together is the simple product of their probabilities. This result can be used in various ways as shown by the following two examples.

EXAMPLE 8.3
Assume that the probability that a certain mRNA fragment incorporates the dye is 0.9, the probability that the fragment hybridizes on a given spot is 0.95 and the probability that we observe it using a microarray experiment is 0.8. Are hybridization and labeling independent?

SOLUTION   If they were independent, the probability of them happening at the same time would be:

$$P(A \cap B) = 0.9 \cdot 0.95 = 0.855$$

Since this probability was given as 0.8, the two protocol steps are not independent. ▯

EXAMPLE 8.4
Assume two unrelated genes are expressed 85% and 75% of the time in the condition under study. What is the probability that both genes are expressed at the same time in a given sample?

SOLUTION   Since the genes are unrelated, the two events are considered to be independent. The probability of both genes being expressed is then:

$$P(A \cap B) = 0.85 \cdot 0.75 = 0.6375$$

▯

8.5.1.3 General multiplication rule

Let us revisit briefly the definition of the conditional probability in (8.47). From this expression, we can extract the probability of the intersection of two events as:

$$P(A \cap B) = P(A) \cdot P(B|A) \qquad (8.52)$$

This can be interpreted as follows: the probability of A and B occurring together is equal to the probability of A occurring times the probability of B occurring given that A has already occurred. This is **the general multiplication rule**.

EXAMPLE 8.5
Assume that the probability that a certain mRNA fragment incorporates the dye is 0.9 and the probability that the labeled fragment hybridizes on a given spot is 0.888. What is the probability to obtain a spot with a non-zero signal?

SOLUTION   The probability of incorporating the dye is the probability of event A (labeling). The probability that the labeled fragment hybridizes is the probability that the hybridization takes place given that the labeling is successful, i.e. the conditional probability of event B (hybridization) given A (labeling). In order to have a non-zero signal, we need both A (labeling) and B (hybridization). The probability of this is:

$$P(A \cap B) = P(B|A) \cdot P(A) = 0.888 \cdot 0.9 = 0.7999$$

▯

8.6   Bayes' theorem

In 1763, the Royal Society published an article entitled "An Essay towards Solving a Problem in the Doctrine of Chances" by the Reverend Thomas Bayes [37, 38]. This article and the approach presented in it has had the most profound effect on the science of the next 250 years. Today, perhaps more than ever, statisticians divide themselves into Bayesians and non-Bayesians, according to how much they use Bayes' approach.

Bayes' theorem can be explained in very technical terms. The truth is that, like many great ideas, Bayes' theorem is really simple. Let us consider two events, *A* and *B*. Let us assume that both events have been observed. This situation could have come into existence in two ways: i) either event *A* happened first followed by event *B* or ii) event *B* happened first followed by event *A*. Let us follow the first alternative. The probability of event *A* happening on its own is $P(A)$. The probability of B happening after A is, by definition, the conditional probability $P(B|A)$ or the probability of B given A. Since we observed both A and B, they both must have happened so the probability of this event can be obtained by multiplying the two probabilities above:

$$P(A) \cdot P(B|A) \tag{8.53}$$

Let us now follow the second possible chain of events: event B happened first, followed by event A. The probability of event B happening on its own is $P(B)$. The probability of A happening after B is again by definition, $P(A|B)$. The probability of observing both A and B in this case is:

$$P(B) \cdot P(A|B) \tag{8.54}$$

From a probabilistic standpoint, it does not matter whether the chain of events was A followed by B or B followed by A. Therefore, the two probabilities must be equal:

$$P(A) \cdot P(B|A) = P(B) \cdot P(A|B) \tag{8.55}$$

From this, we can extract one of the terms, for instance $P(B|A)$:

$$P(B|A) = \frac{P(B) \cdot P(A|B)}{P(A)} \tag{8.56}$$

This is Bayes' theorem. Now that we gained an understanding of it, we can read it the way it is usually read: the probability of an event B occurring given that event A has occurred is equal to the probability of event A occurring given that event B has occurred, multiplied by the probability of event B occurring and divided by the probability of event A occurring.

The event B is usually associated to a cause, whereas A is usually an observed event. Upon the observance of A, one goes back and assesses the probability of the causal hypothesis B. Our goal is to calculate the probability of the cause B, given that we have observed A. The terms $P(A)$ and $P(B)$ are called the a priori[9] probabilities of A and B, respectively. The term $P(B|A)$ is sometimes called the a posteriori probability of B. The same equation may appear more intuitive if written using the notation H for hypothesis, and O for observed event:

$$P(H|O) = \frac{P(H) \cdot P(O|H)}{P(O)} \tag{8.57}$$

Furthermore, the probability of the observed event $P(O)$ can be written as:

$$P(O) = P(H) \cdot P(O|H) + P(nonH) \cdot P(O|nonH) \tag{8.58}$$

which is to say that the event O may happen if the hypothesis H is true (first

---

[9]The terms "a priori" and "a posteriori" actually mean "before the event" and "after the event," respectively.

term) or if the hypothesis H is not true (second term). Using this for the denominator of Eq. 8.57, the Bayes' theorem can now be written as:

$$P(H|O) = \frac{P(H) \cdot P(O|H)}{P(H) \cdot P(O|H) + P(nonH) \cdot P(O|nonH)} \qquad (8.59)$$

The following example will illustrate a typical use of Bayes' theorem.

EXAMPLE 8.6

Let us assume a disease such as cervical cancer, with a prevalence of about 1 case in 5000 women, is to be screened for using a testing procedure that provides positive results for 90% of the women who have the disease (true positives) but also for 0.5% of the women who do not have the disease (false positives). Let O be the event of a positive test outcome and H the event corresponding of having cervical cancer. What is the probability that a woman who tested positive actually has the disease?

SOLUTION   We would like to use Bayes' theorem as expressed in Eq. 8.59. The term $P(H)$ is the probability of a woman having cancer which is

$$P(H) = \frac{1}{5000} = 0.0002 \qquad (8.60)$$

The term $P(O|H)$ is the probability of a positive test given cancer, i.e.

$$P(O|H) = 0.9 \qquad (8.61)$$

The denominator of Eq. 8.59 is the probability of observing a positive test result $P(O)$, which can be expressed as:

$$P(O) = P(H) \cdot P(O|H) + P(nonH) \cdot P(O|nonH) \qquad (8.62)$$

The first two terms in this expression can be calculated using Eq. 8.60 and 8.61. The term $P(nonH)$ is the probability of a woman not having cancer. This is:

$$1 - P(H) = 0.9998 \qquad (8.63)$$

Finally, the term $P(O|nonH)$ is the probability of a positive outcome given that there is no cancer. This is the probability of having a false positive and is:

$$P(O|nonH) = 0.005 \qquad (8.64)$$

From Eq. 8.59, we can now calculate the probability for a woman who tested positive actually to have the disease as:

$$P(H|O) = \frac{0.0002 \cdot 0.9}{0.0002 \cdot 0.9 + 0.9998 \cdot 0.005} = \frac{0.00018}{0.00018 + 0.0049999} = 0.034 \quad (8.65)$$

The conclusion is that fewer than 4% of the women with a positive test result actually have the disease. An examination of Eq. 8.65 reveals that the problem is not the test itself. A test yielding 90% true positives and 0.5% false positives is not necessarily a bad test. The problem is caused by the very low probability of having the disease $P(H) = 0.0002$. It is said that such a disease has a low prevalence (see also Chapter 21). ▯

Interestingly, in problems such as the one in the example above, most intuitive assessments would estimate the probability of having cancer in the case of a positive test result to a value very close to the true positive rate of the test (90% in this example) [174]. This tendency of our intuition to underestimate, sometimes even ignore, certain population characteristics such as the prior probability of an event is known as **base-rate fallacy**. Gigerenzer et al. argue that the people's poor performance in estimating probabilities such as in the case of base-rate fallacy is related to the "natural sampling" approach. The argument is that as humans evolved, the natural format of information was frequency. Thus, the human brain would be very accurate in acquiring and processing information provided as frequencies and much less accurate when the information is provided as probabilities [174].

8.7 Testing for (or predicting) a disease

Let us assume that we have a diagnostic test that was designed to detect a certain condition, such as cancer, based on the analysis of a sample. For any given sample submitted for analysis, this test can usually produce only one of two outcomes: a positive result $(+)$, which will be interpreted as the presence of the disease, or a negative result $(-)$, which will be usually interpreted as the absence of the disease. In a different context, a very similar problem is that of predicting the presence or absence of a given disease in the given sample. In this context, the test is referred to as the **classifier** and the test outcomes are referred to as **predicted classes**. In the following, we will use the terms **diagnostic test** and classifier interchangeably.

It is important to realize that any test or classifier will also make mistakes, or incorrect predictions, from time to time. Hence, one must distinguish between various situations with respect to the relationship between the true disease state and the predictions made. Furthermore, some criteria need to be established in order to allow us to evaluate various classifiers or disease pre-

		True disease state		
		Disease (D)	Healthy (H)	Prevalence $= \frac{D}{D+H} = \frac{D}{N}$
Predicted	Disease (+)	**TP**	FP	PPV $= \frac{TP}{TP+FP}$
	Healthy (−)	FN	**TN**	NPV $= \frac{TN}{FN+TN}$
		Sensitivity $\frac{TP}{TP+FN}$	Specificity $\frac{TN}{FP+TN}$	Accuracy $\frac{TP+TN}{TP+TN+FP+FN}$

FIGURE 8.16: Criteria used to assess the prediction abilities of a classifier or diagnosis device. There are N samples of which D have this disease, and H do not have this disease. The columns indicate the true disease state. The rows contain the predictions of the classifiers: a positive outcome (+) would indicate the presence of the disease, whereas a negative outcome (−) its absence. True positives (TP) are true diseases predicted as disease. False negatives (FN) are true diseases predicted as healthy. False positives (FP) are true healthy predicted as disease and true negatives (TN) are true healthy samples correctly predicted as healthy. The table also shows the definitions of the accuracy, sensitivity, specificity, positive predicted value (PPV), and negative predicted value (NPV).

dictors. In the following, we will discuss the main measures used to evaluate the performance of such devices.

8.7.1 Basic criteria: accuracy, sensitivity, specificity, PPV, NPV

When the true disease state is known, any particular outcome of a test can be characterized with a combination of two qualifiers: positive/negative and true/false. The first one refers to the outcome of the test: a *positive* will be a sample for which the test yielded a positive result, whereas a *negative* will be a sample for which the test yielded a negative result. The second qualifier refers to the relationship between the outcome of the test and the true disease status. For instance, a **true positive** (TP) will be a true cancer predicted as cancer and a **false positive** (FP) will be a true healthy[10] predicted as cancer. Similarly, a **true negative** (TN) will be a healthy predicted as such, whereas a **false negative** (FN) will be a healthy predicted as cancer. These definitions are summarized in Fig. 8.16 where the layout should also be helpful in understanding the terms defined in the following. The matrix shown in this figure is called a **confusion matrix**.

Probably the simplest and most intuitive performance criterion for a classifier is the overall **accuracy**. This is defined as the number of correct prediction

---

[10]In this context and throughout this section, "healthy" is used to describe a situation in which the sample does not have the disease that the test was designed to detect, rather than the absence of any disease.

divided by the total number of samples:

$$accuracy = \frac{TP+TN}{TP+TN+FP+FN} = \frac{TP+TN}{D+H} \quad (8.66)$$

The ideal accuracy is 1 (or 100%), which is obtained when there are no FPs and no FNs, in other words, when the classifier always predicts the true disease state.

The ability of a test to produce a positive outcome in the presence of the disease is referred to as **sensitivity**. This sensitivity of a test can be computed as the fraction of the true diseases detected in a given sample (population):

$$sensitivity = \frac{TP}{TP+FN} = \frac{TP}{D} \quad (8.67)$$

Essentially, the sensitivity characterizes the ability of the test to detect the disease when the disease is present. A sensitive test will produce positive results for a large proportion of the disease samples. Increasing the sensitivity of a test means increasing its ability to detect the disease. The ideal sensitivity is 1 or 100% which is attained when there are no false negatives. This corresponds to a situation in which all true disease samples are correctly classified as such.

The ability of a test to produce a negative result for samples which do not have the targeted disease (in this context labeled with "healthy") is called **specificity**. This is calculated as the ratio between the number of healthy predicted as such and the total number of healthy samples (TN+FP):

$$specificity = \frac{TN}{TN+FP} \quad (8.68)$$

The specificity refers to the ability of a test to produce positive results *only* for those samples that have the targeted results, in other words, the ability of the test to be *specific* to that particular disease. Any false positives will decrease the specificity of the given test. The ideal specificity is 1 (or 100%) which is attained when there are no false positives.

There are two other very important aspects related to the ability of a classifier or diagnostic test to predict the disease. Let us assume that a certain sample comes out as positive after testing. As stressed before, any test can be wrong sometimes and the fact that the test came out positive does not necessarily mean that the person who provided that sample actually has the disease tested for. A very important problem is to assess the probability of that individual actually having the disease when the test result is positive. Given the results of the test on a set of samples for which the true disease state is known as in Fig. 8.16, this probability can be calculated as the fraction of all predicted positives that actually have the disease:

$$PPV = \frac{TP}{TP+FP} \quad (8.69)$$

		True disease state		
		Disease	Healthy	
		100	900	Prevalence = 0.1
Predicted	Disease (+)	100	900	PPV = $\frac{100}{1000}$=0.1
	Healthy (−)	0	0	NPV = $\frac{0}{0}$=?
		Sensitivity	Specificity	Accuracy
		$\frac{100}{100} = 1$	$\frac{0}{900} = 0$	$\frac{100}{1000} = 0.1$

FIGURE 8.17: A (useless) test with 100% sensitivity can easily be obtained by always predicting disease for all samples. The specificity will be 0, NPV will be undefined, and the PPV and accuracy will be equal to the prevalence of the disease.

This quantity is called the **positive predicted value** (**PPV**) of a test, and reflects the ability of a test to actually predict disease in a given sample or population.

While the PPV provides a measure of the probability of having the disease when the outcome of the test was positive, it says nothing about the probability of having the disease (or not) when the outcome of the test was negative. Hence, a similar measure is needed for negative test outcomes. Since, this time we are considering those samples with a negative test outcome, we would be interested in assessing the probability of these negative samples actually not having the disease. This can be calculated as the fraction of the negative test outcomes that are associated with a true negative disease status:

$$NPV = \frac{TN}{TN+FN} \tag{8.70}$$

This quantity is called **negative predictive value** (NPV) and indicates the probability that a sample predicted as healthy is actually such. This value also corresponds to the proportion of the predicted negative samples that are actually true negatives.

8.7.2 More about classification criteria: prevalence, incidence, and various interdependencies

It is interesting to note (and useful never to forget) that an ideal sensitivity can be always and easily obtained by having the test predict disease for all samples, as in the example shown in Fig. 8.17. Always predicting disease will ensure that 100% of the true disease samples will be predicted as disease, thus yielding a sensitivity of 1. However, this will, of course, be completely useless in terms of actually using this test in any diagnostic setup.

Much as for the sensitivity, it is also very easy to bring the specificity of a test to 1 by simply having the test to always predict "healthy," which will ensure that there will be no false positive, and hence that the specificity will

		True disease state		
		Disease	Healthy	
		100	900	Prevalence $= 0.1$
Predicted	Disease $(+)$	0	0	PPV $= ?$
	Healthy $(-)$	100	900	NPV $= \frac{900}{1000} = 0.9$
		Sensitivity	Specificity	Accuracy
		$\frac{0}{100} = 0$	$\frac{900}{900} = 1$	$\frac{900}{1000} = 0.9$

FIGURE 8.18: A (useless) test with 100% specificity can easily be obtained by always predicting the lack of disease for all samples. In this case, the sensitivity will be 0, PPV will be undefined, and the NPV and total accuracy will be equal to 1 minus the prevalence of the disease (in this case 90%).

be 1. However, as shown in Fig. 8.18, this will negatively affect the sensitivity of the test. What is both difficult as well as practically useful and important is to have both high sensitivity *and* high specificity *at the same time*. This can be achieve only by having a small number of false positives together with a small number of false negatives. This is the goal in any practical application.

A very important aspect to underline is that, unlike the sensitivity and specificity of a test which depend on the test itself and nothing else, the PPV also depends on the proportion of the disease cases within the given sample or population. This proportion is called **prevalence** and can be calculated as:

$$prevalence = \frac{D}{D+H} \tag{8.71}$$

The prevalence of a disease in a given population or sample will directly affect the PPV for a given specificity. In other words, our ability to conclude that a certain individual actually has a disease from the positive outcome of a test designed to detect that disease is very much influenced by the prevalence of the given disease in the given population as clearly illustrated in the example 8.6. This also means that exactly the same test can provide very different results that can range from very useful to completely useless depending on the population on which is used.

For instance, ovarian cancer (OVCA) is a disease with a very low prevalence of about 50/100,000 or 0.05% in the population at large. In spite of its low prevalence, the ovarian cancer causes more deaths per year than any other cancer of the female reproductive system, so there are a number of efforts to develop a screening test able to detect this cancer early. Let us consider a diagnostic test with a specificity of 95% and a sensitivity of 95% used to screen a sample of 40,000 people. In general, a test with both specificity and sensitivity values of 95% is a very good test. Indeed, out of the 20 true cancers present in the sample of 40,000 people screened, the test will detect 19, true to its 95% sensitivity. The problem stems from the fact that together with these 19 true cancers, the test will provide positive outcomes for another 1,999

		True disease state		
		Disease	Healthy	Prevalence
		20	39,980	0.0005
Predicted	Disease (+)	19	1999	PPV = 0.009415
	Healthy (−)	1	37,981	NPV = 0.999974
		Sensitivity	Specificity	Accuracy
		0.95	0.95	0.95

FIGURE 8.19: An accurate test with high sensitivity (95%) and specificity (95%) can be practically useless in an application in which the prevalence is very low. In this case, less than 1% of the people who tested positive will actually have the disease.

people for a total of 2,018 positive results. The positive predicted value of a positive outcome, or the proportion of true cancers out of all positive results will be only 0.0094 or less than 1%. The performance of this test in this population is summarized in Fig. 8.19. In practice, such a test is practically useless since 99 of every 100 people who tested positive will actually not have the disease and no clinical decision can be made based on a positive test outcome under these circumstances.

However, let us now consider the situation in which the same test is used in a different way: rather than using it as a screening test in the population at large, such a test could be used to detect recurrence in patients who have been previously diagnosed with ovarian cancer, undertook treatment for it and were declared cancer-free at the end of their treatment. Such a test can be used annually to monitor and detect early the recurrence of the illness. Let us assume that the prevalence of the disease in this population of previously treated cancer patients is 30% and that 400 patients are tested. This situation is illustrated in Fig. 8.20. In this set of 400 patients, there will be 160 patients with a relapse of their cancer and 240 who are still cancer-free. Since it has a sensitivity of 95%, the test will come out positive for 152 out of the 160 cancers. Because of the 95% specificity, the test will also yield 12 false positives out of the 240 cancer-free patients. The 152 true positives and 12 false positives yield a PPV of 92%. In other words, the vast majority of the people with a positive outcome will in fact have the disease and a clinical follow-up can be triggered based on a positive test result. Note that the exact same test was clinically useless in a low prevalence population but very useful in a different setting, when used in a population with higher prevalence.

We have seen above how the specificity of a test influences the PPV for a constant prevalence. The higher the specificity is the higher the PPV will be if the prevalence is constant. A similar relationship exists between sensitivity and NPV. For a given prevalence, the higher the sensitivity is, the higher the NPV will be too.

As a side note, we will mention that the notion of prevalence defined above

		True disease state		Prevalence
		Disease	Healthy	
		160	240	0.40
Predicted	Disease (+)	152	12	PPV = 0.9268
	Healthy (-)	8	228	NPV = 0.9661
		Sensitivity	Specificity	Accuracy
		0.95	0.95	0.95

FIGURE 8.20: The same test with 95% sensitivity and 95% accuracy can be very useful in an application in which the prevalence is high. For instance, if the prevalence of cancer relapses after treatment is 40%, this test can be successfully used to predict such cancer recurrence. In this case, the PPV is 92% meaning that the large majority of the people who tested positive will actually have the disease

is related to, but different from, that of incidence. The **incidence** of a disease is a measure of the risk of developing that disease in a given period of time. If a population at risk has N=1,000 individuals and within one year 50 of those individuals develop a certain disease D, the incidence of D in this population can be calculated as:

$$incidence = \frac{D}{N} = \frac{50}{1000} \tag{8.72}$$

and expressed as "50 in 1,000 individuals" or, even more accurately calculated as:

$$incidence = \frac{D}{N \cdot y} = \frac{50}{1000 \cdot 1} \tag{8.73}$$

where y is the length of the time interval considered expressed in years. In this case, the incidence can be reported as "50 in 1,000 individuals/year." Note that the incidence formula above is numerically identical with the prevalence formula in Eq. 8.72 if we consider $N = D + H$ and the time period $y = 1$. This is to simply say that, over the one year, the initial population of N individuals can be divided into D individuals who developed the disease, and H individuals who did not developed this disease. The difference is that the prevalence looks at a given sample of population *at a given moment* in time, while the incidence looks at a sample or population *over an interval of time*. In the example above, if the prevalence of the disease D at the beginning of that period of time was 0 and no individual died during that interval, the value of the prevalence at the end of the period will be equal to the incidence.

8.8   Summary

This chapter introduced a few necessary statistical terms such as: population, sample, parameters, random variables, etc. Several measures of central tendency (mean, median, mode, percentile) and variability (range, variance, standard deviation) were presented. The discussion emphasized various interesting properties of these quantities, the relationship between them and their utility in the context of microarray data. The chapter also introduced the notions of covariance and correlation, discussed their common usage as well as common mistakes associated to their usage.

The next section of the chapter introduced the notion of probabilities, conditional probabilities and operations with such. Bayes' theorem and a few related issues as well as a typical example of Bayesian reasoning were briefly reviewed. The chapter also discussed the notions of probability distributions, as well as a few specific distributions including the binomial and normal distributions.

Finally, the chapter discussed the basic notions related to predicting or testing for a disease: accuracy, specificity, sensitivity, PPV, NPV, prevalence, etc.

8.9   Solved problems

1. Consider a breast cancer screening test with a sensitivity of 95% and a specificity of 95% applied to a sample of 10,000 subjects. As of 2005, there were an estimated 2.521 million cases of breast cancer in the US population of approximatively 295.560 million. Calculate:

   (a) The expected number of false positives in the given sample.

   (b) The expected number of false negatives in the given sample.

   (c) The PPV.

   (d) The NPV.

   (e) The accuracy of this test.

   SOLUTION   From the data provided, we calculate the prevalence as:
   $$\frac{2.521}{295.560} \approx 0.00853$$

		True disease state		
		Disease	Healthy	Prevalence
		85.3	9914.7	0.0085
Predicted	Disease (+)	81.03	20.26	PPV = 0.8
	Healthy (−)	4.26	9894.45	NPV = 0.9995
		Sensitivity	Specificity	Accuracy
		0.95	0.95	0.95

FIGURE 8.21: The expected results when applying a breast cancer screening test with a sensitivity of 95% and a specificity of 95% on a set of 10,000 people coming from a population in which the breast cancer prevalence is 2.521/295.56 or about 0.85%.

Using this prevalence, we calculate the expected number of true cancers in the given sample:

$$0.00853 \cdot 10,000 \approx 85$$

Using the sensitivity, we then calculate the expected number of true positives as:

$$85 * 0.95 \approx 81$$

and false negatives as:

$$85 \cdot 0.05 \approx 4.25$$

The number of true and false negatives are calculated in a similar manner from the number of true healthy people. The results are summarized in Fig. 8.21.

$\square$

8.10   Exercises

1. Consider a screening test for ovarian cancer that has a sensitivity of 95%. Calculate what the specificity of the test should be for it to have a PPV of 80% given that the prevalence of the ovarian cancer in the given population is 0.05%?

2. Consider a screening test for a disease that has a sensitivity of 90%. Calculate what the specificity of the test should be for it to have a minimum PPV of 86% given that the prevalence of the given disease in the given population is 5%?

3. A classifier is tested on a set of 40 disease samples and 60 healthy samples. The classified yielded a positive outcome for 54 samples which

included 35 of the true disease samples. Calculate the sensitivity, specificity, PPV, NPV, and accuracy for this test.

4. A classifier is tested on a set of 40 disease samples and 60 healthy samples. The classified yielded a negative outcome for 54 samples of which included 52 were true negatives. Calculate the sensitivity, specificity, PPV, NPV, and accuracy for this test.

5. Establish the relationships between the performance criteria used to assess a classifier and the various probabilities used in Bayes theorem. For instance, the a priori probability of a disease, $P(H)$ can be estimated by the prevalence of that disease in a given population. What would the PPV correspond to in terms of probabilities? How about the sensitivity, specificity and NPV?

Chapter 9

Probability distributions

The most important questions of life are, for the most part, really only problems of probability.

—*Pierre Simon Laplace*

9.1 Probability distributions

This section will introduce the notions of random variables and probability distributions. A few classical distributions, including the uniform, binomial, normal, and standard normal distributions will be discussed. These distributions are particularly important due to the large variety of situations for which they are suitable models. The discussion will focus on the general statistical reasoning as opposed to the particular characteristics of any given distribution.

A **random variable** is a variable whose numerical value depends on the outcome of the experiment, which cannot be predicted with certainty before the experiment is run. Examples include the outcome of rolling a die or the amount of noise on a given microarray spot. Random variables can take **discrete** values such as the die roll or the number of spots contaminated by noise on a given microarray, or **continuous** values such as the amount of noise on a given microarray spot. The behavior of a random variable is described by a function called a density function or probability distribution.

9.1.1 Discrete random variables

The **probability density function (pdf)** of a discrete variable X, usually denoted by f, is given by the probability of the variable X taking the value x:

$$f(x) = P(X = x) \tag{9.1}$$

where x is any real number.

The pdf of a discrete variable is also known as **discrete mass function** or **probability mass function**. In this analogy, $f(x)$ is the mass of the point x.

The pdf of a discrete variable is defined for all real numbers but is zero for most of them since X is discrete and can take only integer values. Also the pdf is a probability and therefore $f(x) \geq 0$. Finally, summing f over all possible values has to be 1. This corresponds to the probability of the universal event containing all possible values of X. No matter which particular value X will take, that value will be from this set and therefore this event is certain and has probability 1. In other words:

$$\sum_{all\ x} f(x) = 1$$

EXAMPLE 9.1
A fair (unbiased) 6 face die is rolled. What is the discrete pdf (mass function) of the random variable representing the outcome?

SOLUTION   This variable is uniformly distributed since the die is unbiased. The pdf is zero everywhere with the exception of the following values:

value	1	2	3	4	5	6
probability	1/6	1/6	1/6	1/6	1/6	1/6

$\square$

The **mean** or **expected value** of a variable can be calculated from its pdf as follows:

$$\mu = E(x) = \sum_x x \cdot f(x) \tag{9.2}$$

Furthermore, if a function H depends on the discrete random variable X, the expected (mean) value of H can also be calculated using the pdf of X as follows:

$$E(H(x)) = \sum_x H(x) \cdot f(x) \tag{9.3}$$

The typical example is the computation of the expected value to be gained by buying raffle tickets. Let us assume there is only one prize, the big jackpot, and this prize is worth $1,000,000. Let us also assume that there are 10,000,000 tickets sold. The probability of having the winning ticket is 1/10,000,000 and the expected value is $1,000,000/10,000,000 = $0.10. Clearly, such a raffle would not help saving money for one's retirement if the ticket cost more than 10 cents.

EXAMPLE 9.2
What is the expected value of the sum of two unbiased dice?

SOLUTION The possible outcomes are:

$$1+1, 1+2, 2+1, 2+2, 1+3, 3+1, \ldots$$

The probabilities are shown in the table below. The pdf is shown in Fig. 9.1.

sum	2	3	4	5	6	7	8	9	10	11	12
prob.	1/36	2/36	3/36	4/36	5/36	6/36	5/36	4/36	3/36	2/36	1/36

The expected value can be calculated as:

$$\mu = 2 \cdot \frac{1}{36} + 3 \cdot \frac{2}{36} + 4 \cdot \frac{3}{36} + \cdots + 11 \cdot \frac{2}{36} + 12 \cdot \frac{1}{36} = \frac{252}{36} = 7 \qquad (9.4)$$

Note that the value 7 corresponds to the peak in Fig. 9.1.

☐

The **cumulative distribution function** (or **cdf**) of a discrete random variable X is:

$$F(x) = P(X \leq x) \qquad (9.5)$$

for all real values x.

For a specific real value x_0 and a discrete random variable X, $P[X \leq x_0] = F(x_0)$ can be found by summing the density $f(x)$ over all possible values of X less than or equal to x_0. This can be written as:

$$F(x_0) = \sum_{x \leq x_0} f(x) \qquad (9.6)$$

EXAMPLE 9.3
What is the probability of rolling a value less or equal than 4 using a pair of unbiased dice?

SOLUTION From the table in example 9.2, the values of the pdf are as follows:

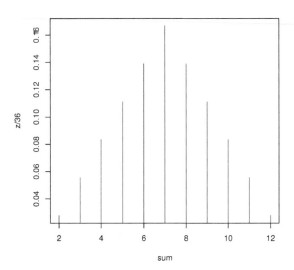

FIGURE 9.1: The pdf of a discrete random variable representing the sum of two fair dice. The horizontal axis represents the possible values of the variable; the vertical axis represents the probability.

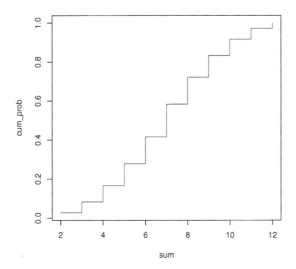

FIGURE 9.2: The cdf of a discrete random variable representing the sum of two fair dice.

sum	2	3	4	5	6	7	8	9
prob.	1/36	2/36	3/36	4/36	5/36	6/36	5/36	...
cum. prob.	0.027	0.083	0.166	0.277	0.416	0.583	0.722	...

The cdf is shown in Fig. 9.2.

□

9.1.2   The discrete uniform distribution

The discrete uniform distribution is a discrete probability distribution that describes a situation in which all values from a finite set of discrete values are equally probable. Examples could include an experiment involving throwing a fair dice like in Example 9.1.

For a discrete uniform distribution on the interval $[a, b]$, all integer values in this interval will be equiprobable with a probability equal to $\frac{1}{b-a+1}$. Fig. 9.3 shows both the pdf (probability mass function) and the cdf for a discrete uniform distribution on the interval $[4, 10]$.

The mean of a discrete uniform distribution taking values in the interval between a and b is:

$$E(X) = \frac{a+b}{2} \tag{9.7}$$

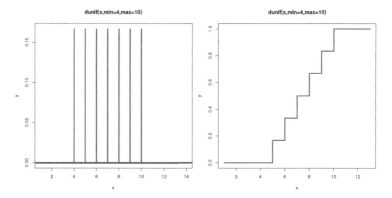

FIGURE 9.3. A discrete uniform distribution on the interval $[a,b] = [4,10]$. The probability of all integer values in this interval is equal to $1/n = 1/(b-a+1)$. The probability mass function (left panel) is zero elsewhere. The cdf (right panel) goes from zero to one in steps corresponding to each of the values in the given interval.

The median of a discrete uniform distribution is equal to its mean. The variance of such a discrete uniform distribution is:

$$Var(X) = \frac{n^2 - 1}{12} \tag{9.8}$$

where $n = b - a + 1$ is the number of integer values in the interval $[a,b]$.

9.1.3 Binomial distribution

The **binomial distribution** is the most important discrete variable distribution in life sciences. This makes it the preferred example distribution in the context of this book. However, the main issues to be learned from this section are related to the general concept of a distribution and the use of a distribution to answer particular statistical questions.

Any distribution will be appropriate for use in a particular statistical framework. Usually, this framework is defined by a set of assumptions. When addressing a specific statistical question, identifying the correct characteristics of the problem is essential in order to choose the appropriate distribution. Furthermore, any distribution will be characterized by a probability density function (pdf) and a cumulative probability density function (cdf). It is important to recall that given a real value x, **the pdf is the function that gives the probability that a variable following the given distribution (e.g., binomial) takes the value x.** Given the same real value x, **the cdf is the function that gives the probability that the variable following the given distribution takes a value less than or equal to x.** According

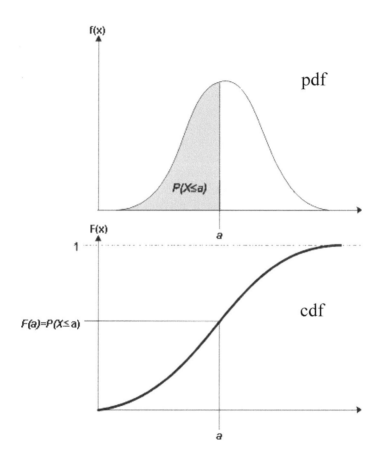

FIGURE 9.4: The relationship between the pdf and the cdf of a given distribution. The value of the cdf is at any point a, equal to the area under the curve of the pdf to the left of point a. This is because the pdf provides the probability that the variable takes precisely the value a, whereas the cdf gives the probability that the variable takes a value less than or equal to a.

to the definition, the cdf can be obtained by calculating the sum of all pdf values from $-\infty$ to the value x. In other words, the value of the cdf in any particular point x is equal to the area under the curve of the pdf to the left of the given point x. This general relationship between the pdf and cdf of a given distribution is illustrated in Fig. 9.4. A few other general properties can be observed in the same figure. Many pdfs will tend to zero towards the extremities of their domain (x axis). This is because in many distributions, really extreme values have a low probability of occurring. A notable example of a distribution for which this is not true is the uniform distribution (see Example 9.1) in which the pdf is constant for the whole range of the variable.

Most cdfs will start at zero and will increase to one. This is because when we are considering an extremely large value (a large value of a in Fig. 9.4) it is very likely to get a value less than or equal to it.

The binomial distribution takes is name from "bi," which means two and "nomen," which means name. Like any other distribution, the binomial distribution describes a phenomenon with certain characteristics. The characteristics of the phenomenon are also assumptions that one makes (or verifies) when one applies the binomial distribution to any specific problem. The results obtained by using the binomial distribution will be accurate only if the phenomenon studied satisfies these assumptions.

The binomial distribution describes a model with the following four assumptions:

1. A **fixed** number of trials are carried out.

2. The trials are **independent**, i.e. the outcome of one trial does not influence in any way the outcome of successive trials.

3. The outcome of each trial can be **only one of two mutually exclusive categories**. Usually, these outcomes are labeled "success" and "failure."

4. The **probability of success is fixed**. If we assume this probability to be p, then the probability of failure will be $1 - p$ since the two categories are mutually exclusive.

A **Bernoulli trial** is an experiment whose outcome is random and can be only one of two possible outcomes. This is essentially condition 3 above. A **Bernoulli process** is a sequence of independent identically distributed Bernoulli trials. The term **independent identically distributed**, or **i.i.d.**, is part of the exotic jargon that usually intimidates non-statisticians. "Independent Bernoulli trials" simply means that the trials in this process are independent of each other, which is condition 2 above. Identically distributed means that the distribution governing the random variables in each trial are all the same; in other words, it means that in every such trial the probability of obtaining a success is the same, which is condition 4 above. If we wanted to use fancy language, we can put together all 4 conditions above, by saying that *the binomial distribution will model the random variable that corresponds to the number of successes in a fixed number of trials of a Bernoulli process.*

Let us assume we carry out n trials and we have x successes. Clearly, x has to be less than or equal to n: $x \le n$. The pdf $f(x)$ is the probability of x successes in n trials. There are several ways in which x successes can occur in n trials. The number of such combinations is given by the combinatorial formula:

$$C_n^x = \binom{n}{x} = \frac{n!}{x! \cdot (n-x)!} \tag{9.9}$$

The notations C_n^x and $\binom{n}{x}$ are both read "n choose x." The notation $n!$ is read "n factorial" and is equal to the product of all integers up to n:

$$n! = 1 \cdot 2 \cdot 3 \cdots (n-1) \cdot n \tag{9.10}$$

We would like to calculate the probability of having x successes and $n - x$ failures. Since the events are independent, the probabilities multiply (see Eq. 8.51). We need to have x successes when each success has probability p and $n-x$ failures when each failure has probability $1 - p$. We obtain the probability of having such a combination by multiplying these values: $p^x \cdot (1-p)^{n-x}$. However, there are several such alternative combinations. Any one of them would be acceptable to us and they cannot occur together, so the probabilities will add up according to the addition rule in Eq. 8.46. The probability of having exactly x successes in n trials is therefore:

$$f(x) = \binom{n}{x} \cdot p^x \cdot (1-p)^{n-x} = \frac{n!}{x! \cdot (n-x)!} \cdot p^x \cdot (1-p)^{n-x} \tag{9.11}$$

This is **the probability distribution function (probability mass function) of the binomial distribution**. The exact shape of the distribution depends on the probability of success in an individual trial p. Fig. 9.5 illustrates the probability of having x successes in 15 trials if the probability of success in each individual trial is $p = 0.5$. Fig. 9.6 shows the cdf of the same distribution. Fig. 9.7 and Fig 9.8 show the pdf and cdf corresponding to having x successes in 15 trials if the probability of success in each individual trial is $p = 0.2$. Note that the mode (the peak) of the pdf corresponds to the expected value. In the first case, the mode is $15 \cdot 0.5 = 7.5$, whereas in the second case, the maximum probability is at $15 \cdot 0.2 = 3$.

In general, if X is a random variable that follows a binomial distribution with parameters n (number of trials) and p (probability of success), X will have the **expected value**:

$$E[X] = np \tag{9.12}$$

and the **variance**:

$$Var(X) = np(1-p) \tag{9.13}$$

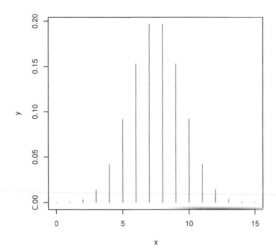

FIGURE 9.5: The pdf of a binomial distribution with the probability of success $p = 0.5$. The horizontal axis represents the possible values of the variable; the vertical axis represents the probability.

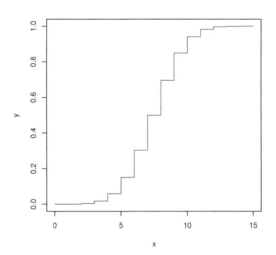

FIGURE 9.6: The cdf of a binomial distribution with the probability of success $p = 0.5$.

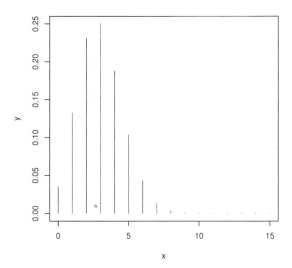

FIGURE 9.7: The pdf of a binomial distribution with the probability of success $p = 0.2$ and a size of 15 trials.

EXAMPLE 9.4

Let us consider the example of an urn containing 60 red balls and 40 black balls. Let us consider an experiment in which we extract a ball, we note the color, and then we replace the ball in the urn. What is the probability of extracting 5 red balls in 10 such trials?

SOLUTION   This type of experiment is called **sampling with replacement** because the ball is returned to the urn after every extraction. Because of this, whenever a ball is picked, there are exactly 60 red balls and 40 black balls to choose from. Hence, the probability of picking a red ball is always constant and equal to $p = \frac{60}{100} = 0.6$. Each extraction is not affected in any way by any of the results of previous trials, so each trial is independent. Finally, the number of trials is fixed $(n = 10)$, and in each trial there are only red and blue balls in the urn so the outcome is binary. Hence, a binomial distribution can appropriately model this situation and the probability of picking $x = 5$ red balls in $n = 10$ trials can be calculated as:

$$f(x) = \binom{10}{5} \cdot 0.6^5 \cdot (1 - 0.6)^{10-5} = 0.2006581 \qquad (9.14)$$

In R, this result could be obtained by simply calling the probability density

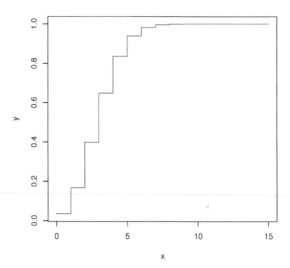

FIGURE 9.8: The cdf of a binomial distribution with the probability of success $p = 0.2$.

function of the binomial distribution, dbinom, with the appropriate parameters:

```
> dbinom(5,10,0.6)
> [1] 0.2006581
```

☐

EXAMPLE 9.5
Let us consider an Affymetrix array featuring 50,000 genes. Let us consider that 2,000 of these genes are known to be involved in a given biological process such as apoptosis. An experiment aims to compare the expression level of all genes represented on this array between a certain type of lung cancer tumors and healthy tissue collected from the same patients (each tumor sample is matched to a healthy tissue sample from the same patient). After the analysis of the microarray data, a set of 1,000 genes are found to be differentially expressed (DE) between the tumor and healthy tissue. Of these 1,000 genes, 50 are involved in apoptosis. Is apoptosis a process that can be significantly involved in this type of lung cancer?

SOLUTION   This situation is very similar to the experiment involved in the previous example. The question is whether the number of genes found to be involved in apoptosis in the set of 1,000 differentially expressed genes is significantly different from what is expected by chance if 1,000 genes were to be picked at random. We can think of the microarray as the urn containing 50,000 balls. The genes on the microarray can be seen as belonging to two disjunct sets: the genes involved in apoptosis, corresponding to the red balls in the previous example, and the genes not involved in the apoptosis, corresponding to the black balls above. Initially, the probability of picking one red ball is:

$$p = \frac{2000}{50000} = 0.04 \tag{9.15}$$

The process of selecting genes is based on their expression changes in the conditions compared. However, we are trying to calculate the probability of obtaining the same number of genes involved in apoptosis in a random set of genes of the same size. Let us assume that we pick a random gene and that this gene is involved in apoptosis, ie. it corresponds to a red ball. Strictly speaking, the binomial model should not be applied here because the probability of picking another apoptosis-related gene, or red ball, has changed since we cannot pick the same gene again, which means that the number of red balls in the urn has changed. Now we only have 1,999 apoptosis-related genes and 48,000 non-apoptosis-related genes, for a total of 49,999 genes. The probability of picking another apoptosis-related genes is:

$$p = \frac{1999}{49999} = 0.0399808 \tag{9.16}$$

However, the difference between the two probabilities is so very, very small (only 0.0000192) that we could still consider this probability as being constant. Under this approximation, we can use the binomial distribution again in order to calculate the probability of finding 50 apoptosis-related genes in a random set of 1,000 genes as:

$$f(x) = \binom{1000}{50} \cdot 0.04^{50} \cdot (1 - 0.04)^{1000-50} = 0.01724183 \tag{9.17}$$

In R, this can be calculated as:

```
> choose(1000,50)*0.04^50*(1-0.04)^950
```

[1] 0.01724183

or simply as the value from the pdf of the binomial distribution corresponding to extracting $x = 50$ red balls in a set of 1,000 extractions, with a constant probability of success of 0.04:

```
> dbinom(50,1000,0.04)
```

[1] 0.01724183

However, this probability corresponds to having *exactly* 50 apoptosis-related genes in a set of 1,000 random genes. In fact, we are interested in the probability of having 50 *or more* such genes in this random set. As it will be explained in Chapter 11, this corresponds to a one-tailed testing in the right tail of the distribution. This probability can be calculated as 1 minus the probability of having 49 of fewer such genes:

$$P(1 - f(49)) = 1 - \sum_{k=1}^{49} \binom{1000}{k} \cdot 0.04^k \cdot (1 - 0.04)^{1000-k} \qquad (9.18)$$

In R, if we wanted to calculate everything explicitly, step by step, we could first define a small function to calculate a single binomial term:

```
> f<- function(n,k,p){
+ f = choose(n,k) * p^k * (1-p)^(n-k)
+ }
```

Then calculate the terms of the sum in Equation 9.18 in an array:

```
> terms = sapply(c(1:49),f,n=1000,p=0.04)
> terms
```

```
 [1] 7.780756e-17 1.619370e-15 2.244627e-14 2.331138e-13
 [5] 1.934845e-12 1.336924e-11 7.910133e-11 4.091022e-10
 [9] 1.878840e-09 7.758043e-09 2.909266e-08 9.990500e-08
[13] 3.163658e-07 9.293246e-07 2.545317e-06 6.529003e-06
[17] 1.574642e-05 3.583039e-05 7.716107e-05 1.576979e-04
[21] 3.066349e-04 5.685521e-04 1.007326e-03 1.708607e-03
[25] 2.779334e-03 4.342709e-03 6.527467e-03 9.451229e-03
[29] 1.319913e-02 1.780049e-02 2.320763e-02 2.928150e-02
[33] 3.578851e-02 4.241113e-02 4.877280e-02 5.447425e-02
[37] 5.913646e-02 6.244344e-02 6.417798e-02 6.424483e-02
[41] 6.267788e-02 5.963104e-02 5.535517e-02 5.016562e-02
[45] 4.440587e-02 3.841268e-02 3.248732e-02 2.687536e-02
[49] 2.175625e-02
```

And then calculate the required probability as:

```
> 1-sum(terms)
```

```
[1] 0.06630587
```

Alternatively, this probabilities can be calculated using the built-in R functions. The *f* function above is nothing but the value of the *k*-th term in a binomial expansion, which is provided by the pdf of the binomial distribution, dbinom:

```
> terms = dbinom(c(1:49),size=1000,prob=0.04)
> terms
```

```
 [1] 7.780756e-17 1.619370e-15 2.244627e-14 2.331138e-13
 [5] 1.934845e-12 1.336924e-11 7.910133e-11 4.091022e-10
 [9] 1.878840e-09 7.758043e-09 2.909266e-08 9.990500e-08
[13] 3.163658e-07 9.293246e-07 2.545317e-06 6.529003e-06
[17] 1.574642e-05 3.583039e-05 7.716107e-05 1.576979e-04
[21] 3.066349e-04 5.685521e-04 1.007326e-03 1.708607e-03
[25] 2.779334e-03 4.342709e-03 6.527467e-03 9.451229e-03
[29] 1.319913e-02 1.780049e-02 2.320763e-02 2.928150e-02
[33] 3.578851e-02 4.241113e-02 4.877280e-02 5.447425e-02
[37] 5.913646e-02 6.244344e-02 6.417798e-02 6.424483e-02
[41] 6.267788e-02 5.963104e-02 5.535517e-02 5.016562e-02
[45] 4.440587e-02 3.841268e-02 3.248732e-02 2.687536e-02
[49] 2.175625e-02
```

```
> 1-sum(terms)
```

```
[1] 0.06630587
```

```
>
```

Furthermore, the required probability can be directly provided by the cdf of the binomial distribution, `pbinom` by specifying that we are looking for the value corresponding to the right tail:

```
> pbinom(49,1000,0.04,lower.tail=FALSE)
```

```
[1] 0.06630587
```

□

9.1.4 Poisson distribution

The **Poisson distribution** is a discrete probability distribution related to the binomial distribution. The Poisson distribution characterizes the probability that a random event[1] occurs in a given interval of time or space. The Poisson is also known as the distribution of rare events. The Poisson distribution describes a phenomenon with the following assumptions:

1. Events occur one at a time; in consequence, two or more events cannot occur exactly at the time moment in time (or in the same place).

---

[1] In this context, the term event is used to denote an occurrence of a given phenomenon such as a birth, the emission of a particle, etc. This usage of the term event is different from the one introduced earlier in this chapter, when the event was defined as a subset of a sample space.

2. The occurrence of an event in any period of time (or region of space) is independent of the occurrence of the event in any other (non-overlapping) period of time (or region of space).

3. The expected number of events during any one period of time (or region of space) is constant. This expected number of event occurrences in the given interval or region is a parameter of the Poisson distribution and is denoted by λ.

A process that satisfies conditions 1 and 2 above is also known as a **Poisson process**. If condition 3 is also true, the process is a **homogeneous** Poisson process.

The classical example of a phenomenon that is well modeled by a Poisson distribution is the decay of a radioactive material. In this example, the event that is observed is the emission of a decay particle. This event satisfies all assumptions of the Poisson distribution: i) only one particle is emitted at any one time, ii) the expected number of particles emitted in each unit of time is constant, and iii) the emission of a particle is independent of any similar event in the past or future. Other examples of phenomena modeled well by a Poisson distribution can include: the number of droplets of rain that fall in a given area, the number of mutations in a given area of a chromosome, etc. The pdf of a Poisson distribution with $\lambda = 2$ is shown in Fig. 9.9. The cdf of the same distribution is shown in Fig. 9.10.

A couple of other remarks can be useful here. The first one regards the relationship between the Poisson distribution and the binomial distribution. In a Poisson process, the time (or space) are continuous. If the time is considered discrete, the appropriate model is the Bernoulli process together with the associated binomial distribution. The relationship between the Poisson and binomial distributions can be illustrated by the following example. Let us consider an experiment in which we flip a coin at a constant average rate of μ flips/minute. If we do not care about *when* the flips happen, but only care about the *number* of heads in a given number of flips, we can model this with a binomial distribution, as discussed in Section 9.1.3. However, if we do care about when the flips happen, and we would rather focus on how many heads occurred in a given period of time instead of the number of heads in a fixed number of flips, then the appropriate model is the Poisson distribution.

The second remark is that the parameter of the Poisson distribution has to be chosen in a meaningful way. In the example above, if the coin is unbiased, the heads are expected to occur at a rate of $\mu/2$ heads/minute. If we are interested in the number of flips in a one minute interval, the appropriate value of the parameter would be $\lambda = \mu/2$. If however, we were interested in the number of flips in a one-hour interval, we would use as a model a Poisson distribution with $\lambda = 30\mu$.

The probability mass function (or pdf) of the Poisson distribution with parameter λ is:

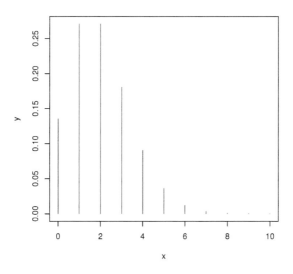

FIGURE 9.9: The pdf of a Poisson distribution with $\lambda = 2$.

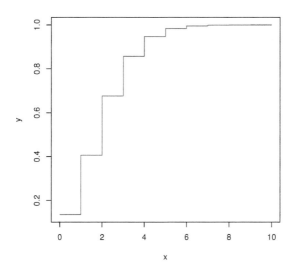

FIGURE 9.10: The cdf of a Poisson distribution with $\lambda = 2$.

	DE	not DE	total
apoptosis	k	m-k	m
not apoptosis	n-k	N-m-(n-k)	N-m
total	n	N-n	N

FIGURE 9.11: A contingency table corresponding to two categorical variables. In this case, one represents the association between a given gene and a specific biological process (apoptosis), while the other variable denotes whether the gene has been selected as differentially expressed (DE) in a given condition.

$$f(x) = \frac{e^{-\lambda} \cdot \lambda^k}{k!} \tag{9.19}$$

The cdf of a Poisson distribution with parameter λ is:

$$F(x) = e^{-\lambda} \sum_i^k \frac{\lambda^i}{i!} \tag{9.20}$$

The mean of the Poisson distribution with parameter λ is:

$$E(X) = \lambda \tag{9.21}$$

The median of the Poisson distribution with parameter λ is also equal to λ.

9.1.5 The hypergeometric distribution

Let us revisit the GO profiling example in which a given GO category is tested for enrichment or depletion in a set of differentially expressed genes. As discussed above, this is similar to an urn experiment in which the ball picked in any one extraction is *not* placed back into the urn after each extraction. This is **sampling without replacement**. The essential difference here is that each ball extracted changes the number of balls of that color remaining in the urn, and hence the probability of picking up another ball of that color.

A convenient way to represent the data associated with this experiment is that of a contingency table as shown in Fig. 9.11. The **hypergeometric distribution** describes the probability that exactly k genes are associated with the given GO category in a sample of n differentially expressed (DE) genes. Generically, such a table can be constructed every time there are two categorical variables with two possible value each. Possible examples could include a variable "outcome" with the two possible values "success" and "failure," a variable "selected" with two possible values "true" and "false," etc.

In general, let us assume there is a set of N objects of which m have a certain property. The hypergeometric distribution describes the probability that exactly k objects have the given property in a sample of n such objects drawn from the given set (see Fig. 9.12).

If a random variable X follows the hypergeometric distribution with parameters N, m, and n, then the probability of having exactly k "successes" is given by:

$$f(k;N,m,n) = \frac{\binom{m}{k} \cdot \binom{N-m}{n-k}}{\binom{N}{n}} \tag{9.22}$$

where the $\binom{N}{n}$ is the usual binomial coefficient:

$$\binom{N}{n} = C_N^n = \frac{N \cdot (N-1) \cdots \cdots (N-n+1)}{n!} \tag{9.23}$$

In order to understand this, let us consider the situation described in Fig. 9.12. There are N objects of which m have a given property A. We are interested to determine the probability of having exactly k objects with property A in a sample of n objects with property B. We can calculate this probability as the ratio of the number of favorable cases over the total number of cases. The total number of different sets of n objects picked up out of the total N objects is, by definition, $\binom{N}{n}$. This will be our denominator. The number of different ways in which we can pick k objects from the m objects having property A is $\binom{m}{k}$. Once we picked these k objects, we need to pick other $n-k$ objects with property B but without property A (see Fig. 9.13). The total number of objects not having property A is $N-m$. Given this, there are $\binom{N-m}{n-k}$ different ways to construct such a set. Therefore, there will be:

$$\binom{m}{K} \cdot \binom{N-m}{n-k} \tag{9.24}$$

favorable cases out of a total of:

$$\binom{N}{n} \tag{9.25}$$

which yield the probability show in Eq. 9.22.

The cdf of the hypergeometric distribution will be:

$$F(x|N,m,n) = \sum_{k=1}^{x} \frac{\binom{m}{k} \cdot \binom{N-m}{n-k}}{\binom{N}{n}} \tag{9.26}$$

The mean of a discrete random variable X that follows a hypergeometric distribution with parameters N, m, and n will be:

$$E(x) = \frac{n \cdot m}{N} \tag{9.27}$$

Note that the contingency table in Fig. 9.12 is symmetrical, i.e. A and B

	B	not B	total
A	k		m
not A			
total	n		N

FIGURE 9.12: The scenario contemplated by the hypergeometric distribution involves two binary variables A and B and a number of N objects. Out of these N objects, m have the property A. The questions is what is the probability to have exactly k objects with property A in a sample of n objects having the property B.

	B	not B	total
A	k		m
not A	n-k		N-m
total	n		N

FIGURE 9.13: There are $\binom{m}{k}$ ways to pick k objects with property B from the set of m objects with property A. For any such choice, there will be $n-k$ objects with property B but not A. The number of different combinations of $n-k$ objects with B but not A picked from $N-m$ objects is $\binom{N-m}{n-k}$. Overall, there are $\binom{m}{k} \cdot \binom{N-m}{n-k}$ favorable cases (with exactly k successes) out of a total number of $\binom{N}{n}$ which yields the probability shown in Eq. 9.22.

	B	not B	total
A	k	m-k	m
not A			
total	n	N-n	N

FIGURE 9.14: There are $\binom{n}{k}$ ways to pick k objects with property B from the set of n objects with property B. For any such choice, there will be $m - k$ objects with property A but not B. The number of different combinations of $m - k$ objects with A but not B picked from $N - n$ objects is $\binom{N-n}{m-k}$. Overall, there are $\binom{n}{k} \cdot \binom{N-n}{m-k}$ favorable cases (with exactly k successes) out of a total number of $\binom{N}{m}$ which yields the probability shown in Eq. 9.28.

can be swapped. In consequence, the hypergeometric probability can also calculated reasoning on the other direction of the table, as illustrated in Fig. 9.14. There are $\binom{n}{k}$ ways to pick k objects with property B from the set of n objects with property B. For any such choice, there will be $m - k$ objects with property A but not B. The number of different combinations of $m - k$ objects with A but not B picked from $N - n$ objects is $\binom{N-n}{m-k}$. Overall, there are $\binom{n}{k} \cdot \binom{N-n}{m-k}$ favorable cases (with exactly k successes) out of a total number of $\binom{N}{m}$, which yields the probability:

$$f(k;N,n,m) = \frac{\binom{n}{k} \cdot \binom{N-n}{m-k}}{\binom{N}{m}} \tag{9.28}$$

9.1.6 Continuous random variables

Continuous random variables are random variables that can take any values in a given interval. The pdf $f(x)$ for a continuous random variable X is a function with the following properties:

1. f is defined for all real numbers.

2. $f(x) \geq 0$ for any value of x.

3. The region between the graph of f and the x axis has an area of 1 (the sum of probabilities of all possibilities is 1). This is expressed mathematically as:

$$\int_{all\ x} f(x) = 1 \tag{9.29}$$

4. For any real a and b, the probability of X being between a and b, $P(a \leq X \leq b)$, is the area between the graph of f, the x axis and vertical lines drawn through a and b (see Fig. 9.15). This is expressed mathematically

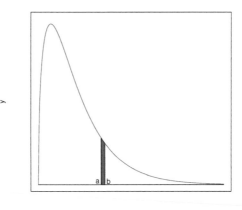

FIGURE 9.15: For any real a and b, the probability of X being between a and b, $P(a \le X \le b)$, is the area between the graph of f, the x axis and vertical lines drawn through a and b.

as:

$$P(a \le X \le b) = \int_a^b f(x)\,dx \qquad (9.30)$$

Note that if we tried to apply the definition of the discrete pdf to a continuous variable (see Eq. 9.1), the result would be zero:

$$f(x) = P(X = x) = 0 \qquad (9.31)$$

since the probability of a continuous variable taking any given precise numerical value is zero.

The cdf for a continuous random variable X is (same as for a discrete random variable) a function F whose value at point x is the probability the variable X takes a value less than or equal to x:

$$F(x) = P(X \le x) \qquad (9.32)$$

The relationship between the pdf and the cdf of a continuous random variable remains the same as the relationship between the pdf and the cdf of a discrete random variable illustrated in Fig. 9.4.

9.1.7 The continuous uniform distribution

The continuous uniform distribution on an interval $[a,b]$ captures a situation in which any (real) value from the given interval is equally probable. The pdf

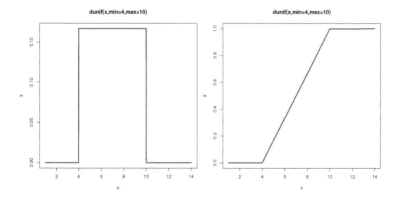

FIGURE 9.16: A continuous uniform distribution on the interval $[a,b] = [4,10]$. The probability of any value in this interval is equal to $1/n = 1/(b-a)$. The pdf is shown in the left panel. The cdf, shown in the right panel, goes from zero to one in a linear fashion. A comparison with Fig. 9.3 emphasizes the differences between the discrete and the continuous uniform distributions.

of the continuous uniform distribution is:

$$f(x) = \begin{cases} 0 & , & x < a \\ \frac{1}{b-a} & , & x \in [a,b] \\ 0 & , & x > b \end{cases} \tag{9.33}$$

The cdf of the continuous uniform distribution is:

$$F(x) = \begin{cases} 0 & , & x < a \\ \frac{x-a}{b-a} & , & x \in [a,b] \\ 1 & , & x > b \end{cases} \tag{9.34}$$

Both pdf and cdf of the continuous uniform distribution are shown in Fig. 9.16. The expected value (mean) of a random variable X that follows a continuous uniform distribution is:

$$E(X) = \frac{a+b}{2} \tag{9.35}$$

9.1.8 The normal distribution

The most important continuous distribution is the **normal distribution**. The normal distribution is important because many natural phenomena follow it. In particular, most biological variables, such as the variation of anatomical and physiological parameters between different individuals of the same species, follow this distribution. It is also reasonable to believe that differences in the

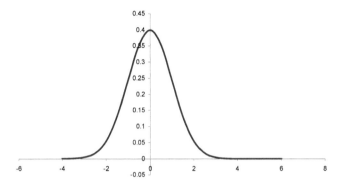

FIGURE 9.17: The probability density function of a normal distribution with a mean of zero and a standard deviation of 1 (standard normal distribution).

gene expression levels between various individuals in the same conditions also follow a normal distribution. Another common assumption is that the noise in a microarray experiment is normally distributed although this may not always be true.

The normal distribution is also called a Gaussian distribution and has a characteristic bell shape shown in Fig. 9.17. Its corresponding cdf is shown in Fig. 9.18. The normal distribution has two parameters: the mean and the standard deviation. The mean is also called the location of the curve since changing this value will move the whole curve to a different location on the x axis. The standard deviation determines the exact shape of the curve. A smaller standard deviation means the values are less dispersed. In consequence, the graph will be narrower and taller, because the total area under the curve has to remain the same (see the definition of the pdf). A larger standard deviation will determine a shorter but wider curve. This is illustrated in Fig. 9.19.

There is also an analytical expression that give the pdf of the normal distribution as a function of its mean and standard deviation. The probability density function for a normal random variable with mean μ and standard deviation σ has the form:

$$f(x) = \frac{1}{\sigma\sqrt{2\pi}} \cdot e^{-\frac{(x-\mu)^2}{2\sigma^2}} \tag{9.36}$$

The normal distribution has several important characteristics as follows:

1. The distribution is symmetrical around the mean.

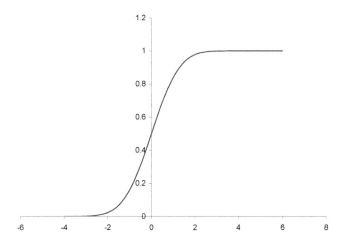

FIGURE 9.18: The cumulative distribution function of a normal distribution with a mean of zero and a standard deviation of 1 (standard normal distribution).

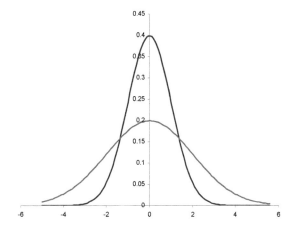

FIGURE 9.19: The probability density function of two normal distributions. The taller one has a mean of zero and a standard deviation of 1. The shorter one has a mean of zero and a standard deviation of 2.

2. Approximatively 68% of the values are within 2 standard deviations from the mean (i.e., less than 1 standard deviation either to the left or to the right of the mean: $x - \mu < \pm\sigma$).

3. Approximatively 95% of the values are within 4 standard deviations from the mean ($x - \mu < \pm 2\sigma$).

4. Approximatively 99% of the values are within 6 standard deviations from the mean ($x - \mu < \pm 3\sigma$).

5. The inflexion points of the curve (where the curve changes from bending up to bending down) occur at $\mu \pm \sigma$.

As an exercise, let us calculate the mean and standard deviation of a new variable, Z, that is obtained from a normally distributed variable X by subtracting the mean and dividing by the standard deviation:

$$Z = \frac{X - \mu}{\sigma} \tag{9.37}$$

The mean of a variable X, μ_X, is actually the expected value of that variable, $E[X]$. Also, the expected value of a constant is equal to the constant $E(c) = c$. Using this and the properties discussed in 8.5.1, the expected value of the new variable Z is:

$$\mu_Z = E\left[\frac{X - \mu_X}{\sigma_X}\right] = \frac{1}{\sigma_X}E\left[X - \mu_X\right] = \frac{1}{\sigma_X}\left(E\left[X\right] - E\left[\mu_X\right]\right) =$$

$$= \frac{1}{\sigma_X}(\mu_X - \mu_X) = 0 \tag{9.38}$$

Therefore, the mean of the new variable Z will be zero. Note that this will happen independently of the mean of the initial variable X.

The standard deviation of the new variable will be:

$$\sigma_Z^2 = E\left[(Z - \mu_Z)^2\right] = E[Z^2] = E\left[\left(\frac{X - \mu_X}{\sigma_X}\right)^2\right] =$$

$$= \frac{1}{\sigma_X^2}E\left[(X - \mu_X)^2\right] = \frac{1}{\sigma_X^2}\sigma_X^2 = 1 \tag{9.39}$$

Therefore, any normal variable can be mapped into another variable distributed with mean zero and standard deviation of one. This distribution is called **the standard normal distribution** and its probability density function can be expressed as:

$$f(x) = \frac{1}{\sqrt{2\pi}} \cdot e^{-\frac{x^2}{2}} \tag{9.40}$$

This distribution is shown in Fig. 9.17.

9.1.9   Using a distribution

Distributions can be used directly to answer a number of simple questions. However, in order to use a given distribution to answer such questions, one has to make sure that the distribution is appropriate by checking the validity of its assumptions. Typical questions that can be asked directly are:

1. What is the probability that the variable has a value higher than a certain threshold t?

2. What is the probability that the variable has values between two thresholds t_1 and t_2?

3. What is the threshold that corresponds to a certain probability?

Tables summarizing various distributions or computer programs able to calculate pdfs and cdfs are used in order to answer such questions. Tables are usually organized by the values of the random variable. For instance, a table for the normal distribution will have rows for different values of the variable Z (e.g., 0.0, 0.1, 0.2, etc.). The columns will correspond to the third digit of the same variable Z (e.g., the value situated at the intersection of row 1.1 and column 0.05 corresponds to the probability of $Z \leq 1.15$). Most general purpose spreadsheets and pocket calculators with statistical capabilities will also be able to provide the necessary values. In Excel, such values can be obtained quickly using the function "normdist(x, mean, standard deviation, TRUE)." If Excel is available, it is not necessary to standardize the variable since the mean and standard deviation can be specified explicitly. The last parameter chooses between pdf ("FALSE") and cdf ("TRUE").

In R, the pdf-s and cdf-s of many useful distributions are readily available with a simple function call. Their use in several contexts is discussed in details in Chapter 10.

EXAMPLE 9.6
The distribution of mean intensities of cDNA spots corresponding to genes that are expressed can be assumed to be normal [276]. Let us assume that the mean of this distribution is 1000 and the standard deviation is 150. What is the probability that an expressed gene has a spot with a mean intensity of less than 850?

SOLUTION   *Firstly, let us note that the problem statement contains two different means: the mean intensity of a spot and the mean of the spot distribution. The mean intensity of the spot is not particularly relevant for the question considered. It merely informs us that given a particular spot, the various pixel intensities of a spot have been combined in a single number by taking their mean. The mean of the pixel intensities is the value chosen to represent the expression level of a gene in this particular example. The value that we* need to focus on is the mean of the spot intensities distribution. If we are to use a table, we need to transform this problem involving a particular normal distribution ($\mu = 1000, \sigma = 150$) into an equivalent problem using the standard normal distribution ($\mu = 0, \sigma = 1$). In order to do this, we apply the transformation in Eq. 9.38 and Eq. 9.39 by subtracting the mean and dividing by the standard deviation:

$$Z = \frac{X - \mu}{\sigma} = \frac{850 - 1000}{150} = -1 \qquad (9.41)$$

The Z transformation above maps problems such as this from any normal distribution to the one standard normal distribution. The normal distribution had its horizontal axis graded in units corresponding to the problem, in our case, pixel intensities. If we were to represent the Z variable, the distribution (see Fig. 9.20) would have the axis graded in standard deviations. Equation 9.41 emphasized that our problem corresponds to determining the probability that a variable following a normal distribution has a value less than 1 standard deviation below its mean:

$$P(X < 850) = P(Z < -1.0) \qquad (9.42)$$

This corresponds to the shaded area in Fig. 9.20. From a table or from a pocket calculator, we can obtain the exact value of this probability: 0.1587. In other words, approximatively 15% of the spots corresponding to expressed genes will have a mean intensity below 850 or below one standard deviation from their mean. □

EXAMPLE 9.7
The distribution of mean intensities of cDNA spots corresponding to genes that are not expressed can be assumed to be normal [276]. Let us assume that the mean of this distribution is 400 and the standard deviation is 150. What is the probability that an unexpressed gene has a spot with a mean intensity of more than 700?

SOLUTION We transform the variable X into a standard normal variable by subtracting the mean and dividing by the standard deviation:

$$Z = \frac{X - \mu}{\sigma} = \frac{700 - 400}{150} = 2 \qquad (9.43)$$

From a table of the standard normal distribution or from a calculator, we obtain the value 0.9772. A mechanical application of the reasoning used in the example above might conclude that the probability of unexpressed genes having spots with an average intensity higher than 700 is 97.72%. Clearly, this cannot be true since it is impossible that 97% of normally distributed values be higher than their own mean of 400.

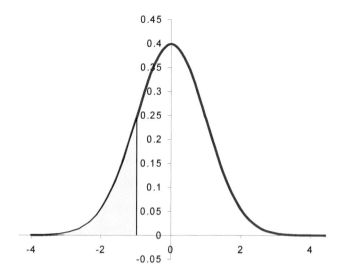

FIGURE 9.20: The shaded area under the graph of the pdf and to the left of the -1 value corresponds to the probability of the X variable having a value less than one standard deviation below its mean.

Fig. 9.21 illustrates the problem: we are interested in the probability of our variable being higher than the given threshold (in the shaded area on the figure), while the cdf provides by definition the probability of the variable Z being less than or equal to the threshold $z = 2$, which corresponds to the unshaded area in the figure. However, the figure also suggests the solution to this problem. Since we know that the area under the curve is equal to 1 (from the definition of a pdf), we can obtain the value of interest by subtracting the value provided by the cdf (unshaded) from 1:

$$P(Z > 2) = 1 - P(Z \le 2) = 1 - 0.9772 = 0.0228 \qquad (9.44)$$

In conclusion, the probability that an unexpressed gene has a high intensity spot (higher than their mean plus 2 standard deviations) is about 2.3%. □

The two examples above showed that it is possible for expressed genes to have lower than usual intensity spots as well as for unexpressed genes to have spots with an intensity higher than usual. Let us now consider an array containing both expressed and unexpressed genes and let us also assume that the parameters of each distribution are not changed by the presence of the other. In other words, we assume the two distributions are independent. This situation is illustrated in Fig. 9.22. As before, the unexpressed genes are normally distributed with a mean of 400 and a standard deviation of 150. The expressed genes are normally distributed with a mean of 1,000 and a standard

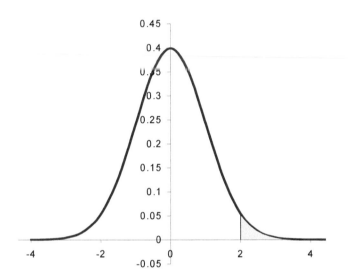

FIGURE 9.21: The shaded area under the graph of the pdf corresponds to the probability that we are interested in $(P(Z > 2))$. However, the cdf of the function provided a value corresponding to the unshaded area to the left of 2 $(F(x) = P(Z \leq 2))$. The value of interest can be calculated as $1 - cdf = 1 - P(Z \leq 2)$.

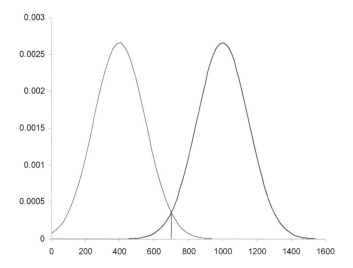

FIGURE 9.22: An array having both expressed and unexpressed genes. The unexpressed genes are normally distributed with a mean of 400 and a standard deviation of 150. The expressed genes are normally distributed with a mean of 1,000 and a standard deviation of 150. Note that there are expressed genes with values less than 700 and unexpressed genes with a value more than 700.

deviation of 150. Note that there are expressed genes with values less than 700 and unexpressed genes with a value less than 700. This shows that in most cases, distinguishing between expressed and unexpressed genes is not as simple as it might seem. The following chapters will discuss in detail how this can be done. However, a conclusion that should already be apparent is that we will have to make some compromises and be prepared to accept the fact that we might make some mistakes. This, in itself, is not very troublesome as long as we know exactly how likely we are to make a mistake when we draw a given conclusion.

9.2 Central limit theorem

Let us revisit the example used to illustrate the concept of mode. We considered the following set of measurements for a given population: 55.20, 18.06, 28.16, 44.14, 61.61, 4.88, 180.29, 399.11, 97.47, 56.89, 271.95, 365.29, 807.80, 9.98, 82.73. We calculated the population mean as 165.570, and we considered two samples from this population. We observed that two different samples can have means very different from each other and also very different from the true population mean. An interesting question is: what would happen if we considered, not only two samples, but *all possible samples of the same size?* In other words, what happens if we consider all possible ways of picking n measurements out of the population?

The answer to this question is one of the most fascinating facts in statistics. If the answer to this question were different from what it is, statistics as a science would probably not exist as we know it. It turns out that if we calculate the mean of each sample, those mean values tend to be distributed as a normal distribution, *independently on the original distribution*. Furthermore, the mean of this new distribution of the means *is exactly the mean of the original population* and the variance of the new distribution *is reduced by a factor equal to the sample size n.*

There are two important observations to be made. First, the distribution of the original values X can have any shape (e.g., the skewed binomial distribution corresponding to a success rate of 0.2 in Fig. 9.7). Second, the result is true for *the means $\overline{X}$ of the subsets picked from the original X* distribution. For each subset picked, the mean $\overline{X}_i$ is calculated; this value is one sample in the new distribution.

This result is known as **the central limit theorem:**

THEOREM 9.1
*When sampling from a population with mean μ_X and variance σ_X^2, the distribution of the sample mean (or the **sampling distribution**) will have the following properties:*

1. *The distribution of $\overline{X}$ will be approximately normal. The larger the sample is, the more will the sampling distribution resemble the normal distribution.*

2. *The mean $\mu_{\overline{X}}$ of the distribution of $\overline{X}$ will be equal to μ_X, the mean of the population from which the samples were drawn.*

3. *The variance $\sigma_{\overline{X}}^2$ of the distribution of $\overline{X}$ will be equal to $\frac{\sigma_X^2}{n}$, the variance of the original population of X's divided by the sample size. The quantity $\sigma_{\overline{X}}$ is called **the standard error of the mean**.*

9.3   Are replicates useful?

There is a joke about an American tourist visiting a remote town in England. Strolling through the village, the tourist gets to the main town square where he notices two large towers, each of them with a large clock. He also notices that the two clocks are not showing the same time. Puzzled, he stops a local person and asks the time. "Can't you see the clocks?" the English man replies. "Yes, but they show different times!" says the confused tourist. "Which one is the correct one?" "Neither of them," replies the local. "If we could build a clock that could show us the exact time, we would only have that one."

Tongue in cheek, we can now use the powerful results of the central limit theorem to investigate this phenomenon through an example.

EXAMPLE 9.8

The distribution of mean intensities of cDNA spots corresponding to genes that are not expressed can be assumed to be normal [276]. Let us assume that the mean of this distribution is 400 and the standard deviation is 150. We are using an array that has each gene spotted in quadruplicates. We calculate the expression value of a gene as the mean of its replicates.

1. *What is the probability that a spot corresponding to an unexpressed gene has a value higher than 700?*

2. *What is the probability that an unexpressed gene has an expression value of 700 or higher?*

SOLUTION   *In order to calculate the probability for one individual spot, we transform the variable X into a standard normal variable by subtracting the mean and dividing by the standard deviation as we did in example 9.7:*

$$Z = \frac{X - \mu}{\sigma} = \frac{700 - 400}{150} = 2 \qquad (9.45)$$

The probability of this happening is:

$$P(Z > 2) = 1 - P(Z \leq 2) = 1 - 0.9772 = 0.0228 \qquad (9.46)$$

This value of 0.0228 or 2.28% may appear to be relatively small and, in consequence, not a cause for concern. However, an array typically contains thousands of spots. Two percent means 20 in 1,000 and 200 in 10,000. This is a good cause for concern. In a typical array, a researcher might look at 10,000 genes with the hope of finding about 50 or 100 genes that are differentially regulated in the condition under study. Two hundred unexpressed genes that have spots with high intensities due to various random causes will mix the 50 or 100 real positives with an additional 200 false positives. The researcher will have no way to distinguish between spots that have high intensity due to a genuinely higher level of expression and spots that have high intensity due to other random factors.

In order to answer the second question, we need to apply the central limit theorem since the second question is referring to the expression level of the gene which is obtained as a mean of four measurements. The standard error of the mean is:

$$\sigma_{\overline{X}} = \frac{\sigma_X}{\sqrt{n}} = \frac{150}{\sqrt{4}} = 75 \tag{9.47}$$

This means that the variance in the expression levels will be four times smaller than the variance of the spot intensities. With the standard error, we can now calculate the Z variable, mapping the problem to a standard normal distribution:

$$Z = \frac{\overline{X} - \mu_{\overline{X}}}{\sigma_{\overline{X}}} = \frac{\overline{X} - \mu_{\overline{X}}}{\frac{\sigma_X}{\sqrt{n}}} = \frac{700 - 400}{\frac{150}{\sqrt{4}}} = \frac{700 - 400}{150} \cdot 2 = 4 \tag{9.48}$$

This is equivalent to asking the probability of a normally distributed variable to take a value 4 standard deviations away from its mean. The probability of this happening is:

$$P(Z > 4) = 1 - F(4) = 1 - 0.999968 = 0.000032 \tag{9.49}$$

This probability is only 0.003%. An array containing 10,000 spots will only contain 2,500 genes (there is no such thing as a free lunch!!), but this array will give 0.000032 · 2500 = 0.079 false positives, i.e. no false positives.

In the given hypotheses, the choices are as follows. On the one hand, an array with 10,000 genes, each spotted once, interrogates 10,000 genes at the same time but probably many of the genes selected as expressed will not be so. On the other hand, an array spotting each gene in quadruplicates interrogates only 2,500 genes at a time, but if the mean of the replicate spots of a gene indicate a high expression value, the genes will probably be truly highly expressed.

□

9.4 Summary

This chapter introduced a few necessary statistical terms such as: population, sample, parameters, random variables, etc. Several measures of central tendency (mean, median, mode, percentile) and variability (range, variance, standard deviation) were presented. The discussion emphasized various interesting properties of these quantities, the relationship between them and their utility in the context of microarray data. The chapter also introduced the notions of covariance and correlation, discussed their common usage as well as common mistakes associated to their usage.

The next section of the chapter introduced the notion of probabilities, conditional probabilities, and operations with such. Bayes' theorem and a few related issues as well as a typical example of Bayesian reasoning were briefly reviewed. The chapter also discussed the notions of probability distributions, as well as a few specific distributions, including the binomial and normal distributions.

Finally, the chapter discussed the central limit theorem and used it to explain through some examples why repeated measurements are more reliable than individual measurements.

9.5 Solved problems

1. We decide to declare a microarray spot as having high confidence if it is either perfectly circular or perfectly uniform (the standard deviation of the signal intensities is less than a chosen threshold) or both. There are 10,000 spots of which 8,500 are circular and 7,000 are uniform. 5,000 spots are both circular and uniform. What is the probability that a randomly chosen spot is good (has high confidence)?

    SOLUTION $P(\text{circular}) = 8500/10000 = 0.85$

    $P(\text{uniform}) = 7000/10000 = 0.70$

    $P(\text{circular and uniform}) = 5000/10000 = 0.50$

    $P(\text{high confidence}) = P(\text{circular}) + P(\text{uniform}) - P(\text{circular and uniform})$

    $P(\text{high confidence}) = 0.85 + 0.70 - 0.5 = 1.05$

    However, a probability cannot be larger than 1. There must be something wrong. Let us calculate the number of spots that are circular and not uniform: $8500 - 5000 = 3500$.

    We can also calculate the number of spots that are uniform but not circular: $7000 - 5000 = 2000$.

    We can now sum up the circular but not uniform spots (3500) with the uniform but not circular spots (2000) and with the spots that are both circular and uniform (5000): $3500 + 2000 + 5000 = 10500$. This number is greater than the number of spots on the array, which clearly indicates that at least one of the given measurements was incorrect.

    This example emphasized the fact that in microarray applications, it is a good idea to put the results of the data analysis in the context of the problem and use them to verify the consistency of the data gathered. $\square$

2. Write code in R to calculate the median of the values of a gene from the

ge data frame using the definition rather than the build in R function median.

SOLUTION   Since this sample has an even number of values, the median value is not one of the values from the set of measurements but rather some in-between value. Let us calculate the median step by step, rather than rely on the built-in function:

```
> ogene1 = sort(gene1)
> ogene1
 [1] -1.269 -0.902 -0.894 -0.580 -0.394
 [6] -0.138 -0.133  0.193  0.903  1.966
> my_median = (ogene1[5]+ogene1[6])/2
> my_modian
 [1] -0.266
> median(gene1)
 [1] -0.266
>
```

The median is defined as the value in the middle of the ordered list of measurements so the code above first orders the values, storing them in an ordered vector, ogene1, then calculates the values halfway in-between the 5th and 6th elements in the ordered vector.   ⬚

9.6   Exercises

1. The distribution of mean intensities of cDNA spots corresponding to genes that are not expressed can be assumed to be normal [276]. Let us assume that the mean of this distribution is 400 and the standard deviation is 150. We are using an array that has each gene spotted in triplicates. We calculate the expression value of a gene as the mean of its replicates.

   (a) What is the probability that a spot corresponding to an unexpressed gene has a value higher than 700?

   (b) What is the probability that an unexpressed gene has an expression value of 700 or higher?

2. The distribution of mean intensities of cDNA spots corresponding to genes that are expressed can be assumed to be normal [276]. Let us assume that the mean of this distribution is 1,000 and the standard deviation is 100. We are using an array that has each gene spotted in

triplicates. We calculate the expression value of a gene as the mean of its replicates.

(a) What is the probability that a spot corresponding to an expressed gene has a value lower than 700?

(b) What is the probability that an expressed gene has an expression value of 700 or lower?

3. The distribution of mean intensities of cDNA spots corresponding to genes that are expressed can be assumed to be normal [276]. Consider an array on which there are both expressed and unexpressed genes. Let us assume that the mean of the distribution corresponding to the expressed genes is 1,000 and the standard deviation is 100; the mean of the distribution corresponding to the unexpressed genes is 400 and the standard deviation is 150 (usually there is higher variability in the low intensity spots). We are using an array that has each gene spotted in triplicates. We calculate the expression value of a gene as the mean of its replicates.

(a) What is the probability that a spot corresponding to an expressed gene has a value lower than 700?

(b) What is the probability that an expressed gene has an expression value of 700 or lower?

(c) Find the threshold T in the intensity range corresponding to the intersection point of the two distributions (the point for which the probability of having an expressed gene is equal to the probability of having an unexpressed gene).

Hint: this point is marked by a vertical bar in Fig. 9.22.

(d) What is the probability that an expressed gene has the mean of its replicates lower than T? (This would be a false negative.)

(e) What is the probability that an unexpressed gene has the mean of its replicates higher than T? (This would be a false positive.)

(f) What is the probability that an individual spot corresponding to an expressed gene has an intensity lower than T? (This would be a false negative on an array not using replicates.)

(g) What is the probability that an individual spot corresponding to an unexpressed gene has an intensity higher than T? (This would be a false positive on an array not using replicates.)

4. Write a more general R function that would correctly calculate the median for a vector with an arbitrary number of element (either even or odd) without using the built-in function `median`.

Chapter 10

Basic statistics in R

I hear, I forget
I see, I remember
I do, I understand

—Chinese proverb

10.1 Introduction

The goal of this chapter is to show how the concepts discussed in Chapter 8 can be used in practice in the R environment. More specifically, this chapter discusses the use of various descriptive statistics such as mean, median, mode, range, variance, standard deviation, etc., as well as how densities, *p*-values, quantiles, and pseudorandom numbers can be obtained from the built-in distributions in R. These capabilities are also used to illustrate some basic properties of certain distributions, as well as the central limit theorem.

10.2 Descriptive statistics in R

The main characteristics of a data set can be captured by a number of measures for central tendency, variability, etc. Each of these measures is a statistic that can be calculated for any given data set. Overall, since these statistics are used to describe the main features of a data set, they are called **descriptive statistics**.

10.2.1 Mean, median, range, variance, and standard deviation

These descriptive statistics can be divided into measures of central tendency, which include the **mean**, **median**, and **mode**; and measures of variability, which include the **range**, **variance**, and **standard deviation**. As a quick refresher, the sample mean is the arithmetical mean of the values: $\frac{\sum_{i=1}^{n} x_i}{n}$, and the median is the value that is positioned at an equal distance from the minimum and maximum values when the values are ordered. A distribution can also be characterized by the presence of one or several modes. The mode can be defined as the most frequent value in a data set. The presence of more than one mode in a given data set is usually taken as a sign that the values are coming from a mixture of unimodal underlying distributions.

The most frequently used descriptive statistics are mean, median, standard deviation, and variance. In R, these can be easily calculated with the built-in functions: `mean`, `median`, `sd`, and `var`, respectively.

In Chapter 6, we have constructed a data set containing the expression values of 5 genes measured across 10 samples. Let us calculate the descriptive statistics for the expression values of gene g1 in this data set.

```
> #setwd("c:/Users/Sorin/Documents/Books/Book revision")
> ge=read.table("./outputfile.txt")
> ge

      g1      g2     g3      g4      g5   group source
1   0.903   0.767 -1.262   0.146   0.971  cancer    DMC
2   0.193   0.515  1.388   0.490   0.207  cancer    KCI
3   1.966   0.354 -0.857  -0.460   0.619  cancer    KCI
4  -0.138   0.448 -1.352  -0.748  -0.899  cancer    DMC
5  -0.894   0.823 -0.523   0.276   1.490  cancer    DMC
6  -0.902   1.838  0.156  -0.942  -0.709 healthy    KCI
```

```
7  -0.133 -3.331  0.377  1.287  1.292 healthy   DMC
8  -0.580  0.300  0.618 -0.243 -1.193 healthy   DMC
9  -1.269  1.392 -0.965  1.395  0.083 healthy   KCI
10 -0.394 -1.165  0.220  0.941  0.707 healthy   DMC

> attach(ge)
> mean(g1)

[1] -0.1248

> median(g1)

[1] -0.266

> sd(g1)

[1] 0.9611124

> var(g1)

[1] 0.9237371
```

The first lines above set R's working directory with `setwd`, and read the data in a data frame `ge`. We then verified that the content of `ge` was what it was supposed to be and attached the data frame in order to avoid the `ge$g1` notation. The mean, median, standard deviation, and variance were then easily calculated with their respective R functions. Let us look at these data and results for a second. The mean is -0.12 which is notably different from zero even though the numbers were originally drawn from a zero-mean random distribution. This illustrates the point that the mean of a small sample (and yes, 10 measurements represent a small sample) can be quite different from the mean of the populations the samples were drawn from. The same goes for the variance and standard deviation, which were supposed to be 1. Let us see, however, what happens if we calculate the mean, variance, and standard deviation across the entire set of 50 values stored in `ge`. Note that if we try `mean(ge[1:10,1:5])`, we do not obtain the result desired in this case:

```
> mean(ge[1:10,1:5])

    g1      g2      g3     g4      g5
-0.1248  0.1941 -0.2200  0.2142  0.2568
```

R actually took into consideration the fact that `ge` is a data frame and did something very sensible: it calculated the mean of every gene, across all samples. In most cases, this is probably more useful than the overall mean we are trying to calculate here. However, the overall mean can be easily calculated as the mean of the means of each gene:

```
>  mean(mean(ge[1:10,1:5]))
```

```
[1] 0.06406
```

Alternatively, we can make a single vector containing all values from ge and then calculate the desired descriptive statistics:

```
> attach(ge)
```

```
The following object(s) are masked from 'ge (position 3)':
```

```
    g1, g2, g3, g4, g5, group, source
```

```
> v=c(g1,g2,g3,g4,g5)
> v
```

```
 [1]   0.903  0.193  1.966 -0.138 -0.894 -0.902 -0.133 -0.580
 [9]  -1.269 -0.394  0.767  0.515  0.354  0.448  0.823  1.838
[17]  -3.331  0.300  1.392 -1.165 -1.262  1.388 -0.857 -1.352
[25]  -0.523  0.156  0.377  0.618 -0.965  0.220  0.146  0.490
[33]  -0.460 -0.748  0.276 -0.942  1.287 -0.243  1.395  0.941
[41]   0.971  0.207  0.619 -0.899  1.490 -0.709  1.292 -1.193
[49]   0.083  0.707
```

```
> mean(v)
```

```
[1] 0.06406
```

```
> median(v)
```

```
[1] 0.2
```

```
> sd(v)
```

```
[1] 1.019253
```

```
> var(v)
```

```
[1] 1.038876
```

The approach above produced the desired result, but we had to manually enumerate all columns of ge in order to construct a vector from the numerical elements of the data frame. This approach cannot be used for instance in a function that takes a data frame as an argument. In this case, the number and names of the columns will only be known at run time and will be different every time the function is called. A more general way of converting the subset of desired values from the given data frame into a vector would use the functions as.matrix and as.vector:

```
> vge = as.vector(as.matrix(ge[1:10,1:5]))
> vge
```

```
 [1]   0.903  0.193  1.966 -0.138 -0.894 -0.902 -0.133 -0.580
 [9]  -1.269 -0.394  0.767  0.515  0.354  0.448  0.823  1.838
[17]  -3.331  0.300  1.392 -1.165 -1.262  1.388 -0.857 -1.352
[25]  -0.523  0.156  0.377  0.618 -0.965  0.220  0.146  0.490
[33]  -0.460 -0.748  0.276 -0.942  1.287 -0.243  1.395  0.941
[41]   0.971  0.207  0.619 -0.899  1.490 -0.709  1.292 -1.193
[49]   0.083  0.707
```

The function `as.matrix` tells R to interpret its argument as a matrix while the function `as.vector` transforms it into a vector.

The ranges of the expression values for each gene can be calculated in various ways. One possibility is to calculate them as follows:

```
> max= sapply(ge[,1:5],max)
> max
```

```
  g1    g2    g3    g4    g5
1.966 1.838 1.388 1.395 1.490
```

```
> min= sapply(ge[,1:5],min)
> min
```

```
   g1     g2     g3     g4     g5
-1.269 -3.331 -1.352 -0.942 -1.193
```

```
> ranges = max -min
> ranges
```

```
  g1    g2    g3    g4    g5
3.235 5.169 2.740 2.337 2.683
```

```
> overall_range = max(max)-min(min)
> overall_range
```

```
[1] 5.297
```

One interesting question is related to the treatment of missing values, NA. Let us assume that some of our values were missing. We can simulate that by replacing a few of the values with NA:

```
> v[c(3,7,10,15,17)]=NA
> v
```

```
[1]    0.903  0.193      NA -0.138 -0.894 -0.902      NA -0.580
[9]  -1.269      NA   0.767  0.515  0.354  0.448      NA  1.838
[17]     NA   0.300   1.392 -1.165 -1.262  1.388 -0.857 -1.352
[25] -0.523  0.156   0.377  0.618 -0.965  0.220  0.146  0.490
[33] -0.460 -0.748   0.276 -0.942  1.287 -0.243  1.395  0.941
[41]  0.971  0.207   0.619 -0.899  1.490 -0.709  1.292 -1.193
[49]  0.083  0.707
```

```
> mean(v)
```

```
[1] NA
```

As shown above, an attempt to directly calculate any of the descriptive statistics would fail because any computation involving a value that is NA (not available) will also be NA (see also Section 6.4.2 in Chapter 6). Fortunately, the functions mean, median, var, and sd allow the user to explicitly ask for the computation to be done without the NA values. This is achieved with the option na.rm=TRUE:

```
> mean(v,na.rm=TRUE)
```

```
[1] 0.09493333
```

This option is always FALSE by default in order to ensure that the user is aware of the presence of missing values.

10.2.2   Mode

The mode is worth some further discussion on the *ge* example above. On the one hand, all values in ge are distinct which means that each of them appears only once. Strictly speaking, this data set does not have a mode since all values appear with the same frequency: once. However, it would be incorrect to conclude that the underlying distribution, from which these values were drawn, does not have a mode. This can be easily shown by plotting a histogram of all values, across all genes in ge. These values have already been extracted in a numeric vector v above so we can plot a histogram by simply calling hist:

```
> v=c(g1,g2,g3,g4,g5)
> plot(hist(v),col="gold")
```

In this command, we allowed the function hist to analyze its input and create its own bins. The resulting plot is shown in the left panel of Fig. 10.1. This plot suggests that the mode of the underlying distribution is between 0 and 1. Also, the distribution appears to be skewed to the left, having a longer and fatter tail to the left of the highest bin. However, these are just artifacts of the binning process undertaken in order to create the histogram. We can show

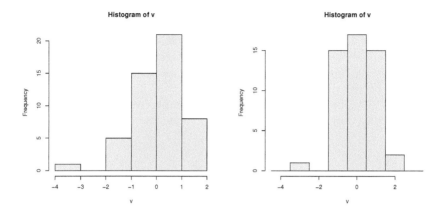

FIGURE 10.1: The histogram of the data shows that the underlying distribution does appear to have a mode even if no value is repeated in the data set. According to the plot in the left panel, the mode appears to be between 0 and 1. According to the plot in the right panel, the mode appears to be between -0.5 and 0.5. This is an artifact of the binning process.

this by redefining the boundaries of the bins to be used in the creation of the histogram. This can be done by providing a vector of specific intervals that should be used to cover the x range. Let us use a number of bins of width=1, equally spaced in the interval from -4.5 to 3.5:

```
> breaks=seq(-4.5,3.5,by=1)
> hist(v,breaks,col="gold")
```

This histogram is shown in the right panel of Fig. 10.1. This plot now suggests that the mode is somewhere in the interval $[-0.5, 0.5]$. A more detailed discussion about the various artifacts that can be introduced by the binning process can be found in Chapter 17. Also note that in these plots the vertical axes shows the raw frequency of the data in each bin. The function **hist** can also be used to produce a probability density histogram in which the area of each bin is proportional with the number of values in the bin (see Section 17.6).

10.2.3 More built-in R functions for descriptive statistics

The built-in function **summary** provides a convenient way to get information about a set of data points:

```
> summary(ge$g1)

   Min. 1st Qu.  Median    Mean 3rd Qu.    Max.
-1.2690 -0.8155 -0.2660 -0.1248  0.1115  1.9660
```

As shown above, the function provides the minimum, maximum, mean, and median as well as two other values 1st Qu. and 3rd Qu., in their appropriate order. The 1st Qu. and 3rd Qu refer to the first and third **quartiles** of the data. These are the values that are greater than one quarter of the data (25%) and three quarters of the data (75%), respectively. The median is also the second quartile.

The function summary also works on more complex objects such as matrices or data frames. For instance, we can summarize the entire ge data frame: summary:

```
> summary(ge)
```

```
      g1                 g2                 g3
Min.   :-1.2690   Min.   :-3.3310   Min.   :-1.3520
1st Qu.:-0.8155   1st Qu.: 0.3135   1st Qu.:-0.9380
Median :-0.2660   Median : 0.4815   Median :-0.1835
Mean   :-0.1248   Mean   : 0.1941   Mean   :-0.2200
3rd Qu.: 0.1115   3rd Qu.: 0.8090   3rd Qu.: 0.3377
Max.   : 1.9660   Max.   : 1.8380   Max.   : 1.3880
      g4                 g5             group     source
Min.   :-0.9420   Min.   :-1.1930   cancer :5   DMC:6
1st Qu.:-0.4057   1st Qu.:-0.5110   healthy:5   KCI:4
Median : 0.2110   Median : 0.4130
Mean   : 0.2142   Mean   : 0.2568
3rd Qu.: 0.8283   3rd Qu.: 0.9050
Max.   : 1.3950   Max.   : 1.4900
```

This function summarizes the data very nicely by calculating the statistics above for any numerical variables. For factors or categorical variable such as the last two columns of the matrix above, the function summarizes the data by counting how many instances there are for each factor level or category.

In this case, the above summary statistics were provided for each gene in the data frame. Note that the columns group and source, which are not numeric, have been summarized by counting the number of instances of each label. Essentially, these have been treated as factors. If NA values are present anywhere in the data frame, the summary also includes the number of such NAs:

```
> summary(v)
```

```
   Min.    1st Qu.    Median     Mean    3rd Qu.     Max.
-3.33100  -0.73820   0.20000   0.06406   0.75200   1.96600
```

10.2.4 Covariance and correlation

The **covariance** is a measure that characterizes the degree to which two variables vary together.[1] The covariance will be positives if both variable tend to be above and below their respective expected values (means) at the same time. This type of behavior provides information about a possible linear relationship between the two variables. In R, the covariance can be calculated with the function cov. For instance, the covariance between the genes g1 and g2 in the data frame ge can be calculated as:

```
> cov(ge$g1,ge$g2)

[1] -0.1959645
```

Very conveniently, the same function can be called on a matrix or on a numerical sub-selection of a data frame. In this case, the result will be the covariance matrix of the genes. The element at position (i, j) in this matrix will be the correlation between genes i and j:

```
> cov(ge[,1:5])

          g1         g2          g3         g4          g5
g1  0.9237371 -0.1959645 -0.17648600 -0.2084118  0.22466038
g2 -0.1959645  2.1487565 -0.39431733 -0.6186605 -0.58636176
g3 -0.1764860 -0.3943173  0.82212044  0.1419104 -0.08364222
g4 -0.2084118 -0.6186605  0.14191044  0.6760186  0.43263727
g5  0.2246604 -0.5863618 -0.08364222  0.4326373  0.87106907
```

The disadvantage of the covariance is that it is dimensional, ie. it depends on the values and units of measurement of the variables measured. If a variable X is measured in meters and another variable Y is measured in kilograms, the covariance between X and Y will be measured in meters $\times$ kilograms. Furthermore, the possible values of the covariance span the entire set of real numbers. The correlation coefficient addresses both these issues by scaling the covariance to the variance of each variable (see Section 8.3.4). The resulting correlation coefficients take values only between -1 and 1 which allows a comparison across different pairs of variables, and is independent of their measurement units. In R, the correlation coefficients can be calculated with the function cor. The function cor calculates the correlation between its first two arguments, which are expected to be vectors, matrices, or data frames, much as for the cov.

```
> cor(ge$g1,ge$g2)

[1] -0.1390944
```

---

[1] See also Section 8.3.4 for a more detailed discussion of covariance and correlation.

The following function call calculates the correlation matrix for the genes in ge:

```
> cor(ge[,1:5])
```

```
            g1          g2          g3          g4          g5
g1   1.0000000 -0.1390944 -0.20252020 -0.2637356  0.25045286
g2  -0.1390944  1.0000000 -0.29667745 -0.5133096 -0.42859365
g3  -0.2025202 -0.2966775  1.00000000  0.1903563 -0.09883973
g4  -0.2637356 -0.5133096  0.19035632  1.0000000  0.56379121
g5   0.2504529 -0.4285937 -0.09883973  0.5637912  1.00000000
```

Note that now the elements of the first diagonal are all equal to 1 (each gene is perfectly correlated with itself).

Both functions cor and cov calculate the Pearson correlation or covariance by default. The Spearman and Kendall tau correlation coefficients can be obtained by using the additional parameter method set to the desired value (method="spearman" or method="kendall").

Since the variance of the variable can be seen as the correlation of the variable with itself, the same basic function can be used to also calculate the variance. Nevertheless, R provides a separate function, var, which is in fact a different interface to the same function.

10.3 Probabilities and distributions in R

10.3.1 Sampling

Random sampling is the basic process of choosing at random a number of values from a larger set. The idea of sampling is central in statistics and used in many situations. Let us consider a vector, x, containing a set of values of any type (numeric, complex, character or logical). In R, sampling some values from a vector can be done with the function:

```
> sample(x, size, replace = FALSE, prob = NULL)
```

The first parameter is the vector containing the values from which the sample should be drawn. The size parameter specifies the size of the desired sample, i.e. how many values will be drawn from x. The parameter replace specifies whether the sampling should be done with or without replacement. In the sampling with replacement, after a value is picked from x, the value is replaced such that it can be chosen again in the future. In consequence, sampling with replacement can yield a sample containing repeated values even if all values in the sampled vector are distinct. In sampling without replacement, the values are not replaced. By default, `size` is equal to the size of `x`. Since `replace` is also FALSE by default, simply calling `sample(x)` will generate a random permutation of `x`. Finally, the parameter `prob` can be used to specify a vector with the probabilities of picking individual elements from `x`.

Let us use this function to construct a sample representing 15 random throws of an unbiased dice:

```
> x=1:6
> y=sample(x,15,replace=T)
> y

 [1] 3 1 4 4 2 1 2 2 4 4 3 6 2 6 6
```

or 10 tosses of a coin that was biased such that the head is 10% more likely than the tail:

```
> sample(c("h","t"),10,replace=T,c(0.6,0.4))

 [1] "t" "t" "h" "h" "t" "t" "h" "h" "t" "h"
```

In practical data analysis, the function `sample` is often used in Monte-Carlo type simulations or in analyses involving resampling.

10.3.2 Empirical probabilities

The empirical probability is the relative frequency of the occurrence of an event. If an outcome A occurred in n_A cases out of n trials, the empirical probability of A is $P(A) = n_A/n$. In many cases, calculating the empirical probability of an event A reduced to counting the number of favorable (A has occurred) and unfavorable (A has not occurred) cases.

R functions useful in this context are `prod(x)`, which calculates the product of the element of a vector, and `choose`, which calculate the number of different ways in which one can choose `k` elements from a set of `n` such elements.

A classical problem involves calculating the number of different permutations of n elements. An example would be the number of different integers that can be written with the digits from 0 to 4 using each digit just once (assuming numbers starting with 0 are acceptable). For the first position, we would have the choice of any one of those numbers, i.e. we would have $n = 5$ different possibilities. Once the first digit is chosen, there will be only $n - 1$ available digits to choose from for the second position. For every one digit chosen for the first position, there are $n - 1$ choices for the second position, so there are $n \times (n - 1)$ choices for the first 2 positions. Overall, there are:

$$n \cdot (n - 1) \cdots 3 \cdot 2 \cdot 1 = n! \tag{10.1}$$

permutations of n numbers. In R, such a factorial can be easily calculated using the `factorial` function, or the `prod` function. For instance, the number of different integers that can be written with the digits from 0 to 4 using each digit just once is:

```
> factorial(5)
```

```
[1] 120
```

```
> prod(1:5)
```

```
[1] 120
```

Another classical problem involves calculating the number of ways in which one can choose k out of n numbers, irrespective of their order. The classical example here is a lottery ticket, which will be a winning ticket if the k chosen numbers are the same as the k numbers extracted, independently of the order in which the numbers appear on the ticket or have been extracted. This number can be calculated as follows: For the first number, there will be n choices. For each such number chosen as the first number, there will be $n - 1$ numbers available to choose from as the second number. For the k-th such choice, there will be only $n - k + 1$ possibilities to choose a number. Overall, this gives $n \cdot (n - 1) \cdots (n - k + 1) \cdot (n - k + 1)$ possible arrangements of k numbers chosen from the available set of n. However, these are individual *arrangements* of k numbers. For the lottery ticket, it doesn't matter whether the ticket shows the numbers 2, 3, 4 or 3, 2, 4. Hence, we need to adjust the computation to account for the fact that many of these possible arrangements contain the same set of numbers. The adjustment is done by dividing by the number of all possible permutations of k numbers. In other words, the quantity needed will be:

$$\frac{n \cdot (n - 1) \cdots (n - k + 1) \cdot (n - k + 1)}{k \cdot (k - 1) \cdots 3 \cdot 2 \cdot 1} = \frac{n!}{(n - k)! \cdot k!} = \binom{n}{k} \qquad (10.2)$$

or "n choose k." In R, we can calculate this quantity either directly implementing the formula above, or by using the built-in function `choose(n,k)`:

```
> n=10;k=5
> n_choose_k = prod(n:(n-k+1))/factorial(k)
> n_choose_k
```

```
[1] 252
```

```
> choose(n,k)
```

```
[1] 252
```

Using nothing more than the above function, we can put R to good use and try to improve our daily life by calculating our chances of winning the lottery, as in the following example.

Matches	Prize
10 of 22	$250,000
9 of 22	$2,500
8 of 22	$250
7 of 22	$25
6 of 22	$7
0 of 22	$1 ticket

TABLE 10.1: In the game of Keno, the player picks 10 numbers from 1 to 80. Subsequently, 22 winning numbers are drawn from the same pool of 1 to 80. The table shows the winning combinations and their respective prizes.

EXAMPLE 10.1

The Keno game[2] available from the Michigan Lottery, allows us to purchase a $1 play slip on which we pick 10 numbers from 1 to 80. Later, the lottery will draw 22 numbers. The various possible winning combinations and their respective prizes are shown in Table 10.1. Calculate: i) the probability of winning each individual prize, ii) the probability of winning anything with a given ticket, and iii) the expected gain for the entire game.

SOLUTION   The idea of investing $1 to make $250,000 does not seem to be a bad *a priori* idea, so let us start by calculating our chances of winning this prize. Since any number can occur on a ticket only once, there are $\binom{80}{10}$ ways to pick the 10 numbers on a ticket. These constitute the set of all possible tickets, which is the set of all possible choices for us. Let us assume that we knew the 22 numbers that will be extracted. Given these 22 "lucky" numbers, a ticket will be winning if all its numbers are "lucky" numbers. The number of winning tickets out there is equal to the number of ways in which we can choose 10 numbers out of the 22 soon-to-be-extracted "lucky" numbers. This is $\binom{22}{10}$. Overall, the probability of winning the big prize can be calculated as the number of favorable outcomes, i.e. the total number of winning tickets, divided by the total number of tickets:

```
> win_prob = choose(22,10)/choose(80,10)
> win_prob

[1] 3.927416e-07

> 1/win_prob

[1] 2546203
```

In other words, the odds of winning are approximatively 1 in 2.5 million.

---

[2]http://www.michigan.gov/lottery/0,1607,7-110-46442_812_872---,00.html

The reality of this is somewhat sweetened by the $250,000 payoff if we win. The expected value for the monetary outcome of this enterprize (the value that we expect to get if we play this game for a very long time) can be calculated as:[3]

```
> 250000*win_prob
```

```
[1] 0.09818541
```

or about 10c for every $1 ticket bought. Consequently, playing this game in the sole hope of winning the big prize does not seem a very profitable enterprise at this time. However, the game also offers prizes for fewer numbers guessed correctly, as shown in Table 10.1, and most certainly, we will not refuse such a lesser prize if entitled to it. Hence, it would be useful to have a function that can calculate the probability of any given outcome. Let us consider there are n numbers available, m numbers on a ticket, k numbers are extracted, and l of them have to match in order to win anything. The denominator here is the total number of possible tickets, which is the number of ways in which we can choose m numbers out of the available n:

$$\text{total number of tickets} = \binom{n}{m} \tag{10.3}$$

In order to calculate the number of winning tickets, we first take the number of all possible ways in which one can choose the required l numbers from the k winning numbers:

$$\text{number of winning combinations (with 1 matches)} = \binom{k}{l} \tag{10.4}$$

This corresponds to the core part that would make any ticket winning (any ticket with l matching numbers is a winning ticket). For each such core winning combination, there are a number of winning tickets that can be obtained by picking the remaining $m - l$ numbers necessary to form a ticket from the remaining $n - k$ numbers which were not picked in the draw. There are:

$$\binom{n-k}{m-l} \tag{10.5}$$

such combinations. The probability can be expressed as:

$$p = \frac{\binom{k}{l} \cdot \binom{n-k}{m-l}}{\binom{n}{m}} \tag{10.6}$$

For instance, the $250 prize will be won by all tickets that have 8 winning numbers. In this case, illustrated in Fig. 10.2, there are $n = 80$ numbers to pick

---

[3] See also Section 9.1.1 in Chapter 8.

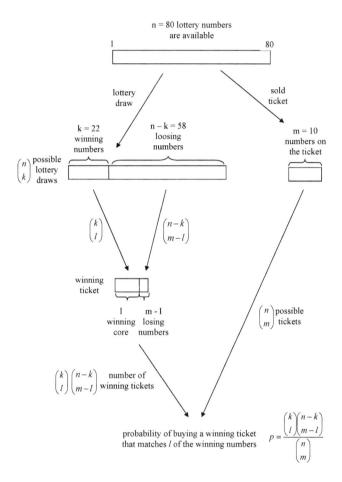

FIGURE 10.2: Calculating the probability of winning a prize in the Keno lottery. In this case, there are $n = 80$ numbers to pick from, $m = 10$ numbers on a ticket, and the lottery extracts $k = 22$ winning numbers. The winning "core" of a ticket can be any combination of $l = 8$ out of the $k = 22$ winning numbers, or $\binom{22}{8}$. For each such combination, one can obtain a winning ticket by adding $m - l = 2$ "losing" numbers from the remaining $n - k = 58$ numbers, which were not picked during the draw. The probability of having a winning ticket is the number of winning tickets divided by the number of possible tickets.

from, $m = 10$ numbers on a ticket, and the lottery extracts $k = 22$ winning numbers. The winning "core" of a ticket can be any combination of $l = 8$ out of the $k = 22$ winning numbers, or $\binom{22}{8}$. For each such combination, one can obtain a winning ticket by adding $m - l = 2$ "losing' numbers from the remaining $n - k = 58$ numbers which were not picked during the draw. The probability will be:

$$p = \frac{\binom{22}{8} \cdot \binom{58}{2}}{\binom{80}{10}} \tag{10.7}$$

The following R function will calculate the probabilities of winning in each case:

```
> keno_prob <- function(n,m,k,l){
+ # n numbers available, m numbers on a ticket,
+ # k numbers extracted, l have to match
+      prob = choose(k,l)*choose(n-k,m-l)/choose(n,m)
+      #odds = 1/prob
+      prob
+ }
```

The probability of winning any of the prizes can be obtained as the sum of the probabilities of winning each individual prize:

```
> allwins = c(10,9,8,7,6,0)
> wp= keno_prob(80,10,22,allwins)
> wp
```

```
[1] 3.927416e-07 1.752232e-05 3.210339e-04 3.196071e-03
[5] 1.922636e-02 3.169130e-02
```

```
> panyprize = sum(wp)
> panyprize
```

```
[1] 0.05445269
```

The vector **wp** was used to stored the probabilities of winning the individual prizes. The probability of winning any prize with a given ticket is the probability of winning the first prize OR the second one, etc. Since these are mutually exclusive events (if we win the big prize we will only collect it even though we also matched 9 out of the 10, etc.), we can calculate the probability of winning anything by adding the probabilities of the individual wins in **panyprize**. This probability is about 5.4%. The overall expected gain can be calculated as the sum of the products between each prize and its probability:

```
> prizes = c(250000,2500,250,25,7,1)
> prizes
```

```
[1] 250000   2500    250     25      7      1
```

```
> wins = wp*prizes
> wins
```

```
[1] 0.09818541 0.04380580 0.08025848 0.07990177 0.13458455
[6] 0.03169130
```

```
> gain = sum(wins)
> gain
```

```
[1] 0.4684273
```

The vector `prizes` stored the prize values while `wins` stored the expected gains for each type of prize. The variable `gain` stores the total gain expected from this game: approximatively 47 cents for every dollar played, which, as far as lotteries go, it is not that bad. ⧠

10.3.3 Standard distributions in R

Most standard distributions likely to be useful in a data analysis context have been implemented in R. For each such distribution, R has built-in functions that can return any one of the following:

1. **Pseudo-random numbers** drawn from the given distribution.

2. The **probability density** (for continuous distributions) or point probability (for discrete distributions) – calculated from the probability density function $f(x)$ (pdf).

3. The **cumulated probability** – calculated from the cumulative distribution function (cdf).

4. The **quantiles** of the given distribution.

Each of the quantities above can be obtained by invoking a function whose name is constructed by prefixing the name of the distribution (e.g. `norm` for normal, `binom` for binomial, etc.) with one of the following letters:

1. `r` for (pseudo-)random numbers – e.g., `rnorm`, `rbinom`, etc.

2. `d` for density – e.g., `dnorm`, `dbinom`, etc.

3. `p` for probability – e.g., `pnorm`, `pbinom`, etc.

4. `q` for quantiles – e.g., `qnorm`, `qbinom`, etc.

10.3.4   Generating (pseudo-)random numbers

Random numbers are often useful to simulate data with certain characteristics. More recently, a whole family of methods has been developed based
on random permutations of the original data. Such methods are very useful
because they allow us to assess the significance of a certain outcome while
taking into account various properties of the original data such as variable
correlations and interdependencies.

In spite of the abundance of random events in nature, genuine random
numbers are not easily obtained on a computer. This is because computers
are designed to be deterministic, i.e. to produce the same results every time a
given program is run with the same input. Hence a fair amount of work was
needed in order to design programs able to generate pseudorandom numbers.
Usually, these pseudo-number generators use a seed in order to generate a
sequence of pseudorandom numbers. The sequence generated will be always
the same for a given seed – which allows debugging and replicating one's
results. If the seed is changed every time (e.g., based on the internal clock
of the machine), every sequence will be different. Even though these are not
genuinely random, the pseudorandom numbers generated by the R built-in
functions will be more than satisfactory for most practical purposes.

For instance, the following code generates a sequence of 1,000 (pseudo-
)random numbers from a normal distribution with mean zero and standard
deviation one:

```
> nums = rnorm(1000,mean=0,sd=1)
> nums[1:10]

 [1] -1.47476195 -0.88889057  0.74446620 -0.07522112  0.70697727
 [6] -1.18309198  1.24421548 -1.31958864  0.96193632  0.36442228

> mean(nums)

[1] -0.02874554

> sd(nums)

[1] 0.9850121
```

10.3.5   Probability density functions

The density values $f(x)$ obtained from the d- built-in functions (e.g., dnorm)
can be intuitively understood as the probability of obtaining a value equal to
(for discrete distributions) or close to (for continuous distributions) the given
value x.

According to recent statistics [328], the mean weight of 19-year-old males
is $\overline{X} = 78.2$ kilograms with a standard error of the mean of $\sigma_{\overline{X}} = 1.3$ (see 9.2

in Chapter 8), measured on a sample of $n = 270$ individuals. Using the central limit theorem, we can calculate an estimate of the population standard deviation, $\hat{\sigma}$ as:

$$\hat{\sigma} = \sigma_{\bar{X}} \cdot \sqrt{n} \qquad (10.8)$$

We can now plot the distribution of the weights of 19-year-old males estimated from these data:

```
> sample=270
> std_err_mean=1.3
> mean_weight=78.2
> est_pop_sd = std_err_mean * sqrt(sample)
> curve(dnorm(x,mean=mean_weight,sd=est_pop_sd),
+ from=mean_weight-3*est_pop_sd,
+ to=mean_weight+3*est_pop_sd,col="red")
```

This distribution is shown in Fig 10.3. In principle, dnorm and similar can be used to calculate the probability of measurement being very close to (or exactly equal to for discrete distributions) a given value. For instance, the probability of a 19-year-old male having a weight close to 70 kilograms, $P(X = 70)$, can be obtained as:

```
> dnorm(70,mean=mean_weight,sd=est_pop_sd)
```

```
[1] 0.01734947
```

However, this is the probability of having a weight *very, very* close to 70 kilograms. For instance, the probabilities of having a weight of 70.001 or 69.999 kilograms would be slightly different (0.01734978 and 0.01734916, respectively) and not included in the number above. This sort of "point" probabilities are not very useful in practice. A more useful probability would be that corresponding to an individual weighing less than or equal to 70 kilograms, $P(X < 70)$, or perhaps somewhere in a larger interval, such as between 69 and 71 kilograms, $P(69 < X < 71)$. Such probabilities correspond to various areas under the curve of the probability density function and can be calculated with the p-functions: pnorm, pbinom, etc.

10.3.6 Cumulative distribution functions

The probability that a random variable X is less than or equal to a given value x, $P(X \leq x)$, can be calculated with the functions pnorm, pbinom, etc. These are the R implementations of the cumulative distribution functions (cdf-s) of their respective distributions (see Chapter 8 for a more in-depth discussion of cdf-s).

These implementations of the cumulative distribution functions provide numbers that are more useful in practice than those provided by the probability density functions. For instance, we can calculate the probability that a 19-year-old boy weighs 70 kilograms or less as:

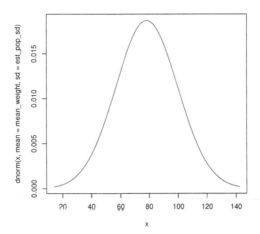

FIGURE 10.3: The distribution of the weights of 19-year-old males in the United States, according to a recent survey [328]. The horizontal axis shows the weight in kilograms.

```
> pnorm(70,mean=mean_weight,sd=est_pop_sd)
```

```
[1] 0.3505359
```

Note that the given probability corresponds to the lower, or left, tail of the distribution $P(X \leq x)$. In principle, the probability corresponding to the right tail $P(X > x)$ can always be obtained by subtracting the left tail probability from 1: $P(X > x) = 1 - P(X \leq x)$. However, R can provide these values directly if the parameter lower.tail is specified with the value FALSE. For instance, the probability that a 19-year-old male weighs more than 70 kilograms is:

```
> pnorm(70,mean=mean_weight,sd=est_pop_sd,lower.tail=F)
```

```
[1] 0.6494641
```

```
> 1-pnorm(70,mean=mean_weight,sd=est_pop_sd)
```

```
[1] 0.6494641
```

The probability that a 19-year-old male weighs more than 69 and less than or equal to 71 kilograms can be calculated as:

```
> pnorm(71,mean=mean_weight,sd=est_pop_sd)-
+ pnorm(69,mean=mean_weight,sd=est_pop_sd)
```

[1] 0.03468813

Note that the in the snippet above, the two probabilities are subtracted; the '+' symbol is just R's way of telling us that the expression on the first line is not syntactically complete and that it is continued on the second line.

10.3.7 Quantiles

The p-th quantile can be defined as the value x with the property that there is a probability p of obtaining a values less than or equal to x:

$$P(X \leq x) = p \tag{10.9}$$

Let us recall that for a random variable X and a given value x, the cumulative distribution function (cdf) can be defined as the probability of X having a value less than or equal to x:

$$F(x) = P(X \leq x) \tag{10.10}$$

Let us assume that the probability of getting a value less than or equal to x is p, or in other words, $F(x) = p$. An examination of the definitions above reveals the relationship between the cdf and the quantiles. For any given value x and its corresponding probability p, the cdf provides the probability p for a given x, while the quantile provides the value x for a given p. In other words, the general quantile function can be understood as the inverse of the cdf. In R, this can be easily verified:

```
> p=pnorm(2)
> p
```

[1] 0.9772499

```
> qnorm(p)
```

[1] 2

In this example, **pnorm** was called with the default values of **mean=0** and **sd=1**. The first line above calculates the probability of having a value less than or equal to 2 from a standard normal distribution. The value obtained is approximately p=0.97725.[4] Conversely, the value that corresponds to this probability can be obtained with **qnorm** and, unsurprisingly, this value is 2.

Quantiles are useful for various purposes. First, quantile functions can be used to directly obtain what they are meant to provide: the values corresponding to a certain probability. For instance, let us say we would like to

---

[4]This means that only about 2.5% of the values are larger than 2 standard deviations, which in turn should remind us that in a normal distribution approximatively 5% of the values are beyond ±2 standard deviations.

know whether a certain individual is substantially overweight and that we define being overweight as weighing more than 90% of the people of same age and sex. For the population above of 19-year-old males, this value can be obtained as:

```
> qnorm(0.9,mean=mean_weight,sd=est_pop_sd)
```

```
[1] 105.5755
```

Similarly, the quartiles for this population can be obtained as:

```
> qnorm(0.25,mean=mean_weight,sd=est_pop_sd)
```

```
[1] 63.7921
```

```
> qnorm(0.5,mean=mean_weight,sd=est_pop_sd)
```

```
[1] 78.2
```

```
> qnorm(0.75,mean=mean_weight,sd=est_pop_sd)
```

```
[1] 92.6079
```

```
> mean_weight
```

```
[1] 78.2
```

```
> est_pop_sd
```

```
[1] 21.36118
```

Note that the second quartile, 78.2, is also the 50th percentile or the median. In this case, because the distribution is normal, this value is also the mean.

By using R's ability to accept vectors as arguments, we can calculate all quartiles at the same time as in:

```
> qnorm(c(0.25,0.5,0.75),mean=mean_weight,sd=est_pop_sd)
```

```
[1] 63.7921 78.2000 92.6079
```

Quantiles are also useful to calculate confidence intervals, the number of replicates necessary in a given experiment (power calculations), and in order to draw quantile-quantile plots (q-q plots).

10.3.7.1   The normal distribution

The probability density function for a normal random variable with mean μ and standard deviation σ has the form:

$$f(x) = \frac{1}{\sigma\sqrt{2\pi}} \cdot e^{-\frac{(x-\mu)^2}{2\sigma^2}} \qquad (10.11)$$

A plot of the pdf and cdf of the standard normal distribution can be easily obtained as:

```
> curve(dnorm(x),from=-3,to=3,col="blue")

> curve(pnorm(x),from=-3,to=3,col="blue")
```

The plots are shown in Fig. 10.4. The function `curve` is very convenient for plotting explicit functions i.e. that can be described by an expression of the form $y = f(x)$. Such functional relationships will only have at most one value of y for any values of x. An alternative way of plotting graphs is to first create a vector of x values, then calculate the corresponding y values, and then plot the (x,y) pairs. In our case, a plot of the cdf and pdf of a standard normal distribution can be obtained with:

```
>  x=seq(-3,3,0.1)
>  y=dnorm(x)
>  plot(x,y,type="l")
```

In the code above, the graph is of `type="l"` which means that R will unite subsequent (x,y) pairs with lines, producing a line graph. Of course, if the x and y values are not subsequently needed, the same result can be obtained with just one line of code:

```
> plot(seq(-3,3,0.1),dnorm(seq(-3,3,0.1)),type="l")
```

A plot overlaying several normal distributions with standard deviations of 1, 1.5 and 2 (see Fig. 10.5) can be obtained by using the argument `add=TRUE` in the function `curve`, as follows:

```
> curve(dnorm(x),from=-7, to=7)
> curve(dnorm(x,mean=0,sd=1.5),from=-7, to=7,add=T,col="red")
> curve(dnorm(x,mean=0,sd=2),from=-7, to=7,add=T,col="blue")
```

Let us try to verify some of the known properties of the normal distribution. For instance, it is known that approximatively 68% of the values are within 1 standard deviation from the mean, and that approximatively 95% of the values are within 2 standard deviations from the mean. For simplicity's sake we will use the standard normal distribution ($\mu = 0$, $sd = 1$). For this distribution, having a value farther than 1 standard deviation to the left of

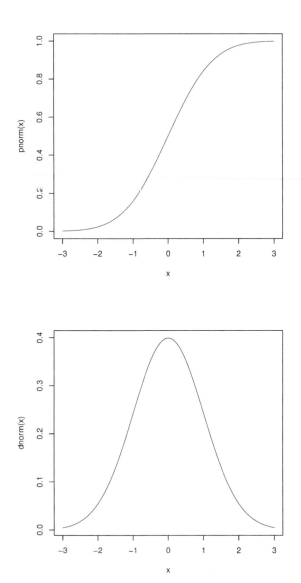

FIGURE 10.4: The pdf and cdf of the standard normal distributions plotted in R.

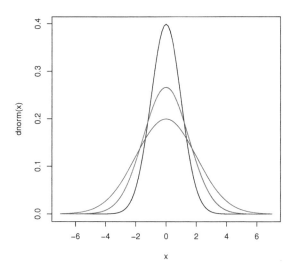

FIGURE 10.5: Normal distributions with zero mean and standard deviations equal to 1, 1.5, and 2.

the mean is equivalent to having a value less than -1. Similarly, a value farther than 1 standard deviation to the right of the mean translates into the value being larger than 1. The probability of getting a value less than -1 can be obtained with `pnorm(-1,lower.tail=TRUE)`.[5] The probability of getting a value larger than 1 can be obtained with `pnorm(1,lower.tail=FALSE)`. If we were interested in the values that are more than 1 standard deviation away either side of the mean, we would need to add these two probabilities (see Fig. 10.6). However, we are interested in how many values are within ± 1 standard deviations, which is the event complementary to the values being outside those limits (unshaded area in Fig. 10.6) so we need to take the complement of this sum of probabilities:

```
> 1- (pnorm(-1,lower.tail=T)+ pnorm(1,lower.tail=F))
```

```
[1] 0.6826895
```

```
> 1- (pnorm(-2,lower.tail=T)+ pnorm(2,lower.tail=F))
```

```
[1] 0.9544997
```

---

[5]The value `TRUE` is the default value for the argument `lower.tail`. In this case, it was made explicit for clarity reasons.

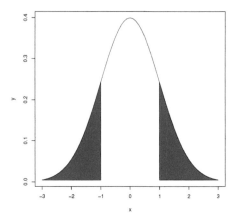

FIGURE 10.6: The shaded areas represent the values outside ±1 standard deviation for a standard normal distribution.

As expected, the first probability shows that the interval of ± 1 standard deviations around the mean include about 68% of the values while extending the interval to ± 2 standard deviations includes about 95% of the values. The same information could have been obtained by asking what is the quantile that corresponds to a probability of 0.16 (which is half of 0.32, for the lower tail of the distribution):

```
> qnorm(.16)

[1] -0.9944579
```

This quantile gives a value that is approximatively one standard deviation to the left of the mean. In order to verify the same for the right tail of the distribution, we have to ask for `qnorm(1-0.16)`, which yields 0.9944579. This shows both that the given interval includes approximatively 68% of the values, and also the symmetry of the distribution ($P(X \leq \mu - x) = 1 - P(X > x)$).

10.3.7.2 The binomial distribution

As previously discussed, the pdf of the binomial distribution is described by the equation:

$$f(x) = \binom{n}{x} \cdot p^x \cdot (1-p)^{n-x} = \frac{n!}{x! \cdot (n-x)!} \cdot p^x \cdot (1-p)^{n-x} \qquad (10.12)$$

In R, this function is implemented as `dbinom(q, size, prob)`, where q is the number of successes, `size` is the size of the trial, and `prob` is the probability of success in each individual trial. For example, the probability density corresponding to a coin toss can be obtained with:

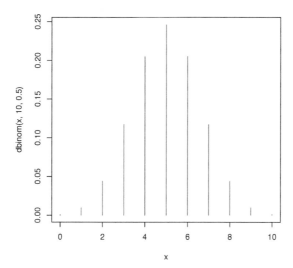

FIGURE 10.7: The pdf of the binomial distribution with the size of the trial equal to 10 and the probability of a success in an individual trial **p=0.5**.

```
> x=0:10
> plot(x,dbinom(x,10,0.5),type="h",col="red")
```

The corresponding plot is shown in Fig. 10.7. Note the **type="h"** parameter used to obtain a histogram-like graph emphasizing that the probabilities are zero for non-integer values.

The function **pbinom** can be used to calculate the probability of various events that can be modeled by a binomial distribution. The same function can be used to plot the graph of the cdf of a binomial distribution. Fig. 10.8 shows the cdf of the binomial distribution corresponding to a trial size of 10 and the probability of success in an individual trial equal to 0.5:

```
> x=0:10
> plot(x,pbinom(x,10,0.5),type="s",col="red")
```

The rather unusual step-like shape of the cdf graph shown in Fig. 10.8 is due to the fact that the outcome of a binomial phenomenon represents the number of successes in a given number of trials, and hence, it can only take discrete values. This is also reflected by the fact that the binomial pdf is zero in-between integer values, as shown in Fig. 10.7. At any point x, the cdf represents the cumulative probability of obtaining a values *less than or equal to* x. For integer values of x, such as 3, there will be a certain probability to

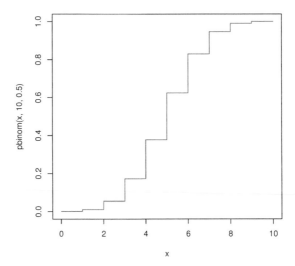

FIGURE 10.8: The cdf of the binomial distribution with the size of the trial equal to 10 and the probability of a success in an individual trial p=0.5.

obtain exactly 3 successes (0.1171875 for 10 trials with 0.5 prob. of success in an individual trial), and a certain probability of obtaining 3 successes or less (0.171875). The former is given by the pdf, while the latter is given by the cdf. For non-integer values of x, such as 3.5, the probability of obtaining exactly 3.5 successes will be zero, while the probability of obtaining less than or equal to 3.5 successes will be the same as the probability of obtaining less than or equal to 3 successes, since there will never be a number of successes between 3 and 3.5. Hence, the cdf of a discrete distribution, such as the binomial in this case, will be constant in-between the discrete values where it will register a jump (discontinuity) as shown in Fig. 10.8.

10.3.8 Using built-in distributions in R

As we have seen in Chapter 8, the typical questions that can be asked directly are:

1. What is the probability that the variable has a value higher than a certain threshold t?

2. What is the probability that the variable has values between two thresholds t_1 and t_2?

3. What is the threshold that corresponds to a certain probability?

In R, any of these questions can be answered very easily, usually with just one or two lines of code. The functions pnorm, pbinom, etc., discussed above, provide direct answers to questions of the first type of questions above. The second type of questions above can be easily answered with simple arithmetical operations on the values provided by the probability functions, while the appropriate quantiles, qnorm, qbinom, etc., provide easy answers to questions of the latter type. These will be illustrated briefly in the following.

EXAMPLE 10.2

The distribution of mean intensities of cDNA spots corresponding to genes that are expressed can be assumed to be normal [276]. Let us assume that the mean of this distribution is 1,000 and the standard deviation is 150. What is the probability that an expressed gene has a spot with a mean intensity of less than 850?

SOLUTION   *The traditional way of addressing such problems was to transform the data in such a way that a standard normal distribution could be used (see Example 9.6). This was required because statistical tables were available only for the standard normal distribution rather than for the normal distribution with the mean and standard deviation required by the problem. In R, exact probabilities and quantiles can be calculated for normal distributions with arbitrary means and standard deviations as required, by simply specifying their values in the function call. Hence, standardizing the values by subtracting the mean and dividing by the standard deviation, is no longer necessary. The answer can be calculated directly as:*

```
> pnorm(850,mean=1000,sd=150)
```

```
[1] 0.1586553
```

▯

EXAMPLE 10.3

The distribution of mean intensities of cDNA spots corresponding to genes that are not expressed can be assumed to be normal [276]. Let us assume that the mean of this distribution is 400 and the standard deviation is 150. What is the probability that an unexpressed gene has a spot with a mean intensity of more than or equal to 700? Plot a graph showing the area under the pdf curve that corresponds to this probability.

SOLUTION   *Here the question is about the probability of obtaining a value higher than or equal to the given threshold. As illustrated in Fig. 10.9, this* *probability corresponds to the right tail of the distribution. Hence, this value can be calculated by subtracting the value providing by* **pnorm**, *corresponding to the area to the left of the threshold, from the value corresponding to the entire area under the curve (which is 1, since the curve is a density function):*

```
> 1-pnorm(700,mean=400,sd=150)
```

```
[1] 0.02275013
```

The graph in Fig. 10.9 can be constructed as follows:

```
>   m=400
>   stdev=150
>   right = 700
>   x=seq(m-4*stdev,m+4*stdev,0.1)
>   y=dnorm(x,mean=m,sd=stdev)
>   plot(x,y,type="l",col="red",axes=T,xlab="X",ylab="probability
+   density")
>   x2=x[x>right]
>   y2=dnorm(x2,mean=m,sd=stdev)
>   x2=c(x2,right)
>   y2=c(y2,y2[length(y2)])
>   polygon(x2,y2,col="yellow")
>   lines(c(-200,1000),c(0,0))
```

In the code above, **x** *and* **y** *store the coordinates of the points that form the graph of the density function. The* **x** *values are obtained as equally spaced samples from the interval given by the mean plus/minus 4 standard deviations. The values of* **y** *are provided by the* **dnorm** *function. The pairs of points (x,y) are plotted by the* **plot** *function, which also sets the color of the plot and adds the desired labels. In order to create the shaded area, those values of* **x** *larger than the desired value are sub-selected in* **x2** *and their corresponding y values are calculated and stored in* **y2**. *The function* **polygon** *can be used to plot an arbitrarily shaped polygon filled with a given color. However, this function closes the polygon drawn by joining the last point with first point. If nothing else is done, the last point in* **x2** *and* **y2** *will be somewhere at the extreme right of the density line. In order to obtain the correct shading, the supplementary point (700,0) needs to be added to the vector of coordinates by concatenating the existing vectors* **x2** *and* **y2** *with the coordinates of this point. The coordinates are specified as* **right** *and* **y2[length(y2)]** *in order to make the code more flexible (usable for other values of* **right**, *for instance).* □

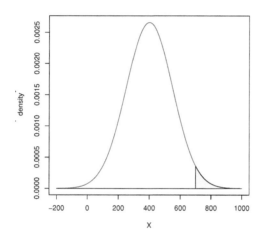

FIGURE 10.9: The shaded area under the graph of the pdf corresponds to the probability that we are interested in, $P(X > 700)$. However, the cdf of the function provided a value corresponding to the unshaded area to the left of 700: `pnorm(700,mean=400,sd=150)`. The value of interest can be calculated as `1-pnorm(700,mean=400,sd=150)`.

10.4 Central limit theorem

As discussed in Section 9.2, the central limit theorem makes some very surprising and rather strong statements about what happens when samples of a certain size are drawn from a given population. Let us try to use R's capabilities in order to investigate this phenomenon in more detail.

Let us recall the setup surrounding the central limit theorem. There is a population characterized by a mean μ_X and a standard deviation σ_X. Usually, these parameters of the population, as well as the distribution governing this population are not known. However, one has the possibility to draw samples from this population. For simplicity, these samples are of equal size, n, and theoretically, an infinite number of samples can be drawn. For each such sample, one calculates the sample mean, $\overline{X}$. The means of these samples follow a distribution which is usually referred to as the sampling distribution. Under these circumstances, the central limit theorem connects the parameters of the sampling distribution with the parameters of the original distribution. This allows us to compute statistics on the sampling distribution and, from them, infer things about the original, unknown, distribution.

The central limit theorem states that (see also Section 9.2): i) the means of the samples will be distributed normally; ii) the mean of this distribution of the means will be the same as the mean of the original population and iii) the standard deviation of these means, $\sigma_{\overline{X}}$, will be smaller by a factor of $\sqrt{n}$ compared to the standard deviation of the original population, σ_X.

Probably the most surprising fact revealed by the central limit theorem is that the distribution of the means of the samples will be approximatively normal, *independently of the shape of the original distribution*. This is nothing short of astonishing. What this means is that we can pick a number of equal samples from any distribution whatsoever, and as long as this distribution has finite mean and variance, the means of the samples will be normally distributed. Let us do some experiments in order to verify these assertions. To make things more interesting, we will take a distribution that is rather different from the normal, such as a χ^2 (chi-square) distribution with 5 degrees of freedom:

```
> chi2=rchisq(1000000,df=5)
> xlim=c(min(chi2),max(chi2))
> plot(density(chi2),main="Random numbers",
+ xlab=paste("mean=",round(mean(chi2),4),"sd=",round(sd(chi2),4)))

> x=seq(0,30,0.1)
> y=dchisq(x,df=5)
> plot(x,y,xlim=xlim,type="l",col="red",axes=T,xlab="X",
+ ylab="Chi-square",main="Theoretical distribution")
```

Here, we first took some equidistant points x in the interval $(0,30)$ and calculated the density of a χ^2 distribution with 5 degrees of freedom in those points. Line 3 above defines the graphical window as a matrix with 1 row and 2 columns, such that we can display the theoretical and simulated distributions side-by-side. The first call to the `plot` function displays the theoretical density function for this χ^2 distribution. The following line of code generates one million random values from this distribution and stores them in the vector v. The distribution of these points is plotted with the subsequent call to the `plot` function. Fig. 10.10 shows the result produced by the code snippet above. Note that this distribution is skewed, having a long tail to the right.

Now, let us try to take some samples from this distribution and construct the distribution of the sample means i.e. the sampling distribution. We would like to see what happens when the sample size increases so we will repeat this process for samples of size 4, 16, and 100. In order to be able to execute the code quickly, we will only construct the sampling distribution with a limited number of samples, 50 in this case. However, it is easy to experiment with the code by changing both the number of samples, as well as the sample size.

```
> set.seed(1012)
> chisq=rchisq(1000000,df=5)
> par(mfrow=c(1,3))
> no_samples = 50
> sizes=c(4,16,100)
> samples=list()
> means=matrix(nrow=length(sizes),ncol=no_samples)
> for( i in 1:length(sizes)){
+     current_sample = matrix(nrow=no_samples,ncol=sizes[i])
+     for(j in 1:no_samples){
+         current_sample[j,]=sample(chisq,sizes[i])
+         means[i,j]=mean(current_sample[j,])
+     }
+     samples<- c(samples,list(current_sample))
+     plot(density(means[i,]),
+     main=paste("sample size=",sizes[i]),
+     xlab=paste("mean=",round(mean(means[i,]),4),
+     "sd=",round(sd(means[i,]),4)),col="red")
+ }
```

The first line above, sets the seed used by the random number generator to a certain value. This value in itself is not important but rather the fact that every time this piece of code will be executed, the numbers drawn will be exactly the same. This is needed here only to make sure that the results are exactly the same between different compilations and executions of this code. The following line draws 1,000,000 random numbers from a chi-square distribution with 5 degrees of freedom. The third line, sets up a graphical window as a matrix with 1 row and 3 columns, one for each sample size. The number of samples is initialized with 50. The `for` loop repeats the experiment

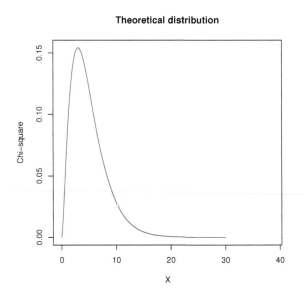

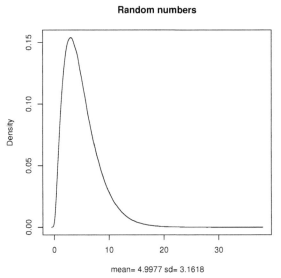

FIGURE 10.10: The chi-square distribution with df=5. The top panel shows the theoretical distribution; the bottom panel shows the distribution of 1,000,000 random numbers drawn from this distribution with `rchisq`.

for the 3 sample sizes we chose: 4, 16, and 100. For each sample size, a matrix `current_sample` is created with one row for each sample and one column for each value in the sample; hence, the matrix will have `no_samples` rows and `sample_size` columns. The inner `for` loop takes the desired number of samples from the data vector `chisq` and calculates the mean of each such sample. These means are values from the distribution of the means. This distribution is plotted in the last call to the `plot` function. Before plotting this, the current sample is attached to the end of the list of samples.

The results of these experiments are shown in Fig. 10.11 which illustrates very well all results predicted by the central limit theorem. First, even though the data sampled originated from the very skewed distribution shown in Fig. 10.10, the sampling distribution nicely gains the symmetry and resemblance of a normal distribution as the sample size increases. The mean of the original data, 4.999, is well approximated by the mean of the sampling distribution right from the beginning. However, the standard deviation is rather large for small sample sizes, but it becomes very small as the sample size increases.

```
> set.seed(1012)
> chisq=rchisq(1000000,df=5)
> sizes=c(4,16,100)
> sd(chisq)/sqrt(sizes)
```

`[1] 1.5798827 0.7899414 0.3159765`

In the line above, we started with the standard deviation of the original data and we calculated the expected standard deviations for each sample size. The values above match well the standard deviations shown in Fig. 10.11. Note that the graphs in this figure are somewhat misleading because the scale of each plot has been automatically adjusted to fit the given graph. The net result of this scaling is that all 3 distributions appear to be equally wide while in reality the distribution for n=100 is extremely narrow compared with the one for n=4. Similarly, the heights of the 3 distributions are very unequal. These differences become very obvious if the 3 graphs are plotted on the same scale (see Fig. 10.12). This figure shows well how an increase in the size of the sample makes the sample means provide more accurate information regarding the mean of the original population. In essence, the mean of a sample of 4 values can be anywhere from 0 to 10 (see also the x axis in Fig. 10.11). At the same time, the mean of a sample of 16 values from the same population is most likely to be somewhere between 3 and 8. However, the mean of a sample of 100 values will essentially be confined to the narrow interval (4,6) around the real mean $\mu = 4.999$.

Finally, the code used above was written in the classical style of a procedural language, such as C/C++/Java, without taking any advantage of the more advanced vector processing capabilities of R. The results obtained with the two nested loops in the previous code fragment can be obtained with the following few lines of code:

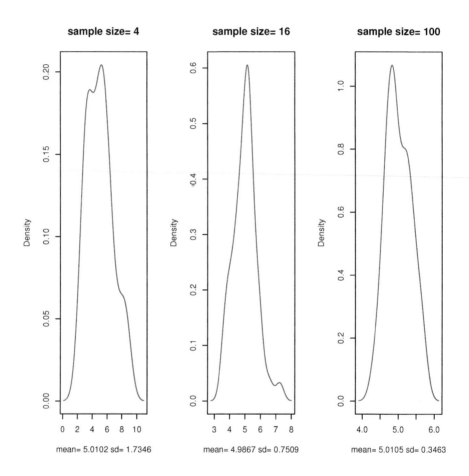

FIGURE 10.11: An illustration of the central limit theorem. The graphs show the sampling distribution for samples of different sizes drawn from the χ^2 distribution with 5 degrees of freedom shown in Fig. 10.10. As the sample size increases from left (n=4), to middle (n=16), to right (n=100), the distribution becomes more similar to the normal distribution. While the mean of the distribution of the means is a reasonable approximation of population mean even for a sample size of 4, the variance of the sampling distribution decreases proportional to the square root of the sample size, $\sqrt{n}$.

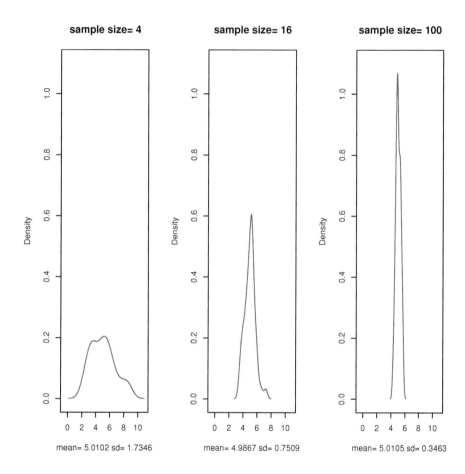

FIGURE 10.12: Another illustration of the central limit theorem. The graphs show the sampling distribution for samples of different sizes drawn from the χ^2 distribution with 5 degrees of freedom shown in Fig. 10.10. By maintaining the same scale between the graphs, this figure shows better how the sampling distribution quickly shrinks around the mean of the original distribution as the sample size increases.

```
> set.seed(1012)
> chisq=rchisq(1000000,df=5)
> par(mfrow=c(1,3))
> no_samples = 100
> sizes=c(4,100,1000)
> samples = lapply(sizes,
+ function(y) {list(values = t(sapply(c(1:no_samples),
+ function(x){sample(chisq,y)}))), size = y)})
> means = lapply(samples,function(x)
+ {list(values = apply(x$values,1,mean), size = x$size)})
> lapply(means, function(x) { mean(x$values)} )
> lapply(means,function(x){plot(density(x$values),
+ xlim=c(-0.5,11),main=paste("sample size",x$size))})
```

In the first line above (after the initializations), a list samples is constructed with `lapply` by executing the function of y for each element of the array sizes. In turn, this function constructs a list with two components: `values`, which stores samples from the appropriate sizes drawn from chisq; and `size`, which stores the size of the samples. The second line calculates the means of the samples drawn by calling a custom function of the samples list. This custom functions takes a matrix and applies the mean for each row using `apply`. Finally, `lapply` is used to construct the density plots for each sample size. While this code is more compact and probably more efficient than the explicit nested loops used above, it is likely to be more difficult to write (and debug) for a beginner in R.

10.5 Summary

This chapter revisited the concept discussed in Chapter 8 in the context of the R environment. Various descriptive statistics available in R, such as mean, median, mode, range, variance, and standard deviation, were illustrated here. At the same time, this chapter demonstrated how how densities, p-values, quantiles, and pseudorandom numbers can be obtained from the built-in distributions in R, thus eliminating the need for the classical statistical tables. R's capabilities were used to illustrate various properties of some classical distributions, as well as the central limit theorem.

10.6 Exercises

1. Write the R code that would illustrate the effects of the central limit theorem. Start with a uniform distribution from which you draw 1,000,000 values. These values form the population to be sampled. Then take 1,000 samples of different sizes: $n = 4, 16, 100$. Plot the sampling distributions obtained in each case. Calculate the population mean, μ and the means of each of the sampling distributions. Calculate the population variance σ^2 and the variances of each of the sampling populations. What is to be expected? How do the results match your expectations?

2. Write the R code that would illustrate the effects of the central limit theorem. Start with an exponential distribution from which you draw 1,000,000 values. These values would form the population to be sampled. Then take 1,000 samples of different sizes: $n = 4, 16, 100$. Plot the sampling distributions obtained in each case. Calculate the population mean, μ, and the means of each of the sampling distributions. Calculate the population variance, σ^2, and the variances of each of the sampling populations. What is to be expected? How do the results match your expectations?

3. Plot the graphs above on the same scale for each problem. Hint: you need to calculate the minimum and maximum values for both x and y axes, and pass these values to the plot function using the formal arguments xlim and ylim.

Chapter 11

Statistical hypothesis testing

The manipulation of statistical formulas is no substitute for knowing what one is doing.

— *Hubert Blalock, Jr., Social Statistics, Chapter 19*

11.1 Introduction

The goal of this chapter is to illustrate the process of testing a statistical hypothesis. Unfortunately, just being familiar with a number of statistical terms and a few distributions is not sufficient in order to draw valid conclusions from the data gathered. A clear formulation of the hypotheses to be tested as well as a clear understanding of the basic *mathematical* phenomena involved are absolutely necessary in order to be able to extract facts from data.

11.2 The framework

In order to illustrate the basic ideas involved in hypothesis testing, we will consider the following example.

EXAMPLE 11.1

The expression level of a gene in a given condition is measured several times. A mean $\overline{X}$ of these measurements is calculated. From many previous experiments, it is known that the mean expression level of the given gene in normal conditions is μ_X. We formulate the following hypotheses:

1. *The gene is up-regulated in the condition under study: $\overline{X} > \mu_X$*

2. *The gene is down-regulated in the condition under study: $\overline{X} < \mu_X$*

3. *The gene is unchanged in the condition under study: $\overline{X} = \mu_X$*

4. *Something has gone awry during the lab experiments and the gene measurements are completely off; the mean of the measurements may be higher or lower than the normal: $\overline{X} \neq \mu_X$*

This is already a departure from the lay thinking. Given two values $\overline{X}$ and μ_X, one could think that there are only 3 possibilities: $\overline{X} > \mu_X$, $\overline{X} < \mu_X$ or $\overline{X} = \mu_X$. This is indeed the case; the two values will necessarily be in only one of these situations. However, the emphasis here is not on the two values themselves but on *our expectations about them.* If we think along these lines, there are clearly four possibilities. The first 3 correspond to the relationship between the two variables: we might expect $\overline{X}$ to be higher than, equal to or lower than μ_X. However, the fourth possibility has nothing to do with the variable but reflects our teleological position: we have no clue!! The mean measurement might be either higher or lower than the real mean: $\overline{X} \neq \mu$. This is the first important thing to remember: *hypothesis testing is not centered on the data; it is centered on our* a priori *beliefs about it.* After all, this is the meaning of the term hypothesis: an *a priori* belief, or assumption, that needs to be tested.

The second important fact in hypothesis testing has to do with the fundamental characteristic of the statistical thinking. Let us consider a normal distribution like the one in Fig. 11.1. Let us consider that this distribution represents the (normalized) distribution of the expression levels of the gene under study. This distribution has the mean $Z = 0$. This corresponds to $\overline{X} = \mu_X$. We know that if the gene is up-regulated in the condition under study, it will have an expression value higher than normal. We would like to set a threshold to the right, such that, if a gene is higher than the given threshold, we can call it up-regulated. This is our goal.

We have seen that 95% or so of the values of a normal distribution will be situated within $\pm 2\sigma$. The fundamental fact that needs to be understood here is that X can actually take *any value whatsoever* even if it follows strictly a normal distribution. Yes, it is unlikely that a variable coming from a normal distribution with zero mean and $\sigma = 1$ takes a value larger than 2. Yes, it is *very* unlikely that it takes a value larger than 3. And it is even more unlikely that it will take a value larger than 5. However, this is not impossible. In fact, the probability of this event can be easily calculated from the expression of the normal distribution. Once this fact is accepted, the conclusion follows: no matter where we set the threshold, a member of the original, normal distribution may end up there just by random chance. Therefore, *independently of the value of the threshold, we can never eliminate the possibility of making a mistake.*

Once we understand this fact, we have to adjust our goals: we should not aim at finding a threshold that allows us to identify correctly the up-regulated genes since this is futile; instead, we should aim at being able to calculate the exact probability of making a mistake for any given threshold. There is a big difference between a value situated 0.5 standard deviations larger than the mean and a value 3 standard deviations larger than the mean. The former will probably occur about 1 in 3 measurements while the latter will probably appear only once in 1,000 measurements. If we were to bet against this, we might feel uncomfortable to do so against the former while betting against the latter would be much safer. However, not all situations are suitable for taking the same amount of risk. For instance betting at roulette with a chance of losing of only 1 in 3 is not bad. This would mean winning 2 out of every 3 turns. Many people would do that.[1] However, jumping from a plane with a parachute that will open only 2 out of 3 jumps would probably not gather many volunteers. Sometimes, even very, very small probabilities of a negative outcome are unacceptable if the cost of such an occurrence is very high. As a real-life example, Ford Motor Co. has recently recalled 13,000,000 Firestone tires because of 174 deaths and 700 injuries related to the tire tread separation. The empirical probability of being injured or dying because of this was only $874/13,000,000 = 0.00006$ or less than 1 in 10,000. However, Ford decided this was unacceptable and recalled the tires for a total cost of approximatively 3,000,000,000 dollars.

We saw earlier that it is not possible to choose a fixed threshold for the measurement values. The examples above illustrated the fact that choosing an a priori, fixed, threshold for the probability of making a mistake is not possible. Therefore, we will focus on what is both possible and necessary, that is calculating the probability of making a mistake for any given situation. Let us consider the situation illustrated in Fig. 11.1. Let us consider that this is the distribution of the gene expression levels in the normal situation (on

---

[1] Heck, many people play roulette anyway, even if the chances of losing are much higher than 0.33.

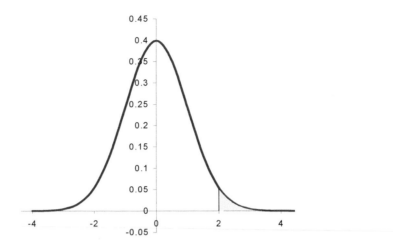

FIGURE 11.1: A normal distribution with zero mean and standard deviation of one. The probability of a measurement being to the right of the marked value is given by the shaded area.

controls). Let us consider that we set the threshold for up-regulated genes at 2 standard deviations away from the mean. If a gene has a measurement of, for instance, $Z = 4$, we will call that gene up-regulated. However, a gene from the normal distribution of controls might have such a high expression value. This is unlikely, of course, but it may happen. If this happens and we call such a gene up-regulated because its value is larger than the threshold, we will be making a mistake: even if the gene has a high expression level, it comes from the normal distribution. The probability of this happening is called the **p-value**. The p-value provides information about the amount of trust we can place in a decision made using the given threshold. A more exact definition of the p-value will be given shortly.

Calculating the p-value for a given threshold is not difficult and can be accomplished as follows. The probability of a value being higher than a threshold x is 1 minus the probability of it being lower than or equal to the given threshold:

$$P(Z > x) = 1 - P(Z \leq x) \tag{11.1}$$

Furthermore, the probability of a value from a distribution being lower than a given threshold x is given directly by its cdf or by the area under its pdf. In Fig. 11.1, the shaded area gives the probability of a value from the given distribution being larger than 2. This is exactly the probability of making a mistake if we used 2 as the threshold for choosing up-regulated genes, or the p-value associated to this threshold.

11.3 Hypothesis testing and significance

Hypothesis testing involves several important steps. The first step is to clearly define the problem. Such a problem might be stated as follows:

EXAMPLE 11.2
The expression level c of a gene is measured in a given condition. It is known from the literature that the mean expression level of the given gene in normal conditions is μ. We expect the gene to be up-regulated in the condition under study, and we would like to test whether the data support this assumption.

The next step is to generate two hypotheses. These are statistical hypotheses and, unlike biological hypotheses, they have to take a certain, very rigid form. In particular, the two hypotheses must be **mutually exclusive** and **all inclusive**. Mutually exclusive means that the two hypotheses cannot be true both at the same time. All inclusive means that their union has to cover all possibilities. In other words, no matter what happens, the outcome has to be included in one or the other hypothesis. One hypothesis will be the null hypothesis. Traditionally, this is named H_0. The other hypothesis will be the alternate or research hypothesis, traditionally named H_a. The alternate hypothesis, or the research hypothesis, has to reflect our expectations. If we believe that the gene should be up-regulated, the research hypothesis will be $H_a : c > \mu$. The null hypothesis has to be mutually exclusive and also has to include all other possibilities. Therefore, the null hypothesis will be $H_0 : c \leq \mu$.

We have seen that the p-value is the probability of a measurement more extreme than a certain threshold occurring just by chance. If the measurement occurred by chance, and we drew a conclusion based on this measurement, we would be making a mistake. We would erroneously conclude that the gene is up-regulated when in fact the measurement is only affected by chance. In the example considered previously (see Fig. 11.1), this corresponds to rejecting the null hypothesis $H_0 : c \leq \mu$ even if this null hypothesis is in fact true. In other words, the **p-value is the probability of drawing the wrong conclusion by rejecting a true null hypothesis**. Choosing a significance level means choosing a maximum acceptable level for this probability. The **significance level is the amount of uncertainty we are prepared to accept** in our studies.

For instance, when we choose to work at a significance level of 10% we accept that 1 in 10 cases our conclusion can be wrong. A significance level of 20% means that we can be wrong 1 out of 5 cases. Usual significance levels are 1%, 5%, 10%, and 15% depending on the situation. At 15% significance level, there might be a real phenomenon, but it may also be just a random effect; a repetition of the experiment is necessary. At 10%, not very many people will doubt our conclusions. At 5% significance level we can start betting money on it, and at 1% other people will start betting money on our claims, as well. The 5% and 1% probability levels are thus standard critical levels.

Returning to the flow of hypothesis testing, once we have defined the problem and made explicit hypotheses, we have to choose a significance level. The next step is to calculate an appropriate statistic based on the data and calculate the p-value based on the chosen statistic. Finally, the last step is to either: i) reject the null hypothesis and accept the alternative hypothesis or ii) not reject the null hypothesis.

To summarize, the main steps of the hypothesis testing procedure are as follows :

1. Clearly define the problem.

2. Generate the null and research hypothesis. The two hypotheses have to be mutually exclusive and all inclusive.

3. Choose the significance level.

4. Calculate an appropriate statistic based on the data and calculate a p-value based on it.

5. Compare the calculated p-value with the significance level and either reject or not reject the null hypothesis.

This process will be now detailed and illustrated on some examples.

11.3.1 One-tailed testing

In the simplest case, we expect the measurement to have a tendency. For instance, if we are working to improve an algorithm or a procedure, we might expect that the results of the improved procedure would be better than the results of the original. In gene expression experiments, we might understand or hypothesize the function of a gene and predict the direction of its changes.

EXAMPLE 11.3
The expression level of a gene is measured 4 times in a given condition. The 4 measurements are used to calculate a mean expression level of $\overline{X} = 90$. It is known from the literature that the mean expression level of the given gene, measured with the same technology in normal conditions, is $\mu = 100$ and the standard deviation is $\sigma = 10$. We expect the gene to be down-regulated in the condition under study and we would like to test whether the data support this assumption.

SOLUTION *We note that the sample mean is 90, which is lower than the normal 100. However, the question is whether this is meaningful since such a value may also be the result of some random factors. We need to choose a* *significance level at which we are comfortable for the given application. Let us say that we would like to publish these results in a peer-reviewed journal and, therefore, we would like to be fairly sure of our conclusions. We choose a significance level of 5%.*

We have a clear problem and a given significance level. We need to define our hypotheses. The research hypothesis is "the gene is down-regulated" or:

$$H_a : \overline{X} < \mu \tag{11.2}$$

The null hypothesis has to be mutually exclusive and the two hypotheses together have to be mutually inclusive. Therefore:

$$H_0 : \overline{X} \geq \mu \tag{11.3}$$

Note that the two hypotheses do cover all possible situations. This is an example of a one-sided, or a one-tail, hypothesis in which we expect the values to be in one particular tail of the distribution.

From the sampling theorem, we know that the means of samples are distributed approximately as a normal distribution. Our sample has size $n = 4$ and mean $\overline{X} = 90$. We can calculate the value of Z as:

$$Z = \frac{\overline{X} - \mu}{\frac{\sigma}{\sqrt{n}}} = \frac{90 - 100}{\frac{10}{\sqrt{4}}} = \frac{-10}{10} \cdot 2 = -2 \tag{11.4}$$

The probability of having such a value just by chance, i.e. the p-value, is:

$$P(Z < -2) = F(-2) = 0.02275 \tag{11.5}$$

The computed p-value is lower than our significance threshold $0.022 < 0.05$. In the given circumstances, we can reject the null hypothesis. The situation is illustrated in Fig. 11.2. The shaded area represents the area corresponding to the critical value. For any value of Z in this area, we will reject the null hypothesis. For values of Z anywhere in the remainder of the area under the graph of the pdf, we will not be able to reject the null hypothesis.

Since the hypotheses were designed to be mutually exclusive and all inclusive, we must now accept the research hypothesis. We state that "the gene is down-regulated at 5% significance level." This will be understood by the knowledgeable reader as a conclusion that is wrong in 5% of the cases or fewer. This is the classical approach to hypothesis testing. The p-value is simply compared to the critical value corresponding to the chosen significance level; the null hypothesis is rejected if the computed p-value is lower than the critical value. However, there is a difference between a p-value barely below the 0.05 threshold and a p-value several orders of magnitude lower. An alternative approach is to merely state the p-value and let the reader assess the extent to which the conclusion can be trusted. If we wanted to follow this approach, we would state that "the gene is down-regulated with a p-value of 0.0228."

□

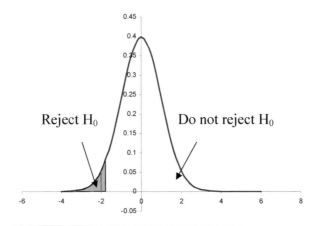

FIGURE 11.2: The computed statistic (-2) is more extreme than the critical value corresponding to the chosen level of significance. In this case, the null hypothesis can be rejected.

EXAMPLE 11.4

The expression level of a gene is measured twice in a given condition. The 2 measurements are used to calculate a mean expression level of $\overline{X} = 90$. It is known from the literature that the mean expression level of the given gene, measured with the same technology in normal conditions, is $\mu = 100$ and the standard deviation is $\sigma = 10$. We expect the gene to be down-regulated in the condition under study and we would like to test whether the data support this assumption.

SOLUTION   *Note that this is exactly the same problem as before, with the only difference that now the gene was measured only twice. We will use the same null and research hypotheses:*

$$H_a : \overline{X} < \mu \tag{11.6}$$

$$H_0 : \overline{X} \geq \mu \tag{11.7}$$

We are still dealing with the mean of a sample. Therefore, independently of the distribution of the gene expression level, we can use the normal distribution to test our hypotheses. We can calculate the value of Z as:

$$Z = \frac{\overline{X} - \mu}{\frac{\sigma}{\sqrt{n}}} = \frac{90 - 100}{\frac{10}{\sqrt{2}}} = \frac{-10}{10} \cdot 1.1414 = -1.414 \tag{11.8}$$

The probability of having such a value just by chance, i.e. the p-value, is:

$$P(Z < -1.41) = F(-1.41) = 0.0792 \tag{11.9}$$

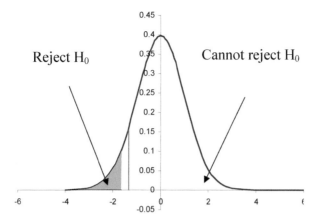

FIGURE 11.3: The computed statistic (-1.414) is less extreme than the critical value corresponding to the chosen level of significance. In this case, the null hypothesis cannot be rejected.

This time, the computed value is higher than the 0.05 threshold, and in this case, we will not be able to reject the null hypothesis. This is illustrated in Fig. 11.3. In this case, the measured difference may be the effect of a random event. □

The two examples above illustrated the essence of a general statistical phenomenon. If a variable seems different from its usual values, we can accept there is a real cause behind it if: i) either the difference is large enough (such as more than 3σ from the mean); or ii) the difference is small but this difference is obtained consistently, over sufficiently many measurements. In principle, the smaller the difference, the more measurements we will need in order to reject the null hypothesis for a given significance level.

The reasoning is the same even if the gene is expected to be up-regulated. This means that the measurement is in the other tail of the distribution. However, this is still a one-sided hypothesis. The following example illustrates this.

EXAMPLE 11.5

The BRCA1 mutation is associated with the over-expression of several genes [453]. HG2855A is one such gene. An experiment involving samples collected from a subject with the BRCA1 mutation measures the expression level of this gene 4 times. The mean of these measurements is 109. It is known from the literature that the standard deviation of the expression level of this gene is 10 and the mean expression level is 100. Does the data support the hypothesis that this gene is up-regulated?

SOLUTION We state the two hypotheses as:

$$H_a : \overline{X} > \mu \qquad (11.10)$$

$$H_0 : \overline{X} \leq \mu \qquad (11.11)$$

Note that the alternative hypothesis reflects our expectations and that the null hypothesis includes the equality.

At this time, that is before doing any analysis, we also choose the significance level. We will work at a significance level of 5%.

The next step is to choose the appropriate statistical model. In this case, we are working with a mean of a sample and we know from the central limit theorem that sample means are normally distributed. Therefore, we can use the normal distribution. We normalize our variable by applying the Z transformation. We subtract the mean and divide by the standard deviation:

$$Z = \frac{\overline{X} - \mu}{\frac{\sigma}{\sqrt{n}}} = \frac{109 - 100}{\frac{10}{\sqrt{4}}} = \frac{9}{10} \cdot 2 = 1.8 \qquad (11.12)$$

We calculate the probability of the Z variable taking a value greater than 1.8:

$$P(Z > 1.8) = 1 - P(Z \leq 1.8) = 1 - F(1.8) = 1 - 0.96409 = 0.0359 \qquad (11.13)$$

This value is lower than the chosen significance level $0.036 < 0.05$. Based on this, we can reject the null hypothesis and accept the research hypothesis: the gene is indeed up-regulated at the 5% significance level. This situation is illustrated in Fig. 11.4.

$\square$

11.3.2 Two-tailed testing

In some situations, however, we might not have any precise expectations about the outcome of the event. For instance, in a gene expression experiment we may not have any knowledge about the behavior of a given gene. Thus, the gene might be either up-regulated or down-regulated. In such cases, the reasoning has to reflect this. This is called a two-side or a two-tailed test. The following example will illustrate such a situation.

EXAMPLE 11.6

A novel gene has just been discovered. A large number of expression experiments performed on controls revealed that the technology used together with consistent normalization techniques measured the mean expression level of this gene as 100 with a standard deviation of 10. Subsequently, the same gene is measured 4 times in 4 cancer patients. The mean of these 4 measurements is 109. Can we conclude that this gene is up-regulated in cancer?

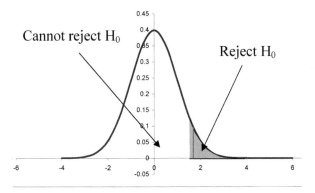

Cannot reject H_0

Reject H_0

FIGURE 11.4: The computed statistic (1.8) is more extreme than the critical value corresponding to the chosen significance level. The probability of this happening by chance (0.036) is lower than the 0.05 significance chosen. In this case, the null hypothesis is rejected: the gene is up-regulated at the 5% significance level.

SOLUTION   *In this case, we do not know whether the gene will be up-regulated or down-regulated. The only research hypothesis we can formulate is that the measured mean in the condition under study is different from the mean in the control population. In this situation, the two hypotheses are:*

$$H_a : \overline{X} \neq \mu \tag{11.14}$$

$$H_0 : \overline{X} = \mu \tag{11.15}$$

We will work at the same significance level of 5%. The significance level corresponds to the probability of calling a gene differentially regulated when the gene has an unusual value in one of the tails of the histogram. In a one-sided test, we expected the gene to be in a specific tail, and we set the threshold in such a way that the area in that tail beyond the threshold was equal to the chosen significance level. However, in this case, the gene can be in either one of the tails. In consequence, we have to divide the 5% probability into two equal halves: 2.5% for the left tail and 2.5% for the right tail. This situation is presented in Fig. 11.5.

We are dealing with a mean of a sample so we can use the normal distribution. We normalize our variable by applying the Z transformation. We subtract the mean and divide by the standard deviation:

$$Z = \frac{\overline{X} - \mu}{\frac{\sigma}{\sqrt{n}}} = \frac{109 - 100}{\frac{10}{\sqrt{4}}} = \frac{9}{10} \cdot 2 = 1.8 \tag{11.16}$$

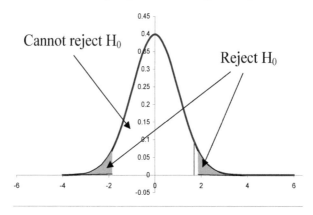

FIGURE 11.5: In a two-tailed test, the probability corresponding to the chosen significance level is divided equally between the two tails of the distribution.

We calculate the probability of the Z variable taking a value greater than 1.8:

$$P(Z > 1.8) = 1 - P(Z \le 1.8) = 1 - F(1.8) = 1 - 0.96409 = 0.0359 \quad (11.17)$$

This value is higher than the chosen significance level 0.036 > 0.025. In this situation, we cannot reject the null hypothesis. ⬚

Note that all the numerical values in examples 11.5 and 11.6 were identical. However, in example 11.5, the null hypothesis was rejected and the gene was found to be differentially regulated at 5% significance while in example 11.6, the null hypothesis could not be rejected at the same significance level. This illustrates the importance of defining correctly the hypotheses to be tested *before the experiment is performed.*

11.4 "I do not believe God does not exist"

The Spanish philosopher Miguel de Unamuno pointed out in one of his writings that one's position with respect to religion can take one of four alternatives. This is somewhat contrary to the common belief that one either believes in God or one does not. According to Miguel de Unamuno, the four different positions that one can adopt are as follows:

1. "I believe God exists"

2. "I do not believe God exists"

3. "I believe God does not exist"

4. "I do not believe God does not exist"

A careful consideration of the 4 statements above shows that all 4 are indeed distinct possibilities, very different from one another. Saying "I believe God exists" expresses a positive conviction about God's existence. This person knows something and that something happens to be about God's existence. "I do not believe God exists" expresses the lack of such a positive conviction. This person expressed their ignorance about something and that something happens to be again about God's existence. The third formulation, "I believe God does not exist" expresses another positive belief, but this time it happens to be about the exact opposite: God's lack of existence. Finally, "I do not believe God does not exist" reflects the lack of a strong opinion regarding the same matter. Statements 1 and 3 reflect a positive attitude, the speaker feels certain about some fact. Statements 2 and 4 reflect a lack of knowledge about the facts of the matter. Let us substitute the "God exists" with x. Clearly, "I do not believe God does not exist" is not equivalent to saying "I believe God exist." The former merely states a lack of information about non-x, whereas the latter makes a strong statement about x. These are not the same and cannot be substituted for each other. The lack of information, or proof, about something can in no circumstances be taken as proof about anything else. In particular, the lack of proof for a fact x cannot, under any circumstances, be taken as proof for the opposite of x. Even Donald Rumsfeld got this right: "the absence of evidence is not the evidence of absence." It is basically saying the same thing in a different way. Simply because you do not have evidence that something does exist does not mean that you have evidence that it doesn't exist.[2]

This applies to the interpretation of the results of the hypothesis testing as described above. In particular, in the hypothesis testing there is a null hypothesis and a research hypothesis. As we have seen, sometimes the data do not allow us to reject the null hypothesis. This means that we do not have enough support to accept the research hypothesis. However, this does not necessarily mean that the research hypothesis is false. As an example, let us consider the null hypothesis: H_0: "Genetically modified foods (GMOs) are safe" and the research hypothesis: H_a: "GMOs make the descendants of humans who consume them grow horns and tails 4 generations later." Since we have started to consume GMOs on a large scale only about 2 generations ago, no study will be able to reject the null hypothesis. The conclusion of such a study may be expressed as: "There is no data to support the idea that GMOs

---

[2] About Iraq's weapons of mass destruction.

are harmful" or "There is no data to support the idea that GMOs are unsafe." However, and this is the big trap, *this does not mean that GMOs are safe.*

As a final, and hopefully convincing example, the tobacco industry claimed for many years that "there was no evidence showing that tobacco is harmful." Millions of people interpreted that as meaning "tobacco is not harmful" and jeopardized their health by smoking. When finally, enough data were gathered to prove that tobacco causes cancer and a plethora of other health problems, people got very upset and the tobacco companies were sued for very large amounts of money. This shows that we should ask for positive evidence of the type "show us it is safe" instead of a lack of evidence of negative effects such as "there is no data showing it is harmful."

Going back to our gene expression experiments, let us assume that the research hypothesis is "the gene is up-regulated" and the null hypothesis is that "the gene is not up-regulated." In these circumstances, only two conclusions are possible:

1. The data are sufficient to reject the null hypothesis or

2. We cannot reject the null hypothesis.

If the null hypothesis is rejected, the research hypothesis has to be accepted because the two are mutually exclusive and all inclusive. Therefore, we can conclude the gene is up-regulated.

If the null hypothesis cannot be rejected, it means we cannot show it is false. However, the statistical argument can never conclude that the null hypothesis is true. In other words, we cannot say "the gene is NOT up-regulated." The gene may or may not be up-regulated. We simply do not have enough data to draw a conclusion at the chosen significance level. It may well happen that the very next experiment will produce more data (e.g., another replicate measurement with the same value), which, together with the already existing data, may become conclusive proof that the gene is up-regulated (see examples 11.5 and 11.6 above).

11.5 An algorithm for hypothesis testing

We can now summarize an algorithm that can be used for hypothesis testing. Each step of the algorithm is illustrated with the corresponding step in Example 11.6.

1. State the problem: Is the gene up-regulated?

2. State the null and alternative hypothesis:

$$H_a : \overline{X} \neq \mu$$

$$H_0 : \overline{X} = \mu$$

Identify whether the situation corresponds to a one-tailed or two-tailed test; in this case, this is a two-tailed test.

3. Choose the level of significance: 5%

4. Find the appropriate test statistic: we were using means of samples; therefore, we used the Z transformation and normal distribution (from the central limit theorem).

5. Calculate the appropriate test statistic: $Z = 1.8$

6. Determine the p-value of the test statistic (the prob. of it occurring by chance):

$$P(Z > 1.8) = 1 - P(Z \leq 1.8) = 1 - F(1.8) = 1 - 0.96409 = 0.0359$$

7. Compare the p-value with the chosen significance level: $0.0359 > 0.025$.

8. Reject or do not reject H_0 based on the test above: here we cannot reject H_0.

9. Answer the question in step 1: the data do not allow us to conclude the gene is up-regulated.

11.6   Errors in hypothesis testing

The hypothesis testing involved defining the H_0 (null) and H_a (research) hypotheses. So far, we have discussed the reasoning that allows us to pass judgment using such hypotheses. A very important issue is to assess the performance of the decisions taken with the algorithm presented in section 11.5. In order to do this, let us consider that the true situation is known. We will apply the algorithm as above then interpret the results in the context of the true situation.

Let us assume that H_0 is actually true and H_a is false. In this case:

- If we accept H_0, we have drawn the correct conclusion. We will call the instances in this category true negatives: they are negatives because they go against our research hypothesis and they are true because H_0 is indeed true and our H_a is false. If H_0 is "the gene is not regulated" and H_a is "the gene is either up or down-regulated," true negatives will be those genes that are not regulated and are reported as such by our algorithm.

- If we reject H_0, we have drawn an incorrect conclusion. We will call the instances in this category false positives. They are positives because they go with our research hypothesis and they are false because they are reported as such by our algorithm while, in fact, they are not. Non-regulated genes reported as regulated by our algorithm would be false positives.

Rejecting a null hypothesis when it is in fact true is called a **Type I error**. The probability of a Type I error is usually denoted by α. Let us consider a normal distribution such as the one presented in Fig.11.4. Using a threshold such as the one shown in the figure will classify as up-regulated any gene expressed at a level higher than the threshold. However, this means all the genes in the shaded area of the graph will be false positives since H_0 will be rejected for them. Therefore, *the probability of a Type I error corresponds directly to the significance level chosen.*

Let us now assume H_0 is false and H_a is true. In this case:

- If we accept H_0, we have drawn an incorrect conclusion. The instances in this category will be false negatives (genes that are in fact regulated but are not reported as such by our algorithm).

- If we reject H_0, we have drawn the correct conclusion. The instances in this category are true positives.

The second type of mistake is called a **Type II error**. The probability of a Type II error is denoted by β. The probability of avoiding a Type II error corresponds to correctly picking the instances that do not belong to the distribution of reference. For our purposes, this corresponds to finding the differentially regulated genes (which do not belong to the distribution representing the normal variation of expression levels). This is exactly the purpose of the hypothesis testing algorithm. The higher this probability, the better will the algorithm be at finding such genes. This probability is called the **power of the test** and can be calculated as $1 - \beta$.

The possible outcomes of hypothesis testing are summarized in Table 11.1. The columns correspond to the true but unknown situation, whereas the rows correspond to the two possible outcomes of the hypothesis testing. Row 1 corresponds to a situation in which H_0 cannot be rejected; row 2 corresponds to a situation in which H_0 can be rejected and, therefore, H_a can be accepted.

A natural tendency is to try to minimize the probability of making an error. The probability of making a Type II error is not directly controllable by the user.[3] However, the probability of making a Type I error, α, is exactly the significance level. Therefore, one might be tempted to use very high standards and choose very low values for this probability. Unfortunately, this is not a good idea since α and β are closely related and using a lower value for α has

---

[3] This probability can be controlled only indirectly, through a careful choice of the number of measurements (sample size).

	True (but unknown) situation	
Reported by the test	H_0 is true	H_0 is false
H_0 was not rejected	true negatives (correct decision) $1 - \alpha$	false negatives (Type II error) β
H_0 was rejected	false positives (Type I error) α	true positives (correct decision) $1 - \beta$

TABLE 11.1: The possible outcomes of hypothesis testing. The probability of making a Type I error corresponds directly to the significance level α. The probability of making a Type II error (β) corresponds to the number of false negatives. The term $1 - \beta$ is the power of the test and corresponds to the number of true positives that can be found.

the immediate consequence of reducing the power of the test. An example will illustrate this phenomenon.

Let us assume that our measurements actually come from two distributions. Fig. 11.6 illustrates this situation. The genes that are not expressed have the distribution to the left (towards lower intensities), while the gene that are expressed will come from the distribution to the right (towards higher intensities). Assume we are using the null hypothesis "the gene is not expressed." We choose the threshold corresponding to a significance level of 5%. For any gene with a measured value to the right of this threshold, we will reject the null hypothesis and conclude the gene is expressed. If a gene from the distribution to the left (not expressed) happens to be in this area, we will be making a Type I error (false positive). If a gene from the distribution to the right is in this area, we will correctly categorize it as expressed. The area to the right of the threshold and belonging to the distribution on the left (horizontal dashing) is equal to the significance level and the probability of a Type I error. The area to the left of the threshold and belonging to the distribution on the right (vertical dashing) is equal to the probability of a Type II error. In this area, we will not be able to reject the null hypothesis. For a gene from the left distribution, this is the correct decision. For a gene from the right distribution, we will be making a Type II error (false negative).

Fig. 11.7 shows what happens if a lower α is used and the threshold is moved to the right. The area corresponding to the Type I error will be reduced, but, at the same time, the area corresponding to a Type II error will be increased. This reduces the power of the test. The genes from the right distribution that fall between the old and the new position of the threshold will now be false negatives while before they were correctly identified as true

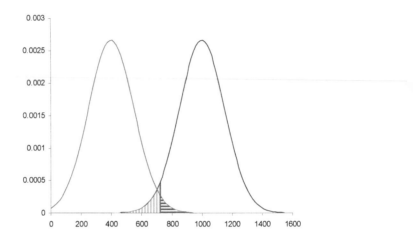

FIGURE 11.6: A problem with measurements from two populations. The distribution to the left represents genes that are not expressed, the distribution to the right corresponds to expressed genes. Assume we are using the null hypothesis "the gene is not expressed." We choose the threshold corresponding to a significance level of 5%. For any gene with a measured value to the right of this threshold we will reject the null hypothesis and conclude the gene is expressed. The area to the right of the threshold and belonging to the distribution on the left (horizontal dashing) is equal to the significance level and the probability of a Type I error. The area to the left of the threshold and belonging to the distribution on the right (vertical dashing) is equal to the probability of a Type II error.

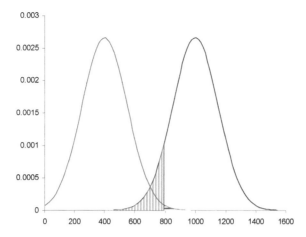

FIGURE 11.7: The same problem as in Fig. 11.6 but with a threshold corresponding to a lower α. The area corresponding to a Type I error is decreased but the area corresponding to a Type II error (β) is increased. In consequence, the power of the test, $1 - \beta$, will be diminished.

positives. The optimal threshold corresponds to the value for which the two distributions yield the same probability (same height on this graph).

11.7 Summary

This chapter presented the framework of the classical statistical hypothesis testing. The key elements introduced here include how to formulate the research and null hypothesis, p-value and its link to the significance level, types of errors and the link between significance and power.

Important aspects discussed in this chapter include:

- The framework of the statistical testing (hypotheses, significance levels, etc.) has to be defined before the experiment is done.

- When using two-tailed testing, the probability corresponding to the significance level is equally distributed between the tails of the histogram. Thus, the computed p-value is compared with a probability of $\alpha/2$ instead of α.

- Not being able to reject the null hypothesis does not mean that the null hypothesis is false. For instance, the fact that there is no evidence that

x is harmful does not mean that x is safe. Similarly, not being able to conclude that a gene is differentially regulated does not mean that the gene is not so.

• There are two types of errors in classical hypothesis testing. The Type I error corresponds to a situation in which a true null hypothesis is rejected. The Type II error corresponds to not rejecting a false null hypothesis. The probability of a Type I error is the significance level chosen. The probability of a Type II error is denoted by β. The quantity $1 - \beta$ is known as the power of the test and is directly proportional to the ability of the test to identify correctly those instances for which the research hypothesis is true (e.g., differentially regulated genes).

• The requirements of decreasing the probability of Type I (α) and Type II (β) errors are contradictory. Lowering α increases β and reduces the power of the test. This is to say that lowering the probability of false positives also reduces the ability of detecting true positives. Conversely, lowering β increases α. This is to say that increasing the ability of the test to detect true positives also increases the probability of introducing false positives.

11.8   Solved problems

1. The expression level of a gene is measured in a particular experiment and found to be 90. A large number of control experiments have been done with the same technology and it is known that:

   (a) The expression level of this gene follows a normal distribution in the control population.

   (b) The standard deviation of the expression level of this gene in controls is 10.

   (c) The mean expression level in controls is 100.

   The experiment is expected to suppress the activity of this gene. Can one conclude that the gene is indeed expressed at a lower level in the experiment performed?

   SOLUTION   We choose a significance level of 5%. We expect the gene to be lower so this is a one-tailed test. We formulate the null and research hypotheses.

$$H_a : \overline{X} < \mu \tag{11.18}$$

$$H_0 : \overline{X} \geq \mu \qquad (11.19)$$

This time we have a single measurement as opposed to a mean of a sample. Therefore, we cannot use the central limit theorem to assume normality. Fortunately, in this case, it is known that the expression level of the given gene follows a normal distribution. If this had not been a known fact, one would have had to study first how the gene is distributed in the control population. However, in the given circumstances we can calculate Z as follows:

$$Z = \frac{X - \mu}{\sigma} = \frac{90 - 100}{10} = -1 \qquad (11.20)$$

The probability of obtaining a value this small or smaller just by chance is:

$$P(Z < -1) = 0.158 \qquad (11.21)$$

This value is higher than 0.05 (5%), and therefore, we cannot reject the null hypothesis. Note that the null hypothesis cannot be rejected even if we lower our standards to 10% or even 15%. Also note that the values involved are exactly the same as in example 11.3: the measured value was 90 and the problem was modeled by a normal distribution with $\mu = 100$ and $\sigma = 10$. However, in example 11.3 there was enough evidence to reject the null hypothesis because the measured value was obtained from 4 replicates. □

2. A particular gene is expected to be up-regulated in a given experiment. The gene is measured 4 times and a mean expression level of 109 is obtained. It is known from the literature that the standard deviation of the expression level of this gene is 10 and the mean expression level is 100. Can one conclude that the gene is indeed up-regulated?

SOLUTION

We choose a significance level of 5%. We expect the gene to be expressed higher so this is a one-tailed test.

We formulate the null and research hypotheses as follows:

$$H_a : \overline{X} > \mu \qquad (11.22)$$

$$H_0 : \overline{X} \leq \mu \qquad (11.23)$$

In this problem, we do not know whether the gene is distributed normally. However, the gene has been measured several times and we know the mean of those measurements. According to the central limit theorem, the distribution of such sample mean values approximates well a normal distribution. We can calculate the value of Z as:

$$Z = \frac{\overline{X} - \mu}{\frac{\sigma}{\sqrt{n}}} = \frac{109 - 100}{\frac{10}{\sqrt{4}}} = 1.8 \qquad (11.24)$$

The probability of obtaining such a value just by chance is (p-value) :

$$P(Z > 1.8) = 1 - P(Z \leq 1.8) = 1 - 0.9647 = 0.0359 \qquad (11.25)$$

We compare the p-value with the significance level $0.0359 < 0.05$ and conclude that we can reject the null hypothesis at the chosen 5% level of significance. We can indeed conclude that, at this significance level, the gene is up-regulated.

□

Chapter 12

Classical approaches to data analysis

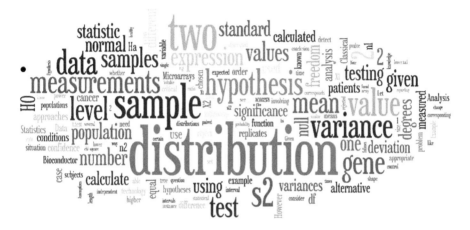

A knowledge of statistics is like a knowledge of foreign languages or of algebra; it may prove of use at any time under any circumstances.

— *Arthur L. Bowley,* Elements of Statistics, *Part I, Chapter 1, p. 4*

12.1 Introduction

This chapter will illustrate the use of classical methods in analyzing microarray data. The reader should be familiar with the basic statistical terms defined in Chapter 8, as well as with the hypothesis testing issues discussed in Chapter 11.

There are two different criteria that are used to distinguish between different types of problems and therefore hypothesis testing situations:

1. The number of samples involved

2. Whether we assume that the data comes from a known distribution or not.

361

According to the first criterion, there are three different types of problems that can arise. These are:

1. Problems involving one sample
2. Problems involving two samples
3. Problems involving more than two samples.

The second criterion divides hypothesis testing into:

1. Parametric testing – where the data are known or assumed to follow a certain distribution (e.g., normal distribution)
2. Non-parametric testing – wherr no a priori knowledge is available and no such assumptions are made.

The two criteria are orthogonal in the sense that each distinct class according to the first criterion can fall into any of the classes according to the second criterion. Thus, there are 6 possible combinations and therefore 6 different approaches to hypothesis testing:

1. Parametric testing

    (a) Involving one sample
    (b) Involving two samples
    (c) Involving more than two samples

2. Non-parametric testing

    (a) Involving one sample
    (b) Involving two samples
    (c) Involving more than two samples

In this text, we will focus on parametric testing. Non-parametric testing does not rely on any particular assumptions about an underlying distribution and often works with the order of the measurements instead of their values. This may be problematic for a small number of measurements as it is often the case in microarray experiments. However, some non-parametric tests (e.g., Wilcoxon's test) are often used in the analysis of microarray data.

12.2 Tests involving a single sample

12.2.1 Tests involving the mean. The t distribution.

Tests involving a single sample may focus on the *mean* or *variance* of the sample. The following hypotheses may be formulated if the testing regards the mean of the sample:

$$H_0 : \mu = c, \ H_a : \mu \neq c \tag{12.1}$$

$$H_0 : \mu \geq c, \ H_a : \mu < c \tag{12.2}$$

$$H_0 : \mu \leq c, \ H_a : \mu > c \tag{12.3}$$

The hypotheses in 12.1 corresponds to a two-tailed testing in which no a priori knowledge is available while 12.2 and 12.3 correspond to a one-tailed testing in which the measured value c is expected to be higher and lower than the population mean, respectively. A typical situation of a test involving a single sample is illustrated by the following example.

EXAMPLE 12.1

The expression level of a gene, measured with a given technology and normalized consistently, is known to have a mean of 1.5 in the normal human population. A researcher measures the expression level of this gene on several arrays using the same technology and normalization procedure. The following values have been obtained: 1.9, 2.5, 1.3, 2.1, 1.5, 2.7, 1.7, 1.2, and 2.0. Is this data consistent with the published mean of 1.5?

This problem involves a single sample, including 9 measurements. It is easy to establish that a two-tailed test is needed since there is no particular expectation for the measured mean: $H_0 : \mu = c, \ H_a : \mu \neq c$. However, we do face an obstacle: the population variance is not known. One way of dealing with the lack of knowledge about the population variances is to estimate it from the sample itself. We have seen that estimating population parameters from a sample, especially a rather small sample like this one, is imprecise since the estimates may be well off the real values. It should be expected that, at some point, we will probably need to give up something but this is the best we can do. Thus, instead of:

$$Z = \frac{\overline{X} - \mu}{\frac{\sigma}{\sqrt{n}}} \tag{12.4}$$

we will use a similar expression in which the population standard deviation σ is substituted by the sample standard deviation s:

$$t = \frac{\overline{X} - \mu}{\frac{s}{\sqrt{n}}} \tag{12.5}$$

However, the Z variable had a special property. If we took many samples from a given population and we calculated the Z value for each sample, the distribution of those Z values would follow a normal distribution. It was because of this reason that we could use the tables or calculated values of the normal distribution in order to calculate p values. If we took again many samples from a given population and calculated the values of the t variable above for all such samples, we would notice that the distribution thus obtained is not a normal distribution. Therefore, we will not be able to use the tables of the normal distribution to solve the given problem. However, the t variable in 12.5 does follow another classical and well known distribution called t **distribution** or **Student's t distribution**.

The name of this distribution has an interesting story. The distribution was discovered by William S. Gossett, a 32-year-old research chemist employed by the famous Irish brewery Guinness. When Gossett arrived in Dublin he found that there was a mass of data about brewing which called for some statistical analysis. Gossett's main problem was to estimate the mean value of a characteristic of various brews on the basis of very small samples. Confronted with this problem, Gossett found the t distribution and developed a method allowing the computation of confidence limits for such small samples. Gossett's findings were of exceptional importance for the scientific community, but Guinness would not allow him to publish the results of his work under his name. Thus, he chose the name "Student" and published his report under this signature in the journal *Biometrika* in 1908 [159, 186]. Since then, this test has been known as Student's t-test. Gossett's name was released long after the publication of his work. He died in 1937 leaving us one of the most useful, simple, and elegant statistical tools as proof that beer research can lead to true advances in scientific knowledge.

The t distribution has zero mean and a standard deviation that depends on the **degrees of freedom**, or **d.f.**, of the data. The notion of degrees of freedom is a perennial cause of confusion and misunderstandings as well as a source of inspiration for many jokes.[1] The notion of degrees of freedom has been extensively discussed in Section 8.4. Here, we will simply recall that the number of d.f. can be defined as the number of independent pieces of information that go into the estimation of a statistical parameter. In general, the number of degrees of freedom of an estimate is equal to the number of independent measurements that go into the estimate minus the number of parameters estimated as intermediate steps in the estimation of the parameter itself. For instance, if the variance, s^2, is to be estimated from a random sample of n independent scores, then the number of degrees of freedom is equal to the number of independent measurements n minus the number of parameters estimated as intermediate steps. In this case, there is one parameter, μ, estimated by $\overline{X}$ and therefore the d.f. is $n - 1$. In a similar way, we can conclude that the number of degrees of freedom in our problem is $n - 1$, where n is the sample size.

The shape of the t distribution is similar to the shape of the normal distribution, especially as n grows larger (see Fig. 12.1). For lower values of n, the t distribution has more values in its tails than the normal distribution. In practical terms, this means that for a given percentage of the distribution,

---

[1]For instance, it has been said that "the number of d.f. is usually considered self-evident – except for the analysis of data that have not appeared in a textbook" [142].

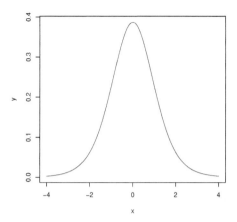

FIGURE 12.1: An example of a t distribution. In this case, the distribution has 8 degrees of freedom. The shape is very similar to that of a normal distribution. For small samples, the t distribution has more values in its tails compared to the normal distribution. For larger sample sizes (degrees of freedom), the t distribution tends to the normal distribution.

we will need to go farther into the tail. For example, given a t distribution with 4 degrees of freedom, if we wanted to set thresholds on the tails such as the central portion includes 95% of the measurements, we need to go approximately 2.78 standard deviations away from the mean. In comparison, for a normal distribution, the interval extends only ±1.96 standard deviations from the mean. Intuitively, this is consistent with the fact that we are estimating a parameter (population standard deviation), and therefore, the data have to show more extreme values, farther from the mean, for us to be able to reject the null hypothesis at any given significance level (5% in the example above). For large values of n, the t distribution tends to the normal distribution. This reflects the fact that as the sample size grows larger, the sample standard deviation will become a more and more accurate estimate of the population standard deviation. For values of n sufficiently large, the t distribution will be indistinguishable from the normal distribution. The t distribution is widely available like any other classical distribution. Its values can be obtained from tables or computer software.

Armed with this new distribution, we can now return to solving our problem.

EXAMPLE 12.1

The expression level of a gene measured with a given technology and normal-

ized consistently is known to have a mean expression level of 1.5 in the normal human population. A researcher measures the expression level of this gene on several arrays using the same technology and normalization procedure. The following values have been obtained: 1.9, 2.5, 1.3, 2.1, 1.5, 2.7, 1.7, 1.2, and 2.0. Is this data consistent with the published mean of 1.5?

SOLUTION

1. State the null and alternative hypothesis:

$$H_0 : \mu = 1.5$$

$$H_a : \mu \neq 1.5$$

Identify whether the situation corresponds to a one-tailed or two-tailed test; in this case, this is a two-tailed test.

2. Choose the level of significance: 5%.

3. Find the appropriate statistical model and test statistic: as we discussed, the appropriate model is the t distribution.

4. Calculate the appropriate test statistic:

$$t = \frac{\overline{X} - \mu}{\frac{s}{\sqrt{n}}} = \frac{1.87 - 1.5}{\frac{0.51}{\sqrt{9}}} = 2.21$$

5. Calculate the p-value corresponding to the calculated statistic: p=0.057. Alternatively, we could have calculated the critical value of the same t distribution corresponding to the 5% significance level: 2.21 < 2.306.

6. Compare the p-value with the chosen significance level: 0.057 > 0.05, or the observed statistic with the critical value corresponding to the chosen significance level.

7. Reject or do not reject H_0 based on the test above. In this example, we see that the calculated p-value (0.057) for the observed t-value (2.21) is higher, or less significant, than our chosen significance threshold (0.05). Alternatively, we can base our conclusion on the fact that the observed t-value (2.21) is less extreme (here higher than) than the critical value corresponding to the significance threshold (2.306). Hence, in this example we cannot reject H_0.

8. Draw the conclusion: the data seem consistent with the published mean.

In R, the function t.test can be used to perform such a one sample two-tailed Student's t-test:

```
> dat=c(1.9,2.5,1.3,2.1,1.5,2.7,1.7,1.2,2.0)
> res=t.test(x=dat,mu=1.5,alternative="two.sided")
> res
```

```
        One Sample t-test

data:   dat
t = 2.2144, df = 8, p-value = 0.05769
alternative hypothesis: true mean is not equal to 1.5
95 percent confidence interval:
 1.484370 2.271186
sample estimates:
mean of x
 1.877778
```

The vector containing the data is passed as the argument x of the `t.test` function, while the `mu` argument specifies the population mean to which we want to compare our sample mean. The `alternative` argument allows us to specify that we want to perform a two-tailed test since we do not have an a priori expectation that our sample mean will be less or greater than the target value 1.5. All the relevant details about our data and test results are returned by the function `t.test` in an object of class *htest*, which is similar to a list object in R. The available fields of this object can be listed using:

```
> names(res)
```

```
[1] "statistic"   "parameter"   "p.value"      "conf.int"
[5] "estimate"    "null.value"  "alternative" "method"
[9] "data.name"
```

and specific values can be retrieved using the name of the elements of the list or by their position in the list:

```
> res$statistic
```

```
       t
2.214385
```

```
> res[[1]]
```

```
       t
2.214385
```

```
> res$p.value
```

```
[1] 0.05768516
```

```
> res[[3]]
```

```
[1] 0.05768516
```

In this case, with the p-value being 0.057, the null hypothesis cannot be rejected at the 0.05 significance level. The 95% confidence interval is also by default provided by the t.test function. The 95% confidence interval is the interval in which the mean will be found 95% of the time. In this case, this interval included the tested value, 1.5, in it, which is consistent with our conclusion that we cannot reject the null hypothesis at the chosen significance level. □

12.2.2 Choosing the number of replicates

The size of the sample has a clear and direct influence on the results. Once a sample is collected, the statistic calculated on it together with the sample size will decide whether the measured change is significant or not. Many times, genes may not vary that much in absolute value while the variation can still be very meaningful from a biological point of view. Taking this into consideration, one could ask the converse question. Given a certain size of the expected variation, what would be the number of measurements that we need to collect in order to be able to draw a statistically meaningful conclusion from the data? Let us consider the following example.

EXAMPLE 12.2

The expression level of a gene measured with a specific technology and normalized in a standard way has a mean of 100 and a standard deviation of 10, in the normal population. A researcher is conducting an experiment in which the expression level of this gene is expected to increase. However, the change is expected to be relatively small, of only about 20%. Assuming that each array provides a single measurement of the given gene, how many arrays does the researcher need to run in order to be able to detect such a change 95% of the time?

The null hypothesis is $H_0 : \mu \leq 100$. We will work at the 1% significance level. Once again, we can use the Z statistic:

$$Z = \frac{\overline{X} - \mu}{\frac{\sigma}{\sqrt{n}}} \tag{12.6}$$

Usually, we know the values of all variables on the right-hand side and use them to calculate the value of Z. In this case, we do not know the value of n, the number of replicate measurements, but we can calculate the critical value of Z that would correspond to the threshold beyond which we can reject the null hypothesis. From a table of the normal distribution, the value of z that corresponds to $\alpha = 0.01$ is $z_\alpha = 2.33$. With this, we can calculate the minimal value of the sample mean that would be significant at this significance level:

$$z_\alpha = \frac{\overline{X}_t - \mu}{\frac{\sigma}{\sqrt{n}}} \tag{12.7}$$

The threshold value of $\overline{X}$ can be extracted as:

$$\overline{X}_t = \mu + z_\alpha \frac{\sigma}{\sqrt{n}} = 100 + 2.33 \frac{10}{\sqrt{n}} \tag{12.8}$$

The problem is that we cannot actually calculate $\overline{X}_t$ because we do not know n yet. However, we have the supplementary requirement of being able to detect a change of a given magnitude a certain percentage of times. This refers to the power of the test. If the true mean is $\mu + 20\% = 120$, then the Z value for the distribution centered at 120 could be written as:

$$z_\beta = \frac{\overline{X}_t - \mu'}{\frac{\sigma}{\sqrt{n}}} = \frac{\overline{X}_t - 120}{\frac{10}{\sqrt{n}}} \tag{12.9}$$

We want to reject the null hypothesis with a probability of $1 - \beta = 0.95$. The Z value that corresponds to $\beta = 0.05$ is -1.645. From this second distribution:

$$\overline{X}_t = \mu' + z_\beta \frac{\sigma}{\sqrt{n}} = 120 + 1.645 \frac{10}{\sqrt{n}} \tag{12.10}$$

Once again, we cannot calculate the value of $\overline{X}_t$ since we do not know the number of replicated measurements n. However, the two conditions ($\alpha = 0.01$ and $\beta = 0.95$) have to be satisfied simultaneously by the same $\overline{X}_t$. This is illustrated in Fig. 12.2. The distribution to the left represents the expression of the given gene in normal individuals ($\mu = 100$, $\sigma = 10$). The distribution to the right represents the expression of the given gene in the condition under study. We assume this distribution has a mean of $\mu = 120$ and the same standard deviation $\sigma = 10$. We would like to detect when a gene comes from the right distribution 95% of the time. The $\overline{X}_t$ has to be chosen such that both conditions on α and β are satisfied at the same time. This condition can be written as:

$$\mu + z_\alpha \frac{\sigma}{\sqrt{n}} = \mu' + z_\beta \frac{\sigma}{\sqrt{n}} \tag{12.11}$$

From this, we can extract the number of replicates as follows:

$$\mu' - \mu = (z_\alpha - z_\beta) \frac{\sigma}{\sqrt{n}} \tag{12.12}$$

and finally:

$$n = \frac{(z_\alpha - z_\beta)^2 \sigma^2}{(\mu' - \mu)^2} \tag{12.13}$$

In R, we can calculate this as:

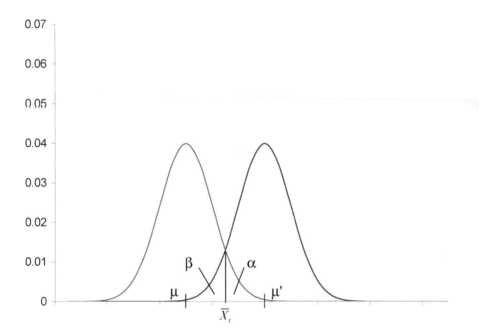

FIGURE 12.2: The distribution to the left represents the expression of the given gene in normal individuals ($\mu = 100$, $\sigma = 10$). The distribution to the right represents the expression of the given gene in the condition under study. We assume this distribution has a mean of $\mu = 120$ and the same standard deviation $\sigma = 10$. We would like to detect when a gene comes from the right distribution 95% of the time. The $\overline{X}_t$ has to be chosen such that it corresponds at the same time to: i) $\alpha = 0.01$ on the distribution to the left and ii) $1 - \beta = 0.95$ on the distribution to the right.

```
> za=qnorm(0.01,lower.tail=F)
> zb=qnorm(0.05)
> s=10
> mu=100
> mup=120
> za
```

[1] 2.326348

```
> zb
```

[1] -1.644854

```
> n=s^2*(za-zb)^2/(mup-mu)^2
> n
```

[1] 3.94261

This formula allows us to calculate the number of replicates needed in order to detect a specific change in the gene expression assessed as a mean of several replicate measurements, when the standard deviation is known and specific thresholds have been chosen for the probability α of making a type I error (false positives) and the power of the test $1 - \beta$ (the ability to detect true positives).

Equation 12.13 is very meaningful and illustrates several important phenomena. First, the number of replicate measurements, or the sample size, depends on the chosen significance level, α, the chosen power of the test, β, the variability of the measurements as reflected by their standard deviation σ and the difference between means that we want to be able to detect. For fixed α and β, the number of replicates is directly proportional to the standard deviation and inversely proportional to the difference between means that we want to detect. A large difference in means can be detected with a smaller number of replicate measurements and a larger variability will require more replicates. Furthermore, the dependence is **quadratic**. In other words, if the standard deviation increases from $\sigma = 1$ to $\sigma = 2$, we will need 4 times more replicates in order to be able to detect the same change $\mu - \mu'$.

EXAMPLE 12.3
Calculate the number of replicate measurements needed in order to detect a difference of at least 20 units between the means of two normally distributed populations having the same standard deviation of 10 units. The test required is a one-tailed test and we would like to be able to detect differences as above 95% of the time, at a significance level of 1%.

SOLUTION Since the data are normally distributed, we can use the R function **power.t.test** to compute the required sample size:

```
> power.t.test(n=NULL,delta=20,sd=10,sig.level=0.01,
+ power=0.95,type ="one.sample",alternative="one.sided")

     One-sample t test power calculation

              n = 6.936908
          delta = 20
             sd = 10
      sig.level = 0.01
          power = 0.95
    alternative = one.sided
```

The results of this function call summarize all input parameters, as well as provide the desired number of replicate measurements. In all practical situations, the number of replicates is obtained by rounding up the value provided by the function call above, in this case n=7. ▯

The relationship explained in Example 12.2 above between n (number of samples per group), delta (the minimum difference to be detected), sd (the standard deviation), sig.level (the significance level), and the power, allows us to calculate any of these 5 parameters, given the other 4. For example, let us assume that we had 7 samples, and we wanted to know what will be the power in the same experimental conditions as above. This can be calculated with the same power.t.test function:

```
> power.t.test(n = 7, delta = 20, sd = 10, sig.level = 0.01,
+ power = NULL, type = "one.sample", alternative = "one.sided")

     One-sample t test power calculation

              n = 7
          delta = 20
             sd = 10
      sig.level = 0.01
          power = 0.953054
    alternative = one.sided
```

The power calculated for 7 replicates is 0.953054, that is 95.30%. As expected, this is slightly higher than the initially required 95% because we used 7 replicates instead of 6.93 replicates calculated above.

12.2.3 Tests involving the variance (σ^2). The chi-square distribution

Example 12.2 illustrated the need for information regarding variance. However, so far, we do not have any tools able to test hypotheses regarding the variance. Let us consider the following example.

EXAMPLE 12.4

The expression level of a gene is measured many times with a certain array technology and found to have a mean expression level of 40 with a standard deviation of 3. A colleague scientist proposes a novel array technology. The same gene is measured 30 times and found to have a standard deviation of 2. Is there evidence that the new technology offers significantly more uniform measurements of the gene expression levels?

This is a typical example of a *one sample test regarding variance*. The null and research hypotheses are as follows:

$$H_0 : \sigma^2 \geq c$$

$$H_a : \sigma^2 < c \qquad (12.14)$$

Unfortunately, none of the distributions discussed so far allows us to test these hypotheses because none of them is concerned with the standard deviation of the sample. We would need a variable similar to the t variable that behaves in a certain standard way. Fortunately, the quantity:

$$\frac{(n-1)s^2}{\sigma^2} \qquad (12.15)$$

behaves like a random variable with the interesting and useful property that if all possible samples of size n are drawn from a normal population with a variance σ^2 and for each such sample the quantity $\frac{(n-1)s^2}{\sigma^2}$ is computed, these values will always form the same distribution. This distribution will be a sampling distribution called a χ^2 (chi-square) distribution. The parametric aspect of the testing using a chi-square distribution comes from the fact that the population is assumed to be normal. The χ^2 distribution is similar to the t distribution inasmuch its particular shape depends on the number of degrees of freedom. If samples of size n are considered, the number of degrees of freedom of the chi-square distribution will be $n - 1$. The χ^2 distributions with degrees of freedom ranging from 1 to 6 can be plotted as:

```
> x=seq(0,15,by=0.2)
> y=sapply(1:6,function(z){dchisq(x,df=z)})
> colors=c("blue","black","yellow","magenta","brown","purple")
> plot(x,y[,1],col="blue",type="l")
> dummy=sapply(2:6,function(df){points(x,y[,df],col=colors[df],
+ type="l")})
> legend(11,.8,paste(rep("df=",6),1:6), text.col="black",
+ lty=1, col= colors)
```

These distributions are shown in Fig. 12.3.

The expected value of the chi-square distribution with ν degrees of freedom is:

$$\mu = E\left(\chi^2\right) = \nu \qquad (12.16)$$

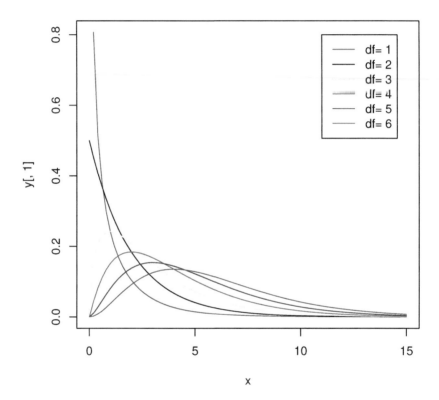

FIGURE 12.3: The shape of the χ^2 (chi-square) distribution depends on the number of degrees of freedom (df). The figure shows the chi-square distributions with degrees of freedom from 1 to 6.

and its variance is:

$$Var(\chi^2) = 2v \tag{12.17}$$

Given that we now know that the chi-square variable behaves in a certain way, we can use this to test our hypotheses. Given a certain sample, we can calculate the value of the chi-square statistic on this sample:

$$\chi^2 = \frac{(n-1)s^2}{\sigma^2} \tag{12.18}$$

If the value of the sample standard deviation s is close to the value of the population standard deviation σ, the value of χ^2 will be close to $n-1$. If the value of the sample standard deviation s is very different from the value of the population standard deviation σ, the value of χ^2 will be very different from $n-1$. Given that the shape of the χ^2 distribution is known, we can use it to calculate a precise value for the probability of the measured sample having a variance different from the known population variance.

Let us use the χ^2 distribution to solve the problem in Example 12.4. In order to do this, we will calculate the value of the χ^2 statistic for our sample and see how unlikely it is for such value to occur if the null hypothesis is true:

$$\chi^2 = \frac{(n-1)s^2}{\sigma^2} = \frac{(30-1)2^2}{3^2} = \frac{29 \cdot 4}{9} = 12.88 \tag{12.19}$$

The value of the χ^2 statistic is expected to be near its mean of $v = n-1 = 29$ if H_0 is true, and different from 29 if H_a is true. The one-tailed probability that corresponds to a χ^2 value of 12.88 for a distribution with 29 degrees of freedom is 0.995. In R, this can be obtained with the function pchisq as follows:

```
> pchisq((30-1)*2^2/3^2,df=29,lower.tail=FALSE)

[1] 0.9957149
```

In other words, if H_0 is true, the measured value of the χ^2 statistic will be less or equal to 12.88 in 99.5% of the cases. In these conditions, we can reject the null hypothesis at the 5% significance level.

As we discussed in Chapter 11, an alternative approach is to compare the calculated value of the χ^2 statistic with the critical values of the χ^2 from a distribution with 29 degrees of freedom. For instance, the critical values for probabilities of 0.05 and 0.95 are 17.70 and 42.55, respectively:

```
> qchisq(0.05,df=29,lower.tail=TRUE)

[1] 17.70837

> qchisq(0.05,df=29,lower.tail=FALSE)

[1] 42.55697
```

Note that this corresponds to a two-tailed testing situation and the sum of the probabilities in both tails is 10% so this corresponds to a 10% significance level. In other words, we will not be able to reject the null hypothesis at 10% significance as long as the calculated χ^2 statistic is in the interval $(17.70, 42.55)$ for a two-tailed test. If the calculated value of the χ^2 statistic is more extreme than these values, the H_0 hypothesis will be rejected.

However, our research hypothesis stated that we expect the new technology to be better and therefore we are in a one-tailed testing situation. In this case, the only interesting critical value is 17.70 and this corresponds to a 5% significance. We note that $12.88 < 17.70$ and we reject the null hypothesis: there is enough data to support the hypothesis that the new microarray technology is significantly better.

We can summarize the testing involving the variance of one sample as follows. The hypothesis can take one of three forms:

$$H_0 : \sigma^2 \geq c \quad H_a : \sigma^2 < c$$

$$H_0 : \sigma^2 \leq c \quad H_a : \sigma^2 > c$$

$$H_0 : \sigma^2 = c \quad H_a : \sigma^2 \neq c \tag{12.20}$$

The first two correspond to a one-tailed test in which the data are expected to be on a specific side of the value c; the last corresponds to a two-tailed test. The value c is chosen or known before the data are gathered. The value c is never calculated from the data itself. In any given experiment, only one set of hypotheses is appropriate.

The testing for variance is done using a χ^2 distribution. The χ^2 distribution is a family of distributions with a shape determined by the number of degrees of freedom. There are two basic alternatives for hypothesis testing. The first alternative is to calculate the p-value corresponding to the calculated value of the χ^2 statistic and compare it with the significance level α. If the p value is smaller than the predetermined alpha level, H_0 is rejected. The second alternative is to compare the calculated value of the χ^2 statistic with the critical value corresponding to the chosen significance level. If the calculated χ^2 value is more extreme than the appropriate critical χ^2 value, the H_0 will be rejected.

12.2.4 Confidence intervals for standard deviation/variance

The χ^2 distribution can be used to calculate confidence intervals for the standard deviation. Let us consider the following example.

EXAMPLE 12.5

The expression level of a specific gene is measured 16 times. The sample variance is found to be 4. What are the 95% confidence limits for the population variance?

We can use the definition of the χ^2 variable:

$$\chi^2 = \frac{(n-1)s^2}{\sigma^2} \tag{12.21}$$

to extract the σ^2:

$$\sigma^2 = \frac{(n-1)s^2}{\chi^2} \tag{12.22}$$

and then use this to set the conditions regarding the probability of making a Type I error, α:

$$C\left[\frac{(n-1)s^2}{\chi^2_{1-\frac{\alpha}{2}}} \leq \sigma^2 \leq \frac{(n-1)s^2}{\chi^2_{\frac{\alpha}{2}}}\right] = 1-\alpha \tag{12.23}$$

We can now extract the square root in order to obtain σ:

$$C\left[\sqrt{\frac{(n-1)s^2}{\chi^2_{1-\frac{\alpha}{2}}}} \leq \sigma \leq \sqrt{\frac{(n-1)s^2}{\chi^2_{\frac{\alpha}{2}}}}\right] = 1-\alpha \tag{12.24}$$

These are the confidence intervals for the standard deviation.

Using this, we can now calculate the 95% confidence interval for the population standard deviation given that the sample variance is 4 estimated from 16 measurements:

```
> low=sqrt((16-1)*4/qchisq(p=1-0.05/2,df=16-1))
> up=sqrt((16-1)*4/qchisq(p=0.05/2,df=16-1))
> c(low,up)

[1] 1.477410 3.095382
```

12.3 Tests involving two samples

12.3.1 Comparing variances. The F distribution.

Let us consider a typical experiment whose goal is to compare the gene expression levels of some cancer patients to the expression levels of the same genes in a group of healthy people. An example of such data is shown in Table 12.1.

Probably, the most interesting question that one could ask refers to whether a given gene is expressed differently between cancer patients and healthy subjects. This is a question that involves the mean of the two samples. However, as we shall see, in order to answer this question we must first know whether the two samples have the same variance. Let us consider the

| Gene ID | Cancer patients | | | | Control subjects | | |
	P 1	P 2	$\cdots$	P n	C 1	C 2	$\cdots$	C n
AC002115	5.2	6.1	$\cdots$	5.9	0.9	0.7	$\cdots$	0.8
AB006782	2.1	1.6	$\cdots$	1.9	-1.3	0.6	$\cdots$	0.2
AB001325	0.5	-2.8	$\cdots$	1.5	1.7	3.8	$\cdots$	0.2
AB001527	0.5	-2.8	$\cdots$	1.5	1.7	3.8	$\cdots$	0.2
AB006190	-2.1	-3.9	$\cdots$	-2.1	-1.9	-1.3	$\cdots$	0.2
AB002086	4.1	3.6	$\cdots$	1.5	-1.3	0.6	$\cdots$	0.2
$\cdots$								
AB000450	-1.4	1.3	$\cdots$	-2.5	2.1	-3.2	$\cdots$	0.8

TABLE 12.1: An experiment comparing the expression levels of some genes in two groups: cancer patients (P) and control subjects (C).

following example, which will help us describe the method used to compare variances of two samples.

EXAMPLE 12.6
The expression level of a gene is measured in a number of control subjects and patients. The values measured in controls are: 10, 12, 11, 15, 13, 11, 12 and the values measured in patients are: 12, 13, 13, 15, 12, 18, 17, 16, 16, 12, 15, 10, 12. Is the variance different between controls and patients?

The problem can be used to state the following pair of hypotheses:

$$H_0 : \sigma_1 = \sigma_2, \quad H_a : \sigma_1 \neq \sigma_2 \tag{12.25}$$

At this point, none of the test statistics used so far is appropriate to study this problem. In order to develop one, let us consider the following situation. Let us consider a normal population with variance σ^2. Let us draw a random sample of size n_1 of measurements from this population and calculate the variance of this sample s_1^2. Let us then draw another sample of size n_2 from the same population and calculate its sample variance s_2^2. Most likely, these two variances will not be the same. We would like to study their relationship, but following two variables at the same time is inconvenient. A good way to combine the two numbers into a single one is to take their ratio $\frac{s_1^2}{s_2^2}$. Since the two sample standard deviations will be, in most cases, slightly different from each other, the ratio $\frac{s_1^2}{s_2^2}$ will be different from one. However, it turns out that this ratio is a random variable that follows a sampling distribution known as the F distribution. Once again, the actual shape of this distribution depends on the degrees of freedom ν_1 and ν_2 of the two samples (see Fig. 12.4). Since the two samples are coming from populations with the same variance, it is intuitive that the expected value of the ratio $\frac{s_1^2}{s_2^2}$ is 1. Values of this

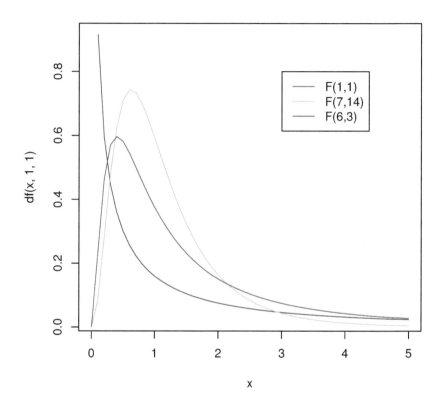

FIGURE 12.4: The F distribution models the behavior of the random variable $\frac{s_1^2}{s_2^2}$, where s_1^2 and s_2^2 are the sample variances of two samples drawn from two normal populations with the same variance σ^2. The shape of the F distribution depends on the degrees of freedom v_1 and v_2 of the two samples considered. The figure shows the shapes of the F distributions for $F_{v_1=v_2=1}$, $F_{v_1=7,v_2=14}$, and $F_{v_1=6,v_2=3}$.

statistic that deviate considerably from 1 will indicate that H_0 is unlikely. This distribution also has the interesting property that the left tail for an F with v_1 and v_2 degrees of freedom is the reciprocal of the right tail for an F with the degrees of freedom reversed:

$$F_{\alpha(v_1,v_2)} = \frac{1}{F_{1-\alpha(v_2,v_1)}} \tag{12.26}$$

We can now address the problem in Example 12.6. In this case, $v_1 = n_1 - 1 = 7 - 1 = 6$ and $v_2 = n_2 - 1 = 13 - 1 = 12$. We can use a spreadsheet to calculate the critical values of F. We choose $\alpha = 0.05$ but we work in a two-tailed framework so we calculate two values, one for the right tail:

$$F_{0.975(6,12)} = 3.73 \tag{12.27}$$

and one for the left tail:

$$F_{0.025(6,12)} = 0.186 \tag{12.28}$$

Thus, if the ratio of the variances (s_1^2/s_2^2) is less than 0.19 or more than 3.72, H_0 can be rejected. In these circumstances, a Type I error of rejecting a true null hypothesis H_0 will happen at most 5% of the time.

The results of a hypothesis testing using an F statistic should (obviously) be independent of the choice of the order of the samples (which one is the numerator and which one is the denominator). The following example will illustrate this.

EXAMPLE 12.7

The expression level of a gene is measured in a number of control subjects and patients. The values measured in controls are: 10, 12, 11, 15, 13, 11, 12 and the values measured in patients are: 12, 13, 13, 15, 12, 18, 17, 16, 16, 12, 15, 10, 12. Is the variance different between controls and patients?

SOLUTION The null hypothesis is: $H_0 : \sigma_A^2 = \sigma_B^2$ and the research hypothesis is: $H_a : \sigma_A^2 \neq \sigma_B^2$. We know that the ratio of the sample variances of two samples drawn from two normal populations with the same population variance will follow an F distribution.

We consider sample 1 coming from population A and sample 2 coming from population B: $v_A = 6$, $s_A^2 = 2.66$, $v_B = 12$, $s_B^2 = 5.74$.

We choose to consider the ratio s_A^2/s_B^2 first. Therefore, we use:

$$F_{0.975(6,12)} = 3.73 \tag{12.29}$$

$$F_{0.025(6,12)} = \frac{1}{F_{0.975(12,6)}} = \frac{1}{5.37} = 0.1862 \tag{12.30}$$

If the statistic s_A^2/s_B^2 is lower than 0.186 or higher than 3.73, we will reject the null hypothesis. If the statistic s_A^2/s_B^2 is in between the two values, we will not be able to reject the null hypothesis:

$$\frac{1}{5.37} \leq \frac{s_A^2}{s_B^2} \leq 3.73 \tag{12.31}$$

Now, let us consider the ratio s_B^2/s_A^2.

$$F_{0.975(12,6)} = 5.37 \tag{12.32}$$

$$F_{0.025(12,6)} = \frac{1}{F_{0.975(6,12)}} = \frac{1}{3.73} = 0.268 \tag{12.33}$$

If the statistic s_B^2/s_A^2 is lower than 0.268 or higher than 5.37, we will reject the null hypothesis. If the statistic s_B^2/s_A^2 is in between the two values, we will not be able to reject the null hypothesis:

$$\frac{1}{3.73} \leq \frac{s_B^2}{s_A^2} \leq 5.37 \tag{12.34}$$

An examination of equations 12.31 and 12.34 reveals that the same conditions are used in both cases, independently of the choice of the populations. For instance, if s_A^2 is to be significantly higher than s_B^2, the statistic s_A^2/s_B^2 has to satisfy:

$$\frac{s_A^2}{s_B^2} > F_{0.975(6,12)} = 3.73$$

or, alternatively, the statistic s_B^2/s_A^2 has to satisfy:

$$\frac{s_B^2}{s_A^2} < F_{0.025(12,6)} = \frac{1}{F_{0.975(6,12)}} = \frac{1}{3.73}$$

Thus, it is clear that the two different choices lead to exactly the same numerical comparison, which is the basis for rejecting or not rejecting the null hypothesis.

□

In R, the same question can be addressed as follows:

```
> x=c(10,12,11,15,13,11,12)
> y=c(12,13,13,15,12,18,17,16,16,12,15,10,12)
```

We need to test whether there is a significant difference between the variances of the two samples. As discussed above, the ratio of the two variances follows an F distribution with $7 - 1 = 6$ and $13 - 1 = 12$ degrees of freedom. If the samples have different variances, the F statistic will be different from 1.

```
> var(x)/var(y)
```

[1] 0.4642857

```
> qf(0.05/2,df1=length(x)-1,df2=length(y)-1,lower.tail=TRUE)
```

[1] 0.1863501

```
> qf(0.05/2,df1=length(x)-1,df2=length(y)-1,lower.tail=FALSE)
```

[1] 3.728292

```
> 1/qf(0.05/2,df1=length(y)-1,df2=length(x)-1,lower.tail=TRUE)
```

[1] 3.728292

Note the last two lines showing alternative ways of calculating the critical value for the right tail. Since `var(x)`/`var(y)` is between the two critical F values, the null hypothesis cannot be rejected. The same conclusion can also be reached by directly calculating the p-value corresponding to the observed F statistic:

```
> pf(var(x)/var(y),df1=length(x)-1,df2=length(y)-1,
+ lower.tail=TRUE)
```

[1] 0.1781315

Since this p-value is higher than our significance threshold of 5%, we conclude that we cannot reject the null hypothesis.

In summary, there are three possible sets of hypotheses regarding variances of two samples:

$$H_0 : \sigma_1^2 \geq \sigma_2^2 \quad H_a : \sigma^2 < \sigma_2^2$$
$$H_0 : \sigma_1^2 \leq \sigma_2^2 \quad H_a : \sigma^2 > \sigma_2^2$$
$$H_0 : \sigma_1^2 = \sigma_2^2 \quad H_a : \sigma^2 \neq \sigma_2^2$$

The first two correspond to a one-tailed test in which one of the samples is expected to have a variance lower than the other. As always, in any experiment only one set of hypotheses is appropriate. The testing for variance is done using an F statistic. The value of the F statistic calculated from the two samples is compared with the critical values corresponding to the chosen α calculated from the F distribution. The F distribution is a family of distributions with a shape determined by the degrees of freedom of the two samples. Alternatively, if the p-value calculated for the F statistic is lower than the chosen significance level, the null hypothesis H_0 can be rejected.

12.3.2 Comparing means

As mentioned in the previous section, given two groups of gene expression measurements, for example, measurements in cancer patients versus measurements in a sample of healthy individuals, probably the most interesting question that one could ask is whether a given gene is expressed differently between cancer patients and healthy subjects. This question refers to the mean of the measurements: we would like to know whether the mean expression level of a gene is different between the two samples. Note that the question refers to *a given gene*. This apparent trivial detail is absolutely crucial for the choice of the approach to follow and the correctness of the final result.

Let us consider for instance the following gene, measured in 6 cancer patients and 6 controls:

```
> AC04219 = c(-5.09,-6.46,-7.46,-4.57,-5.67,-6.84,5.06,
+ 7.25,4.17,6.10, 5.06, 5.66)
> names(AC04219)= c(paste(rep("P",6),1:6),paste(rep("C",6),
+ 1:6))
> AC04219
```

P 1	P 2	P 3	P 4	P 5	P 6	C 1	C 2	C 3	C 4
-5.09	-6.46	-7.46	-4.57	-5.67	-6.84	5.06	7.25	4.17	6.10

C 5	C 6
5.06	5.66

In this case, the answer seems clear: all cancer patients have negative values while all control subjects have positive values. We can be pretty confident that there is a difference between the expression level of this gene in cancer versus healthy subjects. However, in most cases, the numbers are not so very different. For instance, a conclusion is really difficult to draw from the data collected about the following gene:

```
> AC002378 = c(0.66,0.51,1.12,0.83,0.91,0.50,0.41,0.57,-0.17,
+ 0.50,0.22,0.71 )
> names(AC002378 )=c(paste(rep("P",6),1:6),
+ paste(rep("C",6),1:6))
> AC002378
```

P 1	P 2	P 3	P 4	P 5	P 6	C 1	C 2	C 3	C 4
0.66	0.51	1.12	0.83	0.91	0.50	0.41	0.57	-0.17	0.50

C 5	C 6
0.22	0.71

In this case, the range of values for both cancer and controls is much narrower. All but one values are positive and although the values for cancer patients seem a bit larger, there are control cases (e.g., C6) that are larger than some cancer cases (e.g., P2). Furthermore, "pretty confident" may not be

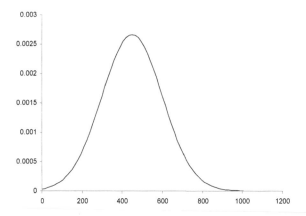

FIGURE 12.5: The null hypothesis H_0: there is no difference in the expression of this gene in cancer patients versus control subjects and all measurements come from a single distribution.

sufficiently accurate. We would like to have a better way to characterize the degree of confidence in our conclusion. As we have seen in Chapter 11, the probability of drawing the wrong conclusion, or the p-value, might be a more useful and more accurate way to characterize the amount of trust that we are willing to put into our conclusion.

Let us approach this problem using the framework defined in Chapter 11:

1. State the problem.

    Given a single gene (e.g., AC002378), is this gene expressed differently between cancer patients and healthy subjects?

2. State the null and alternative hypothesis.

    The null hypothesis, H_0, is that all measurements originate from a single distribution (see Fig. 12.5). The research hypothesis, H_a, is that there are two distributions: the measurements of the cancer patients come from one distribution while the measurements of the healthy patients come from a different distribution (see Fig. 12.6). We summarize the hypotheses as:
    $$H_0 : \mu_1 = \mu_2, \; H_a : \mu_1 \neq \mu_2$$

    In this case, we do not have any expectation regarding in which of the two situations the gene is expressed higher or lower so this is a two-tailed test situation.

    Note that a cursory examination of the data reveals that the measurements corresponding to the cancer patients are in general higher than

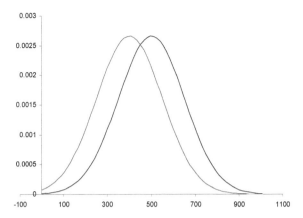

FIGURE 12.6: The research hypothesis H_a: there are two distributions, one that describes the expression of the given gene in cancer patients and one that describes the expression of the same gene in control subjects.

the measurements corresponding to the controls. However, it would be incorrect to use this in order to justify the use of a one-tailed test by saying that we expect the cancer measurements to be higher. The hypotheses must be stated before examining the data and the choice between a one-tailed and a two-tailed test must be made based on our a priori knowledge and expectations and not based on the collected data itself.

3. Choose the level of significance.

   We choose to work at 5% significance level.

4. Find the appropriate statistical model and test statistic.

Returning to the problem at hand, we do not know the variance of the populations these data are drawn from, and therefore, we will need to use the sample variances calculated on the data themselves. Hence, the t distribution is the appropriate statistical model. However, there are several versions of a two sample t test as follows.

1. **One-tailed versus two-tailed.** The one-tailed test is to be used when there is an *a priori* knowledge or expectation that one set of measurements will be either higher or lower than the other. If there is no such knowledge, one should use a two-tailed test.

2. **Paired versus unpaired.** A paired test is used when there is a natural pairing between experiments. A typical example is a situation in which

a set of patients are treated with a given drug. A certain parameter is measured several times before and after the drug is administered and the task is to decide whether the drug has a statistically significant effect. In this case, the data collected from one patient before and after the administration of the drug can be tested with a paired t-test. If there are no known reasons to associate measurements in the two samples, the unpaired test is to be used.

3. **Equal variance versus unequal variance.** As suggested by the names, the equal-variance t-test is appropriate when the two samples are known, or presumed, to come from distributions with equal variances while the unequal variance test is appropriate when the two distributions have different variances.

In our problem, there is no *a priori* expectation that the cancer measurements be either higher or lower than the control measurements. Furthermore, there is no special connection between cancer patients and control subjects. The choice between the equal variance and the unequal variance can be decided by testing the variances as discussed in Section 12.3.1.

12.3.2.1 Equal variances

If the two samples are drawn from normal populations,[2] $(\overline{X}_1 - \overline{X}_2)$ will be distributed normally with an expected value equal to $\mu_1 - \mu_2$. It can be shown that the variance of a difference of two variables is equal to the sum of variances of the two variables. We can write the variance of $\overline{X}_1 - \overline{X}_2$ as:

$$\frac{\sigma_1^2}{n_1} + \frac{\sigma_2^2}{n_2} \tag{12.35}$$

The t-statistic used in the tests for mean involving one sample was (see Eq. 12.5):

$$t = \frac{\overline{X} - \mu}{\sqrt{\frac{s^2}{n}}} \tag{12.36}$$

We can rewrite this t-statistic using the new variable $X_1 - X_2$ with the new mean $\mu_1 - \mu_2$ and the new sample variance $\frac{s_1^2}{n_1} + \frac{s_2^2}{n_2}$:

$$t = \frac{(\overline{X}_1 - \overline{X}_2) - (\mu_1 - \mu_2)}{\sqrt{\frac{s_1^2}{n_1} + \frac{s_2^2}{n_2}}} \tag{12.37}$$

This is the case in which the two variances are assumed to be equal ($\sigma_1^2 = \sigma_2^2$). This assumption is called **homoscedasticity**. In this case, the

---

[2]This assumption makes this test parametric.

best estimate of the overall variability, and hence the new sample variance, will be given by a *pooled sample variance*:

$$s_1 = s_2 = s_p^2 = \frac{(n_1 - 1) \cdot s_1^2 + (n_2 - 1) \cdot s_2^2}{n_1 + n_2 - 2} \tag{12.38}$$

Substituting this in Equation 12.37, we obtain the formula for the t-statistic of two independent samples with equal variances:

$$t = \frac{(\overline{X}_1 - \overline{X}_2) - (\mu_1 - \mu_2)}{\sqrt{s_p^2 \left(\frac{1}{n_1} + \frac{1}{n_2} \right)}} \tag{12.39}$$

Formulae 12.38 and 12.39 can be used to test hypotheses regarding the means of two samples of equal variance. The degrees of freedom are given by the number of measurements minus the number of intermediate values we need to calculate. In this case, there are $n_1 + n_2$ measurements and two intermediate values s_1^2 and s_2^2, so $v = n_1 + n_2 - 2$.

We can now return to the example given at the beginning of this section and answer the question regarding gene AC002378:

```
> AC002378 = c(0.66,0.51,1.12,0.83,0.91,0.50,0.41,0.57,-0.17,
+ 0.50,0.22,0.71 )
> names(AC002378 )=c(paste(rep("P",6),1:6),
+ paste(rep("C",6),1:6))
> AC002378

  P 1   P 2   P 3   P 4   P 5   P 6   C 1   C 2   C 3   C 4
 0.66  0.51  1.12  0.83  0.91  0.50  0.41  0.57 -0.17  0.50
  C 5   C 6
 0.22  0.71
```

For this gene, we calculate the following:

$$\overline{X}_1 = 0.755$$

$$\overline{X}_2 = 0.373$$

$$s_1^2 = 0.059$$

$$s_2^2 = 0.097$$

At this point, we have to decide whether we can assume equal variances or not. For this purpose, we take $H_0 : \sigma_1^2 = \sigma_2^2$ and $H_a : \sigma_1^2 \neq \sigma_2^2$. We choose to work at the 5% significance level and we calculate the F statistic:

$$F = \frac{s_1^2}{s_2^2} = \frac{0.059}{0.097} = 0.60$$

We calculate the probability of obtaining such a value for the F statistic just by chance as 0.29, and we decide that the data do not allow us to reject H_0 at the chosen significance level. We then assume that the two samples have equal variance and calculate the pooled variance:

$$s_p^2 = \frac{(n_1-1)\cdot s_1^2 + (n_2-1)\cdot s_2^2}{n_1+n_2-2} = \frac{(6-1)\cdot 0.059 + (6-1)\cdot 0.097}{6+6-2} = 0.078$$

$$(12.40)$$

Finally, we use formula 12.39 to calculate the value of the t-statistic:

$$t = \frac{(\overline{X}_1 - \overline{X}_2) - (\mu_1 - \mu_2)}{\sqrt{s_p^2\left(\frac{1}{n_1} + \frac{1}{n_2}\right)}} = \frac{(0.755 - 0.373) - 0}{\sqrt{0.078\left(\frac{1}{6} + \frac{1}{6}\right)}} = 2.359 \qquad (12.41)$$

and calculate the p-value, or the probability of having such a value by chance, as 0.04. This value is lower than the chosen significance level of 0.05, and therefore, we reject the null hypothesis: the gene AC002378 is expressed differently between cancer patients and healthy subjects.

In R, all the computation above can be easily performed with two calls to the same pf (for the F- test) and t.test (for the t-test) functions used above:

```
> x=c(0.66,0.51,1.12,0.83,0.91,0.50)
> y=c(0.41,0.57,-0.17,0.50,0.22,0.71)
> pf(var(x)/var(y),df1=length(x)-1,df2=length(y)-1,
+ lower.tail=TRUE)

[1] 0.2993681

> t.test(x,y,alternative="two.sided",var.equal = TRUE)

        Two Sample t-test

data:   x and y
t = 2.3593, df = 10, p-value = 0.04
alternative hypothesis: true difference in means is not equal to 0
95 percent confidence interval:
 0.02122125 0.74211208
sample estimates:
mean of x mean of y
0.7550000 0.3733333
```

Note that it was the outcome of the F-test ($p=0.29$) that allowed us to perform a t-test under the homoscedasticity conditions (equal variance) in the last line above. Had the outcome of the F-test been such that the null hypothesis that the two variances are equal could be rejected, the t.test function would have been called with the parameter var.equal=FALSE.

12.3.2.2 Unequal variances

Sometimes, the two sets of measurements do come from independent normal populations, but those populations do not have the same variance. This situation is called **heteroscedasticity**, and in this case, a modified t-test is necessary. This situation arises if an F test undertaken on the ratio of the sample variances $\frac{s_1^2}{s_2^2}$ provides sufficient evidence to reject the hypothesis that the two variances are equal $H_0 : \sigma_1^2 = \sigma_2^2$. In this case, the t-statistic can still be calculated as:

$$t = \frac{(\overline{X}_1 - \overline{X}_2) - (\mu_1 - \mu_2)}{\sqrt{\frac{s_1^2}{n_1} + \frac{s_2^2}{n_2}}} \tag{12.42}$$

where s_1 and s_2 are the respective sample variances. However, the degrees of freedom need to be adjusted as:

$$\nu = \frac{\left(\frac{s_1^2}{n_1} + \frac{s_2^2}{n_2}\right)^2}{\frac{\left(\frac{s_1^2}{n_1}\right)^2}{n_1-1} + \frac{\left(\frac{s_2^2}{n_2}\right)^2}{n_2-1}} \tag{12.43}$$

Usually, this value is not an integer and needs to be *rounded down*.

A close inspection of formulae 12.39 and 12.42 shows that if $n_1 = n_2$ or $s_1^2 = s_2^2$, the two formulae will yield exactly the same value for the t-statistic.

12.3.2.3 Paired testing

In some situations, the measurements are naturally paired. Examples include:

1. Simultaneous tests – e.g., cells from the same culture are split into two groups and each group is given a different treatment.

2. Before and after tests – e.g., a certain parameter is measured in a given animal before and after the treatment with a given substance.

3. Matched tests – e.g., the subjects are matched in pairs that have similar characteristics (e.g., same age, height, weight, diet, etc.).

In all these cases, the variables measured are dependent. Let us assume that there are n measurements performed before the treatment and n measurements after the treatment. Since these measurements are paired, we can consider that what we want to actually follow is not the values themselves but the differences "after treatment – before treatment." These differences are independent so the t-test developed earlier will be appropriate:

$$t = \frac{\overline{X}_d - \mu_d}{\frac{s_d}{\sqrt{n}}} \tag{12.44}$$

where μ_d is the mean difference, s_d is the sample standard deviation of the differences, etc. Clearly, the degrees of freedom are $n - 1$, where n is *the number of differences*, or paired measurements. Note that in an unpaired t-test there would be $2n - 2$ degrees of freedom as discussed in Section 12.3.2.1.

In conclusion, the only difference between paired and unpaired testing is the computation of the degrees of freedom and variance. The rest of the statistical argument follows the same procedure used in all other hypothesis testing situations.

In R, if the two series of measurements are paired, we can use:

```
> x=c(0.66,0.51,1.12,0.83,0.91,0.50)
> y=c(0.41,0.57,-0.17,0.50,0.22,0.71)
> #t.test(x,y,alternative="two.sided",paired = F, var.equal=T)
> #t.test(x,y,alternative="two.sided",var.equal = TRUE)
> t.test(x,y,alternative="two.sided",paired = TRUE)

        Paired t-test

data:  x and y
t = 1.7153, df = 5, p-value = 0.1469
alternative hypothesis: true difference in means is not equal to 0
95 percent confidence interval:
 -0.1903062  0.9536396
sample estimates:
mean of the differences
              0.3816667
```

which is equivalent to a one sample t-test on a new sample containing the differences between the two series of values:

```
> z=x-y
> t.test(z,alternative="two.sided")

        One Sample t-test

data:  z
t = 1.7153, df = 5, p-value = 0.1469
alternative hypothesis: true mean is not equal to 0
95 percent confidence interval:
 -0.1903062  0.9536396
sample estimates:
mean of x
0.3816667
```

12.3.3 Confidence intervals for the difference of means $\mu_1 - \mu_2$

If the null hypothesis $H_0 : \mu_1 - \mu_2$ can be rejected, a natural question is to ask how different the two means are? This can be translated directly into the goal of establishing confidence intervals for the difference $\mu_1 - \mu_2$. Such intervals can be calculated using the same method used in Section 12.2.4. The intervals are:

$$C\left[(\overline{X}_1 - \overline{X}_2) - t_0 \cdot s_{\overline{X}_1 - \overline{X}_2} \leq \mu_1 - \mu_2 \leq (\overline{X}_1 - \overline{X}_2) + t_0 \cdot s_{\overline{X}_1 - \overline{X}_2}\right] = 1 - \alpha \quad (12.45)$$

where the number of degrees of freedom is $v = n_1 + n_2 - 2$ and the standard error $s_{\overline{X}_1 - \overline{X}_2}$ is:

$$s_{\overline{X}_1 - \overline{X}_2} = \sqrt{s_p^2 \left(\frac{1}{n_1} + \frac{1}{n_2}\right)}$$

12.4 Summary

This chapter discussed a number of statistical techniques that can be used directly to answer questions that arise in the analysis of microarray data. The chapter addressed only parametric testing involving one or two samples. A sample in this context is a set of measurements. Such measurements can be:

- Repeated measurements of the expression level of a gene in various individuals of a biological population (biological replicates)

- Repeated measurements of the expression level of a gene using several spots on a given microarray (spot replicates)

- Repeated measurements of the expression level of a gene using several arrays (array replicates).

Each of these levels allows answering specific questions at that particular level. For instance, spot replicates can provide quality control information for a given array, array replicates can provide quality control information for a given technology, etc. However, the results can be combined. For instance, an analysis at the spot replicate level can be used in order to decide how many replicates should be printed on a custom array and what the confidence limits for the mean of such measurements are. Once this has been decided, the mean of such measurements can be chosen to represent the expression level of a gene on a given array.

The caveats include the fact that the techniques discussed in this chapter do not allow us to test hypotheses about interactions of different levels. ANalysis Of VAriance (ANOVA) methods must be used for such purposes. Also, the techniques discussed here assume that the data was normalized in a consistent way. Applying these techniques on raw data or trying to compare data that was normalized using different techniques will, most probably, produce erroneous results.

The following list includes questions that arise in the analysis of microarray data that can be addressed with the techniques discussed in this chapter.

1. A gene is known to be expressed at level c when measured with a given technology in a given population under given circumstances. A new set of measurements produce the values: $m_1, m_2, \ldots, m_k$. Are these data consistent with the accepted value c?

   - Hypothesis: $H_0 : \mu = c$
   - Suitable distribution: t distribution.
   - Assumptions: The t distribution assumes normality. However, if the variable is a mean of a sample ($\overline{X}$), the central limit theorem allows us to infer valid conclusions even if the original population of X is not normal.
   - Reasoning: classical hypothesis testing.

2. Given that the measurements of a gene in normal conditions has a mean μ and a standard deviation σ, how many replicate measurements are needed in order to detect a change of δ % at least 95% of the time?

   - Hypothesis: there are two null hypotheses involved, one for the initial population and one for the population with an assumed shift in mean that we want to detect.
   - Suitable distribution: Z distribution (normal)
   - Assumptions: The Z distribution assumes normality, but if the variable is a mean of a sample ($\overline{X}$), the central limit theorem allows us to infer valid conclusions even if the original population of X is not normal.
   - Reasoning: Calculate a Z_α value by using the initial distribution and setting conditions for Type I error (α); then use a distribution with a shift in mean equal to the minimum change to be detected, set conditions for the Type II error and calculate a Z_β value. The number of replicates can be calculated by setting the condition: $Z_\alpha = Z_\beta$.

3. The expression level of a gene measured with technology A in a given set of conditions has a standard deviation of σ. The expression level of the same gene in the same conditions is measured n times with technology B and has a sample variance s^2. Does technology B offer a more consistent way of measuring gene expression?

   - Hypothesis: $H_0 : \sigma^2 = c$
   - Suitable distribution: χ^2 distribution
   - Assumptions: normal distribution
   - Reasoning: classical hypothesis testing

4. The expression level of a specific gene is measured n times. The sample variance is found to be s. What are the 95% confidence limits for the population variance?

- Hypothesis: not applicable (not a hypothesis testing situation)
- Suitable distribution: χ^2 distribution
- Assumptions: normal distribution
- Reasoning: Extract the variance from the expression of the χ^2 distribution and set conditions for Type I error (α)

5. Given two sets of measurements of the expression level of a gene (perhaps in two different conditions), are the variances of the two sets equal?

- Hypothesis: $\sigma_1 = \sigma_2$
- Suitable distribution: F distribution
- Assumptions: the two populations are normal
- Reasoning: classical hypothesis testing

6. Given two sets of measurements of the expression level of a gene (perhaps in two different conditions), what is the pooled variance of the combined set of measurements?

- Hypothesis: not applicable (not a hypothesis testing situation)
- Suitable distribution: not applicable (not a hypothesis testing situation)
- Assumptions: the two populations are normal
- Reasoning: see Eq. 12.38

7. Given two sets of measurements of the expression level of a gene in two different conditions, is this gene expressed differently between the two conditions?

- Hypothesis: $\mu_1 = \mu_2$
- Suitable distribution: t distribution
- Assumptions: the two populations are normal or the random variable is a mean of a sample (central limit theorem)
- Reasoning: classical hypothesis testing but care must be taken in using the appropriate formulae for sample variance and degrees of freedom (equal variance versus unequal variance, paired versus unpaired, etc.)

8. If a gene has been found to have different mean measurements between two different conditions, how different are these means? What are the 95% confidence intervals for the difference of the means?

- Hypothesis: not applicable (not a hypothesis testing situation)
- Suitable distribution: t distribution
- Assumptions: normality
- Reasoning: see Eq. 12.45

Note that very often such questions need to be combined. For instance, in order to choose the appropriate test to see whether a gene is expressed differently between two conditions, one usually needs to test whether the two sets of measurements have the same variance. Another example is measuring a gene several times and calculating the sample variance and mean as well as the confidence intervals for population variance and mean. When sufficient data are collected, these can be used as good estimates of the population variance and mean, which in turn can be used to answer questions about the reliability of a new technology or about the expression value in a new condition.

12.5 Exercises

1. Given two independent populations with variances σ_1 and σ_2. Consider the variables X_1 and X_2 drawn from the two populations, respectively. Calculate the variance of the variables:

    (a) $X_1 + X_2$

    (b) $X_1 - X_2$

    (c) $X_1 \cdot X_2$

2. Consider two independent samples for which the null hypothesis of having the same means can be rejected. Calculate the confidence intervals for $\mu_1 - \mu_2$.

Chapter 13

Analysis of Variance – ANOVA

In a nutshell, understanding any system means engineering it.

—Lynn Conway

13.1 Introduction

13.1.1 Problem definition and model assumptions

Let us consider an experiment measuring the expression level of a given gene in a number of k conditions. Each gene i is measured n_i times for a total of $\sum_{i=1}^{k} n_i$ measurements as follows:

		Condition			
1	2	$\cdots$	i	$\cdots$	k
X_{11}	X_{21}	$\cdots$	X_{i1}	$\cdots$	X_{k1}
X_{12}	X_{22}	$\cdots$	X_{i2}	$\cdots$	X_{k2}
$\vdots$	$\vdots$	$\vdots$	$\vdots$	$\vdots$	$\vdots$
X_{1n_1}	X_{2n_2}	$\cdots$	X_{in_i}	$\cdots$	X_{kn_k}

Note that this may look like a matrix with the symbolic notation used but, in reality, is more general because each column can have a different number of measurements n_i.

The typical question asked here is whether there are any differences between the expression level of the given gene between the k conditions. However, this is a rather imprecise formulation. We need to formulate clearly the null and research hypotheses. The null hypothesis seems to be pretty easy to formulate. Under the null hypothesis, the different conditions are not really different and, therefore, all measurements actually come from a single distribution. In these conditions, all means would be the same:

$$H_0 : \mu_1 = \mu_2 = \cdots = \mu_k \qquad (13.1)$$

However, from the initial formulation, it is not clear whether the alternate or research hypothesis would require for *all conditions* or only *a subset of conditions* to be different from each other. In other words, there seem to be three possibilities for H_a:

1. H_a: All means are different from each other.

2. H_a: Several but not all means are different from each other.

3. H_a: There is at least one pair of means that are different from each other.

Let us consider the first choice above as our research hypothesis. The problem would then be formulated as the following pair of hypotheses:

H_0: $\mu_1 = \mu_2 = \cdots = \mu_k$
H_a: All means are different from each other.

It is clear that this pair of hypotheses does not include all possible situations. In other words, it is not all inclusive. For instance, if $\mu_1 \neq \mu_2$ and $\mu_2 \neq \mu_3$ but $\mu_4 = \mu_5 = \cdots = \mu_k$, both hypotheses above would be false, which shows they are not defined correctly. A similar situation happens if the second alternative above is chosen as the research hypothesis since it is always possible to find a situation in which neither hypothesis is true. The only correct combination is to use H_0 as defined above together with its logical negation as the research alternative. From elementary logic, if a statement is of the form "for all x it is true that $P(x)$," its negation will be of the form "there exists at least one x for which $P(x)$ is false." Therefore, the *only* correct problem formulation is:

H_0: $\mu_1 = \mu_2 = \cdots = \mu_k$
H_a: There is at least one pair of means that are different from each other.

This particular data layout and set of hypotheses is characteristic to a **Model I**, or **fixed effects**, **ANOVA**. In Model I ANOVA, the researcher has a specific interest in the conditions or treatments under study. In particular, the question regards whether there are differences between any pair of the specific conditions considered. If such differences exist (at least one), it is of interest to identify which ones are different. Furthermore, in this data layout, each measurement belongs to a single group. In other words, the data were factored one-way. This analysis is called a **one-way ANOVA**.

The other assumptions made here are as follows:

1. The k samples are independent random samples drawn from k specific populations with means $\mu_1, \mu_2, \ldots, \mu_k$ (constant but unknown).

2. All k populations have the same variance σ^2.

3. All k populations are normal.

Since the framework used to develop the ANOVA methodology assumes the populations are normal, ANOVA is a parametric test. ANOVA stands for ANalysis Of VAriance which may seem a bit counterintuitive since the problem was formulated as a test for means. However, the reasoning behind this is still intuitive. It is very likely that two different random samples will have two different means whether they come from the same population or not. Given a certain difference in means, one can reject a null hypothesis regarding the means if the variance of each sample is sufficiently small in comparison to the overall variance. This is illustrated in Fig. 13.1 and Fig. 13.2.

Fig. 13.1 shows two samples with means $\mu_1 = 2$ and $\mu_2 = 6$. In this case, the variability within groups is much smaller than the overall variability. This may allow us to reject the hypothesis that the two samples were drawn from the same distribution and accept that there is a significant difference between the two samples. Fig. 13.2 illustrates an example with two samples in which the variability within groups is comparable to the overall variability. Note that the two sample means are still $\mu_1 = 2$ and $\mu_2 = 6$, as in Fig. 13.1. However, in this situation we are unlikely to reject the null hypothesis that the two samples were drawn from the same distribution. These two examples show how an analysis of *variance* can provide information that allow us to draw conclusions about hypotheses involving *means*.

Another interesting observation is that both the set of hypotheses used here and the examples in Fig. 13.1 and Fig. 13.2 look very similar to hypotheses and examples given in Chapter 12 when the t-test was discussed. This is no coincidence. If there are only two samples, the ANOVA will provide the same

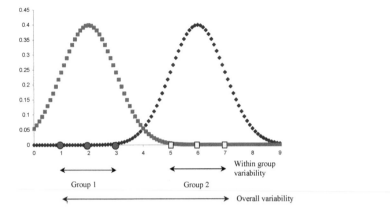

FIGURE 13.1: Two samples with means $\mu_1 = 2$ and $\mu_2 = 6$. The variability within groups is much smaller than the overall variability; this may allow us to reject the hypothesis that the two samples were drawn from the same distribution.

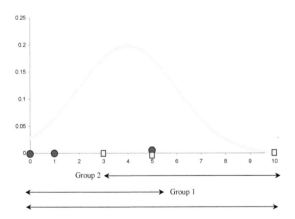

FIGURE 13.2: Two samples with means $\mu_1 = 2$ and $\mu_2 = 6$. The variability within groups is comparable to the overall variability. In this situation, we are unlikely to reject the null hypothesis that the two samples were drawn from the same distribution.

results as a t-test. However, the t-test can be applied to comparisons involving only two samples while ANOVA can be applied to any number of samples. An astute reader could argue that this is not a problem since one could use a divide-and-conquer approach and split the null hypothesis $H_0 : \mu_1 = \mu_2 = \cdots = \mu_k$ into a number of hypotheses involving only two samples $H_0^i : \mu_i = \mu_j$. Subsequently, each of these hypotheses could be tested with an individual t-test. Thus, the given null hypothesis $H_0 : \mu_1 = \mu_2 = \cdots = \mu_k$ will be rejected if at least one of the $H_0^i : \mu_i = \mu_j$ is rejected. Although this may seem as a valid reasoning from a logical point of view, it is invalid from a statistical point of view. The main problem is that the statistical hypothesis testing does not provide absolute conclusions. Instead, each time a null hypothesis is rejected, there is a non-zero probability of the null hypothesis being actually true. This is the probability of a Type I error or the significance level. When many such tests are carried out for the purpose of drawing a single conclusion, as above, a single mistake in each one of the individual tests is sufficient to invalidate the conclusion. Thus, the probability of a Type I error increases with the number of tests even if the probability of a Type I error in each test is bounded by the chosen level of significance α. This will be explained fully in the section dealing with the correction for multiple experiments.

13.1.2 The "dot" notation

A notation very useful when many sums are involved is the "dot" notation. This notation uses a dot instead of the index for which the sum is calculated. Using this notation, the sum of all measurements of a gene in a given condition i can be written as:

$$\sum_{j=1}^{n_i} X_{ij} = X_{i.} \tag{13.2}$$

The same notation can be used for means and other quantities. For instance, the mean of the i-th sample can be written as:

$$\frac{\sum_{j=1}^{n_i} X_{ij}}{n_i} = \frac{X_{i.}}{n_i} = \overline{X}_{i.} \tag{13.3}$$

Furthermore, a summation over two indices can be represented by using two dots as follows:

$$\sum_{i=1}^{k} \sum_{j=1}^{n_i} X_{ij} = X_{..} \tag{13.4}$$

More examples using the dot notation are given in Fig. 13.3.

	Condition 1	$\cdots$	Condition k	
measurement 1	X_{11}	$\cdots$	X_{k1}	$X_{.1} = \sum_{i=1}^{k} X_{i1}$
measurement 2	X_{12}	$\cdots$	X_{k2}	$X_{.2} = \sum_{i=1}^{k} X_{i2}$
$\vdots$	$\vdots$	$\cdots$	$\vdots$	$\vdots$
measurement n	X_{1n}	$\cdots$	X_{kn}	$X_{.n} = \sum_{i=1}^{k} X_{in}$
	$X_{1.} = \sum_{j=1}^{n} X_{1j}$	$\cdots$	$X_{k.} = \sum_{j=1}^{n} X_{kj}$	$X_{..} = \sum_{i,j} X_{ij}$

FIGURE 13.3: The "dot" notation. Summation on an index is represented by a dot substituting the given index. In this example, there are k conditions and each condition is measured n times for a total number of $n \times k$ measurements.

13.2 One-way ANOVA

13.2.1 One-way Model I ANOVA

The general idea behind one-way Model I ANOVA is very simple. The measurements of each condition vary around their mean. This is an *within-group* variability and will be characterized by a corresponding *within-group variance*. At the same time, the means of each condition will vary around an overall mean. This is due to a *inter-group* variability. Finally, as a result of the two above, each individual measurement varies around the overall mean. The idea behind ANOVA is to study the relationship between the inter-group and the within-group variabilities (or variances).

We can start by considering the difference between an individual measurement X_{ij} and the global mean:

$$X_{ij} - \overline{X}_{..} \tag{13.5}$$

Let us add and subtract the mean of the condition i and regroup the terms in a convenient way:

$$X_{ij} - \overline{X}_{..} = X_{ij} - \overline{X}_{..} + \overline{X}_{i.} - \overline{X}_{i.} = \left(\overline{X}_{i.} - \overline{X}_{..} \right) + \left(X_{ij} - \overline{X}_{i.} \right) \tag{13.6}$$

This is a *very* interesting observation. Equation 13.6 tells us that the deviation of an individual measurement from the overall mean can be seen as the sum of the deviations of that individual measurement from the mean of its condition and the deviation of that mean from the overall mean. This is most interesting because the first term $\left(\overline{X}_{i.} - \overline{X}_{..} \right)$ characterizes the deviation of an individual condition mean from the overall mean or the deviations *between groups* while the term $\left(X_{ij} - \overline{X}_{i.} \right)$ characterizes the deviation of an individual measurement from the mean of its condition or the deviation *within group*. Unfortunately, we have seen in Chapter 8 that the deviations are not good estimators of the sample variance because their sum cancels out. The sample variability is better characterized by the variance of the sample which was defined as the sum of squared deviations of individual measurements from the mean $\sum(X_i - \overline{X})^2$ divided by the sample size minus 1. For samples of equal size, the denominator is constant and the variance will be determined by the sum of (deviation) squares. Such sums of squares (SS) turn out to be most important quantities in the ANOVA approach, and we will focus on them.

We can characterize the total variability (using the dot notation defined above) as:

$$\sum_{i=1}^{k}\sum_{j=1}^{n_i}(X_{ij} - X_{..})^2 \tag{13.7}$$

This expression is the sum of the squares of all differences from individual observation to the grand mean of the data set and is called the **Total Sum of Squares** or SS_{Total}. This is a measure of the scatter of all data considered, as a single group, around its mean. Note that if we divide this expression by $N - 1$, where N is the total number of measurements, we obtain exactly the global sample variance.

We need to work out a formula relating the overall variance with the variance between conditions and the variance within conditions. This translates to working out a formula between their respective sum of squares: SS_{Total}, $SS_{between\ groups}$, and $SS_{within\ group}$.

All measurements within a group are measurements of the same gene in the same condition. Any differences between such measurements are due to the experimental error. In consequence, the sum of squares within a group is also called the **Error Sum of Squares** $SS_{within\ group} = SS_{Error}$ or **Residual Sum of Squares**. The sum of squares between groups characterizes the variance of the groups around the overall mean. Therefore, one individual such sum of square will characterize the variance of its condition (or treatment) $SS_{between\ groups} = SS_{Cond}$. This is also called the **Among Treatment Sum of Squares**.

13.2.1.1 Partitioning the Sum of Squares

The **Total Sum of Squares**:

$$SS_{Total} = \sum_{i=1}^{k}\sum_{j=1}^{n_i}(X_{ij} - X_{..})^2 \tag{13.8}$$

can be rewritten using equation 13.6 as:

$$
\begin{aligned}
SS_{Total} &= \sum_{i=1}^{k}\sum_{j=1}^{n_i}\left[(\overline{X}_{i.} - \overline{X}_{..}) + (X_{ij} - \overline{X}_{i.})\right]^2 \\
&= \sum_{i=1}^{k}\sum_{j=1}^{n_i}\left[(\overline{X}_{i.} - \overline{X}_{..})^2 + 2(\overline{X}_{i.} - \overline{X}_{..})(X_{ij} - \overline{X}_{i.}) + (X_{ij} - \overline{X}_{i.})^2\right]
\end{aligned}
$$

$$
= \sum_{i=1}^{k} \sum_{j=1}^{n_i} \left(\overline{X}_{i.} - \overline{X}_{..} \right)^2 + \sum_{i=1}^{k} \sum_{j=1}^{n_i} 2 \left(\overline{X}_{i.} - \overline{X}_{..} \right) \left(X_{ij} - \overline{X}_{i.} \right) +
$$

$$
+ \sum_{i=1}^{k} \sum_{j=1}^{n_i} \left(X_{ij} - \overline{X}_{i.} \right)^2
$$

$$
= \sum_{i=1}^{k} \sum_{j=1}^{n_i} \left(\overline{X}_{i.} - \overline{X}_{..} \right)^2 + \sum_{i=1}^{k} \sum_{j=1}^{n_i} \left(X_{ij} - \overline{X}_{i.} \right)^2 +
$$

$$
+ 2 \sum_{i=1}^{k} \left(\overline{X}_{i.} - \overline{X}_{..} \right) \sum_{j=1}^{n_i} \left(X_{ij} - \overline{X}_{i.} \right) \tag{13.9}
$$

The very last term contains the sum:

$$
\sum_{j=1}^{n_i} \left(X_{ij} - \overline{X}_{i.} \right)
$$

which is the sum of all deviations of measurements in sample i with respect to its mean. We have seen in Chapter 8, equation 8.12, that this sum is zero. In fact, this was an obstacle against using the sum of deviates as a measure of variance. Using this, the total sum of squares becomes:

$$
SS_{Total} = \sum_{i=1}^{k} \sum_{j=1}^{n_i} \left(\overline{X}_{i.} - \overline{X}_{..} \right)^2 + \sum_{i=1}^{k} \sum_{j=1}^{n_i} \left(X_{ij} - \overline{X}_{i.} \right)^2 \tag{13.10}
$$

or

$$
SS_{Total} = SS_{Cond} + SS_{Error} \tag{13.11}
$$

This is the fundamental result that constitutes the basis of the ANOVA approach. The result states that the total sum of squares SS_{Total} or the overall variability can be partitioned into the variability SS_{Cond} due to the difference between conditions (treatments) and the variability SS_{Error} within treatments. This is essentially the same partitioning that can be done on the deviations (Eq. 13.6) only that the SS partitioning is more useful since the SS is directly related to the variance.

We will also introduce here the notation:

$$
T_{i.} = \sum_{j} X_{ij} \tag{13.12}
$$

$$
T_{..} = \sum_{i} \sum_{j} X_{ij} \tag{13.13}
$$

With this notation, and assuming that the data are arranged in a table that contains one column for every condition, the sums of squares can also be calculated with the following, more convenient, formulae:

$$
SS_{Total} = \sum_{i} \sum_{j} X_{ij}^2 - \frac{T_{..}^2}{N} \tag{13.14}
$$

$$SS_{Cond} = \sum_i \frac{T_{i.}^2}{n_i} - \frac{T_{..}^2}{N} \tag{13.15}$$

13.2.1.2 Degrees of freedom

It is interesting to calculate the number of degrees of freedom for each of the sum of squares. The overall variance is reflected by SS_{Total}. For this variance, there are $N-1$ degrees of freedom where $N = \sum_{i=1}^{k} n_i$ is the total number of measurements in the data set. The condition variance is reflected by SS_{Cond}, which has $k-1$ degrees of freedom if there are k different conditions. Finally, there are k error variances, each of them having $n_i - 1$ degrees of freedom. We can calculate the number of degrees of freedom for the errors as:

$$\sum_{i=1}^{k}(n_i - 1) = \sum_{i=1}^{k} n_i - \sum_{i=1}^{k} 1 = N - k \tag{13.16}$$

Now, one can appreciate that the partitioning that is true for the sum of squares:

$$SS_{Total} = SS_{Cond} + SS_{Error} \tag{13.17}$$

also carries over to the degrees of freedom:

$$N - 1 = (k - 1) + (N - k) \tag{13.18}$$

13.2.1.3 Testing the hypotheses

We have seen that the sums of squares are closely related to the variances. The following quantities can be defined:

1. **Error Mean Squares (MS_E)**

$$MS_E = \frac{SS_{Error}}{\sum_{i=1}^{k}(n_i - 1)} = \frac{SS_{Error}}{N - k} \tag{13.19}$$

   It can be shown that the expected value of this quantity is σ^2.

2. **Condition Mean Squares** or **Treatment Mean Squares (MS_{Cond}):**

$$MS_{Cond} = \frac{SS_{Cond}}{k - 1} \tag{13.20}$$

   It can be shown that the expected value of this quantity is $\sigma^2 + \sum_i \frac{n_i(\mu_i - \mu)^2}{k-1}$. The proof of this statement is beyond the scope of this book. However, it can be seen that if the null hypothesis is true and all means are equal, the second term of this expression becomes zero and the expected value of the Condition Mean Squares MS_{Cond} becomes equal to σ^2 as well.

The two statistics above allow us to test the null hypothesis using the techniques developed to test for hypotheses on variance. In particular, the statistic:

$$\frac{MS_{Cond}}{MS_E} \qquad (13.21)$$

is distributed as an F distribution with $v_1 = k - 1$ and $v_2 = N - k$.

We have seen that the expected values for both the denominator and numerator are σ^2 when the null hypothesis is true. Therefore, the F statistic above will have a value of 1 if the null hypothesis is true and a value larger than 1 when some of the means are different. Since the value of this statistic can only be larger or equal to one, the F-test used to possibly reject the null hypothesis will always be one-tailed.

Most statistical software packages allow the user to perform an ANOVA analysis very easily. Furthermore, ANOVA is also available in the Data Analysis Pack that is an option to Microsoft Excel. Let us consider the following example.

EXAMPLE 13.1
The expression level of a gene is measured in two different conditions: control subjects and cancer patients. The measurements in control subjects produced the following values: 2, 3, and 1. The measurements in cancer patients produced the values 6, 7, and 5. Is there a significant difference between the expression level of this gene in the two conditions?

SOLUTION   We can organize the data as follows:

	Control	Cancer
Measurement 1	2	6
Measurement 2	3	7
Measurement 3	1	5

The results of the ANOVA can be summarized as shown in the following table. This is the summary as it is presented by Excel:

Source of variation	SS	df	MS	F	p-value	F critical
Between groups	24	1	24	24	0.00805	7.708
Within groups	4	4	1			
Total	28	5				

Note that the partition of the sum of squares is noticeable in the first column $(24 + 4 = 28)$ and the partitioning of the degrees of freedoms can be observed in the second column $(1 + 4 = 5)$. The third column shows the condition and error mean squares and the fourth column shows the value of the F statistic. The p-value of this observed statistic is 0.008 (column 4), which is smaller than the critical value for the 5% significance level (column

5). In conclusion, the data allow us to reject the null hypothesis and conclude that the given gene is expressed differently in the two conditions. ☐

As an observation, a two-tailed t-test assuming equal variances.[1] will produce exactly the same p-value of 0.008 in this example. As noted above, a comparison that involves *only* two conditions can be analyzed with either ANOVA or a t-test. The following example is a situation in which more than two samples are involved and the t-test cannot be applied directly anymore.

EXAMPLE 13.2

The expression level of a gene is measured in three different conditions: mesothelioma patients who survived more than 1 year after they were operated on (long-term survivors), mesothelioma patients who survived less than 1 year (short-term survivors) and control subjects. Is there a significant difference between the expression level of this gene in the three conditions?

	Control	Short term	Long term
Patient 1	2	6	2
Patient 2	3	7	2
Patient 3	1	5	1

SOLUTION

The results of the ANOVA can be summarized as shown in the following table. The critical value for the F statistic corresponds to a significance level of 5% for $F_{2,6}$.

Source of variation	SS	df	MS	F	p-value	F critical
Between groups	34.88	2	17.44	22.42	0.001	5.14
Within groups	4.66	6	0.77			
Total	39.55	8				

Once again, the data allow us to reject the null hypothesis. However, this time there are three different conditions/treatments, and the only information provided by the model I ANOVA that we have performed is that at least one pair of means is different. Unfortunately, ANOVA does not provide information as to *which* of the three possible pairs of means contradicts the null hypothesis. The three possibilities are: $\mu_1 \neq \mu_2$, $\mu_1 \neq \mu_3$, and $\mu_2 \neq \mu_3$. This can be done with a series of pair-wise t-tests with corrections for multiple experiments as discussed in Chapter 16. ☐

---

[1] This is also known as a homoscedastic test.

EXAMPLE 13.3

There is a conjecture that a certain gene might be linked to ovarian cancer. The ovarian cancer is subclassified into three categories: stage I, stage II, and stage III-IV. There are three samples available, one from each stage. These samples are labeled with three colors and hybridized on a four-channel cDNA array (one channel remains unused). The experiment is repeated five times and the following measurements are collected:

Array	mRNA 1	mRNA 2	mRNA 3
1	100	95	70
2	90	93	72
3	105	79	81
4	83	85	74
5	78	90	75

Is there any difference between the three mRNA samples?

SOLUTION   We will use a one-way Model I ANOVA analysis at the 5% significance level. We start by calculating the sums and averages for the three samples.

Array	mRNA 1	mRNA 2	mRNA 3
1	100	95	70
2	90	93	72
3	105	79	81
4	83	85	74
5	78	90	75
$T_{i.}$	456	442	372
$\overline{X}_i$	91.2	88.4	74.4
$T_{i.}^2$	207936	195364	138384

We calculate the sums of squares:

$$SS_{Treat} = \sum_i \frac{T_{i.}^2}{n_i} - \frac{T_{..}^2}{N} = 810.13 \tag{13.22}$$

$$SS_{Total} = \sum_i \sum_j X_{ij}^2 - \frac{T_{..}^2}{N} = 1557.33 \tag{13.23}$$

$$SS_{Error} = SS_{Total} - SS_{Treat} = 1557.33 - 810.13 = 747.2 \tag{13.24}$$

We calculate the mean squares by dividing the sums of squares by their respective degrees of freedom. There are $k - 1 = 3 - 1 = 2$ degrees of freedom for the treatments and $N - k = 15 - 3 = 12$ degrees of freedom for the error. Hence:

$$MS_{Treat} = \frac{SS_{Treat}}{k - 1} = \frac{810.13}{2} = 405.06 \tag{13.25}$$

$$MS_{Error} = \frac{SS_{Error}}{N-k} = \frac{747.2}{12} = 62.26666667 \qquad (13.26)$$

And finally, we calculate the value of the F statistic:

$$F = \frac{MS_{Treat}}{MS_{Error}} = \frac{405.06}{62.26} = 6.50 \qquad (13.27)$$

These results can be summarized in the usual ANOVA table:

	SS	df	MS	F	p-value	F critical
Treatments	810.13	2	405.06	6.50	0.012	3.885
Error	747.2	12	62.26			
Total	1557.33					

We see that the p-value is lower than the chosen significance level, and we reject the null hypothesis. Implicitly, we accept the research hypothesis that at least one pair of means are different. We now have to test which means are different. We will use a t-test with Bonferroni correction for multiple experiments (see Chapter 16). Our chosen significance level is 0.05 and there are three parallel experiments. The value of α adjusted according to Bonferroni is:

$$\alpha_{Bonferroni} = \frac{0.05}{3} = 0.017 \qquad (13.28)$$

We will use the t-test with the assumption of equal variances (the ANOVA itself assumes that all distributions involved are normally distributed with equal variance). There are $N - k = 15 - 3 = 12$ degrees of freedom. The t statistic is calculated as:

$$t = \frac{\overline{X}_{i.} - \overline{X}_{j.}}{\sqrt{MS_{Error}\left(\frac{1}{n_i} + \frac{1}{n_j}\right)}} \qquad (13.29)$$

The t-values and the corresponding p-values are as follows:

	1 vs. 2	1 vs. 3	2 vs. 3
t	0.561048269	3.687585516	3.07298793
p-value	0.585089205	0.00310545	0.009662999

The null hypothesis can be rejected for the comparisons between 1 versus 3 and 2 versus 3. The null hypothesis cannot be rejected for the comparison 1 versus 2. The conclusion is that 3 is different from both 1 and 2. In other words, this gene seems to be expressed differently in the later stages (III and IV) of ovarian cancer with respect to the earlier stages (stages I and II).

13.2.2   One-way Model II ANOVA

We have seen that in Model I ANOVA, the researcher has a specific interest in the conditions or treatment under study. In particular, the question regards whether there are differences between any pair of the specific conditions considered. If such differences exist (at least one), it is of interest to identify which ones are different. In Model II, or random effects, ANOVA, the particular conditions studied are not of interest and the question focuses on whether there is a significant variability among the conditions. The data layout looks similar to the layout for Model I ANOVA:

	Condition				
1	2	$\cdots$	i	$\cdots$	k
X_{11}	X_{21}	$\cdots$	X_{i1}	$\cdots$	X_{k1}
X_{12}	X_{22}	$\cdots$	X_{i2}	$\cdots$	X_{k2}
$\vdots$	$\vdots$	$\vdots$	$\vdots$	$\vdots$	$\vdots$
X_{1n_1}	X_{2n_2}	$\cdots$	X_{in_i}	$\cdots$	X_{kn_k}

but often the conditions or treatments involved in Model II ANOVA are chosen at random from a larger population. The hypotheses for a Model II can be formulated as follows:

$$H_0: \quad \sigma^2_{Cond} = 0$$
$$H_a: \quad \sigma^2_{Cond} > 0$$

The null hypothesis states that there is no significant difference between conditions (columns) and the research hypothesis states that there is a significant difference between conditions. Once again, in the Model II analysis there is no interest in the individual conditions, which may have been drawn at random from a larger population.

The partitioning of the sums of squares still holds:

$$SS_{Total} = SS_{Cond} + SS_{Error} \tag{13.30}$$

but the expected value for the condition mean square MS_{Cond} is:

$$E(MS_{Cond}) = \sigma^2 + n_0 \sigma^2_{Cond} \tag{13.31}$$

where n_0 is:

$$n_0 = \frac{N - \sum_{i=1}^{k} \frac{n_i^2}{N}}{k - 1} \tag{13.32}$$

and N is the total number of measurements $N = \sum_{i=1}^{k} n_i$.

The expected value of the mean square error MS_{Error} for this test is:

$$E(MS_E) = \sigma^2 \tag{13.33}$$

Once again, if H_0 is true, σ_{Cond} in Equation 13.31 is 0, both numerator and denominator of the ratio involved in the F statistic are σ^2, and the expected value for the F statistic:

$$F = \frac{MS_{Cond}}{MS_{Error}} \tag{13.34}$$

will be 1. If H_0 is not true, the F statistic will be greater than 1 because σ^2_{Cond} is always positive. Again a one-tailed testing can be performed using the F distribution.

EXAMPLE 13.4

The β-actin gene is proposed to be used in the normalization as a housekeeping gene. This gene is assumed to be expressed at approximately constant levels throughout the body. A number of five tissues have been randomly chosen to investigate the expression level of this gene. The expression level was measured five times in each tissue. The measurements of the gene expression levels provided the following data:

	Liver	Brain	Lung	Muscle	Pancreas
Measurement 1	12	14	17	10	32
Measurement 2	15	20	31	15	10
Measurement 3	17	23	19	20	12
Measurement 4	20	21	14	25	13
Measurement 5	12	19	26	30	41

Can the β-actin gene be used as housekeeping gene for normalization purposes?

SOLUTION

In this example, the particular tissues studied are not of interest. These tissues have been picked at random from a larger population. The question of interest is whether there is a significant variability among the tissues for this particular gene. This is a Model II ANOVA situation. The null and research hypotheses are:

$$H_0: \quad \sigma^2_{Tissue} = 0$$
$$H_a: \quad \sigma^2_{Tissue} > 0$$

We can calculate the $T_{i.}$ and $\overline{X}_i$ values as follows:

	12	14	17	10	32
	15	20	31	15	10
	17	23	19	20	12
	20	21	14	25	13
	12	19	26	30	41
$T_{i.}$	76	97	107	100	108
$\overline{X}_{i.}$	15.2	19.4	21.4	20	21.6

We then calculate the basic summary statistics:

$$\sum_i \sum_j X_{ij}^2 = 10980$$

$$T_{..} = \sum_i \sum_j X_{ij} = 488$$

$$N = 25$$

The sum of squares can be calculated using the more convenient formulae:

$$SS_{Total} = \sum_i \sum_j X_{ij}^2 - \frac{T_{..}^2}{N} = 10980 - \frac{488^2}{25} = 1454.24$$

and

$$SS_{Tissue} = \sum_i \frac{T_{i.}^2}{n_i} - \frac{T_{..}^2}{N} = 133.84$$

We can now extract SS_{Error} from the partitioning formula 13.11:

$$SS_{Error} = SS_{Total} - SS_{Tissue} = 1454.24 - 133.84 = 1320.4$$

There are 4 degrees of freedom for the treatments $(k - 1$ where $k = 5)$ and 20 degrees of freedom for the error $(N - k = 25 - 5)$. In consequence, the mean squares will be:

$$MS_{Tissue} = \frac{SS_{Tissue}}{k - 1} = \frac{133.84}{4} = 33.46$$

$$MS_{Error} = \frac{SS_{Error}}{N - k} = \frac{1320.4}{20} = 66.02$$

yielding an F statistic of:

$$F = \frac{33.46}{66.02} = 0.507$$

The p-value corresponding to this F value can be obtained from an F distribution with 4 and 20 degrees of freedom and is $p = 0.73$. Under these circumstances, we cannot reject the null hypothesis. In conclusion, there is no evidence that the chosen gene is expressed differently between tissues, and it is probably[2] safe to use it for normalization purposes. ∎

13.3 Two-way ANOVA

The one-way ANOVA methods allow us to investigate data in which only one factor is considered. However, in many problems, the data might be influenced

---

[2]Note that, strictly speaking, our hypothesis testing did not prove that the gene is expressed uniformly throughout the given tissue but merely showed that there are no obvious inter-tissue differences so far. See Section 11.4 for a more detailed discussion of how to interpret the conclusions of hypothesis testing.

by more than one factor. In fact, even in the data used in the one-way ANOVA there is a second dimension, that of the individual measurements within the same group. These appeared as rows in the data matrix. This section will discuss the methods used to analyze data when several factors might contribute to the variability of the data.

13.3.1   Randomized complete block design ANOVA

Let us consider a microarray experiment in which three mRNA samples are hybridized on a number of arrays. For the sake of the presentation, we will assume this was done with three-channel arrays. On such arrays, the mRNA can be labeled with three colors and three samples can be tested simultaneously on the same array. Each gene will be measured 6×3 times (6 arrays $\times$ 3 mRNA samples). Let us consider the data gathered for one gene:

	Treatments		
	mRNA 1	mRNA 2	mRNA 3
Array 1	100	95	88
Array 2	90	93	85
Array 3	105	79	87
Array 4	83	85	83
Array 5	78	90	89
Array 6	93	75	75

Each of the expression values in this array can be looked at in two ways. First, each value comes from a given mRNA sample. The mRNA sample is the first factor that affects the measurements. Second, each value was obtained using a given microarray. Therefore, the microarray on which the measurement was performed is the second factor that can influence the measurements. All values in any given row share the fact that they have been obtained on the same microarray. Such values represent **a block** of data. The goal of this experiment is to compare the given mRNA samples. However, each microarray will be slightly different from every other microarray, and we would like to remove the variability introduced by the arrays such that we can make a better decision about the mRNA samples. In this situation, we are not interested in the differences between microarrays. We had to measure the three mRNAs at the same time because this is how microarrays work, and we would like to distinguish between the variability introduced by them and the variability due to the different mRNA samples. This experiment design is called **randomized complete block design**. The design is called **randomized** because each mRNA sample was assigned randomly within each block. The term **block** denotes the fact that the data can be partitioned into chunks that are expected to share certain features (e.g., all values obtained from a given microarray will share all characteristics determined by washing or drying). Finally, the term

complete denotes that each mRNA sample is measured exactly once in each block.

The hypotheses will be:

1. H_0: $\mu_{mRNA_1} = \mu_{mRNA_2} = \mu_{mRNA_3}$

2. H_a: There is at least one pair of means that are different from each other.

The data can be organized as follows:

		mRNA sample (Treatment)				Totals	Mean
		1	2	$\cdots$	k		
	1	X_{11}	X_{21}	$\cdots$	X_{k1}	$T_{.1.}$	$\overline{X}_{.1}$
	2	X_{12}	X_{22}	$\cdots$	X_{k2}	$T_{.2.}$	$\overline{X}_{.2}$
Block	$\vdots$	$\vdots$	$\vdots$	$\vdots$	$\vdots$	$\vdots$	$\vdots$
	b	X_{1b}	X_{2b}	$\cdots$	X_{kb}	$T_{.b.}$	$\overline{X}_{.b}$
	Totals	$T_{1.}$	$T_{2.}$	$\cdots$	$T_{k.}$	$T_{..}$	
	Means	$\overline{X}_{1.}$	$\overline{X}_{2.}$	$\cdots$	$\overline{X}_{k.}$		$\overline{X}_{..}$

The notations remain the same as in Section 13.1.2: a dot denotes a summation on the corresponding index, T is the simple sum of the terms, and $\overline{X}$ stands for a mean.

We will make the following assumptions:

1. Each measurement is a random, independent sample from a population. The measurement located in the cell (i, j) is assumed to have mean μ_{ij}. The input data represents samples from $b \times k$ populations.

2. All these populations are normally distributed with a variance σ^2.

3. There is no interaction between the effects of the arrays and those of the samples.

The randomized complete block model assumes that each individual datum is a result of a superposition of various effects:

$$X_{ij} = \mu + B_j + V_i + \varepsilon_{ij} \tag{13.35}$$

where X_{ij} is the value measured for sample i and array j, μ is an overall mean effect, B_j is the effect of the array j, V_i is the effect of the sample i, and ε_{ij} is the random error. In our example, an array is a block and a sample is a variety of mRNA.

The effects above can be quantified as:

1. B_j is the effect of the j-th block. This can be assessed by:

$$B_j = \mu_{.j} - \mu \tag{13.36}$$

which is the difference between the array mean and the overall mean.

This is very intuitive since if we were to describe how different array j is from the overall mean, this is exactly the quantity that we would want to use.

2. V_i is the effect of the i-th variety (or mRNA sample):

$$V_i = \mu_{i.} - \mu \tag{13.37}$$

3. ε_{ij} is the effect of the random error that affects measurement X_{ij} in particular. Therefore, this can be quantified as:

$$\varepsilon_{ij} = X_{ij} - \mu_{ij} \tag{13.38}$$

The rest of the process is exactly as in the one-way ANOVA. The magic happens again with the sum of squares (see Eqs. 13.9 through 13.11), and all the inconvenient terms disappear to leave a very simple and intuitive relation:

$$SS_{Total} = SS_{Treat} + SS_{Blocks} + SS_{Error} \tag{13.39}$$

This partitioning also happens at the level of degrees of freedom. There are k samples so SS_{Treat} will have $k-1$ degrees of freedom, there are b blocks so SS_{Blocks} will have $b-1$ degrees of freedom, and there are $b \times k$ individual measurements so SS_{Error} will have $(b-1)\cdot(k-1)$ degrees of freedom. The partition of the degrees of freedom will be:

$$
\begin{array}{ccccccc}
SS_{Total} & = & SS_{Treat} & + & SS_{Blocks} & + & SS_{Error} \\
bk-1 & = & (k-1) & + & (b-1) & + & (b-1)\cdot(k-1)
\end{array} \tag{13.40}
$$

Following the same steps used in the one-way ANOVA, we then calculate the mean squares MS_{Treat}, MS_{Blocks}, and MS_{Error}:

$$MS_{Treat} = \frac{SS_{Treat}}{k-1} \tag{13.41}$$

$$MS_{Block} = \frac{SS_{Block}}{b-1} \tag{13.42}$$

$$MS_{Error} = \frac{SS_E}{(b-1)\cdot(k-1)} \tag{13.43}$$

and the F statistics for both samples and arrays:

$$F = \frac{MS_{Treat}}{MS_{Error}} \tag{13.44}$$

$$F = \frac{MS_{Block}}{MS_{Error}} \tag{13.45}$$

It can be shown that the F statistics above are 1 when the null hypothesis is true and larger than 1 when the distributions are different. This means that we can test our hypotheses using one tail of the F distribution with the correspondent number of degrees of freedom.

13.3.2 Comparison between one-way ANOVA and randomized block design ANOVA

Before we solve some examples, let us try to identify the differences between the one-way ANOVA and the randomized block design ANOVA. In both cases the data seem to be the same rectangular matrix, so it is important that we clarify when to use which method. Let us reconsider the partitioning of the degrees of freedom for the randomized block design:

$$
\begin{aligned}
SS_{Total} &= SS_{Treat} + SS_{Blocks} + SS_{Error} \\
bk - 1 &= (k-1) + (b-1) + (b-1) \cdot (k-1)
\end{aligned}
\tag{13.46}
$$

This expression is an invitation to some reflections. Let us extract the degrees of freedom corresponding to the error:

$$
(b-1) \cdot (k-1) = bk - 1 - (k-1) - (b-1)
\tag{13.47}
$$

We can now perform the computations and also consider that $bk = N$, which is the total number of measurements. We obtain:

$$
(b-1) \cdot (k-1) = bk - k - b + 1 = N - k - (b-1)
\tag{13.48}
$$

This is a very interesting and meaningful result if we compare it with the number of degrees of freedom of the error in one-way ANOVA (see Eq. 13.18). The degrees of freedom of the error have been reduced by $b-1$ with respect to the one-way ANOVA. It turns out that if everything else is kept constant, reducing the degrees of freedom of the denominator increases the critical values of the F distribution. For any given set of data, it will be harder to produce an F statistic more extreme than the critical value. In other words, we will be able to reject the null hypothesis less often, which means we have decreased the power of the test (see Section 11.6 for the definition of the power of a statistical test). If this were the only difference between the two approaches, the randomized block design would never be used. Fortunately, there is one other important difference. Let us go back to the partitioning of the sum of squares for the randomized complete block design:

$$
SS_{Total} = SS_{Treat} + SS_{Blocks} + SS_{Error}^{two-way}
\tag{13.49}
$$

This equation can be directly compared to the one for the one-way analysis:

$$
SS_{Total} = SS_{Treat} + SS_{Error}^{one-way}
\tag{13.50}
$$

For any given data set, the sum of squares for the total and the treatments are exactly the same. We can now see what the randomized blocking does: **it extracts the variability due to the blocks from the previously unexplained variability**. The quantity that is the sum of squares of the error in the one-way analysis, which represents the variability that our model is unable to account for, is now divided into a variability due to the treatments and an amount of variability that continues to be unexplained:

$$SS_{Error}^{one-way} = SS_{Blocks} + SS_{Error}^{two-way} \tag{13.51}$$

In the above equation, $SS_{Error}^{one-way}$ stands for the sum of squares of the error in one-way ANOVA and $SS_{Error}^{two-way}$ stands for the sum of squares of the error in the randomized block design, which is a particular type of two-way ANOVA. From this equation, it is clear that the SS_{Error} obtained from a two-way analysis will always be smaller than the one obtained from a one-way analysis of the same data set. Since SS_{Error} goes to the denominator of the F statistic ($F = \frac{MS_{Treat}}{MS_{Error}}$), decreasing it will increase the value of F. Hence, if the same data set is analyzed with both one-way and two-way ANOVA, the F value obtained with the two-way analysis will always be more significant than the other one (see Example 13.6). If the variability due to the arrays is large, or if the F statistic obtained with the one-way analysis is close to the critical value, being able to remove the array variability from the error variability may allow us to reject the null hypothesis in situations in which a one-way ANOVA might not.

Furthermore, doing an analysis using blocking can reassure us that the effects of the arrays are not falsifying our conclusions. This should appear as a low value of SS_{Blocks} and a high p-value for the corresponding F statistic. In this case, one can safely conclude that blocking is not necessary and a completely randomized design together with a one-way ANOVA is sufficient. Alternatively, a significant p-value for the blocks in a design as above might signal that the laboratory process that produced the microarrays is inconsistent and the arrays are significantly different from each other.

13.3.3 Some examples

Armed with this new methodology, let us consider the following examples.

EXAMPLE 13.5

A microarray experiment measured the expression level of the genes in the liver of mice exposed to ricin versus control mice. Two mRNA samples are collected from mice exposed to ricin and one is collected from a control mouse. The three samples are labeled and hybridized using a four-channel technology on six arrays. One of the channels is used for normalization purposes so the experiment provides three measurements for each gene. The data gathered for one gene of particular interest are given below. Are there any significant differences between the mRNA samples coming from the three animals?

	Treatments		
	ricin 1	ricin 2	control
Array 1	100	95	88
Array 2	90	93	85
Array 3	105	79	87
Array 4	83	85	83
Array 5	78	90	89
Array 6	93	102	75

SOLUTION

A quick inspection of the data reveals that the last column has consistently lower values than the rest. This column corresponds to the control mouse and it is very likely from a biological point of view to expect the control mouse to be different than the experiment. At this point, it is very tempting to conclude without further ado that this gene does play a role in the defense mechanism triggered by ricin. However, we know better by now so we will analyze the data using ANOVA.

We start by calculating the sums and means for the columns and rows:

		mRNA sample			Totals	Means
		1	2	3		
	Array 1	100	95	88	283	94.33
	Array 2	90	93	85	268	89.33
	Array 3	105	79	87	271	90.33
Block	Array 4	83	85	83	251	83.66
	Array 5	78	90	89	257	85.66
	Array 6	93	102	75	270	90
	Totals	549	544	507	1600	
	Means	91.5	90.66	84.5	88.88	

We then calculate the sums of squares:

$$SS_{Total} = \sum_i \sum_j X_{ij}^2 - \frac{T_{..}^2}{N} = 1141.77$$

$$SS_{Treat} = \sum_i \frac{T_{i.}^2}{b} - \frac{T_{..}^2}{N} = 175.44$$

$$SS_{Blocks} = \sum_j \frac{T_{.j}^2}{k} - \frac{T_{..}^2}{N} = 212.44$$

$$SS_{Error} = SS_{Total} - (SS_{Treat} + SS_{Blocks}) = 1141.77 - (175.44 + 212.44) = 753.88$$

$$MS_{Treat} = \frac{SS_{Treat}}{k-1} = \frac{175.44}{2} = 87.72$$

$$MS_{Block} = \frac{SS_{Block}}{b-1} = \frac{212.44}{5} = 42.48$$

$$MS_{Error} = \frac{SS_E}{(b-1)\cdot(k-1)} = \frac{753.88}{10} = 75.38$$

and the F statistics for both samples and arrays:

$$F = \frac{MS_{Treat}}{MS_{Error}} = \frac{87.72}{75.38} = 1.16 \tag{13.52}$$

$$F = \frac{MS_{Block}}{MS_{Error}} = \frac{42.48}{75.38} = 0.56 \tag{13.53}$$

The p-value for the treatments is obtained from the F distribution with 2 and 10 degrees of freedom, respectively, and the p-value for the blocks is obtained from the F distribution with 5 and 10 degrees of freedom, respectively. The two values are:

$$F_{2,10}(1.16) = 0.35$$

$$F_{5,10}(0.56) = 0.72$$

The conclusion is that the null hypothesis stating that the samples are drawn from the same distribution cannot be rejected. In other words, there are no statistically significant differences between the mice. Furthermore, the null hypothesis that the blocks are the same cannot be rejected either. Therefore, we can conclude that the data offer no indication that there are significant differences between the arrays.

This is usually summarized in an ANOVA results table. The following table has been constructed in Excel and includes both the p-values and the critical values of the F distribution:

	SS	df	MS	F	p-value	F critical
Rows	212.444	5	42.488	0.563	0.726	3.325
Columns	175.444	2	87.722	1.163	0.351	4.102
Error	753.888	10	75.388			
Total	1141.777	17				

□

EXAMPLE 13.6

A microarray experiment measured the expression level of the genes in the liver of mice exposed to ricin versus control mice. Two mRNA samples are collected from mice exposed to ricin and one is collected from a control mouse. The three samples are labeled and hybridized using a four-channel technology on six arrays. One of the channels is used for normalization purposes so the experiment provides three measurements for each gene. The data gathered for one gene of particular interest are given below. Are there any significant differences between the mRNA samples coming from the three animals?

		Treatments	
Array	*ricin 1*	*ricin 2*	*control*
1	110	105	95
2	90	95	83
3	105	79	82
4	85	85	73
5	82	80	74
6	93	102	82

SOLUTION

We will analyze this data as a complete random block design. We start by calculating the sums and means for the columns and rows:

		mRNA sample				
		1	2	3	Totals	Means
	Array 1	110	105	95	310	103.33
	Array 2	90	95	83	268	89.33
	Array 3	105	79	82	266	88.66
Block	Array 4	85	85	73	243	81
	Array 5	82	80	74	236	78.66
	Array 6	93	102	82	277	92.33
	Totals	565	546	489	1600	
	Means	94.16	91	81.5	88.88	

We then calculate the sums of squares:

$$SS_{Total} = \sum_i \sum_j x_{ij}^2 - \frac{T_{..}^2}{N} = 2087.77$$

$$SS_{Treat} = \sum_i \frac{T_{i.}^2}{b} - \frac{T_{..}^2}{N} = 521.44$$

$$SS_{Blocks} = \sum_j \frac{T_{.j}^2}{k} - \frac{T_{..}^2}{N} = 1162.44$$

$$SS_{Error} = SS_{Total} - (SS_{Treat} + SS_{Blocks}) = 2087.77 - (521.44 + 1162.44) = 403.88$$

$$MS_{Treat} = \frac{SS_{Treat}}{k-1} = \frac{521.44}{2} = 260.72$$

$$MS_{Block} = \frac{SS_{Block}}{b-1} = \frac{1162.44}{5} = 232.48$$

$$MS_{Error} = \frac{SS_E}{(b-1) \cdot (k-1)} = \frac{403.88}{10} = 40.38$$

and the F statistics for both samples and arrays:

$$F = \frac{MS_{Treat}}{MS_{Error}} = \frac{260.72}{40.38} = 6.45 \tag{13.54}$$

$$F = \frac{MS_{Block}}{MS_{Error}} = \frac{232.48}{40.38} = 5.75 \tag{13.55}$$

The p-value for the treatments is obtained from the F distribution with 2 and 10 degrees of freedom, respectively, and the p-value for the blocks is obtained from the F distribution with 5 and 10 degrees of freedom, respectively. The two values are:

$$F_{2,10}(6.45) = 0.015$$

$$F_{5,10}(5.75) = 0.009$$

The ANOVA summary will look like this:

	SS	df	MS	F	p-value	F critical
Blocks	1162.44	5	232.48	5.75	0.009	3.325837383
Treatments	521.44	2	260.72	6.45	0.015	4.10
Error	403.88	10	40.38			
Total	2087.77	17				

The conclusion here is that null hypothesis can be rejected for the treatments. There is sufficient evidence that the treatments are different at the 5% significance level.

Let us now analyze the same data using a one-way Model I ANOVA approach. Here we do not consider the rows as blocks and we will disregard their variance. We focus on the inter-group variance as it compares with the within-group variance.

The summary for the one-way Model I is:

	SS	df	MS	F	p-value	F critical
Treatments	521.44	2	260.72	2.49	0.11	3.68
Error	1566.33	15	104.42			
Total	2087.77					

The p-value resulted from the one-way Model I analysis does not allow us to reject the null hypothesis. In this situation, the variability due to the blocks masked the variability due to the treatments. This shows that the two-way analysis, in which the blocks were considered as a source of variance, was more powerful in this case than the Model I analysis, which lumps together the variance due to the blocks and the variance due to the treatments. The two-way analysis showed not only that the treatments are different but also that there is a considerable variation between arrays. □

13.3.4 Factorial design two-way ANOVA

In the randomized block design, it was assumed that the array effect and the sample effect are independent and there are no interactions between the

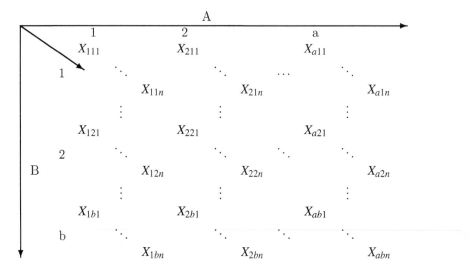

FIGURE 13.4: A factorial design with two factors. The first index represents factor A or the first dimension, the second index represents factor B or the second dimension, and the third index will represent the position of the measurement within the ij cell or the third dimension.

two besides the simple addition of their effects. In principle, two factors can have a **synergetic interaction** in which the result of both factors applied at the same time will be larger than the sum of the results of the two factors applied independently or an **antagonistic interaction**, or interference, in which the two factors applied at the same time have an effect less than the sum of the two factors applied independently. These types of interactions are taken into consideration by the two-way factorial design. A general equation for a factorial design involving two factors would be:

$$X_{ijk} = \mu + \alpha_i + \beta_j + (\alpha\beta)_{ij} + \varepsilon_{ijk} \qquad (13.56)$$

In this equation, α_i is the effect of the i-th unit of factor α, β_j is the effect of the j-th unit of factor β, $(\alpha\beta)_{ij}$ is the effect of the interaction between the i-th factor α and the j-th β factor, and ε_{ijk} is the random noise.

The data for this design will be laid out using three indexes and can be imagined as a three-dimensional matrix. The first index will represent the factor A or the first dimension, the second index the factor B or the second dimension, and the third index will represent the position of the measurement within the ij cell or the third dimension. This layout is shown in Fig. 13.4. The data can also be laid out as shown in Fig. 13.5.

This layout emphasizes a cell at the intersection of every row and column. This cell represents the set of measurements influenced by factor A_i and B_j.

			1		2			a	
							A		
B	1	X_{111}	$\cdot\cdot\cdot$ X_{11n}	X_{211}	$\cdot\cdot\cdot$ X_{21n}	$\cdots$	X_{a11}	$\cdot\cdot\cdot$ X_{a1n}
	2	X_{121}	$\cdot\cdot\cdot$ X_{12n}	X_{221}	$\cdot\cdot\cdot$ X_{22n}	$\cdot\cdot\cdot$	X_{a21}	$\cdot\cdot\cdot$ X_{a2n}
		$\vdots$		$\vdots$			$\vdots$	
	b	X_{1b1}	$\cdot\cdot\cdot$ X_{1bn}	X_{2b1}	$\cdot\cdot\cdot$ X_{2bn}	$\cdots$	X_{ab1}	$\cdot\cdot\cdot$ X_{abn}

FIGURE 13.5: The layout of the factorial design two-way ANOVA.

We can calculate totals and means for each such cell as well as some totals and means for each row and column of cells. This is shown in Fig. 13.6.

Each measurement observed can be thought of as the sum of the effects of factors *A*, *B* as well as their interactions:

$$X_{ijk} = \mu + \alpha_i + \beta_j + (\alpha\beta)_{ij} + \varepsilon_{ijk} \qquad (13.57)$$

These effects can be quantified as :

1. μ is the overall mean effect,

2. $\alpha_i = \mu_{i..} - \mu$ is the effect of the *i*-th factor A,

3. $\beta_j = \mu_{.j.} - \mu$ is the effect of the *j*-th factor B,

4. $(\alpha\beta)_{ij} = \mu_{ij.} - \mu_{i..} - \mu_{.j.} + \mu$ is the effect of the interaction between the *i*-th factor A and the *j*-th factor B, and

5. $\varepsilon_{ijk} = X_{ijk} - \mu_{ij.}$ is the effect of the random noise.

Using these expressions, Equation 13.57 can be rewritten as:

Factor B	Factor A					Level B
	1	2	3	$\cdots$	a	Totals and Means
1	$T_{11.}$ $\overline{X}_{11.}$	$T_{21.}$ $\overline{X}_{21.}$	$T_{31.}$ $\overline{X}_{31.}$	$\cdots$ $\cdots$	$T_{a1.}$ $\overline{X}_{a1.}$	$T_{.1.}$ $\overline{X}_{.1.}$
2	$T_{12.}$ $\overline{X}_{12.}$	$T_{22.}$ $\overline{X}_{22.}$	$T_{32.}$ $\overline{X}_{32.}$	$\cdots$ $\cdots$	$T_{a2.}$ $\overline{X}_{a2.}$	$T_{.2.}$ $\overline{X}_{.2.}$
$\vdots$	$\vdots$	$\vdots$	$\vdots$	$\vdots$	$\vdots$	$\vdots$
b	$T_{1b.}$ $\overline{X}_{1b.}$	$T_{2b.}$ $\overline{X}_{2b.}$	$T_{3b.}$ $\overline{X}_{3b.}$	$\cdots$ $\cdots$	$T_{ab.}$ $\overline{X}_{ab.}$	$T_{.b.}$ $\overline{X}_{.b.}$
Level A Totals and Means	$T_{1..}$ $\overline{X}_{1..}$	$T_{2..}$ $\overline{X}_{2..}$	$T_{3..}$ $\overline{X}_{3..}$	$\cdots$ $\cdots$	$T_{a..}$ $\overline{X}_{a..}$	$T_{...}$ $\overline{X}_{...}$

FIGURE 13.6: Calculating totals and means for each cell as well as totals and means for each row and column of cells.

$$(X_{ijk} - \mu) = (\mu_{i..} - \mu) + (\mu_{.j.} - \mu) - + (\mu_{ij.} - \mu_{i..} - \mu_{.j.} + \mu) + (X_{ijk} - \mu_{ij.})$$

Deviation of individual measurement from the grand mean	=	deviation of the i-th A factor from the grand mean	+	deviation of the j-th B factor from the grand mean	+	interaction of the i-th factor A and j-th factor B	+	unexplaine variability within cell

deviation of cell mean from the grand mean

The assumptions of this model are as follows:

1. Each cell constitutes an independent random sample of size n from a population with the mean μ_{ij}

2. Each such population is normally distributed with the same variance σ^2

The partitioning performed on the model can be also performed on the estimates of the corresponding terms. Each population mean μ can be estimated by the mean of the sample drawn from the corresponding population $\overline{X}$:

Population mean	Sample mean
μ	$\overline{X}_{...}$
$\mu_{i..}$	$\overline{X}_{i..}$
$\mu_{.j.}$	$\overline{X}_{.j.}$
$\mu_{ij.}$	$\overline{X}_{ij.}$

This yields:

$$(X_{ijk} - \overline{X}_{...}) = (\overline{X}_{i..} - \overline{X}_{...}) + (\overline{X}_{.j.} - \overline{X}_{...}) + (\overline{X}_{ij.} - \overline{X}_{i..} - \overline{X}_{.j.} + \overline{X}_{...}) + (X_{ijk} - \overline{X}_{ij.})$$

(13.58)

The sums of squares can be partitioned in:

$$SS_{Total} = \underbrace{SS_A + SS_B + SS_{A \times B}}_{SS_{Cells}} + SS_{Error}$$

(13.59)

where the first three terms correspond to the cell sums of squares:

$$SS_{Cells} = SS_A + SS_B + SS_{A \times B}$$

(13.60)

from which we can extract $SS_{A \times B}$. The same partitioning will hold true for the degrees of freedom, as well:

SS_{Total}	$=$	SS_A	$+$	SS_B	$+$	$SS_{A \times B}$	$+$	SS_{Error}
$abn - 1$	$=$	$(a-1)$	$+$	$(b-1)$	$+$	$(a-1)(b-1)$	$+$	$ab(n-1)$

$$\underbrace{N-1}$$ $$\underbrace{ab-1}$$

Although the details of the formulae are different, the ideas behind the factorial design ANOVA are the same as for every other ANOVA. Sums of squares will be calculated for each of the factors above. Then mean squares will be obtained by dividing each sum of squares by its degrees of freedom. This will produce a variance-like quantity for which an expected value can be calculated if the null hypothesis is true. These mean squares can be compared the same way variances can: using their ratio and an F distribution. Each ratio will have the MS_E as the denominator. Essentially, each individual test asks the question whether a certain component MS_x has the variance significantly different from the variance of the noise MS_E. For instance, in order to test whether there is interaction between the two factors, we can use the ratio:

$$F_{A \times B} = \frac{MS_{A \times B}}{MS_E}$$

(13.61)

with the degrees of freedom

$$\nu_1 = (a-1)(b-1)$$

(13.62)

and

$$\nu_2 = ab(n-1)$$

(13.63)

13.3.5 Data analysis plan for factorial design ANOVA

The data analysis for factorial design ANOVA can be summarized as follows:

1. **Test for interaction between factors.**

   The hypotheses here are:

   (a) H_0: $(\alpha\beta)_{ij} = 0$ for all i, j
   (b) H_a: There is at least one pair i, j for which $(\alpha\beta)_{ij} \neq 0$

   This is tested with by calculating the statistic:

   $$F_{A \times B} = \frac{MS_{A \times B}}{MS_E} \tag{13.64}$$

   with the $v_1 = (a-1)(b-1)$ and $v_2 = ab(n-1)$ degrees of freedom, respectively. The null hypothesis is rejected if the numerator is significantly larger than the denominator.

2. If there are significant interactions, the next step is to **separate the means** by using pair-wise t-tests with correction for multiple experiments (e.g., Bonferroni, see Chapter 16), or Duncan's multiple range test [178].

3. If there are no significant interactions, we need to **test whether there are significant differences in means for each of the factors considered**:

   (a) Test for difference between the A factors. Here, the hypotheses are:

      i. H_0: $\alpha_i = 0$ for all i (or all $\mu_{i.}$ are equal)
      ii. H_a: There exists at least a pair of factors A for which the two corresponding means are not equal.

   This is tested with by calculating the statistic:

   $$F_A = \frac{MS_A}{MS_E} \tag{13.65}$$

   with $a-1$ and $ab(n-1)$ degrees of freedom, respectively.

   (b) Test for difference between the B factors. Here, the hypotheses are:

      i. H_0: $\beta_i = 0$ for all i (or all $\mu_{i.}$ are equal)
      ii. H_a: There exists at least a pair of factors B for which the two corresponding means are not equal.

   This is tested with by calculating the statistic:

   $$F_B = \frac{MS_B}{MS_E} \tag{13.66}$$

   with $b-1$ and $ab(n-1)$ degrees of freedom, respectively.

Source of Variation	SS	df	MS	$E(MS)$
Cells	SS_{Cells}	$ab-1$	MS_{Cells}	$\sigma^2 + n\sum_i\sum_j \frac{(\mu_{ij.}-\mu_{...})^2}{ab-1}$
A factors	SS_A	$a-1$	MS_A	$\sigma^2 + nb\sum_i \frac{(\mu_{i..}-\mu_{...})^2}{a-1}$
B factors	SS_B	$b-1$	MS_B	$\sigma^2 + na\sum_j \frac{(\mu_{.j.}-\mu_{...})^2}{b-1}$
A × B	$SS_{A\times B}$	$(a-1)(b-1)$	$MS_{A\times B}$	$\sigma^2 + n\sum_i\sum_j \frac{(\alpha\beta)_{ij}^2}{(a-1)(b-1)}$
Error	SS_{Error}	$ab(n-1)$	$MS_{A\times B}$	σ^2
Total	SS_{Total}	$abn-1$		

TABLE 13.1: The main quantities involved in a factorial design ANOVA: sums of squares, degrees of freedom and mean squares. The table also shows the expected values for each mean square *MS*.

13.3.6 Reference formulae for factorial design ANOVA

The various sums of squares involved can be calculated as follows:

$$SS_{Total} = \sum_i\sum_j\sum_k \left(X_{ijk}-\overline{X}_{...}\right)^2 = \sum_i\sum_j\sum_k X_{ijk}^2 - \frac{T_{...}^2}{abn} \tag{13.67}$$

$$SS_A = \sum_i\sum_j\sum_k \left(\overline{X}_{i..}-\overline{X}_{...}\right)^2 = \sum_i \left(\frac{T_{i..}^2}{bn}\right) - \frac{T_{...}^2}{abn} \tag{13.68}$$

$$SS_B = \sum_i\sum_j\sum_k \left(\overline{X}_{.j.}-\overline{X}_{...}\right)^2 = \sum_i \left(\frac{T_{.j.}^2}{an}\right) - \frac{T_{...}^2}{abn} \tag{13.69}$$

$$SS_{Cells} = \sum_i\sum_j\sum_k \left(\overline{X}_{ij.}-\overline{X}_{...}\right)^2 = \sum_i\sum_j \left(\frac{T_{ij.}^2}{n}\right) - \frac{T_{...}^2}{abn} \tag{13.70}$$

The expected values of the various mean squares involved are shown in Table 13.1. These formulae are meaningful and instructive from a theoretical point of view, but they will probably never be used in practice in the analysis of microarray data since most of the software available make these computations transparent to the user.

13.4 Quality control

Note that the same data can be analyzed using different models. An example of such a situation was discussed in Example 13.6 in which the same data were analyzed using a complete random block approach and an Model I approach. If this is the case, a natural question is how do we know which model to use? Furthermore, assuming that we have performed the analysis using some model, how can we assess the quality of our work? Recall that each specific ANOVA model makes some specific assumptions (e.g., that the effects can be added, that there are no other effects, etc.) Furthermore, ANOVA itself is a parametric approach since it makes the assumption that the data are distributed normally, the genes are independent, etc. Many of these assumptions are clearly not satisfied in microarray data analysis (e.g., genes do interact with one another) so the results of the ANOVA analysis may be inaccurate. However, ANOVA is so elegant that it provides us with the means to estimate the quality of the results. The key observation is that all ANOVA models make one other common assumption: **all models assume that the noise is random** (and normally distributed like everything else). Furthermore, ANOVA estimates all factors involved in the model used. Thus, at the end of the analysis, one can go back and use these estimates to calculate an estimate of the random error. For instance, if the model:

$$\log\left(y_{ijkg}\right) = \mu + A_i + D_j + G_g + (AD)_{ij} + (AG)_{ig} + (VG)_{kg} + (DG)_{jg} + \varepsilon_{ijkg} \quad (13.71)$$

was used, then the random error can be extracted as:

$$\varepsilon_{ijkg} = \mu + A_i + D_j + G_g + (AD)_{ij} + (AG)_{ig} + (VG)_{kg} + (DG)_{jg} - \log\left(y_{ijkg}\right) \quad (13.72)$$

In this equation, all terms on the right-hand side stand for various effects and their interactions,[3] and can be estimated by using partial sums (see for instance Eq. 13.58). We can then look at the distribution of the residuals and assess the quality of our analysis. If the residuals do not show any particular trends, the conclusion is that the analysis has accounted for all systematic effects and the results are credible. If, on the other hand, the residuals show any trends or substantial deviation from a random normal distribution, the conclusion is that the model did not capture the effects of all factors and interactions. In this case, an alternative model might be sought. If several models show no tendencies in their residual distributions, the model with the lowest MS_E is to be preferred since, for that model, the factors considered managed to explain more of the variability exhibited by the data.

The following example from [256] will illustrate the usage of two ANOVA models to analyze the same data.

EXAMPLE 13.7
An experiment is studying the effect of 2,3,7,8-tetrachlordibenzo-p-dioxin (TCDD) on cells from the human hepatoma cell line HepG2. The experiment used a two-channel cDNA technology. Instead of using replicated spots (the

---

[3]This specific model, as well as other similar models will be discussed in detail in Chapter 15.

same gene spotted several times on every array), the experiment involved per-forming six different labeling and hybridization experiments. The experimental design is shown in the following table:

Array	Cy3	Cy5
1	variety 2	variety 1
2	variety 2	variety 1
3	variety 1	variety 2
4	variety 2	variety 1
5	variety 1	variety 2
6	variety 1	variety 2

There were 1920 genes spotted on the array, but 13 of them were consistently below the detectable level and are discarded from consideration. Calculate the degrees of freedom available to estimate the error for an analysis using the ANOVA models:

$$\log\left(y_{ijkg}\right) = \mu + A_i + D_j + V_k + G_g + (AG)_{ig} + (VG)_{kg} + (DG)_{jg} + \varepsilon_{ijkg} \quad (13.73)$$

and

$$\log\left(y_{ijkg}\right) = \mu + A_i + D_j + G_g + (AD)_{ij} + (AG)_{ig} + (VG)_{kg} + (DG)_{jg} + \varepsilon_{ijkg} \quad (13.74)$$

SOLUTION We will consider there are a arrays, d dyes, v varieties, and n genes. Using this notation, the degrees of freedom for each term in the first model are:

Term	Formula	df
Arrays	$a - 1$	5
Dyes	$d - 1$	1
Varieties	$v - 1$	1
Genes	$n - 1$	$(1920 - 13) - 1 = 1906$
Array $\times$ Gene	$(a-1)(n-1)$	$5 \times 1906 = 9530$
Variety $\times$ Gene	$(v-1)(n-1)$	$(2-1)(1907-1) = 1906$
Dye $\times$ Gene	$(d-1)(n-1)$	$(2-1)(1907-1) = 1906$
Total	$N - 1 = a \times d \times n - 1$	$6 \times 2 \times 1907 - 1 = 22883$

The degrees of freedom for the error can be calculated as the difference between the total degrees of freedom and the sum of degrees of freedom of all known terms:

$$v_{Error} = v_{Total} - \left[v_{arrays} + v_{dyes} + \cdots + v_{Dye \times Gene}\right] \quad (13.75)$$

In this case:

$$v_{Error} = 22883 - (5+1+1+1906+9530+1906+1906) = 22883 - 15225 = 7628$$

For the second model, the degrees of freedom are as follows:

Term	Formula	df
Arrays	$a - 1$	5
Dyes	$d - 1$	1
Genes	$n - 1$	$(1920 - 13) - 1 = 1906$
Array $\times$ Dye	$(a - 1)(d - 1)$	$(6 - 1) \cdot 1 = 5$
Array $\times$ Gene	$(a - 1)(n - 1)$	$5 \times 1906 = 9530$
Variety $\times$ Gene	$(v - 1)(n - 1)$	$(2 - 1)(1907 - 1) = 1906$
Dye $\times$ Gene	$(d - 1)(n - 1)$	$(2 - 1)(1907 - 1) = 1906$
Total	$N - 1 = a \times d \times n - 1$	$6 \times 2 \times 1907 - 1 = 22883$

The degrees of freedom of the error for the second model are:

$$v_{Error} = 22883 - (5 + 1 + 1906 + 5 + 9530 + 1906 + 1906) = 22883 - 15259 = 7624$$

The conclusion is that both models allow a good estimate of the quality of the analysis because there are sufficient degrees of freedom for the error. These models are fully discussed in [256]. □

13.5 Summary

This chapter discussed a set of techniques known as ANalysis Of Variance (ANOVA). In spite of the name, in most cases, ANOVA is used to test hypotheses about the means of several groups of measurements. One-way ANOVA analyzes the data by considering only one way of partitioning the data into groups, i.e. taking into consideration only one factor. Model I ANOVA focuses on the means of the specific groups of data collected. Model I asks the question whether the means of these groups in particular are different from one another. The null hypothesis is that all groups have the same mean; the research hypothesis is that there is at least a pair of groups whose means are different. If the answer is affirmative, the Model I approach proceeds to separate the means and decide which groups in particular are different from one another. Model II ANOVA is not concerned with the specific groups. The null hypothesis is that there is no significant variability between the groups considered, and the research hypothesis is that there is a significant variability. The specific groups are not of interest in Model II ANOVA and the analysis stops by rejecting or not the null hypothesis that the overall variance is zero. Two-way ANOVA considers data that can be grouped according to at least two factors. This chapter also discussed the randomized block design and the factorial design ANOVA. The general idea behind ANOVA is to identify individual sources of variance in such a way that all uninteresting variability can be removed from the test that addresses the research problem. ANOVA is a parametric approach that assumes normality. Moreover, each specific ANOVA model assumes the data are influenced by a certain number of factors interacting in a given way. These assumptions are important and they may not be true in all situations. Once an ANOVA analysis was performed, an inspection of the model's residuals can provide quality control information. If the residuals are normally distributed, it is likely that the model used was able to capture well the phenomenon under study. If the residuals show any noticeable trends, this indicates that the model was not able to explain the data properly. A crucial issue for ANOVA is the experimental design. If the experimental design does not provide enough degrees of freedom for a specific model, ANOVA will not be able to provide an accurate analysis. More details about the design of experiments will be provided in Chapter 15.

13.6   Exercises

1. A set of five genes is proposed as a set of housekeeping genes to be used in the normalization. These genes are assumed to be expressed at approximately constant levels throughout the body. A number of five tissues have been randomly chosen to investigate the expression levels of this set of genes. The measurements of the gene expression levels provided the following data:

	Liver	Brain	Lung	Muscle	Pancreas
Gene 1	12	15	17	10	32
Gene 2	19	20	29	15	10
Gene 3	17	23	19	20	12
Gene 4	21	28	14	25	13
Gene 5	12	19	26	30	41

   Can this set of genes be used as housekeeping genes for normalization purposes?

2. The β-actin gene is proposed to be used in the normalization as a housekeeping gene. This gene is assumed to be expressed at approximately constant levels throughout the body. A number of 5 tissues have been randomly chosen to investigate the expression level of this gene. The expression level was measured five times in each tissue. The measurements of the gene expression levels provided the following data:

	Testis	Brain	Ovary	Muscle	Pancreas
Measurement 1	15	14	20	12	28
Measurement 2	16	20	30	16	12
Measurement 3	14	19	19	20	14
Measurement 4	19	21	15	24	15
Measurement 5	12	20	25	31	41

Can the β-actin gene be used as housekeeping gene for normalization purposes?

3. It is suspected that a certain pollutant present in very low amounts in certain rivers acts as a carcinogen for the aquatic mammals living in those waterways. The mechanism is believed to be related to a certain gene G. A biologist studying gene G collected samples from different rivers. The expression level of this gene was measured in individuals of the same species and recorded in the table below.

	St. Clair	Detroit	St. Lawrence	Hudson	Potomac
Measurement 1	15	14	20	10	32
Measurement 2	16	20	31	15	10
Measurement 3	14	19	19	21	12
Measurement 4	19	21	14	25	13
Measurement 5		20	26		41
Measurement 6		19			

Are there significant differences in the expression of gene G between these geographical areas? If there are differences between these areas, specify which areas are different?

4. An endocrinologist studying genetic and environment effects on the development of the testis, raised three litters of experimental mice. At age two months, he sacrificed the mice, dissected out the testis tissue, and extracted the mRNA. The mRNA was used to measure the expression level of the insulin-like growth factor I (IGF-I) gene. The following table contains the measurements made on the three litters.

Litter 1	Litter 2	Litter3
0.236	0.257	0.258
0.238	0.253	0.264
0.248	0.255	0.259
0.245	0.254	0.267
0.243	0.261	0.262

Are there significant differences in the expression of IGF-1 among the litters?

5. Construct an Excel spreadsheet performing an ANOVA analysis for a randomized complete block design involving four samples hybridized simultaneously (using a four-channel technology) on six arrays. Use random numbers as experimental values. Then change the numbers corresponding to one mRNA sample such that the mean of that sample is lower. See how much the mean can go down without the F test showing a significant difference between the different samples. Settle on a data set such that the p-value is barely below 0.05.

6. Repeat the experiment above by changing the values corresponding to

one array. See how much the array mean can go down without the F-test showing a significant difference.

7. Perform the Model I test on the two data sets obtained above. How do the conclusions of the Model I and Model II (randomized block design) compare? Compare the values of the F-statistic, MS_E, the degrees of freedom, and the power of the two tests.

Chapter 14

Linear models in R

Everything should be made as simple as possible, but no simpler.

—Albert Einstein

If a straight line fit is required, obtain only two data points.

—Velilind's Laws of Experimentation

14.1 Introduction and model formulation

In this chapter, we will illustrate the use of R built-in functions to perform several types of analyses of variance presented in the previous chapter, as well as regression analysis for situations when continuous variables (e.g., age) rather than factors (e.g., disease status) potentially explain the variability in the measured outcome of an experiment (e.g., gene expression).

The typical analysis of gene expression data using linear models involves a continuous variable called response (i.e., the expression level of a given gene) and one or more factors and/or continuous variables which are called explanatory variables or predictors. The goal in this type of analysis is to test the

433

association between the predictors and the response, which in some situations may be equivalent to a simple t-test or a classical analysis of variance.

A linear model can be formulated as

$$y_i = \sum_{j=1}^{p} \beta_j x_{ij} + \beta_0 + \varepsilon_i \qquad (14.1)$$

where y_i is the response for the i-th sample, β_j is the coefficient of the j-th predictor (or explanatory variable), x_{ij} is the value of the predictor x_j for the i-th sample, β_0 is the intercept, and ε_i is an error term or residuals. The residuals ε_i encompass eventual noise in the measurement of the response as well as the effect of all relevant predictors that were left out of the model.

This is a linear model since the equation 14.1 models the behavior of the target variable y_i as a linear function of the predictors x_j. In matrix terms, the equation above can be written as:

$$\mathbf{y} = \mathbf{X} \cdot \beta + \varepsilon \qquad (14.2)$$

where X is referred to as the design matrix, β is the vector of coefficients of the model, $\mathbf{y}$ is the response vector and ε is a vector of residuals. The elements of the first column of the design matrix, $x_{i,0}$, are usually equal to 1, to account for the intercepts β_0.

Building a linear model reduces to finding the values of the βs that would minimize the error. In essence, this means solving the linear system:

$$\mathbf{y} = \mathbf{X} \cdot \beta \qquad (14.3)$$

This would be easily done if $\mathbf{X}$ were a nonsingular square matrix. Then, the solution would simply be:

$$\beta = \mathbf{X}^{-1} \cdot \mathbf{y} \qquad (14.4)$$

However, in the general case, $\mathbf{X}$ is not a non-singular square matrix. However, we can obtain a pretty good approximation as follows. Firstly, we multiply both sides of equation 14.3 with the transpose of $\mathbf{X}$, which always exists:

$$\mathbf{X}^{\mathbf{T}} \cdot \mathbf{y} = \mathbf{X}^{\mathbf{T}} \cdot \mathbf{X} \cdot \beta \qquad (14.5)$$

Now, the matrix $\mathbf{X}^{\mathbf{T}} \cdot \mathbf{X}$ is square and can be inverted. We multiply to the left with the inverse of this matrix:

$$(\mathbf{X}^{\mathbf{T}} \cdot \mathbf{X})^{-1} \cdot \mathbf{X}^{\mathbf{T}} \cdot \mathbf{y} = (\mathbf{X}^{\mathbf{T}} \cdot \mathbf{X})^{-1} \cdot \mathbf{X}^{\mathbf{T}} \cdot \mathbf{X} \cdot \beta \qquad (14.6)$$

which yields:

$$(\mathbf{X}^{\mathbf{T}} \cdot \mathbf{X})^{-1} \cdot \mathbf{X}^{\mathbf{T}} \cdot \mathbf{y} = \mathbf{I} \cdot \beta \qquad (14.7)$$

or:

$$\beta = (\mathbf{X}^{\mathbf{T}} \cdot \mathbf{X})^{-1} \cdot \mathbf{X}^{\mathbf{T}} \cdot \mathbf{y} \qquad (14.8)$$

This is known as the ordinary least square solution because it minimizes the error in the least square sense. In other words, the straight line chosen as the solution is the one that minimizes the sum of the squares of the distances between the line and the data points. The matrix $(\mathbf{X}^\mathrm{T} \cdot \mathbf{X})^{-1} \cdot \mathbf{X}^\mathrm{T}$ is known as the Moore-Penrose pseudo-inverse of the matrix $\mathbf{X}$ and has several important and convenient properties. Among these, the pseudo-inverse exists for all matrices and if the matrix is square and invertible, it reduces to the regular inverse of the matrix.

There are a number of assumptions that are required to obtain accurate and precise estimates for the β_j coefficients and also to justify the use of the ordinary least square method to estimate these coefficients:

1. The ε_i are normally distributed

2. The expected value (mean) of the residuals ε_i is 0 and does not depend on the values of the predictors (mean independence)

3. The variance of ε_i does not depend on the values of the predictors (homoscedasticity)

4. The value of the residual for one sample i is unrelated to the amount of residual for another sample j (no auto-correlation).

14.2   Fitting linear models in R

In this section, we will illustrate how to fit and interpret the results from linear models in R when the predictors are either continuous variables and/or factors. Whenever possible we will link the results and interpretations to the simple t-test or ANOVA model.

Let us use a simple example in which the gene expression level of a particular gene denoted with y is measured in a group of 10 individuals, whose ages vary from 20 to 50 years of age:

```
> y=c(100,120,135,150,160,190,210,260,290,300)
> AGE=c(20, 24, 25,27,30,35,40,41,45,48)
> dat=data.frame(y,AGE)
> dat
```

```
    y  AGE
1  100  20
2  120  24
3  135  25
4  150  27
5  160  30
6  190  35
```

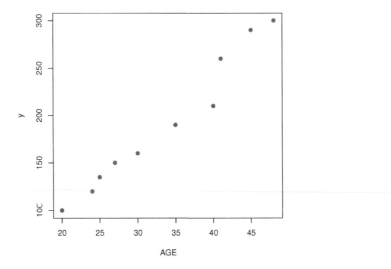

FIGURE 14.1: A regression analysis data set

7	210	40
8	260	41
9	290	45
10	300	48

We would like to test whether the expression of this gene changes with the age of the patients. In this case both the response (y) and the predictor (*AGE*) are continuous variables. The data are shown in Figure 14.1, where the response values are plotted against the age.

The model we will build will express the values of y as a linear function of the *AGE*. Since, we started with the intention to use a linear model and the only thing we need to do is to find the values of the coefficients of this model, the process is also referred to as **fitting** or **regressing**. In this case, we are performing a linear regression because the chosen model is a linear one. Linear regression and regression in general fall under the category of parametric modeling. In **parametric modeling**, the type of model to be used (e.g., the equations) are chosen in advance and the modeling itself reduces to finding the values of the parameters that best fit the given data. In **non-parametric modeling**, both the type of the model as well as the values of all its parameters are to be chosen during the modeling.

The R function we will use here to build this linear model is lm, even

though more general functions such as `glm` can be used as well. We can choose
to pass the arguments required by the `lm` function by naming them, as follows:

```
> mymod=lm(formula="y~AGE",data=dat)
> mymod

Call:
lm(formula = "y~AGE", data = dat)

Coefficients:
(Intercept)          AGE
   -51.549        7.255
```

or by making use of the argument matching mechanism in R, as:

```
> mymod=lm(y~AGE,dat)
```

The first argument in the function call above is of type "formula." This
formula has to contain a tilde $\sim$ sign. To the left of $\sim$ we specify the name of
the response variable while to the right of it, we enumerate the explanatory
variable(s) that we wish to consider in our model, in this case, AGE. The
formula $y \sim AGE$ basically means, that we want to express the variable y as a
function – in this case linear – of the variable AGE. The last argument is the
data frame that contains the values for both variables, `dat`. The end result
of the call to the function lm (the object mymod) is a complex object (of
class lm) representing the fitted model. By default, R includes in any linear
model an intercept term, which can be removed (omitted) in several ways in
the formula: $y \sim AGE - 1$ or $y \sim AGE + 0$.

```
> mymod=lm(y~AGE-1,dat)
> mymod

Call:
lm(formula = y ~ AGE - 1, data = dat)

Coefficients:
  AGE
5.824
```

In this particular example, the absence of an intercept ($\beta_0 = 0$) will force
the estimated (predicted) $\hat{y}$ value to be 0 at $AGE = 0$, in other words, the fitted
expression level is forced to pass through the origin. This leads to a different
estimate for the age coefficient compared to the situation when the intercept
term is allowed, and very often this means a worse fit of the data. This is
illustrated in Fig. 14.2.

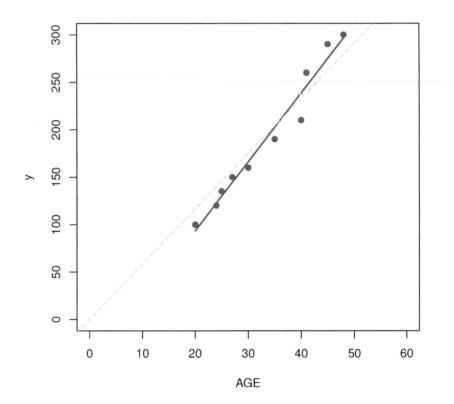

FIGURE 14.2: A comparison between a linear model with and without an intercept for the same data. The solid line corresponds to a model with an intercept term. The dashed line corresponds to a model without intercept.

14.3 Extracting information from a fitted model: testing hypotheses and making predictions

In many cases, the goal of the linear modeling is to test hypotheses about the association between the explanatory variables (such as the age in our example), and the response. When the explanatory variable (predictor) is continuous, as is the age in our example, then the null hypothesis (there is no association between the age and the response) can be simply formulated as $\beta_1 = 0$. The estimated value of the true unknown β_1, denoted with $\hat{\beta}_1$, as well as its standard deviation, can be extracted from the fitted model using the summary function:

```
> mymod=lm(y~AGE,dat)
> summary(mymod)

Call:
lm(formula = y ~ AGE, data = dat)

Residuals:
    Min      1Q  Median      3Q     Max
-28.659  -5.224   4.234   6.248  15.065

Coefficients:
            Estimate Std. Error t value Pr(>|t|)
(Intercept) -51.5490    16.6847   -3.09   0.0149 *
AGE           7.2552     0.4803   15.10 3.65e-07 ***
---
Signif. codes:  0 '***' 0.001 '**' 0.01 '*' 0.05 '.' 0.1 ' ' 1

Residual standard error: 13.94 on 8 degrees of freedom
Multiple R-squared: 0.9661,       Adjusted R-squared: 0.9619
F-statistic: 228.1 on 1 and 8 DF,  p-value: 3.652e-07
```

In the output above, under the "Coefficients" heading, we can find all relevant information regarding the estimated coefficients. First, we can observe that there are two coefficients that were estimated: one for the "(Intercept)" and one for the "AGE" variable. The column "Estimate" in this table gives the estimated values based on an ordinary least squares (OLS) approach, which basically finds that unique pair of values $\hat{\beta}_0 = -51.5490$ and $\hat{\beta}_1 = 7.2552$ (corresponding to the intercept and AGE terms, respectively) that minimize the sum of squared errors between the observed y_i and predicted $\hat{y}_i$ values. The estimated response values for all values of the predictors can be obtained by replacing the true β_j coefficients with the estimated value:

$$\hat{y}_i = \sum_{j=0}^{p} \hat{\beta}_j x_{ij} \qquad (14.9)$$

A t-score and corresponding p-values for these coefficients are also given in the output from the function summary. We can see that the probability to observe such a coefficient $\beta_1 = 7.2552$ when the null hypothesis is true, is very low ($3.65e-07$). The null hypothesis that the gene expression is not linearly associated with the age of the patients can be rejected at 0.05 confidence level.

The general function predict can be used to apply the fitted model on any data frame that contains a numeric column named "AGE," including the original data frame that was used to estimate the model's parameters:

```
> yhat=predict(mymod,dat)
> yhat
```

```
      1        2        3        4        5        6        7
93.5549 122.5757 129.8309 144.3412 166.1068 202.3828 238.6588
      8        9       10
245.9139 274.9347 296.7003
```

The real and predicted data can be both plotted against the AGE values using:

```
> plot(y~AGE,dat,col="red",pch=19)
> points(dat$AGE,yhat,type="b",lwd=2,col="blue",pch=19)
```

Fig. 14.3 shows the original data together with the predicted values. The predicted values are all situated on the straight line that defined the fitted model.

14.4   Some limitations of linear models

A limitation to keep in mind is that a parametric modeling approach only finds the best values for the parameters *available in the model*. It is the responsibility of the user to choose a model that is appropriate for the given phenomenon. Fig. 14.4 shows a set of data as well as the best linear model that can be found for these data. It is clear that in this case a linear model does not capture the nature of the phenomenon, which is of a higher order.

```
> x=seq(-10,9,by=1)
> y=x^2+x-33
> newdat=data.frame(x,y)
```

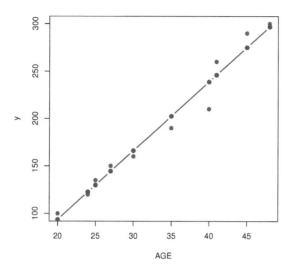

FIGURE 14.3: The real(red) versus predicted values (blue). The predicted values are all situated on the straight line that defined the fitted model.

```
> mymod=lm(y~x,newdat)
> summary(mymod)

Call:
lm(formula = y ~ x, data = newdat)

Residuals:
    Min     1Q Median     3Q    Max
    -33    -27     -8     23     57

Coefficients:
               Estimate Std. Error t value Pr(>|t|)
(Intercept) -5.619e-16   7.010e+00       0        1
x           -6.882e-17   1.211e+00       0        1

Residual standard error: 31.23 on 18 degrees of freedom
Multiple R-squared: 8.088e-34,       Adjusted R-squared: -0.05556
F-statistic: 1.456e-32 on 1 and 18 DF,  p-value: 1
```

In this case, the data were samples from the function $y = x^2 + x - 33$, which is a second-degree function. The linear model fit to these data is essentially a horizontal line with an intercept of approximately zero. In essence, if this

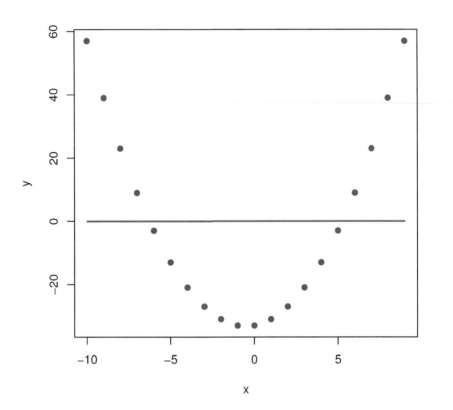

FIGURE 14.4: A limitation worth keeping in mind is that a parametric modeling approach only finds the best values for the parameters available in the model. The figure shows the best linear model found for the given data. It is clear that in this case, a linear model does not capture the nature of the phenomenon, which is of a higher order.

model were to be used in practice to predict the values of the variable *y*, it will predict always a value close to zero, for all values of *x*.

Note that there is no error or warning message of any kind since the function `lm` dutifully found the linear model that fits the data with the smallest squared error. The problem in this case is that the fitted model does not really capture the nature of the phenomenon. However, this is not because the search for the optimal model parameters was not successful (as noted above, the pseudo-inverse always exists), but because the type of model (linear) is not able to capture well the behavior of a quadratic funtion. The *p*-values associated with both coefficients (for the intercept and *x*) are 1, meaning that the null hypothesis cannot be rejected. This, together with the larger values of the residuals indicate the fact that the fit is poor and the fitted model is not a good explanation for the data.

14.5 Dealing with multiple predictors and interactions in the linear models, and interpreting model coefficients

So far we have seen how to fit a linear model when we have only one predictor and how to test if the association between the predictor and the response is significant or not. In this section, we will consider the situations when multiple explanatory variables are used as predictors in the model.

Let us consider the same hypothetical gene expression data set that we used above but adding an additional explanatory variable, namely, the body mass index (BMI) of the patients.

```
> y=c(100,120,135,150,160,190,210,260,290,300)
> AGE=c(20, 24, 25,27,30,35,40,41,45,48)
> BMI=c(40,45,24,20,26,28,21,24,20,25)
> dat=data.frame(y,AGE,BMI)
> dat
```

	y	AGE	BMI
1	100	20	40
2	120	24	45
3	135	25	24
4	150	27	20
5	160	30	26
6	190	35	28
7	210	40	21
8	260	41	24

```
9   290   45   20
10  300   48   25
```

To fit a model that includes both the main effect of the AGE and BMI together with a default intercept term, we have several options regarding the formula argument in the lm function: a) either we specify both predictors at the right of the $\sim$ sign separated by an + sign: $y \sim AGE + BMI$, or b) we tell R that we want all the existing variables in the data frame, except y to be included as predictors using $y \sim .$ as formula argument in the lm function call, where the "." stands for everything except the response variable, in this case y. Obviously, the latter formulation is more convenient when there are several predictors since one does not need to enumerate them.

```
> mymod2=lm(y~AGE+BMI,dat)
> summary(mymod2)

Call:
lm(formula = y ~ AGE + BMI, data = dat)

Residuals:
    Min      1Q  Median      3Q     Max
-28.703  -5.212   4.216   6.272  15.046

Coefficients:
            Estimate Std. Error t value Pr(>|t|)
(Intercept) -50.87064   37.22262  -1.367    0.214
AGE           7.24727    0.63982  11.327 9.36e-06 ***
BMI          -0.01512    0.72837  -0.021    0.984
---
Signif. codes:  0 '***' 0.001 '**' 0.01 '*' 0.05 '.' 0.1 ' ' 1

Residual standard error: 14.9 on 7 degrees of freedom
Multiple R-squared: 0.9661,      Adjusted R-squared: 0.9564
F-statistic: 99.82 on 2 and 7 DF,  p-value: 7.156e-06
```

In the "Coefficients" table above, we can see that the AGE is still a significant variable after adjusting for the BMI. This "adjustment" for additional covariates is similar in concept with a two-way ANOVA in which the variance is partitioned between the two variables and the significance of each variable is influenced by the amount of variability explained by the other variable.

The meaning of the coefficient for AGE, in this case $\beta_1 = 7.247$, is that there will be an increase of 7.247 units in the response for each unit increase in the age (which is expressed in years), while adjusting for the BMI of the person. Moreover, the meaning of the coefficient for BMI, in this case $\beta_2 = -0.015$, is that there will be a decrease of 0.015 units in the response for each unit increase in the BMI, while adjusting for the age. However, while the p-value for the

AGE is highly significant at the 10^{-6} level allowing us to comfortably reject the null hypothesis for this factor, the *p*-value for the BMI is 0.984, indicating that the association between BMI and our outcome is not significant.

Sometimes it is useful to study the effect of an interaction between the independent variables on the response variable. For our problem, this would mean testing if the effect of age changes with the BMI or equivalently is the effect of the BMI changes with the age. To specify an interaction between two or more variables, we list the names of the variables separated by a $*$ sign:

```
> mymod3=lm(y~AGE*BMI,dat)
> summary(mymod3)

Call:
lm(formula = y ~ AGE * BMI, data = dat)

Residuals:
    Min      1Q  Median      3Q     Max
-30.841  -3.861   4.367   6.009  14.857

Coefficients:
              Estimate Std. Error t value Pr(>|t|)
(Intercept) -88.78460   88.94795  -0.998   0.3567
AGE           8.75884    3.24961   2.695   0.0358 *
BMI           1.50578    3.28956   0.458   0.6632
AGE:BMI      -0.06232    0.13103  -0.476   0.6512
---
Signif. codes:  0 '***' 0.001 '**' 0.01 '*' 0.05 '.' 0.1 ' ' 1

Residual standard error: 15.8 on 6 degrees of freedom
Multiple R-squared: 0.9674,      Adjusted R-squared: 0.951
F-statistic: 59.26 on 3 and 6 DF,  p-value: 7.517e-05
```

The interaction term is displayed in the table above as "AGE:BMI," and we can see that it does not have a significant effect. The interpretation of the coefficient for the interaction term, $\beta_3 = -0.0623$, is that the the effect of BMI is smaller for older people, even though as said, this effect is not significant in this case.

14.5.1 Details on the design matrix creation and coefficients estimation in linear models

Now that we have an idea of how we can fit linear models and test linear association between continuous variables, we will explain how R constructed the design matrix X and how it estimated the coefficients.

The design matrix $\mathbf{X}$ that appears in Equation 14.2 of this chapter is automatically generated by R for a given problem based on the formula used in the lm function call, for instance $y \sim AGE * BMI$, and the data frame "dat" that we passed as arguments to the function lm. In this case, the design matrix **X** had 4 columns: the intercept, AGE, BMI, and the interaction between AGE and BMI. Even though we did not specify that we want an intercept in the model, lm included one by default. The column corresponding to the intercept is filled with values of 1. The columns corresponding to the AGE and BMI are filled with corresponding AGE and BMI columns taken as they are from the dat data frame. However, the last column corresponding to the interaction term was filled in automatically as the element by element product between the AGE and BMI columns for the 10 patients. We can manually construct this matrix as follows:

```
> X=as.matrix(cbind("Intercept"=rep(1,10),dat[,c("AGE","BMI")],
+ "Interaction"=dat$AGE*dat$BMI))
> X
```

```
       Intercept AGE BMI Interaction
 [1,]          1  20  40         800
 [2,]          1  24  45        1080
 [3,]          1  25  24         600
 [4,]          1  27  20         540
 [5,]          1  30  26         780
 [6,]          1  35  28         980
 [7,]          1  40  21         840
 [8,]          1  41  24         984
 [9,]          1  45  20         900
[10,]          1  48  25        1200
```

We can now verify that this is indeed the model matrix used by the function lm. In order to do this, we will call again the function lm but this time, we will request that the function return the model matrix used. This can be done with the parameter x=T. If this parameter is true, the function lm will also include the model matrix, x, as one of the elements of the list returned as result. We can then print this model matrix as follows:

```
> mymod3=lm(y~AGE*BMI,dat,x=T)
> mymod3$x
```

```
  (Intercept) AGE BMI AGE:BMI
1           1  20  40     800
2           1  24  45    1080
3           1  25  24     600
4           1  27  20     540
5           1  30  26     780
6           1  35  28     980
7           1  40  21     840
```

```
8              1  41  24     984
9              1  45  20     900
10             1  48  25    1200
attr(,"assign")
[1] 0 1 2 3
```

We can use the manually constructed matrix X to directly solve Equation 14.8 in order to obtain ordinary least squares (OLS) estimates of the beta coefficients and the response value y from the "dat" data frame as:

```
> solve(t(X) %*% X) %*% t(X) %*% matrix(dat$y)

                     [,1]
Intercept    -88.78459731
AGE            8.75884284
BMI            1.50577969
Interaction  -0.06232126
```

The function `solve` above returns the inverse of the square matrix resulting from the matrix multiplication of the transpose of X with itself. The "matrix" function above was used to convert the vector or response values into a one-column matrix. The formula above is a direct implementation in R of Equation 14.8. Note that the coefficients obtained by directly solving the equation are the same as those returned by the function `lm`:

```
> mymod3

Call:
lm(formula = y ~ AGE * BMI, data = dat, x = T)

Coefficients:
(Intercept)          AGE          BMI      AGE:BMI
  -88.78460      8.75884      1.50578     -0.06232
```

14.5.2 ANOVA using linear models

In the sections above, we discussed the use of linear models when the explanatory or independent variables were quantitative (continuous). However, linear models can also be used when the predictors are factors with two or more levels. Most of the aspects discussed above for continuous predictors also apply to the case in which the predictors are factors, except for the way in which R creates the design matrix used in the coefficient estimation, and the interpretation of the coefficients in the model. The ANOVA models presented in Chapter 13 for factor-like explanatory variables can be handled with linear models as well, with improved versatility since both continuous and factor variables can be used together in the same model.

14.5.2.1 One-way Model I ANOVA

We will start with an example of a one-way Model I ANOVA. Let us consider Example 13.3 discussed in Chapter 13. In this example, the expression level of a gene is measured in three types of mRNA sources (mRNA1, mRNA2, mRNA3) using five different arrays:

Array	mRNA 1	mRNA 2	mRNA 3
1	100	95	70
2	90	93	72
3	105	79	81
4	83	85	74
5	78	90	75

We are interested in any difference between the three mRNA sources, henceforth called groups. The preferred way to organize the data for linear modeling is within a data frame, in which the gene expression level y, as well as the factor describing the different groups, denoted with Group, appear both as columns. The array number is dropped since it is not relevant in this analysis.

```
> mydat=data.frame(y=c(100,90,105,83,78,95,93,79,85,90,70,72,81,
+ 74,75),Group=factor(rep(c("mRNA1","mRNA2","mRNA3"),each=5)))
> mydat
```

```
      y Group
1   100 mRNA1
2    90 mRNA1
3   105 mRNA1
4    83 mRNA1
5    78 mRNA1
6    95 mRNA2
7    93 mRNA2
8    79 mRNA2
9    85 mRNA2
10   90 mRNA2
11   70 mRNA3
12   72 mRNA3
13   81 mRNA3
14   74 mRNA3
15   75 mRNA3
```

In order to compute the F statistic that would allow us to test whether there is any difference between the three groups, we can use a linear model in which the response variable y is the gene expression level and predictor variable is the Group factor, having three levels: mRNA1, mRNA2, and mRNA3. The model we need to specify as the argument of the lm function is the same as it was with continuous predictors:

```
> my3=lm(y~Group,data=mydat)
> summary(my3)

Call:
lm(formula = y ~ Group, data = mydat)

Residuals:
   Min     1Q Median     3Q    Max
 -13.2   -3.9   -0.4    5.6   13.8

Coefficients:
             Estimate Std. Error t value Pr(>|t|)
(Intercept)    91.200      3.529  25.844 6.87e-12 ***
GroupmRNA2     -2.800      4.991  -0.561  0.58509
GroupmRNA3    -16.800      4.991  -3.366  0.00561 **
---
Signif. codes:  0 '***' 0.001 '**' 0.01 '*' 0.05 '.' 0.1 ' ' 1

Residual standard error: 7.891 on 12 degrees of freedom
Multiple R-squared: 0.5202,        Adjusted R-squared: 0.4402
F-statistic: 6.505 on 2 and 12 DF,  p-value: 0.0122
```

In the coefficients table above there are three terms: an intercept, "GroupmRNA2," and "GroupmRNA3." By default, for a factor with k levels (in our case 3) R estimates $k-1$ coefficients. Each coefficient captures the effect of of a given level in the factor compared to a reference group chosen alphabetically. Since in the mRNA1 is the first among the levels of the factor Group, it is chosen as reference. The mean of the y values for the mRNA1 level is the estimated coefficient in the Intercept term. The remaining two coefficients are the difference in means between the remaining levels (mRNA2, and mRNA3) and the reference level (mRNA1). The coefficient for mRNA3 level is negative (-16.8) and it has a significant p-value ($p = 0.00561$), meaning that the mean response in mRNA3 group is lower than the one in the reference group (mRNA1) by 16.8 units.

We can verify this by calculating the means of the three groups:

```
> m=tapply(mydat$y, mydat$Group, mean)
> m

mRNA1 mRNA2 mRNA3
 91.2  88.4  74.4

> m[3]-m[1]
```

```
mRNA3
-16.8
```

```
> m[2]-m[1]
```

```
mRNA2
 -2.8
```

The F statistic produced by the summary function from the fitted model allows us to test the hypothesis that there is at least one significant difference between any of the groups. The p-value corresponding to this global test is 0.0122.

The design matrix used by R to obtain the coefficients in the model my3 above was:

```
> X=cbind(Intercept=rep(1,15),mRNA2=c(rep(0,5),rep(1,5),rep(0,5)),
+ mRNA3=c(rep(0,10),rep(1,5)))
> X
```

	Intercept	mRNA2	mRNA3
[1,]	1	0	0
[2,]	1	0	0
[3,]	1	0	0
[4,]	1	0	0
[5,]	1	0	0
[6,]	1	1	0
[7,]	1	1	0
[8,]	1	1	0
[9,]	1	1	0
[10,]	1	1	0
[11,]	1	0	1
[12,]	1	0	1
[13,]	1	0	1
[14,]	1	0	1
[15,]	1	0	1

which as described previously can be used to obtain the estimates of the coefficients via the OLS method:

```
> solve(t(X) %*% X) %*% t(X) %*% matrix(mydat$y)
```

```
             [,1]
Intercept   91.2
mRNA2       -2.8
mRNA3      -16.8
```

R also has a function that performs a classical ANOVA. This function is called anova and takes as an input argument the model found by lm (or glm):

```
> anova(my3)

Analysis of Variance Table

Response: y
          Df Sum Sq Mean Sq F value Pr(>F)
Group      2 810.13  405.07  6.5054 0.0122 *
Residuals 12 747.20   62.27
---
Signif. codes:  0 '***' 0.001 '**' 0.01 '*' 0.05 '.' 0.1 ' ' 1
```

This is exactly the output format used in Chapter 13 with the exception that the columns for degrees of freedom (Df) and sum of squares (Sum Sq) are swapped, and that the total sum of squares is missing (but please recall that is redundant since this value is the sum of the two sum of squares shown). The p-value shown here is the one corresponding to the F-statistic shown which can be used to test the hypothesis that there is at least one pair of groups that are significantly different. Note that the number of degrees of freedom, the value of the F-statistic and the corresponding p-value are the same as the ones reported in the last part of the output of the linear model function call.

Returning to linear models, so far we have seen how to interpret a linear model's coefficients when the predictor is a factor with more than two levels. We have obtained a p-value on the global test for any difference between all levels of the factor Group, as well as individual tests for two out of three possible comparisons between the three groups, namely, mRNA2 versus mRNA1, and mRNA3 versus mRNA1. However, no comparison was done by default between mRNA2 and mRNA3 groups. There are at least two ways to perform other comparisons rather than the ones chosen by lm as default. One possibility is to simply rename the factor levels so a different group will become the reference and we will obtain direct comparisons between the new reference group and the remaining ones. A second option is based on the idea that any comparison not performed already can be obtained by a simple subtraction of two coefficients already obtained. We now illustrate the first option, while the second alternative will be illustrated later this chapter using the limma library.

Since the missing comparison is between mRNA2 and mRNA3, we will rename mRNA2 with AmRNA2, so that it becomes the first level in alphabetical order:

```
> mydat2=mydat
> levels(mydat2$Group)[2]<-"AmRNA2"
> mydat2$Group<-factor(as.vector(mydat2$Group))
> my4=lm(y~Group,data=mydat2)
> summary(my4)

Call:
```

```
lm(formula = y ~ Group, data = mydat2)

Residuals:
   Min    1Q Median    3Q    Max
  -13.2  -3.9   -0.4   5.6   13.8

Coefficients:
            Estimate Std. Error t value Pr(>|t|)
(Intercept)  88.400      3.529  25.050 9.93e-12 ***
GroupmRNA1    2.800      4.991   0.561  0.5851
GroupmRNA3  -14.000      4.991  -2.805  0.0159 *
---
Signif. codes:  0 '***' 0.001 '**' 0.01 '*' 0.05 '.' 0.1 ' ' 1

Residual standard error: 7.801 on 12 degrees of freedom
Multiple R-squared: 0.5202,        Adjusted R-squared: 0.4402
F-statistic: 6.505 on 2 and 12 DF,  p-value: 0.0122
```

The new GroupmRNA1 and GroupmRNA3 coefficients that were obtained represent the effect of the group mRNA1 and mRNA3 compared to mRNA2 that was renamed to "AmRNA." Note that the coefficient GroupmRNA3 obtained with my4 model represents the difference in means between mRNA3 and mRNA2 while in my3 model it represented the difference between mRNA3 and mRNA1.

The p-values for the individual comparisons obtained from the linear model are somewhat different than the ones shown in the solution for this problem in Chapter 13. This is because in that analysis a equal variance was assumed, which is not the case with linear models. However, both analyses yielded the same qualitative result: the null hypothesis could be rejected between 1 and 3 and 2 and 3, but it could not be rejected between 1 and 2.

Note that the last line of the lm output shows exactly the same values for the F-statistic and its associated p-value. Clearly, the F-statistic for the overall test should not depend on the choice of the reference condition, as long as the same data for the three groups remain the same. Hence, the ANOVA tables obtained from my3 and my4 should be exactly the same:

```
> anova(my3)

Analysis of Variance Table

Response: y
          Df Sum Sq Mean Sq F value Pr(>F)
Group      2 810.13  405.07  6.5054 0.0122 *
Residuals 12 747.20   62.27
---
Signif. codes:  0 '***' 0.001 '**' 0.01 '*' 0.05 '.' 0.1 ' ' 1
```

```
> anova(my4)

Analysis of Variance Table

Response: y
          Df Sum Sq Mean Sq F value Pr(>F)
Group      2 810.13  405.07  6.5054 0.0122 *
Residuals 12 747.20   62.27
---
Signif. codes:  0 '***' 0.001 '**' 0.01 '*' 0.05 '.' 0.1 ' ' 1
```

14.5.2.2   Randomized block design ANOVA

Let us revisit Example 13.5 in Chapter 13. In this example, there are three mRNA samples, mRNA1 (ricin 1), mRNA2 (ricin 2), and mRNA3 (control), and six arrays, each mRNA sample being measured once on each array:

| | Treatments | | |
	ricin 1	ricin 2	control
Array 1	100	95	88
Array 2	90	93	85
Array 3	105	79	87
Array 4	83	85	83
Array 5	78	90	89
Array 6	93	102	75

In R, these data can be organized as follows:

```
> Group=factor(rep(paste("mRNA",1:3,sep=""),each=6))
> Block=factor(rep(paste("array",1:6,sep=""),3))
> y=c(100,90,105,83,78,93,95,93,79,85,90,102,88,85,87,83,89,75)
> md2<-data.frame(Group,Block,y)
> md2

   Group  Block   y
1  mRNA1 array1 100
2  mRNA1 array2  90
3  mRNA1 array3 105
4  mRNA1 array4  83
5  mRNA1 array5  78
6  mRNA1 array6  93
7  mRNA2 array1  95
8  mRNA2 array2  93
9  mRNA2 array3  79
10 mRNA2 array4  85
```

```
11 mRNA2 array5  90
12 mRNA2 array6 102
13 mRNA3 array1  88
14 mRNA3 array2  85
15 mRNA3 array3  87
16 mRNA3 array4  83
17 mRNA3 array5  89
18 mRNA3 array6  75
```

We want to know whether there are significant differences between the three groups while accounting for the block effect, in this case the array number. The `anova` method in R applied to the fitted model `my5` provides a classical ANOVA table for example. An overall test for the significance of the Group and Sample factors can be obtained using:

```
> my5=lm(y~Group+Block,data=md2)
> anova(my5)
```

```
Analysis of Variance Table
```

```
Response: y
          Df Sum Sq Mean Sq F value Pr(>F)
Group      2 175.44  87.722  1.1636 0.3513
Block      5 212.44  42.489  0.5636 0.7264
Residuals 10 753.89  75.389
```

14.5.3 Practical linear models for analysis of microarray data

There are two main formats of microarray data: single channel and two channel. The single-channel data format contains one measurement for each probe(set) on the array and there is only one biological replicate (sample) per array. Two-channel experiments use two samples per array labeled with two different dyes, where one of the samples on each array may be a common reference sample. A matrix with G rows and N columns can represent the results of the single-channel experiment, where G is the number of genes (probe(sets)) and N is the number of samples. Data used in the analysis are typically log2 transformed and quantile normalized so that the distribution of values in each array after normalization matches the one of a artificial "average" sample. In most of the cases, the two-channel data can be summarized in the same format, except that the values stored in the matrix are the log2 ratios between the two channels for each array. In general, from this point further, the two cases of data formats are treated in the same way in terms of identifying genes whose expression is associated with important explanatory variables such as disease status or treatment dose amount. Note that, at this point, the effect of each individual array is already corrected for in the normalization step, but there is still room to correct for eventual batch effects if needed.

14.5.4 A two-group comparison gene expression analysis using a simple *t*-test

In this section, we illustrate how to use R to analyze a real gene expression data set. A single channel technology (Affymetrix) was used to profile the expression level of 12 samples which are RNA pools from two or three placenta tissues. Each of the 12 samples is described by two factors: a) the Labor variable, indicating whether the women were at term in labor (TIL) or term not in labor (TNL), and b) the Region factor, indicating the tissue the sample was extracted from: placental amnion (PA) or reflected amnion (RA).

To load the data into R and display it in part, we use:

```
> load("placenta.RData")
> head(eset)
```

	PRB_CJ.Kim_1.CEL	PRB_CJ.Kim_2.CEL	PRB_CJ.Kim_3.CEL
1007_s_at	10.139246	10.180721	10.220019
1053_at	7.036337	7.280764	7.288650
117_at	5.791837	5.511455	5.960270
121_at	7.736955	7.758362	7.887433
1255_g_at	4.271423	3.613137	3.873642
1294_at	5.973166	5.664603	5.971478

	PRB_CJ.Kim_4.CEL	PRB_CJ.Kim_5.CEL	PRB_CJ.Kim_6.CEL
1007_s_at	10.570528	10.276756	10.469487
1053_at	7.254231	7.319713	7.084867
117_at	6.182366	5.610828	6.152365
121_at	8.048690	8.154927	8.317754
1255_g_at	5.448111	4.394194	4.333071
1294_at	6.509880	5.973398	6.456714

	PRB_CJ.Kim_7.CEL	PRB_CJ.Kim_8.CEL	PRB_CJ.Kim_9.CEL
1007_s_at	10.298518	10.332854	10.125108
1053_at	7.251156	7.337214	7.224029
117_at	5.759165	5.856644	6.119106
121_at	7.685909	7.579158	7.832778
1255_g_at	3.693009	4.306265	4.146974
1294_at	5.683163	5.653184	6.175387

	PRB_CJ.Kim_10.CEL	PRB_CJ.Kim_11.CEL	PRB_CJ.Kim_12.CEL
1007_s_at	10.507251	10.349308	10.290880
1053_at	7.436952	7.493823	7.389693
117_at	6.028664	7.135994	5.894144
121_at	7.927043	7.696860	8.193026
1255_g_at	4.995780	4.627351	4.563757
1294_at	6.018250	6.009788	5.725767

```
> Region
```

```
[1] "PA" "PA" "PA" "RA" "RA" "RA" "PA" "PA" "PA" "RA" "RA" "RA"
```

```
> Labor
```

```
 [1] "TNL" "TNL" "TNL" "TNL" "TNL" "TNL" "TIL" "TIL" "TIL" "TIL"
[11] "TIL" "TIL"
```

```
> eset=eset[1:1000,]
```

In the code above, the function **head** was used to extract and display the first part of the data structure **eset** containing the data.

Let us assume that we want to study the effect of the Region regardless of the Labor status of the patients, using a simple two-group *t*-test. First, we create a function that performs the *t*-test and computes the difference between the PA and RA regions, then we apply it on the rows of the *eset* matrix that contains the expression data.

```
> f<-function(x)
+ {
+ g1=x[Region=="PA"]
+ g2=x[Region=="RA"]
+ p=t.test(g1,g2,var.equal=TRUE)$p.value
+ logFC=mean(g1)-mean(g2)
+ c(logFC,p)
+ }
> res=t(apply(eset,1,f))
> rownames(res)<-rownames(eset)
> colnames(res)<-c("logFC","p")
> res<-res[order(res[,2]),]
> res[1:5,]
```

```
                    logFC            p
1405_i_at      -1.1769736 4.899162e-05
1552546_a_at   -0.8844603 5.947453e-05
1553630_at      1.0191412 1.078619e-04
1552774_a_at   -0.4863275 1.337571e-04
1553452_at      0.7406848 1.434401e-04
```

The transpose of the result from the function apply was needed to obtain the results in a column-wise format, and then ordering of the results were done in increasing *p*-values order, displaying the top five genes (probes). Note that the logFC column gives the difference in means of the log_2 transformed gene expression data, which is equivalent to the log_2 of the ratio between region PA and region RA. The ratio *PA/RA* can be obtained simply as:

```
> RatioPAvsRA<-2^res[,"logFC"]
```

The most widely used approaches for gene expression data analysis do not analyze each probe/gene data independently, but information is borrowed across the different genes to obtain more reliable estimates for gene expression variance. One of the most popular approaches is a moderated t-test implemented in the `limma` package of Bioconductor. We will show here a few examples using the `limma` package functionalities, assuming that the data are already available in R and were properly normalized and processed.

14.5.5 Differential expression using the `limma` library of Bioconductor

14.5.5.1 Two group comparison with single-channel data

We will now illustrate the same example as above, as well as several others, using the `limma` package of Bioconductor. Given that the expression data are available in the `eset` matrix, and the `Region` variable specifies which column of the matrix `eset` corresponds to a either PA or RA regions, we can use the following commands to fit a linear model (equivalent to a simple unpaired t-test) but we use instead a moderation of the gene standard deviation:

```
> require(limma)
> REG <- factor(Region)
> design <- model.matrix(~0+REG)
> colnames(design)<-levels(REG)
> fit <- lmFit(eset, design)
>   cont.matrix <- makeContrasts(
+   PAvsRA=PA-RA,
+   levels=design)
> fit2 <- contrasts.fit(fit, cont.matrix)
> fit2 <- eBayes(fit2)
> limres<-topTable(fit2,coef=1, number=5, adjust="fdr")
> head(limres)
```

```
                ID      logFC   AveExpr          t      P.Value
9        1405_i_at -1.1769736 4.637881 -7.522874 3.161368e-06
205 1552546_a_at -0.8844603 6.915289 -7.108195 5.926628e-06
986    1553630_at  1.0191412 4.900675  6.800008 9.593640e-06
853    1553452_at  0.7406848 3.938416  6.305395 2.134663e-05
72   1552343_s_at  1.2585929 5.315067  6.200640 2.539602e-05
        adj.P.Val        B
9     0.002963314 4.822607
205   0.002963314 4.244405
986   0.003197880 3.797572
853   0.005079204 3.049305
72    0.005079204 2.885863
```

Note the the ranking of the Affymetrix probesets changed between the regular t-test analysis we did before and the new moderated t-test results. The explanation of the syntax above, goes as follows. The `model.matrix` function specifies a linear model without an intercept based on the levels of the factor REG, which in this case has only two levels: PA and RA.

```
> head(design)

  PA RA
1  1  0
2  1  0
3  1  0
4  0  1
5  0  1
6  0  1
```

This design matrix has two columns, one for each level of the factor, with values of 1 for the samples whose REG value have the level given by the column name of the design matrix. This way of specifying the design matrix allows the user to specify later which comparison (s)he want to make, i.e. PA versus RA or RA versus PA. This is very useful when the factor has more than two levels. The `lmFit` function fits the model to the data in each row in the `eset` matrix. Then the `makeContrasts` function allows to create a matrix `cont.matrix` (standing for the contrast matrix) that allows us to specify the comparisons we are interested in, as well as the reference we want, rather than a default based on some alphabetical order, as we saw that it was the case for the basic function `lm`. In the `makeContrasts` function call, the "PAvsRA" is an arbitrary name chosen by us to be suggestive for the comparison. However, the text at the right of the '=' sign has to contain only names of the columns of the design matrix, in this case PA and RA and "+" or "−" signs. In this case "*PA − RA*" means that what we estimate as effect from the model is the log2 ratio for each gene between the PA samples and RA samples.

```
> head(cont.matrix)

       Contrasts
Levels PAvsRA
    PA      1
    RA     -1
```

Our `cont.matrix` in this case has only one column since we had only one comparison to make, and there is one row for each column of the design matrix. The effect we want to extract will be simply the coefficient of the PA group multiplied with 1 and summed up with the coefficient for the RA group multiplied with −1. The function `eBayes` adjusts the standard deviations for each probe set allowing to obtain the moderated t-test statistics and p-values, which are shown by calling the function `topTable`. The `coef` argument in this function call refers to the first (and only) comparison or contrast that we have PAvsRA. Note the the data frame produced by the function `topTable` also contains the corrected *p*-value using the method specified by the "adjust" argument.

14.5.5.2   Multiple contrasts with single-channel data

To illustrate how multiple contrasts can be easily obtained using the `makeContrasts` function in `limma`, we generate a new indicator variable by merging the Region and Labor variables:

```
> REGLAB=factor(paste(Region,Labor,sep=""))
> REGLAB

 [1] PATNL PATNL PATNL RATNL RATNL RATNL PATIL PATIL PATIL RATIL
[11] RATIL RATIL
Levels: PATIL PATNL RATIL RATNL
```

Let us say that we are interested in the effect of labor in each specific region. In this case, we can use:

```
> design <- model.matrix(~0+REGLAB)
> colnames(design)<-levels(REGLAB)
> fit <- lmFit(eset, design)
>  cont.matrix <- makeContrasts(
+  LaborvsNoLaborPA=PATIL-PATNL,
+  LaborvsNoLaborRA=RATIL-RATNL,
+  levels=design)
> fit2 <- contrasts.fit(fit, cont.matrix)
> fit2 <- eBayes(fit2)
> limres1<-topTable(fit2,coef="LaborvsNoLaborPA", number=5,
+ adjust="fdr")
> head(limres)

            ID      logFC AveExpr          t    P.Value
9       1405_i_at -1.1769736 4.637881 -7.522874 3.161368e-06
205 1552546_a_at -0.8844603 6.915289 -7.108195 5.926628e-06
986    1553630_at  1.0191412 4.900675  6.800008 9.593640e-06
853    1553452_at  0.7406848 3.938416  6.305395 2.134663e-05
72  1552343_s_at  1.2585929 5.315067  6.200640 2.539602e-05
       adj.P.Val         B
9    0.002963314 4.822607
205  0.002963314 4.244405
986  0.003197880 3.797572
853  0.005079204 3.049305
72   0.005079204 2.885863

> limres2<-topTable(fit2,coef="LaborvsNoLaborRA", number=5,
+ adjust="fdr")
> head(limres2)
```

```
            ID      logFC   AveExpr          t     P.Value
63     1552327_at -0.8359371 5.042075 -6.000833 6.793283e-05
956 1553587_a_at  0.5384325 7.246628  3.970331 1.936097e-03
627    1553145_at -1.1095219 6.125600 -3.809565 2.581020e-03
230 1552583_s_at  0.4396128 4.080053  3.479960 4.687573e-03
392 1552804_a_at  0.4821649 5.245231  3.472069 4.755509e-03
      adj.P.Val          B
63   0.06793283 -1.049805
956  0.61276342 -2.274284
627  0.61276342 -2.398234
230  0.61276342 -2.664512
392  0.61276342 -2.671079
```

In the `topTable` function call above, the results for the contrasts of interest are obtained using the names of the contrasts "LaborvsNoLaborPA" and
"LaborvsNoLaborRA" rather than the number of the contrast, as we showed
in the example before. Note that the effect of labor in PA region is significant
for several genes, but it is not so in the RA tissue.

More examples of microarray analysis with the `limma` package can be found
in `limma`'s user's guide.

14.6 Summary

The goal of this chapter was to introduce linear models, make the connection with the previously discussed ANOVA, and introduce the reader to some
real-world data analyses in the R environment. The chapter introduced the
idea of a linear model using some simple examples, discussed the general solution provided by a linear model on continuous variables, and its most important limitation – that it only constructs a linear model, no matter what
the data are. Basic R functions commonly used in this context were described
and illustrated on some examples: `lm`, `predict`, `solve`, and `anova`. In general, the coefficients associated with various explanatory variables represent
the amount of variation in the response caused by a change of one unit in
the given explanatory variable. The p-value associated with each explanatory
variable tests the null hypothesis that the associated coefficient is zero. The
size of the residuals and the overall p value provide information about the
goodness of fit. High values of the residuals and p-values that are not significant may indicate that the phenomenon is not well modeled by a linear model.
Linear models are also capable to deal with several explanatory variables at
the same time, as well as with interactions between them. When applied on
categorical variables, the `lm` functions choose the first group (in alphabetical
order) as the reference group. Differences with respect to this group will be

calculated for all the other groups. However, no differences between groups that do not involve the reference are calculated. The coefficients calculated for categorical variables have a different meaning, representing the difference in means between the group associated with the given coefficient and the reference group. The `anova` function can be called on a model obtained with `lm` (or `glm`) to perform the classical ANOVA analysis as discussed in Chapter 13. Examples of one-way model I ANOVA and randomized block design ANOVA performed in R are included. This chapter also includes a few examples of finding differentially expressed genes using the `limma` packages in Bioconductor. These examples include a two group comparison with single-channel data and a multiple contrast with single-channel data.

Chapter 15

Experiment design

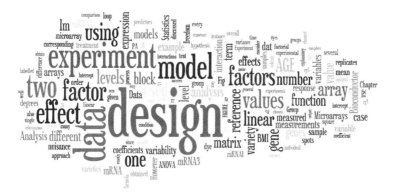

One day when I was a junior medical student, a very important Boston surgeon visited the school and delivered a great treatise on a large number of patients who had undergone successful operations for vascular reconstructions. At the end of the lecture, a young student at the back of the room timidly asked, "Do you have any controls?" Well, the great surgeon drew himself up to his full height, hit the desk, and said, "Do you mean did I not operate on half the patients?" The hall grew very quiet then. The voice at the back of the room very hesitantly replied, "Yes, that's what I had in mind." Then the visitor's fist really came down as he thundered, "Of course, not. That would have doomed half of them to their death." It was absolutely silent then, and one could scarcely hear the small voice ask, "Which half?"

—*E.E. Peacock*, Medical World News, *Sept. 1, 1972*

It is often said that experiments must be made without preconceived ideas. That is impossible. Not only would it make all experiments barren, but that would be attempted which could not be done.

— *Henri Poincare:* The Foundations of Science, Science and Hypothesis

To consult a statistician after an experiment is finished is often merely to ask him to conduct a post-mortem examination. He can perhaps say what the experiment died of.

—*Ronald A. Fisher: Indian Statistical Congress, 1938, vol. 4, p. 17*

15.1   The concept of experiment design

Chapter 13 discussed in detail some of the classical statistical analysis methods able to process data from experiments influenced by several factors. Very conveniently, in Chapter 13 the data happened to be such that these methods could be applied. Measurements were available for all needed factor interactions. However, this does not happen automatically in every experiment involving data collection. In order for the experiment to provide the data necessary for the analysis, the experiment needs to be *designed*. The design of the experiment is a crucial but often neglected phase in microarray experiments. A designed experiment is a test or a series of tests in which a researcher makes purposeful changes to the input variables of a process or a system such that one may observe and identify the reasons for changes in the output response [315]. If the experiments are not designed properly, no analysis method will be able to obtain valid conclusions. It is very important to provide data for a proper comparison for every major source of variation.

Unfortunately, many a data analyst has been called to analyze the data only after the experiment was carried out and the data collected. It is only by sheer luck that an experiment will provide all required data if it was not designed to do so. Therefore, in many such instances, the statistician or data analyst will have no other choice but to require that a new set of data be collected.

In a designed experiment, the factors thought to contribute to the noise of the system are identified, some of them controlled, and the statistical methods for analysis are chosen from the very beginning of the experiment. A laboratory experiment in which these elements have not been identified from the very beginning is a data collection study rather than an experiment.

15.2   Comparing varieties

Ronald Fisher was a statistician who pioneered the field of experiment design. He is currently considered the "grandfather of statistics." His focus at that time was on agricultural experiments. The development of new varieties of crops is long-term and tedious work. The evaluation of several varieties in a study may need a vast area of land. Fisher's classical experiment involves comparing the yield of two strains of corn x and y. In order to compare the two strains, the researcher plans to seed one acre of land with each strain and compare the amount of corn produced from each strain. Unfortunately, the land available for this is divided into two lots A and B situated a certain distance away from each other. The researcher decides to seed lot A with strain

Corn variety

Field		x	y
	A	M_{Ax}	
	B		M_{By}

FIGURE 15.1: A confounding experiment design. The data are influenced by two factors. As performed, this experiment does not allow us to decide whether the difference between the two measurements is due to the different strains of corn or to the different soil in the two lots.

x and lot B with strain y. One crop season later, the researcher will be able to compare the two yields only to realize there is no way to assign the difference to a specific cause. Assuming that the crop on lot A, planted with seed x yielded more corn, the researcher will never know whether this was because x is indeed more productive or perhaps because lot A had a more fertile soil. This is a typical example of a confounding experiment design. In most cases, the data provided by a confounded experiment design simply do not allow the researcher to answer the question posed, and no data analysis method or approach can change this.[1] This situation is illustrated in Fig. 15.1.

Let us recall how the number of degrees of freedom is partitioned in ANOVA:

$$v_{Total} = \sum v_{factors} + v_{error}$$

Here the total number of degrees of freedom is $N - 1 = 2 - 1 = 1$. Facing this hardship, we can reduce our demands and be willing to perhaps ignore the influence of the different lots and consider the two measurements as two measurements of the two strains in the same conditions. This would collapse the rows of the matrix in Fig. 15.1 producing a matrix with a single row and two columns. Even in this case, the degrees of freedom for the variety would then be $2 - 1 = 1$ and there are no degrees of freedom available to estimate the error.

$$
\begin{array}{ccccc}
v_{Total} & = & v_{Corn} & + & v_{error} \\
1 & = & 1 & + & 0
\end{array}
$$

Alternatively, one can ignore the difference between the varieties and calculate the difference between the lots. This would collapse the columns in Fig. 15.1 producing a matrix with a single column and two rows. Even in this case, the number of measurements is insufficient:

$$
\begin{array}{ccccc}
v_{Total} & = & v_{Lots} & + & v_{error} \\
1 & = & 1 & + & 0
\end{array}
$$

---

[1] However, in some situations, an experiment can be designed in such a way that certain uninteresting variables are confounded.

Corn variety

Field		x	y
	A	M_{Ax}	M_{Ay}
	B	M_{Bx}	M_{By}

FIGURE 15.2: A better experiment design. There are $4-1=3$ total degrees of freedom, $2-1=1$ block (lot) degrees of freedom and $2-1=1$ variety (corn) degrees of freedom. The error can be estimated using $3-(1+1)=1$ degree of freedom.

A better experiment design would seed each strain on both lots available. This time the matrix of the experiment design would look like the one in Fig. 15.2. There are $4-1=3$ total degrees of freedom, $2-1=1$ block (lot) degrees of freedom, and $2 \quad 1 = 1$ variety (corn) degrees of freedom. The error can now be estimated using $3-(1+1)=1$ degree of freedom. The ANOVA approach described in Chapter 13 can now be used to test the hypothesis that the two strains are different while adjusting for the variability introduced by the different soil in the two lots. Furthermore, the same approach can even test whether the two lots are significantly different. Fisher's landmark contribution was the development of the ANOVA methods discussed in Chapter 13 that simultaneously estimate the relative yield of the crop varieties and the relative effects of the blocks of land.

15.3 Improving the production process

Experiment design methods have been used extensively to improve performance in production plants. In this context, the objective of the experiment design is to troubleshoot a process and transform it into a "robust" process with minimal influence of external sources of variability [315].

Let us consider the example of a microarray production process in a microarray core facility. A major problem in a microarray facility is establishing the correct protocols for printing the arrays, hybridization, and scanning. The arrays produced in such a facility need to be reproducible and meet high-quality standards. The factorial design described in Chapter 13 can be used to ensure the robustness of the process and assess the effects of different experimental factors [461]. In general, the target preparation step is a major contributor to the overall experimental variability. The target preparation involves two factors: i) reverse transcription of mRNA (which is dependent on the reverse transcriptase enzyme), and ii) incorporation of fluorescently labeled nucleotides during this reverse transcription. Other factors that add to

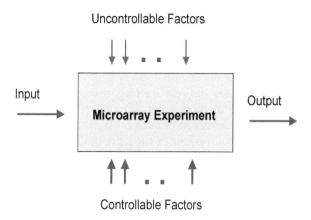

FIGURE 15.3: The output of a process is affected by the input of that process as well as by a number of controllable and uncontrollable factors.

the variability of the targets include the RNA type used in the experiment, age and type of the fluorescent label, dNTP age, and the enzyme type.

Printing the arrays is a process that involves many sources of variance: different protocols, different people performing various steps, different consumable materials (such as glass slides, buffers, enzymes, etc.), instrumentation, etc. This combination of factors transforms the input into an output that has one or more observable responses. This is illustrated in Fig. 15.3. There are many variables that have an influence on the output. Some of these are controllable $(x_1, x_2, \ldots, x_n)$ and some uncontrollable $(z_1, z_2, z_3, \ldots, z_n)$. The questions posed by the experimenter may include [315]:

1. Which variables are most influential on the response, y?

2. Where can we set the influential x-s so that y is almost always near the desired nominal value? Some of the sources of noise are understandingly unavoidable. What can we do to minimize them?

3. Where to set the influential x-s so that variability in y is as small as possible?

4. Where to set the influential x-s so that the effects of the uncontrollable variables $z_1, z_2, \ldots, z_3$ are minimized?

For example, let us consider an experiment studying the effect of a drug on the expression levels of certain genes. In this case, the input is the amount of drug administered, and the output is the expression level of the genes or, more precisely, the difference between the expression levels of the genes without the drug and the expression levels of the genes when the drug is administered. There are situations in which the variance introduced by the various sources is so great that the interesting effect, the gene regulation due to the drug, is completely covered by the variation in the output determined by the other sources of variance. In such circumstances, a preliminary study can be undertaken in order to assess the relative importance of the various sources of variance as well as find ways of minimizing the undesired variability [461]. Once the levels of these factors have been optimized, the experiment can then proceed to vary the concentrations of the drug and measure the differences of the output signals.

15.4 Principles of experimental design

There are three basic principles of experimental design:

1. replication

2. randomization

3. blocking

15.4.1 Replication

In a strict linguistic sense, to **replicate** means to duplicate, to repeat, or to perform the same task more than once. Replication allows the experimenter to obtain an estimate of the experimental error. This estimate of error can become the basis for drawing conclusions whether the observed differences in the data are significant.

Replication is a widely misunderstood term in the microarray field. Often, the misunderstanding is related to the definition of the task to be performed. Thus, if the purpose is to understand and control the noise introduced by the location of the spot on the slide, one can replicate spots by printing exactly the same DNA at different locations on the same slide. If the purpose is to understand and control the noise introduced by the hybridization stage, for instance, one can print several exact copies of a given slide (with all other parameters and DNA sources exactly the same) and hybridize several times with exactly the same mRNA in exactly the same conditions. Finally, if the purpose is to control the biological variability, different mRNA samples can be collected from similar specimens and the microarray should be used in exactly the same conditions from all other points of view. The common misunderstanding is related to the fact that often researchers refer to replicates without specifying which one factor was varied while keeping everything else constant. Even more misleading, sometimes the term "replicates" is used to describe results obtained by varying several factors at the same time. For instance, combining two different expression values of the same gene obtained on different arrays, in different hybridization conditions, and perhaps with different mRNA, will certainly contain more information than a single expression value, but many people would not think of these two measurements as being replicates. Strictly speaking, these measurements *can* be considered replicates only that the unexplained variability (or noise) will be so large that, in most cases, these data will not provide sufficient support to reject a null hypothesis. An alternative point of view could give up the strict hypothesis-testing approach used so far in favor of a Bayesian approach that make inferences based on prior knowledge and accumulated experience.

Spot replication of DNA sequences by printing them adjacent to one another has been used in the field, but is not always the best choice. The very purpose of spotting the same sequence more than once is to have several independent pieces of information. If the spots are printed next to one another, any defect that affects the slide locally, any time during the whole process, is likely to affect all such spots and thus defeat the purpose of having independent measurements. For filter microarrays using radioactive labeling, if a gene is expressed at a high level, its spot can be extremely large[2] and can cover part or all of its neighbors. Since overlapping spots cannot be distinguished from one another, this is a source of experimental error. As another example, printing all replicates of a gene in a limited area on the array means that if that particular area is affected by a local washing or drying problem, all spots corresponding to the given gene will be affected and no reliable information about this gene will be available. A better choice is to distribute the spots randomly on the entire surface of the array. Clearly, spots that are replicated on the same array share all sources of variability related to the array and the process that the array was subject to. Thus, the only remaining source of variability between such measurements is the location of the spots on the slide.

For spotted cDNA arrays, there is a non-negligible probability that the hybridization of any single spot containing complementary DNA will not reflect the presence of the mRNA. This probability is about 5% [276]. This means that if a cDNA array contains 5,000 spots corresponding to expressed genes, approximately 250 such spots will not appear to have a high signal. The converse situation is also true. It turns out that a spot can provide a substantial signal even if the corresponding mRNA is not present. This probability is even higher, of about 10% [276]. These probabilities are rather large and they should make us think twice before deciding to print single spots.

If we accept that printing each spot just once is not a very good idea, the next question is how many replicates do we need? Would two spot replicates be

---

[2]This phenomenon is known as bleeding.

sufficient? How about three? And what if we wanted to have really good data and are prepared to pay more for it? Shall we go for five? Several approaches are possible here in order to answer this question. An analytical approach would try to estimate the population mean and variance, choose a minimum detectable change, and calculate the number of replicates as in Example 12.2. However, this approach is not always feasible for a large variety of reasons. The fundamental phenomenon in action here is that the variance of the mean of a set of measurements is reduced by the number of replicates as follows (see Theorem 9.1 in Chapter 8):

$$\sigma_{\bar{X}}^2 = \frac{\sigma^2}{n} \tag{15.1}$$

where $\sigma_{\bar{X}}^2$ is the variance of the sample mean, σ^2 is the variance of the measurements, and n is the number of replicates.

An empirical approach would just choose a "reasonable" number knowing that any number of replicates is better than not having replicates at all. If this approach is chosen, a good minimum for the number of spot replicates is three. This is because if two replicate spots yield two very different values such as 10 and 100, one would not know which value is more likely to be the outlier. However, if there are three spot replicates and the values read are 10, 90, and 100, it is more likely that the value 10 is the outlier. In such cases, the software used for the analysis can be instructed to disregard the outliers. Of course, it is still possible to have ambiguous situations such as three measurements of 10, 50, and 100 but such a situation is far less probable than having an ambiguous situation when using only two measurements. Indeed, Lee and colleagues [276] used three replicates and showed that both false positives (spots showing signal in the absence of the correspondent mRNA) and false negatives (spots not showing signal in the presence of the correcpondent mRNA) can be reduced considerably by combining the data provided by only three replicate spots.

Another perennial question related to replication regards the practice of pooling biological samples in microarray experiments. This method has been proposed as a way to reduce costs and time in such studies. Let us consider, for example, a situation in which a certain treatment is applied to a number of 5 animal subjects. There are two approaches. A first approach would combine, or pool, the mRNA coming from the five animals and use the pooled sample to hybridize a number, let us say, five arrays. The second approach would use each individual sample to hybridize a different array. One of the advantages of pooling is that it can reduce the data collection effort. In this case, the pooled sample could have been hybridized only on three arrays instead of five. However, the classical question is: assuming that the number of arrays is the same, which way is better: hybridize a pooled sample five times or hybridize each of the five individual samples separately?

Arguments can be brought for both choices, but, overall, pooling has the drawback of averaging without control. As Claude Bernard put it: "*If we collect a man's urine during twenty-four hours and mix all this urine to analyze the average, we get an analysis of a urine which simply does not exist; for urine, when fasting, is different from urine during digestion. A startling instance of this kind was invented by a physiologist who took urine from a railroad station urinal where people of all nations passed, and who believed he could thus present an analysis of* average *European urine!"* [48]. Conversely, the advantage of not pooling is that an average can always be calculated from the individual values whereas the individual values, cannot be extracted from the average or pooled sample.

As another example, consider the case in which four of the animals above have a certain gene expressed consistently at a low level. However, the fifth animal, due to some individual characteristics (e.g., some illness), has a very high level of expression of this gene. When pooling the samples, the large amount of mRNA coming from this last animal can increase the overall level such that the gene appears expressed higher in any hybridization with the pooled sample. In this situation, individual hybridizations would have made us aware of the fact that four out of five measurements were consistently low while still giving us the opportunity to calculate an average should we wish to do so.

There are several arguments in favor of pooling, as well. The most compelling of them all is that sometimes it is not possible to extract enough mRNA from a single individual. In such cases, of course, having some measurement is better than having no measurement at all and pooling is the only way to go.

15.4.2 Randomization

Randomization has been proclaimed to be the cornerstone underlying statistical methods [315]. **Randomization** requires the experimenter to use a random choice for every factor that is not of interest but might influence the outcome of the experiment. Such factors are called **nuisance factors**. The simplest example is the printing of replicate spots on the array. If such replicates are printed next to one another, a localized defect of the array will affect all of them making it impossible to distinguish the interesting gene effect from the uninteresting effect of the defect. Randomization requires that the replicate spots be printed at random locations throughout the array. Another example is the use of microarray slides from different batches in an experiment comparing a treatment group versus a control group. If all control animals are tested using slides from one batch and all treated animals are tested using slides from a different batch, it will be impossible to distinguish between the uninteresting variability introduced by the slides and the interesting variability introduced by the treatment. These two factors would be confounded in such an experiment design. However, if the slides are assigned randomly between the controls and the treated animals, the bias is eliminated and the influence of this nuisance factor reduced. Randomization may not always be possible but should always be attempted.

15.4.3   Blocking

Blocking is a design technique used to increase the accuracy with which the influence of the various factors is assessed in a given experiment. A **block** is a subset of experimental conditions that are expected to be more homogeneous than the rest. **Blocking** refers to the method of creating homogeneous blocks of data in which the nuisance factor is kept constant and the factor of interest is allowed to vary. Blocking is used to eliminate the variability due to the difference between blocks (see Section 13.3.1 in Chapter 13 for more details). A typical example of a block in microarrays is the microarray slide itself. Since all spots on a given slide are subject to the same factors during the slide processing (hybridization, washing, drying, etc.), it is expected that the measurements of the spots coming from a single slide will be more homogeneous (have a lower variance) than the measurements across the whole experiment. This can be observed particularly well on the control spots if such spots are being used. The two-channel cDNA process deals with this very elegantly by hybridizing both control and treatment samples on the same array. Unfortunately, this is not currently possible with the Affymetrix technology, which requires each sample to be hybridized on its own array. However, the cDNA process introduces the supplementary nuisance factor of the dyes, which, in turn, would require blocking.

Both blocking and randomization deal with nuisance factors. The difference is that blocking can only be used when the nuisance factor is under our control. Examples include any choice of materials or substances. If the nuisance factor considered is not under our control (e.g., drying marks on the surface of the microarray), randomization remains the only tool available. This is summarized by the general rule: "**block what you can, randomize what you cannot**."

15.5   Guidelines for experimental design

The following guidelines for designing experiments can be taken into account when planning an experiment:

1. **Describe exactly what your research problem is.** Write the questions you want answered in a laboratory book or your LIMS system. Spell out every detail. The fact is that you may ask many questions from a single experiment. It is necessary to make an effort and foresee what the problems in the experiment might be, what can affect it in a negative or positive way, what the sources of variation might be, what the goals are, and what the final results could be. There is an abundance of literature on every subject and sometimes a researcher can gather a lot of information a long time before the actual experiment starts. Stat-

ing the objectives is extremely important for going through the whole process of designing the experiment, planning it, implementing it, and analyzing the data.

2. **Choose the technology to be used.** Make the choices between cDNA or oligo arrays, commercial or custom arrays, and the specific brand and type of microarray if commercial arrays are to be used. A very important issue is to choose an array that is appropriate for the biological question being asked. In an exploratory research, in which no hypothesis has been yet formulated, large, comprehensive arrays may be best. In a hypothesis-driven research, in which phenomena involving certain specific gene regulatory pathways are hypothesized, more focused arrays may be more useful and convenient. An important issue is to choose a commercial array that has a good representation of the genes conjectured to be relevant. The same issue applies to custom arrays printed in-house. These issues are discussed in more detail in Chapter 26.

3. **Involve a collaborator who has experience in experiment design and data analysis**. Try to obtain a firm commitment from them for the data analysis part since experiment design and data analysis are parts of the same thought process, and it is not a good idea to change collaborators in-between. Be aware of the fact that communication across field boundaries can be very challenging. In spite of a growing number of people with interdisciplinary interests, it was not too long ago that the word "mitochondria" seemed offensive to some statisticians, and even today, "heteroscedasticity" will sound a bit scary to most biologists! The goal of this chapter is not to allow the life scientist to design their own experiments but help them understand the issues involved, such that they can communicate effectively with a statistician or computer scientist.

4. **Choose the factors that can influence your output in a significant way and their corresponding levels of interest.** Identify which are the major factors you want to follow. In most cases, the major inputs will be the ones directly related to your scientific question: the effect of the drug, differences between illness and healthy, etc. In most case, the interesting outputs of a microarray experiment will include the expression of various genes or, in ANOVA terms, the variety-gene interaction (see Chapter 13).

5. **Identify the nuisance factors that you would like to consider.** The nuisance factors could be the types of enzyme, types of dyes, nucleotides, who prepared the mRNA sample, etc. Each nuisance factor you choose to follow will require extra work; so the rule of thumb is to keep the number of factors as low as possible by choosing only the ones that are expected to influence considerably the outcome. Divide the nuisance factors into controllable and uncontrollable.

6. **Choose the significance level and desired power.** Consider the issues of Type I (rejecting a true null hypothesis, e.g., concluding that a gene is differentially regulated when the gene is in fact unchanged) and Type II errors (not rejecting a false null hypothesis, e.g., not detecting a true differentially regulated gene). Recall from Chapter 11 that there is always a compromise between false positives and false negatives. For instance, having stringent requirements for when a gene is differentially regulated (low alpha) means that many truly differentially regulated genes may not be detected (low power).

7. **Design your experiment.** Block the controllable nuisance variables and randomize the others. Calculate the number of replicates at every level from the power requirements.

8. **Perform the experiment and collect the data.** Record every detail and check the quality at every step. Any error in experimental procedure will destroy the experimental plan and validation! This author has seen a situation in which a PhD student performed about 100 hybridizations as part of a randomized block design, without ever checking the scanned images. At the end of the experiment, it turned out that the large majority of arrays provided no signal whatsoever on one of the channels. The lack of mRNA prevented repeating the arrays and the experiment was seriously compromised.

9. **Perform the data analysis.** The data analysis is still a challenge but at least the necessary conditions are met. Tools useful at this stage are discussed in Chapters 17, 18, and 21.

10. **Extract the biological meaning from the results of the data analysis**. Statistical methods are not the untouchable proof – it is the biological meaning of your experiments that validates the work! Translate the lists of differentially regulated genes into biological knowledge by mapping differentially regulated genes to the biological processes involved, affected pathways, etc. This step is discussed in detail in Chapters 23 and 28.

15.6 A short synthesis of statistical experiment designs

This section will review in a very concise manner several important experiment designs. In this section, a source of variability under our control will be considered a factor. Factors can include: source of mRNA sample, treatment applied to various patients of experiment animals, cell culture, etc. Each factor will have several levels, that is, possible values. For instance, a factor such as

Factor

1	2	$\cdots$	i	$\cdots$	a
X_{11}	X_{21}	$\cdots$	X_{i1}	$\cdots$	X_{a1}
X_{12}	X_{22}	$\cdots$	X_{i2}	$\cdots$	X_{a2}
$\vdots$	$\vdots$	$\vdots$	$\vdots$	$\vdots$	$\vdots$
X_{1n}	X_{2n}	$\cdots$	X_{in}	$\cdots$	X_{an}

FIGURE 15.4: The data layout for a fixed effect design with one factor.

the dye will have two values (or levels): cy3 and cy5. A factor such as a drug concentration will have as many levels as required by the study undertaken. A factor such as mRNA source will have as many levels as there are mRNA samples, etc.

15.6.1 The fixed effect design

In this design, there is only one factor and data are collected repeatedly, at the various levels of the factor. In this design, the data can be laid out as shown in Fig. 15.4. This is very similar to the data layout used in the discussion of the model I ANOVA in Chapter 13, only that the conditions have been now been replaced by the more general factor levels. In this example, there are a different levels for the factor under study, and there are n different measurements for each level. Because the number of observations within each treatment level is the same, we say this is a **balanced design**. If the number of observations within each treatment were different, the design would be **unbalanced**. A balanced design has two advantages: i) the model is less sensitive to departures from the equal variance assumption and ii) has a better power with respect to an unbalanced design.

This design, either balanced or unbalanced, has the general model:

$$X_{ij} = \mu + \tau_i + \varepsilon_{ij} \tag{15.2}$$

where μ is the overall mean, τ_i is the effect of the factor level (treatment) i and ε_{ij} is the term corresponding to the random noise or unexplained variability. In this equation, i takes values from 1 to a and corresponds to the various levels of the factor considered; j takes values from 1 to n and corresponds to the various measurements for each factor level. This experiment design can be analyzed as discussed in Section 13.2.1.

	Factor					
Block	1	2	$\cdots$	i	$\cdots$	a
1	X_{11}	X_{21}	$\cdots$	X_{i1}	$\cdots$	X_{a1}
2	X_{12}	X_{22}	$\cdots$	X_{i2}	$\cdots$	X_{a2}
$\vdots$	$\vdots$	$\vdots$	$\vdots$	$\vdots$	$\vdots$	$\vdots$
b	X_{1b}	X_{2b}	$\cdots$	X_{ib}	$\cdots$	X_{ab}

FIGURE 15.5: The data matrix for a randomized complete block design. Every treatment is measured in each block; the distribution of the treatments to specific blocks is random.

15.6.2 Randomized block design

Recall that the data are influenced by: i) the factors studied and ii) nuisance factors. In turn, nuisance factors can be a) controllable and b) uncontrollable. A fully randomized design would assign treatments to experimental units (e.g., hybridizations) in a completely random manner. A **block design** considers the individual groups of measurements that are expected to be more homogeneous than the others. Such groups are called blocks, and various treatments are assigned randomly to such blocks. If all treatments are present on every block, the design is a **randomized complete block design**. If some blocks do not include some treatments, the design is an **incomplete block design**. The layout of the data matrix for a randomized complete block design is illustrated in Fig. 15.5.

The model for the block design is:

$$X_{ij} = \mu + \tau_i + \beta_j + \varepsilon_{ij} \tag{15.3}$$

where μ is the overall mean, τ_i is the effect of treatment i, β_j is the effect of the block j, and ε_{ij} is the effect of the random noise. The index i takes values from 1 to the number of factor levels a, whereas the index b takes values from 1 to the number of blocks b. This experiment design can be analyzed as discussed in Section 13.3.1.

15.6.3 Balanced incomplete block design

Sometimes it is not possible to run all treatment combinations in each block as the randomized complete block design requires. For instance, if more than two treatments are compared using a two-channel cDNA microarray (e.g., using cy3-cy5), an array, which is a block, will only be able to provide information about two samples at any given time. A design that does not include all treatment combinations on every block is an **incomplete block design**. A

balanced incomplete block design makes sure that any pair of treatments occurs together the same number of times as any other pair.

The data layout of the incomplete block design is the same as the one for the randomized complete block design shown in Fig. 15.5 with the only difference that the matrix will have missing elements since not all treatment combinations are available on each block. The statistical model for the incomplete block design is the same as for the complete block design (Eq. 15.3).

15.6.4   Latin square design

The randomized block design above takes into consideration one factor studied and one nuisance factor. However, many times there is more than one nuisance factor. For instance, if two such factors exist, one would need to measure the value of each level of the treatment for each combination of the two nuisance factors. Let us assume that the factor to be studied has four levels denoted by the four Latin letters: A, B, C, and D. Furthermore, let us assume that there are two nuisance factors, each of them having four levels. A Latin square design for such an experiment can be illustrated as in Fig. 15.6. The Latin square has the property that each row and each column contain each treatment exactly once. If numbers are used instead of the Latin letters, the sum of the elements of every row and every column would be the same. The popular Sudoku game is an example of a Latin square. The matrix for a Latin square design can be obtained easily by starting with a random first row containing each symbol once. Each subsequent row can be obtained from the one above by shifting the elements by one position. An alternative approach would apply the same procedure to the columns.

The statistical model for a Latin square design is:

$$X_{ijk} = \mu + \alpha_i + \tau_j + \beta_k + \varepsilon_{ijk} \tag{15.4}$$

where μ is the overall mean, α_i is the effect of the i-th row (the i-th level of the first nuisance factor), τ_j is the effect of the j-th treatment, β_k is the effect of the k-th column (or k-th level of nuisance factor 2) and ε_{ijk} is the random noise. Note that this model is a strictly additive model, i.e. it does not take into consideration potential interactions between the factors considered.

15.6.5   Factorial design

The factorial design is an experiment design that takes into consideration all possible combinations of the levels considered. Furthermore, a factorial design allows us to analyze the interactions between factors. An experiment with two factors A and B, with factor A having a levels and factor B having b levels, will require $a \cdot b$ measurements. If there are n replicates for each combination, the experiment will require a total of $a \cdot b \cdot n$ measurements. The data layout for a factorial design with two factors is shown in Fig. 15.7.

The statistical model for a factorial design with two factors is:

Nuisance factor 1	Nuisance factor 2			
	1	2	3	4
1	A	B	C	D
2	B	C	D	A
3	C	D	A	B
4	D	A	B	C

FIGURE 15.6: A 4×4 Latin square design. Each treatment A, B, C, and D is measured once for each combination of the nuisance factors.

B	A			
	1	2		a
1	X_{111} X_{11n}	X_{211} X_{21n}	$\cdots$	X_{a11} X_{a1n}
	$\vdots$	$\vdots$		$\vdots$
2	X_{121} X_{12n}	X_{221} X_{22n}		X_{a21} X_{a2n}
	$\vdots$	$\vdots$		$\vdots$
b	X_{1b1} X_{1bn}	X_{2b1} X_{2bn}	$\cdots$	X_{ab1} X_{abn}

FIGURE 15.7: The data layout for a factorial design with two factors. Factor A has a levels, factor B has b levels, and there are n replicates for each combination of factor levels. This design requires $a \cdot b \cdot n$ measurements.

Source of Variation	Sum of Squares	Degrees of Freedom	Mean Squares	Expected Mean Square	F_0
A	SS_A	$a-1$	MS_A	$\sigma + \frac{bcn\sum\tau_i^2}{a-1}$	$F_0 = \frac{MS_A}{MS_E}$
B	SS_B	$b-1$	MS_B	$\sigma^2 + \frac{acn\sum\beta_j^2}{b-1}$	$F_0 = \frac{MS_B}{MS_E}$
C	SS_C	$c-1$	MS_C	$\sigma^2 + \frac{abn\sum\gamma_k^2}{c-1}$	$F_0 = \frac{MS_C}{MS_E}$
AB	SS_{AB}	$(a-1)(b-1)$	MS_{AB}	$\sigma^2 + \frac{cn\sum\sum(\tau\beta)_{ij}^2}{(a-1)(b-1)}$	$F_0 = \frac{MS_{AB}}{MS_E}$
AC	SS_{AC}	$(a-1)(c-1)$	MS_{AC}	$\sigma^2 + \frac{bn\sum\sum(\tau\gamma)_{ik}^2}{(a-1)(c-1)}$	$F_0 = \frac{MS_{AC}}{MS_E}$
BC	SS_{BC}	$(b-1)(c-1)$	MS_{BC}	$\sigma^2 + \frac{an\sum\sum(\beta\gamma)_{jk}^2}{(b-1)(c-1)}$	$F_0 = \frac{MS_{BC}}{MS_E}$
ABC	SS_{ABC}	$(a-1)(b-1)(c-1)$	MS_{ABC}	$\sigma^2 + \frac{n\sum\sum\sum(\tau\beta\gamma)_{ijk}^2}{(a-1)(b-1)(c-1)}$	$F_0 = \frac{MS_{ABC}}{MS_E}$
Error	SS_E	$abc(n-1)$	MS_E	σ^2	
Total	SS_T	$abcn-1$			

TABLE 15.0: The ANOVA table for the general factorial design with 3 factors.

$$X_{ijk} = \mu + \tau_i + \beta_j + (\tau\beta)_{ij} + \varepsilon_{ijk} \tag{15.5}$$

where μ is the overall mean, τ_i and β_j are the main effects of the two factors, $(\tau\beta)_{ij}$ is the interaction between the two factors, and ε_{ijk} is the random noise.

The equations Eq. 15.5 and Eq. 15.3 might seem similar inasmuch they both use only two factors β and τ. However, in the model 15.3, β is a nuisance factor, whereas in Eq. 15.5 it is a factor under study. We can control the levels for a factor we study while we can only block the level for a nuisance factor, i.e. make sure that all measurements in a block are affected by the same level of the nuisance factor. Furthermore, the model in Eq. 15.5 also takes into consideration the interaction between the two factors, interaction which is not considered in Eq. 15.3.

The factorial model can be generalized to a situation in which there are a levels for factor A, b levels for factor B, c levels for factor C, etc. In general, if the complete experiment is to be replicated n times, there will be $a \cdot b \cdot c \cdots n$ measurements. Note that it is necessary to have at least two replicates $(n \geq 2)$ of the complete experiment in order to determine a sum of squares for the error if all possible interactions are included in the model.

As discussed in Chapter 13, the test statistic for each main effect or interaction can be calculated by dividing the corresponding mean square for the given effect or interaction by the mean square error. The number of degrees of freedom for any main effect is the number of levels of the factor minus one. The number of degrees of freedom for any interaction is the product of the degrees of freedom of the component. For instance, the three factor ANOVA model for a factorial design is:

$$X_{ijkl} = \mu + \tau_i + \beta_j + \gamma_k + (\tau\beta)_{ij} + (\tau\gamma)_{ik} + (\beta\gamma)_{jk} + (\tau\beta\gamma)_{ijk} + \varepsilon_{ijkl} \quad (15.6)$$

with $i = 1, 2, \ldots, a$, $j = 1, 2, \ldots, b$, $k = 1, 2, \ldots, c$, $l = 1, 2, \ldots, n$. The degrees of freedom, mean squares, the expected mean squares, and the formulae for the computation of the F test are given in Figure 15.8.

15.6.6 Confounding in the factorial design

The complete factorial design requires a very large number of experimental runs. For instance, factors that might be taken into consideration in a cDNA microarray experiment can include [461]:

1. type of fluorescent label,

2. age of the fluorescent label,

3. type of RT enzyme,

4. age of dNTPs,

5. incubation time for transcription, and

6. the use of total RNA or poly(A) RNA.

If each of these factors is considered at two levels, a complete factorial experiment would require $2 \times 2 \times 2 \times 2 \times 2 \times 2 = 64$ runs.

In most microarray applications, it would be impossible to perform a complete replicate of a factorial design in one block. As a consequence, we will not be able to obtain information about all factors and all interactions. However, not all factors and interactions are equally important. **Confounding** is a technique that allows the designer to group the various runs of an experiment in blocks, where the block size is smaller than the number of treatment combinations in one replicate [315]. Because of this, the effect of certain factors or factor interactions will be indistinguishable from the effect of the blocks. However, if this is done by design, we can make sure that the confounded factors are factors in which we are not interested. Note that the variability due to these confounded factors is still subtracted from the total variability. Thus, the confounded factors are not lost from under control and the confounding does not necessarily decrease the reliability of our conclusions about the factors of interest. In principle, any factor or any interaction can be confounded. This is done by grouping together all the measurements corresponding to the chosen factor and assigning them to a block. For instance, if a variety (mRNA sample) is always assigned to a dye, the two effects A_i and D_j will be confounded. Confounding the main variety effect with the main dye effect may not be necessarily bad. Recall that in most cases, we are particularly interested in the variety-gene interaction $(VG)_{kg}$, which is to say we are interested in how specific genes change from one mRNA sample to another. As another example of purposeful confounding design, if a variety appears always in the same array-dye combination, the variety will be confounded with the array-dye interaction (see for instance Eq. 15.10).

15.7 Some microarray specific experiment designs

This section will use the notation and methodology developed in Chapter 13 to discuss and compare a few experiment designs and the associated data analysis methods.

15.7.1 The Jackson Lab approach

Let us consider some real-world examples of experiment design in the analysis of microarray data. Kathleen Kerr and Gary Churchill were the first to recognize the suitability of the ANOVA approach for studying microarray data. They also did pioneering work in data analysis and associated experimental design. One of their first papers on this topic proposed the following model [259]:

$$\log(y_{ijkg}) = \mu + A_i + D_j + V_k + G_g + (AG)_{ig} + (VG)_{kg} + \varepsilon_{ijkg} \qquad (15.7)$$

The model assumes that these effects can be summed in a linear way. Several researchers showed that this is a reasonable assumption in logarithmic scale. This means the y_{ijkg} are the *logs of* the intensities read from the microarray. The particular base of the logarithm is not very important but base 2 is convenient because it makes the interpretation easier.

In order to illustrate how such an additive model works, let us consider the data in Table 15.1. In this example, there are four main effects: the mouse effect, dye effect, array effect, and overall mean effect. The mouse effect adds 5 to all values measured on Mouse 1, and 3 to all values measured on Mouse 2. The dye effect adds 3 to all values measured with cy3 and 2 to all values measured with cy5. The array effect adds 4 to array 1 and 6 to array 2. Finally, the overall mean effect is 50. The value measured in each condition is the result of the addition of the corresponding effects. For instance, in Table 15.1, the value measured for Mouse 1, on array 1, on the cy5 channel is:

$$50 + 5 + 2 + 4 = 61$$

	Mouse 1		Mouse 2			
	5		3			Mouse effect
	cy3	cy5	cy3	cy5		
	3	2	3	2		Dye effect
Array 1	62	61	60	59	4	
Array 2	64	63	62	61	6	Array effect
		50				Overall mean effect

TABLE 15.1: An example of a superposition of effects in a linear model. There are 4 main effects: the mouse effect, dye effect, array effect and overall mean effect. The mouse effect adds 5 to all values measured on Mouse 1, and 3 to all values measured on Mouse 2. The dye effect adds 3 to all values measured with cy3 and 2 to all values measured with cy5. The array effect adds 4 to array 1 and 6 to array 2. Finally, the overall mean effect is 50. The value measured in each condition is the result of the addition of the corresponding effects.

In this situation, there are no interactions between factors. This means that if two factors are applied simultaneously the result would simply be the sum of the effects of the individual factors. If the result of the two factors acting together were larger than the sum of the individual effects, we would have a **synergy**. If the result of the two factors acting together were smaller than the sum of the individual effects, we would have an **interference**. Also, in this case, there is no noise since the values measured are exactly the values obtained by adding the individual effects. The noise is sometimes referred to as the unexplained variability and is the difference between the actual value and the sum of the factors considered.

In light of this example, let us revisit Eq. 15.7:

$$\log\left(y_{ijkg}\right) = \mu + A_i + D_j + V_k + G_g + (AG)_{ig} + (VG)_{kg} + \varepsilon_{ijkg} \tag{15.8}$$

In this model, there are five terms describing main effects and two terms describing interactions. The term μ corresponds to the overall mean effect. This is a quantity that is present in all measurements, independent on what array, channel or gene the value was measured on. The term A_i represents the effect of array i. This quantity accounts for a change at the array level such as a longer hybridization time. Such an array effect would affect all genes on the given array. The two dyes used in cDNA microarray experiments have different chemical properties, which may be reflected in their differential incorporation. One dye may be consistently brighter than the other one. The term D_j represents the effect of the dye j, such as a higher overall efficiency of one of the dyes. Also, some genes have been observed to have consistently higher or lower expressions than others. Additionally, some sequences show differential labeling and hybridization efficiencies. The term G_g represents the contribution of such gene specific effects. Finally, the last main effect, V_k, is the effect of variety k. In this context a variety is an mRNA sample. Since Fisher's pioneering work on ANOVA compared crop varieties, the term variety has remained in the terminology denoting different subspecies to be compared.

Besides the main factors above, the model in Eq. 15.8 also considers certain factor interactions. The variety-gene interactions $(VG)_{kg}$ reflect differences in expression for particular genes under the effect of a given variety. This is the effect that we need in order to answer questions such as: "What are the genes differentially regulated in condition A versus condition B?" The last interaction term considered is the $(AG)_{ig}$ interaction. This term accounts for the interaction between arrays and genes. When several arrays are printed, the spots for a gene on the different arrays will vary from one another due to differences in cDNA available for hybridization, printing differences, or other causes. Therefore, the location of a gene on an array will have its effect on the final outcome. This is the so-called spot effect represented by the interactions term $(AG)_{ig}$.

Finally, ε_{ijkg} is the effect of the random noise. This is assumed to be normally distributed and have a zero mean. As explained in Chapter 13, once the analysis is done, the effects above should be used to calculate the residuals, which are estimates of effects of the ε_{ijkg} term. The inspection of these residuals will validate or invalidate the data analysis and implicitly the experiment design. If the distribution of the residuals shows non-random features, the model used did not capture all sources of systematic variability. In some cases, the data are such that an alternative model may be used to reanalyze the data. However, many times, a non-random distribution of the residuals can indicate an inadequate experiment design.

The model in Eq. 15.7 does not take into account the interactions between the dyes and genes. If one desires to consider this interaction, the model can be augmented by adding the (DG) term [255]:

$$\log\left(y_{ijkg}\right) = \mu + A_i + D_j + V_k + G_g + (AG)_{ig} + (VG)_{kg} + (DG)_{jg} + \varepsilon_{ijkg} \quad (15.9)$$

Yet another model can be constructed by adding a term for the interaction between arrays and dyes [258]:

$$\log\left(y_{ijkg}\right) = \mu + A_i + D_j + G_g + (AD)_{ij} + (AG)_{ig} + (VG)_{kg} + (DG)_{jg} + \varepsilon_{ijkg} \quad (15.10)$$

A careful analysis of this model shows that it is complete inasmuch as all combinations of arrays, dyes, and varieties are directly or indirectly accounted for. The missing term corresponding to the variety (V_k) is indirectly accounted for by the (AD) term since a variety appears only once for every array-dye combination. This design will not be able to distinguish between the variability due to the variety alone and the variability due to the array-dye interaction, but this is not important since these are normalization issues that are not interesting. The important fact is that the variability due to all

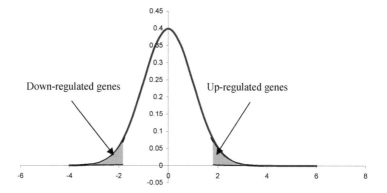

FIGURE 15.0: Two varieties can be compared directly using a cy3 cy5 cDNA microarray. A ratio can be calculated between the values measured on the two channels for each gene and represented graphically as a histogram (frequency versus ratio value). Most genes are expected to be expressed at similar levels, and therefore, most ratios will be around the value 1 (0 if log values are used). Two thresholds can be used to select the tails of the ratio distribution and hence select the differentially regulated genes.

these factors is subtracted from the overall variability allowing us to test more accurately the hypotheses related to the $(VG)_{kg}$ term, which corresponds to the differential regulation between varieties (mRNA samples). More detailed comparisons between the various models above can be found in papers such as [256] and [257].

15.7.2 Ratios and flip-dye experiments

A common and simple problem is to compare two conditions A and B looking for the genes that are expressed differently. The simplest experiment design could compare these varieties directly. If a two-channel technology is used (e.g. cDNA using cy3 and cy5 labels), the two samples A and B can be compared using a single array. A common approach is to calculate the ratio of the values measured on the two channels. This is illustrated in Fig. 15.9. Most genes are expected to be expressed at similar levels, and therefore, most ratios will be around the value 1. The thinking is that the differentially regulated genes will be the ones with unusually large or unusually small ratios. Therefore, two thresholds can be used to select the tails of the ratio distribution and hence select the differentially regulated genes.

Let us consider this experiment in more detail. We can write the expression corresponding to the measurements on the two channels cy3 and cy5 according to the model Eq. 15.7. The values for cy3 are collected from array 1, channel 1, variety 1, and gene g and can be written as:

$$y_{111g} = \mu + A_1 + D_1 + V_1 + G_g + (VG)_{1g} + (DG)_{1g} + (AG)_{1g} + \varepsilon_{111g} \qquad (15.11)$$

The values for cy5 are collected from the same array1, but channel 2 and variety 2. These values can be written as:

$$y_{122g} = \mu + A_1 + D_2 + V_2 + G_g + (VG)_{2g} + (DG)_{2g} + (AG)_{1g} + \varepsilon_{222g} \qquad (15.12)$$

Let us recall that these are log values. Taking the ratios corresponds to subtracting the two expression above since:

$$\log \frac{a}{b} = \log a - \log b \qquad (15.13)$$

Subtracting equations 15.11 and 15.12, we obtain:

$$y_{111g} - y_{122g} = (D_1 - D_2) + (V_1 - V_2) + (DG)_{1g} - (DG)_{2g} + (VG)_{1g} - (VG)_{2g} + \varepsilon_g \qquad (15.14)$$

Let us consider the remaining terms in the above equation. The term $(D_1 - D_2) + (V_1 - V_2)$ is the average-log-ratio bias term. Normalization can eliminate the above term, since this is not gene related. The term $(VG)_{1g} - (VG)_{2g}$ is the interaction between the variety and the genes, i.e. the gene regulation due to the treatment. This is the effect of interest. However, the result also contains another term, $(DG)_{1g} - (DG)_{2g}$, which is the dye-gene interaction or the gene specific dye effect. In this experiment design, we cannot separate the effect of the uninteresting dye-gene interaction from the effect of the interesting variety-gene interaction. This is an example of *confounded effects* very similar to the example in which the two varieties of corn were planted on two different fields and the field effect could not be distinguished from the variety effect (see Fig. 15.1).

In order to be able to separate these two effects, more data are necessary. However, simply repeating the hybridization with another array will not help. The two effects would still remain undistinguishable from each other. In order to be able to separate them, we must flip the dyes. This experiment design is sometimes called the dye swap experiment, or the flip fluor experiment.

This experiment design will provide two measurements for every variety with the property that each variety is measured on each dye exactly once. This experiment design provides data allowing us to calculate the sum of squares corresponding to the gene-dye interaction and therefore subtract it from the overall variability in order to estimate more precisely the gene-variety interaction.

It should be noted that the VG effects are orthogonal to all other effects, which means that other factorial effects will not bias estimates of VG effects.

	A_1	A_2	A_3
Red	R	R	R
Green	V_1	V_2	V_3

FIGURE 15.10: A classical reference design. In this design, each condition is compared with the reference. R and G denote the red and green channels, respectively. A variety is a condition such as treated/not treated or a time point in a time series analysis such as sporulation. In this design, the reference is measured n times, while each of the n varieties is measured only once.

If the experiment is designed properly, the precision of the VG estimates is not affected by other sources of variation. In the dye swap experimental design, VG effects are orthogonal to gene-specific dye effects DG.

15.7.3   Reference design versus loop design

The pioneers of the microarray technology started by using microarrays in what Gary Churchill calls a **reference design** [95]. In this approach, a number of conditions or time points in a time series $c_1, c_2, \ldots, c_n$ are pair-wise compared to a reference r.

Researchers using the reference design would use one dye to label the reference variety, and the other dye to label the varieties of interest. This would have the advantage that there are only $n + 1$ labeling reactions, one for each variety (mRNA sample) and one for the reference. Thus, n two-channel experiments would compare each condition to the reference as follows: experiment 1 compares c_1 labeled with cy3 to r labeled with cy5, experiment 2 compares $c_2/cy3$ to $r/cy5$, ... , experiment n compares $c_n/cy3$ to $r/cy5$. This is illustrated in Fig. 15.10. The same design is represented in Fig. 15.11 using the Kerr-Churchill notation. In this notation varieties, or mRNA samples, are represented by squares. An arrow linking two squares represents an array (a hybridization). A direction can be chosen arbitrarily (e.g., from red to green in Fig. 15.11) to represent the two channels used (cy3/cy5).

Kerr and Churchill made the very interesting observation that this design collects the most information exactly on the least interesting variety, namely, the reference, since this reference appears on every array. Thus, the reference is measured n times while each interesting variety is measured only once.

Another and more serious criticism of the reference design is the fact that the dye effects are completely confounded with variety effects. Thus, the variability introduced by the dyes cannot be subtracted from the overall variability and erroneous conclusions may be drawn. For instance, if something happens in the second labeling reaction that reduces the amount of dye incorporated in the target by 10%, all values measured for variety 2 will be 10% lower. No matter what analysis method is used, there will be no way to distinguish between a gene that is 10% lower due to the dye and a gene that is 10% lower

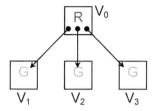

FIGURE 15.11: A classical reference design in the Kerr-Churchill notation. *R* and *G* denote the red and green channels, respectively. Arrays are represented by arrows. For instance, the arrow from V_0 to V_1 represents the array in which variety V_0 was labeled with red and variety V_1 with green.

	A_1	A_2	A_3
Red	V_1	V_2	V_3
Green	V_2	V_3	V_1

FIGURE 15.12: A loop design. This design compares each condition with every other condition. Furthermore, each condition is measured once on every channel (dye).

due to the label. In order to provide data allowing this separation between dye effects and array effects, one needs to swap the dyes. This would require $2n$ arrays and would measure each variety twice and the reference $2n$ times.

One might try to optimize the information collected by measuring the reference fewer times in favor of more measurements for the various varieties while still swapping dyes. This is achieved by the **loop design** illustrated in Fig. 15.12. This design uses the same number of arrays as the reference design but collects twice as much data on the varieties of interest. The same design is shown in Fig. 15.13 in the Kerr-Churchill notation. In a loop design, experiment 1 would compare r labeled with cy3 versus c_1 labeled with cy5, experiment 2 would compare c_1 labeled with cy3 versus c_2 labeled with cy5, experiment 3 would compare c_2 labeled with cy3 versus c_3 labeled with cy5, etc. until the last experiment which closes the loop by comparing c_n labeled with cy3 versus r labeled with cy5. In this way, the loop experiment design maximizes the amount of information that can be extracted from the data with a given number of arrays.

Although the loop design is an elegant theoretical solution, problems might appear in the practical use. Each sample must be labeled with both Cy5 and Cy3 dyes, which doubles the number of labeling reactions and thus increases the time and cost of the experiment. Furthermore, if many varieties are involved and the loop becomes very large, the data will not provide direct comparisons between all pairs. For instance, in Fig. 15.14, comparing varieties

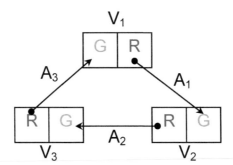

FIGURE 15.13: A loop design in the Kerr-Churchill notation. In the classical reference design, each condition is compared with the reference. In the loop design, conditions are compared with each other, flipping dyes at every comparison. R and G denote the red and green channels, respectively. For instance, the arrow $A1$ represents the array in which variety $V1$ was labeled with red and variety $V2$ with green. In this design, each variety is measured twice. This is also a balanced design since all combinations variety-dye appear together the same number of times.

	A_1	A_2	A_3	A_5	A_6	A_7
Red	V_1	V_2	V_3	V_4	V_6	V_7
Green	V_2	V_3	V_4	V_5	V_7	V_1

FIGURE 15.14: A design involving many varieties creates a large loop. In this example, comparing V_3 and V_6 can only be done indirectly by comparing $V_3 \to V_4 \to V_5 \to V_6$ or $V_3 \to V_2 \to V_1 \to V_7 \to V_6$.

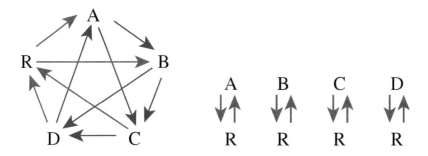

FIGURE 15.15: A comparison between the full loop design with reference and a flip-dye reference design. Each letter represents a condition (such as treatments with various drugs). An arrow represents a two-channel array on which the source condition in labeled with cy3 and the destination condition is labeled with cy5.

V_3 and V_6 can only be done indirectly by comparing $V_3 \rightarrow V_4 \rightarrow V_5 \rightarrow V_6$ or $V_3 \rightarrow V_2 \rightarrow V_1 \rightarrow V_7 \rightarrow V_6$. Either path would involve four comparisons with three other arrays. In order to alleviate this problem, the loop design can be improved by adding direct comparisons between varieties that are not neighbors on the loop. Furthermore, the loop design in Fig. 15.12 does not provide any information about the reference. Although the focus is on studying the varieties, sometimes it is useful to compare the reference directly with the conditions studied. If nothing else, the inclusion of the reference provides an internal control and the basis for the alignment of the data gathered with the rest of the data from the literature. A loop design in which the control is included as another variety is called a **loop with reference**. Fig. 15.15 shows a comparison between a full loop with reference and a flip-dye reference. The loop design with reference requires 10 arrays as opposed to the 8 arrays required by the classical reference with flip-dye design.

These and many other issues related to the statistical design of cDNA microarray experiments are discussed in detail in the work of Kerr and Churchill [257, 258, 259].

15.8 Summary

This chapter presented the main concepts related to experiment design. Replication, randomization, and blocking have been discussed as the main tools used in experiment design. Some guidelines for experimental design were outlined. The chapter also included a discussion of some classical experiment designs, such as the fixed effect design, the randomized block design, the balanced incomplete block design, the Latin square design, and the factorial design. Confounding was discussed in the context of increasing the efficiency of the factorial design by reducing the number of experimental runs. Finally, the chapter discussed in detail several of the ANOVA models proposed by Kerr and Churchill for cDNA microarrays, as well as a related ratio-based experiment design. The reference design, loop design, and loop-with-reference design were also discussed. More details about microarray specific experiment design issues can be found in the work of Kerr and Churchill [259, 258, 259]. A very complete treatment of the experiment design topic can be found in [315].

Chapter 16

Multiple comparisons

Quantitative accumulations lead to qualitative changes.

—Karl Marx

16.1   Introduction

The problem of multiple comparisons is probably the most challenging topic for the typical life scientist. We will introduce this problem through an example that illustrates the dangers that lurk behind multiple comparisons. We will then approach the problem from a statistical perspective and prove that multiple comparisons need to be treated in a special way. Once this is clear, the chapter will present a few classical solutions to the problem as well as the best choices from a microarray data analysis perspective.

| Gene | Tumor | | | | | Controls | | | | |
	T1	T2	T3	T4	T5	C1	C2	C3	C4	C5
g 1	0.340	0.232	0.760	0.610	0.224	0.238	0.075	0.624	0.978	0.198
g 2	0.155	0.724	0.163	0.100	0.143	0.257	0.833	0.062	0.578	0.796
g 3	0.659	0.273	0.003	0.202	0.332	0.752	0.010	0.585	0.694	0.201
g 4	0.034	0.918	0.749	0.748	0.643	0.807	0.760	0.520	0.930	0.638
g 5	0.887	0.532	0.091	0.254	0.487	0.380	0.075	0.936	0.730	0.362
g 6	0.630	0.177	0.352	0.638	0.555	0.151	0.765	0.619	0.833	0.593
g 7	0.676	0.243	0.673	0.289	0.066	0.494	0.553	0.277	0.159	0.962
g 8	0.374	0.334	0.619	0.095	0.287	0.831	0.952	0.077	0.802	0.601
g 9	0.591	0.771	0.407	0.473	0.647	0.937	0.804	0.881	0.394	0.524
g 10	0.200	0.761	0.681	0.960	0.734	0.005	0.439	0.355	0.745	0.852
g 11	0.342	0.220	0.025	0.149	0.045	0.500	0.222	0.091	0.899	0.828
g 12	0.794	0.122	0.279	0.311	0.046	0.507	0.714	0.963	0.941	0.879
g 13	0.455	0.083	0.409	0.756	0.268	0.868	0.442	0.008	0.610	0.104
g 14	0.239	0.913	0.990	0.754	0.558	0.971	0.444	0.253	0.674	0.948
g 15	0.332	0.569	0.065	0.956	0.543	0.510	0.842	0.851	0.800	0.307
g 16	0.473	0.817	0.076	0.239	0.054	0.154	0.432	0.582	0.396	0.525
g 17	0.282	0.296	0.448	0.801	0.871	0.672	0.532	0.278	0.575	0.774
g 18	0.778	0.212	0.170	0.624	0.790	0.372	0.468	0.611	0.255	0.017
g 19	0.670	0.397	0.767	0.159	0.909	0.798	0.258	0.080	0.904	0.930
g 20	0.594	0.517	0.078	0.336	0.802	0.077	0.964	0.059	0.751	0.207

TABLE 16.1: Expression data from two groups of subjects: cancer patients and healthy controls. The data are already normalized.

16.2 The problem of multiple comparisons

Let us consider an experiment comparing the gene expression levels in two different conditions such as healthy tissue versus tumor. Let us consider that we have 5 tumor samples, 5 healthy tissue samples, and we are following 20 genes. The data have been pre-processed and normalized in such a way that all values are between 0 and 1. The data can be organized as shown in Table 16.1. The task is to find those genes that are differentially expressed between cancer and healthy subjects.

An examination of Table 16.1 shows that each gene has a number of measurements for both cancer and healthy subjects. A simple approach would be to consider each gene independently and perform a test for means involving two samples, as discussed in Chapter 12. As it was discussed there, this test for means should be preceded by a test for variance. For simplicity, we will assume that the variance of the cancer population is equal to the variance of the healthy population. Using this assumption, we can calculate the value of the t-statistic and the associated p-value. Recall that the p-value is the prob-

Gene	Tumor					Controls					p-value
	T1	T2	T3	T4	T5	C1	C2	C3	C4	C5	
g 1	0.340	0.232	0.760	0.610	0.224	0.238	0.075	0.624	0.978	0.198	0.959
g 2	0.155	0.724	0.163	0.100	0.143	0.257	0.833	0.062	0.578	0.796	0.230
g 3	0.659	0.273	0.003	0.202	0.332	0.752	0.010	0.585	0.694	0.201	0.419
g 4	0.034	0.918	0.749	0.748	0.643	0.807	0.760	0.520	0.930	0.638	0.522
g 5	0.887	0.532	0.091	0.254	0.487	0.380	0.075	0.936	0.730	0.362	0.825
g 6	0.630	0.177	0.352	0.638	0.555	0.151	0.765	0.619	0.833	0.593	0.438
g 7	0.676	0.243	0.673	0.289	0.066	0.494	0.553	0.277	0.159	0.962	0.604
g 8	0.374	0.334	0.619	0.095	0.287	0.831	0.952	0.077	0.802	0.601	0.115
g 9	0.591	0.771	0.407	0.473	0.647	0.937	0.804	0.881	0.394	0.524	0.323
g 10	0.200	0.761	0.681	0.960	0.734	0.005	0.439	0.355	0.745	0.852	0.366
g 11	0.342	0.220	0.025	0.149	0.045	0.500	0.222	0.091	0.899	0.828	0.072
g 12	0.794	0.122	0.279	0.311	0.046	0.507	0.714	0.963	0.941	0.879	0.014
g 13	0.455	0.083	0.409	0.756	0.268	0.868	0.442	0.098	0.619	0.194	0.787
g 14	0.239	0.913	0.990	0.754	0.558	0.971	0.444	0.253	0.674	0.948	0.870
g 15	0.332	0.569	0.065	0.956	0.543	0.510	0.842	0.851	0.800	0.307	0.382
g 16	0.473	0.817	0.076	0.239	0.054	0.154	0.432	0.582	0.396	0.525	0.607
g 17	0.282	0.296	0.448	0.801	0.871	0.672	0.532	0.278	0.575	0.774	0.864
g 18	0.778	0.212	0.170	0.624	0.790	0.372	0.468	0.611	0.255	0.017	0.342
g 19	0.670	0.397	0.767	0.159	0.909	0.798	0.258	0.080	0.904	0.930	0.953
g 20	0.594	0.517	0.078	0.336	0.802	0.077	0.964	0.059	0.751	0.207	0.815

TABLE 16.2: Expression data from two groups of subjects: cancer patients and healthy controls. The last column shows the p-values of a t-test done gene by gene, assuming the populations are normal with equal variance. The null hypothesis that the cancer and healthy measurements are coming from the same distribution can be rejected for genes 11 and 12 at 10% significance level. In other words, genes 11 and 12 differ significantly between cancer and healthy.

ability of rejecting a true null hypothesis or the probability associated with a false positive (a gene that is declared to be differentially expressed although it is not). We do not have any a priori expectations so we will use a two-tailed test. We choose to work at a 10% significance level.

Now that we have clearly formulated our hypotheses and assumptions, we can use R or a simple spreadsheet to quickly generate these p-values.[1] The data together with the computed p-values are shown in Table 16.2. An examination of the last column shows that the null hypothesis that the cancer and healthy measurements are coming from the same distribution can be rejected for genes 11 and 12 at our chosen 10% significance level. In other words, genes 11 and

---

[1]The appropriate Excel function is "ttest(array1, array2, 2, 2)." The third parameter uses the value 2 for a homoscedastic test (equal variance), and the fourth parameter indicates we are performing a two-tailed test.

12 differ significantly between cancer and healthy. Also, gene 8 is really close to the threshold with a p-value of 0.115, so we might want to take a closer look at it, as well. These values came from 10 different subjects, 10 different mRNA preparations, 10 different hybridizations, etc. Let us have a close look at gene 11:

			Tumor					Controls		
Gene	T1	T2	T3	T4	T5	C1	C2	C3	C4	C5
g 11	0.342	0.220	0.025	0.149	0.045	0.500	0.222	0.091	0.899	0.828

The mean of the cancer values is 0.156 and the mean of the healthy values is 0.508.[2] Taking into consideration that the values are all normalized to the (0,1) interval, this is a considerable difference. Furthermore, even the t-test tells us that the change is significant for this gene at a significance level of 10%. There seems to be a lot of statistical evidence that this is a gene that is down-regulated in cancer. Right?

Wrong! In fact, there is little if any evidence that this gene is indeed differentially expressed. The gene may or may not be so, but this conclusion does not follow from these data. The problem here is that multiple comparisons have been done in parallel. The significance level of 5% is the probability of a Type I error that we are prepared to accept. The definition of the significance level tells us that it is likely that we will make one mistake (a false positive) every 20 times we apply the test. Well, it turns out that there are exactly 20 genes in the table. By calculating the p-value from the t-test for each gene, we have, in fact, applied the test 20 times. Sure enough, one gene appeared to have a p-value lower than the threshold of 0.05. In fact, using the 10% threshold, we expect a type I error to occur once in every 10 tests. In this case, we have performed 20 such tests so there should be no surprise that there are two genes (g11 and g12) with p-values lower than 0.1.

In fact, the data in Table 16.1 were obtained from a random number generator as illustrated by the code below:

```
> set.seed(100)
> ngenes = 20
> ncases = 5
> rownames = paste("g",1:ngenes,sep="")
> colnames = c(paste("T",1:ncases,sep=""),
+ paste("C",1:ncases,sep=""))
> l=list(rownames,colnames)
> data = matrix(rnorm(ngenes*2*ncases),nrow=ngenes,
+ ncol=2*ncases,dimnames=l)
> data
```

---

[2]Note that there is no special link between any cancer patient and any control subject. Thus, a regular t-test was used instead of a paired one. A paired t-test would have produced a p-value of 0.006. A paired t-test would have been appropriate for instance if the data had been collected before and after treatment for the same subjects.

	T1	T2	T3	T4	T5
g1	-0.50219235	-0.43808998	-0.10162924	-0.26199577	0.89682227
g2	0.13153117	0.76406062	1.40320349	-0.06884403	-0.04999577
g3	-0.07891709	0.26196129	-1.77677563	-0.37888356	-1.34534931
g4	0.88678481	0.77340460	0.62286739	2.58195893	-1.93121153
g5	0.11697127	-0.81437912	-0.52228335	0.12983414	0.70958158
g6	0.31863009	-0.43845057	1.32223096	-0.71302498	-0.15790503
g7	-0.58179068	-0.72022155	-0.36344033	0.63799424	0.21636787
g8	0.71453271	0.23094453	1.31906574	0.20169159	0.81736208
g9	-0.82525943	-1.15772946	0.04377907	-0.06991695	1.72717575
g10	-0.35986213	0.24707599	-1.87865588	-0.09248988	-0.10377029
g11	0.08988614	-0.09111356	-0.44706218	0.44890327	-0.55712229
g12	0.09627446	1.75737562	-1.73859795	-1.06435567	1.42830143
g13	-0.20163395	-0.13792961	0.17886485	-1.16241932	-0.89295740
g14	0.73984050	-0.11119350	1.89746570	1.64852175	-1.15757124
g15	0.12337950	-0.69001432	-2.27192549	-2.06209602	-0.53029645
g16	-0.02931671	-0.22179423	0.98046414	0.01274972	2.44568276
g17	-0.38885425	0.18290768	-1.39882562	-1.08752835	-0.83249580
g18	0.51085626	0.41732329	1.82487242	0.27053949	0.41351985
g19	-0.91381419	1.06540233	1.38129873	1.00845187	-1.17868314
g20	2.31029682	0.97020202	-0.83885188	-2.07440475	-1.17403476

	C1	C2	C3	C4	C5
g1	-0.3329234	-0.77371335	-0.40842503	-0.242269499	-1.23972284
g2	1.3631137	0.42400240	-2.13649386	0.059031382	0.58987389
g3	-0.4691473	-0.58394698	0.15682192	-0.177271868	0.12401929
g4	0.8428756	0.41503568	0.66004890	0.794680268	-0.52370779
g5	-1.4579937	-1.54526166	-0.98183441	0.006737787	0.62022800
g6	-0.4003059	-0.51874950	-1.11364370	-0.629790293	0.70822158
g7	-0.7764173	-0.27979155	-0.43734768	-0.252489783	-0.09319835
g8	-0.3692965	1.00745738	-0.51611125	-0.690422163	-0.29519670
g9	1.2401015	-0.46956995	0.41899599	0.202542145	-1.08581523
g10	-0.1074338	0.29789704	0.13415544	0.846381438	-0.62481506
g11	0.1725935	-0.41779443	1.03468645	0.632074062	-0.23300654
g12	0.2546013	-0.85038078	1.65350323	0.201413525	-0.25081686
g13	-0.6145338	0.68904619	-0.01794682	-0.091070644	0.95389534
g14	-1.4292151	-0.46019619	-0.02420332	0.289484125	-0.26597251
g15	-0.3309754	1.34818438	0.25024690	-0.054684939	1.89527595
g16	0.1283861	0.44307138	-0.33712454	-2.041849854	-0.42999083
g17	1.0181200	-0.15092619	-0.11335370	0.358369241	1.57554699
g18	-0.2555737	0.45554886	-0.09888291	-0.372600852	0.16194120
g19	-0.3025410	-0.04015468	0.26408682	1.268308841	-1.08545291
g20	1.6151907	0.45612104	0.13898369	2.168600317	0.57693730

```
> labels=c(rep("tumor",ncases),rep("control",ncases))
> mytest = function (x){
+ tumors    = x[labels=="tumor"]
```

```
+ controls = x[labels=="control"]
+ p=t.test( tumors, controls,var.equal=TRUE)$p.value
+ foldchange=mean(tumors)-mean(controls)
+ c(foldchange,p)
+ }
> results=t(apply(data,1,mytest))
> rownames(results)<-rownames(data)
> colnames(results)<-c("logFC","p")
> results<-results[order(results[,2]),]
> results
```

```
          logFC           p
g17 -1.2425105 0.02109769
g15 -1.7077999 0.02664108
g8   0 8294332 0 05353653
g18  0.7093357 0.05961968
g13 -0.6270931 0.13584092
g16  1.0850587 0.13736632
g1   0.5179938 0.13742521
g14  0.9814332 0.16142621
g20 -1.1525251 0.22673566
g10 -0.5467774 0.25337128
g5   0.5955697 0.26985110
g3  -0.4736879 0.28613640
g11 -0.3490123 0.31418421
g6   0.4571497 0.35606553
g7   0.2056308 0.48526338
g2   0.3760856 0.58124677
g19  0.2516817 0.71517331
g4   0.1489743 0.85050648
g9  -0.1176411 0.85820715
g12 -0.1058645 0.89747021
```

This was just a small-scale demonstration of what happens when multiple comparisons are performed without any special care. Applying the t-test to a list of 10,000 genes (a number comparable with the number of genes on the Affymetrix HG95Av2 array) will produce approximatively 500 genes that appear to be differentially expressed even if they are in fact random. Of course, real genes interact in complex ways and their values are not random. However, the example does show it is perfectly possible for tens or hundreds of genes to appear as being significantly differentially expressed even if they are not. Let us perform a similar experiment but on a larger scale as follows.

EXAMPLE 16.1
Generate a matrix of 10,000 × 20 random numbers. Now, let us pretend that

the first 10 columns are measurements from tumors, while the remaining 10
columns are measurements from healthy tissue samples. Use a t-test to cal-
culate a p-value for every gene and sort by p-values. Count the number of
"genes" that have a p-value less than the chosen significance threshold α. The
top $10000 \cdot \alpha$ "genes" should have p-values lower than α.

```
> set.seed(100)
> ngenes = 10000
> ncases = 10
> rownames = paste("g",1:ngenes,sep="")
> colnames = c(paste("T",1:ncases,sep=""),
+ paste("C",1:ncases,sep=""))
> l=list(rownames,colnames)
> data = matrix(rnorm(ngenes*2*ncases),nrow=ngenes,
+ ncol=2*ncases,dimnames=l)
> labels=c(rep("tumor",ncases),rep("control",ncases))
> mytest = function (x){
+ tumors   = x[labels=="tumor"]
+ controls = x[labels=="control"]
+ p=t.test( tumors, controls,var.equal=TRUE)$p.value
+ foldchange=mean(tumors)-mean(controls)
+ c(foldchange,p)
+ }
> results=t(apply(data,1,mytest))
> rownames(results)<-rownames(data)
> colnames(results)<-c("logFC","p")
> results<-results[order(results[,2]),]
> alpha=0.001
> sum(results[,2]<alpha)

[1] 11

> sum(results[,2]<alpha*10)

[1] 94

> sum(results[,2]<alpha*100)

[1] 991
```

Here for obvious reasons we did not show the entire data matrix of 10,000
by 20 elements. However, the number of "genes" that have a raw *p*-value lower
than 0.001 is 11, which is very close to the expected number of $10,000*0.001 =$
10. Similarly, the number of raw *p*-values lower than 0.01 is 94 (compared with
the 100 expected) and the number of those lower than 0.1 is 991 (compared
with the 1,000 expected).

Perhaps an even more convincing example can be made with one's own real data. If you have data involving two groups of samples, merely change the labels of the sample or, equivalently, move them randomly from one group to another. A subsequent t-test performed at a significance level of 5% will always provide approximately 5% "differentially expressed" genes for any labeling and hopefully many more when the real labels are used.

Now that we have established that this approach will produce a list of "differentially expressed" genes no matter what the data look like, let us try to discuss the reasons for this very unpleasant outcome. The typical objection brought by the young statistical mind is that the gene should not care whether it is tested in the presence of the other 9,999 genes or by itself. Thus, this one gene was measured five times, using five different mRNA samples, five different mRNA preparations and five different hybridizations for the cancer. The same was also done for the controls. This gene has exhibited a behavior *consistently different* between the two conditions. The means of the two conditions are considerably different, and even when the variance is taken into consideration by calculating the t-test, the two means appear to be significantly different, as well. It seems that everything is in place and the conclusion cannot be denied. And yet, there is a problem. The problem is related to the fact that this gene *has been selected* from a very, very large number of similar genes. Had this experiment been performed on this one gene alone, the same data would have been very convincing because, it is very, very unlikely that a single gene exhibits such consistently different behavior between two conditions in 10 hybridizations. However, the microarray experiment involved perhaps 10,000 such genes. In such a very large number of genes, the probability that was negligible for any one gene suddenly becomes large enough for things to happen. With so many genes, it will be likely that a few genes are affected by random effects that will make them appear to be consistently lower in a condition and consistently higher in the other condition. "Yes, the young statistician can object, but the gene didn't know that it was hybridizing in the presence of another 9999 genes and, therefore, it could not have used this information in order to alter its values." This is entirely true. This individual gene did not know about the other genes. However, *we did!* This gene was selected based on the very fact that its p-value was low. In performing this selection, we have used the fact that there were many genes. If there had been fewer, we probably would not have been able to find such an outrageous false positive that was affected so consistently by random factors to make it appear significant.

At this point, the balance starts to tip and our young statistician starts to believe that the gene under scrutiny is, indeed, nothing more than a false positive. However, the balance goes all the way and now our young statistician has adopted the opposite point of view: "If this gene cannot be trusted in spite of its consistent behavior across samples, then no individual gene coming from a large microarray can. After all, a gene can hardly be more consistent than having low values for all patients and higher values for all controls." That is not the case. In fact, any individual gene can be trusted as long as it is chosen *before* the experiment and not based on the p-values coming from the experiment. "Hold on!" my young opponent jumps. "What if I had chosen precisely the same gene as before? You know, the one that had a low p-value after selecting it from the many other genes?" Well, this would have been a Type I error and we would have been wrong. However, the probability of a gene exhibiting such consistent behavior due exclusively to random chance is extremely small. In fact, this is similar to a lottery with extremely low odds of winning, only that the outcome is reversed. In the lottery, the rare event is winning. In our experiment, the rare event is being wrong. Picking up a gene from the beginning and then being so unlucky that precisely this particular gene is the one in many thousands that is perturbed by random factors in a mischievous way is similar to being so lucky to win the lottery. However, it is not difficult at all to pick the unlucky gene[3] *after* the experiment in the same way it is not difficult at all to pick the winning lottery number after the draw. No weight should be given to such a gene, the same way no prize is associated to picking the winning lottery ticket after the draw.

16.3   A more precise argument

Let us study this phenomenon using a more rigorous point of view. The significance level α was defined as the acceptable probability of a Type I error. This corresponds to a situation in which the null hypothesis is rejected when it is in fact true. The genes that are called differentially expressed when in fact they are not will be **false positives**.

Let us now think in the terms of hypothesis testing. When the t statistic for a gene is more extreme than the threshold t_α, we will call this gene differentially expressed. However, the gene may be so just due to random effects. This will happen with probability α. If this happens and we call this gene differentially expressed, we will be making an erroneous decision. Therefore, the probability of making a mistake of this kind is exactly α. If we do not make a mistake, we will be drawing the correct conclusion for that given gene. This will happen with probability:

$$\text{Prob}(\text{ correct }) \quad = \quad 1 - p$$

Now we have to take into consideration the fact that there are many such genes. Let us consider there are R such genes. For each of them, we will follow the same reasoning. However, at the end, we would like to draw the correct conclusion from all of them. This means, we have to have the correct conclusion for the first gene AND for the second gene AND ... AND for the last gene. We have seen that the probability of such an event is the multiplication of the

---

[3]It is unlucky because it has a low p-value just by chance.

Number of genes	significance level used for individual genes			
	0.01	0.05	0.1	0.15
10	0.095617925	0.401263061	0.65132156	0.803125596
20	0.182093062	0.641514078	0.878423345	0.961240469
50	0.394993933	0.923055025	0.994846225	0.999704235
100	0.633967659	0.994079471	0.999973439	0.999999913
500	0.993429517	1	1	1
1000	0.999956829	1	1	1
5000	1	1	1	1
10000	1	1	1	1

TABLE 16.3: The probability of making a Type I error (at least one false positive) in a multiple comparison situation. An array with as few as 20 genes has a probability of 87.84% of having at least one false positive if the gene level test is performed at a gene significance value of 0.1. For an array with 100 genes, the same probability becomes 99.99%.

probabilities corresponding to the individual events (see Chapter 8). Therefore, the probability of drawing the correct conclusion from all experiments is:

$$\text{Prob(globally correct)} \quad = \quad (1-p) \cdot (1-p) \cdots (1-p) = (1-p)^R$$

We can now calculate the probability of being wrong somewhere. This would be 1 minus the probability of being correct in all experiments:

$$\text{Prob(wrong somewhere)} \quad = \quad 1 - \text{Prob(globally correct)} \quad = 1 - (1-p)^R$$

In this situation, being wrong means drawing the wrong conclusion for at least one gene. This is in fact the α value for the whole experiment. Table 16.3 shows the values of this probability for various significance levels and various sizes of the array. An array with as few as 20 genes has a probability of 87.84% of having at least one false positive if the gene level test is performed at a gene significance value of 0.1. For an array with 100 genes, the same probability becomes 99.99%. Although this is worrisome, the table does not paint the whole picture. After all, having a false positive from time to time may be deemed to be acceptable. Microarrays cannot be trusted completely anyway.[4] One might argue that any gene found as differentially expressed using microarrays should be confirmed with alternative assays such as quantitative real-time polymerase chain reactions (Q-RT-PCR) and further biological experiments. The question then becomes how many such false positives are expected for a given array size and gene level significance? Table 16.4 shows these numbers. For instance, a small array with 5,000 genes on which the gene-level analysis is performed at

---

[4]First, individual gene hybridizations are inherently unreliable [276]. Second, they only reflect the phenomena at the mRNA level and completely ignore the translation and post-translational modifications.

	Gene significance level			
Number of genes	0.01	0.05	0.1	0.15
10	< 1	< 1	1	1.5
20	< 1	1	2	3
50	< 1	2.5	5	7.5
100	1	5	10	15
500	5	25	50	75
1000	10	50	100	150
5000	50	250	500	750
10000	100	500	1000	1500

TABLE 16.4: The expected number of false positives for a given gene significance level and size of the array if no correction for multiple comparison is performed.

0.05 significance level is expected to produce about 250 false positives mixed up with whatever true positives there are in the given condition. For such numbers, performing alternative assays in order to sort out the true positives from the false positives is not an option anymore.

An experiment involving multiple comparisons is a good example of a situation in which small quantitative changes accumulate until a qualitative change occurs. In the multiple comparison, a hypothesis testing approach that was perfectly valid for a single test (e.g., any one gene chosen before the experiment is performed) or a small number of such tests, becomes inadequate for analyzing data coming from large arrays. Our task is to control the global or experiment level significance level. This is the probability of having a Type I error anywhere. This probability is also known as the **family-wise error rate** (FWER).

16.4 Corrections for multiple comparisons

16.4.1 The Šidák correction

After discussing the problem both from an intuitive perspective and from a statistical one, let us try to focus on how we can address the problem. The issue here is that we would like to control the overall probability of making a Type I error. This probability is equal to the probability of making at least one such mistake, calculated above:

$$\text{Prob(wrong somewhere)} \quad = \quad 1 - (1-p)^R$$

This can be rewritten as:

$$\alpha_e = 1 - (1 - \alpha_c)^R \qquad (16.1)$$

where α_e is the probability of a Type I error at the experiment level and α_c is the probability of a Type I error at the gene level (single comparison). The task is to calculate the α level that we need to use for individual genes (α_c) in order to ensure that the global, or experiment-level Type I error is less or equal to α_e, which was chosen. Using simple algebraic manipulations, we can extract α_c from the equation above:

$$\alpha_c = 1 - \sqrt[R]{1 - \alpha_e} \qquad (16.2)$$

This is the so-called **Šidák correction** for multiple comparisons [439].

16.4.2 The Bonferroni correction

Bonferroni [60, 61] noted that for small p, Eq. 16.1 can be approximated by taking only the first two terms of the binomial expansion of $(1 - p)^R$:

$$\alpha_e = 1 - (1 - \alpha_c)^R = 1 - (1 - R \cdot \alpha_c + \cdots) \approx R \cdot \alpha_c \qquad (16.3)$$

Using this approximation, we can calculate the experiment level α_c value as:

$$\alpha_e = \alpha_c \cdot R \Rightarrow \alpha_c = \frac{\alpha_e}{R} \qquad (16.4)$$

This is the **Bonferroni correction** for multiple comparisons. This is a very simple formula but it is only an approximation of the exact value given by Eq. 16.2. Bonferroni starts to depart from the exact values even for as few as 20 genes (see Table 16.5).

However, this is the least of our problems. Unfortunately, both Bonferroni and Šidák corrections are unsuitable for gene expression analysis because for large number of genes R, the required significance at the gene level becomes very small, very quickly. Table 16.5 shows the significance levels that need to be used at the individual gene level in order to ensure an overall significance level of 0.05. For arrays involving more than 1,000 genes, the technology is simply not able to provide values precise enough such that the genes will appear significant at those levels. At such stringent significance levels, the hypothesis testing approach will not be able to reject the null hypothesis for many genes. It is said that Bonferroni and Šidák are conservative methods in the sense that if a gene is significant after either Bonferroni or Šidák adjustments,[5] then the gene is truly different between the groups. However, if a gene is not significant according to these adjustments, then it may still be truly different. In other words, Bonferroni and Šidák are sufficient but not necessary conditions.

---

[5] Between the two of them, Bonferroni is slightly more conservative than Šidák (see Table 16.5).

Genes	Šidák	Bonferroni
1	0.05	0.05
10	0.005116197	0.005
20	0.002561379	0.0025
100	0.000512801	0.0005
1000	0.000051292	0.00005
5000	0.0000102586	0.00001
10000	0.00000512932	0.000005
20000	0.00000256466	0.0000025

TABLE 16.5: The significance levels that need to be used at individual gene level in order to ensure an overall significance level of 0.05. Both Šidák and Bonferroni corrections require that tests at gene level be performed with extremely high significance, which is unfeasible in gene expression experiments.

16.4.3 Holm's step-wise correction

A family of methods that allow less conservative adjustments of the p-values is the Holm step-down group of methods [211, 215, 217, 381]. These methods order the genes in increasing order of their p-value and make successive smaller adjustments.

Let us consider we have a set of R genes. Each gene is measured in two groups, for example, patients and controls. For a given gene, g_i, the null hypothesis is that the mean of the values of gene i measured in controls is the same as the mean of the values measured in patients, i.e., $H_i: \mu_{ic} = \mu_{ip}$. For each gene, we will use an independent test statistic Y_i (e.g. a t-test between the patient's group and the control's group) to generate a p_i value. The p_i value will be the probability of the corresponding test statistic to have the observed value just by chance, that is, when the null hypothesis is true. Holm's step-wise correction proceeds as follows:

Holm's step-wise correction procedure:

1. Choose the experiment-level significance level α_e.

2. Order the genes in the increasing order of individual p-values:

Genes	g_{i_1}	g_{i_2}	$\cdots$	g_{i_k}	$\cdots$	g_{i_R}
Increasing p-values	p_1	p_2	$\cdots$	p_k	$\cdots$	p_R

3. Compare the p-values of each gene with a threshold that depends on the position of the gene in the list of ordered values. The thresholds are as follows: $\frac{\alpha_e}{R}$ for the first gene, $\frac{\alpha_e}{R-1}$ for the second gene, etc.

Genes	g_{i_1}	g_{i_2}	$\cdots$	g_{i_k}	$\cdots$	g_{i_R}
p-values	p_1	p_2	$\cdots$	p_k	$\cdots$	p_R
Test	$p1 < \frac{\alpha_e}{R}$	$p_2 < \frac{\alpha_e}{R-1}$	$\cdots$	$p_k < \frac{\alpha_e}{R-k+1}$	$\cdots$	$p_R < \frac{\alpha_e}{1}$

4. Let k be the largest i for which $p_i < \frac{\alpha_e}{R-i+1}$. Reject the null hypotheses H_i for $i-1,2,\ldots,k$. These genes are indeed different between the two groups at chosen α_e significance level.

Now it should be clear why Holm's procedure is called step-wise. Unlike Šidák and Bonferroni, where the corrected threshold was unique for all genes and calculated in a single step procedure, the Holm's thresholds are different for every gene and they depend on their order in the ordered list of uncorrected p-values.

Note that the correction procedure in itself does not bring any additional information about the relative significance of one gene versus another gene. All that the procedure does is to adjust all values such that they reflect what happens at the global level of the entire set of tests. Since no additional information about any of the genes is brought into play here, the correction procedure should not change the ranking of the genes. If two genes g_i and g_j are such that g_i is more significant than g_j, they should remain in this relative order of significance even after the correction. Sometimes however, the correction formula yields a corrected value for g_j that is more significant than the corrected value for g_i. If this were taken **ad literam**, it would change the significance order of the two genes. In practice, when this happens, g_j will be given the same corrected p-value as g_i and be maintained in the same relative order. This is why sometimes several genes share the same corrected p-values. This phenomenon will be illustrated in Section 16.5.

16.4.4 The false discovery rate (FDR)

Bonferroni, Šidák, and Holm's step-down adjustment are statistical procedures that assume the variables are independent. However, the genes of an organism are actually known to be involved in complex dependencies and regulatory mechanisms [116, 117]. The False Discovery Rate (FDR) correction procedure was initially proven for independent variables [45] but was recently extended to allow for some dependencies [46].

The FDR procedure adjusts the p-values in a manner similar to Holm's. The genes are ordered in increasing order of the p-values provided by the individual independent tests. However, the threshold for the i-th p-value will be: $p_i < \frac{i}{n} \frac{\alpha_e}{p_0}$ where α_e is the chosen experiment-level significance level and p_0 is the proportion of the null hypotheses H_i that are actually true. Since this proportion is not known (if we knew which of the genes are not differentially expressed we wouldn't need to do this), the p_0 can be conservatively estimated as being 1. This assumes that all null hypotheses are actually true and there are no differentially expressed genes. The conservative aspect means that if the null hypothesis can be rejected for any particular gene in these circumstances, then the null hypothesis will still be rejected if some of the null hypotheses are actually false and $p_0 < 1$.

False Discovery Rate (FDR) correction procedure:

1. Choose the experiment-level significance level α_e.

2. Order the genes in the increasing order of individual p-values:

Genes	g_{i_1}	g_{i_2}	$\cdots$	g_{i_k}	$\cdots$	g_{i_R}
Increasing p-values	p_1	p_2	$\cdots$	p_k	$\cdots$	p_R

3. Compare the p-values of each gene with a threshold that depends on the position of the gene in the list of ordered values. The thresholds are as follows: $\frac{1}{R}\alpha_e$ for the first gene, $\frac{2}{R}\alpha_e$ for the second gene, etc.

Genes	g_{i_1}	g_{i_2}	$\cdots$	g_{i_k}	$\cdots$	g_{i_R}
p-values	p_1	p_2	$\cdots$	p_k	$\cdots$	p_R
Test	$p_1 < \frac{1}{R}\alpha_e$	$p_2 < \frac{2}{R}\alpha_e$	$\cdots$	$p_k < \frac{k}{R}\alpha_e$	$\cdots$	$p_R < \alpha_e$

4. Let k be the largest i for which $p_i < \frac{i}{R}\alpha_e$. Reject the null hypotheses H_i for $i = 1, 2, \ldots, k$. These genes are indeed different between the two groups at the chosen α_e significance level.

Note that in this procedure, we are looking for the *largest* index i for which the given condition holds. In order to do this, we can test the individual genes backwards, from the largest p-value to the smallest one. If we do so, the first gene for which the relationship holds will have the largest index sought by the procedure. If we start from the most significant p-value, p_1, theoretically we will have to test the condition above for all genes, since finding one that fails the test does not guarantee that all other p-values with larger indices will also fail the test.

16.4.5 Permutation correction

The Westfall and Young (W-Y) step-down correction [457] is a more general method that adjusts the p-value while taking into consideration the possible correlations.

Let us consider the following data set.

Gene	Tumor					Controls					t
	T1	T2	T3	T4	T5	C1	C2	C3	C4	C5	
g 1	0.340	0.232	0.760	0.610	0.224	0.238	0.075	0.624	0.978	0.198	t_1
g 2	0.155	0.724	0.163	0.100	0.143	0.257	0.833	0.062	0.578	0.796	t_2
g 3	0.659	0.273	0.003	0.202	0.332	0.752	0.010	0.585	0.694	0.201	t_3
g 4	0.034	0.918	0.749	0.748	0.643	0.807	0.760	0.520	0.930	0.638	t_4
g 5	0.887	0.532	0.091	0.254	0.487	0.380	0.075	0.936	0.730	0.362	t_4
$\vdots$	$\vdots$	$\vdots$	$\vdots$	$\vdots$	$\vdots$	$\vdots$	$\vdots$	$\vdots$	$\vdots$	$\vdots$	$\vdots$

This procedure starts by changing the measurements randomly between the patient's and control's groups. Alternatively, the same result can be achieved by randomly assigning the "patient" and "control" labels to the various measurements. A first such permutation may be:

Gene	C5	C3	T3	C1	T5	T2	C2	T1	T4	C4	t
g 1	0.340	0.232	0.760	0.610	0.224	0.238	0.075	0.624	0.978	0.198	t_{11}
g 2	0.155	0.724	0.163	0.100	0.143	0.257	0.833	0.062	0.578	0.796	t_{12}
g 3	0.659	0.273	0.003	0.202	0.332	0.752	0.010	0.585	0.694	0.201	t_{13}
g 4	0.034	0.918	0.749	0.748	0.643	0.807	0.760	0.520	0.930	0.638	t_{14}
g 5	0.887	0.532	0.091	0.254	0.487	0.380	0.075	0.936	0.730	0.362	t_{15}
⋮	⋮	⋮	⋮	⋮	⋮	⋮	⋮	⋮	⋮	⋮	⋮

New p-values are calculated using the chosen gene level test (e.g. t-test) for this permutation and the values are corrected for multiple experiments using Holm's step-down method discussed above. Then, a new permutation is done and new p-values resulting from this permutation are calculated:

Gene	C3	C5	C1	T5	T3	T4	C2	T1	T2	C4	t
g 1	0.340	0.232	0.760	0.610	0.224	0.238	0.075	0.624	0.978	0.198	t_{21}
g 2	0.155	0.724	0.163	0.100	0.143	0.257	0.833	0.062	0.578	0.796	t_{22}
g 3	0.659	0.273	0.003	0.202	0.332	0.752	0.010	0.585	0.694	0.201	t_{23}
g 4	0.034	0.918	0.749	0.748	0.643	0.807	0.760	0.520	0.930	0.638	t_{24}
g 5	0.887	0.532	0.091	0.254	0.487	0.380	0.075	0.936	0.730	0.362	t_{25}
⋮	⋮	⋮	⋮	⋮	⋮	⋮	⋮	⋮	⋮	⋮	⋮

This whole process (random labeling + testing) is repeated thousands or tens of thousands of times. Finally, the p-value for a gene i will be the proportion of times the value of t calculated for the real labels t_i is less or equal to the value of t calculated for a random permutation:

$$p\text{-value for gene } i: \quad \frac{\text{number of permutations for which } u_j^{(b)} \geq t_i}{\text{total number of permutations}}$$

where $u_j^{(b)}$ are the values corrected as in Holm's step-down method for permutation b. More details and an example of applying this method to microarray data can be found in [138].

The main important advantage of the W-Y approach is that it fully takes into consideration all dependencies between genes. This is extremely important for tightly correlated genes such as those being involved in the same pathways. Disadvantages include the fact that it is an empirical process lacking the elegance of a more theoretical approach. Also, the label permutation process is computationally intensive and, therefore, inherently slow.

This method is a refinement of a more general approach known as bootstrapping [155, 257, 457]. The method samples with replacement the pool of observations to create new data sets and calculates p-values for all tests. For each data set, the minimum p-value on the resampled data sets is compared with the p-value on the original test. The adjusted p-value will be the proportion of resampled data where the minimum pseudo-p-value is less than or equal to an actual p-value. Bootstrap used with sampling without replacement is known as the permutation method [74, 208].

16.4.6 Significance analysis of microarrays (SAM)

Tusher et al. have reported that the step-down adjustment method of Westfall and Young was still too stringent for their microarray data [427]. In response to this, they have developed another method called **significance analysis of microarrays** (SAM). SAM assigns a score to each gene taking into consideration the relative change of each gene expression level with respect to the standard deviation of repeated measurements. The basic statistic used is similar to that of the t-test used in Chapter 12, section 12.3.2.1. The basic idea of a t-test is to calculate a difference between means divided by an estimate of the standard deviation. In other words, the purpose is to express the difference between means in units of standard deviations. How exactly the estimate of the standard deviation is calculated depends on whether the two populations are known (or assumed) to have the same variance, etc. SAM calculates the following statistic which is very similar to the t-statistic discussed in Chapter 12:

$$d_i = \frac{\bar{x}_{i1} - \bar{x}_{i2}}{s_i + s_0} \tag{16.5}$$

This is, again, a difference between means over a standard deviation. The second term of the denominator, s_0, is a "fudge" term. Its purpose is to prevent the computed statistic d_i from becoming too large when the estimated variance s_i is close to zero. The estimated variance is calculated based on the specific problem. In many cases, the assumption of equal variance is reasonable and the pooled variance is calculated as for the t-test involving two samples, equal variance in section 12.3.2.1:

$$s_i = \sqrt{\frac{(n_1 - 1) \cdot s_1^2 + (n_2 - 1) \cdot s_2^2}{n_1 + n_2 - 2} \left(\frac{1}{n_1} + \frac{1}{n_2} \right)} \tag{16.6}$$

Note that this variance is exactly the same as the one used in Eq. 12.39. If the assumptions of the experiment change, the test statistic above can be modified accordingly. However, the general approach remains the same.

Note that SAM calculates a gene-by-gene variance which will allow for the selection of the appropriate genes independently of their expression levels. However, the issue of multiple comparisons still remains since many thousands of genes are analyzed at the same time. SAM uses the same permutation idea to estimate the percentage of genes identified just by chance (the false discovery rate). Many permutations of the labels are done. For each permutation

i, the value of the test statistic t_i is calculated. Each such t_i is actually an observation when the null hypothesis is true (the labels are actually random). These values are used to construct an empirical distribution for t_i values. In practice, the algorithm is as follows:

False discovery rate in SAM:

1. Fix a threshold for differentially expressed genes

2. Count how many genes are reported as differentially expressed in each permutation (false positives)

3. Calculate the median number of false positives across all permutations

4. Calculate FDR as the number of false positives divided by the number of genes in the original data.

SAM is available (freely for nonprofit use) as an Excel macro that can be downloaded from: http://www-stat.stanford.edu/ tibs/SAM/index.html.

16.4.7   On permutation-based methods

A word of caution should be added for all methods using permutations. In order for these methods to work, a large number of random permutations is necessary. However, the total number of distinct permutations is limited by the number of samples in each group. For instance, if there are only three patients and three controls, there are only

$$C_6^3 = \binom{6}{3} = \frac{6 \cdot 5 \cdot 4}{3 \cdot 2 \cdot 1} = 20$$

distinct permutations, which is completely insufficient in order to construct a good re-sampling distribution. If the number of patients and controls is increased to six each, the number of distinct permutation becomes:

$$C_{12}^6 = \binom{12}{6} = \frac{12 \cdot 11 \cdot 10 \cdot 9 \cdot 8 \cdot 7}{6 \cdot 5 \cdot 4 \cdot 3 \cdot 2 \cdot 1} = 924$$

which is more acceptable. In general, the number of permutations should be at least 1,000 or so. One order of magnitude more permutations ($\approx 10,000$) will make the results much more trustworthy.

16.5   Corrections for multiple comparisons in R

R has implemented many p-value adjustment methods. The main idea is to start with a vector of raw p-values and generate a new vector with corrected

p-values and use the desired threshold on the corrected p-value. This method of dealing with multiple testing does not apply to SAM or Westfall and Young step-down correction methods.

The main function in R used to obtain corrected p-value from raw p-value is `p.adjust`. The methods available are:

```
> p.adjust.methods
```

```
[1] "holm"        "hochberg"    "hommel"      "bonferroni"
[5] "BH"          "BY"          "fdr"         "none"
```

As an example, consider that the p-values for 10 genes were computed using a two-group t-test:

```
> rawp=c(0.1,0.8,0.05,0.001,0.2,0.011,0.012,0.5,0.89,0.9)
> names(rawp)<-paste("gene",1:length(rawp),sep="")
> rawp
```

```
 gene1   gene2   gene3   gene4   gene5   gene6   gene7   gene8   gene9
 0.100   0.800   0.050   0.001   0.200   0.011   0.012   0.500   0.890
gene10
 0.900
```

If we want to select differentially expressed genes so that the false discovery rate is about 10%, we can compute the FDR corrected p-values and chose those genes with the adjusted p-values less than 0.1.

```
> adjustedp=p.adjust(rawp,method="fdr")
> adjustedp[adjustedp<0.1]
```

```
gene4 gene6 gene7
 0.01  0.04  0.04
```

Note that both genes 6 and 7 got the same corrected p-value 0.04 even though their nominal p-values are not identical. Other correction methods can be used by changing the "method" argument in the p.adjust function call to: "holm," "hochberg," "hommel," "bonferroni," "BH," or "BY." Details on algorithms to which these options correspond can be found in the help of the p.adjust function.

Armed with this, let us revisit the example above in which several out of the 10 random genes appeared to be differentially expressed:

```
> set.seed(100)
> ngenes = 20
> ncases = 5
> rownames = paste("g",1:ngenes,sep="")
> colnames = c(paste("T",1:ncases,sep=""),
```

```
+ paste("C",1:ncases,sep=""))
> l=list(rownames,colnames)
> data = matrix(rnorm(ngenes*2*ncases),nrow=ngenes,
+ ncol=2*ncases,dimnames=l)
> data
```

	T1	T2	T3	T4	T5
g1	-0.50219235	-0.43808998	-0.10162924	-0.26199577	0.89682227
g2	0.13153117	0.76406062	1.40320349	-0.06884403	-0.04999577
g3	-0.07891709	0.26196129	-1.77677563	-0.37888356	-1.34534931
g4	0.88678481	0.77340460	0.62286739	2.58195893	-1.93121153
g5	0.11697127	-0.81437912	-0.52228335	0.12983414	0.70958158
g6	0.31863009	-0.43845057	1.32223096	-0.71302498	-0.15790503
g7	-0.58179068	-0.72022155	-0.36344033	0.63799424	0.21636787
g8	0.71453271	0.23094453	1.31906574	0.20169159	0.81736208
g9	-0.82525943	-1.15772946	0.04377907	-0.06991695	1.72171575
g10	-0.35986213	0.24707599	-1.87865588	-0.09248988	-0.10377029
g11	0.08988614	-0.09111356	-0.44706218	0.44890327	-0.55712229
g12	0.09627446	1.75737562	-1.73859795	-1.06435567	1.42830143
g13	-0.20163395	-0.13792961	0.17886485	-1.16241932	-0.89295740
g14	0.73984050	-0.11119350	1.89746570	1.64852175	-1.15757124
g15	0.12337950	-0.69001432	-2.27192549	-2.06209602	-0.53029645
g16	-0.02931671	-0.22179423	0.98046414	0.01274972	2.44568276
g17	-0.38885425	0.18290768	-1.39882562	-1.08752835	-0.83249580
g18	0.51085626	0.41732329	1.82487242	0.27053949	0.41351985
g19	-0.91381419	1.06540233	1.38129873	1.00845187	-1.17868314
g20	2.31029682	0.97020202	-0.83885188	-2.07440475	-1.17403476

	C1	C2	C3	C4	C5
g1	-0.3329234	-0.77371335	-0.40842503	-0.242269499	-1.23972284
g2	1.3631137	0.42400240	-2.13649386	0.059031382	0.58987389
g3	-0.4691473	-0.58394698	0.15682192	-0.177271868	0.12401929
g4	0.8428756	0.41503568	0.66004890	0.794680268	-0.52370779
g5	-1.4579937	-1.54526166	-0.98183441	0.006737787	0.62022800
g6	-0.4003059	-0.51874950	-1.11364370	-0.629790293	0.70822158
g7	-0.7764173	-0.27979155	-0.43734768	-0.252489783	-0.09319835
g8	-0.3692965	1.00745738	-0.51611125	-0.690422163	-0.29519670
g9	1.2401015	-0.46956995	0.41899599	0.202542145	-1.08581523
g10	-0.1074338	0.29789704	0.13415544	0.846381438	-0.62481506
g11	0.1725935	-0.41779443	1.03468645	0.632074062	-0.23300654
g12	0.2546013	-0.85038078	1.65350323	0.201413525	-0.25081686
g13	-0.6145338	0.68904619	-0.01794682	-0.091070644	0.95389534
g14	-1.4292151	-0.46019619	-0.02420332	0.289484125	-0.26597251
g15	-0.3309754	1.34818438	0.25024690	-0.054684939	1.89527595
g16	0.1283861	0.44307138	-0.33712454	-2.041849854	-0.42999083
g17	1.0181200	-0.15092619	-0.11335370	0.358369241	1.57554699
g18	-0.2555737	0.45554886	-0.09888291	-0.372600852	0.16194120

```
g19 -0.3025410 -0.04015468  0.26408682  1.268308841 -1.08545291
g20  1.6151907  0.45612104  0.13898369  2.168600317  0.57693730

> labels=c(rep("tumor",ncases),rep("control",ncases))
> mytest = function (x){
+ tumors   = x[labels=="tumor"]
+ controls = x[labels=="control"]
+ p=t.test( tumors, controls,var.equal=TRUE)$p.value
+ foldchange=mean(tumors)-mean(controls)
+ c(foldchange,p)
+ }
> results=t(apply(data,1,mytest))
> rownames(results)<-rownames(data)
> colnames(results)<-c("logFC","p")
> results<-results[order(results[,2]),]
> results

          logFC           p
g17 -1.2425105 0.02109769
g15 -1.7077999 0.02664108
g8   0.8294332 0.05353653
g18  0.7093357 0.05961968
g13 -0.6270931 0.13584092
g16  1.0850587 0.13736632
g1   0.5179938 0.13742521
g14  0.9814332 0.16142621
g20 -1.1525251 0.22673566
g10 -0.5467774 0.25337128
g5   0.5955697 0.26985110
g3  -0.4736879 0.28613640
g11 -0.3490123 0.31418421
g6   0.4571497 0.35606553
g7   0.2056308 0.48526338
g2   0.3760856 0.58124677
g19  0.2516817 0.71517331
g4   0.1489743 0.85050648
g9  -0.1176411 0.85820715
g12 -0.1058645 0.89747021

> adjustedp=p.adjust(results[,2],method="fdr")
> results=cbind(results,adjustedp)
> results

          logFC           p adjustedp
g17 -1.2425105 0.02109769 0.2664108
g15 -1.7077999 0.02664108 0.2664108
```

```
g8    0.8294332 0.05353653 0.2980984
g18   0.7093357 0.05961968 0.2980984
g13  -0.6270931 0.13584092 0.3926434
g16   1.0850587 0.13736632 0.3926434
g1    0.5179938 0.13742521 0.3926434
g14   0.9814332 0.16142621 0.4035655
g20  -1.1525251 0.22673566 0.4768940
g10  -0.5467774 0.25337128 0.4768940
g5    0.5955697 0.26985110 0.4768940
g3   -0.4736879 0.28613640 0.4768940
g11  -0.3490123 0.31418421 0.4833603
g6    0.4571497 0.35606553 0.5086650
g7    0.2056308 0.48526338 0.6470178
g2    0.3760856 0.58124677 0.7265585
g19   0.2010017 0.71517331 0.8413904
g4    0.1489743 0.85050648 0.8974702
g9   -0.1176411 0.85820715 0.8974702
g12  -0.1058645 0.89747021 0.8974702
```

Note that now, the *p*-values corrected for multiple comparisons correctly show that none of the random genes is significantly different between the two groups.

The `multtest` package of Bioconductor also contains a function that performs *p*-value adjustments:

```
> require(multtest)
> mymethods <- c("Bonferroni", "Holm", "Hochberg",
+ "SidakSS", "SidakSD", "BH", "BY")
> res <- mt.rawp2adjp(rawp, mymethods)
> res$adjp
```

	rawp	Bonferroni	Holm	Hochberg	SidakSS	SidakSD
[1,]	0.001	0.01	0.010	0.010	0.00995512	0.00995512
[2,]	0.011	0.11	0.099	0.096	0.10471169	0.09475398
[3,]	0.012	0.12	0.099	0.096	0.11372307	0.09475398
[4,]	0.050	0.50	0.350	0.350	0.40126306	0.30166270
[5,]	0.100	1.00	0.600	0.600	0.65132156	0.46855900
[6,]	0.200	1.00	1.000	0.900	0.89262582	0.67232000
[7,]	0.500	1.00	1.000	0.900	0.99902344	0.93750000
[8,]	0.800	1.00	1.000	0.900	0.99999990	0.99200000
[9,]	0.890	1.00	1.000	0.900	1.00000000	0.99200000
[10,]	0.900	1.00	1.000	0.900	1.00000000	0.99200000

	BH	BY
[1,]	0.0100000	0.02928968
[2,]	0.0400000	0.11715873
[3,]	0.0400000	0.11715873

```
 [4,]  0.1250000 0.36612103
 [5,]  0.2000000 0.58579365
 [6,]  0.3333333 0.97632275
 [7,]  0.7142857 1.00000000
 [8,]  0.9000000 1.00000000
 [9,]  0.9000000 1.00000000
[10,]  0.9000000 1.00000000
```

To illustrate the use of this package for computing the Westfall and Young step-down multiple testing procedure [457], we will use the Golub data set contained in this package. This data set contains 3,051 genes, 38 samples, and 2 groups: group 0 with 27 genes and group 1 with 11 genes.

The function `mt.maxT` takes as argument the raw gene expression data matrix, called "golub," a vector with the class level of the samples (columns) called golub.cl and the number of permutations (B=1000) used in the algorithm.

```
> data(golub)
> golub[1:5,1:5]
> dim(golub)
> table(golub.cl)
> getOption("width")
> options(width=40)
> resT <- mt.maxT(X=golub, classlabel=golub.cl, B = 1000)
> head(resT)
```

The result from this function includes the row number for each gene (the column index) and the adjusted p-value (column adjp).

16.6   Summary

This chapter discussed the general issue of correcting for multiple comparisons. An example and an informal discussion introduced the problem. The Šidák and Bonferroni corrections were presented as single-step methods. Both methods are conservative, and, therefore, they can be used only when few multiple comparisons are used. The two methods are not appropriate for microarray data when many thousands of genes are compared since, in these cases, the adjusted p-values are too small and the methods yield many false negatives. Holm's correction method is less conservative but assumes the variables are independent. In many cases, this assumption does not hold for microarray experiments involving interacting genes. The false discovery rate (FDR) is able to cope with some degree of interaction and is computationally efficient. Bootstrapping-based methods such as permutation (Westfall and Young) and

SAM are very powerful inasmuch they take into account any dependencies and correlations between variables but can be very computationally intensive. Significance Analysis of Microarrays is another method recently proposed for selecting differentially expressed genes while also correcting for multiple experiments. The chapter concludes with a section discussing various procedures for multiple comparison corrections available in R.

Chapter 17

Analysis and visualization tools

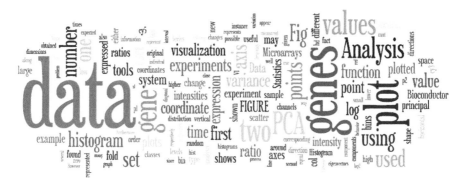

Here is the answer that I will give to President Roosevelt.... Give us the tools and we will finish the job.

—*Sir Winston Churchill, British Prime Minister, Radio Broadcast, 9 Feb. 1941*

17.1   Introduction

This chapter will present briefly a few basic analysis and visualization tools used in gene expression analysis. All tools with the exception of gene pies are equally effective in the analysis of both Affymetrix and cDNA data. Gene pies are most useful for the visualizing of two-channel cDNA array data.

17.2   Box plots

A **box plot** is a plot that represents graphically several descriptive statistics[1] of a given data sample. The box plot usually has a box including a central line and two tails (see Fig. 17.1). The central line in the box shows the position of the median (the value located halfway between the largest and smallest data value). The upper and lower boundaries of the box show the location of the **upper quartile** (UQ) and **lower quartile** (LQ), respectively. The upper and lower quartiles are the 75th and 25th percentiles, respectively. Thus, the box will represent the interval that contains the central 50% of the data. The interval between the upper and the lower quartiles is called the **interquartile distance** (IQD). The length of the tails is usually $1.5 \cdot IQD$. Data points that fall beyond $UG + 1.5 \cdot IQD$ or $LQ - 1.5 \cdot IQD$ are considered outliers.

Box plots can be easily created in R using the function `boxplot`. Assume we have a vector x containing the data, we can use the following code to draw a box plot:

```
> set.seed(10)
> x=rnorm(2000)
> boxplot(x,col="yellow")
```

The result of this call is shown in Fig. 17.1. In this case, the data was drawn from a normal distribution.

Sometimes it is useful to compare side by side the distribution of several sets of values in separate box plots. One way to obtain such a representation is to enumerate each set one after the other in the box plot function call and specify names for each of them.

```
> set.seed(10)
> x1=rnorm(100)
> x2=rnorm(2000,sd=0.5, mean=-1)
> x3=rnorm(200,sd=0.2, mean=1)
> boxplot(x1,x2,x3,names=c("x1","x2","x3"),col=c("red","purple",
+ "magenta"))
```

Fig. 17.2 shows the three box plots side by side. Note how the number of outliers increases with the number of points drawn from the normal distribution: no outliers for the data set with 10 values, only 2 outliers for the data set with 200 values, and multiple outliers for the data set with 2,000 values. Also, note how the different colors for each box plot were specified in an array.

On many occasions, the data may be available in a data frame with in which interesting subsets are defined by the level of a given factor. In this

---

[1] See Chapter 8 for a discussion of several descriptive statistics, including mean, median, percentiles, etc.

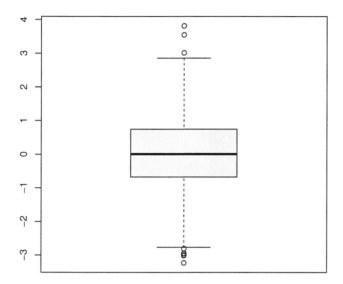

FIGURE 17.1: A box plot in R. The central line in the box shows the position of the median (the value located halfway between the largest and smallest data value). For these settings, the upper and lower boundaries of the box show the location of the upper quartile (UQ) and lower quartile (LQ), respectively. The upper and lower quartiles are the 75th and 25th percentiles, respectively. The data outside the ends of the tails are considered outliers.

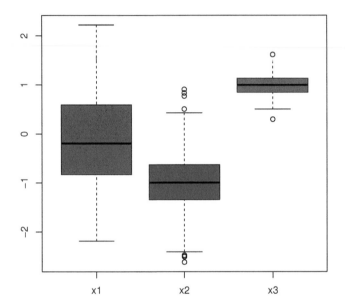

FIGURE 17.2: Multiple box plots in R. Drawing box plots of several data sets next to each other allows a quick visual comparison between these data sets.

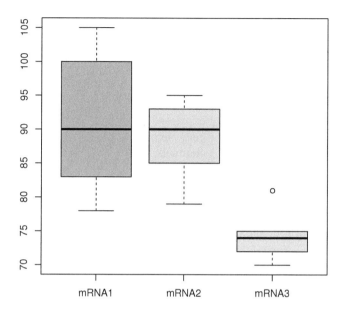

FIGURE 17.3: Box plot of a data frame showing the data grouped by the various levels of a given factor.

case, the call to the function `boxplot` can be made in the same style as for a linear model, in which a $\sim$ sign separates the name of the variable containing the numerical values (at the left) and the name of the factor that defines the different groups (subsets) of values (at the right). The name of the data frame containing both the values and the group factor is provided via the "data" argument of the function `boxplot`:

```
> mydat=data.frame(y=c(100,90,105,83,78,95,93,79,85,
+ 90,70,72,81,74,75),Group=factor(rep(c("mRNA1","mRNA2",
+ "mRNA3"),each=5)))
> boxplot(y~Group,data=mydat,col=c("darksalmon","deepskyblue1",
+ "darkseagreen"))
```

The result is shown in Fig. 17.3.

Intensities		Magnitude of cy3/cy5 ratio	Biological meaning	Interesting gene
cy3	cy5			
low	low	meaningless; can be anything	the gene is not expressed in either condition	no
low	high	meaningless; close to zero	gene expressed on cy5, not expressed on cy3	yes
high	low	meaningless; very large	gene expressed on cy3, not expressed on cy5	yes
high	high	meaningful; can be anything	gene expressed on both channels, useful ratio	depends on the ratio

TABLE 17.1: The meaning of the ratio cy3/cy5 depends in an essential way on the absolute intensities cy3 and cy5. Furthermore, a gene may be interesting from a biological point of view even if it has a meaningless ratio.

17.3 Gene pies

Gene pies are nice visualization tools most useful for cDNA data obtained from two color experiments. This type of data is characterized by two types of information: absolute intensity and the ratio between the cy3 and cy5 intensities. In general, the meaning of the ratio between the channels depends essentially on the absolute intensity levels on the two channels, cy3 and cy5. If the intensities of the two channels are close to the background intensity, the ratio is completely meaningless since the ratio of two numbers close to zero can be almost anything. Furthermore, from a biological point of view, the gene is not particularly interesting since it seems to be shut off in both mRNA samples. If the intensity is high on one channel and close to the background level on the other channel, the ratio will be either close to zero or a very large number, depending on which intensity is the denominator. For instance, if the values are 0.1 and 100, the ratio can be either $100/0.1 = 1000$ or $0.1/100 = 0.001$. In either case, this is a biologically significant change: the gene was expressed in a sample and not expressed in the other sample. Even though this is an interesting gene, the specific value of the ratio is not trustworthy since the value close to the background level is most probably just noise. Finally, the ratio is most informative if the intensities are well over background for both cy3 and cy5 channels. For instance, if the values are 100 and 200, the ratio is either 0.5 or 2 and is highly informative either way. These observations are summarized in Table 17.1

A gene pie conveys information about both ratios and intensities. The

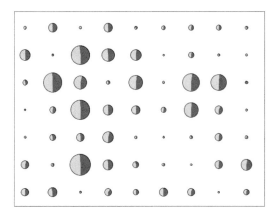

FIGURE 17.4: A gene pie plot. Gene pies convey information about both ratios and intensities. The maximum intensity is encoded in the diameter of the pie chart while the ratio is represented by the relative proportion of the two colors within any pie chart. The net effect is that the genes that might be biologically relevant are made more conspicuous independently on their ratio (which may not always be meaningful).

maximum intensity is encoded as the diameter of the pie chart, while the ratio is represented by the relative proportion of the two colors within any pie chart. From Table 17.1, it follows that the genes that may be biologically interesting are those with at least one large individual channel intensity. Since the gene pies encode the maximum intensity as the diameter of the pie, the net effect is that the genes that might be biologically relevant are made more conspicuous independently on their ratio. An example of a gene pie plot is given in Fig. 17.4.

17.4 Scatter plots

The **scatter plot** is probably the simplest tool that can be used to for a comparative analysis of DNA expression levels. In a scatter plot, each axis corresponds to an experiment and each expression level corresponding to an individual gene is represented as a point. Scatter plots are very useful to convey information about two-dimensional data, and they have been used in many a research paper to show various features of the data [354, 355, 409, 421]. If a gene G has an expression level of e_1 in the first experiment and that of e_2 in the second experiment, the point representing G will be plotted at coordinates (e_1, e_2) in the scatter plot. Fig. 17.5 presents a scatter plot.

In such a plot, genes with similar expression levels will appear somewhere on the first diagonal (the line $y = x$) of the coordinate system. A gene that has an expression level that is very different between the two experiments will appear far from the diagonal. Therefore, it is easy to identify such genes very quickly. For instance, in Fig. 17.5, gene A has higher values in the experiment represented on the horizontal axis (below the diagonal $y = x$) and gene B has a higher value in the experiment represented on the vertical axis (above the diagonal). The further away the point is from the diagonal, the more significant is the variation in expression levels from one experiment to another. In principle, the scatter plot could be used as a tool to identify differentially regulated genes.

Simple as it may be, the scatter plot does allow us to observe certain important features of the data. Fig. 17.6 presents a scatter plot of the logs[2] of the background corrected values in a typical cy5/cy3 experiment. The cy3 was used to label the control sample and is plotted on the horizontal axis; the cy5 was used to label the experiment sample and is plotted on the vertical axis. The data appear as a comma (or banana) shaped blob. Note that in this plot, most genes appear to be down-regulated in the experiment versus control since most of the genes are plotted below the $x = y$ diagonal. However, in the huge majority of experiments involving living organisms, most genes are expected to be expressed at roughly the same expression levels. If too many genes change too much at any given time, it is likely that the functioning of the organism would be disrupted so badly that it would probably die. In other words, we expect most genes to be placed somewhere around the diagonal. The fact that we do not see this indicates that the data on one channel are consistently lower than the data on the other channel. This shows the need for some data preprocessing and normalization. Furthermore, the specific comma shape of the data is not an accident. This specific shape indicates that the dyes introduce a nonlinear effect distorting the data. Again, this should be corrected through an appropriate normalization procedure as will be discussed in Chapter 20.

Another type of plot commonly used for two-channel cDNA array data is the ratio-versus-intensity plot. In this plot, the horizontal axis represents a quantity directly proportional with the intensity. This quantity may be the intensity of one of the channels, or a sum of the intensities on the two channels $\log(cy3) + \log(cy5)$. Note that if log values are used, it follows from the properties of the logarithmic function that the sum of the log intensities is equal to the log of the product of the intensities:

$$\log(cy3) + \log(cy5) = \log(cy3 \cdot cy5) \qquad (17.1)$$

The vertical axis represents the ratio of the two channels $\log(\frac{cy5}{cy3})$. Once again, one can note the typical nonlinearity introduced by the dyes (the comma

---

[2]Chapter 20 will explain in detail the role of the logarithmic function.

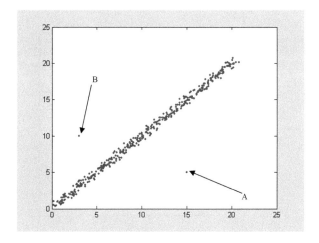

FIGURE 17.5: Expression levels in two experiments visualized as a scatter plot (synthetic data). The control is plotted on the horizontal axis, while the experiment is plotted on the vertical axis. Points above the diagonal $y = x$ represent genes with expression levels higher in the experiment (e.g., gene B), whereas points below the diagonal represent genes with expression levels higher in control (gene A).

or banana shape). Note that according to the same properties of the logarithmic function:

$$\log\left(\frac{cy5}{cy3}\right) = \log(cy5) - \log(cy3) \tag{17.2}$$

Therefore, if most of the genes have equal expression values in both control and experiment, the expression above will be zero. In other words, we expect the points to be grouped around the horizontal line $y = 0$. The fact that most of the points are off this line shows that there is a systematic tendency for the values measured on one of the two channels to be higher than the values measured on the other channel.

Many times, the data will exhibit a high variance at low intensities and a lower variance at high intensities. This is because there is an error inherent to the measurement of the fluorescent intensity. In general, this error is higher for low intensities. Even if the error were constant, at high intensities the error will represent a small part of the signal read while at low intensities, the same absolute value of the error will represent a much larger proportion of the signal. In consequence, the same absolute value will affect more genes expressed at low levels.

For example, let us consider two genes g_1 and g_2, for which the intensity

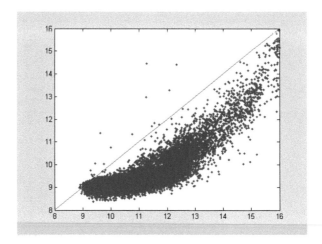

FIGURE 17.6: Typical graph of the cy5 (vertical axis) versus cy3 (horizontal axis) scatter plot of the raw data obtained from a cDNA experiment (real data). The cy3 was used to label the control sample; the cy5 was used to label the experiment sample. The straight line represents the first diagonal cy3=cy5.

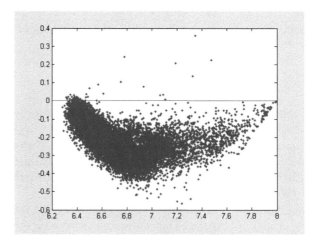

FIGURE 17.7: In this ratio versus intensity plot, the horizontal axis is used to plot the log of the product of the intensities (which is the sum of the log intensities) of the two channels. Sometimes, the mean of the log intensities is used instead of the sum on the horizontal axis. The vertical axis plots the ratio between the channels. The typical raw data set will exhibit the same banana shape visible in the cy3 versus cy5 plot.

read on the control array is 1,000 and 30,000, respectively.[3] Let us assume that in a treated animal, both genes go up 1.5 times to 1,500 and 45,000, respectively. Let us also assume that the error in reading the intensity values is ± 100. The following table shows how the individual intensities will vary.

Untreated sample		Treated sample	
actual	read	value	read
1000	$900-1100$	1500	$1400-1600$
30000	$29900-30100$	45000	$44900-45100$

However, if we calculate ratios between such intensity, the variance will increase (see also Chapter 8). The lowest possible ratios will be obtained by combining the lowest possible value of the numerator with the highest possible value of the denominator. Similarly, the highest possible ratio will be obtained by combining the highest value of the numerator with the lowest value of the denominator. The following table shows what happens to the ratios as well as the log ratios.

values read		range for ratios	range for log ratios
$\frac{1400}{1100}$	$\frac{1600}{900}$	$1.272-1.777$	$0.104-0.249$
$\frac{44900}{30100}$	$\frac{45100}{29900}$	$1.491-1.508$	$0.173-0.178$

Note that the same absolute error in reading the intensity leads to wildly different ranges for the log ratios of the two genes. The ratio of the low-intensity gene can more than double: the highest possible ratio is 0.249, while the lowest possible ratio is 0.101. However, the ratio of the high-intensity gene remains within a very reasonable range. This means that at the low end of the intensity range, the data points will appear to be more scattered than at the high end. This phenomenon is the cause of the funnel shape of a scatter plot of the ratios versus intensities. An example of this is shown in Fig. 17.8.

17.4.1 Scatter plots in R

Scatter plot can be easily obtained using the function **plot** by providing two arrays, one with the x values and the other one with the y values. The function will plot a point using each combination of $(x[i], y[i])$.

We can illustrate this using the placenta data set from Section 14.5.4. Let us say we want to plot the expression values of all genes in sample 5 versus the expression values of the same genes in sample 1. The two samples belong to two different regions of the placental tissues. Note that the original **eset** object contains the log2-transformed gene intensities. This plot can be constructed as follows:

```
> load("placenta.RData")
```

---

[3] Recall that the image file is a 16-bit tiff file that can capture values from 0 to 65,535.

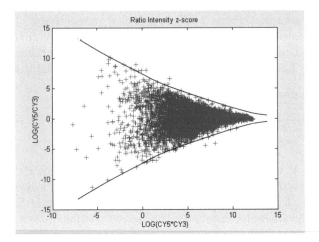

FIGURE 17.8: Typically, the data will exhibit high variance at low intensities and lower variance at higher intensities. This will produce a funnel shape. This funnel shape is more accentuated in the ratio plots such as this one. Note that here the data have been normalized and most of the points lie near the $y = 0$ reference.

```
> intens=2^eset
> plot(intens[,1],intens[,5],cex=1,pch=19,
+ col="mediumaquamarine")
```

The M-A plot of the same data (the log of the ratio as a function of the average log intensity) can be constructed as follows:

```
> A=1/2*log2(intens[,1]*intens[,5])
> M=log2(intens[,1]/intens[,5])
> plot(A,M,cex=0.75,pch=19,col="steelblue")
```

The graph produced by the code above is shown in Fig. 17.10.

17.4.2 Scatter plot limitations

The main disadvantage of scatter plots is that they can only be applied to data with a very small number of components since they can only be plotted in two or three dimensions. In practice, this means that one can only analyze data corresponding to two or three different arrays at the same time. Data involving hundreds or thousands of individuals or time points in a time series cannot be analyzed with scatter plots. Dimensionality reduction techniques such as Principal Component Analysis (PCA) are usually used to extend the usefulness of scatter plots.

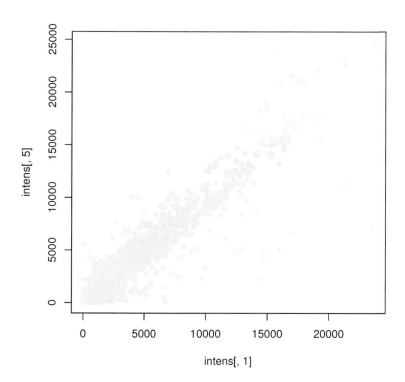

FIGURE 17.9: A scatterplot in R. The plot shows the log values measured in sample 5 of the placenta data (see Section 14.5.4), as a function of the log values of the same genes measured in sample 1. In this graph, each point corresponds to a gene. The x coordinate of the point is the value of the gene measured in sample 1, while the y coordinate is the value of the same gene measured in sample 5.

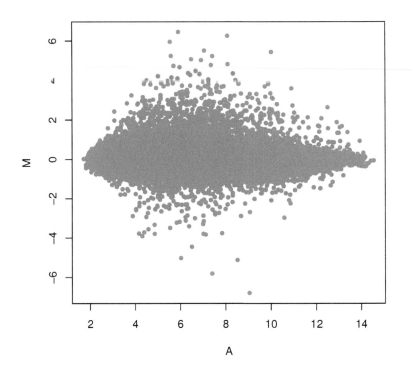

FIGURE 17.10: An MA plot in R. The plot shows the log of the ratio $\log_2(sample1/sample5)$, as a function of the average log intensity $\log_2(sample1 * sample5)/2$.

17.4.3   Scatter plot summary

1. The scatter plot is a two- or three-dimensional plot in which a vector is plotted as a point having the coordinates equal to the components of the vector.

2. Scatter plots are highly intuitive and easy to understand. Limitations include the reduced number of dimensions in which data can be plotted.

3. In a scatter plot of two expression experiments, including many genes or whole genomes, in which each experiment is represented on an axis, most data points are expected to lie around the line $y = x$ (or $z = y = x$). The assumption here is that most genes will not change. Points far from the diagonal are differentially expressed in the experiments plotted. The assumption may not hold if the set of genes has been preselected in some relevant way.

4. In a scatter plot of two expression experiments, including many genes or whole genomes in which the ratio is plotted against an intensity, most data points are expected to lie around the line $y = 0$.

5. A consistent departure from the reference lines above may indicate a systematic trend in the data and the need for a normalization procedure.

6. Many scatter plots of microarray data will exhibit a funnel shape with the wider variance towards low intensities.

17.5   Volcano plots

In principle, the selection of differentially expressed genes can be made using two types of criteria: fold changes and *p*-values. Traditionally, life scientists have favored fold changes because they have an intuitive interpretation and are easy to understand and conceptualize. Also traditionally, statisticians have preferred *p*-values because they capture information about both the number of measurements, as well as about the difference between groups with respect to the variance. The two types of evidence really capture slightly different aspects. It is possible to have a gene that has a relatively large fold change but an insignificant *p*-value. A life scientist would argue that a gene that doubles or triples its value should be studied further. A statistician, on the other hand, could argue that such a change can happen quite often just by chance. It is also possible to have a gene that has a significant *p*-value (e.g., p=0.001) but a small fold-change (e.g., 1.3). A statistician would argue that the change in this gene is much larger than expected by chance, and therefore, the gene should be studied further. A life scientist would argue that this gene has such a small fold change that it probably will not be possible to confirm it with an independent assay. The arguments presented by both sides have their merits in each case. Because the two criteria capture slightly different phenomena, it is useful to use at those genes that have large fold changes, *as well as* significant *p*-values. The volcano plots are a very useful tool to visualize such simultaneous requirements.

A volcano plot has the fold changes on the horizontal axis and the *p*-values on the vertical axis. In this plot, the horizontal axis can use a log2 scale on which one unit corresponds to a two-fold change, and the vertical axis can use a negative log10 scale on which a significance threshold such as 0.01 would appear as 2 ($-\log_{10}(0.01) = 2$). Such a plot is shown in Fig. 17.11.

As shown in this figure, thresholds can be set on both axis. Interesting genes will be situated in the upper left and right quadrants, as defined by the thresholds. Note the V-shaped empty area in the middle of the plot. This area would correspond to genes that have a fold change close to zero and highly significant *p*-values. Clearly, this is not possible. Hence, such a distinctive V-shaped area will appear in all volcano plots. Furthermore, the areas in the lower left and right quadrants correspond to very high fold changes associated with insignificant *p*-values. This is also not possible so these areas will also be empty in all volcano plots. The combination of these two features gives this plot the distinctive shape suggested by its name.

17.5.1   Volcano plots in R

Let us consider that the data are available in a matrix that contains the probes or genes as rows, and the conditions as columns. Let us further assume that we already performed an analysis that calculates a fold change and a *p*-value for each probe/gene. These are available as additional columns. The data would look like this:

```
[1] "GSM239309.CEL" "GSM239310.CEL" "GSM239311.CEL"
[4] "GSM239312.CEL" "GSM239313.CEL" "GSM239314.CEL"

Background correcting
Normalizing
Calculating Expression

> head(peDat)

            GSM239309.CEL GSM239310.CEL GSM239311.CEL
209140_x_at    12.622693     12.644389     12.671890
201641_at      11.122038     10.953200     11.187770
201163_s_at     2.109063      2.046446      1.976826
208791_at       6.669523      6.740302      6.648747
209969_s_at     8.009124      7.936813      7.912223
218723_s_at     6.144639      6.268876      6.180939
```

	GSM239312.CEL	GSM239313.CEL	GSM239314.CEL
209140_x_at	11.026198	11.060462	11.028351
201641_at	4.553666	4.564932	4.784074
201163_s_at	6.257076	6.150340	6.310329
208791_at	9.504053	9.600896	9.613349
209969_s_at	5.091673	5.056348	5.122406
218723_s_at	3.113827	3.200814	3.172252

	p.value	FC
209140_x_at	2.067896e-07	1.1456729
201641_at	4.051030e-07	2.3925621
201163_s_at	4.353157e-07	0.3276215
208791_at	5.801449e-07	0.6984597
209969_s_at	8.254649e-07	1.5623767
218723_s_at	1.138986e-06	1.9600150

We define a function that will create a volcano plot. In this plot, we want the probes/genes that meet both criteria to be plotted with red, while the rest of the probes/genes are plotted with black. Furthermore, we would like to draw the thresholds used for both fold changes, as well as for the *p*-value. Such a function could look like this:

```
> myvolcano  = function(myx,myy,vt1,vt2,ht,myxlab,myylab,
+ cex.lab=1.2,...){
+   plot(myx,myy, pch=18, col=colors()[552], xlab=myxlab,
+                        ylab=myylab, cex.lab=cex.lab,...)
+   points(myx[myx<vt2 & myx>vt1 | myy<ht], myy[myx<vt2 &
+                   myx>vt1 | myy<ht], pch=18, cex=1.05)
+   abline(v=vt1, lty=2)
+   abline(v=vt2, lty=2)
+   abline(h=ht, lty=2)
+ }
```

This function takes the fold changes and *p*-values as the first two arguments. The next three arguments represent the vertical thresholds, for the fold changes, and the horizontal threshold, for the *p*-values. The following two parameters can be used for the labels, while the last parameters allows us to increase the size of the character used to plot the data points. With this function, a volcano plot can be created as:

```
> myvolcano(log2(peDat[,"FC"]),-log10(peDat[,"p.value"]),-1,1,3,
+ "log2(fold changes)","-log10(p-values)")
```

This creates a volcano plot with the vertical thresholds located at ± 1, which corresponds to a fold change of ± 2, and the horizontal threshold at 3, which corresponds to a significance threshold $\alpha = 0.001$. This volcano plot is show in Fig. 17.11.

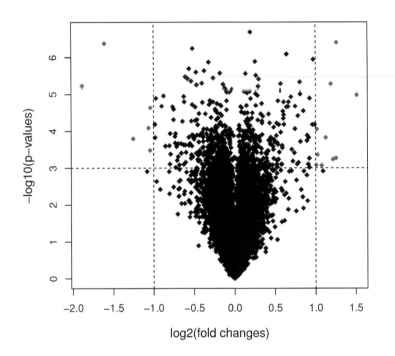

FIGURE 17.11: A volcano plot shows the fold changes on the horizontal axis, and the p-values on the vertical axis, for every probe or gene. Thresholds for both fold changes and p-values can be applied simultaneously. In this plot, the red probes meet both criteria: a fold change more than two-fold (± 1 in log2 scale), and a p-value more significant than 0.001 ($-\log_2(p) > 3$).

17.6 Histograms

A **histogram** is a graph that shows the frequency distribution of the values in a given data set. The horizontal axis of a histogram spans the entire range of values encountered in the data set. The vertical axis shows the frequency of each value. Usually, the histogram is represented as a bar graph. If this is the case, a bar of height y is drawn at position x to represent the fact that the value x appears y times in the data set. A histogram may also be drawn as a graph $y = f(x)$ with the same meaning: a point plotted at coordinates (x, y) means that the value x appears y times in the given data set. Fig. 17.12 shows an example of a histogram of the normalized log ratios of the background subtracted cy3/cy5 ratios[4] from the publicly available yeast cell cycle data set [4, 95]. This data set includes the expression of 6,118 genes of the YSC328 yeast strain collected over 7 times points. The sporulation was induced by transferring the cells into a nitrogen-deficient medium. Data were collected at 0.5, 2, 5, 6, 7, 9, and 11.5 hours. The histogram included only the data corresponding to the first time point at 0.5 hours.

In general, histograms provide information about the shape of the distribution that generated the data. The histogram may be used as an empirical probability density function (pdf) since the frequency of a certain value will be directly proportional to the probability density for that value.

Let us recall that most expression data are real numbers. Real numbers may have a large number of decimal places, depending on the computer used to process the data. Currently, most computers store floating point numbers using 64 or 128 bits which gives the ability to represent a staggering number of decimal places. This makes it likely that almost all values extracted from the image will be distinct even if they are very close to each other. For instance, the numbers 2.147, 2.148, and 2.151 are different even though they are all close approximations of 2.15. Furthermore, even in log scale, it is almost sure that the least significant digits are completely irrelevant to the gene expression which can only be effectively measured with a much lower accuracy. If the frequency of each distinct real value is computed, it is likely that the histogram will be a very wide graph with as many bars as data points and all bars having the height equal to 1 since each real data point is likely to be unique. In order to avoid this and make the histogram more meaningful, one must define some **bins**. A bin is an interval that is used as an entity in counting the frequencies. The histogram will contain a bar for each bin and the height of the bar will be equal to the number of values falling in the interval represented by the bin. In order to illustrate the importance of the binning and the phenomena related to it let us consider the following example.

---

[4]A full discussion of the preprocessing and normalization techniques can be found in Chapter 20.

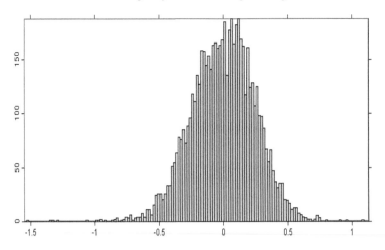

FIGURE 17.12: A histogram is a graph showing the frequency of various values in the data set. The horizontal axis spans the range of values while the vertical axis shows the frequency. A bar of height y is drawn at position x to represent the fact that the value x appears y times in the data set.

EXAMPLE 17.1

Let us consider the following data set:

```
> hd = read.table("histdata.csv")
> hd$V1
```

```
 [1] 0.23970217 0.90357686 0.32631680 0.96863302 0.15623996
 [6] 0.64392895 0.23757565 0.44174449 0.58400632 0.73742638
[11] 0.47582837 0.32175581 0.17516263 0.48354816 0.13785403
[16] 0.07578331 0.47921325 0.56467984 0.93206682 0.03959783
[21] 0.10162151 0.23614322 0.37389189 0.20355521 0.32704396
[26] 0.11007081
```

As we have seen, if we calculate the frequency of each value, we notice that all values are unique. A histogram with no binning will contain all values on the x axes and a number of bars equal to the number of values, all bars having height equal to 1. This is not very informative. In order to obtain more information about the data, we divide the range of values (0.04, 0.97) into a number of equal bins. Fig. 17.13 shows from the top down the histograms obtained with 100, 40, 20, and 10 bins, respectively. Note that the histogram obtained for 100 bins is very similar to the one that would be produced with no binning: almost all bar heights are equal to 1. As the number of bins decreases, the intervals corresponding to each bin increase and the number of

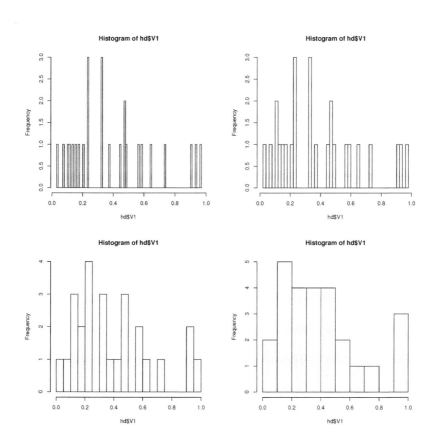

FIGURE 17.13: The effect of the bin size on the shape of the histogram. From top to bottom and left to right, the data are exactly the same, but there are 100, 40, 20 and 10 bins, respectively. If bins are very narrow, few values fall into each bin; as bins get wider, more values fall into each bin, and the shape of the histogram changes.

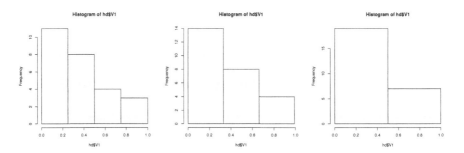

FIGURE 17.14: An artifact of the binning process. From left to right, the same data are shown with bin sizes of 0.25, 0.33, and 0.5, respectively. Note that all three histograms suggest a clear tendency for higher frequencies at lower values. The data are in fact uniformly distributed, and this tendency is entirely an artifact of the binning process and the relatively small sample size (26 data points).

values falling into each such interval starts to increase and also vary from bin to bin.

Should we continue the process by reducing the number of bins further, the heights of the bars will continue to increase. Fig. 17.14 shows what happens when histograms are constructed for the same data with bin sizes of 0.24, 0.33, and 0.5, respectively. Interestingly, these three graphs appear to show a clear tendency of the data to have a higher frequency at lower values. In reality, this data set was drawn from a uniform distribution across the interval (0,1). This is a good example of a **binning artifact**. A binning artifact is an apparent property of the data that is in fact due exclusively to the binning process. In general, binning artifacts can be easily detected because they disappear as the number of bins is changed. This is why it is very important to plot several histograms across a large range of bin numbers and sizes before trying to extract any data features. For instance, the histogram in Fig. 17.12 exhibits a gap right at the peak of the histogram. One would be well advised to construct several other histograms with different bin numbers and sizes before concluding that the data really have such a feature at that point. Indeed, another histogram constructed from the same data using only 88 bins (see Fig. 17.15) does not exhibit any gap at all. Most software packages implementing this tool will give the user the ability to change the number of bins used to create the histogram.

In R, one can plot a histogram for a given data set using the function `hist`. In this function, we can specify the number of equal bins in which the range will be split into using the parameter `breaks`:

```
> par(mfrow=c(2,2))
> hist(eset[,1],breaks=5,col="peachpuff")
```

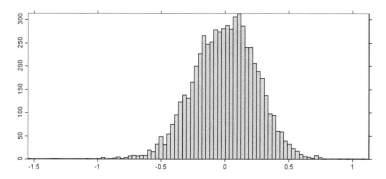

FIGURE 17.15: The histogram of the first experiment in the yeast sporulation data constructed using 88 bins. The data are the same as in Fig. 17.12, but this histogram does not have the artifactual gap near the peak.

```
> hist(eset[,1],breaks=10,col="seagreen")
> hist(eset[,1],breaks=20,col="yellow2")
> hist(eset[,1],breaks=50,col="sienna")
> par(mfrow=c(1,1))
```

In the code above, we first set up our graphical display into an array of 2×2 plots. Then, we plotted the same data using histograms with 5, 10, 20, and 50 bins, respectively. The resulting plot is shown in Fig. 17.16.

In fact, the parameter **breaks** can also be used to specify a vector giving arbitrary "custom" breakpoints between the bins (e.g. "Scott," "FD," or "Sturges" which is the default), or a character string naming one of the predefined algorithms available for the computation off the number of bins, or even a custom, user-defined function to be used for this computation.

```
> par(mfrow=c(2,2))
> hist(eset[,1],breaks="Scott",col="violetred")
> hist(eset[,1],breaks=c(min(eset[,1]),6,6.5,7,max(eset[,1])),
+ col="yellowgreen")
> hist(eset[,1],breaks=function(x){(max(x)-min(x))/2},
+ col="turquoise")
> hist(eset[,1],breaks="FD",col="wheat")
> par(mfrow=c(1,1))
```

The results of this are shown in Fig. 17.17.

Another useful feature of the function hist, is the parameter **freq**. If the value of this parameter is TRUE, the resulting histogram is a representation of frequencies of the values in each bin. If the value is FALSE, the histogram will plot the probability density for each bin (obtained from the raw counts

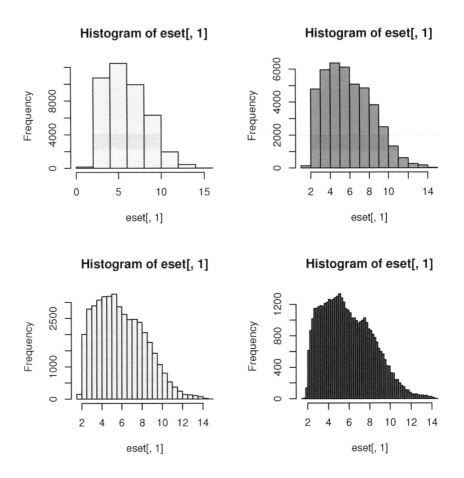

FIGURE 17.16: Histograms in R. The same data are plotted in histograms with 5 (top left), 10 (top right), 20 (bottom left), and 50 (bottom right) equally spaced bins.

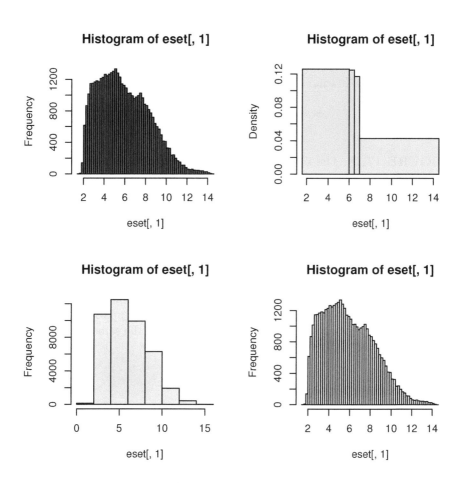

FIGURE 17.17: Histograms in R. The same data are plotted in histograms in which the number of bins was calculated with the "Scott" algorithm (top left), arbitrary custom bins defined by the user (top right), calculated with a user-defined function (bottom left), and calculated with the "FD" built-in algorithm (bottom right).

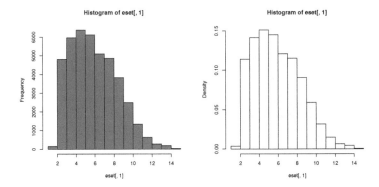

FIGURE 17.18: The same function hist can be use to plot either the raw frequency (left) or the probability density (right) for each bin. The shapes of the two histograms are the same, but note the units on the vertical axis and the meaning of the respective heights of each bin. In the frequency plot, the total area of the bins will sum up to the number of values; in the probability density plot, the total area will be 1.

by dividing by the total number of values). The total area under the plot of frequencies will sum up to the total number of data points. Conversely, the total area under the plot of probability densities will sum up to one. Fig. 17.18 shows the frequency and probability density plots for the same data obtained with:

```
> hist(eset[,1],col="royalblue1")
> hist(eset[,1],freq=F,col="papayawhip")
```

When analyzing expression data, histograms are often constructed for the log ratios of expression values measured in two different experiments (either two different Affymetrix arrays or the two channels of a cDNA array). For such data, the peak of the histogram is expected to be in the vicinity of value 0 if the experiment has been performed on the whole genome or on a large number of randomly selected genes. This is because most genes in an organism are expected to remain unchanged, which means that most ratios will be around 1, and therefore, most log ratios will be around 0. A shift of the histogram away from these expected values might indicate the need for data prepro-cessing and normalization. Again, caution must be exercised in assessing such shifts since small shifts can also appear as the result of the binning process. Both histograms in Fig. 17.12 and Fig. 17.15 are centered around zero. This is because these data have been normalized.[5] Again, for experiments involving many genes or for whole genome experiments, the histogram is expected to

---

[5]Various normalization procedures are discussed in Chapter 20.

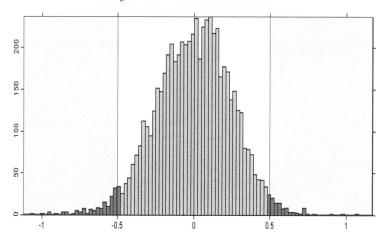

FIGURE 17.19: The histogram of the normalized log ratios can be used to select the genes that have a minimum desired fold change. In this histogram, the tails beyond ± 0.5 contain the genes with a fold change of more than 1.4 ($\log_2 x = 0.5 \Rightarrow x = 2^{0.5} = 1.41$).

be symmetrical, with a shape more or less similar to that of a normal distribution. Although the distribution can often depart from normality inasmuch as the tails are fatter or thinner than those of a normal distribution most such differences will not be visible at a visual inspection. In most cases, large departures from a normal-like shape indicate either problems with the data, strong artifacts, or the need for some normalization procedures.

In general, differentially expressed genes will have ratios either considerably larger than 1 or considerably lower than 1. In consequence, their log ratios will be either positive or negative, relatively far from 0. In principle, one could find the differentially regulated genes in a given data set by calculating the histogram of the ratios or log-ratios and subsequently selecting the tails of the histogram beyond a certain threshold. For instance, if the values are normalized first and then the log is taken in base 2, selecting the tails beyond the values ± 2 will select the genes with a fold change of at least 4 ($\log_2 \frac{ex_1}{ex_2} > 2 \Rightarrow \frac{ex_1}{ex_2} > 2^2 = 4$). The use of the histogram as a tool for selecting differentially regulated genes is illustrated in Fig. 17.19. Note that selecting genes based on their fold change **is not** a very good method as will be discussed in detail in Chapter 21.

17.6.1   Histograms summary

- Histograms are plots of data counts as a function of the data values. Usually, they are drawn using bars. A value x that occurs y times in the data set will be represented by a bar of height y plotted at location x.

- The histogram provides information about the distribution of the data and can be used in certain situations as an empirical approximation of a probability density function (pdf).

- The exact shape of a histogram depends on the number of data collected and the size of the bins used. A small sample size and/or the binning process may create gross artifacts distorting the nature of the data distribution. Such artifacts may be detected by comparing the shapes of several histograms constructed using different number of bins.

- Two experiments can be compared by constructing the histogram of the ratios of the corresponding values. If the experiments involve a large number of genes and the data are suitably preprocessed and normalized, the histogram of the ratios is expected to be centered on either zero (if logs are used) or 1 (no logs are used) and be approximatively symmetrical.

- Differentially expressed genes will be found in the tails of such a histogram.

17.7   Time series

A time series is a plot in which the expression values of genes are plotted against the time points when the values were measured. Fig. 17.20 shows an example of the time series of the normalized log ratios of the background subtracted R/G ratios[6] from the publicly available yeast cell cycle data set [4, 95]. The horizontal axis is the time. The vertical axis represents the measured expression values. In most cases, the time axis will plot the time points at equal intervals. Since the time collection points are usually chosen in a manner that is suitable to the biological process under study and are not equidistant, this graphical representation will introduce a distortion. For instance, the data in the sporulation data set were collected at $t = 0, 0.5, 2, 5, 6, 7, 9$, and 11.5 hours after transfer to the sporulation medium. This means that the first segment on the plot represents 0.5 hours while the last segment represents 2.5 hours. A hypothetical gene that would increase at a constant rate will be represented

---

[6]A full discussion of the preprocessing and normalization techniques can be found in Chapters 19 and 20.

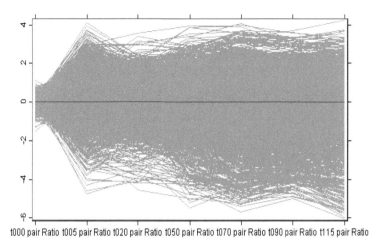

t000 pair Ratio t005 pair Ratio t020 pair Ratio t050 pair Ratio t070 pair Ratio t090 pair Ratio t115 pair Ratio

FIGURE 17.20: A time series plots the expression values of the genes as a function of the time points when the measurements were made. The horizontal axis is the time although in most cases the time will not be represented to scale. The vertical axis represents the measured expression values, in this case the normalized log ratios of the two channels, measured at $t = 0$, 0.5, 2, 5, 6, 7, 9, and 11.5 hours.

by line segments with two different slopes in the two segments. An illustrative example is gene YLR107W shown in Fig. 17.21. This gene seems to exhibit a small increase between the first two time points and a large increase between the last two time points. This seems to suggest that the gene is much more active in the last interval than it was during the first interval. In fact, the rate of change is higher during the first interval. During the first interval, the gene changed from 0 to approximately 0.5 over 0.5 hours, i.e. change at a rate of approximately 1 log/hour. During the last interval, the gene changed from 0.5 to 2.5 over 2.5 hours, i.e. changed at a rate of approximately 0.8 log/hour. This example illustrates why mere data visualization may sometimes be misleading.

17.8 Time series plots in R

Time series plots can be obtained using one call to the function `plot` followed by multiple calls to the function `points`, going iteratively across all data points to be plotted. To illustrate this, we will create an artificial data set of 50 genes,

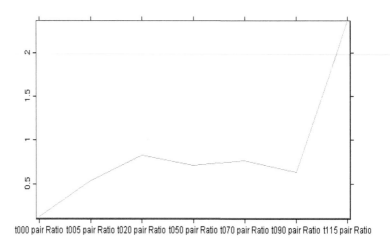

t000 pair Ratio t005 pair Ratio t020 pair Ratio t050 pair Ratio t070 pair Ratio t090 pair Ratio t115 pair Ratio

FIGURE 17.21: An example of a distortion introduced by a non-uniform time scale. The yeast gene YLR107W seems to exhibit a small increase between the first two time points and a large increase between the last two time points of the sporulation time course. This seems to suggest that the gene is more active in the last time segment. In fact, the rate of change is higher in the first interval. The distortion is because the first segment on the time scale represents 0.5 hours while the last segment represents 2.5 hours.

in which sets of 10 genes can have one of 5 different behaviors described below. This example was originally created by John Quackenbush.[7]

EXAMPLE 17.2

This artificial time series data set will include "genes" from each of the following categories:

- *Are relatively constant:*
    - *around zero*
    - *at some positive value*
    - *at some negative value*
- *Increase steadily*
- *Decrease steadily*
- *Oscillate*
    - *up then down then up around a positive, zero or negative value*
    - *down then up then down around a positive, zero or negative value*

In total, there are nine such typical gene profiles shown in Fig. 17.22. A number of other genes are created by adding random noise to each of these profiles. Typical genes are created by adding random noise to each profile for a total of 900 prototype-based genes. Furthermore, an additional 150 genes are created with totally random values.

In order to achieve the goal outlined above, we first create a matrix of 90 rows (genes), with 10 columns (times points) with random normal values, with a 0.1 standard deviation. Then, to each consecutive block of 10 rows(genes) we add the mean values defined by each prototype of gene respectively: type 1 - steady around 0, type 2 – steady around 1, type 3 – steady around -1, type 4 – steady increasing from -5 to 4, type 5 – steady decreasing from 5 to -4, etc. This will produce a set of 10 genes that roughly follow each prototype behavior while still having some random fluctuations, as expected from the real genes.

```
> set.seed(10)
> sg=100
> mat <- matrix(rnorm(9*sg*10,sd=0.2),9*sg,10)
> t <- matrix(0,nrow=9,ncol=10)
> t[1,] <- rep(0,10)
```

---

[7]We modified the data set by adding 150 completely random genes to make it more realistic, but the credit of creating such a wonderful example remains entirely John's.

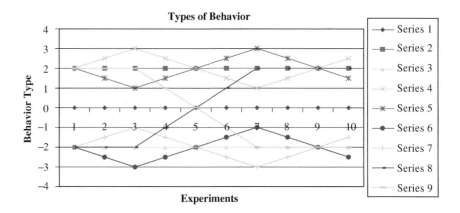

FIGURE 17.22: Typical gene profiles. The prototypes include genes that are constant at -2 (Series 3), 0 (Series 1), and 2 (Series 2), genes that decrease over the 10 experiments (Series 9), genes that increase (Series 8) and genes that oscillate in various ways (Series 4, 5, 6, and 7).

```
> t[2,] <- rep(2,10)
> t[3,] <- -t[2,]
> t[4,] <- c(2,2,2,1,0,-1,-2,-2,-2,-2)
> t[5,] <- -t[4,]
> t[6,] <- c(2,1.5,1,1.5,2,2.5,3,2.5,2,1.5)
> t[7,] <- c(2,2.5,3,2.5,2,1.5,1,1.5,2,2.5)
> t[8,] <- t[6,]-4
> t[9,] <- t[7,]-4
> t_mat <- NULL
> for (i in 1:dim(t)[1]){
+    t_mat <- rbind(t_mat,matrix(rep(t[i,],sg),nrow=sg,
+    byrow=TRUE))
+ }
> mat <- mat + t_mat
```

In the code snippet above, we first generate the noise from a normal distribution with standard deviation 0.2 and zero mean. The number of such noise samples is equal to the number of prototype behaviors we would like to model (9) times the number of genes following each profile (sg = 100) times the number of time points (10). The matrix t is used to store the behavior prototypes: there is one row for each type and one column for each time point. We then create a t_mat matrix by adding the appropriate number of additional functions for each profile (matrix(rep(t[i,],sg)). These rows are bound together in t_mat using rbind. In the last line, the random noise is added to the profiles created above. This is noise that affects the value of each gene at each time point.

To plot the expression of each gene as a function of time, we can use:

```
> cols=c("red","green","blue","magenta","darkgrey","papayawhip",
+ "navyblue","pink","yellow")
> mycols=rep(cols,each=sg)
> plot(1:10,mat[1,],ylim=c(-4,8),type="l",col=mycols[1],lwd=2)
> for(i in 2:(9*sg)){
+ points(1:10,mat[i,],type="l",col=mycols[i],lwd=2)
+ }
> legend(8,8,legend=paste("type ",1:dim(t)[1]),fil=cols,cex=0.8)
```

In order to model a real experiment, in addition to these 90 genes which vary in a controlled way, consistent with the underlying phenomenon, we will also have a large number of genes whose values are purely random and completely independent from the genes above and from the underlying phenomenon.

```
> set.seed(100)
> nrg=150
> rndgenes = matrix(rnorm(nrg*10,sd=1.5),nrow=nrg,ncol=10)
> gdata=NULL
> gdata = rbind(mat,rndgenes)
> plot(1:10,gdata[1,],ylim=c(-4,4),type="l",col="gray",lwd=2)
> for(i in 2:dim(mat)[1]){
+ lines(1:10,gdata[i,],type="l",col="gray",lwd=2)
+ }
> for(i in (dim(mat)[1]+1):dim(gdata)[1]){
+ lines(1:10,gdata[i,],type="l",col="gray",lwd=2)
+ }
```

By changing the color used in the first **for** loop above to another color (e.g., "red") one can obtain a figure in which all genes derived from prototypes are plotted in red while all the purely random genes are plotted in gray. As is, the code generates the plot in Fig. 17.24 showing the entire set of genes across the set of 10 experiments. This graph looks fairly typical in the sense that there is no particular behavior that can be distinguished for any of the genes. Furthermore, this graph does not show that there are some typical profiles, nor that there are several genes behaving in a similar way. We will use this example shortly to illustrate the capabilities of the principal component analysis discussed in the next section.

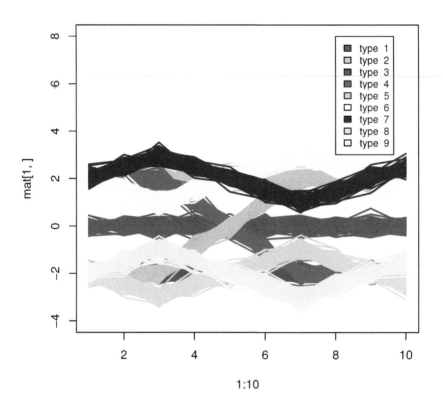

FIGURE 17.23: Some noisy genes following the behavior of the nine proto-types defined in Fig. 17.22.

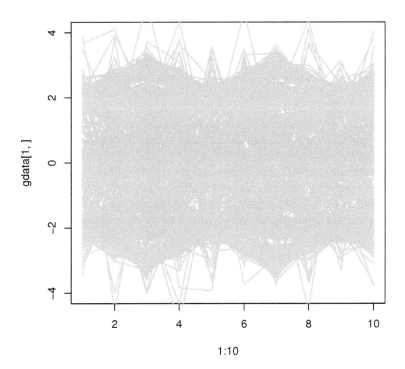

FIGURE 17.24: The data used for the PCA example. These genes were constructed by adding random noise to each prototype in Fig. 17.22 as well as by adding 150 completely random genes. Each prototype was the base for constructing 50 genes of similar behavior. The set shown here includes a total of 1,050 genes.

17.9 Principal component analysis (PCA)

One very common difficulty in many problems is the large number of dimensions. In gene expression experiments, each gene and each experiment may represent one dimension. For instance, a set of 10 experiments involving 20,000 genes may be conceptualized as 20,000 data points (genes) in a space with 10 dimensions (experiments) or 10 points (experiments) in a space with 20,000 dimensions (genes). Both situations are well beyond the capabilities of current visualization tools and frankly, probably well beyond of the visualization capabilities of our brains.

A natural approach is to try to reduce the number of dimensions and, thus, the complexity of the problem, by eliminating those dimensions that are not "important." Of course, the problem now shifts to defining what an important dimension is. A common statistical approach is to pay attention to those dimensions that account for a large variance in the data and to ignore the dimensions in which the data do not vary much. This is the approach used by Principal Component Analysis (PCA).

PCA works by calculating a new system of coordinates. The directions of the coordinate system calculated by PCA are the **eigenvectors of the covariance matrix of the patterns** (see Chapter 8 for a definition of the covariance matrix). An **eigenvector** of a matrix A is defined as a vector $\mathbf{z}$ such as:

$$A\mathbf{z} = \lambda\mathbf{z} \tag{17.3}$$

where λ is a scalar called **eigenvalue**. Each eigenvector has its own eigenvalue although it is possible that the eigenvalues of different eigenvectors have the same numerical value. For instance, the matrix:

$$A = \begin{bmatrix} -1 & 1 \\ 0 & -2 \end{bmatrix} \tag{17.4}$$

has the eigenvalues $\lambda_1 = -1$ and $\lambda_2 = -2$ and the eigenvectors $z_1 = \begin{bmatrix} 1 \\ 0 \end{bmatrix}$ and $z_2 = \begin{bmatrix} 1 \\ -1 \end{bmatrix}$.

It can be verified that:

$$A\mathbf{z_1} = \begin{bmatrix} -1 & 1 \\ 0 & -2 \end{bmatrix}\begin{bmatrix} 1 \\ 0 \end{bmatrix} = (-1)\cdot\begin{bmatrix} 1 \\ 0 \end{bmatrix} = \lambda_1\mathbf{z_1} \tag{17.5}$$

and

$$A\mathbf{z_2} = \begin{bmatrix} -1 & 1 \\ 0 & -2 \end{bmatrix}\begin{bmatrix} 1 \\ -1 \end{bmatrix} = (-2)\cdot\begin{bmatrix} 1 \\ -1 \end{bmatrix} = \lambda_2\mathbf{z_2} \tag{17.6}$$

In intuitive terms, the covariance matrix **captures the shape of the set**

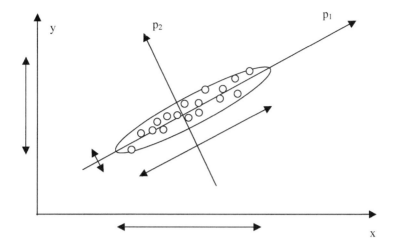

FIGURE 17.25: In spite of the fact that each point has two coordinates, this data set is essentially one dimensional: most of the variance is along the first eigenvector p_1 with the variance along the second direction p_2 being probably due to the noise. The PCA will find a new coordinate system such as the first coordinate is the direction on which the data have maximum variance (the first eigenvector), the second coordinate is perpendicular on the first and captures the second largest variance, etc.

of data points. If one imagines an n-dimensional hyper-ellipsoid including the data, the eigenvectors of the covariance matrix, or the directions found by the PCA, will be the directions of the main axes of the ellipse. This is illustrated in Fig. 17.25.

The essential aspect of the PCA is related to the fact that the absolute value of the eigenvalues are directly proportional to the dimension of the multidimensional ellipse in the direction of the corresponding eigenvector. Since the eigenvectors are obtained from the covariance matrix that captures the shape occupied by the points in the original n-dimensional space, it follows that the absolute values of the eigenvalues will tell us how the data are distributed along these directions. In particular, **the eigenvalue with the largest absolute value will indicate that the data have the largest variance along its eigenvector** (which is a particular direction in space). Second largest eigenvalue will indicate the direction of the second largest variance, etc. It turns out that many times, only few directions manage to capture most of the variability in the data. For instance, Fig. 17.25 shows a data set that lies essentially along a single direction although each particular data point will have two coordinates. This is a set of data points in a two-dimensional space, and each data point is described by two coordinates. However, most of the variability in the data lies along a one-dimensional space that is described by

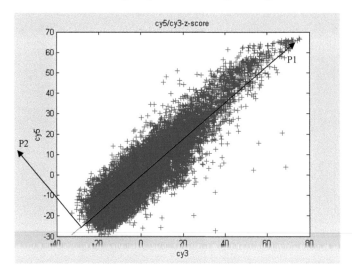

FIGURE 17.26: The PCA can also be used to separate the within-experiment variation from the inter-experiment variation. The new axis P1 will be aligned with the direction of the within-experiment variation. The within-experiment variation stems from the fact that genes are expressed at all levels in either sample. The new axis P2 will be perpendicular on P1 and will capture the inter-experiment variation. The inter-experiment variation is the interesting variations if we want to find those genes that differ between the two samples.

the first principal component. PCA will be able to discover the relevant directions as well as indicate the amount of variance that each new axis captures. In this example, the first principal component is sufficient to capture most of the variance present in the data and the second principal component may be discarded.

Note that the direction of the highest variance may not always be the most useful. For instance, in Fig. 17.26, PCA is used to distinguish between the within-experiment variation and the inter-experiment variation. Fig. 17.26 shows the new axes found by the PCA applied to a typical cy3-cy5 data set. Note that the longer axis, *P*1, corresponds to the within-experiment variation. In either sample, genes will be expressed at various levels, and they will appear at various positions along this axis: genes expressed at low levels will be closer to the origin while genes expressed at high levels will be further up the axis. This variation is not particularly interesting since we know *a priori* that genes will be expressed at all levels. However, the other axis, *P*2, will be aligned with the direction of the inter-experiment variation. The genes that are differentially regulated will appear far from the origin of this axis while genes that are expressed at the same level in both mRNA samples will be

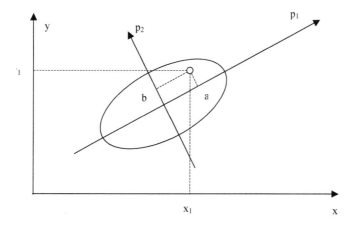

FIGURE 17.27: The principal components of the data are the projections of the data on the new coordinate system. In this figure, the data point shown has the coordinates (x_1, x_2) in the original coordinate system xOy. The ellipse represents the area in which most of the data points lie. The first eigenvector, p_1, will show the direction of the largest variance. The second eigenvector will be perpendicular on the first one. The principal components of the data point (x_1, x_2) will be (a, b) with a being the coordinate along the first eigenvector and b the coordinate along the second eigenvector.

projected close to the origin of this axis. Note that this happens independently of the level of expression of the given gene. Therefore, in principle, the second principal component would be sufficient to look for genes expressed differentially between the two experiments.

The nice aspect of the PCA is that the computation involved is always the same, independently of whether we are interested in the directions of the large variances or low variances. If the input data is n dimensional, i.e. each input point is described by n coordinates, the PCA will provide a complete set of n directions together with the associated eigenvalues and will leave the choice of what to use up to us. Most packages implementing PCA can provide an explicit list of the eigenvectors and associated eigenvalues and allow the user to select a small subset on which to project the data.

The projections of the data points in the new coordinate system found by the PCA are called **the principal components of the data** (see Fig. 17.27). These can be easily obtained by multiplying each data point (which is an n-dimensional vector) by the matrix of the eigenvectors.

Depending on the problem, we will select a small number of directions (e.g., 2 or 3) and look at the projection of the data in the coordinate system formed with only those directions. By projecting the n-dimensional input data

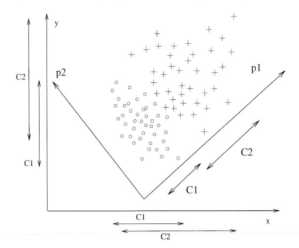

FIGURE 17.28: Principal Component Analysis (PCA) – If one of the two original axes x or y is eliminated, the classes cannot be separated. The coordinate system found by PCA ($p1$, $p2$) allows the elimination of $p2$ while preserving the ability to separate the given classes. It is said that PCA performs a dimensionality reduction.

into a space with only 2 or 3 coordinates, one achieves a **dimensionality reduction**.

However, one should note that the dimensionality reduction obtained through PCA may not always be a true reduction of the number of variables we need to consider. For instance, in the example shown in Fig. 17.27, the PCA found the axis p_1 that captures most of the variance in the data. Under these circumstances, one could discard the second principal component p_2 and describe the data only through its first component. However, each component is a linear combination of the original variables. For instance, the direction of the first component p_1 can be described by an equation such as $a \cdot x + b \cdot y + c = 0$, where x and y are the original variables. In order to conclude that any of the original variables may be discarded, one needs to look at the coefficients of the equations describing the chosen PCA directions. If a particular variable (such as y) has coefficients (such as b) that are close to zero in all chosen PCA directions, then one could indeed conclude that this original variable is not important under the given circumstances. Since the original coordinates correspond to either genes or experiments, this is a way of selecting those genes or experiments that are truly important for the phenomenon under study.

An example of using PCA is shown in Fig. 17.28. The data, which include patterns from two classes (red crosses and blue circles), are given in the original coordinate system with axes x_1 and x_2. If the data are projected on each of the two axes, the clusters corresponding to the two classes overlap and the classes cannot be separated using any single dimension. Therefore, a dimensionality reduction is not possible in the original coordinate system. In order to separate the two classes, one needs both x_1 and x_2. However, the PCA approach can analyze the data and extract from it a new coordinate system with axes p_1 and p_2. The direction of the new axes will be the direction in which the data have the largest and second largest variance. If the two clusters are projected on the axes of the new coordinate systems one can notice that the situation is now different. The projections of the two classes on p_2 overlap completely. However, the projections of the two classes on p_1 yield two clusters that can be easily separated. In these conditions, one can discard the second coordinate p_2. In effect, we have achieved a dimensionality reduction from a space with two dimensions to a space with just one dimension while retaining the ability to separate the two classes.

PCA also provides good results in those cases in which the data have certain properties that become visible in the principal component coordinate system while they may not be so in the original coordinate system. In order to illustrate this, we will use the Multiple Experiment Viewer (MEV) software from TIGR on the data constructed in Example.

We can now apply PCA, construct the new coordinate system and select from the new axes the ones corresponding to the three-largest variances. Fig. 17.29 shows the genes in this three-dimensional space. In this case, PCA has been used to effectively reduce the dimensionality of the experiment space from 10 to 3. This PCA space reveals a lot of information about the data. We can see from the plot that there are nine distinct types of gene behavior represented by the small spherical clusters at the extremities of the axes and the small cluster centered in the origin. It can be seen that, for each such behavior, there are a number of genes that behave in that particular way. Finally, one can also conclude that there are some genes that behave in a completely random way since they are distributed uniformly in a large area of the space. The nine clusters in Fig. 17.29 can be distinguished more easily by comparing it with Fig. 17.30 from which the completely random genes have been removed.

In the previous example, PCA was applied to reduce the dimensionality of the experiment space. The result was a space in which the axes were the eigenvectors of the correlation matrix of the experiments or "principal experiments." In this space, each data point is a gene. Such a plot provides information about the structure contained in the behavior of the genes. As shown in Example 17.2, this can be used to see whether there are genes that behave in similar ways, whether there are distinct patterns of gene behavior, etc. However, PCA can also be applied to the gene expression vectors. In this case, PCA will calculate the eigenvectors of the correlation matrix of the genes or "principal genes." In this space, each point is an experiment. Such a plot provides information about the structure contained in the various exper-

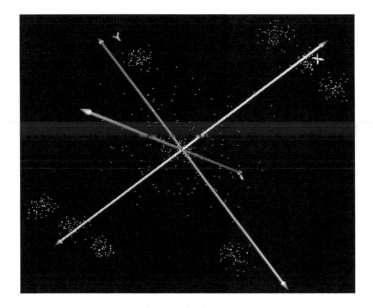

FIGURE 17.29: Using PCA for visualization. The figure shows the genes from Fig. 17.24 plotted in the space of the first three principal components of the 10 experiments (principal experiments). In this plot, each dot represents a gene. The coordinates of each dot are equal to the first three principal components of the respective gene. The plot shows clearly that there are nine distinct types of gene behavior (the tight spherical clusters at the extremities of the axes and the tight cluster centered in the origin) as well as some random genes (the more dispersed collection of points around the origin).

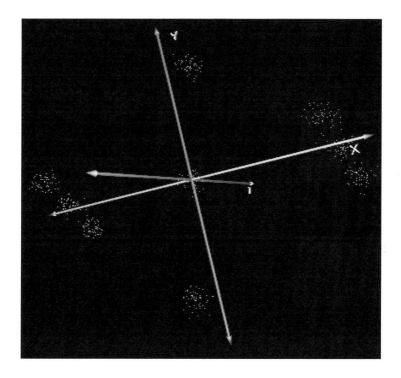

FIGURE 17.30: The PCA plot of the non-random genes. The plot shows clearly nine clusters corresponding to each of the profiles present in the data.

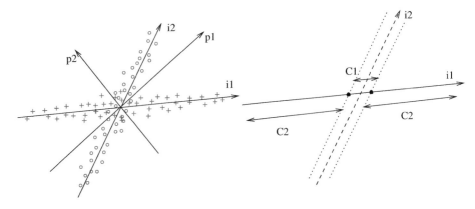

FIGURE 17.31: The difference between principal component analysis (PCA) and independent component analysis (ICA). The coordinate system $\{p_1, p_2\}$ found by PCA is less useful than the $\{i_1, i_2\}$ found by ICA. The classes cannot be separated using any one axis from the coordinate system found by PCA. The right panel shows how the classes can be separated with minimal misclassification using just the projections on i_1.

iments: whether there are similar experiments, whether various experiments tend to have a certain profile, etc.

PCA has been shown to be extremely effective in many practical problems, including gene expression data [147, 210, 350] and is currently available in a number of software tools.

17.9.1 PCA limitations

In spite of its usefulness, PCA has also drastic limitations. Most such limitations are related to the fact that PCA only takes into consideration the variance of the data, which is a first-order statistical characteristic of the data. Furthermore, the eigenvectors of the correlation matrix are perpendicular on one another which means that any axes found by the PCA will also be perpendicular on one another. In effect, this means that the transformation provided by the change of coordinate system from the original system to the principal components system is a rotation perhaps followed by a scaling proportional to the eigenvalues of the same covariance matrix. There are cases in which such a system is not suitable. For instance, in Fig. 17.31, the coordinate system (p_1, p_2) found by PCA is not useful at all since both classes have the same variance on each of the p_1 and p_2 axes.

Furthermore, the direction of the new axes found by PCA is determined exclusively based on the variance of the data. The class of each data point, that is whether the point is a blue circle or red cross, is not taken into consideration by the PCA algorithm. Therefore, PCA may not always be as useful as in the

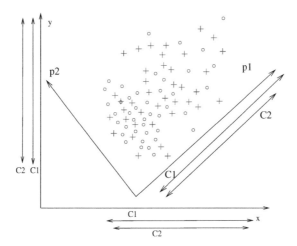

FIGURE 17.32: An example in which the Principal Component Analysis (PCA) is not very useful. The coordinate system found by PCA (*p*1, *p*2) is based exclusively on the spatial distribution of the data points, i.e. on their position, and disregards completely their type (blue circles or red crosses). This figure illustrates a problem in which the data are placed in the same positions as in Fig. 17.28 but have different labels (e.g., red circles are cancer and blue crosses are healthy); in this case, the coordinate system found by the PCA algorithm is not helpful in separating the two classes.

example shown in Fig. 17.28. This point is illustrated in Fig. 17.32 which shows a data set in which the coordinates of the points are the same as in Fig. 17.28 but their classes are different. The class is represented by the color/shape of the individual data points with red crosses representing cancer patients and blue circles representing healthy controls. Since the numerical values are the same, PCA will calculate the same coordinate system based on the variance of the data and disregarding their class. In turn, this will produce the same principal components as in Fig. 17.28. However, in this case, this principal components coordinate system is not particularly useful.

17.9.2 Principal component analysis in R

R provides a good set of tools for performing a PCA analysis. As explained above, in a space with 10 coordinates corresponding to each of the 10 times points in the example above, a gene will appear as as a point with 10 coordinates. The principal components can be calculated with the **pca** function from the **pcurve** package in R.

```
> library(pcurve)
> library(lattice)
```

```
> specpca <- pca(gdata)
> md<-data.frame(PC1=specpca$pcs[,1],PC2=specpca$pcs[,2],
+ PC3=specpca$pcs[,3])
> mynewcols=c(mycols,rep("black",150))
> print(cloud(PC3~PC1+PC2,data=md,col=mynewcols,pch=19,
+ cex=0.5,bg=cols,cex.main=0.5))
```

In this example, we plotted each gene as a point in the three-dimensional space corresponding to the first three principal components, PC1, PC2, and PC3 using the function cloud of the lattice package.

Note that the pca function calculates and returns the coordinates of each data point from gdata in a space with 10 dimensions corresponding to the new coordinate system calculated by the PCA. Each data point has 10 coordinates from which in this example we decided to use only the first 3.

As another example, the same pcurve and lattice are used here to compute the principal components and produce a plot of the 79 samples in the bfust data. The results are shown in Figure 17.34.

```
library(lattice); library(pcurve)
 pc<- pca(t(exprs(bfust)))
cloud(pc$pcs[,3]~ pc$pcs[,1]+
pc$pcs[,2],col=mycols,pch=19,xlab="PC1", ylab="PC2", zlab="PC3")
```

17.9.3 PCA summary

Principal Component Analysis is a technique that uses first-order statistical properties of the data in order to construct a new coordinate system. The direction of the new axes will be the eigenvectors of the correlation matrix of the data. The PCA has the following properties:

1. The directions are chosen in decreasing order of the amount of data variance they explain.

2. The directions are perpendicular on each other.

3. PCA can be performed on either genes or experiments. PCA performed on genes will produce a graph in which each point is an experiment. PCA performed on experiments will produce a graph in which each point is a gene.

4. PCA can be used to achieve dimensionality reduction by choosing either high variance axes or low variance axes, depending on the goal of the analysis. In most cases, one selects the directions that explain most of the variance in the data (high variance axes).

5. The dimensionality reduction achieved through PCA can be useful in visualization and classification.

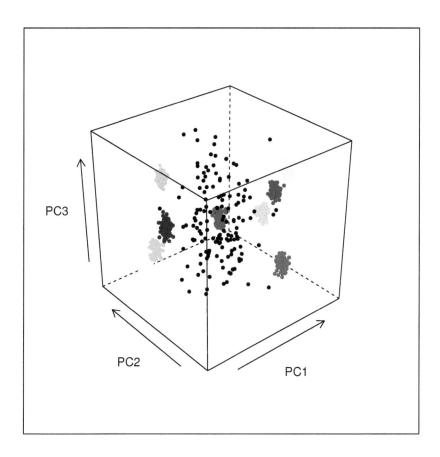

FIGURE 17.33: A PCA plot of the data shown in Fig. 17.24. Each group of points shown corresponds to one of the nine prototypes used to define the general behavior of the gene. The black random points distributed approximately uniform around the center are the 150 completely random genes added to the set.

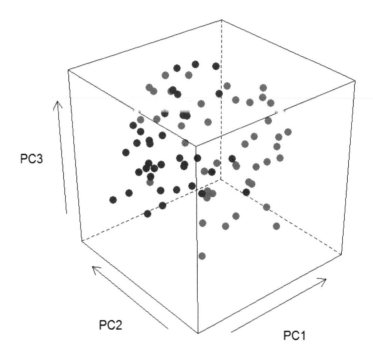

FIGURE 17.34: A PCA plot: the 79 samples are projected on the first three principal components (PC) derived from the 50 original features. The colors are used to denote the two classes.

17.10   Independent component analysis (ICA)

Independent Component Analysis (ICA) [41] is a technique that is able to consider higher-order statistical dependencies like skew and kurtosis. In Fig. 17.31, i_1 and i_2 are the two axes of the coordinate system found by ICA. The blue circles class has a very large variance along i_1 and a very low variance along i_2, whereas the red crosses class has a very large variance along i_2 and a very low variance along i_1. Therefore, the two classes may be separated reasonably well using just one of the two dimensions.

ICA has been used very successfully in different variations of the blind source separation problem. A particular example of the blind source separation is the "cocktail party problem." In this problem, the n original signals are n people speaking in the same room. The available information is a set of recordings coming from n microphones placed somewhere in the room. Each microphone captures a mixing of the n voices speaking at the same time. Thus, each individual recording is a jumbled superposition of all voices in the room. The task is to use the n recordings to recover the speech of each individual person. This is possible because the individual sources (speakers) are statistically independent. This problem has been tackled very successfully by the ICA approach. Currently, there are software programs that are able to process such recordings and produce signals corresponding to individual speakers [41].

In the blind source separation problem, there are a number of n sources $s_1, s_2, \ldots, s_n$ that are mixed by an unknown mixing matrix A. Nothing is known about the sources or about the mixing process. All that we observe is a set of n mixed signals that form a vector x:

$$s = \begin{bmatrix} s_1 s_2 \ldots s_n \end{bmatrix} \tag{17.7}$$

$$\begin{bmatrix} x_1 \\ x_2 \\ \vdots \\ x_n \end{bmatrix} = As = A \begin{bmatrix} s_1 \\ s_2 \\ \vdots \\ s_n \end{bmatrix} \tag{17.8}$$

The task is to recover the original sources by finding a square matrix W, which is a permutation and rescaling of the inverse of the unknown matrix A:

$$\begin{bmatrix} \overline{s_1} \\ \overline{s_2} \\ \vdots \\ \overline{s_n} \end{bmatrix} = W \begin{bmatrix} x_1 \\ x_2 \\ \vdots \\ x_n \end{bmatrix} \tag{17.9}$$

The same approach can be used in the DNA microarray data analysis.

Here, the n genes expressed on a particular array can be seen as the n signals to be separated. One array corresponds to one microphone in the cocktail party problem. A series of experiments done over the range of a given variable (e.g., time for sporulation data, dosage for drug tests, etc.) will provide the signals analogous to the microphone recordings in the cocktail party problem. Given a large number of genes, arrays and time points (as usually available with microarray data), the task is to unveil "functional signals" of each individual gene, i.e., to be able to follow the evolution in time of each individual gene.

17.11 Summary

This chapter presented a number of general purpose tools useful in the analysis and visualization of microarray data. For each tool, the discussion included a detailed explanation of the tool as well as a discussion of its advantages, limitations, as well as several potential problems that can occur in its practical use.

Box plots are informative graphical tools that present in a concise manner several statistics of the data: mean, top and bottom quartiles (or other chosen percentiles), and the interquartile distance. Box plots afford a convenient way of identifying the outliers in a given data set.

Gene pies are tools particularly useful in the analysis of two-channel cDNA data. Gene pies represent each gene as a pie chart divided into two-colored regions. The ratio of the two channels corresponds to the ratio of the areas of the two regions. A useful feature is the ability to code the maximum absolute intensity on one of the two channels as the diameter of the pie chart. This allows an easy identification of those ratios that correspond to genes expressed at higher levels in at least one of the two channels. Such genes are usually biologically meaningful.

The scatter plot is a two- or three-dimensional plot in which a vector is plotted as a point having the coordinates equal to the components of the vector. Limitations include the reduced number of dimensions in which data can be plotted. In a scatter plot of two expression experiments including many genes or whole genomes, in which each experiment is represented on an axis, most data points are expected to lie around the line $y = x$ (or $z = y = x$). The assumption here is that most genes will not change. Points far from the diagonal are differentially expressed in the experiments plotted. The assumption may not hold if the set of genes has been preselected in some relevant way. In a scatter plot of two expression experiments including many genes or whole genomes in which the ratio is plotted against an intensity, most data points are expected to lie around the line $y = 0$. A consistent departure from the reference lines above may indicate a systematic trend in the data and the need for a normalization procedure. Many scatter plots of microarray data will exhibit a funnel shape with the wider variance towards low intensities.

Histograms are plots of data counts as a function of the data values. Usually, they are drawn using bars. A value x that occurs y times in the data set will be represented by a bar of height y plotted at location x. The histogram provides information about the distribution of the data and can be used in certain situations as an empirical approximation of a probability density function (pdf). The exact shape of a histogram depends on the number of data collected and the size of the bins used. A small sample size and/or the binning process may create gross artifacts distorting the nature of the data distribution. Such artifacts may be detected by comparing the shapes of several histograms constructed using different number of bins. Two experiments can be compared by constructing the histogram of the ratios of the corresponding values. If the experiments involve a large number of genes and is suitably preprocessed and normalized, the histogram is expected to be centered on either zero (if logs are used) or 1 (no logs are used) and be approximatively symmetrical. Differentially regulated genes will be found in the tails of a histogram of log ratios.

Time series are plots in which expression values are plotted against the time they were measured at. A nonuniform time scale is often used because of the nature of the biological experiments. Such as scale may produce very misleading results.

Principal Component Analysis is a technique that uses first-order statistical properties of the data in order to construct a new coordinate system. The direction of the new axes will be the eigenvectors of the correlation matrix of the data. The directions are chosen in decreasing order of the amount of data variance they explain and they are perpendicular on each other. PCA can be used to achieve dimensionality reduction by choosing either high variance axes or low variance axes depending on the goal of the analysis. In most cases, one selects the directions that explain most of the variance in the data (high variance axes).

Independent Component Analysis is a technique that takes into consideration higher-order statistical properties of the data. Furthermore, the directions of the new axes found by ICA are not necessarily perpendicular on each other. ICA has been used successfully to solve the blind source separation problem and may be used in a similar way in genomics to separate the signals from individual genes.

Chapter 18

Cluster analysis

Birds of a feather flock together.

—*Unknown*

18.1 Introduction

Cluster analysis is currently the most frequently used multivariate technique to analyze gene sequence expression data. Clustering is appropriate when there is no a priori knowledge about the data. In such circumstances, the only possible approach is to study the similarity between different samples or experiments. In a machine learning framework, such an analysis process is known as unsupervised learning since there is no known desired answer for any particular gene or experiment.

Clustering has become so popular in this field that most authors presenting results obtained with microarrays feel the need to include some type of clustering diagram in their papers [5, 44, 66, 99, 147, 150, 171, 204, 206, 342, 413, 426, 430, 445, 459, 486, 489]. In fact, the popularity of the clustering techniques is so great that sometimes clustering is mistakenly taken as a very fuzzy and all-inclusive ultimate goal of microarray data analysis. This author

567

has been approached by several accomplished life scientists seeking help in order to "do their clustering." Subsequent probing revealed that "doing the clustering" could mean anything from selecting a subset of differentially regulated genes to identifying gene interactions to building classifiers based on gene expression data. In fact, clustering is **the process of grouping together similar entities**. Clustering can be done on any data: genes, samples, time points in a time series, etc. The particular type of input makes no difference to the clustering algorithm. The algorithm will treat all inputs as a set of n numbers or **an n-dimensional vector**.

If one is to group together things that are similar, one should start by defining the meaning of similarity. In other words, we need a very precise **measure of similarity**. Such a measure of similarity is called a **distance** or a **metric**. A distance is a formula that takes two points in the input space of the problem and calculates a positive number that contains information about how close the two points are to each other. The input space of the problem is an n-dimensional space, so the two points can be, for instance, two genes measured across n experiments or two experiments, each represented by the expression values of n genes. There are many different ways in which such a measure of similarity can be calculated. The final result of the clustering depends in a very essential way on the exact formula used. In the following, we will discuss a number of distances used in the analysis of gene expression data.

18.2   Distance metric

A **distance metric** d is a function that takes as arguments two points x and y in an n-dimensional space $\mathbb{R}^n$ and has the following properties:

1. **Symmetry**. The distance should be symmetric, i.e.:

$$d(x,y) = d(y,x) \qquad (18.1)$$

This means that the distance from x to y should be the same as the distance from y to x.

2. **Positivity**. The distance between any two points should be a real number greater than or equal to zero:

$$d(x,y) \geq 0 \qquad (18.2)$$

for any x and y. The equality is true if and only if $x = y$, i.e. $d(x,x) = 0$.

3. **Triangle inequality**. The distance between two points x and y should

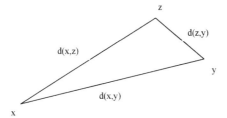

FIGURE 18.1: Triangle inequality: the distance between two points x and y should be shorter than or equal to the sum of the distances from x to a third point z and from z to y. The equality holds true only when z is on the line between x and y.

be shorter than or equal to the sum of the distances from x to a third point z and from z to y:

$$d(x,y) \leq d(x,z) + d(z,y) \qquad (18.3)$$

This property reflects the fact that the distance between two points should be measured along the shortest route (see Fig. 18.1).

It turns out that many different distances can be defined. The only properties shared by all distances are the three properties above. Other properties may intuitively appear to be associated with distances, but they may only hold true for certain ways of defining the distance. Furthermore, certain familiar concepts such as that of a circle are strongly influenced by the implicit distance used to define it. Distances are discussed extensively in the literature in the context of clustering and classification [136, 242].

18.2.1 Euclidean distance

The Euclidean distance between two n-dimensional vectors $\mathbf{x} = (x_1, x_2, \ldots, x_n)$ and $\mathbf{y} = (y_1, y_2, \ldots, y_n)$ is:

$$d_E(\mathbf{x}, \mathbf{y}) = \sqrt{(x_1 - y_1)^2 + (x_2 - y_2)^2 + \cdots + (x_n - y_n)^2} = \sqrt{\sum_{i=1}^{n} (x_i - y_i)^2} \qquad (18.4)$$

This is the usual distance that we use for most practical purposes. Its numerical value comes from the Pythagorean theorem (see Fig. 18.2). This distance will have all properties extrapolated from our common-sense concept of distance. For instance, if both points are translated in space, the distance between them will remain the same, a circle will be the familiar round and symmetrical shape, etc.

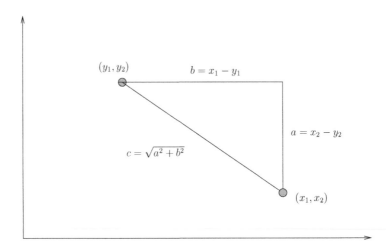

FIGURE 18.2: The Euclidean distance is computed in accordance to the Pythagorean theorem.

As an example, let us calculate the distance from the origin $O(0,0)$ to the point $A(3,4)$ using the Euclidean metric:

$$d_E(O,A) = \sqrt{3^2 + 4^2} = \sqrt{25} = 5 \qquad (18.5)$$

Let us now assume that the coordinates of point A are measured incorrectly due to some experimental error. Let us assume that the measured point is $A'(4,4)$. The distance between the origin and the measured A' is:

$$d_E(O,A) = \sqrt{4^2 + 4^2} = \sqrt{32} = 5.65 \qquad (18.6)$$

which represents a change of $\frac{5.65}{5} = 1.13$. A change of one unit in one of the coordinates determined a change of 13% with respect to the true distance.

18.2.2 Manhattan distance

The Manhattan distance between two n-dimensional vectors $\mathbf{x} = (x_1, x_2, \ldots, x_n)$ and $\mathbf{y} = (y_1, y_2, \ldots, y_n)$ is:

$$d_M(\mathbf{x}, \mathbf{y}) = |x_1 - y_1| + |x_2 - y_2| + \cdots + |x_n - y_n| = \sum_{i=1}^{n} |x_i - y_i| \qquad (18.7)$$

where $|x_i - y_i|$ represents the absolute value of the difference between x_i and y_i.

The Manhattan distance, or city-block distance, is named after the well-known New York borough because it represents the distance that one needs to

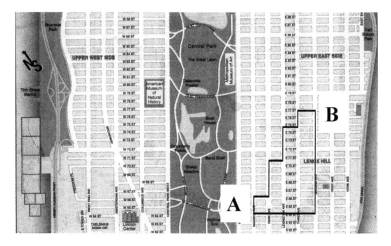

FIGURE 18.3: The Manhattan (or city-block) distance is the distance that one needs to travel in an environment in which one can move only along directions parallel to the x and y axes (no diagonal movements). The Manhattan distance is independent on the path traveled between the two points.

travel in an environment in which one can move only along directions parallel to the x and y axes (no diagonal movements). This is similar to the Manhattan borough where most streets are straight and cross one another at right angles. Note that in the city-block world, as expected, the distance between two points depends only on the points themselves and not on the path followed to travel between the points (see Fig. 18.3).

As an illustration of the fact that the definition of the distance changes dramatically the properties of any object or concept that uses distances even in an implicit way, Fig. 18.4 shows a comparison between a circle in a space using a Manhattan distance and the usual circle in a Euclidean space.

Let us calculate the same distance OA using the Manhattan distance:

$$d_M(O,A) = 3 + 4 = 7 \tag{18.8}$$

The first observation is that the city-block distance provided a larger absolute value. If one of the measurements is incorrect and A is measured as $A'(4,4)$, the distance between the origin and the measured A' is:

$$d_M(O,A) = 4 + 4 = 8 \tag{18.9}$$

which represents a change of $\frac{8}{7} = 1.14$. A change of one unit in one of the coordinates leads to a change of 14% with respect to the true measurement. We can see that, in comparison to the Euclidean distance, the city-block distance tends to yield a larger numerical value for the same relative position of the points. If the two given points are an outlier and the center of a given cluster,

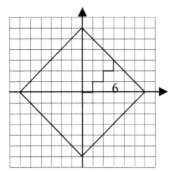 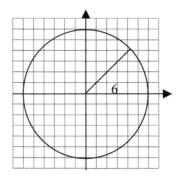

FIGURE 18.4: The distance confers essential properties to the space and objects therein. For instance, a circle is the locus of all points situated at a constant distance from a fixed point called the center. The right panel shows a circle of radius 6 in a Euclidean metric space. The left panel shows the same circle of radius 6 in a Manhattan metric space.

we can say that the city-block distance slightly emphasizes the outliers of a data set: an outlier will appear a bit farther away when using the Manhattan distance.

18.2.3 Chebychev distance

The Chebychev distance between two n-dimensional vectors $\mathbf{x} = (x_1, x_2, \ldots, x_n)$ and $\mathbf{y} = (y_1, y_2, \ldots, y_n)$ is:

$$d_{\max}(\mathbf{x}, \mathbf{y}) = \max_i |x_i - y_i| \tag{18.10}$$

The Chebychev distance simply picks the largest difference between any two corresponding coordinates. For instance, if the vectors $\mathbf{x} = (x_1, x_2, \ldots, x_n)$ and $\mathbf{y} = (y_1, y_2, \ldots, y_n)$ are two genes measured in n experiments each, the Chebychev distance will pick the one experiment in which these two genes are most different and will consider that value the distance between the genes. The Chebychev distance is to be used when the goal is to reflect any big difference between any corresponding coordinates. The Chebychev distance behaves inconsistently with respect to outliers since it only looks at one dimension. If any or all other coordinates are changed due to measurement error without changing the maximum difference, the Chebychev distance will remain the same. In this situation, the Chebychev distance is resilient with respect to noise and outliers. However, if any one coordinate is affected sufficiently such that the maximum distance changes, the Chebychev distance will change. Thus, this distance is in general resilient to small amounts of noise even if they affect several coordinates but will be affected by a single, large change.

18.2.4 Angle between vectors

The angle distance between two n-dimensional vectors $\mathbf{x} = (x_1, x_2, \ldots, x_n)$ and $\mathbf{y} = (y_1, y_2, \ldots, y_n)$ is:

$$d_\alpha(\mathbf{x}, \mathbf{y}) = \cos(\theta) = \frac{\mathbf{x} \cdot \mathbf{y}}{\|\mathbf{x}\| \, \|\mathbf{y}\|} \tag{18.11}$$

where $\mathbf{x} \cdot \mathbf{y}$ is the dot product of the two vectors:

$$\mathbf{x} \cdot \mathbf{y} = x_1 y_1 + x_2 y_2 + \cdots + x_n y_n = \sum_{i=1}^{n} x_i y_i \tag{18.12}$$

and $\|\cdot\|$ is the norm, or length, of a vector:

$$\|\mathbf{x}\| = \sqrt{x_1^2 + x_2^2 + \cdots + x_n^2} = \sqrt{\sum_{i=1}^{n} x_i^2} \tag{18.13}$$

Note that, if this distance is used, if a point A is moved anywhere on the line that goes through its original position and the origin, in a new location A', the distance $d(OA')$ will be the same as $d(OA)$. In particular, if a point is shifted by scaling all its coordinates by the same factors, the angle distance will not change.

18.2.5 Correlation distance

The Pearson correlation distance between two n-dimensional vectors $\mathbf{x} = (x_1, x_2, \ldots, x_n)$ and $\mathbf{y} = (y_1, y_2, \ldots, y_n)$ is:

$$d_R(\mathbf{x}, \mathbf{y}) = 1 - r_{xy} \tag{18.14}$$

where r_{ik} is the Pearson correlation coefficient of the vectors $\mathbf{x}$ and $\mathbf{y}$:

$$r_{xy} = \frac{S_{xy}}{\sqrt{S_x}\sqrt{S_y}} = \frac{\sum_{i=1}^{n}(x_i - \bar{x})(y_i - \bar{y})}{\sqrt{\sum_{i=1}^{n}(x_i - \bar{x})^2}\sqrt{\sum_{i=1}^{n}(y_i - \bar{y})^2}} \tag{18.15}$$

Note that since the Pearson correlation coefficient r_{xy} varies only between -1 and 1, the distance $1 - r_{xy}$ will take values between 0 and 2.

The Pearson correlation focuses on whether the coordinates of the two points change in the same way (e.g., corresponding coordinate increase or decrease at the same time). The magnitude of the coordinates is less important since the denominator will be proportional to the magnitudes of the vectors. If the vector is, for instance, a set of measurements of given genes in a particular experiment and two such experiments are compared, the Pearson distance will be high if the genes vary in a similar way in the two experiments even if the magnitude of the change differs greatly.

A related issue is the problem of outliers. If a gene is measured incorrectly in one of the experiments, its coordinate along that particular dimension can be very different. This can produce a low overall correlation. In order to address this, the **jackknife correlation** calculates the correlation n times, each time leaving one dimension out and calculating the correlation only on the remaining $n-1$. This will produce n different correlation values for any two vectors (e.g., genes). The distance between the two vectors is taken to be the minimum correlation distance between the n different values. If one of the measurements is wrong, the correlations will be low every time the incorrect measurement is taken into consideration. However, since the jackknife discards a different dimension each time, the incorrect measurement will be eventually discarded. For that particular computation, the correlation between the given genes will be much higher, and therefore, their distance $(1-r)$ will be lower. This is the value that will represent the jackknife distance between the given genes. This distance can be written as:

$$d_J(\mathbf{x}, \mathbf{y}) = \min \left\{ d_R^1(\mathbf{x}, \mathbf{y}), d_R^2(\mathbf{x}, \mathbf{y}), \ldots, d_R^n(\mathbf{x}, \mathbf{y}) \right\} \qquad (18.16)$$

where $d_R^k(\mathbf{x}, \mathbf{y})$ is the correlation distance between $\mathbf{x}$ and $\mathbf{y}$ calculated disregarding the k-th component of the vectors $\mathbf{x}$ and $\mathbf{y}$.

Sometimes the jackknife correlation is too radical since it takes the least value of these correlation coefficients as the measure of the similarity. Furthermore, the method works by discarding data, which is not always a good idea. Finally, the jackknife correlation is only robust to a single outlier. For more outliers, a more general definition of jackknife correlation is needed, which makes the method more computationally intensive and more dangerous (for n outliers, n data points will be ignored at every step).

18.2.6 Squared Euclidean distance

The squared Euclidean distance between two n-dimensional vectors $\mathbf{x} = (x_1, x_2, \ldots, x_n)$ and $\mathbf{y} = (y_1, y_2, \ldots, y_n)$ is:

$$d_{E^2}(\mathbf{x}, \mathbf{y}) = (x_1 - y_1)^2 + (x_2 - y_2)^2 + \cdots + (x_n - y_n)^2 = \sum_{i=1}^{n} (x_i - y_i)^2 \qquad (18.17)$$

When compared to the Euclidean distance, the squared Euclidean distance tends to give more weights to the outliers due to the lack of the square root. The squared Euclidean distance from the origin $O(0,0)$ to the point $A(3,4)$ is:

$$d_{E^2}(O, A) = 3^2 + 4^2 = 25 \qquad (18.18)$$

The distance between the origin and the measured A' is:

$$d_{E^2}(O, A) = 4^2 + 4^2 = 32 \qquad (18.19)$$

which represents a change of $\frac{32}{25} = 1.28$. A change of one unit in one of the coordinates leads to a change of 28% with respect to the true measurement. When we compare this with the change of 13% for Euclidean and 14% for

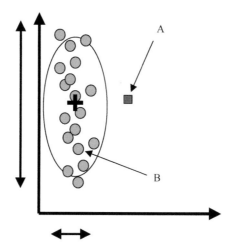

FIGURE 18.5: Data with unequal coordinate variances. A point A that is a certain distance away from the center along x may be very unusual, and therefore interesting, while a point B that is the same distance away from the center along the y-axis may be well within the usual distribution of the patterns. The ellipse represents a contour of equal probability: the points outside the ellipse are unlikely to come from the same distribution as the points inside.

Manhattan, we see why an outlier will be overemphasized by the squared Euclidean distance.

18.2.7 Standardized Euclidean distance

All distances discussed so far give exactly the same importance to all dimensions. The idea behind standardized Euclidean is that not all directions are necessarily the same. For instance, a data set may be known to have a nonhomogeneous distribution that is characterized by a larger variance across the y-axis and a smaller variance across the x-axis. In these conditions, a point that is a certain distance away from the center along x may be very unusual, and therefore interesting, while a point that is the same distance away from the center along the y-axis may be well within the usual distribution of the patterns (see Fig. 18.5). In this situation, the border marking a region of constant probability will be an ellipse instead of a circle. The standardized Euclidean distance takes this into consideration by dividing with the standard deviation of each dimension. The standardized Euclidean distance between

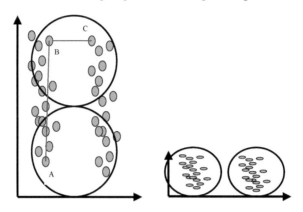

FIGURE 18.6: A data set that has a larger variance along the y-axis. Even though the points B and A belong to the same cluster, they are farther apart than B and C. A clustering algorithm using the usual Euclidean distance will group B and C in the same cluster as shown in the left panel. A clustering algorithm using a standardized Euclidean distance will effectively equalize the variances on each axis and produce the clusters in the right panel which reflect better the structure of the data. Image courtesy of BioDiscovery Inc.

two n-dimensional vectors $\mathbf{x} = (x_1, x_2, \ldots, x_n)$ and $\mathbf{y} = (y_1, y_2, \ldots, y_n)$ is:

$$d_{SE}(\mathbf{x}, \mathbf{y}) = \sqrt{\frac{1}{s_1^2}(x_1 - y_1)^2 + \cdots + \frac{1}{s_n^2}(x_n - y_n)^2} = \sqrt{\sum_{i=1}^{n} \frac{1}{s_i^2}(x_i - y_i)^2} \quad (18.20)$$

Because of this, the standardized Euclidean distance may provide better results in certain situations. Fig. 18.6 shows an example in which the data are distributed as two long and thin clusters. In this case, the usual Euclidean distance would not be able to form the correct clusters while the standardized Euclidean has no difficulty in doing so. In Fig. 18.6, the points B and A belong to the same cluster. However, the Euclidean distance between B and A is larger than the Euclidean distance between B and C. Therefore, a clustering algorithm using the usual Euclidean distance will group B and C in the same cluster as shown in the left panel. A clustering algorithm using a standardized Euclidean distance will effectively equalize the variances on each axis and produce the clusters in the right panel, which reflect better the structure of the data.

The standardization can be done using the sample variance, as in Eq. 18.20, or by using the range:

$$d_{SE}(\mathbf{x}, \mathbf{y}) = \sqrt{\frac{1}{R_1^2}(x_1 - y_1)^2 + \cdots + \frac{1}{R_n^2}(x_n - y_n)^2} = \sqrt{\sum_{i=1}^{n} \frac{1}{R_i^2}(x_i - y_i)^2} \quad (18.21)$$

where R_i is the range of the data along dimension i.

18.2.8   Mahalanobis distance

The standardized Euclidean distance used the idea of weighting each dimension by a quantity inversely proportional to the amount of variability along that dimension. This is equivalent to distorting the space by shrinking it along the axes of large variance and expanding it along the axes of low variance. This can be generalized. One might want to distort the space in an arbitrary way, not necessarily along the axes. This is achieved by the Mahalanobis distance.

The Mahalanobis distance between two n-dimensional vectors $\mathbf{x} = (x_1, x_2, \ldots, x_n)$ and $\mathbf{y} = (y_1, y_2, \ldots, y_n)$ is:

$$d_{MI}(\mathbf{x}, \mathbf{y}) = \sqrt{(\mathbf{x} - \mathbf{y})^T S^{-1} (\mathbf{x} - \mathbf{y})} \qquad (18.22)$$

where S is any $n \times n$ positive definite matrix and $(\mathbf{x} - \mathbf{y})^T$ is the transposition of $\mathbf{x} - \mathbf{y}$.

The role of the matrix S is to distort the space as desired. Usually, this matrix is the covariance matrix of the data set. This achieves the same purpose as the standardized Euclidean only that the directions of distortion can be arbitrary, as best suited to the data. This is in contrast to the distortion introduced by the standardized Euclidean which was limited to scaling along the axes. If the space warping matrix S is taken to be the identity matrix, the Mahalanobis distance reduces to the classical Euclidean distance;

$$d_{MI}(\mathbf{x}, \mathbf{y}) = \sqrt{(\mathbf{x} - \mathbf{y})^T (\mathbf{x} - \mathbf{y})} = \sqrt{\sum_{i=1}^{n} (x_i - y_i)^2} \qquad (18.23)$$

18.2.9   Minkowski distance

The Minkowski distance is a generalization of the Euclidean and Manhattan distance. The Minkowski distance between two n-dimensional vectors $\mathbf{x} = (x_1, x_2, \ldots, x_n)$ and $\mathbf{y} = (y_1, y_2, \ldots, y_n)$ is:

$$\begin{aligned} d_{Mk}(\mathbf{x}, \mathbf{y}) &= \{|x_1 - y_1|^m + |x_2 - y_2|^m + \cdots + |x_n - y_n|^m\}^{\frac{1}{m}} = \\ &= \left\{ \sum_{i=1}^{n} |x_i - y_i|^m \right\}^{\frac{1}{m}} \qquad (18.24) \end{aligned}$$

Recalling that $x^{\frac{1}{m}} = \sqrt[m]{x}$, we note that for $m = 1$ the Minkowski distance reduces to Manhattan, i.e. a simple sum of absolute differences. For $m = 2$, the Minkowski distance reduces to Euclidean distance.

18.2.10   When to use what distance

With so many distances, a natural question is when to use what? This section will discuss briefly the issues and criteria that can be taken into consideration when choosing the distance to be used in clustering.

Sometimes, different types of variables need to be mixed together. To do this, any of the distances above can be modified by applying a weighting scheme. For instance, mixing clinical data with gene expression values can be done by assigning different weights to each type of variable in a way that is compatible with the purpose of the study.

In many cases, it is necessary to normalize and/or preprocess the data.[1] One possible step in the normalization procedure is to standardize genes or arrays. This may be necessary or desirable in order to compare the amount of variation of two different genes or arrays from their respective central locations. Standardizing genes can be done by applying a z-transform, i.e. subtracting the mean and dividing by the standard deviation (see Eqs. 9.38 and 9.39 in Chapter 8). For a gene g and an array i, standardizing the gene means adjusting the values as follows:

$$x_{gi} = \frac{x_{gi} - \bar{x}_{g.}}{s_{g.}}$$
(18.25)

where $\bar{x}_{g.}$ is the mean of the gene g over all arrays and $s_{g.}$ is the standard error of the gene g over the same set of measurements. The values thus modified will have a mean of zero and a variance of one across the arrays.

Standardizing the arrays means adjusting the values as follows:

$$x_{gi} = \frac{x_{gi} - \bar{x}_{.i}}{s_{.i}}$$
(18.26)

where $\bar{x}_{.i}$ is the mean of the array and $s_{.i}$ is the standard error of the array across all genes. Similar standardization can be performed using median instead of mean as a more robust estimator of central tendency (see Chapter 8) and the median absolute deviation as an estimator of the amount of variability.

In some sense, gene standardization makes all genes similar. A gene that is affected only by the inherent measurement noise will be indistinguishable from a gene that varies 10-fold from one experiment to another. Although there are situations in which this is useful, gene standardization may not necessarily be a wise thing to do every time. Standardizing the arrays, on the other hand, is applicable in a larger set of circumstances. However, standardizing the arrays is rather simplistic if used as the only normalization procedure. More normalization issues will be discussed in Chapter 20. Table 18.1 shows how various distances behave with respect to gene or experiment standardization.

There is an important interaction between the choice of the distance and the type of values being compared. It is commonly believed that the Affymetrix

---

[1] See Chapter 20 for a more complete discussion of normalization issues.

Distance	Variable standardization	Observation standardization
Euclidean	different	different
Standardized Euclidean	same	different
Correlation	different	different
Mahalanobis	different	different

TABLE 18.1: The effect of standardization on several distances.

Distance	Clustering genes	Clustering samples
Euclidean	different	same
Manhattan	different	same
Correlation	same	different

TABLE 18.2: The effect of using absolute or relative expression values with several distances. If the distances are the same, the clustering will also be the same.

technology measures the absolute abundance of mRNA and hence is an absolute measure of the expression level of a gene. On the other hand, cDNA arrays are used most commonly with two channels, one of which is the reference.[2] Furthermore, the spots are often characterized by a ratio or log-ratio of the values measured on the two channels. For these reasons, the cDNA technology is often thought to measure a relative expression of the expression level in a condition with respect to a reference. In reality, the difference is only superficial since one could, in principle, pair each Affymetrix array exploring a condition with another Affymetrix array hybridized with a control and then take the ratio of the corresponding genes. This would provide relative measurements with respect to that condition. Alternatively, one could measure different conditions on the two channels of the same cDNA microarray and analyze the values without calculating ratios with respect to the reference (see for instance the ANOVA loop model discussed in Chapter 15). This approach would extract "absolute" expression levels from cDNA arrays. In conclusion, whether we are using absolute or relative expression values has little to do with the technology itself. However, as noted before, there is a strong link between the type of values measured (absolute or relative) and the distances. Specifically, the question is whether the distance between two given genes or two given experiments depends on whether the values are relative or absolute for a given distance. The answers are summarized in Table 18.2 for the most commonly used distances.

---

[2]We have seen in Chapter 15 that this may not be the best practice.

18.2.11 A comparison of various distances

1. Euclidean distance – the usual distance as we know it from our environment. It will be used as a reference when summarizing the other distances.

2. Squared Euclidean – tends to emphasize the distances. Same data clustered with squared Euclidean might appear more sparse and less compact.

3. Angle between vectors – takes into consideration only the angle, not the magnitude. For instance, a gene g_1 measured in two experiments, $g_1 = (1,1)$, and a gene $g_2 = (100,100)$ will have the distance (angle):

$$\cos(A) = \frac{\mathbf{x} \cdot \mathbf{y}}{\|\mathbf{x}\| \|\mathbf{y}\|} = \frac{[100 \ 100] \begin{bmatrix} 1 \\ 1 \end{bmatrix}}{\sqrt{100^2 + 100^2} \cdot \sqrt{1^2 + 1^2}} = \frac{100 + 100}{100 \cdot \sqrt{2} \cdot \sqrt{2}} = 1 \quad (18.27)$$

Therefore, the angle between these two vectors is zero. Clustering with this distance will place these two genes in the same cluster although their absolute expression levels are very different.

4. Correlation distance – will look for similar variation as opposed to similar numerical values. Let us consider a set of five experiments and a gene g_1 that has an expression of 1, 2, 3, 4, and 5 in the 5 experiments, respectively. This gene can be represented as $g_1 = (1,2,3,4,5)$. Let us also consider the genes $g_2 = (100,200,300,400,500)$ and $g_3 = (5,4,3,2,1)$. The correlation distance will place g_1 in the same cluster with g_2 and in a different cluster from g_3 because $(1,2,3,4,5)$ and $(100,200,300,400,500)$ have a high correlation $(d(g_1,g_2) = 1 - r = 1 - 1 = 0)$, whereas $(1,2,3,4,5)$ and $(5,4,3,2,1)$ are anticorrelated $(d(g_1,g_3) = 1 - (-1) = 2)$. However, the Euclidean distance will place g_1 in the same cluster with g_3 and in a different cluster from g_2 because $d_E(g1,g2) = 734.20$ while $d_E(g1,g3) = 6.32$.

5. Standardized Euclidean – eliminates the variance information. All directions will be equally important. If genes are standardized, genes with a small range of variation (e.g. affected only by noise) will appear the same as genes with a large range of variation (e.g., changing several orders of magnitude).

6. Manhattan – the set of genes or experiments being equally distant from a reference does not match the similar set constructed with Euclidean distance (see Fig 18.4).

7. Jackknife – robust with respect to one or few erroneous measurements.

8. Chebychev – focuses on the most important difference: $(1,2,3,4)$ and $(2,3,4,5)$ have distance 2 in Euclidean and 1 in Chebychev. $(1,2,3,4)$ and $(1,2,3,6)$ have distance $\sqrt{2}$ in Euclidean and 2 in Chebychev.

9. Mahalanobis – can warp the space in any convenient way. Usually, the space is warped using the correlation matrix of the data.

10. Minkowski – a generalization of Euclidean and Manhattan.

18.3   Clustering algorithms

Before we start to discuss specific clustering approaches and algorithms, it is useful to clarify the terminology as well as make a few general observations. Any clustering algorithm can be used to group genes or experiments or any set of homogeneous entities described by a set of numbers usually arranged as a vector. We shall refer to such entities as patterns or instances. Similar patterns grouped together by the algorithm form **clusters**. A set of clusters including all genes or experiments considered form a **clustering**, **cluster tree** or **dendrogram**.

There are a few general observations that can be made about clustering. Contrary to the popular belief, clustering is not a goal in itself and, by itself, is seldom convincing. **Anything can be clustered**. In order to prove this, let us consider the well known leukemia data set [183]. This data set contains measurements corresponding to acute lymphoblastic leukemia (ALL) and acute myeloid leukemia (AML) samples from bone marrow and peripheral blood. There are 11 AML samples and 27 ALL samples for a total of 38 samples. These samples were analyzed with Affymetrix 6,800 arrays containing approximately 6,800 genes. The genes that appeared to be different between the two data sets can be found using a t-test. We ordered the genes in the increasing order of their p-values and we selected the top 35 genes. These genes are expected to be most different between the two groups. The top panel in Fig. 18.7 shows a clustering of these top 35 genes.

In order to illustrate the fact that anything can be clustered, we generated a number of fake expression values using a random number generator. These numbers were organized in a matrix with 6,800 rows and 38 columns. Each row represents a fake gene and each column represents a fake experiment. We divided the experiments in two classes, labeled ALL and AML. Respecting the structure of the original data, we picked 11 experiments for the AML group and 27 experiments for the ALL group. We then selected the genes that appeared to be different in both data sets using a t-test. We ordered the genes by p-values and picked the top 35 genes. These genes were then clustered. The bottom panel in Fig. 18.7 shows the results for the random data set. In both panels, one can observe two distinct groups of genes: i) genes with lower expression values in the ALL group (green) and higher values in the AML (red) and ii) genes with lower values in AML and higher values in ALL.

The astute reader will notice that the two dendrograms are not *exactly* the

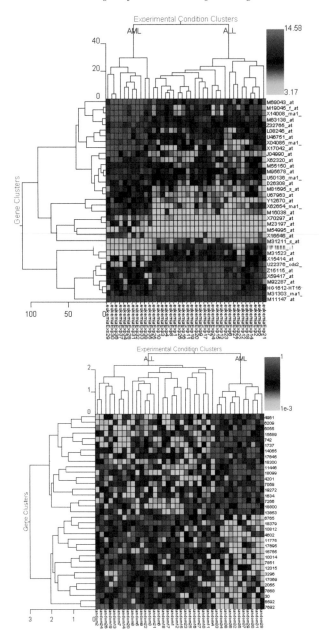

FIGURE 18.7: Anything can be clustered. The top panel shows the top 35 genes in the Leukemia data set [183]; the bottom panel shows the top 35 genes in a random data set. In both panels, one can observe two distinct groups: i) genes with lower expression values in the ALL group (green) and higher values in the AML (red) and ii) genes with lower values in AML and higher values in ALL. Image obtained with GeneSight 3.5.2.

same: the one constructed from the random data has a more grainy appearance whereas the dendrogram of the real data is smoother. This is in fact due to the completely random data. However, a similar experiment can be done by using real expression data coming from some other experiment or even using the same leukemia data set in which the experiment labels have been assigned randomly between the ALL and AML groups. Once this random assignment has been done, a t-test can be used to select the genes that are different between the two groups. A clustering obtained from the top 35 genes will be equally smooth and real-looking as the clustering obtained using the correct labels.

In conclusion, the first major observation is that given enough genes, the **genes will always cluster**. Given the large number of genes in most organisms, there is no surprise and therefore no scientific value in the fact that there are genes that behave in a similar way. The scientific value should always come from what can be said about the genes that fall in the same cluster and what can be done with said genes. In the leukemia data set, the interesting result was that a classifier constructed based on these genes could distinguish correctly between ALL and AML on a different data set used for validation.[3] In other data sets, the interesting result may be related to the functional analysis of the genes that are clustered together (see Chapter 23) or other biologically meaningful relationships between the members of a given cluster.

The second general observation is that, in most cases, the clustering produced by a given algorithm is **highly dependent on the distance metric used**. Changing the distance metric may affect dramatically the number and membership of the clusters as well as the relationship between them. Looking at a clustering without knowing the distance used to generate it is uninformative and can be very misleading. Let us consider for instance the three genes discussed in Section 18.2.11: $g_1 = (1, 2, 3, 4, 5)$, $g_2 = (100, 200, 300, 400, 500)$, and $g_3 = (5, 4, 3, 2, 1)$. The correlation distance will place g_1 in the same cluster with g_2 and in a different cluster from g_3 while the Euclidean distance will place g_1 in the same cluster with g_3 and in a different cluster from g_2. The mere observation that two genes are close, and therefore, the clustering itself can be interpreted only if the distance is clearly specified.

Another observation is that the clustering is not necessarily deterministic. This means that **the same clustering algorithm applied to the same data may produce different results**. Many clustering algorithms have an intrinsically nondeterministic component. For instance, the initialization of the clusters in both k-means and self-organizing feature maps (SOFMs) is done entirely randomly. The stochastic aspects may involve, for instance, a random

---

[3]It turns out that the two types of samples were collected from two different sources in a biased way: most ALL samples came from one source and most AML samples came from a second source. Thus, it is not clear whether the classifier learned to distinguish between ALL/AML or between the data sources. However, no matter what the class corresponds to, the paper showed it is possible to construct a classifier based on gene expression and this in itself remains a landmark result.

choice of the initial cluster centers or a random choice of the patterns to be used as initial clusters. It follows that the membership of any particular gene to any particular cluster should be taken with a grain of salt and analyzed carefully. For instance, one should always check whether the given gene would fall into the same cluster if the same algorithm were applied again, etc.

Finally, one must note that in most clustering algorithms (e.g., k-means and hierarchical clustering) **the position of the patterns within the clusters does not reflect their relationship in the input space.** For instance, in Fig. 18.8 genes M11147_at[4] and M55150_at (in the second gene cluster from the top) are right next to each other in the dendrogram. However, their expression profiles are probably as different as they get between any two genes within that cluster. The gene M55150 has a profile more similar to that of gene U50136_rna1_at which is plotted 5 genes away.

Furthermore, the fact that two patterns belong to a given cluster does not necessarily mean that they are close to one another. In fact, a pattern belonging to a cluster A may be closer to some patterns from a different cluster B than it is to other patterns in its own cluster A. In Fig. 18.8, the experiment Exp35 (rightmost experiment in the rightmost cluster) is closer to some experiments in the AML cluster than it is to the other experiments in its own cluster. This is hardly surprising since this *is* an AML sample. Contrary to the appearances, this is neither a software bug nor an algorithmic mistake. Fig. 18.9 shows the same phenomenon in two dimensions. Note that the cluster assignment of each pattern is correct: each pattern is assigned to the closest cluster center. However, the patterns that appear to be far from each other in the clustering are in fact very close to each other. Furthermore, the pattern in the upper cluster is closer to the patterns indicated by arrows than it is to other patterns in its own cluster.

18.3.1 k-means clustering

The k-means algorithm is one of the simplest and fastest clustering algorithms. In consequence, it is also one of the most widely used algorithms. The k-means clustering algorithm takes the number of clusters, k, as an input parameter. This is usually chosen by the user. The program starts by randomly choosing k points as the centers of the clusters (see upper-left panel in Fig. 18.10). These points may be just random points in the input space, random points from more densely populated volumes of the input space, or just randomly chosen patterns from the data itself.

Once some cluster centers have been chosen, the algorithm will take each pattern and calculate the distance from it to all cluster centers. Each pattern will be associated with the closest cluster center. A first approximate clustering

---

[4]For convenience, we are using here the Affymetrix probe IDs as gene names. Gene identifiers can be easily converted between Affymetrix probe ID, GenBank accession IDs and UniGene cluster IDs using the Onto-Convert software available at http://vortex.cs.wayne.edu/Projects.html.

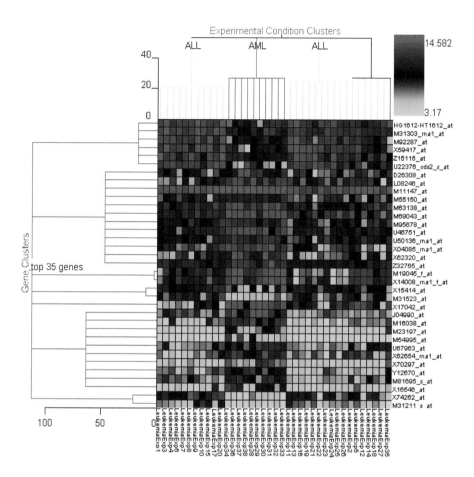

FIGURE 18.8: The way the clusters are plotted may be misleading. The data represent 35 genes that can be used to separate the ALL/AML classes in the leukemia data set [183]. M11147_at and M55150_at are next to each other in the dendrogram even though their expression profiles are not very similar. Gene M55150 has a profile more similar to that of gene U50136_rna1_at which is plotted five genes away. Image obtained with GeneSight 3.5.2.

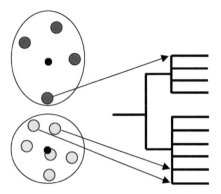

FIGURE 18.9: A two-dimensional example in which the pattern indicated by the top arrow is closer to patterns from another cluster than it is to other patterns in its own cluster. Note that the clustering is correct inasmuch each pattern is assigned to the closest cluster center.

is obtained after allocating each pattern to a cluster. However, since the cluster centers were chosen randomly, it is not said that this is the correct clustering. The second step starts by considering all patterns associated to one cluster center and calculating a new position for this cluster center (upper-right panel in Fig. 18.10). The coordinates of this new center are usually obtained by calculating the mean of the coordinates of the points belonging to that cluster (i.e., the center is calculated as the centroid of the group of patterns). Since the centers have moved, the pattern membership needs to be updated by recalculating the distance from each pattern to the new cluster centers (in the bottom-left panel in Fig. 18.10 three patterns move from one cluster to the other). The algorithm continues to update the cluster centers based on the new membership and update the membership of each pattern until the cluster centers are such that no pattern moves from one cluster to another. Since no pattern has changed membership, the centers will remain the same and the algorithm can terminate (bottom right in Fig. 18.10).

18.3.1.1 Characteristics of the k-means clustering

The k-means algorithm has several important properties. First of all, the results of the algorithm, i.e. the clustering or the membership of various patterns to various clusters, *can change* between successive runs of the algorithm (see Fig. 18.11). Furthermore, if some clusters are initialized with centers far from all patterns, no patterns will fall into their sphere of attraction and they will produce empty clusters. In order to alleviate these problems, care should be taken in the initialization phase. A common practice initializes centers with

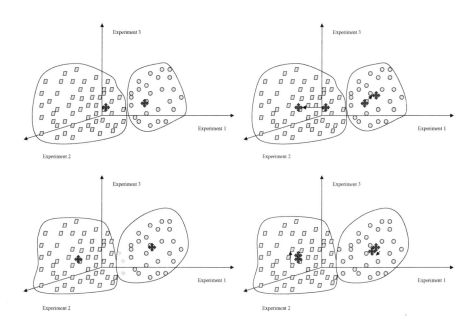

FIGURE 18.10: The k-means algorithm with $k = 2$. Upper-left: two cluster centers are chosen randomly and patterns are assigned to each cluster based on their distance to the cluster center. Upper-right: new centers are calculated based on the patterns belonging to each cluster. Bottom-left: patterns are re-assigned to the new clusters based on their distance to the new cluster centers. Three patterns move from the left cluster to the right cluster. Bottom-right: new cluster centers are calculated based on the patterns assigned to each cluster. The algorithm continues to try to reassign patterns, but no pattern will need to be moved between clusters. Since no patterns are moved from one cluster to another, the cluster centers remain the same and the algorithm stops.

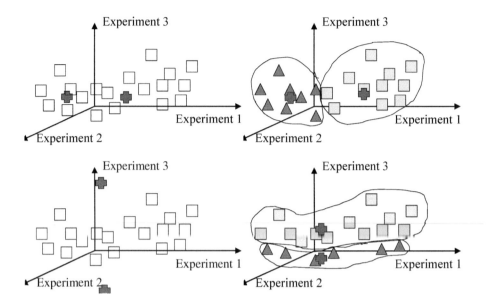

FIGURE 18.11: The k-means algorithm can produce different results in different runs depending on the initialization of the cluster centers. An initialization as in the upper left panel will lead to a clustering as in the upper right panel. An initialization as in the bottom left panel will lead to a clustering as in the bottom right panel.

k points chosen randomly from the existing patterns. This ensures that i) the starting cluster centers are in the general area populated by the given data and ii) each cluster will have at least one pattern. This is because if a pattern is initialized as a center of a cluster, it will probably remain in that cluster.

In the k-means example shown in Fig. 18.9, a different initialization might produce a different clustering in which the top cluster has only three patterns and the bottom cluster has seven patterns. A natural question arises regarding the meaning of the k-means clustering results: if k-means can produce different clusters every time, what confidence can one have in the results of the clustering? This issue can be refined into a number of questions that will be briefly considered in the following.

18.3.1.2 Cluster quality assessment

Given a particular clustering, how good is a particular cluster? Can one assess the quality of a specific cluster? One way to assess the goodness of fit of a given clustering is to compare **the size of the clusters versus the distance to the nearest cluster**. If the inter-cluster distance is much larger than the size of the clusters, the cluster is deemed to be more trustworthy (see Fig. 18.12).

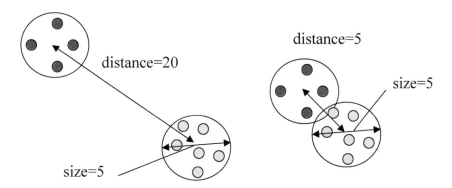

FIGURE 18.12: Cluster quality assessment. The quality of a cluster may be assessed by the ratio between its diameter and the distance to the nearest cluster. Image courtesy of BioDiscovery, Inc.

The ratio between the distance D to the nearest cluster center and its diameter d can be calculated for each cluster and can be used as an indication of the cluster quality.

Another possible quality indicator is the average of the **distances between the members of a cluster and the cluster center**. In the dendrogram shown in Fig. 18.14, the length of the branches of the tree are proportional to the average square distance of the members of the clusters to the cluster centroid. Thus, shorter clusters are better than taller clusters. Fig. 18.14 shows a clustering and the two-dimensional PCA plot of the same data. The height of the clusters reflects the average distance from the gene to the center of the cluster.

The **diameter of the smallest sphere** including all members of a given cluster may also be used as a cluster quality measure. Such as example is shown in Fig. 18.13. In this figure, the cluster to the left represents a genuine similarity of the patterns while the cluster to the right was formed mainly due to the imposed number of clusters (two). A potential disadvantage of this measure of cluster confidence is the fact that the diameter of the smallest sphere including all members of the cluster is determined by the farthest pattern from the cluster. In consequence, this measure is sensitive to cluster outliers. A tight cluster with only one pattern far from the center will have a large diameter.

Another interesting question is how confident can one be that a gene that fell into a cluster will fall into the same cluster if the clustering is repeated? This question can be addressed by repeating the clustering several times and following the particular gene of interest. Certain software packages offer the possibility of creating partitions based on the cluster membership. Such partitions will have all genes in a cluster coded with the same color (see Fig. 18.14).

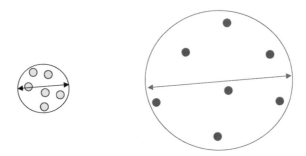

FIGURE 18.13: Cluster quality assessment. The quality of a cluster may be assessed by calculating the average distance from the members of the cluster to the centroid or the diameter of the sphere including all members of the cluster. The cluster to the left represents a genuine similarity of the patterns, while the cluster to the right was formed mainly due to the imposed number of clusters (two).

Repeating the clustering several times will reveal whether the colors remain grouped together. Those genes that are clustered together repeatedly are more likely to be genuinely similar.

Fig. 18.14 shows a clustering of the top 35 genes from the leukemia data set [183]. The left panel shows the result of a k-means clustering with $k = 5$. The genes are colored by their cluster membership. The right panel shows the same genes in a two-dimensional PCA plot of the first two principal components. Note that the patterns are not grouped in any meaningful way. In spite of this, the k-means dutifully produced five clusters as requested. Maintaining the color coding and repeating the clustering a few times shows that the same genes are grouped differently every time.

The question of how confident one can be that a given pattern belongs to a given cluster can also be addressed using a **bootstrapping** approach. Bootstrapping is a general technique that allows the computation of some goodness-of-fit measure based on many repeats of the same experiment on slightly different data sets all constructed from the available data [146]. The bootstrapping idea was discussed in Chapter 16 as a mean of correcting for multiple experiments. In the context of cluster confidence, this idea was investigated by a number of researchers [155, 257]. In particular, one could use the replicate measurement data in combination with a bootstrap approach in order to address the issue of cluster confidence.

Fig. 18.15 illustrates this approach. The idea is to use the fact that each gene expression value is usually the result of several measurements. In cDNA arrays, the gene expression may be the mean of the background corrected mean intensities of several spots. In Affymetrix arrays, the gene expression may be

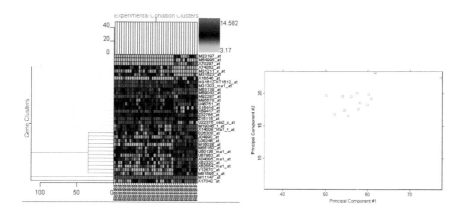

FIGURE 18.14: A clustering of the top 35 genes from the leukemia data set [183]. The left panel shows the result of a k-means clustering with $k = 5$. The genes are colored by their cluster membership. The right panel shows the same genes in a two-dimensional PCA plot. Note that the patterns are not grouped in any meaningful way. In spite of this, the k-means dutifully produced five clusters as requested. Maintaining the color coding and repeating the clustering a few times shows that the same genes are grouped differently every time. Note how the height of the clusters reflects the average distance from the gene to the center of the cluster. Image obtained with GeneSight 3.5.2.

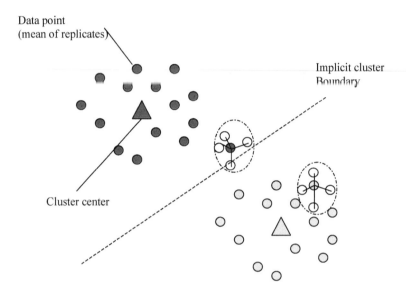

Data point
(mean of replicates)

Implicit cluster
Boundary

Cluster center

FIGURE 18.15: The residual-based cluster confidence approach [155, 257]. Each gene is represented by a mean of some individual measurements. The confidence that a given gene belongs to a cluster is inversely proportional to the number of times the gene falls into a different cluster during the bootstrapping process. The gene shown in red remains in the red cluster 3 out of 4 times, which translates into a confidence of 0.75. Image courtesy of BioDiscovery Inc.

the mean of the values measured on several arrays. ANOVA can be used as explained in Chapter 13 to fit a model suitable to the problem. The ANOVA will have some residuals that will correspond to the unexplained variation in the data. These residuals form a population. The bootstrapping approach proposed by Kerr and Churchill [257] draws n samples from this population and uses them to construct fictitious data that preserve the characteristics of the original data. In particular, the variance of the noise will be exactly the same as that of the original data. The clustering is performed on the original data. The clusters and boundaries between clusters are stored. Subsequently, the same clustering is performed n times using the data constructed from the model and the population of residuals. For each gene, the number of times that the gene moves from one cluster to another is inversely proportional with the confidence of that gene-cluster assignment. A gene that will remain in the same cluster independently of the noise will have a confidence of 1 or 100%. A gene that moves to a different cluster 50% of the time will have a confidence of 0.5. In Fig. 18.15, the number of fake measurements is $n = 4$. For 3 out of these 4 bootstrap runs, the gene close to the border between clusters remained in its cluster. However, in one of the 4 runs, the noise was such that the gene appeared to be in the adjacent cluster. This particular gene will have a confidence value of 0.75.

18.3.1.3   Number of clusters in k-means

The choice of the number of clusters is another issue that needs careful consideration. If it is known in advance that the patterns to be clustered belong to several different classes (e.g., cancer and healthy), one should cluster using the known number of classes. Thus, if there are features that clearly distinguish between the classes, the algorithm might use them to construct meaningful clusters. Note that it is not necessary to know which pattern belongs to each class but only that there are two different classes. If the analysis has an exploratory character and the number of existing classes is not known, one could repeat the clustering for several values of k and compare the results, i.e. track the genes that tend to fall in the same cluster for different values of k. This approach is heuristic in nature, and its utility will vary widely, depending on the particular problem studied.

18.3.1.4   Algorithm complexity

The complexity of the k-means algorithm must also be considered. It can be shown that the k-means algorithm is linear in the number of patterns, e.g. genes, N. This means that the number of computations that need to be performed can be written as $c \cdot N$, where c is a value that does not depend on N. The value c does depend on the number k of clusters chosen by the user as well as the number of iterations. However, the number of clusters is generally very small in comparison with the number of patterns. Typically, hundreds or thousands of genes are clustered in a handful of clusters. Overall, one can conclude that k-means has a very low computational complexity, which translates directly into a high speed.

18.3.2   Hierarchical clustering

Hierarchical clustering has been used since the very beginning of the microarray field [147, 206, 489]. Hierarchical clustering aims at the more ambitious task of providing the definitive clustering that characterizes a set of patterns in the context of a given distance metric. The result of k-means clustering is a set of k clusters. All these clusters, as well as all elements of a given cluster, are on the same level. As we have seen, no particular inferences can be made about the relationship between members of a given cluster or between clusters. In contrast, the result of a hierarchical clustering is a complete tree with individual patterns (genes or experiments) as leaves and the root as the convergence point of all branches.

The diagrams produced by the hierarchical clustering are also known as dendrograms. A dendrogram is a branching diagram representing a hierarchy of categories based on degree of similarity. Different genes and/or experiments are grouped together in clusters using the distance chosen. Different clusters are also linked together based on a cluster distance such as the average distance between all pairs of objects in the clusters. A combined dendrogram with gene clustering plotted horizontally and experiment clustering plotted vertically is presented in Fig. 18.16.

Unlike the real trees, hierarchical trees are usually drawn with the root on top and the branches developing underneath. The tree can be constructed in a **bottom-up** fashion, starting from the individual patterns and working upwards towards the root or following a **top-down** approach, starting at the root and working downwards towards the leaves. The bottom-up approach is sometimes called **agglomerative** because it works by putting smaller clusters together to form bigger clusters. Analogously, the top-down approach is sometimes called **divisive** because it works by splitting large clusters into smaller ones.

Unlike k-means, a hierarchical clustering algorithm should be completely deterministic. Applied on a given data set and using a chosen distance, the same hierarchical clustering algorithm should always produce the same tree. However, different hierarchical clustering algorithms, e.g. a bottom-up approach and a top-down approach, may produce different trees.

The bottom-up method works as follows. It starts with n clusters, each consisting of a single pattern. The pattern can be either a gene or an experiment, depending on what the algorithm is applied to. The algorithm calculates a table containing the distances from each cluster to every other cluster. For n points, this computation will require on the order of n^2 arithmetical operations. Then, the bottom-up method repeatedly merges the two most similar clusters into a single super-cluster until the entire tree is constructed. A distance able to assess the similarity of two clusters is required.

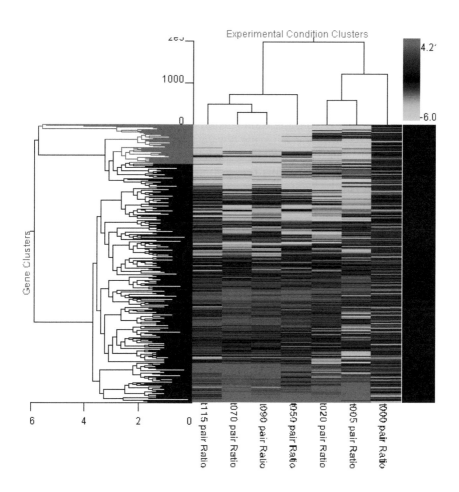

FIGURE 18.16: A hierarchical clustering of both genes and experiments in the yeast sporulation data set [4, 95]. Image obtained with GeneSight 3.5.2.

The top-down approach starts by considering the whole set of patterns to be clustered. Subsequently, the algorithm uses any of a large number of non-hierarchical clustering algorithms to divide the set into two clusters. A particular choice of such a non-hierarchical algorithm can be the k-means with $k = 2$. Subsequently, the process is recursively repeated on each of the smaller clusters as they are obtained. The process stops when all small clusters contain a single pattern. The top-down approach tends to be faster than the bottom-up approach.

Finally, another approach to building a hierarchical clustering uses **an incremental method**. This approach can be even faster than the top-down approach. Such methods build the dendrogram by adding one pattern at a time, with minimal changes to the existing hierarchy. In order to add a new gene, the gene under consideration is compared with each cluster in the tree, starting with the root and following always the most similar branch according to the distance used. When finding a cluster containing a single gene, the algo rithm adds a branch containing the gene under consideration. As mentioned, this approach can be lightning fast compared with the others. However, the weakness is that the results can depend not only on the distance metric (as any clustering) or the distance metric and some random initialization (as the top-down approach) but also on the *order* in which the points are considered.

18.3.2.1 Inter-cluster distances and algorithm complexity

The distance between clusters can be taken to be the distance between the closest neighbors (known as **single linkage** clustering), farthest neighbors (**complete linkage**), the distance between the centers of the clusters (**centroid linkage**), or the average distance of all patterns in each cluster (**average linkage**). The centroid of a group of patterns is the point that has each coordinate equal to the mean of the corresponding coordinates of the given patterns. For instance, the set of experiments: $Exp_1 = (1,2,3)$, $Exp_2 = (2,3,4)$, and $Exp_3 = (3,4,5)$ has the centroid in:

$$\left(\frac{1+2+3}{3}, \frac{2+3+4}{3}, \frac{3+4+5}{3} \right) \tag{18.28}$$

Clearly, the total complexity of the algorithm and therefore its speed is very much dependent on the linkage choice. Single or complete linkages require only choosing one of the distances already calculated, while more elaborated linkages, such as centroid, require more computations. Such further computations are needed *every time two clusters are joined*, which greatly increases the total complexity of the clustering. However, much like always, cheaper is not always better. Simple and fast methods such as single linkage tend to produce long, stringy clusters, e.g. if using a Euclidian distance. More complex methods such as centroid linkage or neighbor joining [398] tend to produce clusters that reflect more accurately the structure present in the data but are extremely slow. The complexity of a bottom-up implementation can vary between n^2 and n^3,

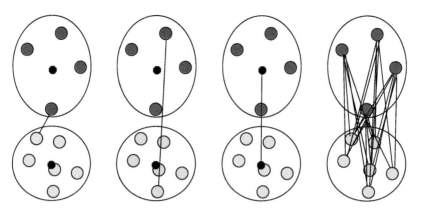

FIGURE 18.17: Linkage types in hierarchical clustering. Left to right: single linkage, complete linkage, centroid linkage and average linkage.

depending on the linkage chosen. In the context of gene expression, one should try to prune as much as possible the set of genes of interest before attempting to apply a bottom-up clustering with a more complex linkage.

18.3.2.2 Top-down versus bottom-up

In general, algorithms working by division require less computation and are therefore faster. However, obtaining the results quicker may not necessarily be a reason for joy because a hierarchical clustering algorithm working by division, or top-down, may produce results inferior to the results of an algorithm working by agglomeration. This can happen because in dividing the clusters the most important splits, affecting many patterns, are performed at the beginning before accumulating enough information and two patterns inadvertently placed in different clusters by an early splitting decision will never be put together again.

The top-down clustering tends to be faster. but the clusters produced tend to reflect less accurately the structure present in the data. Furthermore, theoretically, the results of a hierarchical clustering should only depend on the data and the metric chosen. However, a top-down approach will rely essentially on the qualities of the partitioning algorithm chosen. For instance, if k-means is chosen to divide clusters into sub-clusters, the overall result may be different if the algorithm is run twice with the same data. This can be due to the random initialization of the cluster centers in the k-means division. Fig. 18.18 shows two different clusterings obtained by running the same top-down hiearchical clustering algorithm twice on the top 35 genes selected from the ALL-AML data set [183].

The complexity of the top-down approach can require between $n \log n$ and

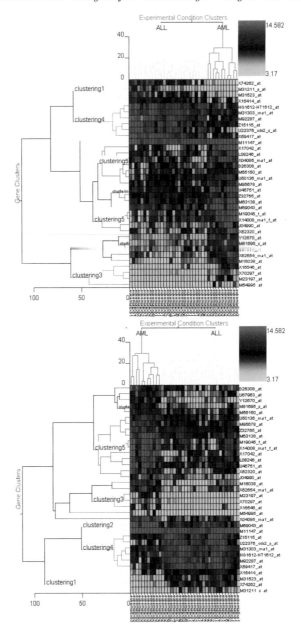

FIGURE 18.18: Two different hierarchical clusterings constructed by division (top-down) using k-means to split the larger clusters. The deterministic aspect of the hierarchical clustering is lost due to the usage of the non-deterministic k-means. Both diagrams above have been obtained using the same distance (Euclidean) on the same data (35 top genes in the ALL-AML data set [183]). Image obtained with GeneSight 3.5.2.

n^2 computations and is therefore intrinsically faster than the bottom-up approach, especially when a complex linkage is involved.

Fig. 18.19 shows the effect of the various combinations between the linkage type and the approach used. All diagrams in this figure use exactly the same data (35 top genes in the ALL-AML data set [183]) and the same distance (Euclidean). The upper left panel shows the results of the clustering using division; upper right: agglomeration with single linkage; bottom left: agglomeration with complete linkage and bottom right: agglomeration with average linkage. For this data, the centroid linkage produces the same clustering as the average linkage.

18.3.2.3 Cutting tree diagrams

A hierarchical clustering diagram may be used to divide the data into a predetermined number of clusters. This division may be done by cutting the tree at a certain depth (distance from the root). For instance, the tree in Fig. 18.20 can be cut to generate two clusters (left panel) or five clusters (right panel). The tree can also be cut at different depths on different branches in order to reflect better the structure of the data. The lowest possible cut is at the individual pattern level. This cut will always generate as many clusters as patterns with only one pattern in each cluster. The highest possible cut is at the root level. This is equivalent to saying that there is a single cluster containing all available data.

As another example, let us consider the set of genes following the nine profiles shown in Chapter 17, Example 17.2, Fig. 17.22. The data for the 900 genes following these profiles were stored in the matrix `mat`. We can build a dendrogram showing the hierarchical clustering of these genes as follows:

```
> d=dist(mat)
> hcgdata=hclust(d)
> palette("default")
> plot(hcgdata,main="Hierarchical clustering of 900
+  genes from the PCA example")
> yh=1.8
> lines(c(0,900),c(yh,yh))
> x <- palette(adjustcolor(palette(), 0.25))
> junk=sapply((1:8*100),function(x){
+ polygon(c(x+0,x+0,x+100,x+100),c(0,yh,yh,0),col=((x+1)/100))
+ })
> palette("default")
```

This dendrogram (shown in Fig. 18.21) was produced with the Euclidean distance and complete linkage, which are the default choices of `dist` and `hclust`, respectively. The vertical distance between the joining points of various branches are proportional to the distance between their corresponding subclusters. Note how all subclusters lower in the dendrogram (towards the leafs) are rather close to one another. If we cut the diagram horizontally at the

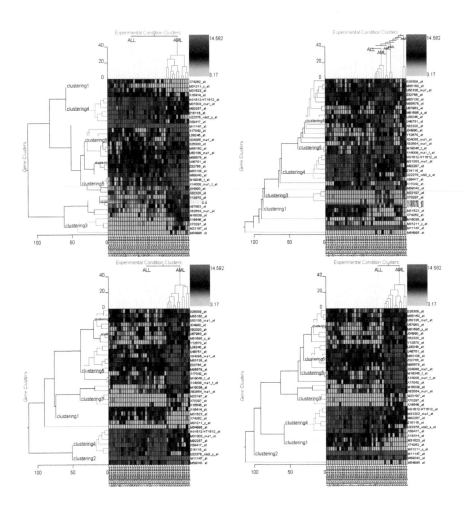

FIGURE 18.19: The effect of the various combinations between the linkage type and the approach used. All diagrams use the same data (35 top genes in the ALL-AML data set [183]) and the same distance (Euclidian). Upper left: division clustering; upper right: agglomeration with single linkage; bottom left: agglomeration with complete linkage and bottom right: agglomeration with average linkage. For this data, the centroid linkage produces the same clustering as the average linkage.

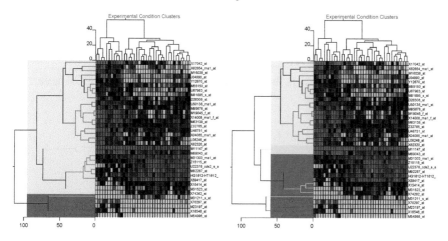

FIGURE 18.20: A complete hierarchical tree structure can be cut at various depths to obtain a different number of clusters. For instance, in the left panel, the dendrogram is cut at the depth of 1 to generate two clusters. The same dendrogram is cut at a depth of 4 to generate five clusters in the right panel. Hierarchical trees can also be cut at different depths on different branches in order to reflect better the structure of the data. The data used here are a subset of 35 genes from the AML-ALL data [183].

first level that shows a larger distance between the cluster, we obtain exactly nine sub-clusters, corresponding to the nine different sub-sets of genes, each following one of the nine profiles. The subcluster to the left, rather dissimilar to the other 8, corresponds to the group of genes shown in red near the origin of the PCA plot in Fig. 17.33. The two subclusters to the right of this diagram correspond to the two subclusters along the second principal component axis, PC2. Finally, the two groups of three subclusters in the middle correspond to the two groups of three clouds of points towards the extreme values of the first principal component in Fig 17.33.

18.3.2.4 An illustrative example

Armed with all the necessary knowledge, let us construct manually a very simple hierarchical clustering for the following example.

EXAMPLE 18.1

Let us consider a gene measured in a set of five experiments: A, B, C, D, and E. Let us consider that the values measured in the five experiments are: $A = 100$, $B = 200$, $C = 500$, $D = 900$, and $E = 1100$. We will construct the hierarchical clustering of these values using Euclidean distance, centroid linkage, and an agglomerative approach.

Hierarchical clustering of 900 genes from the PCA example

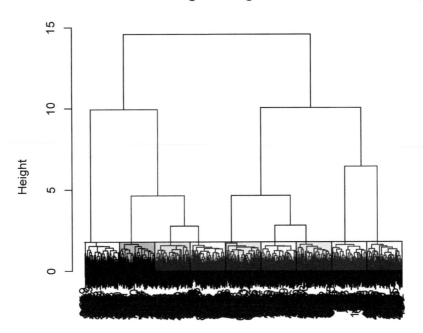

d
hclust (*, "complete")

FIGURE 18.21: A dendrogram showing the results of the hierarchical clustering of the 900 genes following the nine profiles used in the PCA example 17.2 from Chapter 17. The time profiles of these genes are shown in Fig. 17.22. This dendrogram was produced with the Euclidean distance and complete linkage. The vertical distance between the joining points of various branches are proportional to the distance between their corresponding subclusters. Note how all subclusters lower in the dendrogram (towards the leaves) are rather close to each other. If we cut the diagram horizontally at the first level that shows a larger distance between the cluster, we obtain exactly nine subclusters, corresponding to the nine different subsets of genes, each following one of the nine profiles.

SOLUTION   *The closest two values are 100 and 200. The centroid of these two values is 150. Now, we are clustering the values 150, 500, 900, 1100. The closest two values are 900 and 1100. These values are joined, and the centroid is calculated. The centroid of 900 and 1100 is 1000. The remaining values to be joined are 150, 500 and 1000. The closest values are 150 and 500. These values are joined together. Finally, the two resulting subtrees are joined in the root of the tree. The resulting dendrogram is shown in the left panel of Fig. 18.22.*   □

We obtained this result because we started with the experiments ordered by the measurements of the gene considered. However, when more than one gene is involved, the experiments cannot be ordered anymore. Therefore, it is important to see what happens when the values are considered in a different, arbitrary, order. The same clustering process applied to an initial arbitrary ordering will produce a different looking tree as shown in the right panel of Fig. 18.22. It is important to notice that the two trees shown in Fig. 18.22 are *exactly the same*. Indeed, tree are judged by their topology i.e. the way their branches converge. In both trees in Fig. 18.22, $A = 100$ is most similar to $B = 200$, $C = 500$ is most similar to the group (A, B), and $D = 900$ is most similar to $E = 1100$. However, the tree in the left panel shows the experiment A furthest from E while the tree in the right panel shows the experiment A closest to E. It clear from this example that **nothing can be inferred from the fact that two genes or experiments are plotted next to each other in a hierarchical dendrogram**.

18.3.2.5   Hierarchical clustering summary

A few conclusions can be drawn from this discussion of various hierarchical clustering methods. A first conclusion is that various hierarchical clustering implementations using the same data and the same metric can still produce different dendrograms if they use different approaches. Another important conclusion is that merely obtaining a clustering is not an issue and that the dendrogram itself is almost never the answer to the research question. A dendrogram connecting various genes in a graphically pleasant way can be obtained relatively quickly from any data set. The real problem is to obtain a clustering that reflects the structure of the data. A clustering that reflects well the properties of the data may require more work. Finally, various implementations of hierarchical clustering should not be judged simply by their speed. Many times, slower algorithms may simply be trying to do a better job of extracting the data features. Finally, and most importantly, hierarchical diagrams convey information only in their topology. The order of the genes within a given cluster and the order in which the clusters are plotted do not convey useful information and can be misleading.

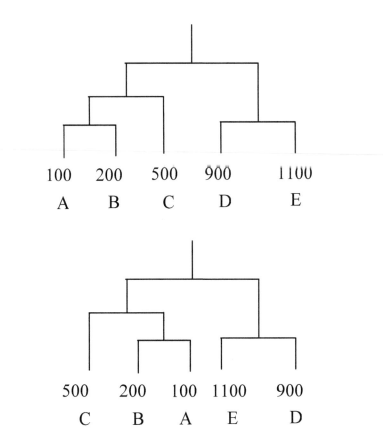

FIGURE 18.22: Two hierarchical clusters of the expression values of a single gene measured in five experiments. The dendrograms are identical: both diagrams show that A is most similar to B, C is most similar to the group (A,B), and D is most similar to E. In the top panel, pattern A and E are plotted far from each other. In the bottom panel, A and E are immediate neighbors. This example shows that the proximity in a hierarchical clustering (e.g., A and E in the bottom panel) does not necessarily correspond to similarity.

18.3.3 Kohonen maps or self-organizing feature maps (SOFM)

The **Kohonen map**, also called **self-organizing feature map** (SOFM), was proposed by Teuvo Kohonen in the late 1980s [268, 269]. The SOFM is a type of clustering. As any clustering algorithm, the SOFM will divide the input patterns into groups of similar patterns. In this respect, it is similar to k-means and hierarchical clustering. However, unlike the clustering produced by k-means and hierarchical clustering, the relationship between the clustered patterns actually conveys information about the relationships and reciprocal positions of the patterns in the original input space.

As we have seen, in k-means, the relative position of the patterns in the resulting clustering is not only uninformative but can also be misleading. This was illustrated in Fig. 18.9 in which two patterns very close in the input space were assigned to different clusters and appeared very far from one another in the resulting clustering. The hierarchical clustering is more informative than the k-means inasmuch as large groups of patterns that are indistinguishable in k-means are further divided and organized into subtrees that provide more information regarding the relative relationships between the respective patterns. However, even in a hierarchical clustering diagram, the elements of any particular subtree can still be plotted swapped around, bringing dissimilar patterns to be drawn nearby in the dendrogram (Fig. 18.22).

Unlike k-means and hierarchical, the SOFM clustering is designed to create a plot in which similar patterns are plotted next to one another. It is said that the SOFM maps the input space into a feature space in which the neighborhood relationship reflects the degree of similarity between patterns. Plotting the patterns in such a space creates a **feature map**. A feature map has the property that distances and relationships measured on the feature map are proportional to distances and relationships between patterns according to the similarity metric chosen.

The SOFM is actually a **neural network** technique [192, 203]. Neural networks are a class of techniques inspired by the brain. The fundamental paradigm is to perform complex computations using networks of very simple elements. These simple elements are called units or neurons. In such networks, much like in the brain, the processing abilities come from the pattern and strength of the connections between units. A SOFM is usually a grid of such very simple elements. A SOFM can use a one-dimensional grid (like a string), a two-dimensional grid (an array), or a three-dimensional grid (a cube or parallelepiped). One-dimensional and two-dimensional SOFMs are usually the most widely used.

Fig. 18.23 illustrates a two-dimensional SOFM with four inputs. Each unit of the SOFM is connected to all inputs. Each such connection is characterized by a **weight** or **connection strength**. For any unit in the SOFM, all its weights form a vector. The size of the vector is equal to the number of dimensions of the input space since each unit has a link from each input. If

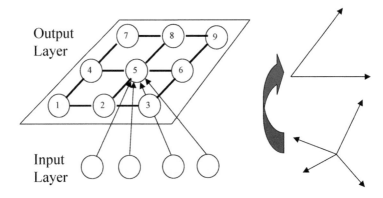

FIGURE 18.23: A two-dimensional self-organizing feature map (SOFM). Each unit in the SOFM is connected to all inputs. The map uses a distance measure and a neighborhood. The neighborhood of a unit is a set of nearby units. For instance, the neighborhood of unit 5 can include units 2, 4, 6, and 8 (a 4-neighborhood) or units 1, 2, 3, 4, 6, 7, 8, and 9 (an 8-neighborhood). This SOFM implements a dimensionality reduction by projecting a space with four dimensions into a space with two dimensions.

the SOFM is used to analyze experiments, each experiment will be a pattern and the number of inputs (or the number of input features or the dimensionality of the input space) will be equal to the number of genes. If the SOFM is used to analyze genes, each gene will be a pattern and the dimensionality of the input space will be equal to the number of experiments. The values of the weight vector determine a point in the input space. It is said that each unit represents a **prototype**. The weights are initialized with random values, which is equivalent to saying that the units of SOFM are randomly distributed in the input space.

The SOFM is constructed by **training**. The neural network training process is similar to the learning process that happens in the brain. The input patterns are presented repeatedly. Every such presentation of an input pattern modifies slightly the strength of some of the connections in the network. If the connections are modified randomly, nothing useful happens. However, if the connections are modified according to some **training rule**, the process leads to a gradual adaptation that is known as **learning**. The purpose of the learning in the training of the self-organizing feature map is to extract from the data the most important features and, based on them, to group the data into meaningful clusters that share such important features. In SOFM, this is achieved using two ideas: a **neighborhood** and a **winner-take-all approach**.

Let us assume that we are clustering 10,000 genes. Let us assume that each gene was measured in each of four experiments. The input space has four dimensions and there are 10,000 patterns, each pattern being described by a vector with four elements. The SOFM is trained as follows. Each gene is presented to the input of the network. Using the chosen distance, each unit in the SOFM calculates the distance between its weight vector (a vector of 4 numbers) and the current gene (another vector of 4 numbers). The unit which is found to be closest to the current gene is declared to be the winner. The weights of the winner are modified in such a way that the weight vector becomes more similar to the current gene. The exact amount of modification brought is determined by a small positive number called the **learning rate**. Essentially, the input winner is moving in the direction of the current input pattern with the magnitude of the move being determined by the learning rate. At the same time, all units in the neighborhood of the winner are also changed in the same way but to a lesser extent. The winning unit is practically pulling its neighbors closer to the current gene. This process can be visualized by imagining all SOFM units being connected to each other by rubber bands. When a particular unit is moved, its neighbors will also be moved a little bit. The process then continues for the next gene until all 10,000 genes have been processed. This represents one iteration of the algorithm. In order to ensure a convergence, the learning rate and sometimes the size of the neighborhood is gradually reduced over a number of iterations.

The result of this training algorithm is that different units in the SOFM become prototype genes representing profiles most often encountered in the genes analyzed. The SOFM provides three benefits. First, each unit of the SOFM will contain a prototype. This prototype will represent the typical behavior of the genes triggering that particular unit. More generally, the prototype will represent the set of common features extracted from the input patterns by a given unit. Second, the SOFM yields a set of clusters. All input patterns activating a unit will be clustered together. Finally, the relationship between the units activated by specific genes will be closely related to the relationships between the genes. A gene will always be more similar to another gene in its immediate neighborhood than to a gene farther apart. Also the placement of the clusters themselves will have the same feature. The landscape of the feature map constructed by the Kohonen network will represent faithfully the properties and features of the data. For this reason, it is said that the SOFM implements a dimensionality reduction: the n-dimensional input space is projected into a space with one, two or three dimensions depending on the network used (see Fig. 18.23).

SOFMs can be used as clustering and/or visualization tools. One could simply perform the Kohonen training and then plot the clusters generated from the SOFM. An example of this usage is shown in Fig. 18.24.

Another way to use the SOFM is to plot the prototypes in the input space together with the topology of the network showing the links between the units. For example, Fig. 18.25 represents a two-dimensional self-organizing feature map trained on three-dimensional data. Each dot represents a gene. The three

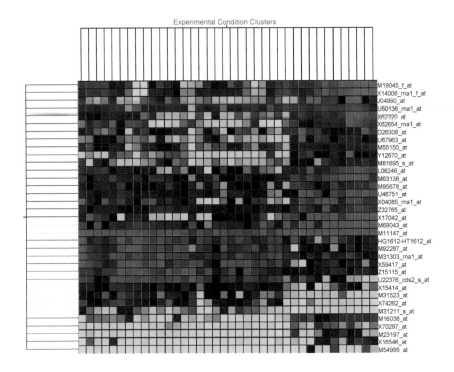

FIGURE 18.24: The results of a one-dimensional clustering in Euclidian distance on both genes and experiments for 35 selected genes from the ALL-AML data set [183]. Unlike the results of a k-means or hierarchical clustering, two genes that are plotted next to each other are necessarily similar according to the chosen distance.

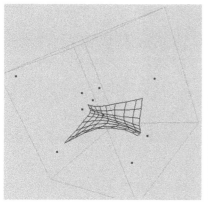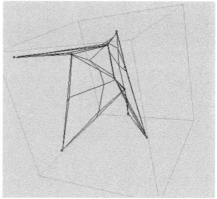

FIGURE 18.25: A two-dimensional self-organizing feature map trained on 3-dimensional data. Each dot represents a gene. The three coordinates of each dot are the expression values measured in three different mRNA samples. The left panel represents an early stage during the training. The feature map has started to bend and stretch in order to capture the features of the data. The right panel represents the feature map at the end of the training. The feature map is bent and stretched such that is covers the entire data set. In this case, there were more units than genes so each unit models a gene and there are unused units. Note that similar genes are represented by neighboring nodes in the network.

coordinates of each dot are the expression values measured in three different mRNA samples. The left panel represents an early stage during the training. The feature map has started to bend and stretch in order to capture the features of the data. The right panel represents the feature map at the end of the training. The feature map is bent and stretched such that it covers the entire data set. In this case, there were more units than genes so each unit models a gene and there are unused units. Note that similar genes are represented by neighboring nodes in the network. Since the plot is drawn in the input space, it is clear that this approach can be used directly only when the dimensionality of the input space is small (2-3 dimensions). However, this technique can be coupled nicely with PCA. The PCA can be used first to project the problem into a space with two or three dimensions, and then SOFM can be used to create a feature map that would indicate similarities and relationships between the various data points.

Finally, SOFMs can be used for visualization. In this case, the plot will show the units of the trained Kohonen network as they are activated by various input patterns. This is truly a dimensionality reduction since any n-dimensional pattern can be seen through the activation determined by it on the one-, two-, or three-dimensional feature map. Such an example is shown

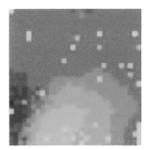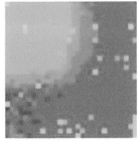

FIGURE 18.26: Examples of responses of a trained 20×20 Kohonen map when patterns from three different clusters were used as inputs. The patterns are structures of several HIV proteases resistant to Indinavir. Each HIV protease is described by a vector with 22 elements. The 22-dimensional input space is conveniently projected on the two-dimensional space of the feature map while maintaining the relationships between patterns. Left to right: high resistance, medium resistance, and low resistance patterns [133]. Images obtained with SNNS.

in Fig. 18.26. The patterns are three-dimensional atomic structures of several HIV proteases resistant to Indinavir. Each HIV protease is described by a vector with 22 elements. The 22-dimensional input space is conveniently projected on the two-dimensional space of the feature map while maintaining the relationships between patterns. From left to right, the figure shows the activation of the map when a high resistance, medium resistance and low resistance pattern is presented at the inputs [133].

Another type of visualization available with SOFM is to plot together all input patterns that activate a given unit from the Kohonen map. Fig. 18.27 shows a two-dimensional self-organizing feature map constructed on a 4×4 network. Each profile represents the expression level of a yeast gene over the sporulation time series [95]. Each of the 4×4 cells corresponds to a unit from the Kohonen network. The profiles plotted in each cell are those of the genes that are mapped into that particular unit. Note that the profiles in any two neighboring cells share certain characteristics.

As for any other algorithm using a distance to assess similarity, the results obtained with the SOFM depend essentially on the choice of the distance. To illustrate this, we will consider again the data shown in Fig. 17.24. These data contain several genes having distinct profiles clearly visible in a three-dimensional PCA plot such as that in Fig. 17.29. A self-organizing feature map of these data obtained using the Euclidean distance is shown in Fig. 18.28. The Euclidean distance groups together profiles with small differences in values regardless whether they occur at the same or different time points. The clusters group together profiles with expression values in the same range (e.g., row 1, column 1 and row 1, column 3) without distinguishing between the

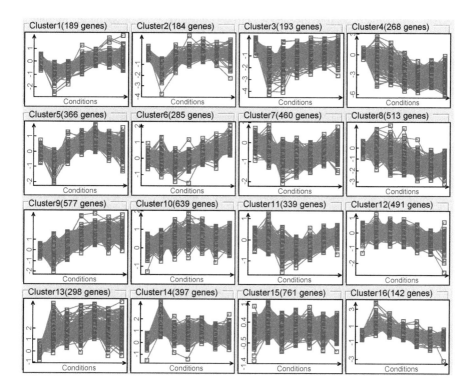

FIGURE 18.27: A two-dimensional self-organizing feature map on a 4×4 network. Each profile represents the expression level of a yeast gene over the sporulation time series [95]. Each of the 4×4 cells corresponds to a unit from the Kohonen network. The profiles plotted in each cell are those of the genes that are mapped into that particular unit. Note that the profiles in any two neighboring cells share certain characteristics.

different shapes within the given range. Although informative, the clustering is not perfect: three of the resulting clusters are empty. A different initialization and a different range of parameters (neighborhood radius and learning rate) might produce a better clustering. Given the data, the results of the clustering are good but not exceptional. The SOFM was able to identify 3 of the profiles present in the data: row 1, column 2, row 2, column 2 and row 3, column 1. However, the "up-down-up" and "down-up-down" profiles were grouped together at low intensities (row 1, column 1). The same happened for the "up-down-up" and "down-up-down" profiles at high intensities (row 1, column 3). We would have probably preferred to see that the algorithm identifies each individual shape as well as their ranges.

Fig. 18.29 shows the same data clustered using the same two-dimensional SOFM but with a correlation distance instead of the Euclidean distance. As expected for the correlation distance, the clusters group together the profiles with a similar shape over time regardless of their position on the vertical axis. For instance, the cell on row 2, column 1 contains all genes with an "up-down-up" behavior regardless of whether they vary around -2, 0, or 2. Like any microarray data set with many genes, this data set contains a lot of random noise. Many genes affected by random noise happen to have profiles similar to the meaningful genes and are picked up by various clusters. The cell on row 1, column 2 found the "low – steady increase – high" behavior but also captured some noisy genes.

Finally, Fig. 18.30 shows the self-organizing feature map obtained from the same data using the Chebychev distance. As discussed in Section 18.2, the Chebychev distance uses the maximum differences between any pair of coordinates. In this case, the Chebychev distance is most useful. The SOFM was able to extract in an unsupervised manner 8 out of the 9 different gene profiles present in the data. The network was able to distinguish between the similar "up-down-up" profiles in row 2, column 1 and row 3, column 3.

The self-organizing feature maps also have drawbacks. The random initialization makes the results nondeterministic: the same algorithm applied twice on the same data may produce different results. The initialization of the Kohonen network may be tricky. If some initial prototypes are initialized with values very unlike any of the data, they will never win the competition, will never be modified, and will remain unused. To avoid this, one may always initialize the network with some random patterns from the data. This will ensure that all units will win at least once (for the pattern they have been initialized with), which means they will all be used.

Another drawback of SOFM is related to the size (number of units) and layout (string, array or cube) of the network. Choosing the size and layout of the network has an heuristic component. In most cases, this is done by trial and error. Usually, many very similar clusters may indicate a network that is too large for the given data and indistinct clusters lumping together many inhomogeneous patterns may indicate that the network is too small.

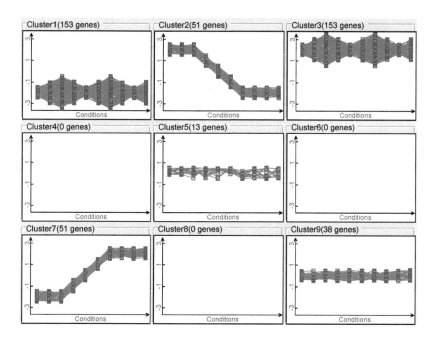

FIGURE 18.28: A two-dimensional SOFM clustering of the TIGR sample data shown in Fig. 17.24 using the Euclidean distance. The Euclidean distance groups together profiles with small differences in values regardless whether they occur at the same or different time points. The clusters group together profiles with expression values in the same range (e.g., row 1, column 1 and row 1, column 3) without distinguishing between the different shapes within the given range. Image obtained with GeneSight 3.5.2.

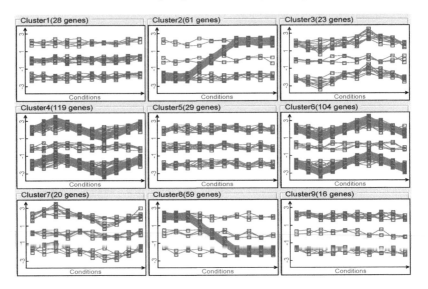

FIGURE 18.29: A two-dimensional SOFM clustering of the TIGR sample data shown in Fig. 17.24 using the correlation distance. The correlation distance groups together the profiles with the same shape disregarding the range. Image obtained with GeneSight 3.5.2.

18.4   Partitioning around medoids (PAM)

Another approach to clustering is called **Partitioning Around Medoids** (PAM) [251]. Similarly to k-means and hierarchical clustering, PAM starts with computing a dissimilarity matrix ($n \times n$) from the original data structure (the $n \times p$ matrix of measurements). Hence, any distance measure discussed in Section 18.2 can be used in conjunction with PAM. The algorithm maps the resulting distance matrix into a specified number of clusters. The medoids are representations of the cluster centers, which are robust with respect to outliers. The robustness is particularly important in the common situation in which many elements do not have a clear-cut membership to any specific cluster [429]. A measure of cluster distinctness is the *silhouette* computed for each observation in a data set, relative to a given partition of the data set into clusters. The silhouette measure contrasts the average proximity of an observation to other observations in the partition to which it is assigned with the average proximity to observations in the nearest partition to which it is not assigned. This quantity tends to one for a "well-clustered" observation and can be negative if an observation seems to have been assigned to the wrong cluster.

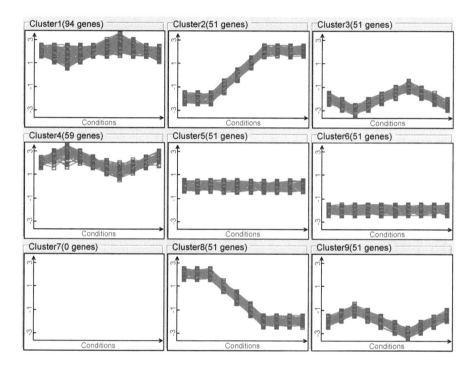

FIGURE 18.30: A two-dimensional SOFM clustering of the TIGR sample data shown in Fig. 17.24 using the Chebychev distance. In this case, the Chebychev distance is most useful. The SOFM was able to extract in an unsupervised manner 8 out of the 9 different gene profiles present in the data. Note how the use of the Chebychev distance allowed the SOFM to separate between the similar up-down-up profiles in row 2, column 1 and row 3, column 3. Image obtained with GeneSight 3.5.2.

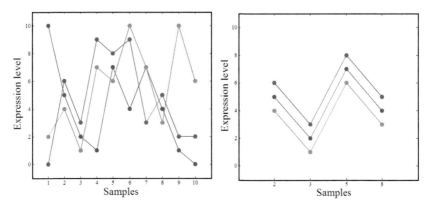

FIGURE 18.31: An example of biclustering. The expression levels of three genes over 10 samples are shown. (a) The genes are not correlated across all of the 10 samples. (b) The genes are strongly correlated in a subset of the samples (2,3,5,8).

18.5 Biclustering

In analyzing microarray expression data, one can observe that the activities of genes are not independent of each other.[5] Hence, it is important to study groups of genes rather than to perform a single gene analysis. However, traditional clustering techniques, such as k-means and hierarchical clustering, assume that related genes should have similar expression profiles across all the samples [288]. This assumption does not hold in all of the experiments. From the biological perspective, not all the genes are involved in each biological pathway, and some of these pathways may be active under a subset of the samples [293]. **Biclustering** was proposed to overcome these limitations of traditional clustering algorithms [91].

Biclustering refers to simultaneously clustering both rows and columns in a given matrix, which helps in discovering local patterns that cannot be identified by the standard one-way clustering algorithms. A bicluster can be defined as a subset of genes that are correlated under a subset of samples. Biclustering has been used in several applications such as clustering microarray data [293], identifying protein interactions [280], collaborative filtering [170], and text mining [76].

To illustrate the concept of biclustering, a simple example is shown in Figure 18.31. In this figure, the expression levels of three genes across 10 samples

---

[5]The section on biclustering was contributed by Chandan Reddy, Assistant Professor, Department of Computer Science, Wayne State University.

are shown. Considering all of the samples, it is difficult to observe any correlation between the three genes as shown in Figure 18.31(a). However, a strong correlation between the three genes can be found in a subset of the samples, namely {2,3,5,8}, as shown in Figure 18.31(b). Traditional clustering techniques cannot capture such correlations. However, biclustering has emerged as a powerful tool to simultaneously cluster both dimensions of a data matrix by using the relationship between the genes and the samples [326].

There are several issues that should be considered while searching for biclusters in gene expression data. A subset of genes can be correlated only across a small subset of conditions due to the heterogeneity of the samples, which could be taken from different patients. Moreover, a gene can be involved in more than one biological pathway; therefore, there is a need for a given biclustering algorithm to allow for overlapping between the biclusters [111], i.e., the same gene can be a member of more than one bicluster (Fig. 18.32(b)). In addition, since genes can be positively or negatively correlated [238], it is important to allow both types of correlations in the same bicluster. Furthermore, the biclusters can be arbitrarily positioned in the gene expression data.

18.5.1 Types of biclusters

There are several interesting types of biclusters that can be biologically relevant [293]:

- Biclusters with constant values. All the elements in this type have the same value. An example of this type is shown in red in Figure 18.32(a).

- Biclusters with constant values on rows. An example of this type is shown in green in Figure 18.32(a).

- Biclusters with constant values on columns. An example of this type is shown in blue in Figure 18.32(a).

- Biclusters with coherent values. In this type, a subset of genes are co-regulated in a subset of samples without considering their actual expression levels. An example of this type is shown in Figure 18.31(b).

Also, there are different types of correlations between genes in any cell. Examples of such relationships are positive and negative correlations [469]. Figure 18.33 shows an example of these correlations. In a positive correlation, genes show similar patterns while in a negative correlation, genes show opposite patterns [326]. Since it is possible that genes with both types of correlations exist in the same biological pathway [238], new algorithms for simultaneously capturing both types of correlations are being proposed.

In addition to the variations in the types of the biclusters, there are other important sources of variations, such as the variations in the size and the position of the biclusters in the gene expression data. Figure 18.32(b) shows

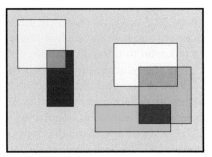

5	1	4	9	2	5	6	7	7	7
1	10	9	6	0	6	4	7	7	7
2	2	3	4	5	3	8	7	7	7
9	2	3	4	5	7	5	7	7	7
2	2	3	4	5	0	4	0	5	9
8	2	3	4	5	3	9	8	8	1
5	1	9	5	1	2	9	2	4	3
10	4	6	4	4	4	4	4	5	8
1	3	0	5	5	5	5	5	2	4
4	8	1	6	6	6	6	6	8	1

FIGURE 18.32: Examples of bicluster types and structures. The bicluster in red includes genes that are constant for that subgroup of samples. The bicluster in green includes genes that are constant on rows. The bicluster in blue includes genes that are constant on columns. The right panel shows possible overlapping and arbitrary placement of biclusters.

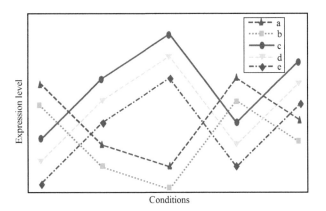

FIGURE 18.33: Different types of relationships between genes in one co-cluster. The genes $\{a,b\}$ are positively correlated with each other, and the genes $\{c,d,e\}$ are positively correlated with each other. However, the genes $\{a,b\}$ are negatively correlated with the genes $\{c,d,e\}$.

2	3	4	6	5	5	1	5	6
2	3	4	6	5	5	9	1	4
2	3	4	2	2	2	6	9	5

FIGURE 18.34: Three biclusters with their errors computed using the MSR function defined in Equation (18.29). The errors are 0 in the left cluster, 0.05 in the middle cluster, and 6.1 in the right cluster.

overlapping and arbitrarily positioned biclusters. Hence, extracting such biclusters from large gene expression data becomes an important challenge. It has been proved that the problem of finding all the significant biclusters is an NP-hard problem [91]. However, there are many biclustering algorithms that have been proposed to discover at least a subset of biclusters from gene expression data.

18.5.2 Biclustering algorithms

In the first biclustering algorithm, Cheng and Church (CC) proposed the mean-squared residue (MSR) score as a measurement of the coherence between genes [91]. Given a gene expression submatrix X that has I genes and J conditions, the residue is computed as follows:

$$H(I,J) = \frac{1}{|I||J|} \sum_{i \in I, j \in J} (x_{ij} - x_{Ij} - x_{iJ} + x_{IJ})^2 \qquad (18.29)$$

where $x_{iJ} = \frac{\sum_{j \in J} x_{ij}}{|J|}$ is the row mean, $x_{Ij} = \frac{\sum_{i \in I} x_{ij}}{|I|}$ is the column mean and $x_{IJ} = \frac{\sum_{i \in I, j \in J} x_{ij}}{|I||J|}$ is the overall mean of the matrix X. A perfect bicluster will have MSR=0. Figure 18.34 shows examples of three biclusters and their corresponding MSR values. The MSR function has been used in many other biclustering algorithms as well [91, 468, 111, 325].

The CC algorithm starts with the original data matrix; then a set of row/column deletions and additions are applied to produce one bicluster, which will be replaced with random numbers. This procedure is repeated until a certain number of biclusters is obtained. The algorithm has two main limitations: (i) It finds only one bicluster at a time, and (ii) random interference (masking the discovered biclusters with random numbers) reduces the quality of the biclusters and obstructs the discovery of other biclusters.

Coupled two-way clustering (CTWC) technique was proposed in [171]. In this technique, a subset of genes (conditions) are used to cluster the conditions (genes), while the Order-Preserving Submatrices (OPSMs) [43] algorithm finds local patterns in which the expression levels of all genes induce the same linear ordering of the experiments. However, the OPSM algorithm finds only one bicluster at a time and captures only positively correlated genes. The XMOTIF algorithm finds a subset of genes that is simultaneously conserved across a subset of samples [317].

Iterative Signature Algorithm (ISA) [232] is a statistical biclustering algorithm which defines a transcription module (bicluster) as a co-regulated set of genes under a set of experimental conditions. ISA starts from a set of randomly selected genes (or conditions) that are iteratively refined until they are mutually consistent. At each iteration, a threshold is used to remove noise and to maintain co-regulated genes and the associated co-regulating conditions.

Robust Overlapping Biclustering (ROCC) [111] is a biclustering algorithm that works with several Bregman divergence measures. This model allows overlapping biclusters and finds k row clusters and l column clusters simultaneously. ROCC performs biclustering through two steps. In the first step, the Bregman biclustering algorithm [33] is used to find biclusters arranged in a grid structure. In the second step, biclusters with large errors are pruned; then similar biclusters are merged. This algorithm does not handle the negative correlation among the rows.

POsitive and NEgative correlation based Overlapping biclustering (PO-NEOCC) was proposed in [325] to efficiently extract significant biclusters from gene expression data. This algorithm uses a ranking-based objective function to find the biclusters with minimum errors and is able to extract large and arbitrarily positioned overlapping biclusters simultaneously. Furthermore, both of positively and negatively correlated genes are allowed to be members of the same bicluster. Capturing both of the positive and negative correlations has been shown to be a biologically important requirement when searching for biclusters in gene expression data [469].

Some of the above biclustering algorithms, such as CC, ISA and OPSM, are available in the Biclustering Analysis Toolbox (BicAT) http://www.tik.ee.ethz.ch/sop/bicat/ (www.tik.ee.ethz.ch/sop/bicat/). BicAT is a graphical user interface tool that provides some of the popular biclustering and other standard clustering algorithms [34].

18.5.3   Differential biclustering

Microarray studies are used to measure the expression level of thousands of genes under different conditions. These conditions could be different tissue types (normal vs. cancerous), different subject types (male vs. female), different group types (African-American vs. Caucasian American) [252], different stages of cancer (early stage vs. late stage) [326] or different time points [175]. Differential Biclustering [326, 329, 154] aims to find gene sets that are correlated under a subset of conditions in one class of conditions but not in the other class. Identifying such differential biclusters can provide valuable knowledge for understanding the roles of genes in various diseases [326].

As an example, the results of a biclustering on the ALL-AML data are shown in Figure 18.35.

18.5.4   Biclustering summary

Compared to traditional one-dimensional clustering, biclustering is considered more informative and more practical [33] because it simultaneously measures the degree of coherence in the samples and the attributes of a given matrix [171]. There had been a lot of interest in extracting biclusters from gene expression data and several researchers have developed different biclustering algorithms for this purpose. This chapter provided an overview of different types of biclusters and various biclustering algorithms.

18.6   Clustering in R

Let us revisit the example used in Section 18.2.11 in order to illustrate some of R's clustering capabilities. In this example, there were 5 experiments and a gene g_1 that has an expression of 1, 2, 3, 4, and 5 in the 5 experiments, respectively. This gene can be represented as $g_1 = (1,2,3,4,5)$. Let us also consider the genes $g_2 = (100,200,300,400,500)$ and $g_3 = (5,4,3,2,1)$. The function dist calculates the distances between the data points represented as rows in a data matrix.

```
> genes=matrix(c(1,2,3,4,5,100,200,300,400,500,5,4,3,2,1),
+ byrow=T,ncol=5)
> genes

     [,1] [,2] [,3] [,4] [,5]
[1,]    1    2    3    4    5
[2,]  100  200  300  400  500
[3,]    5    4    3    2    1

> d=dist(genes)
> d

           1          2
2 734.203650
3   6.324555 736.922655
```

The various distances currently available can be selected using the parameter method. This parameter can take one of the values: "euclidean," "maxi-

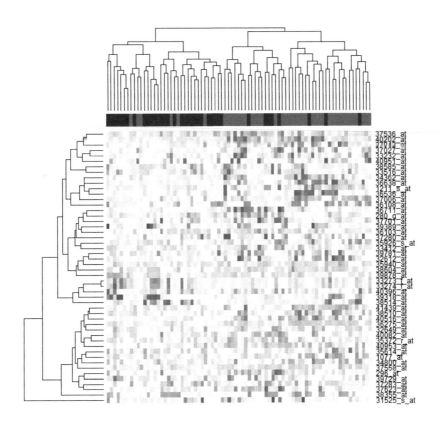

FIGURE 18.35: Biclustering the ALL data after filtering. Class membership is indicated by blue (fusion) or magenta (fusion negative) stripe at top of plot region. Rows correspond to data features (genes) while columns to data points (samples). Hierarchical clustering is applied simultaneously to both rows (genes) and columns (samples) of the expression matrix.

mum," "manhattan," "canberra," "binary," or "minkowski". In fact any unambiguous substring of the above can also be used. The "maximum" here is the Chebychev distance discussed above. For the Minkowski distance, the exponent can be specified using the parameter p, equal to 2 by default. By default, printing a distance matrix will only show the minimum number of values, i.e. only half of the matrix and without the diagonal. Since a distance should be symmetric, the other half of the matrix is redundant. Furthermore, the distance $d(x,x) = 0$ so the diagonal should be zero. However, the function dist has options for plotting the diagonal and/or the upper half of the matrix.

```
> dist(genes,method="max",diag=T,upper=T)

    1   2   3
1   0 495   4
2 495   0 499
3   4 499   0

> dist(genes,method="man",diag=T,upper=T)

     1    2    3
1    0 1485   12
2 1485    0 1485
3   12 1485    0

> dist(genes,method="euc",diag=T,upper=T)

          1          2          3
1  0.000000 734.203650   6.324555
2 734.203650   0.000000 736.922655
3   6.324555 736.922655   0.000000

> dist(genes,method="can",diag=T,upper=T)

         1        2        3
1 0.000000 4.900990 2.000000
2 4.900990 0.000000 4.831802
3 2.000000 4.831802 0.000000

> dist(genes,method="min",diag=T,upper=T,p=1)

     1    2    3
1    0 1485   12
2 1485    0 1485
3   12 1485    0

> dist(genes,method="min",diag=T,upper=T,p=2)
```

```
           1          2          3
1    0.000000 734.203650   6.324555
2 734.203650   0.000000 736.922655
3   6.324555 736.922655   0.000000

> dist(genes,method="min",diag=T,upper=T,p=3)

           1          2          3
1    0.000000 602.137998   5.241483
2 602.137998   0.000000 605.396935
3   5.241483 605.396935   0.000000
```

The code above also illustrates the relationship between Minkowski, Euclidean and Manhattan distances. Arbitrary distances, including 1 - correlation, can be implemented easily as follow:

```
> dd = as.dist((1 - cor(t(genes)))/2)
> dd
```

```
              1             2
2 5.551115e-17
3 1.000000e+00 1.000000e+00
```

Note the use of the function as.dist, which is used to make the matrix produced by 1- cor(t(genes)) look like a distance matrix.

The function hclust can be used to produce a hierarchical clustering. Let us use this function to illustrate the different ways in which the three genes above are clustered by Euclidean and correlation distances. The correlation distance is expected to place g_1 in the same cluster with g_2 and in a different cluster from g_3 because $(1,2,3,4,5)$ and $(100,200,300,400,500)$ have a high correlation $(d(g_1,g_2) = 1 - r = 1 - 1 = 0)$, whereas $(1,2,3,4,5)$ and $(5,4,3,2,1)$ are anticorrelated $(d(g_1,g_3) = 1 - (-1) = 2)$. However, the Euclidean distance will place g_1 in the same cluster with g_3 and in a different cluster from g_2 because $d_E(g1,g2) = 734.20$ while $d_E(g1,g3) = 6.32$.

```
> par(mfrow=c(2,1))
> plot(hclust(d),main="Euclidean distance",sub=NULL)
> plot(hclust(dd), main="correlation distance")
```

The two clusters are shown in Fig. 18.36.

Let us now build a cluster diagram for the ALL/AML data used previously. First, we will take advantage of the popularity of this data set and load the preprocessed data from an existing package, multtest:

```
> library(multtest); data(golub)
> dimnames(golub)[[1]]<-golub.gnames[,3]
> dimnames(golub)[[2]]<-c(paste("ALL", 1:27, sep=""),
+ paste("AML", 28:38, sep=""))
```

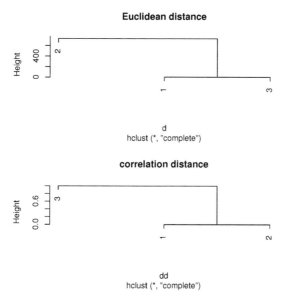

FIGURE 18.36: The same data can be clustered in different ways depending on the distance used. The correlation distance places g_1 in the same cluster with g_2 and in a different cluster from g_3 because $(1,2,3,4,5)$ and $(100,200,300,400,500)$ have a high correlation $(d(g_1,g_2) = 1 - r = 1 - 1 = 0)$, whereas $(1,2,3,4,5)$ and $(5,4,3,2,1)$ are anticorrelated $(d(g_1,g_3) = 1 - (-1) = 2)$. However, the Euclidean distance placed g_1 in the same cluster with g_3 and in a different cluster from g_2 because $d_E(g1,g2) = 734.20$, while $d_E(g1,g3) = 6.32$.

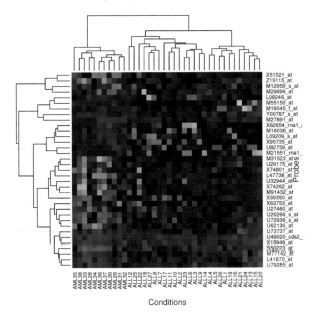

FIGURE 18.37: A clustering of the leukemia data set obtained with the function `heatmap`.

In this data sets, the samples from 1 to 27 are ALL, while the ones from 28 to 38 are AML. We will now perform a *t*-test and order the probe sets using the *p*-values from the *t*-test, then select the top 35 genes as being differentially expressed (DE):

```
> ttestfun<-function(x) t.test(x[1:27],x[28:38])$p.value
> p.value<-apply(golub, 1, ttestfun)
> golub.order<-golub[order(p.value),]
> DEgene<-golub.order[1:35,]
```

We can now plot a hierarchical clustering dendrogram with the function `heatmap`, as follows:

```
> #library(gdata,quietly=T,verbose=F,warn.conflicts=F)
> library(gplots)
> heatmap(DEgene, col=greenred(75),xlab="Conditions",
+ ylab="Probes")
```

The package `gplots` was only needed for the `greenred` function that sets the color palette used for the heatmap. The dendrogram produced by heatmap is shown in Fig. 18.37.

There are also other functions that are able to produce the same type of map. One such function in `heatmap.2` which can be used to show a histogram

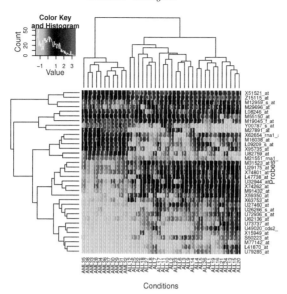

FIGURE 18.38: A clustering of the leukemia data set obtained with the function `heatmap.2`. Note the addition of a color key and histogram showing the distribution of the values in the heatmap.

of the values across the color map used. Fig. 18.38 shows the same top 35 genes from the leukemia data set clustered with `heatmap.2`.

Fig. 18.39 shows the same data clustered and plotted with a customized function that also added the coloring of the conditions clusters, as well as a scale showing the heights of the trees.

Finally, we can plot the leukemia data in the space of the first two principal components:

```
> library(cluster)
> km.gene<-kmeans(DEgene, centers=5)
> km.sample<-kmeans(t(DEgene), centers=2)
> clusplot(t(DEgene), km.sample$cluster, color=TRUE, shade=TRUE,
+ labels=2, lines=0)
```

Let us now build a SOFM of the yeast sporulation data. We will use a two-dimensional rectangular Kohonen network with a size of 4×4. This can be easily achieved with the function `som`. This function takes as arguments a matrix with each row representing an object; a `grid` describing the desired topology for the Kohonen map; `rlen`, which is the number of training iterations to be performed; the learning rate `alpha`; the radius of the neighborhood, `radius`, which can be either a value or a vector of values; and a parameter `keep.data`, which tells the function whether to save the data in the return

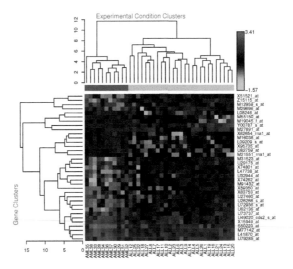

FIGURE 18.39: A custom plot of the clustering of the leukemia data set. Note the coloring of the condition clusters, the color scale in the top right of the heatmap and the scale for the heights of the trees.

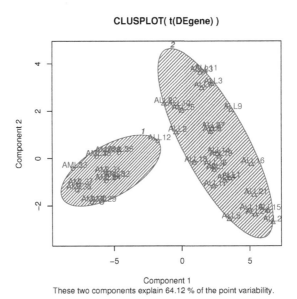

FIGURE 18.40: Leukemia data clustered in a two-dimensional PCA plot.

object. One can also specify whether the edges of the map should be wrapped over (toroidal=TRUE) as well as the initial values for each node (init). If the initial values are not given, they are chosen randomly without replacement from the input data.

```
> library(kohonen)
> data(yeast)
> gridx=4; gridy=4
> dat=na.omit(yeast$elu)
> # "som" function cann't handle missing values
> dat.som=som(scale(dat), grid=somgrid(gridx, gridy,"rectangular"),
+ rlen=500,keep.data=T)
> par(mfrow=c(4,4))
> for (i in 1:(gridx*gridy)) {
+ tempdat<-dat[dat.som$unit.classif == i,,drop=F]
+ if (nrow(tempdat) == 1) {
+ plot(tempdat, ylim=range(dat), xlab="time",
+ ylab="gene expression", Type="b", main=paste("unit", i))
+ }
+ else if (nrow(tempdat) > 1) {
+ plot(tempdat[1,], ylim=range(dat), xlab="time",
+ ylab="gene expression",type="b", main=paste("unit", i))
+ for (j in 1:nrow(tempdat[-1,])) {
+ lines(tempdat[1+j,], ylim=range(dat), type="b")
+ }
+ }
+ else plot(NULL, ylim=range(dat), xlab="time",
+ ylab="gene expression", main=paste("unit", i))
+ }
```

The unit.classif values indicate the winning node for each data object. In the code above, we used them to select and plot the objects belonging to each unit in its corresponding graph.

18.6.1 Partition around medoids (PAM) in R

The Partitioning Around Medoids algorithm can be applied to the bfust ExpressionSet using the brokering code in the MLInterfaces:

```
> #source("http://www.bioconductor.org/biocLite.R")
> #biocLite("ALL")
> library(ALL)
> data(ALL)
> bio = which( ALL$mol.biol %in% c("BCR/ABL", "NEG"))
> isb = grep("^B", as.character(ALL$BT))
> bfus = ALL[, intersect(bio,isb)]
```

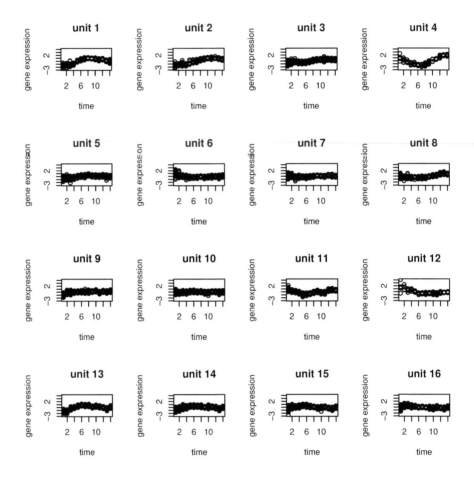

FIGURE 18.41: SOFM of the yeast sporulation cycle. There are 16 units organized in a 4×4 grid. Each of the 16 plots shows the genes that were clustered together in each of these units. Note how the profiles in each location share some similarities with all of their neighbors.

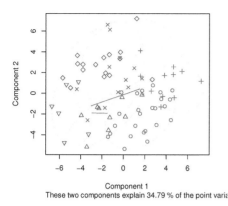
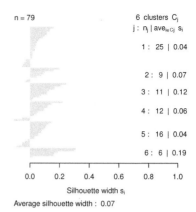

These two components explain 34.79 % of the point varia

Average silhouette width : 0.07

FIGURE 18.42: Two views of the partition obtained by PAM: left, principal component display; right, silhouette display. The ellipses plotted on the left are cluster-specific minimum volume ellipsoids for the data projected into the principal components plane. These should be regarded as two-dimensional representations of the robust approximate variance-covariance matrix for the projected clusters. The silhouette display is composed of a single horizontal segment for each observation, ordered by clusters and by object-specific silhouette value within cluster. Large average silhouette values for a cluster indicate good separation of most cluster members from members of other clusters; negative silhouette values for objects indicate instances of indecisiveness or error of the given partition.

```
> bfust = bfus[ apply(exprs(bfus),1,mad) > 1.43, ]
> bfust$mol.biol = factor(bfust$mol.biol)
> library(MLInterfaces)
> dopam = pam(t(exprs(bfust)), k=6)
> par(mfrow=c(1,2))
> plot(dopam, main="")
```

The graphical output shown in Figure 18.42 is obtained using the R command `plot(RObject(dopam))`. On the left, the the smallest cluster-specific ellipsoids containing all the data in each cluster are displayed in a two-dimensional principal components projection; on the right, the *silhouette* display is presented. High silhouette values indicates "well-clustered" observations while negative values indicate that an observation might have been assigned to the wrong cluster.

18.6.2   Biclustering in R

We will illustrated the biclustering using the top 50 DE genes from the `golub` data set used earlier in the chapter. We will use the Cheng and Church (CC) biclustering algorithm described in Section 18.5.2. This type of biclustering can be peformed with the function `biclust`. This function implements a number of different biclustering methods including the CC method, which can be specified with the parameter `method=BCCC()`. The argument "delta" specifies the maximum threshold for MSR score, "`alpha`" is a scaling factor multiplied by the overall matrix score to give a minimum score value in the deleting of rows and columns, and "`number`" designates the number of biclusters to be found. The `drawHeatmap` function can be used to visualize the heatmap for the identified bicluster.

```
> #install.packages("biclust", repos="http://cran.mtu.edu/",
> #lib="C:/Program Files/R/R-2.13.0/library")
> library("biclust")
> library(multtest); data(golub)
> dimnames(golub)[[1]]<-golub.gnames[,3]
> dimnames(golub)[[2]]<-c(paste("ALL", 1:27, sep=""),
+ paste("AML", 28:38, sep=""))
> ttestfun<-function(x) t.test(x[1:27],x[28:38])$p.value
> p.value<-apply(golub, 1, ttestfun)
> golub.order<-golub[order(p.value),]
> DEgene<-golub.order[1:50,]
> res1<-biclust(DEgene, method=BCCC(), delta=0.1, alpha=0.5,
+ number=1)
> drawHeatmap(DEgene, res1, 1, local=T)

> drawHeatmap(DEgene, res1, 1, local=F)
```

The results of this biclustering are shown in Fig. 18.43. One bicluster of 21 genes and 26 samples is identified from the input data matrix of 50 genes and 38 samples.

18.7   Summary

Two frequently posed problems related to microarrays are: i) finding groups of genes with similar expression profiles across a number of experiments and ii) finding groups of individuals with similar expression profiles within a population. This is the task of the cluster analysis or clustering.

A distance measure is necessary in order to assess the degree of similarity between patterns. This chapter discussed various distances, their definitions

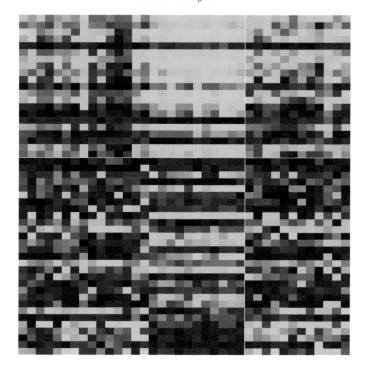

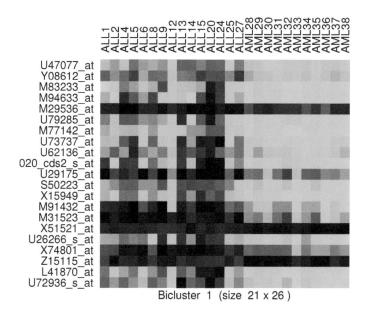

Bicluster 1 (size 21 x 26)

FIGURE 18.43: The Cheng and Church (CC) biclustering algorithm on the 50 genes × 38 samples data results in a 21 genes × 26 samples bicluster. This bicluster appears in the top-left corner of the overall heat map in the top panel, and is shown on its own in the bottom panel.

and their properties. The following distances have been discussed: Euclidean, standardized Euclidean, squared Euclidean, angle, correlation, jackknife, Manhattan (city block), Chebychev, Minkowski, and Mahalanobis. Each distance has certain properties that can be used to emphasize certain characteristics of the data.

The distance is usually used to cluster the genes and/or experiments. The chapter presented the k-means clustering, hierarchical clustering and self-organizing feature maps (Kohonen maps). The k-means algorithm clusters patterns in a given number of groups. The relationship or ordering between different groups in a given k-means cluster diagram is meaningless. The relationship or ordering between the elements of the same group is also meaningless. The k-means clustering uses a random initialization and is therefore nondeterministic (different runs on the same data may produce different results).

The hierarchical clustering constructs a complete tree structure between patterns. It is more informative than k-means since each cluster is further divided into subtrees showing the internal structure of the clusters. There are several approaches: top-down (by division), bottom-up (by agglomeration), and incremental. Each approach has different properties. The bottom-up may be slower but may provide a more informative tree structure. The information extracted by the hierarchical clustering algorithm is contained in the topology of the dendrogram (how the branches are connected). A particular dendrogram or tree diagram may not necessarily reflect the true relationships between patterns. For instance, patterns drawn next to each other may be more different than patterns drawn at large distances from each other.

The self-organizing feature map (SOFM) or Kohonen map is a regular grid of elements that store data prototypes. Usually, the Kohonen map is a one-, two- or three-dimensional array, i.e. a string, a rectangular grid, or a parallelepiped. Each element in the grid is connected directly to its neighbors. The prototypes are a summation of certain common features shared by several pieces of data and are extracted from the data without user intervention. The process of extracting this information from the data is known as learning or training. Unlike the diagrams constructed by k-means or hierarchical clustering, the Kohonen feature map has the property that if the images of two data points are close on the feature map, those two data points are similar according to the distance used. Kohonen feature maps also have some drawbacks. They tend to be sensitive to the way they are initialized, are nondeterministic and the size and layout involve heuristic choices.

The existing literature is very rich in papers concerned with clustering methods and algorithms, as well as their applications [171, 196, 202, 257, 269, 310, 369, 413, 423, 467, 476, 475]. A particularly interesting approach was used by Cho et al. to discern whether hierarchical clusters are enriched in specific functional categories [93].

Chapter 19

Quality control

Sometimes I lie awake at night, and I ask, "Where have I gone wrong?" Then a voice says to me, "This is going to take more than one night."

—Charles M. Schulz

19.1 Introduction

An assessment of the quality of the data obtained in a given experiment or from a specific array is absolutely crucial for the further analysis and the correct interpretation of microarray data. Even a single or few abnormal arrays can completely compromise the results of the analysis of a large data set. Hence, no analysis should be performed before first making sure that the data are of acceptable quality.

In most cases, problems with the quality of the array come from the sample itself. Unlike DNA, the mRNA is susceptible to rapid degradation if the sample is not processed properly immediately after collection. Laboratory methods exist and should be used to assess the quality of an mRNA sample. Other laboratory methods and checks exist for many of the processing steps involved in the microarray process. All these checks and quality assessments should be

635

performed. The methods discussed here are not meant to substitute in any way the laboratory quality control procedures. They are meant to be used in addition to the laboratory procedures, in order to catch anything that may have escaped detection up to this point. Any sample and/or array that does not meet the necessary quality standards should be discarded. This is sometimes difficult to accept since some of these samples may be extremely difficult to obtain and/or very expensive; hence, one is tempted to continue the processing and analysis of a sample even if the quality may not be satisfactory in the hope that they will still add to the scientific value of the experiment. In fact, the truth is exactly the opposite. No valid results can be obtained from a degraded sample or a compromised array. Furthermore, the inclusion of data coming from such samples or arrays will make the analysis of the other samples more difficult in the best case and will usually provide erroneous results.

19.2 Quality control for Affymetrix data

In this section, we will illustrate some tools and concepts relevant to the quality control of Affymetrix data. Although we will focus on tools and packages available in R and Bioconductor, the principles are general and similar quality control checks may be performed in various other ways.

19.2.1 Reading raw data (.CEL files)

The library *affy* contains many relevant functions so we start by loading it. We then list the .CEL files available in the directory and display the content of this list to make sure we are in the correct directory and have the files we expected:

```
> library(affy)
> CELfiles<-list.celfiles()
> CELfiles
```

```
 [1] "1LCX.CEL"       "1LFU.CEL"          "1LLS.CEL"
 [4] "2NLCX.CEL"      "2NLFU.CEL"         "2NLLS.CEL"
 [7] "3LCX.CEL"       "3LFUFundus.CEL"    "3LS.CEL"
[10] "4LCX.CEL"       "4LFU.CEL"          "4LLS.CEL"
[13] "5LCX.CEL"       "5LFU.CEL"          "5LLS.CEL"
[16] "6LCX.CEL"       "6LFU.CEL"          "6LLS.CEL"
[19] "7NLCX.CEL"      "7NLFU.CEL"         "7NLLS.CEL"
[22] "8LCX.CEL"       "8LFU.CEL"          "8LLS.CEL"
[25] "9PTLNLCX.CEL"   "9PTLNLFU.CEL"      "9PTLNLLS.CEL"
```

This time there are nine patients. Three samples were collected from each patient, one from each of three different tissues for a total of 27 arrays. Since everything seems to be fine so far, we proceed to read these CEL files with the function *ReadAffy* (see Section 20.7.1 for details):

```
> rawdata=ReadAffy(filenames=CELfiles)
```

Now we can examine the raw data through a number of exploratory graphs and tools that provide information about the quality of the arrays: array intensity images, intensity distribution curves, box plots, and other diagnostic plots.

19.2.2 Intensity distributions

Let us start with the intensity distributions. A plot of the intensity distributions can be easily obtained with the function *hist*:

```
> library("RColorBrewer")
> usr.col=brewer.pal(9, "Set1")
> mycols=rep(usr.col,each=3)
> hist(rawdata, lty=rep(1,length(CELfiles)), col=mycols)
> legend("topright", rownames(pData(rawdata)),
+ lty=rep(1,length(CELfiles)), col=mycols,
+ cex=0.6)
```

The first three lines above are used only to set up some nice colors. The function *brewer.pal*, part of the *RColorBrewer* package, selects some nice colors and returns them in a character vector that can be used by R. The third line prepares an array of colors by repeating each color three times, such that the three arrays corresponding to one patient are all plotted with the same color. The function *legend* adds a legend in the top right corner of the image using the same colors, and the name of the arrays picked up as the row names of the *pData* slot in *rawdata*. Alternatively, the array names can be accessed through *sampleNames(rawdata)*.

Fig. 19.1 shows the plot of the intensity distributions of these 27 arrays obtained with the code above. This plot shows the presence of two different problems. First, three of the arrays (in blue in Fig. 19.1) exhibit something that looks like saturation. This phenomenon is indicated by the presence of a second mode in the intensity distributions showing that a large number of probes have values at the high end of the detection limit. In this particular plot, we have chosen to plot the three samples coming from one patient with the same color. This allows us to see that all three arrays that exhibit saturation involve biological samples coming from the same patient. These arrays must be discarded from further analysis since no type of normalization can recover the correct values for those probes.

A second problem that can be observed on this graph is the group of three arrays (in green in Fig 19.1) that are shifted considerably towards the

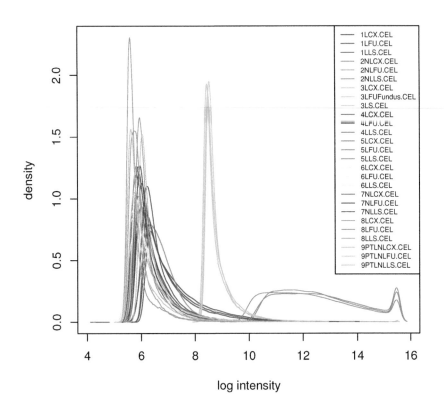

FIGURE 19.1: Intensity distribution plots of a data set that includes some defective arrays. Saturation is indicated by the presence of a second mode in the blue intensity distributions corresponding to patient 2 (arrays 2NLCX, 2NLFU and 2NLLS). These distributions show that a large number of probes have values at the high end of the detection limit. The arrays corresponding to patient 3 (in green) also appear to be very different from the other 21 arrays.

right with respect to all other 21 arrays. This indicates that these arrays are substantially different from the others. In principle, such a shift may be taken care by the normalization but further analysis is needed in order to decide whether these three arrays should be accepted or not in this data set.

19.2.3 Box plots

Let us have a look at the box plots of these arrays:

```
> boxplot(rawdata,col=mycols,las=3,cex.axis=0.5,
+ names=sampleNames(rawdata))
```

The box plots produced with the line above are shown in Fig. 19.2. The *las=3* parameters instructs R to write the labels of the x axis, vertically while the `cex.axis=0.5` makes the font size smaller (50% of the default size) so the labels fit below the axis. The colors in this figure are the same as in Fig 19.1. The information provided by this plot is very similar with that provided by the intensity distributions. The blue box plots corresponding to the saturated arrays from patient 2 are situated at much higher intensities than all the others and have much wider distributions. The green distributions from patient 3 are much more similar to the other 21 distributions but they are placed at considerably higher intensities. The intensity distributions provide more information about the actual shape of the distributions. On the other hand, the box plots allow an easy assessment of the relative position of the medians with respect to one another. One should also note that in general, quality assessment should be done on raw data, as we are doing here. It is clear for instance that if an array normalization that brings all array to a common median is used, the box plots will not be able to point out the defective arrays, especially if the widths of the distributions are comparable. In this particular example, the saturated arrays have a distribution shape very different from those of the normal arrays so probably these arrays would stand out even after certain types of array normalization. However, yet again, quality control should be done on raw rather than normalized data since it may be much more difficult to find abnormal arrays if the quality control is attempted after normalization.

19.2.4 Probe intensity images

The array **probe intensity image** is derived from mapping probe intensities to a gray scale, and can be used to examine the data for potential spacial artifacts on the array. In this type of image, array defects such as scratches, hairs, or other contaminants should be visible as spacial artifacts involving the values read from the array in the areas affected by the defect. Other defects such as surface defects or hybridization problems may involve large areas of a given array. Let us have a look at one typical array from this batch, let

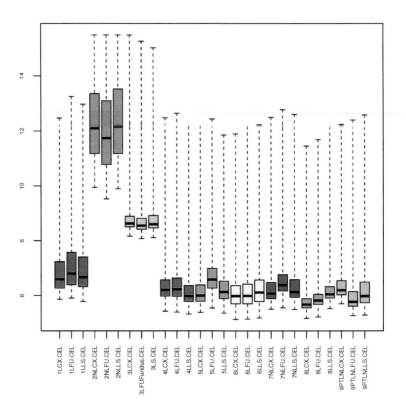

FIGURE 19.2: Box plots of the intensity data for the same arrays as in Fig 19.1. The colors also correspond to the same arrays. Patients 1 through 9 are plotted from left to right. The blue box plots correspond to the saturated arrays from patient 2. Not only that they are situated at much higher intensities than all the others but they also have much wider distributions. The green distributions from patient 3 are much more similar to the other 21 distributions, but they are placed at considerably higher intensities.

us say 1LCX.CEL, and compare it with one of the arrays that appear to be saturated, let us say, 3LCX.CEL.

```
> image(rawdata[,1])
> image(rawdata[,4])
```

The images of the two arrays shown in Fig. 19.3 clearly shows that there is a problem with the saturated array. As expected from the analysis of the intensity distributions and box plots, the array from patient 2 has a large number of probes with maxed out intensities. The other two arrays hybridized with samples from the same patient show a similar saturation. The visual inspection of the images of the arrays hybridized with samples from patient 3 do not show any obvious problem even though their intensity histograms and box plots show a distribution shifted towards higher intensities. An inspection of the arrays coming from patients 4 through 9 does not show anything abnormal either.

Based on these the quality assessment made so far, we conclude that we should eliminate the three arrays corresponding to patient 2. The arrays corresponding to patient 3 need further analysis before a decision is made whether to include or exclude them. The other arrays appear acceptable so far.

19.2.5 Quality control metrics

In order to reach a conclusion about patient 3's arrays, we need to look at some of the quality control probes provided by Affymetrix. Currently, such quality control probes are available to monitor background intensity, array variation and RNA degradation. These probes allow the user to check whether the arrays hybridized correctly, and whether the sample quality was acceptable. Summary statistics from these control probes can be computed using the same "simpleaffy" package. The *simpleaffy* function *qc* generates the following metrics used to assess the quality of an array:

- Average background

- Number of genes called present (P)

- Scale factor

- $3'$ to $5'$ ratios for two control genes β-actin and GAPDH

- Values for spiked-in control transcripts (hybridization controls)

The β-actin and GAPDH genes are used as controls because they are assumed to be more or less constant in various tissues and conditions, although this is not always true [275, 363, 380].

The standard recommendations from Affymetrix regarding the above metrics are as follows [462]:

1LCX.CEL

2NLCX.CEL

FIGURE 19.3: The intensity images of two Affymetrix arrays. Array 1LCX.CEL is one of the 21 arrays with a very similar intensity distribution in Fig. 19.1. Array 2NLCX.CEL is one of the arrays that appeared to be saturated from the distribution of its intensities. Indeed, in this synthetic intensity image, 2NCX.CEL appears to be much, much brighter than the typical array and is likely to be saturated.

- The **average background** is expected to be similar across all arrays. Different levels of the background can be produced by different amounts of cDNA, different hybridization efficiencies, etc. Consistency within the experiment is very much the target here. Also, when trying to combine data sets from different experiments, considerable differences in the average background can generate problems.

- The **number of genes called Present** is expected to also be consistent within an experiment. Variations can be explained by variations in the amount of starting material, differences in processing pipelines, etc. Much like for the average background, absolute values are not informative since they are expected to vary from one experiment to another. However, within a given set of arrays associated with a condition and processed in the same batch, the percentage of present genes is not expected to vary much.

- The **scale factor** measures the overall expression level on an array with respect to a reference. The assumption here is that the vast majority of transcripts will not change much their expression level in any given experiment (otherwise the cells would probably not be alive). In consequence, the trimmed mean intensity should be constant for all arrays. By default, the Affymetrix analysis software MAS 5.0 scales the intensities for every sample so that the means are the same. The scale factor indicates the amount of scaling needed in order to achieve this. Affymetrix recommends that the scale factor within a set of arrays that are to be compared in a given analysis not exceed three-fold between any two arrays.

- The **3′ to 5′ ratios for β-actin and GAPDH** are there to indicate the presence of truncated transcripts or unsatisfactory RNA quality. If the RNA quality was good and if the in vitro transcription was performed well, the amount of signal coming from different probes should be the same no matter whether the probes were close to the 3′ end, in the middle of the gene, or close to the 5′ end of the gene. Hence, ratios values close to 1 signal a good experiment whether high values indicate degraded RNA or problems in the sample processing. If the Affymetrix Small Sample protocol was used, the ratio between the 3′ end probes and the middle probes of the genes should be used instead of the ratio between the 3′ and the 5′ probes (which is to be used in conjunction with the Affymetrix Standard Protocol). However, due to the inherent 3′ bias in the trascripts, a ratio 3′:5′ of less than 3 is acceptable.

- The **values for the spiked-in controls** are there to indicate the quality of the hybridization. Some transcripts for a different organism (*Bacillus Subtilis* for the human Affymetrix arrays) are added during the latter stages of the sample preparation protocol. These are not expected to be perturbed by the actual sample material and are expected to be called

Present for all arrays. An experiment in which such a positive control is not called Present on more than 30% of the arrays is probably performing at suboptimal sensitivity.

The function *avbg* calculates the average background intensity for each array. As explained above, this average background intensity is expected to be comparable among arrays. The function *sfs* computes the scale factor. As detailed above a scale value within three-fold is considered normal. The function *percent.present* gives the percentage of present calls on each array. Let apply these functions to our data:

```
> library("simpleaffy")
> dat.qc<-qc(rawdata)
> avbg(dat.qc)
```

1LCX.CEL	1LFU.CEL	1LLS.CEL	2NLCX.CEL
63.02117	67.14067	64.19655	1229.57884
2NLFU.CEL	2NLLS.CEL	3LCX.CEL	3LFUFundus.CEL
914.85329	1180.11088	312.68777	294.66348
3LS.CEL	4LCX.CEL	4LFU.CEL	4LLS.CEL
305.01759	50.77590	48.25695	44.35516
5LCX.CEL	5LFU.CEL	5LLS.CEL	6LCX.CEL
46.02232	55.30487	46.84669	40.07431
6LFU.CEL	6LLS.CEL	7NLCX.CEL	7NLFU.CEL
39.85145	40.65796	49.10808	57.82419
7NLLS.CEL	8LCX.CEL	8LFU.CEL	8LLS.CEL
50.85542	41.60108	44.96122	52.03011
9PTLNLCX.CEL	9PTLNLFU.CEL	9PTLNLLS.CEL	
56.05221	41.45675	43.28261	

```
> sfs(dat.qc)
```

```
 [1]  4.77482271  2.77074534  3.51041549  0.03927639  0.03861967
 [6]  0.03014661  3.94692237  3.03983430  2.99136965  8.33865625
[11]  7.69351750 13.29079614 11.80067399  9.59636113 18.22558881
[16] 19.13564029 14.51176912 11.93364154 10.97835955  8.42757026
[21]  8.47438672 53.70747134 28.07549220 30.50822128 10.01336288
[26] 15.02504000  7.81468503
```

```
> percent.present(dat.qc)
```

1LCX.CEL.present	1LFU.CEL.present
31.450974	37.438619
1LLS.CEL.present	2NLCX.CEL.present
34.888326	50.887058
2NLFU.CEL.present	2NLLS.CEL.present
54.371931	53.627435

```
          3LCX.CEL.present 3LFUFundus.CEL.present
                 11.341676                17.844131
           3LS.CEL.present        4LCX.CEL.present
                 16.782829                28.639316
          4LFU.CEL.present        4LLS.CEL.present
                 29.360051                20.426105
          5LCX.CEL.present        5LFU.CEL.present
                 24.996040                19.293521
          5LLS.CEL.present        6LCX.CEL.present
                 14.565183                14.018692
          6LFU.CEL.present        6LLS.CEL.present
                 19.404404                19.642009
         7NLCX.CEL.present       7NLFU.CEL.present
                 25.756376                26.287027
         7NLLS.CEL.present        8LCX.CEL.present
                 27.364169                 5.037225
          8LFU.CEL.present        8LLS.CEL.present
                 13.701885                 8.252812
       9PTLNLCX.CEL.present    9PTLNLFU.CEL.present
                 26.540472                23.031839
       9PTLNLLS.CEL.present
                 32.369713
```

In the statistics above, we see seriously abnormal average background values for the arrays hybridized with the sample from patient 2. This is to be expected for saturated arrays and confirms our decision to discard them from the data set. The scale factors go from 0.03 for array 6 to 53.70 for array 22. This also shows that a serious problem exists in this data set if we continue to include the saturated arrays. The percentages of P genes are abnormally high for the saturated arrays, but since Affymetrix does not provide guidelines for the upper limit of this statistic, no decision can be taken based on this observation alone. Of course, in the context of the intensity distributions, box plots, and array images shown above, the high percentage of Present genes on the saturated arrays makes perfect sense.

The function `ratios` can be used to compute $3'/5'/$Middle probe intensity ratios for control genes. As explained above, a $3'/5'$ ratio over 3 may signal RNA degradation.

```
> ratios(dat.qc)
```

	actin3/actin5	actin3/actinM	gapdh3/gapdh5
1LCX.CEL	0.599622548	0.03131228	0.05583826
1LFU.CEL	0.427585026	-0.10765543	0.18231705
1LLS.CEL	0.492858422	-0.13172847	0.12134049
2NLCX.CEL	-0.783270489	4.21952899	-0.99770527
2NLFU.CEL	0.260719866	1.20496516	-0.31751770

2NLLS.CEL	-1.126211296	0.37548934	-0.09691680
3LCX.CEL	3.488295046	6.38291961	4.70204341
3LFUFundus.CEL	6.298310954	3.22290611	4.84990950
3LS.CEL	5.159639973	3.24461258	5.04729422
4LCX.CEL	0.584127789	-0.19097262	0.75984893
4LFU.CEL	0.701688527	-0.06362691	1.21025451
4LLS.CEL	1.115842759	0.13755247	1.73143217
5LCX.CEL	1.075486629	-0.10376354	1.02123897
5LFU.CEL	0.830662520	-0.03650162	0.93235992
5LLS.CEL	0.804011079	-0.34840200	1.65811077
6LCX.CEL	0.629294101	-0.15362147	0.72974601
6LFU.CEL	0.656390049	0.01319634	0.34412578
6LLS.CEL	0.533328954	-0.10230868	0.50586920
7NLCX.CEL	0.748563395	-0.05990448	0.83183152
7NLFU.CEL	0.002704417	0.51872886	0.68176151
7NLLS.CEL	0.358340555	-0.37691258	0.81189525
8LCX.CEL	-0.406535774	-1.82087671	1.74283297
8LFU.CEL	-0.048927496	-0.68034950	1.13540486
8LLS.CEL	0.172543504	-1.30798362	1.59150473
9PTLNLCX.CEL	1.579622236	0.30982025	1.96374745
9PTLNLFU.CEL	0.341146549	-0.36417807	0.87008648
9PTLNLLS.CEL	0.221689343	-0.32251139	0.79464880

	gapdh3/gapdhM
1LCX.CEL	-0.035358493
1LFU.CEL	-0.006849236
1LLS.CEL	0.130174408
2NLCX.CEL	-0.554373581
2NLFU.CEL	0.287995886
2NLLS.CEL	0.008321588
3LCX.CEL	5.146882295
3LFUFundus.CEL	2.902612886
3LS.CEL	2.841049935
4LCX.CEL	0.298461336
4LFU.CEL	0.399150728
4LLS.CEL	0.751715655
5LCX.CEL	0.607756951
5LFU.CEL	0.317647625
5LLS.CEL	0.500375594
6LCX.CEL	0.575925941
6LFU.CEL	0.491566674
6LLS.CEL	0.309347649
7NLCX.CEL	0.244476998
7NLFU.CEL	0.249026553
7NLLS.CEL	0.085161071
8LCX.CEL	1.224636896

8LFU.CEL	0.115965950
8LLS.CEL	0.667692612
9PTLNLCX.CEL	0.849318518
9PTLNLFU.CEL	0.503704884
9PTLNLLS.CEL	0.484833728

This particular quality assessment is the first one so far that provides some indication about the causes of the abnormal intensity distributions for patient 3. Indeed, all beta actin ratios for the three arrays coming from patient 3 have values above the recommended threshold of 3 indicating that the degradation of the RNA may be an issue.

19.2.6 RNA degradation curves

The RNA degradation slopes and p-values for the degradation curves can be further summarized using the function *summaryAffyRNAdeg*.

```
> dat.deg=AffyRNAdeg(rawdata)
> summaryAffyRNAdeg(dat.deg)
```

	1LCX.CEL	1LFU.CEL	1LLS.CEL	2NLCX.CEL	2NLFU.CEL	2NLLS.CEL
slope	0.017	0.116	0.0332	0.266	-0.147	-0.254
pvalue	0.871	0.338	0.7760	0.134	0.382	0.151

	3LCX.CEL	3LFUFundus.CEL	3LS.CEL	4LCX.CEL	4LFU.CEL
slope	1.19e+00	1.35e+00	1.26e+00	0.077	0.231
pvalue	9.06e-11	1.16e-10	6.35e-10	0.629	0.181

	4LLS.CEL	5LCX.CEL	5LFU.CEL	5LLS.CEL	6LCX.CEL	6LFU.CEL
slope	0.2570	0.00465	-0.147	-0.4770	-0.4550	-0.3350
pvalue	0.0862	0.97200	0.419	0.0145	0.0155	0.0586

	6LLS.CEL	7NLCX.CEL	7NLFU.CEL	7NLLS.CEL	8LCX.CEL	8LFU.CEL
slope	-0.3450	-0.0115	-0.0597	0.055	-0.347000	-0.160
pvalue	0.0591	0.9300	0.6920	0.686	0.000636	0.114

	8LLS.CEL	9PTLNLCX.CEL	9PTLNLFU.CEL	9PTLNLLS.CEL
slope	-0.39600	0.3860	0.0212	0.0703
pvalue	0.00696	0.0107	0.8680	0.6250

Again, there is significant evidence of overall RNA degradation according to the degradation curve slope and p-values for beta actin for arrays 3xxx.CEL. The function *plotAffyRNAdeg* produces an RNA degradation plot that displays the probe intensity shifting with respect to the probe position:

```
> par(mar=c(5,4,4,7) + .1, xpd=T)
> plotAffyRNAdeg(dat.deg, col=mycols)
> legend(15.5,34,sampleNames(rawdata),cex=0.7,lty=1,col=mycols)
```

This function will produce the plot shown in Fig. 19.4. This RNA degradation plot shows the average probe intensities (on the y-axis) across probe sets

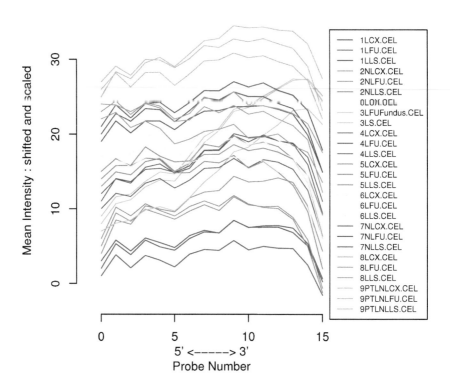

FIGURE 19.4: An RNA degradation plot. The y-axis gives average probe intensities across probe sets for a given probe number (shown on x-axis). The color assignments in these plot are the same as the ones in Fig 19.1 and Fig. 19.2. From this plot, it is clear that the three arrays corresponding to patient 3 (and plotted in green here) show a much higher slope than the others, indicating significant RNA degradation.

for a given probe number (on the x-axis). From this plot, it is clear that the 3 arrays corresponding to patient 3 (and plotted in green here) show a much higher slope than the others, indicating significant RNA degradation. This is in agreement with the conclusion drawn from the slopes and p-values provided by the *summaryAffyRNAdeg*. Based on this, we conclude that the three arrays hybridized with the samples from patient 3 should also be removed from further analysis.

19.2.7 Quality control plots

A summary of many of the above quality control metrics can be obtained in a single plot with the function *plot(dat.qc)*, where *dat.qc* is the object that contains the data returned by the *qc* function:

```
> plot(dat.qc)
```

The image plotted by this call is shown in Fig. 19.5. The blue area spans the interval $[center - 1.5, center + 1.5]$, where center is the mean of the log_2, transformed scale factors. Essentially, this is the permissible area, the area in which we wish to see all scale factors in any given collection of arrays that are to be analyzed together. A point outside the blue region means the scale factor of this area is more than $2^{1.5} = 2.8$-fold changes away from the geometric mean of all scale factors. The scale factors are plotted as horizontal lines starting from the middle and ending with a small circle. If the maximum pair-wise log fold change of scale factors exceeds 3, *all* scale factors are plotted in red, indicating that there is an issue with this set of arrays. In other words, all scale factors will be red if there exists at least one pair of them such that their fold change is greater than 3. If all scale factors are located within blue region, they will be colored in blue since they are within maximum ± 1.5 fold changes of each other and therefore, a pair with a fold change of more then 3 cannot possibly exist; if all scale factors are plotted in red, some of them must have been outside the blue region. The stand-alone triangles represent the ratios between the $3'$ and $5'$ probes for beta-actin. The stand-alone circles represent the same ratios for GAPDH. Individual values outside the acceptable limits (default 1.25 for GAPDH and 3 for beta-actin) and therefore indicating quality problems, would be plotted in red. The plot also shows the percentage of present genes and the average background for each array. Both of these are colored in blue for all arrays unless the maximum pair-wise difference exceeds 20 for background, or 10% for present percentage, in which case all will be colored red instead. The red text "bioB" means that the detection PMA (present/marginal/absent) call of the bioB spike-in control is not "P". bioB is a hybridization control gene from *E.coli*, expected to be flagged as "P" under good hybridization conditions. More details about the utilization of the *qc* metrics can be found in the "QC and Affymetric data" document accompanying the *simpleaffy* package [462], as well as in the "Expression Analysis Fundamentals" manual available from Affymetrix.

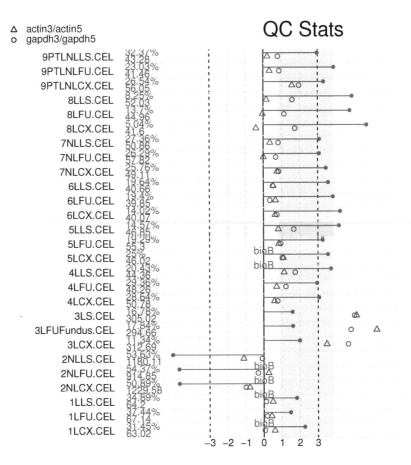

FIGURE 19.5: A summary of the QC data for Affymetrix. The blue area spans the interval [center −1.5, center +1.5], where center is the mean of the log2 transformed scale factors. Essentially, this is the permissible area, the area in which we wish to see all scale factors in any given collection of arrays that are to be analyzed together. The triangles represent the ratios between the 3′ and 5′ probes for beta-actin. The circles represent the same ratios for GAPDH. Values outside the acceptable limits and therefore indicating quality problems, would be plotted in red. The plot also shows the percentage of present genes and the average background for each array. The arrays are represented by their CEL file names. In this particular data set, we can see that the arrays corresponding to patient 2, the saturated arrays already identified by the previous analyzes, are very different from the others: their scaling factors are not only considerably larger but also in the opposite direction in comparison with the others. Also, the average backgrounds for these three arrays are about 2 orders of magnitude higher than the average backgrounds of all other arrays.

In this particular data set, we can see that the arrays corresponding to patient 2, the saturated arrays already identified by the previous analyzes are very different from the others, not only in magnitude but also in direction. In this plot, we can also see that the circles and triangles corresponding to the arrays belonging to patient 3 are red, indicating RNA degradation. Both sets of arrays need to be removed from this data set between any attempts to identify differentially expressed (DE) genes are made.

Let us remove first the saturated arrays corresponding to patient 2. Fig 19.6 shows the same quality plot of the data set from which patient 2 was removed. Note that the scaling factors are still plotted in red because some of them are outside the blue area. The beta action ratios between the 5′ and the 3′ probes (triangles) are all within acceptable margins with the exception of the arrays corresponding to patient 3. The same arrays also show unacceptable rates for the GAPDH ratios indicating RNA degradation. Let us now also remove these arrays and replot the same quality measures for the remaining arrays.

```
> CELfiles<-list.celfiles()
> saturatedarrays = 4:6
> cleanerrawdata<-ReadAffy(filenames=CELfiles[-saturatedarrays])
> cleanerdat.qc<-qc(cleanerrawdata)
> setwd("c:/Users/Sorin/Documents/Books/Book revision/")

> plot(cleanerdat.qc)
```

Fig. 19.6 shows the summary of the QC data for the data set from which the saturated arrays corresponding to patient 2 have been removed. The scaling factors are still plotted in red because some of them are outside the blue area. The beta action ratios between the 5′ and the 3′ probes (triangles) are all within acceptable margins, with the exception of the arrays corresponding to patient 3. Let us now also remove the arrays corresponding to patient 3, recalculate all statistics, and plot the figure again.

```
> CELfiles<-list.celfiles()
> badarrays=4:9
> cleanestrawdata<-ReadAffy(filenames=CELfiles[-badarrays])
> cleanestdat.qc<-qc(cleanestrawdata)
> setwd("c:/Users/Sorin/Documents/Books/Book revision/")

> plot(cleanestdat.qc)
```

The QC plot of the data without patients 2 and 3 is shown in Fig 19.7. Some of the scaling factors are still outside the range recommended by Affymetrix and four of the GAPDG ratios are a bit high but overall this data set is now of almost acceptable quality. The intensity distributions and the box plots of the intensities of these arrays are shown in Fig. 19.8.

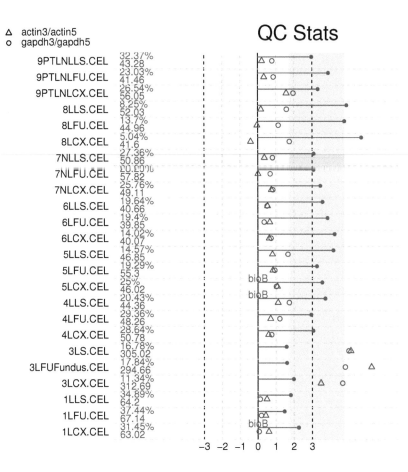

FIGURE 19.6: A summary of the QC data for the data set from which the saturated arrays corresponding to patient 2 have been removed. The scaling factors are still plotted in red because some of them are outside the blue area. The beta action ratios between the 5′ and the 3′ probes (triangles) are all within acceptable margins, with the exception of the arrays corresponding to patient 3.

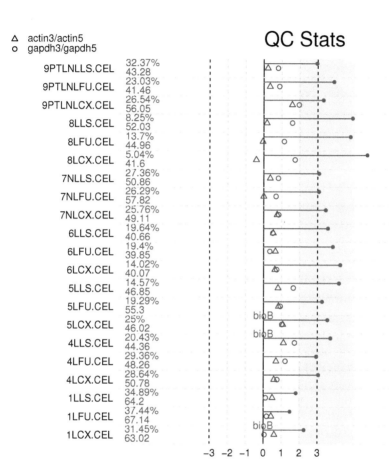

FIGURE 19.7: A summary of the QC data for the data set from which the bad arrays corresponding to patient 2 (saturated) and 3 (poor RNA quality) have been removed. The arrays from patients 1 and 8 still have scaling factors slightly outside the recommended limits, but overall, this set of arrays is almost acceptable.

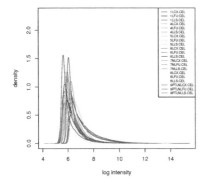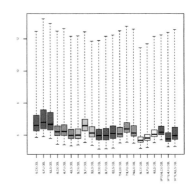

FIGURE 19.8: Intensity distributions and box plots of the Affymetrix data set shown in Figs. 19.1 and 19.2 after the removal of the six bad arrays corresponding to patients 2 and 3. The medians of the arrays are still different, and the arrays corresponding to patient 1 appear to be a bit higher and wider than the rest.

19.2.8 Probe-level model (PLM) fitting. RLE and NUSE plots

The presence of probe effects in raw data may elude the above quality assessment. Probe-level model (PLM) can compensate for such effects to produce robust gene expression values. To this end, the function *fitPLM* from the package "affyPLM" can be employed to fit a model and output a number of model statistics useful for quality diagnostics. Weights are used to down-weight outlier probes in the model, hence small weights are associated with outliers. Since the probe effects have been taken into consideration, the residuals should therefore be randomly distributed on the chip.

Fig. 19.9 shows the PLM weights and residuals images. The weights and residuals for one array from patient 4 and one array from patient 2 are shown in the top and bottom row, respectively. The left panels show the weights while the right panels show the residuals for both arrays. The array from patient 4 appears normal with only 2 small areas with some outliers. The array from patient 2 has a large number of outliers (bottom left) and large residuals (bottom right).

```
> library("affyPLM")
> dat.PLM=fitPLM(rawdata, background=F, normalize=F)
> par(mfrow=c(2,2))
> image(dat.PLM, type="weights", which=10)
> image(dat.PLM, type="resids", which=10)
```

```
> image(dat.PLM, type="weights", which=6)
> image(dat.PLM, type="resids", which=6)
```

Furthermore, a Relative Log Expression (RLE) plot can be constructed from the PLM model. For each array, the RLE plot is a box plot of the relative expression for each gene, which is defined as the difference of the PLM estimated gene expression and the median expression across arrays.

Yet another useful plot is the Normalized Unscaled Standard Error (NUSE) plot. The standard error from the PLM fit for each gene on each array is divided by the median standard error for that gene across all arrays. The NUSE plot is a box plot of this normalized standard error. An abnormal array may stand out in both RLE and NUSE plots.

```
> par(mfrow=c(1,2))
> RLE(dat.PLM, ylim=c(-0.5,7), col=mycols, las=3, cex.axis
+ =0.6, names=sampleNames(rawdata))
> NUSE(dat.PLM, ylim=c(0.95,1.35), col=mycols, las=3,
+ cex.axis=0.6, names=sampleNames(rawdata))
```

The RLE and NUSE box plots for this data set are shown in Fig. 19.10. The plots show that the arrays from patients 2 (already identified as saturated) and 3 (already identified as having RNA quality problems) are very different from the others. So far, so good. However, the three arrays from patient 1 appear to be slightly different from the remaining ones in both RLE and the NUSE plots, even though they did not exibit anything abnormal in any of the analyses and tools deployed so far (intensity distributions, box plots, QC plots, etc.). An examination of the CEL files corresponding to these arrays[1] reveals that the first three experiments used the HG95av1 array while all the other experiment used HG95av2 arrays. These explains pretty much everything. These three arrays look normal in the intensity distribution plots and box plots because they *are* normal.[2] They are perfectly good Affymetrix arrays, that appear to have been processed in accordance to the manufacturer's specifications and protocols and provided perfectly good data in the normal intensity range (hence, the box plots are normal) and having the typical number of probes in each intensity range (hence, their intensity distributions were normal). However, in the perpetual quest for better products, Affymetrix changed some of the probes that represent some of the genes from the version 1 of this array, which was the one used for the first 3 arrays, to the version 2 of the same array, used for the other patients. Hence, probe level analyses such as RLE and NUSE, were able to detect that there is a difference between these arrays and the rest.

---

[1]Let us recall that the CEL files are simple text files that can be opened with any text editor from Notepad, to Word, to emacs.

[2]Normal is meant here in the sense of not being defective, not in the sense of the data being distributed as a normal distribution.

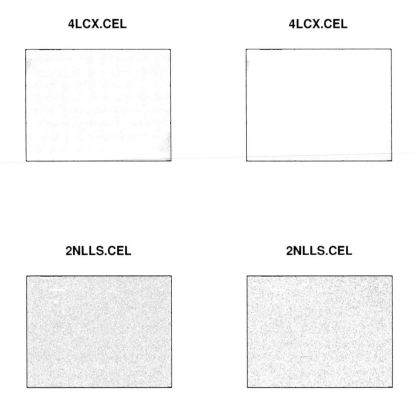

FIGURE 19.9: PLM weights and residuals images: in the left panels, darker green represents smaller weights (outliers); in the right panels, blue stands for negative residuals, while red stands for positive residuals. The weights and residuals for one array from patient 4 and one array from patient 2 are shown in the top and bottom row, respectively. The array from patient 4 appears normal with only two small areas with some outliers. The array from patient 2 has a large number of outliers (bottom left) and large residuals (bottom right).

Issues related to the fact that the same gene may be represented by different probes on the two versions of the array may create problems in the later analysis and interpretation of the data so at this point, eliminating these three arrays should also be considered. This however, is not absolutely necessary. If this is not done, one should do all subsequent analysis at the probe level and take into consideration that the probes for any given gene could be different between the first 3 arrays and the other arrays.

Let us now construct the same RLE and NUSE plots for the data set from which we eliminate the arrays corresponding to patient 1 (different array version), patient 2 (saturated arrays), and patient 3 (poor RNA quality).

```
> #exclude arrays from sample 2 and 3
> badarrays=1:9
> rawdata.QC<-rawdata[,-badarrays]
> library("RColorBrewer")
> usr.col<-rep(brewer.pal(7, "Set1"), each=3)
> library("affyPLM")
> dat.PLM<-fitPLM(rawdata.QC)

> RLE(dat.PLM, ylim=c(-1,1), col=usr.col, las=3, cex.axis
+ =0.6, names=sampleNames(rawdata.QC), main="RLE")

> NUSE(dat.PLM, ylim=c(0.95,1.1), col=usr.col, las=3,
+ cex.axis=0.6, names=sampleNames(rawdata.QC), main="NUSE")
```

The two plots are shown in Fig. 19.11. Although the plots show a much greater consistency, there is still one group of arrays that stand out. This time, it is patient 8. The left panel of Fig. 19.11 shows the RLE plot, in which the arrays from sample 8 stand out from others in terms of wider spread of the relative expression. The right panel shows the NUSE plot, in which the arrays from sample 8 not only have a wider spread but also have higher standard errors for the PLM estimated gene expression. Clearly, there is an issue with the arrays from patient 8, as well. Let us look again at the quality metrics for these three arrays. These arrays will be arrays 22, 23 and 24 in the set of 27 arrays we started with. We can interrogate the dat.qc object that we have already constructed for the particular values that correspond to these three arrays for any particular quality metric. Let us have another look at the percent present:

```
> percent.present(dat.qc)[22:24]

8LCX.CEL.present 8LFU.CEL.present 8LLS.CEL.present
        5.037225        13.701885         8.252812
```

Let us recall that the manufacturer's expectations are that about 30% of the genes should be present in any one particular condition. For whatever reasons, these three arrays have extremely low percentages of Present genes.

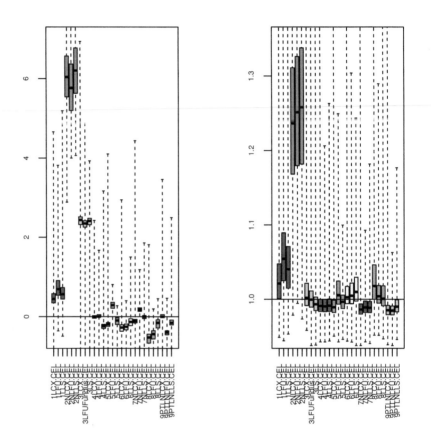

FIGURE 19.10: RLE (left) and NUSE (right) box plots for this data set. The plots shows that the arrays from patients 2 (already identified as saturated) and 3 (already identified as having RNA quality problems) are very different. In addition, the arrays from patient 1 appear to be different in both RLE and the NUSE plots.

According to Affymetrix, such an event indicates that these arrays are below the minimum acceptable sensitivity. If not eliminated from further analysis, these three arrays are likely to introduce false positives in terms of genes would appear to be down-regulated. Notably, this information was available to us earlier, even from the first QC plot or the first time we looked at the *dat.qc* results. However, at that time, we were focused on the more obvious problems associated with patients 2 and 3 pointed out by the intensity distributions and the box plots. Luckily, as we gradually eliminated arrays, the arrays with large and obvious abnormalities, these three arrays eventually made their way back and reentered the scope of our attention. As large abnormalities were eliminated, their smaller but still important differences became noticeable and were eventually discovered.

It should be pointed out that it is not easy to identify the reason for which these three arrays have such a low percentage of Present genes. The fact that the very same phenomenon affected all three arrays coming from patient 8 and none of the other arrays certainly suggests that this may be a sample (or sample processing) issue, rather than a set of three random defective arrays that happen to affect only the samples coming from patient 8. Nevertheless, the fact remains that a percentage of Present genes at or below 10% is only a third of what Affymetrix recommends as a minimum. Whatever the cause was, one would be well advised to discard these three arrays from further consideration.

Finally, let us now look at the QC plot of the data set from which we eliminated all problematic arrays. This plot is shown in Fig. 19.12. This time all scaling factors are blue, indicating that none of them falls outside the acceptable range. Also, the great majority of the $3'/5'$ ratios are within the limits indicating good quality. Overall, this is now a good quality data set.

```
> badarrays=c(1:9,22:24)
> rawdata.QC=rawdata[,-badarrays]
> cleanestdat.qc=qc(rawdata.QC)

> plot(cleanestdat.qc)
```

The intensity distributions and the box plots of the intensities of these arrays are shown in Fig. 19.13. This plot shows that even for a set of data that includes only good quality arrays a normalization step is still necessary in order to align the medians and make the data comparable between arrays. Without this further normalization, the selection of differentially expressed genes can produce incorrect results. Various methods for normalization will be discussed in the next chapter.

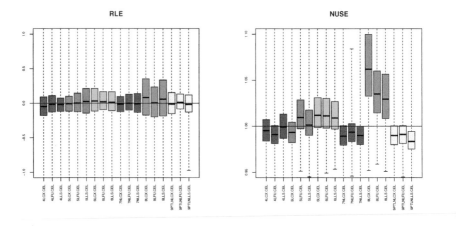

FIGURE 19.11: RLE and NUSE plots based on PLM after excluding arrays from samples 2 and 3: Here the PLM model is fitted after background adjustment and normalization. The left panel shows the RLE plot, in which the arrays from sample 8 stand out from others in terms of wider spread of the relative expression. The right panel shows the NUSE plot, in which the arrays from sample 8 not only have a wider spread but also have higher standard errors for the PLM estimated gene expression.

19.3 Quality control of Illumina data

19.3.1 Reading Illumina data

In this section, we will discuss some functions that are useful when reading and assessing the quality of Illumina data. To summarize the information detailed in Chapter 3, Ilumina's BeadArray technology uses randomly arranged arrays of beads that have DNA probes attached to them. The beads that carry the same sequence are collectively known as a bead-type. BeadArrays are placed on either a matrix of 8 by 12 hexagonal arrays, known as the Sentrix Array Matrix (SAM), or on a rectangular chip known as BeadChip. The surface of the BeadChip is further divided into strips known as sections. The Ilumina software that scans these arrays produces one image for each section. The number of sections assigned to a given biological sample varies between 1 (for HumanHT12 arrays) and 10 or more on SNP arrays.

In order to illustrate the usage of the R and Bioconductor functions that are available for Illumina data, we will use some data from NCBI's Gene Expression Omnibus (GEO) repository. Illumina expression data can be found in GEO in multiple formats: i) bead-level data produced from BeadScan/iScan

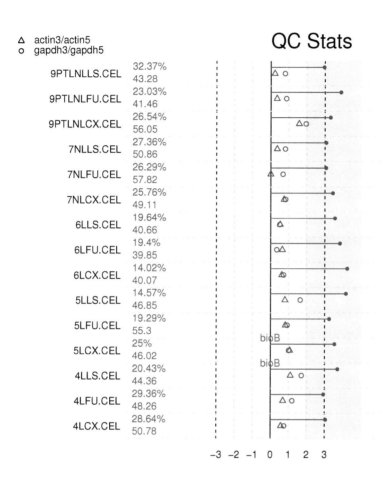

FIGURE 19.12: A summary of the QC data for the data set from which all problematic arrays corresponding to patients 1 (different array type), 2 (saturated arrays), 3 (poor RNA quality), and 8 (insufficient sensitivity) have been removed. This time all scaling factors are blue, indicating that none of them falls outside the acceptable range. Also, the great majority of the $3'/5'$ ratios are within the limits, indicating good quality. Note the correlation between the scaling factors and the percentage of Present genes. Overall, this is now a good quality data set. Nevertheless, a normalization step is still necessary as demonstrated by the box plots in Fig. 19.13.

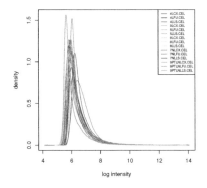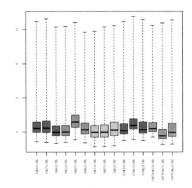

FIGURE 19.13: Intensity distributions and box plots of the Affymetrix data set shown in Fig. 19.1 and Fig. 19.2 after the removal of the 12 problematic arrays corresponding to patients 1, 2, 3, and 8.

(raw TIFF images, text files, etc.), ii) bead-summary data exported from BeadStudio/GenomeStudio (files containing summarized probe-level intensities, etc), and iii) normalized data (preprocessed by the data submitter). Here we illustrate how to use package "beadarray" (for both bead-summary and bead-level) and "lumi" (only for bead-summary) to import and preprocess such data in R.

If the goal is to import data that is already normalized for the purpose of further analysis (e.g., finding differentially expressed genes), a convenient way is to use the function *getGEO* from the package "*GEOquery*." This function takes an argument that represents the identifier of a data set which can be either a GDS or a GSE number. The matrix of expression values across experiments can be extracted through the function *Table* or *exprs*, as follows:

```
> require(GEOquery)
```

```
> prepData=Table(getGEO("GDS3599"))
> prepData=exprs(getGEO("GSE17439", GSEMatrix=T)[[1]])
```

```
> head(prepData)
```

However, as already explained above quality assessment is best done on raw data either on bead-level or probe-level since the presence of many artifacts may be hidden by certain normalization steps. Hence, in the following, we will download and use raw data.

19.3.2   Bead-summary data

19.3.2.1   Raw probe data import, visualization, and quality assessment using "beadarray"

The function *getGEOSuppFiles* can be used to download the raw data from the "Supplementary file" section from GEO repository into a local directory. In the case of the particular data set, we will use here ("GSE17439"), the supplementary data consists of probe-level data as summarized by Illumina's BeadStudio software. After extracting the data files into a local working directory, the function *readBeadSummaryData* can be used to read these data. The argument "*ProbeID*" is used to specify the column of unique probe identifier, while "*columns*" gives corresponding data column header for average expression value and its standard error, number of replicates, and detection p value. "skip" designates the number of rows to ignore from top when data is imported. Note these arguments have to be specified according to the version of BeadStudio and the format of the file containing the bead-summary data. The function *readBeadSummaryData* creates an object of class "*ExpressionSetIllumina*" storing expression matrix, which can be accessed with function *exprs*. In addition, the functions *se.exprs*, *fData*, *nObservations*, *Detection* and *pData* are useful to explore various other stored components.

```
> getGEOSuppFiles("GSE17439", makeDirectory=F)

> datDir<-"./QualityControl"
> rawfile<-paste(datDir, "GSE17439_EsetKnockDownRawData.txt",
+ sep="/")
> library(beadarray)
> BSData<-readBeadSummaryData(dataFile=rawfile, ProbeID=
+ "TargetID", skip=7, columns=list(exprs="AVG_Signal",
+ se.exprs="BEAD_STDEV", nObservations="Avg_NBEADS",
+ Detection="Detection"))
> dim(BSData)

Features  Samples Channels
   46120        6        1

> exprs(BSData)[1:2,]
```

	1488790103_A	1488790103_B
10181072_239_rc-S	183.9	163.5
10181072_290-S	197.9	182.4
	1488790103_C	1488790103_D
10181072_239_rc-S	162.1	166.0
10181072_290-S	190.0	166.5
	1488790103_E	1488790103_F
10181072_239_rc-S	163.9	151.3
10181072_290-S	159.0	175.4

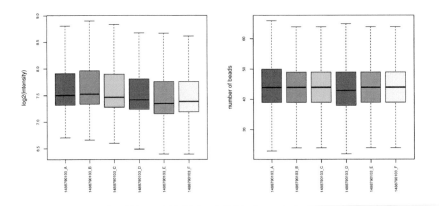

FIGURE 19.14: Box plots of bead-summary data: intensities (left panel) and number of beads (right panel). The average number of beads for a good data set should be 30 or higher.

```
> nObservations(BSData)[1:2,]
```

	1488790103_A	1488790103_B
10181072_239_rc-S	44	40
10181072_290-S	45	39
	1488790103_C	1488790103_D
10181072_239_rc-S	58	44
10181072_290-S	34	49
	1488790103_E	1488790103_F
10181072_239_rc-S	54	43
10181072_290-S	45	38

Box plots can be used to visualize the spread of probe intensity and number of beads for each array. Note the average number of beads per type should to be around 30, which is well exceeded in this data set.

```
> sumraw<-as.data.frame(exprs(BSData))
> library("RColorBrewer")
> usr.col<-brewer.pal(ncol(sumraw), "Set1")

> boxplot(log2(sumraw), las=3,outline=F,col=usr.col,
+ ylab="log2(intensity)", cex.axis=0.9)

> boxplot(as.data.frame(nObservations(BSData)),las=3,outline=F,
+ col=usr.col, ylab="number of beads", cex.axis=0.9)
```

The function *plotMAXY* can be used to build MA-plots for both Illumina and Affymetrix data. In addition, the functions *plotXY* and *plotMA* are available separately for XY plots (scatter plots) and MA plots, respectively.

```
> plotMAXY(sumraw, arrays=1:ncol(sumraw), pch=16)
```

Array probe intensity distribution can be plotted as follows:

```
> plot(density(log2(sumraw[,1]),na.rm=TRUE), col=usr.col[1],
+ xlab="log2(intensity)", main=NA)
> for (i in 2:ncol(sumraw)) {
+ lines(density(log2(sumraw[,i]),na.rm=TRUE), col=usr.col[i])
+ }
> legend("topright",colnames(sumraw), lty=1, col=usr.col)
```

19.3.2.2 Raw probe data import, visualization, and quality assessment using "lumi"

There is an alternative package "lumi"[135] that includes functions designed for preprocessing bead-summary level data. This package offers even more visualization and assessment methods. Here, we briefly outline some basic functions available for step-wise preprocessing. A wrapper function able to combine all preprocessing in one step is available as well and will be discussed later.

```
> library(lumi)
> rawfile = "GSE17439_EsetKnockDownRawData.txt"
> rawsum.lm <- lumiR(rawfile, sep = "\t",
+     columnNameGrepPattern = list(exprs = "AVG_Signal",
+         se.exprs = "BEAD_STDEV", beadNum = "Avg_NBEADS",
+         detection = "Detection", verbose = F))
```

```
Annotation columns are not available in the data.
Perform Quality Control assessment of the LumiBatch object ...
```

```
> head(exprs(rawsum.lm))
```

	1488790103_A	1488790103_B
10181072_239_rc-S	183.9	163.5
10181072_290-S	197.9	182.4
10181072_290_rc-S	163.5	178.0
10181072_311-S	155.0	140.0
10181072_311_rc-S	165.9	178.2
10181072_418-S	120.4	129.4
	1488790103_C	1488790103_D
10181072_239_rc-S	162.1	166.0
10181072_290-S	190.0	166.5
10181072_290_rc-S	177.2	169.8
10181072_311-S	148.7	158.2
10181072_311_rc-S	151.2	157.1

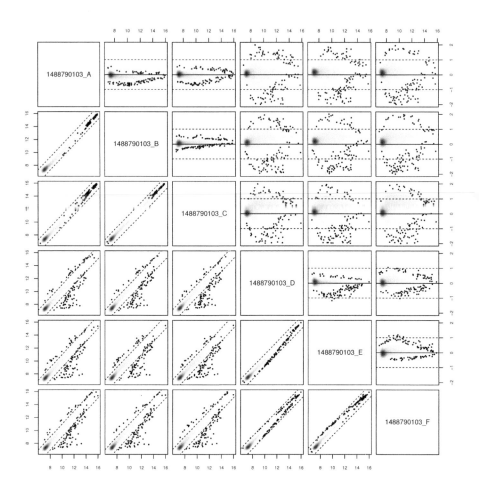

FIGURE 19.15: A MAXY plot of bead-summary data. M versus A plots are shown above the diagonal while XY scatter plots of the same pairs of arrays are shown in the corresponding positions below the diagonal. The dashed lines show the ±2 fold changes. Note the differences between the first three arrays and the other three. In particular, the width of the distribution is different much narrower between any two arrays from the first three (A, B, C), wider for the plots involving any two of the last three arrays (D, E. F), and very wide for the plots involving one array from the first group and one array from the second group.

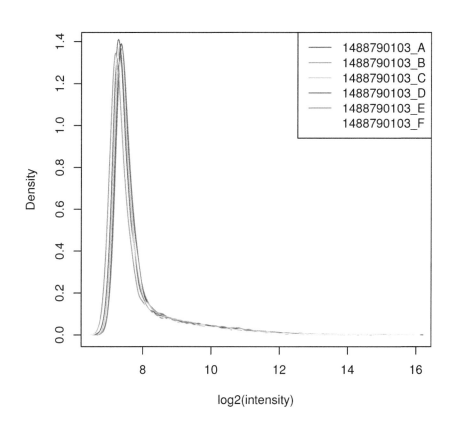

FIGURE 19.16: Array probe intensity distribution.

```
10181072_418-S              118.2          115.8
                    1488790103_E 1488790103_F
10181072_239_rc-S           163.9          151.3
10181072_290-S              159.0          175.4
10181072_290_rc-S           160.9          174.4
10181072_311-S              145.0          147.6
10181072_311_rc-S           139.0          145.2
10181072_418-S              113.2          121.1

> summary(rawsum.lm, "QC")

Data dimension:  46120 genes x 6 samples

Summary of Samples:
                        1488790103_A 1488790103_B
mean                        7.9110       7.9580
standard deviation          1.1000       1.1420
detection rate(0.01)        0.2369       0.2408
distance to sample mean    30.5800      38.5400
                        1488790103_C 1488790103_D
mean                        7.8880       7.8110
standard deviation          1.1210       1.0710
detection rate(0.01)        0.2381       0.2334
distance to sample mean    30.3500      30.0100
                        1488790103_E 1488790103_F
mean                        7.7440       7.7510
standard deviation          1.0840       1.0500
detection rate(0.01)        0.2348       0.2194
distance to sample mean    35.7700      39.3100


Major Operation History:
            submitted              finished
1 2011-10-28 21:33:00 2011-10-28 21:33:04
2 2011-10-28 21:33:00 2011-10-28 21:33:04
3 2011-10-28 21:33:00 2011-10-28 21:33:04
4 2011-10-28 21:33:04 2011-10-28 21:33:04
                                                          command
1                          lumiR("GSE17439_EsetKnockDownRawData.txt",
2    se.exprs = "BEAD_STDEV", beadNum = "Avg_NBEADS", detection = "Detection",
3                                                      verbose = F))
4          lumiQ(x.lumi = x.lumi, detectionTh = detectionTh, verbose = verbose)
  lumiVersion
1      2.4.0
2      2.4.0
3      2.4.0
4      2.4.0
```

```
> #the following plots are similar to those produced by "affy"
> #and "beadarray"
> density(rawsum.lm)     #probe intensity distribution
> plotCDF(rawsum.lm)     #probe intensity CDF
> boxplot(rawsum.lm)     #boxplot of intensity spread
> pairs(rawsum.lm, smoothScatter=T)    #pairwise array probe
>                                      #intensity scatterplot
> MAplot(rawsum.lm, smoothScatter=T, cex=0.9)   #pairwise MAplot
```

Besides the above functions, "*lumi*" provides additional visualization tools such as density plot of coefficient of variance, sample clustering, and sample distance plot after multidimensional reduction.

```
> plot(rawsum.lm, what='cv', main=NA)


> plot(rawsum.lm, what='sampleRelation')

> plotSampleRelation(rawsum.lm, method='mds',col=rep(c(1,2),
+ each=3),xlim=c(-25,25))
```

19.3.3   Bead-level data

19.3.3.1   Raw bead data import and assessment

Bead-level raw data is produced by Ilumina's BeadScan/iScan software and can be read using the function *readIllumina*, which is part of the *beadarray* package [141]. Depending on version and setup of BeadScan/iScan, the bead-level raw data files may include .txt files containing the location and background subtracted intensities for decoded beads, .tif files containing scanned array images, .locs files containing location information for all beads, .idat files which are proprietary binary files storing intensity data to be read in by BeadStudio/GenomeStudio, .xml files containing scanning settings and information used to extract intensities, .bgx files containing probe annotations, Metrics.txt containing a summary of scanning quality parameters, .sdf files containing sample/array structure and layout, and IBS (Illumina Bead Summary) files, saved as .csv, and containing a summary of the number of beads/mean/standard error of intensity for each beadtype. The function *read-Illumina* can read bead intensities either from .txt files or .tif images. If the parameter *useImages* is false, the intensities read from the .txt files (corrected for local background as mentioned above) will be used as the starting point for all subsequent analysis.

The text files contain decoded bead intensities reading that are already corrected for the local background. This local background correction is performed using the pixel intensities surrounding each bead and is somewhat

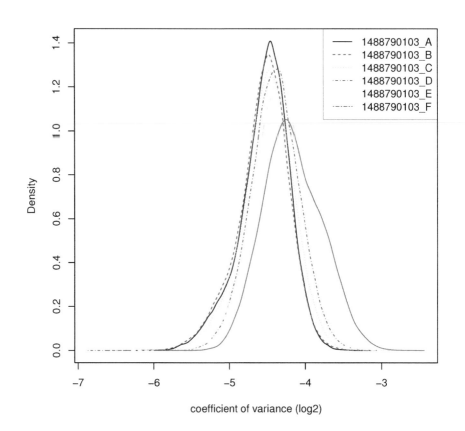

FIGURE 19.17: Density plot of coefficients of variance (ratio of standard deviation to the mean). After log transformation, this ratio will be negative for most probes. Also note that this statistic is sensitive to mean values close to zero. Note that in this plot, the arrays appear to be different, indicating that the variance differs between the arrays. In particular, the array 1486790103_F has a distribution of its coefficients of variance a bit dissimilar to the others.

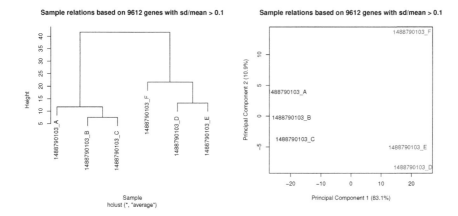

FIGURE 19.18: A hierarchical clustering of samples and PCA plot after multi-dimensional scaling (MDS). The clustering and PCA plots show that samples A, B, and C are rather similar. Samples D, E, and F appear to be a little different. This is indicated by the larger vertical distance in between the clusters (A,B,C) and (D,E,F) in the dendrogram, and the positioning in the PCA plot. Sample F appears to be somewhat different from all the others, as indicated by the PCA plot.

similar to the previously discussed local background correction for cDNA arrays. However, this background correction does not correct for nonspecific hybridization and other similar issues. These corrections are done later on in the normalization step using negative controls. Alternatively, the intensities can be calculated directly from the images, when the data is read. Together with .locs file, all beads intensities (including the ones that failed to be decoded) can be recovered. If available, both .sdf and Metrics.txt files can be read by *readIllumina* as well.

Reading bead-level data is not trivial due to the flexible format of the bead data files and the existence of different versions of both Illumina software and "beadarray" R package. For instance, some bead-level .txt files may use commas instead of decimal points in bead coordinate values, which causes problem when importing with **readIllumina**. Sometimes the .txt file storing decoded beads information is present in the form of .csv file. The correct naming convention of files is also essential. For instance, when the argument "sectionNames" is used to specify the array file names to be read in, the file extensions should be eliminated. Some of these issues will be encountered when reading some of the sample data sets used here.

Illumina has two microarray platforms, Sentrix BeadChip and Sentrix Array Matrix (SAM). As an illustration of bead-level data visualization and physical defect identification, we will download and extract two bead-level raw data sets from those two platforms.

The first data set we will use includes eight customized Mouse-6 version 1 BeadChips from a spike-in experiment [140] that is available at http://www.compbio.group.cam.ac.uk/Resources/spike/index.html. After downloading, extraction, elimination of two un-useful columns in .csv and conversion of its file format to .txt, we will only read in two strips from two arrays, 1377192001_A_1 and 1377192002_A_1.

```
> library(beadarray)

> CSVname=dir(path=SPKdir, pattern=".*csv")
> arrayName=gsub(".csv", "", CSVname)
> for (i in 1:length(CSVname)) {
+ tmpCSV=read.csv(paste(SPKdir, "/", CSVname[i], sep=""))
+ write.table(tmpCSV[,!colnames(tmpCSV) %in% c("Index", "Red")],
+ paste(SPKdir,"/", arrayName[i], ".txt", sep=""), quote=F,
+ sep="\t", row.names=F)
+ }
> BLspk=readIllumina(dir=SPKdir,sectionNames=c("1377192001_A_1",
+ "1377192002_A_1"), illuminaAnnotation="Mousev1")
```

The function *imageplot* can be used to obtain a pseudo-image of the bead chip, image that can be inspected for the presence of spatial artifacts. This type of quality control can only be performed with bead-level data. Such images are not possible when using summarized output from BeadStudio because the summary values are obtained by averaging over spatial coordinates. This averaging loses the spatial information, making it impossible to construct pseudo-images from summary data.

By default, the function *imageplot* displays the image for log2 transformed bead intensities. Note that this is not an "exact" reproduction of the original .tif image due to the absence of .locs file that contains the locations of all beads for this data set. Furthermore, the image is not a faithful reproduction of the array also because *imageplot* does a "compression" by mapping a small window (specified by the argument "squareSize") of original pixels to a single pixel in the pseudo-image.

```
> imageplot(BLspk, array=1, low="lightgreen", high="darkgreen",
+ horizontal=T, main=names(BLspk@beadData)[1], squareSize=16)

> imageplot(BLspk, array=2, low="lightgreen", high="darkgreen",
+ horizontal=T, main=names(BLspk@beadData)[2], squareSize=16)
```

The second data set used here (from http://www.compbio.group.cam.ac.uk/Resources/illumina/) includes ten SAM arrays, compressed in a single file "SAMExample.zip." A similar sequence is used to read the data in R:

1377192001_A_1

z−range 2.6 to 15.9 (saturation 2.6, 15.9)

1377192002_A_1

z−range 2.7 to 15.1 (saturation 2.7, 15.1)

FIGURE 19.19: Pseudo-images of bead intensities from two strips. The BeadChip here has nine segments in each strip. These two strips are supposed to be replicates of each other on two different arrays. However, the array 1377192001 drawn on top shows a blank area due to scanning errors. These errors produced negative bead coordinates, which caused the intensities of these beads to be set to zero. In fact, all 12 strips from the chip 1377192001 have such physical defects-.

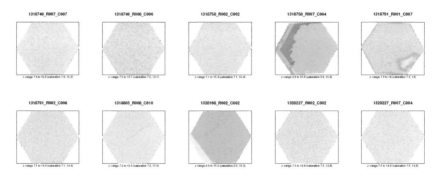

FIGURE 19.20: Pseudo-images of ten SAM arrays. Arrays 4, 5, and 8 (counting left to right and top to bottom) have abnormal intensity pattern that might be attributed to hybridization and/or scanning errors.

```
> SAMtmp<-dir(path=DirSAM, pattern="^[0-9].*csv")
> SAMfile<-SAMtmp[SAMtmp != "1318782_R004_C012.csv"]
> #exclude an unused file
> arraySAM<-gsub(".csv", "", SAMfile)
> for (i in 1:length(SAMfile)) {
+ tmpCSV<-read.csv(paste(DirSAM, "/", SAMfile[i], sep=""))
+ write.table(tmpCSV[, !colnames(tmpCSV) %in% c("Index",
+ "Red")], paste(DirSAM, "/", arraySAM[i], ".txt", sep=""),
+ quote=F, sep="\t", row.names=F)
+ }
> BLsam<-readIllumina(dir=DirSAM, sectionNames=arraySAM)
```

The pseudo-images for these SAM arrays are shown in Fig. 19.20. This figure shows that 3 of the 10 arrays have abnormal intensity patterns. These may be due to hybridization and/or scanning errors.

```
> par(mfrow=c(2,5))
> for (i in 1:length(SAMfile)) {
+ imageplot(BLsam, array=i, low="lightgreen", high="darkgreen",
+ horizontal=F, main=names(BLsam@beadData)[i], squareSize=16)
+ }
```

A third example will use a data set from GEO: GSE13974. Interestingly, these data look normal after assessing the pseudo-images of bead intensity. To save processing time, we only choose the first six arrays (out of a total of 24) to be imported. The platform used in this experiment is "GPL6885 Illumina MouseRef-8 v2.0 expression beadchip."

```
> getGEOSuppFiles("GSE13974", makeDirectory=T)
```

```
> #unzip the file to extract the raw data files
> #before proceeding further; also set the dir with
> #DATdir="./GSE13974"

> require(beadarray)
> txtFile=dir(path=DATdir, pattern="^GSM.*txt")
> arrayName=gsub(".txt","",txtFile) #get rid of file extension
> getwd()

[1] "c:/Users/Sorin/Documents/Books/Book revision"

> BLdata<-readIllumina(dir=DATdir, sectionNames=arrayName[1:6],
+ illuminaAnnotation="Mousev2")

> slotNames(BLdata)  # the slots available in this data structure

[1] "beadData"        "sectionData"
[3] "phenoData"       "experimentData"
[5] "history"

> head(BLdata[[1]]) #accessing the first array

      ProbeID       GrnX       GrnY Grn
[1,]    10008   197.0973   4639.391  91
[2,]    10008  1808.9160   9162.940  42
[3,]    10008   447.4015   7614.163  48
[4,]    10008  1019.5700  14905.310  58
[5,]    10008   337.8100   7189.099  50
[6,]    10008   691.3507  14082.210  62

> #"GrnX", "GrnY", "Grn" give coordinates and bead intensity
```

In the above code data are read from .txt files. The bead intensity values have already been corrected for the local background. To import raw bead data directly from .tif images, we need to rename the downloaded .tif files by adding "_Grn" to the end of file name, since this is what the *read-Illumina* function is expecting. For example, "GSM351169.tif" is renamed as "GSM351169_Grn.tif". After this renaming, the images can be read as follows:

```
> BLimg<-readIllumina(dir=DATdir, sectionNames=arrayName[1:6],
+ useImages=T, illuminaAnnotation="Mousev2")
> head(BLimg[[1]])

      ProbeID       GrnX       GrnY      Grn
[1,]    10008   197.0973   4639.391  91.71893
[2,]    10008  1808.9160   9162.940  42.76772
[3,]    10008   447.4015   7614.163  48.38862
[4,]    10008  1019.5700  14905.310  58.59826
```

```
[5,]    10008   337.8100   7189.099 50.44994
[6,]    10008   691.3507 14082.210 62.97472
          GrnF    GrnB
[1,] 764.7189 673.0
[2,] 713.9677 671.2
[3,] 721.5886 673.2
[4,] 730.3983 671.8
[5,] 719.6499 669.2
[6,] 729.9747 667.0


> #now we have two extra columns of foreground and background
> #intensity; bead intensity is computed by  subtracting these
> #two columns
```

The data can be accessed with the function *getBeadData*, which takes as arguments a data structure of class *beadLevelData*, the index of the desired array, *array*, and a parameter specifying what column we need, *what*. For instance, the intensity values for the green channel can be obtained as:

```
> x=getBeadData(BLdata,array=1,what="Grn")
> head(x)
> par(mfrow=c(1,2))
> y=log2(x)
> hist(y)
> boxplot(y)
```

Box plots can be used to compare intensities between arrays. Such comparisons can be done both on foreground and background intensities. The function *boxplot* as implemented in this package, takes as an argument a object of class beadLeveldata, and a transformation function (that defaults to $\log_2$) that is applied to the data before plotting. A simple box plot comparing the $\log_2$ values of the six arrays in this experiment can be plotted with:

```
> boxplot(BLdata, las=3)
```

As already mentioned, the pseudo-images of background-corrected bead intensity do not show any problems for this data set. We can further construct pseudo-images for log2 transformed background and foreground bead intensity, by specifying a different function to the argument "transFun" in *imageplot*. Here, we will first define two custom functions that apply a log transformation to the green foreground (GrnF) and green background (GrnB), respectively:

```
> transFunF<-function(BLData, array) log2(getBeadData(BLData,
+ what="GrnF", array=array))
> transFunB<-function(BLData, array) log2(getBeadData(BLData,
+ what="GrnB", array=array))
```

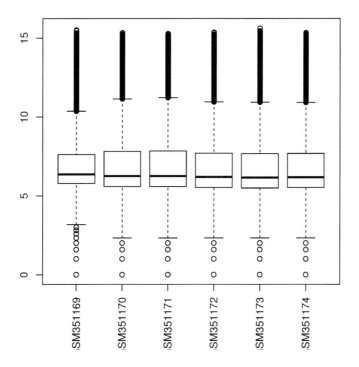

FIGURE 19.21: The box plots of the six Illumina arrays in this sample data set.

Then, these functions are specified as the transformation to be applied in *imageplot*:

```
> par(mfrow=c(1,3))
> imageplot(BLimg, array=1, low="lightgreen",high="darkgreen",
+ horizontal=F, main="intensity")
> imageplot(BLimg, array=1, transFun=transFunF,low="lightgreen",
+ high="darkgreen", horizontal=F, main="foreground")
> imageplot(BLimg, array=1, transFun=transFunB,low="lightgreen",
+ high="darkgreen", horizontal=F, main="background")
```

Fig. 19.22 shows the images for bead intensity, foreground intensity and background intensity for array GSM351169. The image corresponding to the background shows an abnormality affecting a large area of the array. Luckily, in this case a further inspection of the data shows that the values in this region do vary but the variation is limited to a very narrow interval from 9.27 to 9.51. In essence, this is an array defect that was detected using this approach but the defect is so small that is unlikely to seriously affect later analysis stages. This also explains why this defect was not seen in the background-corrected bead intensity images.

More subtle physical defects are hard to identify through a visual examination of the pseudo-images. BASH (BeadArray Subversion of Harshlight) is a more sophisticated package that implements for bead arrays the methodology available in the "Harshlight" package for Affymetrix data. This approach is able to discern three types of physical defects on illumina arrays: i) extended (gradual but significant intensity shifts across surface), ii) diffuse (area with high density of outliers), and iii) compact (large connected clusters of outliers). This function assigns weights to beads to facilitate removal of outliers, where weight of zero indicates the bead located within identified defect area.

```
> BLimgBASH<-BASH(BLimg, array=1:6, bgcorr="medianMAD")

> BLimgBASH$QC
```

	BeadsMasked	ExtendedScore
1	247	0.1137852
2	84	0.1045524
3	174	0.1110771
4	180	0.1110628
5	182	0.1103662
6	378	0.1067582

```
> #record outliers from BASH in a list
> bashoutProbe<-vector("list", 6)
> names(bashoutProbe)<-names(BLimg@beadData)
> for (i in 1:6) {
+ bashoutProbe[[i]]<-BLimg[[i]][,1][BLimgBASH$wts[[i]]==0]
+ }
```

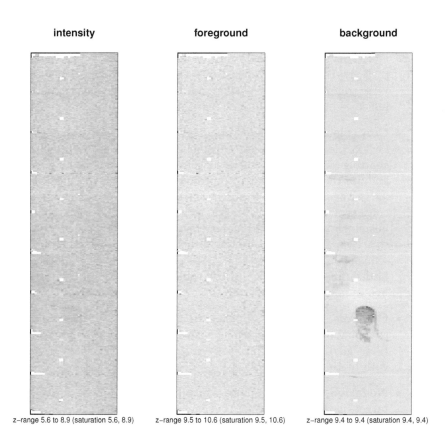

FIGURE 19.22: Pseudo-images of bead intensity, foreground, and background intensity for array "GSM351169" after a log transformation. An abnormality that affects a large area of the array can be found in the background image of this array. However, an examination of the data itself (the log2 background intensity), shows that in this area, the values only vary within a narrow interval from 9.27 to 9.51. Hence, this defect is not as serious as it looks in the image.

Much like the Affymetrix arrays, the Illumina arrays also include various positive and negative controls that should signal if something goes wrong during the processing pipeline.

- **Controls for biological specimen:** These are housekeeping genes that should be expressed in all samples.

- **Controls for sample labeling:** These are RNA spikes lysA, pheA, thrB, trpF. These are spiked into samples immediately before the reverse transcription and labeling reactions. Low signals from these probes indicates problems with these reactions. This type of control is optional.

- **Controls for hybridization:** There are three types of such controls: i) Cy3-labeled hyb (the target oligonucleotides for six Cy3 hybridization controls are present at three concentrations of low, medium, and high, and are therefore expected to produce gradually increasing signals); ii) low-stringency hyb (four probes corresponding to the medium and high concentration Cy3 hybridization targets, except that each probe has two mismatch bases); and iii) high-stringency hyb (one probe corresponding to a Cy3-labled oligonucleotide target, with very high GC content in its sequence). A high signal from low-stringency hyb suggests that the hybridization stringency is too low, while a low signal from high-stringency hyb implies that the hybridization stringency is too high.

- **Controls for signal generation:** biotin controls (two probes corresponding to biotin-tagged oligonucleotide targets). A positive hybridization signal from these probes reveals successful secondary Cy3 staining with biotin-labeled RNA.

- **Negative controls:** hundreds of probes of random sequences without targets in the genome, reflecting background signal from imaging system, nonspecific binding of dye, and cross-hybridization.

Generally, information about the control probes can be obtained in one of three ways: i) specifying an annotation package in the "readIllumina" function, ii) from a .bgx file, and iii) using a control probe profile exported from Bead-Studio/GenomeStudio. For this data set, the first two methods can be used without resorting to the Illumina software. If we want to use the annotations built in the "beadarray" package, the control probe information for different expression arrays are available in the "ExpressionControlData" object.

```
> data(ExpressionControlData)
> names(ExpressionControlData)  #available expression arrays

[1] "Humanv3"   "Humanv2"   "Humanv1"
[4] "Mousev2"   "Mousev1"   "Mousev1p1"
[7] "Ratv1"     "Humanv4"
```

```
> ctrMV2<-ExpressionControlData[["Mousev2"]]
> #for Mouse version 2 array
> ctrMV2[1:3,]

  Array_Address Reporter_Group_Name
1      5860278             negative
2       610201             negative
3      4670735             negative
  Reporter_Group_Identifier
1         permuted_negative
2         permuted_negative
3         permuted_negative
                                    Probe_Sequence
1 GCGTATTGGCTGCTGGTCTTGACCAGTGCCGGAATTCCGCTCTGATATAG
2 TGAATGAGAACTCTTGGCCCCGGCTCCTTTCACAAAGACGGTTAGCTTGG
3 GGAGGCATGCCACCTCTTCCTACGAACAAGTCAGGAAACGGTTCGAAGCC

> dim(ctrMV2)

[1] 974    4

> length(unique(ctrMV2[,1]))   #number of unique control probes

[1] 970
```

Alternatively, a control profile can be built from the .bgx file of this data set, by copying the control probe information at the end of the .bgx to a new .txt file named "CtrProbes.txt." The built-in control annotation contains 974 probes, of which 970 are unique and 4 are redundant. In the control probe section of the .bgx file there are 832 probes, of which 829 are unique and 3 are redundant. Furthermore, all of the 832 from the .bgx file are contained in the 974 probe set from the package. In fact, all those extra probes are not present in all the six arrays, indicating these probes might be lost when BeadScan/iScan was processing the image scanning and intensity extraction to .txt file. So the two sources of control information (built-in annotation in the package, and the annotation extracted from .bgx) are consistent with each other. In the following, we will use the one constructed from the .bgx file.

```
> ctrBGX<-read.delim(paste(DATdir, "/CtrProbes.txt", sep=""))
> ctrBGX[1:3, 1:4]

      Probe_Id Array_Address_Id
1 ILMN_1380403          5860278
2 ILMN_1379274           610201
3 ILMN_1380434           270672
  Reporter_Group_Name Reporter_Group_id
1            negative permuted_negative
2            negative permuted_negative
3            negative permuted_negative
```

```
> dim(ctrBGX)

[1] 581    6

> length(unique(ctrBGX[,2])) #number of unique control probes

[1] 580

> table(ctrBGX[,3])    #probes in general groups

            biotin                    cy3_hyb
                 1                          3
      housekeeping                   labeling
                 8                          7
low_stringency_hyb                      negat
                 3                          1
          negative
               558

> table(ctrBGX[,4])    #probes in detailed subgroups

                                    housekeeping
                 1                             8
              lysA             permuted_negative
                 2                           558
    phage_lambda_genome phage_lambda_genome:high
                 1                             1
 phage_lambda_genome:low  phage_lambda_genome:mm2
                 2                             2
  phage_lambda_genome:pm                     pheA
                 1                             2
              thrB                          trpF
                 1                             2

> all(ctrBGX[,2] %in% ctrMV2[,1])

[1] TRUE

> #a subset relation of the two control profiles
> extraProbe<-setdiff(ctrMV2[,1], ctrBGX[,2])
> tmp<-NULL
> for (i in 1:6) {
+ tmp[i]<-any(extraProbe %in% BLimg[[i]][,1])
+ }
> tmp    #extra probes are not present in all 6 arrays

[1] TRUE TRUE TRUE TRUE TRUE TRUE
```

As shown above, controls can be categorized in 6 general groups or in 12 detailed subgroups. A convenient way to summarize all controls in a single array is provided by the function *quickSummary*. This function can display the results either categorized into general groups or detailed subgroups. This capability is extended to multiple arrays through the function *makeQCTable* (which calls *quickSummary* internally). Clearly, the summary results using either source of control information should be the same.

```
> unlist(quickSummary(BLimg, array=1, reporterIDs=ctrBGX[,2],
+ reporterTags=ctrBGX[,4]))

                                        housekeeping
                5.966244                    11.818181
                    lysA            permuted_negative
                5.833929                     5.846513
    phage_lambda_genome phage_lambda_genome:high
               11.820705                    13.217455
 phage_lambda_genome:low  phage_lambda_genome:mm2
                7.878502                     9.269581
  phage_lambda_genome:pm                         pheA
                     NaN                     5.843918
                    thrB                         trpF
                5.775448                     5.971034

> makeQCTable(BLimg, controlProfile=ctrBGX[,c(2,4)])

              Mean: Mean:housekeeping Mean:lysA
GSM351169 5.966244            11.81818  5.833929
GSM351170 5.787945            12.01865  5.514149
GSM351171 5.668962            11.89779  5.541340
GSM351172 5.397896            11.43984  5.538593
GSM351173 5.480645            11.87888  5.519945
GSM351174 5.657313            11.73736  5.540875
              Mean:permuted_negative
GSM351169                  5.846513
GSM351170                  5.623936
GSM351171                  5.612572
GSM351172                  5.580264
GSM351173                  5.536259
GSM351174                  5.559644
              Mean:phage_lambda_genome
GSM351169                   11.82071
GSM351170                   11.92588
GSM351171                   11.82436
GSM351172                   11.84500
GSM351173                   11.84044
```

```
GSM351174                        11.84604
              Mean:phage_lambda_genome:high
GSM351169                            13.21746
GSM351170                            13.11180
GSM351171                            13.15822
GSM351172                            13.16453
GSM351173                            13.03350
GSM351174                            13.03750
              Mean:phage_lambda_genome:low
GSM351169                             7.878502
GSM351170                             7.783525
GSM351171                             7.721204
GSM351172                             7.559452
GSM351173                             7.590117
GSM351174                             7.599044
              Mean:phage_lambda_genome:mm2
GSM351169                             9.269581
GSM351170                             9.165325
GSM351171                             8.854066
GSM351172                             8.533891
GSM351173                             8.886149
GSM351174                             8.806598
              Mean:phage_lambda_genome:pm Mean:pheA
GSM351169                             NaN   5.843918
GSM351170                             NaN   5.595276
GSM351171                             NaN   5.565484
GSM351172                             NaN   5.517902
GSM351173                             NaN   5.571459
GSM351174                             NaN   5.488599
              Mean:thrB Mean:trpF      Sd:
GSM351169  5.775448  5.971034 0.5669962
GSM351170  5.520517  5.560926 0.6042703
GSM351171  5.460121  5.583488 0.6489057
GSM351172  5.453239  5.479814 0.8462814
GSM351173  5.436420  5.425572 0.6559612
GSM351174  5.428891  5.510098 0.4388993
              Sd:housekeeping   Sd:lysA
GSM351169         1.917233 0.5810514
GSM351170         1.423721 0.6602595
GSM351171         1.486399 0.5402085
GSM351172         2.412083 0.6364740
GSM351173         2.073567 0.6045428
GSM351174         2.108097 0.6787372
              Sd:permuted_negative
GSM351169             0.5688399
```

```
GSM351170              0.6121027
GSM351171              0.6293511
GSM351172              0.6118337
GSM351173              0.6150189
GSM351174              0.6270178
           Sd:phage_lambda_genome
GSM351169              0.3435849
GSM351170              0.3033237
GSM351171              0.3847869
GSM351172              0.3243948
GSM351173              0.3707864
GSM351174              0.3816634
           Sd:phage_lambda_genome:high
GSM351169               0.3741886
GSM351170               0.4493577
GSM351171               0.4667655
GSM351172               0.4277884
GSM351173               0.4357164
GSM351174               0.4337470
           Sd:phage_lambda_genome:low
GSM351169               0.5279084
GSM351170               0.5616879
GSM351171               0.6149477
GSM351172               0.5727639
GSM351173               0.5783215
GSM351174               0.6231966
           Sd:phage_lambda_genome:mm2
GSM351169               1.471274
GSM351170               1.461730
GSM351171               1.512979
GSM351172               1.460425
GSM351173               1.434835
GSM351174               1.459912
           Sd:phage_lambda_genome:pm   Sd:pheA
GSM351169                           NA 0.5687383
GSM351170                           NA 0.6215551
GSM351171                           NA 0.6205162
GSM351172                           NA 0.5512531
GSM351173                           NA 0.6073493
GSM351174                           NA 0.5719235
             Sd:thrB     Sd:trpF
GSM351169 0.5985980 0.6473715
GSM351170 0.6200778 0.5946134
GSM351171 0.6620173 0.6667549
GSM351172 0.5782457 0.6453126
```

```
GSM351173 0.5607409 0.5894422
GSM351174 0.6129117 0.6661350
```

These summaries are calculated from the raw data and may therefore be different from the analogous quantities provided by Illumina's BeadStudio software which first eliminates the outliers. A first observation about the control data displayed above is that the values for the subgroup "phage_lambda_genome:pm" (corresponding to perfect match in Cy3-labeled hyb) are missing. This appears to be caused by the redundancy of probes in the annotation file and the method implemented in function *quickSummary*. The three redundant probes we mentioned earlier are exactly the three probes for "phage_lambda_genome:pm." These three probes are also annotated as "phage_lambda_genome:high" or "phage_lambda_genome:med" (corresponding to high or medium concentration in Cy3-labeled hyb), which is as expected from the array control design. This redundancy, however, corrupts the calculation in the function *quickSummary*, because this function relies on the *match* function to link the control subgroup in the annotation to the probes in the array data. For example, the probe "6450180" in the array data will be linked to the subgroup "phage_lambda_genome:high" instead of "phage_lambda_genome:pm," because the former subgroup is the first record that *match* finds for probe "6450180" in the annotations.

However, the missing values can be calculated explicitly as follows:

```
> pmCtr<-ctrBGX[ctrBGX[,4]=="phage_lambda_genome:pm",][,2]
> pm<-matrix(rep(NA,12), nrow=2)
> rownames(pm)<-c("Mean", "Sd")
> colnames(pm)<-names(BLimg@beadData)
> for (i in 1:6) {
+ pm[1,i]=mean(log2(BLimg[[i]][BLimg[[i]][,1] %in% pmCtr,4]))
+ pm[2,i]=sd(log2(BLimg[[i]][BLimg[[i]][,1] %in% pmCtr,4]))
+ }
> pm
```

```
          GSM351169   GSM351170   GSM351171   GSM351172
Mean 13.2174551 13.1117968 13.1582151 13.1645291
Sd    0.3741886  0.4493577  0.4667655  0.4277884
          GSM351173 GSM351174
Mean 13.0335020 13.037496
Sd    0.4357164  0.433747
```

And we can now see the redundancy that caused the problem:

```
> ctrBGX[ctrBGX[,2] %in% pmCtr,][,1:4]
```

```
          Probe_Id Array_Address_Id
211 ILMN_2038770          6450180
212 ILMN_2038770          6450180
```

	Reporter_Group_Name	Reporter_Group_id
211	cy3_hyb	phage_lambda_genome:high
212	low_stringency_hyb	phage_lambda_genome:pm

Now we can take a look at the summary result for all controls. Overall, there is no single array that is much different from others. We can make the following observations for each type of controls:

- **Controls for biological specimen:** Housekeeping genes are well expressed across all samples.

- **Controls for sample labeling:** RNA spikes (lysA, pheA, thrB, trpF) signals are as low as the background level. But this doesn't necessarily imply labeling problems, since this control is optional. In fact, if these controls have not been used in this experiment, it is expected for these controls to remain at the level of the background.

- **Controls for hybridization:** Cy3-labeled hyb (at three levels: low, med, high) show gradual increasing of signal as expected. Low-stringency hyb (mm2) intensity is one order of magnitude lower than perfect match probes (pm). High-stringency hyb is not available from this annotation.

- **Controls for signal generation:** Biotin controls (phage lambda genome) having high-intensity values suggests a good Cy3 staining reaction.

- **Negative controls:** The signal is at a low level as desired.

The function *poscontPlot* can plot intensity spread of housekeeping/biotin (or other specified groups) bead types in each array. This offers a visualization tool to examine each probe within a group, rather than averaged statistics for each group as provided by the above summary functions.

```
> par(mfrow = c(2, 3), mar = c(5, 2, 1, 2))
> for (i in 1:6) {
+ poscontPlot(BLimg, array = i, ylim = c(4, 15),
+ main=names(BLimg@beadData)[i])
+ }
```

Before finally summarizing bead-level data, Illumina removes any bead observation with an intensity more than 3 median absolute deviations (MAD) from its median for the same beadtype. The function *outlierplot* implements this criterion and also plots the outliers found on the given array. The indices of outlier beads can be returned through function *illuminaOutlierMethod*.

```
> par(mfrow=c(2,3))
> for (i in 1:6) {
+ outlierplot(BLimg, array=i, horizontal=F, main=
+ names(BLimg@beadData)[i])
+ }
```

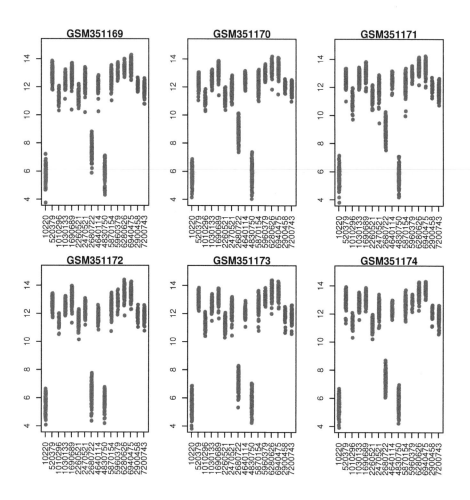

FIGURE 19.23: Plot of housekeeping/biotin controls on each array. Red and blue stand for housekeeping and biotin controls, respectively. There are 14 housekeeping probes and two biotin probes. The spread pattern looks similar among arrays. Note the two housekeeping probes that have an intensity at the level of the background.

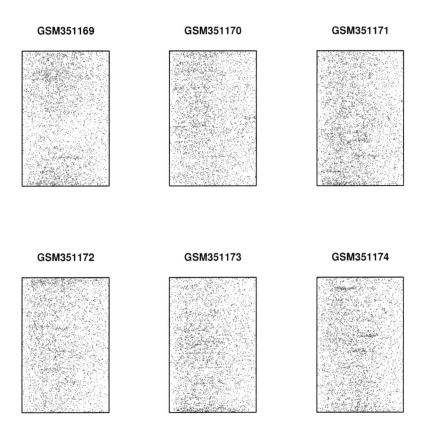

FIGURE 19.24: Outlier visualization. The outliers are shown as points on the array. The colors indicate whether the given outlier is 3 MAD below (blue) or above (red) median intensity for a given bead type.

A natural question is how do the outliers identified by *BASH* compare with those identified by *illuminaOutlierMethod*? It turns out there are no overlapping between the two sets of outliers, suggesting that the two approaches may be complementary rather than redundant. The outliers from *illuminaOutlierMethod* will be removed later on when summarizing the data. Eliminating the outliers identified by *BASH* can be accomplished through storing BASH weights in the "beadLevelData" object. The QC result from BASH can be appended to the object as well, if desired.

```
> outlierProbe<-vector("list", 6)
> names(outlierProbe)<-names(BLimg@beadData)
> for (i in 1:6) {
+ outlierProbe[[i]]=illuminaOutlierMethod(
+ logGreenChannelTransform(BLimg, i),
+ getBeadData(BLimg, array=i, what="ProbeID"))
+ }
> NinCommon<-function(x,y) {x<-as.character(x);
+ y<-as.character(y); length(intersect(x,y))}
> mapply(NinCommon, bashoutProbe, outlierProbe)
```

```
GSM351169 GSM351170 GSM351171 GSM351172 GSM351173
        0         1         0         1         1
GSM351174
        0
```

```
> #storing BASH weights
> for (j in 1:6) {
+ BLimgBS<-setWeights(BLimg, wts=BLimgBASH$wts[[j]], array=j)
+ }
> BLimgBS<-insertSectionData(BLimgBS, what="BASHQC",
+ data=BLimgBASH$QC)
```

19.3.3.2 Summarizing from bead-level to probe-level data

The final summarization of bead-level data to probe-level can be performed with the function summarize. This function takes as arguments a bead-level data object and a list of objects of "illuminaChannel" class defining the sequential summarization procedures for each array. In the following snippet of code, we first do a log2 transform on bead intensity, remove outliers, and then summarize mean and standard deviation for each bead type. The summarized object "BSimg" is of class "ExpressionSetIllumina." Note that in the summarization function the argument "useSampleFac" is set to "FALSE" since the "SampleGroup" variable in the slot "sectionData" of object "BLimgBS" has missing values due to the absence of .sdf file storing sample description. After summarization, the quality assessment tools introduced in the previous section on bead-summary data, can be used to visualize and diagnose the summarized object "BSimg," which was constructed here from bead-level data.

```
> MEANfun<-function(x) mean(x, na.rm=T)
> SDfun<-function(x) sd(x, na.rm=T)
> #creat an object of class "illuminaChannel"
> #with five slots "transFun",
> #"outlierFun", "exprFun", "varFun" and "name"
> greenChannel<-new("illuminaChannel", logGreenChannelTransform,
+ illuminaOutlierMethod, MEANfun, SDfun, "GCh")
> BSimg<-summarize(BLimgBS, list(greenChannel), useSampleFac=F)

> head(exprs(BSimg))

             GSM351169 GSM351170 GSM351171
ILMN_1250052  6.185174  6.097674  5.658175
ILMN_3122480  5.863684  5.382607  5.559047
ILMN_2599935  7.384543  7.752884  7.516926
ILMN_2675543  5.882986  5.686702  5.690595
ILMN_2686883  5.972513  6.013517  6.071102
ILMN_2751818  5.826776  5.710805  5.762353
             GSM351172 GSM351173 GSM351174
ILMN_1250052  6.034787  5.782647  5.721739
ILMN_3122480  5.478990  5.499477  5.402247
ILMN_2599935  7.456536  7.463074  7.474087
ILMN_2675543  5.550885  5.568104  5.572028
ILMN_2686883  5.716214  5.670689  5.699867
ILMN_2751818  5.355358  5.601275  5.431417
```

19.4 Summary

This chapter illustrated some of the quality control steps that can be taken at the data analysis level in order to detect various types of problems that can affect a data set. The quality control performed on the data should be seen as an additional step rather than a substitute for the quality control that is performed in the laboratory. Various tools such as intensity distribution plots, box plots, etc., have been used in this chapter to look at some Affymetrix and Ilumina data sets. The Affymetrix data set included 27 arrays coming from nine patients. The quality problems detected in this data set include intensity saturation, degraded mRNA, usage of inconsistent array types, and arrays with low sensitivity. The Illumina data included several examples both at bead level as well as bead summary level. Different methods to read and check the quality of these data (both bead level and bead summary data) were discussed and illustrated with examples. The chapter discussed various quality control measures and indicators and uses the sample data set to illustrate how these diagrams, tools, and quality measures look like for good arrays versus arrays with the types of problems mentioned above. Although the chapter focused on Affymetrix and Illumina data, the same concepts, and many of the same techniques and tools can be applied to any type of data.

Chapter 20

Data preprocessing and normalization

If at first you don't succeed, transform your data set.

—*Unknown* [165]

20.1 Introduction

Preprocessing is a step that extracts or enhances meaningful data characteristics. Sometimes, preprocessing prepares the data for the application of certain data analysis methods. A typical example of preprocessing is taking the logarithm of the raw values. Normalization is a particular type of preprocessing done to account for systematic differences across data sets. A typical example of normalization is modifying the values in order to compensate for the different dye efficiency in the two channel microarray experiments using cy3 and cy5.

This chapter will discuss the main preprocessing and normalization procedures currently used, emphasizing their motivations, effects and limitations.

Certain preprocessing steps such as the logarithmic transformation are equally applicable to cDNA and Affymetrix data while others such as background correction and probe level preprocessing are specific to a given technology.

```
> dat=c(999,2.5,1.3,2.1,1.5,2.7,1.7,1.2,2.0)
> res=t.test(x=dat,mu=1.5,alternative="two.sided")
> res

        One Sample t-test

data:  dat
t = 1.0034, df = 8, p-value = 0.3451
alternative hypothesis: true mean is not equal to 1.5
95 percent confidence interval:
 -142.8197  368.1530
sample estimates:
mean of x
  112.6667
```

20.2 General preprocessing techniques

20.2.1 The log transform

The logarithmic function has been used to pre-process microarray data from the very beginning [470, 472]. There are several reasons for this. First, the logarithmic transformation provides values that are more easily interpretable and more meaningful from a biological point of view. Let us consider two genes that have background corrected intensity values of 1000 in the control sample. A subsequent measurement of the same two genes in a condition of interest registers background corrected intensity values of 100 and 10,000 respectively for the two genes (a 16-bit tiff file can contain values between 1 and 65,536). If one considers the absolute difference between the control values and the two experiment values, one would be tempted to consider that one gene is much more affected than the other since:

$$10000 - 1000 = 9000 >> 1000 - 100 = 900$$

This effect is illustrated in Fig. 20.1. However, from a biological point of view the phenomenon is the same, namely, both genes registered a 10-fold change. The only difference between the genes is that the 10-fold change was an increase for one gene and a decrease for the other one. It is very convenient to transform the numbers in order to eliminate the misleading disproportion between these two relative changes. The logarithmic transformation (henceforth log) accomplishes this goal. Using a log transform, in base 10 for instance,

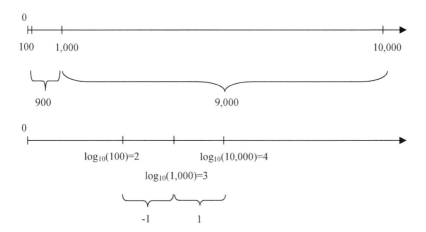

FIGURE 20.1: The effect of the logarithmic transform. The top panel shows the relative position of the values 100, 1000, and 10000. If one considers the absolute difference between these values, one is tempted to consider the difference to the right more important than the difference to the left since $10000 - 1000 = 9000 \gg 1000 - 100 = 900$. The bottom panel shows the relative position of the log-transformed values 2, 3, and 4. The transformed numbers reflect the fact that both genes increased and decreased in a similar way (10 times).

will transform the values into:

$$\log_{10}(100) = 2$$

$$\log_{10}(1000) = 3$$

and

$$\log_{10}(10000) = 4$$

reflecting the fact that the phenomena affecting the two genes are the same only that they happen in different directions. This time, the genes are shown to vary by:

$$2 - 3 = -1$$

for one gene and:

$$4 - 3 = 1$$

for the other gene. Note how the values now reflect the fact that the two genes change by the same magnitude in different directions. It is said that the log partially decouples the variance and the mean intensity, i.e. makes changes such as ± 10 fold in the example above more independent of where they happen. Thus, fold changes happening around small intensities values will be comparable to similar fold changes happening around large intensities values.

A second and very strong argument in favor of the log transformation is related to the shape of the distribution of the values. The log transformation makes the distribution symmetrical and almost normal [291, 399, 472]. This is illustrated in Fig. 20.2. In this figure, the top panel shows the histogram of the background corrected intensity values. Note that the intensity range spans a very large interval, from zero to tens of thousands. The distribution is very skewed having a very long tail towards high intensity values. The bottom panel in Fig. 20.2 shows the distribution of the same values after the log transformation.

Finally, a third argument in favor of using the log transformation is convenience. If the log is taken in base 2, the later analysis and data interpretation are greatly facilitated. For instance, selecting genes with a 4-fold variation can be done by cutting a ratio histogram at the value $\log_2(ratio) = 2$. Henceforth in this text, the base of the logarithm will be assumed to be equal to 2.

20.2.2 Combining replicates and eliminating outliers

As we have seen, due to the large amount of noise typically associated with microarray data, one must make repeated measurements. Chapter 13 showed how such repeated measurements can be used to estimate the amount of noise and compare the inter-experiment and within experiment variations. However, in certain situations it is convenient to combine the values of all replicates in order to obtain a unique value, representative for the given gene/condition combination. Such repeated measurements may in fact be, depending on the

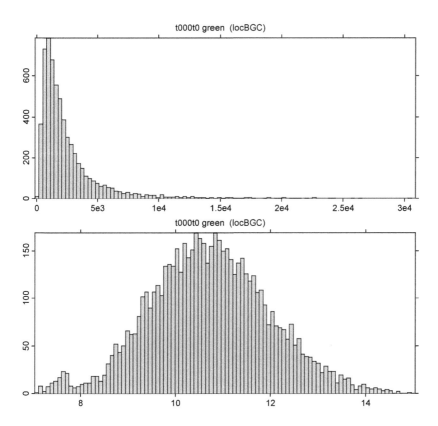

FIGURE 20.2: The effect of the log transform on the distribution of the intensity values. The top panel shows the histogram of the background corrected intensity values. The distribution is very skewed (elongated) towards high-intensity values. The bottom panel shows the distribution of the same values after the log transformation.

situation, different spots in cDNA arrays or different values measured on different arrays in both cDNA or Affymetrix arrays. Typically, such values are combined by calculating a measure of central tendency such as mean, median, or mode. However, substituting a set of values by a unique value does imply a loss of information and must be done with care. Chapter 8 discussed several examples in which various measures of central tendency are misleading and represent poorly the set of values. Nevertheless, there are strong incentives to calculate a unique expression value for a given gene in a unique condition. Such incentives may include the ability to compare various genes across conditions or tissues, storage and retrieval in expression databases, etc.

Two approaches may be attempted in order to somehow alleviate the loss of information associated to the compression of many repeated measurements into condition representative values. The first approach is to store several parameters of the distribution of original values besides the measure of central tendency. Such values may include: the number of values, standard deviation as well as other parameters of the original distribution. Such additional parameters may be used to asses the confidence into a particular value such as a mean. For instance, a mean value obtained from 10 replicates with a low variance will be much more trustworthy than a mean value obtained from 3 replicates with a high variance. The second approach is to try to clean the data by eliminating the outliers. This can be done by calculating a mean and standard deviation σ from the original data and eliminating the data points situated outside some given interval (e.g. $\pm 3\sigma$). The remaining data are reprocessed in the same way by calculating a new mean and standard deviation. The process is repeated until no more outliers are detected. Finally, the representative value may be taken as the mean of the remaining values.

20.2.3 Array normalization

The fundamental driving force behind the extensive use of microarrays is the hope that arbitrary comparisons between the gene expression levels in various conditions and/or tissues will be possible eventually. A crucial requirement before such arbitrary comparisons are possible is to normalize the data in such a way that the data are independent of the particular experiment and technology used [62, 63]. In an ideal world, the technology would allow the computation of the gene expression level in some universal reference system using some standard units such as number of mRNA copies per cell. At this time, there is no agreed upon way of normalizing microarray data in such a universal way. It is also still controversial whether data collected with different technologies such as oligonucleotide and cDNA arrays can be compared directly. For instance, Yuen et al. reported a fairly good correlation between expression data measured using the GeneChip with data measured using a cDNA chip [480] while Kuo et al. found no correlation [271]. The main difference between these two studies is that the latter compared data from two different labs (i.e., data based on the same 60 cell lines that were cultured in different labs), whereas the other group used data based on identical biological material. Another study by Li et al. showed a good correlation between the data obtained with the two technologies [281]. Interestingly, Yuen et al. found that the bias observed with the commercial oligonucleotide arrays was less predictable and calibration was unfeasible. The same study concluded that fold-change measurements generated by custom cDNA arrays were more accurate than those obtained by commercial oligonucleotide arrays after calibration. At the same time, Li et al. conclude that the data from oligonucleotide arrays is more reliable when compared to long cDNA array (Incyte Genomics) [281]. With such a lack of consensus in the literature, the next best thing to arbitrary cross-technology comparisons is to be able to compare data obtained with a given technology. Fortunately, this is a simpler problem and various approaches allow such comparisons.

The difficulty is related to the fact that various arrays may have various overall intensities. This can be due to many causes, including different protocols, different amounts of mRNA, different settings of the scanner, differences between individual arrays and labeling kits, etc. For Affymetrix arrays, there is a difference between the overall mean of each individual array. For cDNA arrays, there can also be a difference between each individual channel (dye) on the same array. Fig. 20.3 illustrates the need for an array normalization. In this figure, the top panel shows a comparison of the green channels on the arrays in the yeast sporulation data [95]. An examination of the box plots shows that there are important differences between the various arrays corresponding to different time points. A comparison of the uncorrected values would produce erroneous results. For instance, a gene expressed at an average level on array 5 (left to right) would appear as highly expressed when compared with values coming from array 6. If this gene is also expressed at an average level on array 6, the comparison of its un-normalized values between the two arrays might conclude that the gene is up-regulated in 5 with respect to 6. In fact, the difference in values may be due exclusively to the overall array intensity. The bottom panel shows the same comparison after all values have been divided by the mean of the array from which they were collected. Exactly the same phenomenon can be observed in the case of Affymetrix data.

For both oligonucleotide and cDNA arrays, the goal is to normalize the data in such a way that values corresponding to individual genes can be compared directly from one array to another. Generally, array normalization methods can be divided into two broad categories: i) methods using a **baseline array** or **reference**, and ii) methods that combine information from all arrays in a given data set. The latter groups are also known as **complete methods**. The methods that use a baseline array include **scaling methods** and **nonlinear methods**. A common array used as the baseline is the array having the median of the median intensities. Scaling methods are generally equivalent to fitting a linear-relationship with zero intercept between the baseline array and each of the arrays to be normalized. The standard Affymetrix normalization, which will be discussed in the following sections is an example of a scaling

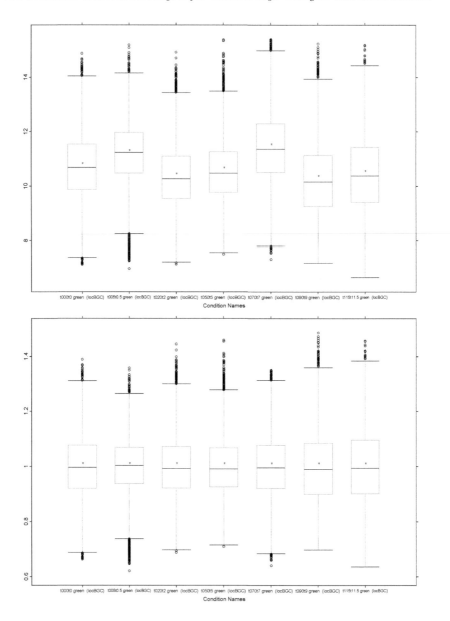

FIGURE 20.3: The need for array normalization. The top panel shows a comparison of the green channels on the arrays in the yeast sporulation data [95]. The graph shows that there are important differences between the various arrays corresponding to different time points. A comparison of the uncorrected values would produce erroneous results. The bottom panel shows the same comparison after all values have been divided by the array mean.

method. Non-linear methods do something very similar but use a nonlinear relationship instead. The dChip method proposed by Li and Wong described in Section 20.5.2 is an example of a nonlinear method. Complete data normalization methods include **cyclic Loess** and the **quantile normalization** described in Section 20.5.4.

Specific normalization methods will be described in the following.

20.2.3.1 Dividing by the array mean

The mean can be substituted with the median, mode, or percentile if the distribution is noisy or skewed. One could also apply a Z transformation by subtracting the mean and divide by the standard deviation [377]. This method can be seen as a scaling method that does not use a baseline array. For cDNA arrays, this approach adjusts overall intensity problems but does not address dye non-linearity. This will be addressed with specific color normalization techniques, such as LOWESS, discussed below.

Variations of this approach eliminate the values in the upper and lower 10% of the distribution (thought to be regulated) and then divide by the mean of the remainder. The motivation behind this is that the number of genes regulated and the amount of regulation should not influence the normalization.

20.2.3.2 Subtracting the mean

Subtracting the mean is usually used in the context of the log transformed data. A well-known property of the logarithmic function is:

$$\log_a \frac{x}{y} = \log_a x - \log_a y \qquad (20.1)$$

or log of ratio is the difference of the logs of the numerator and denominator. Therefore, dividing all values by the mean of the array and then taking the log of the normalized value should be analogous to applying the log of all values, then calculating the mean of the log values on the array and normalizing by subtracting this mean. This is indeed the case; the effect of the two processing sequences is similar. However, they are not quite the same. Let us consider an array A containing n genes. The values x_i are read from this array (e.g., background corrected values for cDNA arrays or average differences for Affymetrix arrays). We can divide the values by the mean of the array to obtain:

$$x_i' = \frac{x_i}{\overline{x_i}} = \frac{x_i}{\frac{\sum_{i=1}^{n} x_i}{n}} \qquad (20.2)$$

For reasons explained in Section 20.2.1, we can now take the log to obtain the normalized values:

$$x_{in} = \log\left(x_i'\right) = \log\left(\frac{x_i}{\overline{x_i}}\right) \qquad (20.3)$$

where x_{in} are the normalized values of x_i according to the "divide by mean – log" sequence. Alternatively, we can choose to take the log of the raw values first and then to subtract the mean. Taking the log would provide the following values:

$$x_i' = \log(x_i) \tag{20.4}$$

We now subtract the mean of the log values on the whole array:

$$x_{in} = \log(x_i) - \overline{\log(x_i)} = \log(x_i) - \frac{\sum_{i=1}^{n} \log(x_i)}{n} \tag{20.5}$$

We can use the properties of the log function to rewrite the last term as:

$$\frac{\sum_{i=1}^{n} \log(x_i)}{n} = \frac{1}{n}\log(x_i \cdot x_2 \cdots x_n) = \log \sqrt[n]{x_i \cdot x_2 \cdots x_n} \tag{20.6}$$

This last form emphasizes the fact that this is the geometric mean of the x_i values. Using the notation:

$$\overline{x_i} = \frac{\sum_{i=1}^{n} x_i}{n} \tag{20.7}$$

for the arithmetic mean of the values x_i and the notation:

$$\overline{x_{ig}} = \sqrt[n]{x_i \cdot x_2 \cdots x_n} \tag{20.8}$$

for the geometric mean of the values x_i, we see that the equations 20.3 and 20.5 are in fact exactly the same but for the type of the mean used. Thus, the sequence "divide by mean – log" uses the arithmetic mean:

$$x_{in} = \log\left(\frac{x_i}{\overline{x_i}}\right) \tag{20.9}$$

while the sequence "log – subtract mean" uses the geometric mean:

$$x_{in} = \log\left(\frac{x_i}{\overline{x_{ig}}}\right) \tag{20.10}$$

The geometric mean is the same (when all x_i are equal) or lower than the arithmetic mean. In general, the arithmetic mean is used when finding the average of numbers that are added to find the total, and the geometric mean is used when the items of interest are multiplied to obtain the total. For instance, if a company has a profit of 7, 9, and 11 million over 3 consecutive years, the average yearly profit would be calculated appropriately using the arithmetic mean. However, if a company's track record is described in terms of yearly growth such as 3%, 2.5%, and 2%, the average growth would be calculated appropriately using the geometric mean. Since the intensities corresponding to various genes are additive, the arithmetic mean is probably more suitable than the geometric mean for array normalization purposes.

20.2.3.3 Using control spots/genes

This involves modifying the values of each channel/experiment such that certain control spots/genes in both experiments have the same or similar values. The control spots should span the whole intensity range. If the intensity of a control gene on array A is found to be c times higher than the intensity of the same control gene on array B, all genes with intensity values in the same range on array A should be divided by c in order to make them comparable with the values read from array B.

20.2.3.4 Iterative linear regression

This approach was proposed in the context of normalizing the two channels of a cDNA array [19]. However, the method can be used to align any two sets of values. The basic idea is that of an iterative linear regression. Performing a linear regression means fitting a straight line of the form $y = m \cdot x + n$ through the data in such a way that the errors (the differences between the y values predicted by using the straight line model above and the real values) are minimal in the least square sense. At each step, those genes that are not modeled well by the linear model are thrown out. Essentially, the approach assumes that there is a linear correspondence between the two sets of values and tries to fit the best slope and fit that would make the two sets match for those genes that are unchanged.

If we consider the sets of values x_{1i} and x_{2i} as coming from either two channels of a cDNA array, or two different oligonucleotide arrays the processing can be described in the following steps.

1. Apply a log transform to obtain $\log(x_{1i})$ and $\log(x_{2i})$.

2. Perform a simple linear regression. This means fitting a straight line of the form: $\log(x_{1i}) = \log(x_{2i}) \cdot m + b$ through the data in such a way that the errors (residuals) are minimal.

3. Find the residuals $e = \log(x_{1i}) - \log(x_{1i})_c$, where $\log(x_{1i})_c$ is the value calculated as a function of $\log(x_{2i})$.

4. Remove all genes that have residuals greater than 2σ from 0.

5. Repeat the steps above until the changes between consecutive steps are lower than a given threshold.

6. Normalize $\log(x_{2i})$ using the regression line found above: $\log(x_{2i})_n = log(x_{2i}) \cdot m + b$

Note again that this normalization does not take into consideration, nor does it correct for any nonlinear distortion introduced by the dyes, if such distortion exists. Normalization techniques able to correct for this are discussed in Section 20.3.3.

20.2.3.5   Other aspects of array normalization

An explicit array normalization is not always necessary. For instance, if an ANOVA approach is used (see Chapter 13) together with a suitable experiment design, the systematic difference between arrays will be extracted as the array bias term. Also, other techniques such as LOWESS achieve also the array normalization as part of their nonlinear correction. Thus, if two channels of the same cDNA arrays are to be compared and LOWESS is used to compensate for the dye non-linearity, an explicit step of array normalization (e.g., divide by mean) may not be necessary.

20.3   Normalization issues specific to cDNA data

The cDNA experiment is discussed in detail in Chapter 3 but, for convenience, the process will be summarized here. In a typical multichannel cDNA experiment, various samples are labeled with different dyes. The most usual experiment uses two dyes, or colors, such as cy3 and cy5. For instance, the control sample will be labeled with cy3 and the experiment sample will be labeled with cy5. Once labeled, the two samples will be mixed and hybridized on the array. For a gene expressed in both samples, the hybridization will involve a competitive aspect since cDNA from both samples will have to compete for the few complementary strands spotted on the array. We assume that each spot is represented by a value computed from the pixel intensities as it was discussed in Chapter 5. The process and phenomena are essentially the same even if more or fewer dyes are used.

20.3.1   Background correction

A first preprocessing step is the background correction. The idea behind the background correction is that the fluorescence of a spot is the effect of a summation between the fluorescence of the background and the fluorescence due to the labeled mRNA. Thus, the theory goes, in order to obtain the value proportional to the amount of mRNA, one needs to subtract the value corresponding to the background. This can be done in several ways as follows.

20.3.1.1   Local background correction

The intensity of the background is calculated in a local area around the spot. A measure of central tendency (e.g., mean, median or mode) is calculated and subtracted from the spot intensity. This method is preferred when the background intensity varies considerably from spot to spot. This method is to be avoided when the local neighborhood of the spots does not contain sufficiently many pixels. This may happen on very high density arrays when the spots may be separated by only a few pixels.

20.3.1.2 Sub-grid background correction

A measure of central tendency is calculated for all spots in a subgrid. This is an useful approach in high-density arrays. A subgrid includes sufficiently many pixels to allow a more reliable estimate of a measure of central tendency while it is still smaller than the whole array and may be flexible enough to compensate for local variations in the background intensity. Furthermore, most current robots print a sub-grid using the same pin so a sub-grid should be homogeneous as far as the shape and size of the spots are concerned.

20.3.1.3 Group background correction

This is similar to the subgrid correction but uses a smaller number of spots. This method would use a neighborhood of the given spot (e.g., all spots situated in a circle of radius 3) and calculate a measure of local tendency using the background pixels in this neighborhood. This is more flexible than the subgrid, allowing a better adaptation to a non-uniform background while still estimating the background value using more than the few pixels around a single spot.

20.3.1.4 Background correction using blank spots

This method can be used when the design of the array included a few blank spots, i.e. spot locations where no DNA was deposited. Again, a measure of central tendency is calculated on a number of such blank spots.

20.3.1.5 Background correction using control spots

A particular criticism of the approach above is related to the assumption that the spot intensity is the result of a simple summation between the background intensity and the labeled DNA intensity. In fact, the background intensity depends on the properties of the interaction between the labeled target in the solution and the substrate. However, the spot intensity depends on the properties of the interaction between the labeled target and the DNA deposited in the spot. Some researchers have studied this and concluded that the labeled target may be more likely to stick to the substrate in the background of a spot than to hybridize non-specifically on a spot containing some DNA. If this is the case, subtracting any value characterizing the target-substrate interaction may be an over-correction. A possibility is to use some control spots using exogenous DNA and use the intensity of the nonspecific hybridization on such spots as a better background correction. Such spots are called control spots.

20.3.2   Other spot level preprocessing

One may choose to discard those spots that have found to be unreliable in the image processing stage. This can be done by flagging those spots in the image processing stage and by discarding the values coming from those spots in the data preprocessing stage. If values are discarded, the missing values may or may not be substituted with some estimates. If the missing values are to be estimated, common estimates may be the mean, median, or mode of other spots representing the same gene on the given arrays (if replicate spots are used, as they should) or the mean, median, or mode of the same gene in other experiments. Care should be exercised if missing values are estimated because this affects the number of degrees of freedom of the sample (see Chapter 12).

20.3.3   Color normalization

One important problem is that various dyes have slightly different biochemical properties that may affect the data collected. The purpose of the color normalization is to eliminate the data artifacts introduced by the dyes. Fig. 20.4 shows an example of a scatter plot cy3 versus cy5 of the raw data in a real cDNA experiment. As discussed in Section 17.4, in any given experiment, most of the genes of an organism are expected to be unchanged. In consequence, most of the points in a cy3-cy5 scatter plot are expected to appear along the diagonal cy3=cy5. A brief examination of the scatter plot in Fig. 20.4 shows that this is hardly the case. In this plot, most of the points are found below the diagonal suggesting that for most of the points, the values measured on the cy3 channel are higher than the values measured on the cy5 channel. In principle, this can have two possible causes: either the mRNA labeled with cy3 was more abundant for most of the genes or the cy3 dye is somehow more efficient and, for the same amount of mRNA, the average intensities read are higher.

The idea of a flip-dye experiment was introduced in order to control such phenomena. In a flip dye experiment, the two samples of mRNA, A and B, are labeled first with cy3 (A) and cy5 (B) and then with cy5 (A) and cy3 (B). Subsequent hybridization and image analysis will produce two sets of data which represent the same biological sample. A plot of the expression levels measured on the two channels for the same mRNA, for instance, A labeled with cy3 versus A labeled with cy5, should produce a straight line $cy3 = cy5$. One is prepared to accept small random variations from the reference but any general trend will indicate a non-negligible influence that should be corrected.

In the example above, the very same mRNA is labeled with two different colors so any differences will be exclusively due to the labeling process. Thus, we expect the graph to be symmetric independently of the number of genes involved. Similar expectations hold even if each sample is labeled with a different dye *if the experiments involve all or most genes in a genome*. In such situations, it is assumed that most genes will not change since across the board

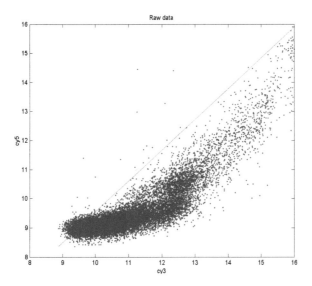

FIGURE 20.4: Typical graph of the cy3/cy5 scatter plot of the raw data obtained from a cDNA experiment. The straight line represents the first diagonal cy3=cy5.

expression changes would probably kill the organism. However, such expectations are **not** warranted if a small subset of genes is considered, especially if the genes are known to be functionally related.

The dyes used may have different overall efficiencies due to many possible reasons. A nonlinear dye effect with a stronger signal provided by one of the two dyes is certainly present if the commonly used cy3 and cy5 fluorescent dyes are used. However, the problem should be investigated and addressed for any dyes used, even if the claim of the manufacturer is that the dyes are equally efficient and perfectly linear. For convenience, we will assume that only two dyes are used, and we will discuss the color normalization referring to these dyes as cy3 and cy5. However, any other dyes can be used in practice. Furthermore, the same methods can be used for any number of dyes (e.g., if performing a 4 color experiment).

The best way to control and compensate for the dye effect is perform a flip-dye experiment as described above. If the experiment design includes flipping the dyes (see Chapter 15), no extra hybridizations are needed. The essential aspect is to collect data from the same mRNA sample using both dyes. Once such data are available, one could draw a scatter plot of the same mRNA labeled with cy5 and cy3. Such a plot is shown in Fig. 20.4. Since the data were obtained from the same mRNA labeled differently, we know that any departure from the $x = y$ line is due either to the inherent random noise or to the dyes. The plot exhibits two major, non-random, features: i) the data are consistently off the diagonal and ii) the "cloud" of data points has a specific shape often described as "banana-like" or "comma-like." This particular shape is the effect of the nonlinear distortion introduced by the dyes and is the main target of the color normalization. The color distortion can be corrected using one of the following approaches:

1. Curve fitting and correction

2. Lowess normalization

3. Piece-wise linear normalization

20.3.3.1   Curve fitting and correction

The color normalization may be achieved through a curve fitting followed by a corresponding data correction. It has been noted [219] that the color distortion introduced by the dyes has an exponential shape in the ratio-intensity scatter plot. Based on this observation, one can use the data to find the parameters (base and shift) of the exponential function that fits the data. This process is illustrated in Fig. 20.5. The data represent a single time point from the yeast cell cycle expression data [400]. In all graphs, the horizontal axis represents $\log(cy5)$, while the vertical axis represents the log of the ratio $\log(cy3/cy5)$. The data in the top left panel are normalized by subtracting the mean but not corrected for color distortion. The distribution exhibits a clear nonlinear distortion. The data are divided into intensity intervals on the horizontal axis $\log(cy5)$. This is equivalent to dividing the graph into vertical slices corresponding to such intervals. A centroid of the data is calculated for each interval. An exponential curve of the form:

$$y = a + b \cdot e^{-cx} \tag{20.11}$$

is fitted through the centroids. The purpose of this fit is to find the best combination of values a, b, and c that would make the exponential curve above to represent best the color distortion present in the data. Once this function is found, it is used to adjust the data: the vertical coordinate of each data point will be shifted by a value given by the fitted exponential curve at that location in such a way that the color distortion is compensated for. The bottom right panel shows the color corrected plot.

The example above used an exponential curve because this was suitable for the distortion introduced by the commonly used dyes cy3 and cy5. Different dyes may introduce distortions that are better compensated by using functions different from the exponential. However, the general idea behind this approach will remain the same: fit a suitable function through the data and thus obtain a model of the distortion; use this model to adjust the data and eliminate the dye effect. A related idea is used by the LOWESS transformation.

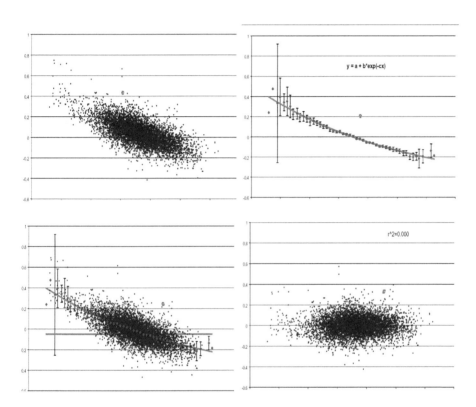

FIGURE 20.5: Exponential normalization. Top left: ratio-intensity plot of the raw data. Top right: the data are divided into groups based on the signal intensity $\log(cy5)$ and a centroid of the data is calculated for each such group. Bottom left: an exponential curve is fitted through the centroids. Bottom right: each data point is shifted with a value given by the fitted exponential. The normalized data is straight and centered around $\log \frac{cy3}{cy5} = 0$. The data are a subset (a single time point) of the yeast cell cycle [400]. In all graphs, the horizontal axis represents $\log(cy5)$, while the vertical axis represents the log of the ratio $\log(cy3/cy5)$.

20.3.3.2 LOWESS/LOESS normalization

The LOWESS transformation, also known as LOESS, stands for LOcally WEighted polynomial regreSSion [100, 101]. In essence, this approach divides the data into a number of overlapping intervals and fits a function in a way similar to the exponential normalization discussed above. However, the function fitted by LOWESS is a polynomial of the form:

$$y = a_0 + a_1 x + a_2 x^2 + a_3 x^3 + \cdots \qquad (20.12)$$

Polynomials are very nice mathematical objects in the sense that they can approximate a large category of functions.[1] However, the polynomial approximation has two general problems. First, the approximation is good only in a small neighborhood of the chosen point and the quality of the approximation gets worse very quickly as one gets farther away from the point of approximation. Second, the polynomial approximation is very prone to overfitting if higher-degree polynomials are used. Overfitting produces highly nonlinear functions that attempt to match closely the target function around the known data points but "wiggle" excessively away from them. Both problems are illustrated in Fig. 20.6. The figure shows a function and several polynomial approximations using polynomials of increasing complexity. The first-order polynomial is a straight line (1), the second-order polynomial is a parabola (2), etc. One can note that, as the degree of the polynomial increases, a better approximation is obtained in the neighborhood of the chosen point, but the approximation degrades quickly far from the given point as the chosen polynomial has more inflexion points (bends).

The approach used by LOWESS/LOESS deals with both issues in an elegant way. Firstly, the degrees of the polynomials used are limited to 1 (in LOWESS) or 2 (in LOESS) in order to avoid the over-fitting and the excessive twisting and turning. Secondly, since the polynomial approximation is good only for narrow intervals around the chosen point, LO(W)ESS will divide the data domain into such narrow intervals using a sliding window approach. The sliding window approach starts at the left extremity of the data interval with a window of a given width w. The data points that fall into this intervals will be used to fit the first polynomial in a weighted manner. The points near the point of estimation will weigh more than the points further away. This is achieved by using a weight function such as:

$$w(x) = \begin{cases} \left(1 - |x|^3\right)^3 & , \quad |x| < 1 \\ 0 & , \quad |x| \geq 1 \end{cases} \qquad (20.13)$$

where x is the distance from the estimation point. Other weighting functions can also be used as long as they satisfy certain conditions [100].

---

[1]For the mathematically minded reader, we can recall that any differentiable function can be approximated around any given point by a polynomial obtained by taking a finite number of terms from the Taylor series expansion of the given function.

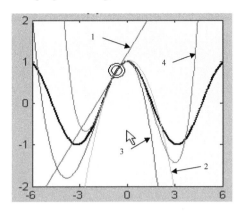

FIGURE 20.6: Polynomial approximation as used in LOWESS/LOESS. The sinusoid function is the target function to be approximated. Several approximations are shown: first order (1), second order (2), third order (3) and fourth order (4). As the degree of the polynomial increases, a better approximation is obtained in the neighborhood of the chosen point but the approximation degrades quickly far from the given point as the chosen polynomial "wiggles" more.

The procedure continues by sliding the window to the right, discarding some data points from the left but capturing some new data points from the right. A new polynomial will be fitted with this local data set and the process will continue sliding the window until the entire data range has been processed. The result is a smooth curve that provides a model for the data. The smoothness of the curve is directly proportional to the number of points considered for each local polynomial, i.e. proportional with the size of the sliding window. If there are n data points and a polynomial of degree d is used, one can define a smoothing parameter q as a user-chosen parameter between $\frac{d+1}{n}$ and 1. The LO(W)ESS will use $n \cdot q$ (rounded up to the nearest integer) points in each local fitting. Large values of q produce smooth curves that wiggle the least in response to variations in the data. Smaller values of q produce more responsive curves that follow the data more closely but are less smooth. Typically, useful values of q for microarray data range between 0.05 and 0.5 for most data sets. These values tend to be lower than the values of q used in other applications.

The effects of the LOWESS normalization are illustrated in Fig. 20.7 and Fig. 20.8. Fig. 20.7 shows the ratio-intensity plot before (left panel) and after (right panel) the LOWESS correction. In this plot, the horizontal axis represents the sum of the log intensities $\log_2(cy3 \cdot cy5) = \log_2(cy3) + \log_2(cy5)$, which is a quantity directly proportional to the overall intensity of a given spot. The vertical axis represents $\log_2(cy3/cy5) = \log_2(cy3) - \log_2(cy5)$, which is the usual log-ratio of the two samples. Similar plots are obtained if the

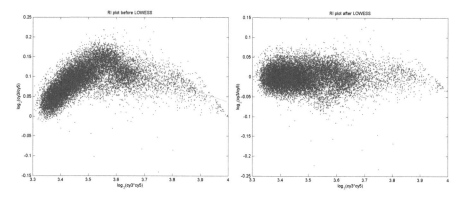

FIGURE 20.7: LOWESS normalization. A ratio-intensity plot before (left panel) and after (right panel) the LOWESS correction. The horizontal axis represents $\log_2(cy3) + \log_2(cy5)$ which is a quantity proportional to the overall signal intensity. The vertical axis represents the log-ratio of the two samples $\log_2(cy3/cy5)$. The left panel also shows the nonlinear regression curve.

horizontal axis is chosen to represent $1/2(\log_2(cy3) + \log_2(cy5))$ or $\log_2(cy3)$. Note the strong nonlinear dye distortion in the left panel and how this is corrected by LOWESS in the right panel. The left panel also shows the nonlinear regression curve calculated by LOWESS. Fig. 20.8 shows the same data as a scatter plot cy5 versus cy3. The horizontal axis represents $\log_2(cy3)$, and the vertical axis represents $\log_2(cy5)$. In this case, the overall mean of the data was normalized before applying LOWESS.

The biggest advantage of LO(W)ESS is that there is no need to specify a particular type of function to be used as a model (e.g., exponential function in the exponential normalization). The only parameters that need to be specified by the user are the degree of the polynomials d and the smoothing factor q. Furthermore, LO(W)ESS methods use least squares regression, which is very well studied. Existing methods for prediction, calibration, and validation of least squares models can also be applied to LO(W)ESS.

Disadvantages of LO(W)ESS include the fact that it does not produce a regression function, or model, that is easily representable as a mathematical formula. In particular, the color distortion model found on a particular data set cannot be transferred directly to another data set or group of researchers. LO(W)ESS needs to be applied every time, on every data set and will produce a slightly different model in each case. Another disadvantage is related to the fact that the procedure is very computationally intensive. As an example, a set of 10,000 genes processed with $q = 0.1$ means that each individual least square curve fitting will be done on a subset of 1,000 data points. Furthermore, there will be approximately 10,000 such computations, as the sliding window scrolls through the entire data range. However, this is a relatively minor problem in the context of all other challenges related to microarray data analysis and tak-

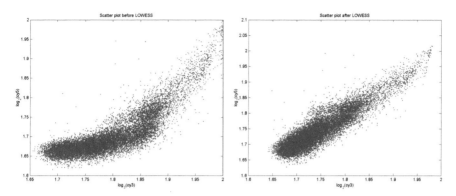

FIGURE 20.8: LOWESS normalization. A scatter plot before (left panel) and after (right panel) the LOWESS correction. The horizontal axis represents $\log_2(cy3)$ and the vertical axis represents $\log_2(cy5)$.

ing into consideration the availability of powerful and cost effective computer hardware. A more important disadvantage is related to LO(W)ESS's susceptibility to noise and outliers. More robust versions of LO(W)ESS have been proposed [100], but even the robust version may be disrupted by extreme outliers. This is one of the reasons for which outliers (e.g., flagged spots) should be removed from the data before attempting a LO(W)ESS normalization.

20.3.3.3 Piece-wise normalization

The piece-wise normalization is closely related to LO(W)ESS attempting to preserve its advantages while improving on the computational aspects [122]. The idea is that LO(W)ESS performs a lot of computations, which, for microarray data, are often redundant. Let us consider again the example of a set of 10,000 genes on which the size of the sliding window is chosen to be $nq = 1,000$. The classical LO(W)ESS requires approximatively 10,000 curve fittings. In turn, each such curve fitting is an iterative process requiring of the order of 1,000 computations. In practice, for microarray data, the difference between two adjacent models each calculated on 1,000 points of which 999 are common is very likely to be minimal. The piece-wise normalization substitutes the sliding window approach with a fixed set of overlapping windows. In each such interval, the data are approximated by a linear function. A quadratic function could also be used as in LOESS. The user controls the smoothness of the resulting curve by choosing the number of such intervals and the degree of overlap.

The results of the piece-wise normalization are very comparable to the results of the LO(W)ESS normalization. Fig. 20.9 shows a comparison of the two methods on the same data set. The top-left panel in this figure shows the scatter plot of the uncorrected average log values cy5 versus cy3. The

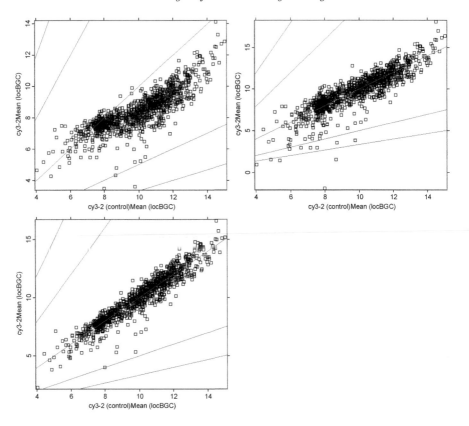

FIGURE 20.9: A comparison between piece-wise linear and LOWESS normalization. Top-left: scatter plot of the uncorrected average log values cy5 versus cy3. Top-right: same plot after piece-wise linear normalization. Bottom-left: same data corrected with LOWESS. Note that both the piece-wise linear and the LOWESS use linear functions for the local regression.

top-right panel shows the same plot after piece-wise linear normalization. The bottom-left panel shows the same data corrected with LOWESS. Note that both the piece-wise linear and the LOWESS use linear functions for the local regression.

Advantages of the piece-wise normalization include a much better speed.[2] Perhaps more importantly, the piece-wise normalization can produce a compact mathematical description of the non-linearity that can be stored and used on different data sets. This description will be a piece-wise linear or quadratic function with a number of pieces equal to the number of intervals specified by the user. In principle, this could be done with LOWESS as well, only that

---

[2]For the computer scientist reader, the algorithm complexity is reduced from $O(n^2)$ to $O(n)$.

the number of individual functions to be stored would be the same order of magnitude as the size of the data set which is unfeasible.

As LOWESS and perhaps a bit more so, the piece-wise normalization is susceptible to the effect of the outliers. A more robust version uses adaptive bins that take into consideration the change in parameters between adjacent intervals [122].

20.3.3.4   Other approaches to cDNA data normalization

It is important to note that the measured variance is dependent on the mean intensity having high variance at low-intensity levels and low variance at high intensities. This can be corrected by using an iterative algorithm that adjusts gradually the parameters of a probabilistic model [89]. This approach can be further refined using a Gamma-Gamma-Bernoulli model [321]. Various other approaches to the normalization of cDNA data are discussed in the literature [53, 138, 156, 471, 479, 199, 254, 446, 470, 472].

20.4   Normalization issues specific to Affymetrix data

The Affymetrix technology as well as the data processing techniques implemented in the Data Mining Tool software package and other related software packages provided by the company are proprietary. Because of this, there is relatively little variation among the way Affymetrix data are processed in different laboratories. Also, there is relatively little research on the preprocessing and normalization of such data [277, 278, 371, 454]

The normalization of the oligonucleotide arrays designed for gene expression is slightly different from the normalization of cDNA data. Here, a gene is represented by a number of probe pairs (short oligonucleotide sequences) with each pair containing a perfect match and a mismatch (the same sequence with a different nucleotide in the middle). The terminology is illustrated in Fig. 20.10. A brief description of the major preprocessing steps and a discussion of the most important issues will be included here for reference. However, as with any proprietary technology, the particulars are subject to changes. For most up-to-date algorithms and preprocessing details, we refer the reader to the Affymetrix technical documentation [7, 8, 9, 233].

20.4.1   Background correction

The image of the chip is captured in .CEL files. Also the .CEL file contains the raw intensities for all probe sets. Unlike cDNA, the Affymetrix array does not have a background as such. Therefore, the background correction of the probe sets intensities is performed using the neighboring probe sets. For this purpose, the array is split into a number of K rectangular zones (at the moment $K = 16$). Control cells and masked cells are ignored in this computation. The cells in each zone are ranked and the lowest 2% are chosen to represent the background value in the given zone. The background value is computed as a weighted sum of the background values of the neighboring zones with the weight being inversely proportional to the square of the distance to a given zone (see Fig. 20.10). Specifically, the weighting of zone k for a cell situated at the chip coordinates (x,y) is:

$$w_k(x,y) = \frac{1}{d_k^2(x,y)+c} \tag{20.14}$$

where c is a smoothing constant that ensures the denominator never gets too close to zero. For every cell, the weighted background of the cell at coordinates (x,y) will be:

$$b(x,y) = \frac{1}{\sum_{k=1}^{K} w_k(x,y)} \sum_{k=1}^{K} w_k(x,y) b_{Z_k} \tag{20.15}$$

where b_{Z_k} is the background of zone Z_k.

As for cDNA arrays, one would like to correct the values by subtracting the background intensity value from the cell intensity values. However, it is very possible for a particular cell to have an intensity lower than the background value calculated according to Eq. 20.15. This would produce a negative background-corrected value, which in turn would produce problems in the subsequent processing (e.g., the log function is not defined for negative values). In order to address this problem, one can use the same approach to calculate a local noise value:

$$n(x,y) = \frac{1}{\sum_{k=1}^{K} w_k(x,y)} \sum_{k=1}^{K} w_k(x,y) n_{Z_k} \tag{20.16}$$

where n_{Z_k} is the local noise value in zone Z_k calculated as the standard deviation of the lowest 2% of the background in that zone.

The individual cell intensities can be adjusted using a threshold and a floor:

$$I'(x,y) = \max(I(x,y), 0.5) \tag{20.17}$$

where $I(x,y)$ is the raw intensity at location (x,y). Finally, the background corrected intensity values can be calculated as follows:

$$I_c(x,y) = \max\left(I'(x,y) - b(x,y), NF \cdot n(x,y)\right) \tag{20.18}$$

where NF is a selected fraction of the global background variation (the default is currently 0.5).

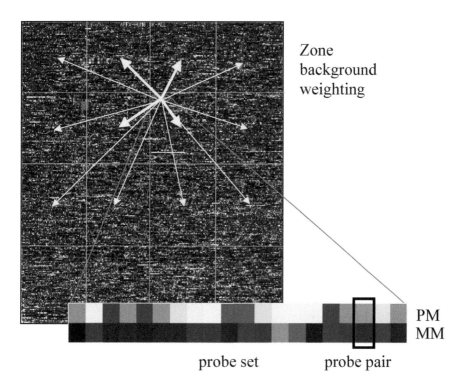

FIGURE 20.10: Affymetrix terminology and background zone weighting computation. A gene is represented by a number of perfect match (PM) and mismatch (MM) probe pairs. Each pair is composed of two cells. The background value is computed as a weighted sum of the nearby zone background values. The weight of a zone is inversely proportional to the distance to it. The background value for a given zone is calculated from the cells with the lowest 2% intensities in that zone.

20.4.2 Signal calculation

As for cDNA arrays, the goal of the preprocessing is to modify the gene-specific intensity values in order to obtain a value that reflects as accurately as possible the amount of transcript in the solution used. As noted before, Affymetrix arrays use several probes with both perfect matches (PM) and mismatched (MM) cells. The challenge is to combine these values in a unique, meaningful value, proportional to the transcript. The amount of hybridization on the mismatched sequence is thought to be representative for nonspecific hybridization, and its intensity is usually subtracted from the intensity of the perfect match. Subsequently, such differences are averaged for all probes corresponding to a gene and an average difference (AD) is calculated. The higher this AD, the higher the expression level of the gene. The software provided by Affymetrix also calculates a *call*, i.e. a ternary decision about the gene: absent (A), marginally present (M) and present (P). A conundrum can occur if genes with an A call have AD higher than P genes. Options include ignoring the calls and using only the AD, considering only P genes, considering only genes with the AD larger than a threshold [183], or calculating the expression values according to some other model (e.g., dChip [277]). Furthermore, a typical problem often encountered with earlier versions of the preprocessing software is related to obtaining negative values for the average difference. This is rather difficult to interpret since usually the PM sequence should hybridize stronger than the MM sequence.[3] Recent versions of the Affymetrix software use a modified algorithm that eliminates the possibility of obtaining negative signal values.

20.4.2.1 Ideal mismatch

The very reason for including a MM probe is to provide a value that estimates the background nonspecific hybridization as well as any other stray signals affecting the PM probe. There are three cases as follows:

1. If the MM value is lower than the PM value, the MM is considered a reasonable estimate of the background and the ideal mismatch (IM) will be taken to be equal to MM:

$$IM_{i,j} = MM_{i,j} \tag{20.19}$$

2. If the MM is larger than PM for a given probe, clearly it cannot be used to estimate the background. In this case, the ideal mismatch IM is estimated from the differences between PM and MM for the other probes in the same set (corresponding to the same gene):

$$SB_i = T_{bi}(\log_2(PM_{i,j}) - \log_2(MM_{i,j})) \quad , \quad j = 1, \ldots, n \tag{20.20}$$

where T_{bi} is a one-step Tukey's biweight estimate similar to a weighted

---

[3]The sequences have been chosen to be specific to the target gene.

mean. Full details about the computation of this estimate can be found in [10]. If the SB_I value thus calculated is greater than a threshold τ_c called *contrast tau*, the IM is calculated as:

$$IM_{i,j} = \frac{PM_{i,j}}{2^{SB_i}} \tag{20.21}$$

3. Finally, if the SB_i value calculated above is smaller than the contrast, one cannot consider it an accurate estimate of MM. In these situations, the IM will be estimated by a value slightly lower than the PM:

$$IM_{i,j} = \frac{PM_{i,j}}{2^{1+\frac{\tau_c}{\tau_c - SB_i}}} \tag{20.22}$$

where τ_s is another threshold called *scale tau*. The current values for these thresholds are $\tau_c = 0.03$ and $\tau_s = 10$ [10].

20.4.2.2 Probe values

Once the ideal match value IM has been calculated for every probe, the probe value is calculated as the difference between the PM and the IM:

$$V_{i,j} = PM_{i,j} - IM_{i,j} \tag{20.23}$$

This is the classical idea of calculating the difference between the PM and MM values for every probe pair in the probe set. The log function is again useful and probe values PV can be calculated as:

$$PV_{i,j} = \log(V_{i,j}), \quad j = 1, \ldots, n \tag{20.24}$$

Finally, the value proportional to the amount of transcript in the solution is the **signal log value** which is calculated again using the one-step Tukey's biweight estimate T_{bi}:

$$SLV = T_{bi}(\log V_{i,j}, \ldots, \log V_{i,n_i}) \tag{20.25}$$

Once again, the Tukey's biweight estimate is just a weighted mean of the values $\log V_{i,j}, \ldots, \log V_{i,n_i}$. Here, we considered a probe set with n_i probes.

It is possible that the values $PM_{i,j}$ and $IM_{i,j}$ are really close for a given probe pair. This will make their difference $V_{i,j} = PM_{i,j} - IM_{i,j}$ very close to zero and the log of the difference will be a very large negative value that may disrupt the weighted mean T_{bi}. In order to avoid this, the software uses another threshold $\delta = 2^{-20}$ for the $V_{i,j}$ values. Any $\log V_{i,j}$ value lower than δ will be set equal to it. This will set a floor of -20 for all values in the weighted mean.

20.4.2.3 Scaled probe values

A trimmed mean is calculated in order to improve the reliability of the values obtained. A trimmed mean is obtained by first eliminating the lowest and highest 2% of the values, and then calculating the mean of the remaining values. However, since the signal values obtained above were logs, this computation needs to be performed on the antilog (or exponential) values: 2^{SLV}. We will use TM to denote the trimmed mean:

$$TM(2^{SLV}, 0.02, 0.98) \tag{20.26}$$

where the parameters 0.02 and 0.98 specify where the tails of the distribution have been cut (lowest and highest 2%).

Finally, the values RV_i reported by the software are calculated as:

$$RV_i = nf \cdot sf \cdot 2^{SLV} \tag{20.27}$$

where nf is a normalization factor and sf is a scaling factor.

The scaling factor sf is calculated such as a target signal TS value is obtained for the trimmed mean:

$$sf = \frac{TS}{TM(2^{SLV}, 0.02, 0.98)} \tag{20.28}$$

The normalization factor nf is calculated in order to allow the direct comparison of two arrays, usually called baseline (or reference) and experiment. The normalization factor is computed as a simple ratio of the two trimmed means corresponding to the two arrays:

$$nf = \frac{TM(2^{SLV_{baseline}}, 0.02, 0.98)}{TM(2^{SLV_{experiment}}, 0.02, 0.98)} \tag{20.29}$$

Since a comparison analysis is done at the probe pair level, the individual probe pair values are also modified by the scaling and normalization factors. The scaled probe values SPV are calculated as:

$$SPV_{i,j} = PV_{i,j} + \log_2(nf \cdot sf) \tag{20.30}$$

This rather complex sequence of steps has been shown to provide good results. The main ideas of this preprocessing sequence can be summarized as follows:

1. The cell intensities are corrected for background using some weighted average of the backgrounds in the neighboring zones.

2. An ideal mismatch value IM is calculated and subtracted from the PM intensity. If the mismatch MM is lower than the PM, the IM will be taken to be the difference $PM - MM$.

3. The adjusted PM values are log-transformed.

4. A robust mean of these log-transformed values is calculated. The signal value is calculated as the exponential of this robust mean.

5. The signal value obtained is scaled using a trimmed mean.

20.4.3 Detection calls

The idea of the detection call is to characterize the gene as either present (P), which means that the expression level is well above the minimum detectable level, absent (A), which means the expression level is below this minimum, or marginal (M), which means that the expression level of the gene is somewhere near the minimum detectable level. The MAS software package from Affymetrix uses a classical hypothesis testing approach using a Wilcoxon rank test, which is a nonparametric test based on the rank order of the values. The p-value reported by the software is the probability of the null hypothesis being true (see Chapter 11).

In order to calculate the calls, the software first removes the saturated probe pairs. If all probe pairs corresponding to a gene are saturated, the gene is reported as present and the p-value is set to zero. The probe pairs for which the PM and MM values are very close (within a limit τ) are also discarded. A discrimination score R is calculated for the remaining probe pairs:

$$R_i = \frac{PM_i - MM_i}{PM_i + MM_i} \tag{20.31}$$

The hypotheses are:

$$H_0 : median(R_i - \tau) = 0$$

$$H_a : median(R_i - \tau) > 0$$

The default value of τ is 0.015. Clearly, increasing τ will reduce the number of false positives (false P calls) but will also reduce the number of true detected calls. This is a clear situation in which reducing the probability of a Type I error also reduces the power of the test. The calls are decided based on the p-value using two thresholds α_1 and α_2:

Present	Marginal	Absent
$p < \alpha_1$	$\alpha_1 \leq p < \alpha_2$	$p \geq \alpha_2$

It is important to note that in this case, the true probability of a Type I error does not correspond directly to the chosen alpha level. This is due to two different reasons. First, the hypotheses formulation involves a constant τ. For a given data set, changing the value of τ can change the calls made and thus affect the false-positive rate. Second, the results will also depend on the number of probe-pairs used. This number will be constant for a given array type but may change from one array type to another.

20.4.4 Relative expression values

When two conditions are to be compared directly, it is best to make the comparison at the individual probe level. If there are differences between the hybridization efficiencies of various probes selected to represent the same gene, these differences will be automatically cancelled out in a probe level comparison. In order to do this, the software first calculates a probe log ratio PLR for each probe j in the probe set i on both baseline and experiment arrays:

$$PLR_{i,j} = SPV_{i,j}^{experiment} - SPV_{i,j}^{baseline} \tag{20.32}$$

In this equation, the SPV values are calculated as in Eq. 20.30. Once these probe log ratios are available, they can be combined using the same one-step Tukey's biweight estimate in a unique, gene-specific, signal log ratio:

$$SLR_i = T_{bi}(PLR_{i,1}, \dots, PLR_{i,n_i}) \tag{20.33}$$

The fold change can now be calculated as the exponential of the log ratio:

$$FC = \begin{cases} 2^{SLR_i}, & SLR_i \geq 0 \\ -2^{-SLR_i}, & SLR_i < 0 \end{cases} \tag{20.34}$$

20.5 Other approaches to the normalization of Affymetrix data

20.5.1 Cyclic Loess

The cyclic Loess normalization is a complete data normalization method, i.e. it takes into consideration information from all arrays in a data set. This approach uses the Loess method described in Section 20.3.3.2, but rather than applying it to the two color channels of the same array, it applies it to probe intensities from two arrays at a time. For any two arrays i and i with probe intensities x_{ki} and x_{kj} where k denotes the probes, the cyclic Loess uses the M and A values defined as follows:

$$M_k = \log_2 \left(\frac{x_{ki}}{x_{kj}} \right) \tag{20.35}$$

$$A_k = \frac{1}{2} \log_2 \left(x_{ki} \cdot x_{kj} \right) \tag{20.36}$$

As noted before, the A values correspond to the mean of the two log values, while the M corresponds to the difference of the two log values. A Loess normalization curve is fitted to this M versus A plot. The normalization curve will provide an adjustment $M'_k = M_k - \widehat{M}_k$ where $\widehat{M}_k$ is the fit provided by the curve. The adjusted probe intensities will be:

$$x'_{ki} = 2^{A_k + \frac{M'_k}{2}} \tag{20.37}$$

$$x'_{kj} = 2^{A_k - \frac{M'_k}{2}} \tag{20.38}$$

The preferred method is to calculate the normalization curves using rank invariant sets of probes [59].

Since a typical Affymetrix experiment involves more than two arrays, the method applies the above normalization for every pair of two arrays from the set. If there are n arrays in the data set, every array k will require $n - 1$ adjustments: $1, 2, \ldots, k - 1, k + 1, \ldots n$. These adjustments are all weighted equally and applied to the set of arrays. This entire process is repeated until the calculated adjustments are below a certain threshold, at which point the data are considered stable (and normalized).Because of the need to repeat the process a couple of times, with a complete iteration through all possible pairs in each repetition, the amount of computation is rather large and this process can be slow.

20.5.2 The model-based dChip approach

Another approach for the preprocessing and normalization of high-density oligonucleotide DNA arrays relies on the idea that the probe-specific biases are significant, but they are also reproducible and predictable. Thus, one could develop a statistical model of the phenomena happening at the probe level and use such model to calculate estimates of the gene expression. Furthermore, the same model can be used to detect cross-hybridizations, contaminated regions, and defective regions of the array [277, 278].

The model proposed by Wong et al. assumes that the intensity value of a probe j increases linearly with the expression of a gene in the i-th sample, θ_i. Furthermore, it is assumed that this happens for both PM and MM, only that the PM intensity will increase at a higher rate than the MM intensity. These ideas can be formalized in the following linear model (see Chapter 14):

$$MM_{ij} = v_j + \theta_i \alpha_j + \varepsilon \tag{20.39}$$

and

$$PM_{ij} = v_j + \theta_i \alpha_j + \theta_i \phi_j + \varepsilon \tag{20.40}$$

where PM_{ij} and MM_{ij} denote the PM and MM intensity values for the i-th array and the j-th probe pair for the given gene, v_j is the baseline response of the j-th probe pair due to nonspecific hybridization, α_j is the rate of increase of the MM response of the j-th probe pair, ϕ_j is the additional rate of increase in the corresponding PM response, and ε is the random error. It has been shown that this model fits the data well yielding a residual sum of squares (see Chapter 13) of only 1%, much better than a simple additive model. From

Equations 20.39 and 20.40 one can extract an even simpler model for the *PM − MM* differences:

$$y_{ij} = PM_{ij} - MM_{ij} = \theta_i \phi_j + e_{ij} \tag{20.41}$$

This model can be extended to all genes in the sample and used to detect and handle cross-hybridizations, image contaminations and outliers from other causes [277, 278]. This technique has been implemented in the package dChip which is available free of charge for academic use.[4]

20.5.3 The Robust Multi-Array Analysis (RMA)

A very popular normalization method is the log-scale Robust Multiarray Analysis (RMA) [59, 234, 235]. The method uses a linear model of the form:

$$T(PM_{ij}) = e_i + a_j + \varepsilon_{ij} \tag{20.42}$$

where T represents the transformation that background corrects, normalizes, and logs the PM intensities, e_i represents the $\log_2$ scale expression value found on arrays $i = 1,...I$, a_j represents the log scale affinity effects for probes $j = 1,...J$, and ε_{ij} represents the error for array i and probe j. A robust linear-fitting procedure is used to estimate the log scale expression values e_i. The resulting summary statistic is referred to as RMA.

Note that this model does not use at all the intensity values of the mismatch (MM) probes available on the Affymetrix arrays. Irizzary et al argue that although the MM probes might contain vinformation regarding the nonspecific hybridization that they are supposed to capture, the simple subtraction is not the appropriate way of taking that into consideration [235].

Fig. 20.11 shows a comparison of the three most popular Affymetrix normalization approaches. Each panel shows a scatter plot of the normalized values obtained after hybridizing two different amounts from the same sample (1.25 μg on the horizontal axis and 20 μg on the vertical axis) and normalizing the data with one of the three normalization methods described above. Since these are different amounts from the same biological sample, there should be no differentially expressed (DE) genes. The results obtained with the Affymetrix MAS 5 package (section 20.4), dChip (section 20.5.2, and RMA normalizations 20.5.3 are shown in panels A, B, and C, respectively. Genes reported as having two- to three-fold changes are shown in yellow. Genes reported as having fold changes larger than three-fold are shown in red.

20.5.4 Quantile normalization

The stated goal of the quantile normalization is to align the distributions of the probe intensities in a set of array. In other words, after this normalization,

---

[4]The dChip software was available at http://www.dchip.org at the time of going to press.

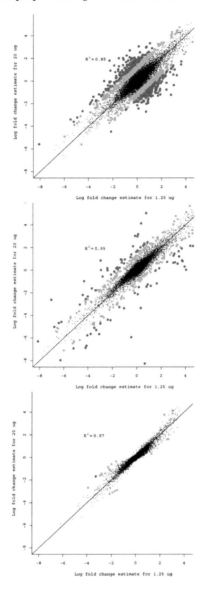

FIGURE 20.11: A comparison of the three most popular Affymetrix normalization approaches. Each panel shows a scatter plot of the normalized values obtained after hybridizing two different amounts from the same sample. Since these are different amounts from the same biological sample, there should be no differentially expressed (DE) genes. The results obtained with the Affymetrix MAS 5 package (section 20.4), dChip (section 20.5.2), and RMA (section 20.5.3) are shown in panels A, B, and C, respectively. Genes reported as having two- to three-fold changes are shown in yellow. Genes reported as having fold changes larger than three fold are shown in red. Figure reprinted with permission from [235].

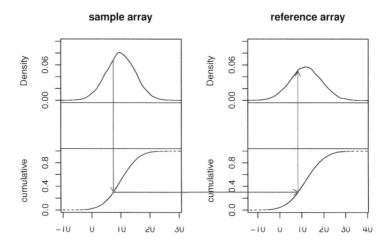

FIGURE 20.12: Quantile normalization.

all arrays should have the same distribution of their probe intensities. This is done as follows [59]:

1. Given a set of n arrays with p probes, construct a matrix X with p rows and n columns, in which each column is an array of raw intensities

2. Sort each column to construct the matrix X_{sort}

3. Calculate the mean across each row of X_{sort} and assign this mean to each element in the row to get X'_{sort}

4. Construct $X_{normalized}$ by rearranging each column of X'_{sort} in the ordering of the original X.

The idea of the quantile normalization is illustrated in Fig. 20.12. In essence, this method calculates the empirical distribution of the averaged sample quantiles and uses it to index the empirical distribution of each original array. This method has been shown to be both very computationally efficient, as well as very effective in terms of achieving the normalization goals by reducing the variance of the data caused by non-biological reasons. In a comparison of several normalization methods applied to Affymetrix data, the quantile normalization was shown to provided the smallest distance between arrays corresponding to same biological samples [59].

20.6 Useful preprocessing and normalization sequences

It is useful to try to discuss the preprocessing as a whole as opposed to focusing on the details of every single preprocessing step. In order to do this, we may imagine the preprocessing as a pipeline, with raw data entering at one end and the normalized data coming out at the other end. As discussed, the specific preprocessing and normalization steps may be combined in a large number of ways to produce useful sequences. For instance, the normalization for the overall array level can be achieved by performing a division of the values by the mean of each array before taking the logs or by subtracting the mean after the log is taken. Also, the values of the replicate spots may be combined at various stages during the preprocessing.

A preprocessing and normalization sequence that might be useful for cDNA data would include the following steps:

1. Background correction. This can be done using:

    (a) Local background correction

    (b) Subgrid background correction

    (c) Local group background correction

    (d) Background correction using blank spots

2. Eliminate the spots flagged in the image processing stage.

3. Substitute missing values (optional).

4. Normalize overall array intensity. This may be optional if the normalization for color distortion equalizes the array levels, as well. The alternatives include:

    (a) Divide by mean

    (b) Iterative regression

    The normalization of overall array intensity may be performed after eliminating the outliers and/or differentially regulated genes.

5. Normalize for color distortion. This can be done using:

    (a) Curve fitting and correction

    (b) LO(W)ESS

    (c) Piece-wise normalization

6. Calculate the ratio of the two channels cy3/cy5 or cy5/cy3.

7. Apply a logarithmic transformation (base 2 recommended).

8. Combine the replicate values to calculate a mean/median/mode value and the standard deviation.

For Affymetrix data, we recommend the use of the *simplyaffy* package in Bioconductor for the preprocessing and normalization. Examples illustrated how to use various functions from this and other related packages follow later in this chapter, as well as the next chapter. A viable alternative for the preprocessing is the dChip package implementing the model proposed by Wong et al. (Eq. 20.41).

If the data are available as average differences (as from older versions of the Affymetrix DMT software), the following sequence may be useful:

1. Apply a floor function to bring all negative values to a small positive value.

2. Apply a logarithmic transformation

3. Normalize overall array intensity.

In many cases, it is useful to use the detection calls provided by the Affymetrix package. However, it is not a good idea to eliminate completely the genes that are absent (A) or marginally detected (M). Table 20.1 shows that the only genes that can probably be safely ignored are those that are absent (A) or only marginally detected (M) in both arrays. All other combinations are or may be interesting from a biological point of view. For instance, a gene that is absent in the baseline and present in the experiment is probably a gene whose activity may be up-regulated in the condition studied. Even though it may not be possible to report a meaningful fold change for this gene, its further study may be well warranted.

It should be said however, that there is another school of thought whose adopters propose to completely ignore the Affymetrix calls and base the entire analysis on the probe-level intensities, as found in the CEL files. Furthermore, certain approaches also ignore the values of the mismatch (MM) probes and perform both the normalization, as well as later analysis steps such as the selection of differentially expressed genes based entirely on the intensities corresponding to the perfect match probes.

20.7 Normalization procedures in R

20.7.1 Normalization functions and procedures for Affymetrix data

In this section, we will illustrate the main steps of a possible preprocessing and normalization procedure for Affymetrix data executed in R. We start by

Detection calls		Magnitude of exp/base ratio	Biological meaning	Interesting gene
exp	base			
M/A	M/A	meaningless; can be any-thing	the gene is probably not expressed in either condition	no
M/A	P	meaningless; close to zero	gene expressed on the baseline array, not expressed in the experiment	yes
P	M/A	meaningless; very large	gene expressed in the experiment, not expressed in the baseline array	yes
P	P	meaningful; can be any-thing	gene expressed on both channels, useful ratio	depends on the ratio

TABLE 20.1: The interplay between the A/M/P calls and the expression values when an experiment array is compared to a baseline array. The only genes that can be safely ignored are those that are absent (A) or only marginally detected (M) in both arrays. All other combinations are or may be interesting from a biological point of view.

downloading some Affymetrix data from GEO (http://www.ncbi.nlm.nih.gov/geo/query/acc.cgi?acc=GSE9412). Henceforth, we will assume that the uncompressed .CEL files containing the raw data are available in the current directory.

The library *affy* contains many relevant functions so we start by loading it. We then list the .CEL files available in the directory and display the content of this list to make sure we are in the correct directory and have the files we expected:

```
> library(affy)
> CELfiles<-list.celfiles()
> CELfiles

[1] "GSM239309.CEL" "GSM239310.CEL" "GSM239311.CEL"
[4] "GSM239312.CEL" "GSM239313.CEL" "GSM239314.CEL"
```

Everything seems to be fine so we proceed to read these CEL files with the function *ReadAffy*. The function *ReadAffy* creates a "rawdata" object of "AffyBatch" class, storing probe-level raw data and phenotypic information for the given samples and experiments. The main slots of this object type are:

- *cdfName*: Object of class character representing the name of CDF file associated with the arrays in the AffyBatch.

- *nrow*: Object of class integer representing the physical number of rows in the arrays.

- *ncol*: Object of class integer representing the physical number of columns in the arrays.

- *assayData*: Object of class AssayData containing the raw data, which will be at minimum a matrix of intensity values. This slot can also hold a matrix of standard errors if the "sd" argument is set to TRUE in the call to ReadAffy.

- *phenoData*: Object of class AnnotatedDataFrame containing phenotypic data for the samples.

- *annotation* A character string identifying the annotation that may be used for the ExpressionSet instance.

- *protocolData*: Object of class AnnotatedDataFrame containing protocol data for the samples.

- *featureData* Object of class AnnotatedDataFrame containing feature-level (e.g., probeset-level) information.

- *experimentData*: Object of class "MIAME" containing experiment-level information.

Trying to display the content of "rawdata" will trigger "affy" to automatically download the chip definition file (CDF) if this file is not already available in your library:

```
> rawdata<-ReadAffy(filenames=CELfiles)
> rawdata

AffyBatch object
size of arrays=732x732 features (11 kb)
cdf=HG-U133A_2 (22277 affyids)
number of samples=6
number of genes=22277
annotation=hgu133a2
notes=
```

The function *intensity* can be used to extract the expression matrix from the rawdata object.

```
> head(intensity(rawdata))
```

	GSM239309.CEL	GSM239310.CEL	GSM239311.CEL	GSM239312.CEL
1	53	51	69	64
2	8137	4728	5301	7173

3	70	92	78	78
4	8513	6102	5723	7426
5	81	73	98	62
6	58	55	64	54

	GSM239313.CEL	GSM239314.CEL
1	72	68
2	6605	6284
3	95	75
4	6431	6175
5	67	71
6	67	58

Now we can examine the raw data through exploratory graphs: array intensity images, intensity distribution curves, box plots, and MA-plots of log-intensity ratios (M-values) versus log-intensity averages (A-values).

The array probe intensity image is derived from mapping probe intensities to a gray scale, and can be used to examine the data for potential spacial artifacts on the array. Note that probe intensities may be transformed on a log scale before mapping to a gray scale image, in order to alleviate the effect of extreme probe intensities:

```
> par(mfrow=c(2,3))
> image(rawdata, transfo=log)
```

The images of the arrays produced with by the fragment of code above are shown in Fig. 20.13. In this type of image, array defects such as scratches, hairs, or other contaminants should be visible as spacial artifacts involving the values read from the array in the areas affected by the defect. Other defects such as surface defects or hybridization problems may involve large areas of a given array, or even entire arrays as discussed in Chapter 19. No particular defects appear to be present on the arrays included in this data set.

Another useful exploratory graph is the M-A plot (see Chapter 17 and section 20.5.1). As explained previously, the M-A graph shows M-values versus A-values, standing for log-intensity differences (i.e., log fold change) and log-intensity averages, respectively. As the M-A plot is a particular type of scatter plot, it can only be used to plot data from two arrays at any one time. Hence, if a set of arrays are to be explored with M-A plots, a number of such plots equal to the number of possible pairs of arrays are necessary. However, plotting both array *i* versus array *j* and array *j* versus array *i* would be redundant so only half of the pairs are actually needed. Furthermore, no plots are needed for the pairing of an array with itself since all those log-ratios will be zero. M-A plots can be obtained with the function *MAplot*:

```
> MAplot(rawdata, pairs=T, plot.method="smoothScatter",
+ cex=0.9, cex.main=0.8, cex.lab=0.8, pch=20)
```

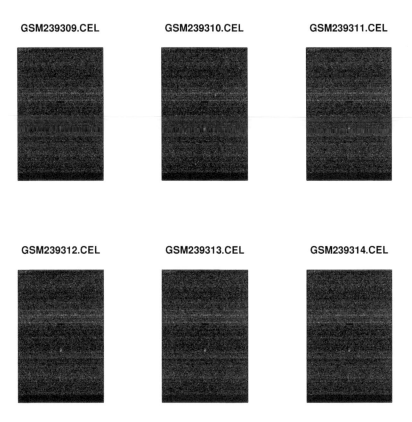

FIGURE 20.13: The intensity images of the Affymetrix arrays included in this data set.

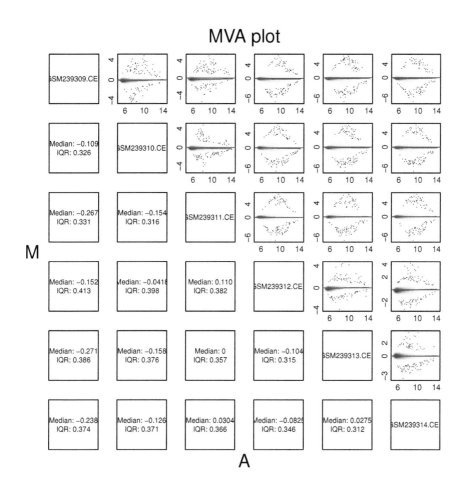

FIGURE 20.14: MA-plots of the Affymetrix arrays in this sample data set. The smoothed blue color density representations are derived from kernel density estimations for each array. Data points located in low-density area are depicted as black dots. The red line is the fitted Loess curve. The two summary statistics, median and IQR (interquantile range), for M-values are also shown.

In fact, as seen above this function is smart enough to take an entire data set, including several arrays, construct all possible pairs from the available arrays, and plot an M-A plot for each such pair. The plot shown in Fig. 20.14 shows the distribution of the data in the graphs above the diagonal, and the corresponding statistics in the symmetric positions below the diagonal. In other words, the graph shown in position (i, j) will correspond to the statistics shown in the position (j, i). The blue color represents the density of the data (darker levels of blue correspond to higher data densities). Data points situated in low-density areas are shown explicitly as black dots. The red line is the fitted Loess curve (see Section 20.3.3.2). The summary statistics shown in the position (i, j) below the diagonal refer to the M values, or the log-ratio of the same probes between arrays i and j. An alternative to plotting the M-A plots for all pairs as we did here, is to construct a "reference" array and construct the M-A plot of each array with respect to this reference. Thus, the number of plots is reduced from $(n^2 - n)/2$ to n. Such a reference could be an abstract "median array" constructed by taking the median for each probe.

20.7.2 Background adjustment and various types of normalization

In the *simpleaffy* package, the quantile normalization is implemented by the function *normalize*. The same function can also apply a simple background normalization, as well as the dChip model-based normalization described in Section 20.5.2. Affymetrix's MAS processing pipeline described in Section 20.4 is implemented in the *mas* function, while the Robust Multiarray Average (RMA) described in Section 20.5.3 is available as *rma*. Other normalization methods can be applied as well, such as scaling in MAS5, but note that Affymetrix applies normalization after summarization rather than at this stage.

Let us correct out data for background and then normalize our data with three different methods: quantiles, invariant set (dChip), and scaling to a reference array, and compare the results:

```
> data.bg<-bg.correct(rawdata, method="rma")
> data.qtl<-normalize(data.bg, method="quantiles")
> data.scl<-normalize(data.bg, method="constant")
> data.dChip<-normalize(data.bg, method="invariantset")
> usr.col<-rep(c("red", "blue"), each=3)
> usr.line<-rep(1:3, 2)

> hist(data.bg, lty=usr.line, col=usr.col, main="background
+ corrected")
> legend("topright", sampleNames(rawdata), lty=usr.line,
+ col=usr.col)

> hist(data.scl, lty=usr.line, col=usr.col, main="scaling")
```

```
> legend("topright", sampleNames(rawdata), lty=usr.line,
+ col=usr.col)

> hist(data.qtl, lty=usr.line, col=usr.col, main="quantiles")
> legend("topright", sampleNames(rawdata), lty=usr.line,
+ col=usr.col)

> hist(data.dChip,lty=usr.line,col=usr.col,main="invariantset")
> legend("topright", sampleNames(rawdata), lty=usr.line,
+ col=usr.col)


> boxplot(data.bg, col=usr.col, las=3, names=
+ sampleNames(rawdata),ylim=c(4,8),cex.axis=0.6,main=
+ "background corrected")

> boxplot(data.scl, col=usr.col, las=3, names=sampleNames
+ (rawdata), ylim=c(4,8), cex.axis=0.6, main="scaling")

> boxplot(data.qtl, col=usr.col, las=3, names=sampleNames
+ (rawdata), ylim=c(4,8), cex.axis=0.6, main="quantiles")

> boxplot(data.dChip, col=usr.col, las=3, names=sampleNames
+ (rawdata), ylim=c(4,8), cex.axis=0.6, main="invariantset")
```

Fig. 20.15 shows the comparison of the probe intensity distributions after the application of these three methods. Fig. 20.16 shows the comparison between the box plots of the same data. The quantile normalization provides the best results, closely followed by the invariant set method implemented by dChip. This is how the data should look like before attempting to find differentially expressed genes or attempting to apply any further analysis methods.

20.7.3 Summarization

After normalization, one may want to do a probe-specific correction (e.g., subtracting MM) and a summarization to probe set-level data on log scale. Various methods are available in *expresso* from the *affy* package using the following functions: *bgcorrect.methods*, *pmcorrect.methods*, *normalize.methods*, *express.summary.stat.methods*. The function *threestep* from the package "affy-PLM" can also be used for this purpose and is usually faster than using *expresso*. Sample function calls would look like this:

```
> eset<-expresso(data.norm, bg.correct=F, normalize=F,
+ pmcorrect.method="pmonly", summary.method="medianpolish")
> eset<-threestep(data.norm, background=F, normalize=F,
+ summary.method="median.polish")
```

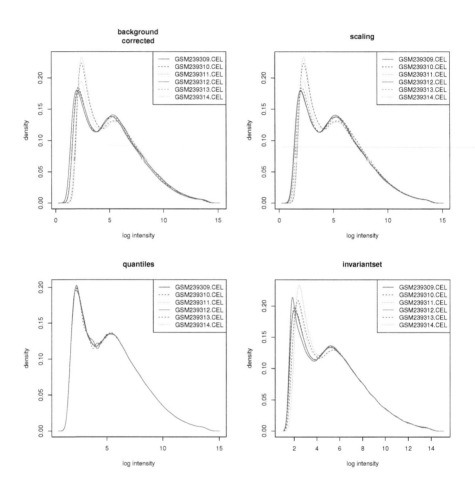

FIGURE 20.15: A comparison of three normalization methods. The figure shows the probe intensity distributions of the background corrected data (upper left), and the same data after three different normalization methods: scaling (upper right), quantiles (bottom left), and invariant set (bottom right).

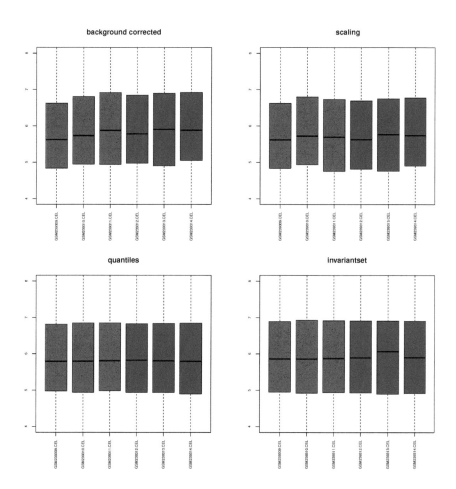

FIGURE 20.16: A comparison of three normalization methods. The figure shows the box plot of probe intensities of the background corrected data (upper left) and after three different normalization methods: scaling (upper right), quantiles (bottom left), and invariant set (bottom right). Note the almost perfect alignment of the distributions provided by the quantile normalization, and the very good alignment provided by the invariant set method used by dChip. The red and blue colors show the class membership of that particular sample.

20.8   Batch preprocessing

In most cases, once one understands the characteristics of one's data and establishes a standardized set of preprocessing and normalization steps, it is desirable to execute all such steps in a batch mode that minimizes the human intervention. This allows for off-line processing of larger sets of arrays. Several preprocessing steps can be wrapped into a sequence that can be used in a batch execution mode using one of the following four functions: *expresso*, *threestep*, *rma*, and *mas5*.

```
> eset<-expresso(rawdata, bgcorrect.method="rma",
+ normalize.method="quantiles", pmcorrect.method="pmonly",
+ summary.method="medianpolish")
> eset<-threestep(rawdata, background.method="RMA.2",
+ normalize.method="quantile", summary.method=
+ "median.polish")
> eset<-rma(rawdata)
> eset<-mas5(rawdata)
```

Another common issue related to the analysis of large Affymetrix data sets is the large amount of memory required. It is a rather frequent occurence to run out of memory during the analysis even for relatively small experiments involving only a few tens of arrays. In order to address this, one can use a convenient and fast wrapper *justRMA*, which takes raw .CEL files as input and produces preprocessed data without creating an "AffyBatch" object. This turns out to be advantageous for experiments involving a large number of arrays because *justRMA* uses much less memory compared to the above methods.

```
> eset<-justRMA(filenames=CELfiles)
```

No matter which of the specific methods described above are used, the preprocessed data will be stored in an object named "eset." The normalized probe set level log intensity expression value matrix can be simply extracted as following:

```
> MA<-exprs(eset)
```

Methods for the selection of differentially expressed genes, as well as other more sophisticated types of analysis, can be now applied to these data.

20.9   Normalization functions and procedures for Illumina data

The normalization should be performed after some basic quality control checks have been performed. For Illumina, the data can be either bead-summary data exported from BeadStudio/GenomeStudio, or summarized data obtained after explicit processing in R, as discussed in Chapter 19. Here we will illustrate some functions and procedures using the bead-summary data from Chapter 19.

Note that the bead-summary data is assumed to have been corrected for the local background by the BeadStudio software. Thus, we can directly go to the normalization step using function `normaliseIllumina`:

```
> datDir<-"./Normalization"
> rawfile<-paste(datDir, "GSE17439_EsetKnockDownRawData.txt",
+ sep="/")
> library("RColorBrewer")
> mycol<-brewer.pal(6, "Set1")
> myline<-rep(1, 6)
> library(beadarray)
> BSData<-readBeadSummaryData(dataFile=rawfile, ProbeID=
+ "TargetID", skip=7, columns=list(exprs="AVG_Signal",
+ se.exprs="BEAD_STDEV", nObservations="Avg_NBEADS",
+ Detection="Detection"))
> BSData.norm<-normaliseIllumina(BSData, method="quantile",
+ transform="log2")
```

The first few lines above do some housekeeping: setting the directory, preparing the full path of the data file, preparing a vector of colors and lines types to be used later, etc. The last two commands are more substantive: reading the data with *readBeadSummaryData* and normalizing it with *BSData.norm*.

However, for Illumina data some preprocessing has been shown to be very useful. Lin et al. have proposed a variance-stabilizing transformation (VST) method [284] that takes advantage of the larger number of technical replicates available on the Illumina platform. The variance stabilization of microarray data is useful because in this type of data there is often an increased variance associated with high-intensity values. Much like the high variance at low-intensity values usually associated with cDNA data, the high variance at high-intensity values violates the constant variance assumption used by most statistical tools from hypothesis testing to linear models. A VST can be performed either by specifying the argument `transform="vst"` in the above *normaliseIllumina* function, or with the function `lumiT` from the package "lumi.":

```
> library(lumi)
```

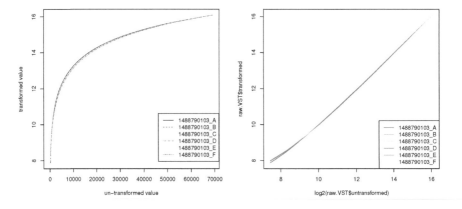

FIGURE 20.17: VST and log transformations. The left panel shows a graph of the VST transformed data as a function of the raw data. The right panel shows the graph of the save VST transformed data as a function of the log2 of the raw data.

```
> rawsum.lm<-lumiR(rawfile, sep="\t", columnNameGrepPattern=
+ list(exprs="AVG_Signal", se.exprs="BEAD_STDEV", beadNum=
+ "Avg_NBEADS", detection="Detection", verbose=F))

> rawsum.lmT<-lumiT(rawsum.lm)  #apply VST

> raw.VST<-plotVST(rawsum.lmT, addLegend=T)

> matplot(log2(raw.VST$untransformed), raw.VST$transformed,
+ type='l', col=mycol, lty=myline)
> legend("bottomright", legend=pData(rawsum.lmT)$sampleID, col=
+ mycol, lty=myline)
```

To visualize the effect of variance stabilization through VST transformation, we can use function `meanSdPlot` to plot row standard deviations versus row means or rank of row means.

```
> library(vsn)
> meanSdPlot(rawsum.lm[,1:3], main="raw")

> meanSdPlot(rawsum.lmT[,1:3], main="VST")

> meanSdPlot(rawsum.lm[,1:3], ranks=F, main="raw")

> meanSdPlot(rawsum.lmT[,1:3], ranks=F, main="VST")
```

The results of this variance stabilization are shown in Fig. 20.18. In this figure, the left column shows the standard deviation (sd) of the data before the VST is applied. The top left panel shows the sd as a function of the ranks of the means. The bottom left panel shows the sd as a function of the means themselves. The top right panel shows the sd after the VST transformation as a function of the ranks of the means. The bottom right panel shows the sd after the VST transformation as a function of the means themselves. Note how the largest standard deviation was reduced from 5,000 to 0.4. The data include only the first three arrays corresponding to one of the two phenotypic groups in order to avoid mixing up into the picture the variance due to the differences between the groups.

After this transformation, the function `lumiN` from package "lumi" can be used to apply various normalization methods as well. We can compare these methods using probe density plots and box plots, as we did for Affymetrix data.

```
> rawsum.lmQTL=lumiN(rawsum.lmT, method="quantile")
> rawsum.lmVSN=lumiN(rawsum.lm, method="vsn") #apply on raw data
> rawsum.lmRIV=lumiN(rawsum.lmT, method=
+ "rankinvariant") #similar to the one used in GenomeStudio

> density(rawsum.lmT, lty=myline, col=mycol, main="VST
+ transformed")

> density(rawsum.lmQTL, lty=myline, col=mycol, main="quantile")

> density(rawsum.lmVSN, lty=myline, col=mycol, main="vsn")

> density(rawsum.lmRIV, lty=myline, col=mycol, main=
+ "rankinvariant")
```

The intensity distributions of the VST transformed bead-summary data before and after applying three different normalization methods are shown in Fig. 20.19. The top left panel shows the data after the VST transformation. This is the data used as input for all three normalization methods compared here. The top right panel shows these data after the quantile normalization. The bottom left panel shows the data after the VSN normalization, and the bottom right panel shows the data after the rank invariant normalization. All three normalization methods perform well. After the normalization the modes of the distributions of the six arrays are perfectly aligned. For the quantile normalization, the distributions after the normalization are practically identical, to the extent that the last drawn array (yellow) actually covers entirely all existing plots.

Also in Fig. 20.15, note the apparent departure of array 1488790103_F after the VST transformation. In essence, this array is not different in terms of distribution of intensities but in terms of their variances. Hence, the original density plot shown in Fig. 19.16 in Chapter 19 did not show that array as being

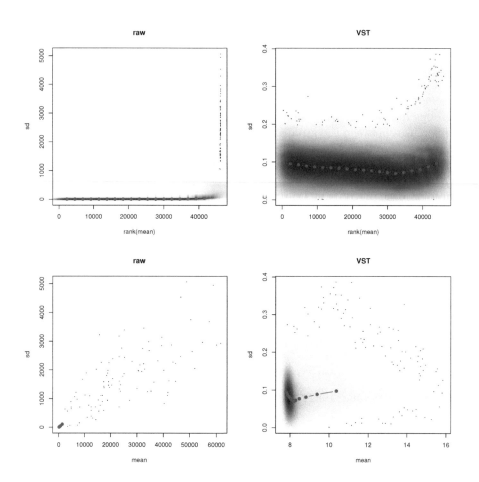

FIGURE 20.18: The variance stabilization effect of the VST transform. There is a significant dependence between the standard deviation of the raw data and its mean. This is much alleviated after the VST transformation. The top left panel shows the sd as a function of the ranks of the means. The bottom left panel shows the sd as a function of the means themselves. The top right panel shows the sd after the VST transformation as a function of the ranks of the means. The bottom right panel shows the sd after the VST transformation as a function of the means themselves. Note how the maximum standard deviation was reduced from 5,000 to 0.4.

different, while the plot shown here in Fig. 20.15, as well as the coefficient of variance plot in Fig. 19.17 in Chapter 19 did show this array as being different.

```
> boxplot(rawsum.lmT, col=mycol, main="VST transformed")

> boxplot(rawsum.lmQTL, col=mycol, main="quantile")

> boxplot(rawsum.lmVSN, col=mycol, main="vsn")

> boxplot(rawsum.lmRIV, col=mycol, main="rankinvariant")
```

The box plots of the same data are shown in Fig. 20.20. In this figure, the top left panel shows the VST transformed data, which is used as the input for all three normalization methods. As noted in the intensity distribution plot, the array 1488790103_F appears to be a bit different after the VST transformation. The top right panel shows the results of the quantile normalization. The bottom left panel shows the data after the VSN normalization, and the bottom right panel shows the data after the rank invariant normalization. All three normalization methods aligned the arrays very well.

As a final note, there is also a wrapper function `lumiExpresso` to allow the batch preprocessing of Illumina data, a counterpart similar to the function `expresso` in the package "affy."

```
> rawsum.lmE<-lumiExpresso(rawsum.lm, bg.correct=F)
```

20.10 Summary

The aim of the normalization is to account for systematic differences across different data sets (e.g., quantity of mRNA) and eliminate artifacts (e.g., non-linear dye effects). The normalization is crucial if results of different experimental techniques are to be combined. This chapter discussed several issues related to the preprocessing and normalization of microarray data. A few general preprocessing techniques useful for both cDNA and Affymetrix data include the logarithmic transformation, combining replicates and eliminating outliers and array normalization. The log transform partially decouples the variance and the mean intensity and makes the distribution almost normal. Combining the replicates should be done with care since in many cases this involves losing some information. Outliers can be eliminated in an iterative process involving calculating the mean and standard deviation σ and eliminating values outside $\pm 3\sigma$. Array normalization can be done by dividing by the mean before the log or by subtracting the mean after the log. The former is equivalent to a correction using the arithmetical mean while the latter is equivalent to a correction using the geometrical mean.

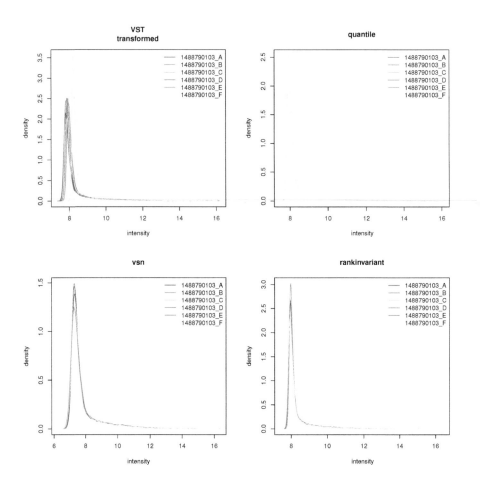

FIGURE 20.19: Probe intensity distribution of VST transformed bead-summary data with three different normalization methods. The top left panel shows the data after the VST transformation. This is the data used as input for all three normalization methods compared here. The top right panel shows these data after the quantile normalization. The bottom left panel shows the data after the VSN normalization, and the bottom right panel shows the data after the rank invariant normalization. All three normalization methods perform well. After the normalization, the modes of the distributions of the six arrays are perfectly aligned. For the quantile normalization, the distributions after the normalization are practically identical, to the extent that the last drawn array (yellow) actually covers entirely all existing plots.

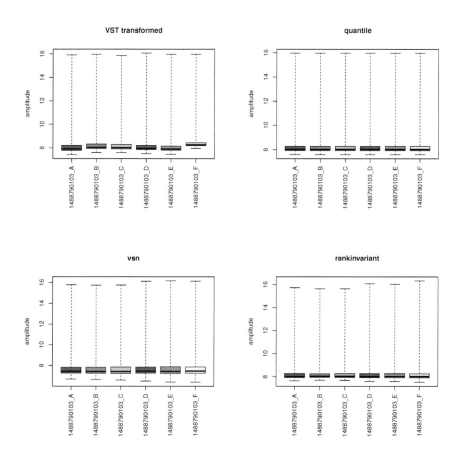

FIGURE 20.20: Box plots of VST transformed bead-summary data with three different normalization methods. In this figure, the top left panel shows the VST transformed data, which is used as the input for all three normalization methods. The top right panel shows the results of the quantile normalization. The bottom left panel shows the data after the VSN normalization, and the bottom right panel shows the data after the rank invariant normalization. All three normalization methods aligned the arrays very well.

Many other preprocessing techniques are specific to a given technology and they were discussed separately. The cDNA data are usually preprocessed by correcting for the background, correcting for color nonlinearities or using more sophisticated statistical models. The background correction can be done locally (only the background local to the spot is considered), using a group of spots such as a sub-grid (suitable for high-density arrays where there are not enough pixels around a single spot to produce a reliable value) or blank spots. The color normalization can be performed using a model-based approach (e.g., the exponential normalization), a LO(W)ESS approach or a piece-wise linear normalization. The exponential normalization has been shown to work well for the usual cy3/cy5 dyes. The LO(W)ESS approach should work well for any dyes but is computationally intensive and does not provide a normalization model that can be stored for reference or used for other data sets. The piece-wise linear normalization is similar to LO(W)ESS but is faster and does provide a normalization model.

The Affymetrix data are preprocessed as follows. The cell intensities are corrected for background using some weighted average of the backgrounds in the neighboring zones. An ideal mismatch value IM is calculated and subtracted from the PM intensity. If the mismatch MM is lower than the PM, the IM will be taken to be the difference *PM − MM*. The adjusted PM values are log-transformed and a robust mean of these log-transformed values is calculated. The signal value is calculated as the exponential of this robust mean and scaled using a trimmed mean. Alternatively, model-based normalizations approaches also exist (e.g., dChip) and have been shown to perform very well. Quantile normalization is arguably one of the best methods available both in terms of results, as well as computational efficiency.

Various packages and functions available for the normalization and preprocessing of Illumina data have also been discussed and illustrated in this chapter. Illumina users have the choice to use raw bead-level data or bead summaries. The normalization of Illumina data follows the same principles and aims to achieve the same goals as for the other types of data.

20.11 Appendix: A short primer on logarithms

The log is the inverse of the exponential function. Thus, $y = \log_a x$ is defined as the power to which one needs to raise a in order to obtain x: $a^y = x$. For instance, $\log_2 8 = 3$ since $2^3 = 8$ and $\log_{10} 100 = 2$ since $10^2 = 100$. The quantity a is called the base of the logarithm. Since the exponential and the logarithm are each other's inverse, the following identities hold true:

$$a^{\log_a x} = \log_a a^x = x \tag{20.43}$$

Fig. 20.21 shows the graphic of the logarithmic function $y = \log_2 x$.

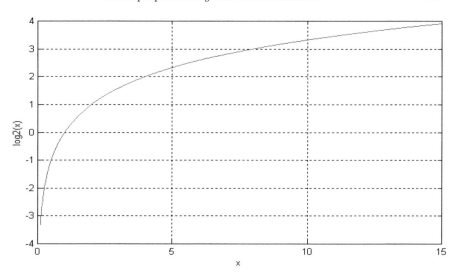

FIGURE 20.21: The logarithmic function: $y = \log_2 x$.

The main properties of the logarithms are as follows:

1. $\log_a a^x = x$, which is to say that the power to which one needs to raise a in order to obtain a^x (a to the power x) is the very same x.

2. $a^{\log_a x} = x$, which is to say that a raised to the power to which one needs to raise a in order to obtain x is indeed x.

3. $\log_a (x \cdot y) = \log_a x + \log_a y$ or log of a product is the sum of the logs of the factors.

4. $\log_a \frac{x}{y} = \log_a x - \log_a y$ or log of ratio is the difference of the logs of the numerator and denominator.

5. $\log_a x^y = y \cdot \log_a x$. This can be easily proven by repeatedly applying the product rule above.

Chapter 21

Methods for selecting differentially expressed genes

Everything should be made as simple as possible, but not simpler.

—Albert Einstein

21.1   Introduction

In many cases, the purpose of the microarray experiment is to compare the gene expression levels between two different phenotypes.[1] In most cases, one sample is considered the reference or control and the other one is considered the experiment. Obvious examples include comparing healthy versus disease, treated versus untreated, drug A versus drug B, before and after treatment, etc. Sample comparison may be done using different arrays (e.g., oligonucleotide arrays) or multiple channels on the same array (e.g., cDNA or Illumina arrays). In all such comparative studies, a very important problem is to

---

[1]Some of the material in this chapter is reprinted from S. Draghici, "Statistical intelligence: effective analysis of high-density microarray data" published in Drug Discovery Today, Vol. 7, No. 11, p. S55-S63, Copyright (2002), with permission from Elsevier.

determine the genes that are differentially expressed (DE) in the two samples being compared.

Although simple in principle, this problem becomes more complex in reality because the measured intensity values are affected by numerous sources of fluctuation and noise [131, 377, 461]. For spotted cDNA arrays, there is a non-negligible probability (about 5%) that the hybridization of any single spot containing complementary DNA will not reflect the presence of the mRNA. Furthermore, the probability that a single spot will provide a signal even if the mRNA is not present is even greater (about 10%) [276]. Most cDNA arrays address this problem by having several replicate spots for each gene.

Illumina addresses this challenge by having a large number of beads carrying the same DNA sequence. In practice, these many beads constitute technical replicates that allow us to estimate the variance introduced by the process for every mRNA species. As shown in Chapter 19, this information can be used to detect various types of problems.

The Affymetrix technology tries to respond to the challenge of a poor reliability for single hybridizations by representing a gene through a set of probes. The probes correspond to short oligonucleotide sequences thought to be representative for the given gene. Each oligonucleotide sequence is represented by two probes: one with the exact sequence of the chosen fragment of the gene (perfect match or PM) and one with a mismatch nucleotide in the middle of the fragment (mismatch or MM). For each gene, the value that is often taken as representative for the expression level of the gene is the average difference between PM and MM (see Chapters 3, 5 and 20). In principle, this value is expected to be positive because the hybridization of the PM is expected to be stronger than the hybridization of the MM. However, many factors, including nonspecific hybridizations and a less than optimal choice of the oligonucleotide sequences representing the gene may determine an MM hybridization stronger than the PM hybridization for some probes. In such cases, older versions of the software reported negative average differences. The latest versions have corrected this by using a more sophisticated preprocessing method (see Chapter 20). However, the numerical manipulations performed in order to avoid such negative values may still introduce nonlinearities in the expression values and make the gene selection task difficult even for Affymetrix data.

In this context, distinguishing between genes that are truly differentially regulated and genes that are simply affected by noise becomes a real challenge. All methods discussed here are completely independent of the technology used to obtain the data (e.g., cDNA, Illumina, Affymetrix, etc.). The main difference between the different types of data is the preprocessing as discussed in Chapter 19 and 20. The Affymetrix data can be pre-processed by combining the fluorescence levels of individual probes between match and mismatch to yield average differences (or expression indexes) and detection calls (present, absent, marginal, etc.). Another approach is to treat the probes as completely independent and postpone their mapping to genes until after the selection process identified those that are differentially expressed. The cDNA data are minimally processed by subtracting the background from the fluorescence values of the spots and correcting for color non-linearity. Furthermore, regardless of which technology is used, when data from different arrays are compared, such comparisons must be first made meaningful by bringing the arrays at comparable levels of intensity. This is usually done by some global normalization as discussed in Chapter 20. Finally, in most cases, one would like to apply a log or a variance stabilization transformation in order to improve the characteristics of the distribution of expression values. Such transformations can be applied at various stages of the preprocessing and normalization, as illustrated in various examples throughout Chapter 19 and Chapter 20.

In the following, we will exemplify several approaches for the selection of differentially expressed (DE) genes or probes using the very simple example of a comparison between two conditions: experiment and control. Consistently with the approach followed in the rest of the book, we will approach the task at hand starting with the simplest approaches, discuss each approach's advantages and disadvantages, and gradually progress towards more sophisticated techniques.

21.2 Criteria

In order to assess the performance of a gene selection method, we need a set of criteria able to quantify the outcome of the selection process. The performance of a gene selection method can be calculated in terms of accuracy, positive predicted value (PPV), negative predicted value (NPV), specificity, and sensitivity. In general, as discussed in Chapter 8, Section 8.7.1, for any diagnosis or classification method, one could compare the truth with the results reported by the method. Here, we will discuss the use of the same criteria for the assessment of gene selection methods. In a binary decision situation, such as changed/unchanged, the results can always be divided into four categories: truly changed that are reported as changed (**true positives**), unchanged that are reported as changed (**false positives**), truly changed that are reported as unchanged (**false negatives**), and truly unchanged that are reported as such (**true negatives**). Based on these, one can define the four quantitative criteria: **positive predicted value** (PPV), **negative predicted value** (NPV), **specificity**, **sensitivity**, and **accuracy**. Let us recall the definitions of these quantities from Section 8.7.1:

$$PPV = \frac{TP}{TP + FP} \tag{21.1}$$

$$NPV = \frac{TN}{TN + FN} \tag{21.2}$$

$$Specificity = \frac{TN}{TN+FP} \tag{21.3}$$

$$Sensitivity = \frac{TP}{TP+FN} \tag{21.4}$$

$$Accuracy = \frac{TP+TN}{N} \tag{21.5}$$

where TP is (the number of) true positives, TN is true negatives, FP is false positives, FN is false negatives, and N is the total number of instances $N = TP+FN+TN+FN$. These criteria are shown in Table 21.1. These quantities range from 0 to 1. Sometimes, they are expressed as percentages with 1 being equal to 100%. A perfect method would yield no false positives and no false negatives. In this case, the accuracy, specificity, sensitivity, PPV, and NPV would all be equal to 1 or 100%. These measures also depend on the proportion of truly changed $(TP+FN)$ with respect to the total number of instance N. The proportion $\frac{TP+FN}{N}$ is called **prevalence**. The prevalence does influence the other measures as well as the usefulness of any classification or diagnosis method. Example 8.6 in Chapter 8 showed that a diagnosis method with a sensitivity of 90% and a PPV of 99.5% may still be insufficient for practical purposes for a disease with a prevalence of only 1 in 5,000 individuals (0.02%).

The most important difference between the general use of these criteria in a classification problem and their use in this application is related to the specific meaning given to the terms. For instance, if the numbers are reported for up-regulated genes, TP would be the number of genes truly up-regulated and TN would be the number of genes not up-regulated. It has to be noted, that in this case, the set of genes that are not up-regulated includes the unchanged genes, as well as the down-regulated genes. In consequence, FP would be the number of genes reported as up-regulated where in fact they are either not regulated or down-regulated, and FN would be the number of genes that are truly up-regulated and are not reported as such. Similarly, if the numbers are reported for down-regulated genes, TP would correspond to those genes that are truly down-regulated and TN will include the genes that are not differentially expressed, as well as the genes that are up-regulated. In essence, this comes down to the correct formulation of the hypothesis to be tested in the hypothesis testing process. The specific research hypothesis used in each test determines the meaning associated with each of the terms above, as well as how they are calculated.

Reported	True			
	changed	unchanged		
changed	TP	FP	Positive predicted value	$\frac{TP}{TP+FP}$
unchanged	FN	TN	Negative predicted value	$\frac{TN}{TN+FN}$
	Sensitivity $\frac{TP}{TP+FN}$	Specificity $\frac{TN}{TN+FP}$	Accuracy $\frac{TP+TN}{TP+TN+FP+FN}$	

TABLE 21.1: The definitions of positive predicted value (PPV), negative predicted value (NPV), specificity, sensitivity, and accuracy. A perfect method will produce PPV = NPV = specificity = sensitivity = accuracy = 1. Note that it is possible to obtain perfect individual values for sensitivity, specificity, etc. in trivial ways (see discussion in Chapter 8, Section 8.7.1).

21.3   Fold change

21.3.1   Description

The simplest and most intuitive approach to finding the genes that are differentially regulated is to consider their fold change between control and experiment. Typically, an arbitrary threshold such as two- or three-fold is chosen and the difference is considered as significant if it is larger than the threshold [239, 114, 113, 421, 409, 455, 459]. Sometimes, this selection method is used in parallel on expression estimates provided by several techniques such as radioactive and fluorescent labeling [354]. A convenient way to select by fold change is to calculate the ratio between the two expression levels for each gene. This is more traditional for the cDNA data but can be easily done for Affymetrix data, as well. Such ratios can be plotted as a histogram (Fig. 21.1).

As discussed previously, in a screening experiment involving many genes, most genes will not change. Thus, the experiment/control ratio of most genes will be grouped around 1, which means that their logs will be grouped around 0. The horizontal axis of such a plot represents the log ratio values. In consequence, selecting differentially regulated genes can be simply done by setting thresholds on this axis and selecting the genes outside such thresholds. For instance, in order to select genes that have a fold change of 4 and assuming that the log has been taken in base 2, one would set the thresholds at $+/-2$. Note the strong resemblance between this figure and Fig. 11.5 in Chapter 11. This indicates already that the process used by the fold selection method is in principle the same as that used in a classical hypothesis testing situation. The difference is that in a hypothesis testing situation, the thresholds are chosen very precisely in order to control the probability of the Type I error (calling a

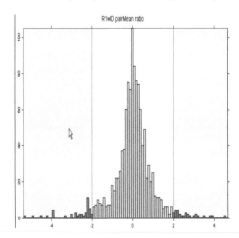

FIGURE 21.1: Fold change on a histogram. Experiment-control ratios can be plotted as a histogram showing the number of genes (vertical axis) for every ratio value (horizontal axis). The horizontal axis is graded in fold-change units. Selecting differentially regulated genes based on fold change corresponds to selecting the genes in the tails of the histogram (blue areas) by setting thresholds at the desired minimum fold change. Note the resemblance of this figure with Fig. 11.5 in Chapter 11.

gene differentially regulated by mistake) while in the fold change method, the thresholds are chosen arbitrarily.

If the log expression levels in the experiment are plotted against the log expression levels in control in a scatter plot (left panel in Fig. 21.2), the genes selected will be at a distance of at least 2 from the diagonal that corresponds to the expression being the same in control and experiment. The fold change method reduces to drawing lines parallel to the diagonal at a distance corresponding to the chosen threshold and selecting the genes outside the central area defined by these thresholds. In a ratio-intensity plot, the reference line corresponding to unchanged genes will be the horizontal line $y = 0$ (right panel in Fig. 21.2). Similarly, the method works by establishing selection boundaries parallel to the reference line and selecting the genes outside the boundaries, away from the reference.

21.3.2 Characteristics

The fold change method is often used because it is simple and intuitive. However, the method has important disadvantages. The most important drawback is that the fold threshold is chosen arbitrarily and may often be inappropriate. For instance, if one is selecting genes with at least two-fold change and the condition under study does not affect any genes to the point of inducing a

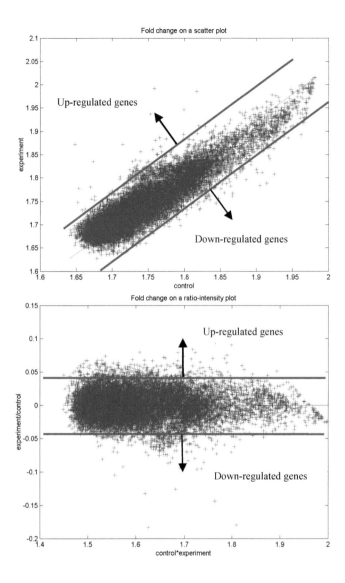

FIGURE 21.2: Fold change on a scatter plot of log values (left) and ratio-intensity plot (right). The fold method reduces to drawing two linear selection boundaries at a constant distance from the *exp = control* line and selecting the genes as shown. The line *exp = control* is the diagonal $y = x$ in a scatter plot and the line $y = 0$ in a ratio-intensity plot.

two-fold change, no genes will be selected resulting in zero sensitivity. Reciprocally, if the condition is such that many genes change dramatically (or if a lower threshold is chosen), the method will select too many genes and will have a low specificity. In this respect, the fold-change method is nothing but a blind guess.

Another important disadvantage is related to the fact that the microarray technology tends to have a bad signal/noise ratio for genes with low expression levels. On a scatter plot and ratio-intensity plot, this is illustrated by a funnel shape of the distribution. This is caused by a larger variance of the values measured at the low end of the scale (to the left) and a low variance of values measured at the high end of the scale (to the right) and by certain numerical phenomena discussed in Chapter 17 (see Fig. 17.8). A gene that is closer to the diagonal at a high expression level might be more reliable than a gene that is a bit farther from the diagonal at a low level. Since the fold change uses a constant threshold for all genes, it will introduce false positives at the low end, thus reducing the specificity, while missing true positives at the high end and thus reducing the sensitivity.

A recent *PNAS* paper studying the effect of the threshold choice [333] showed that: i) the biological conclusion drawn from a specific microarray experiment may or may not be supported by the data, depending on the choice of the threshold, and ii) the functional categories significantly represented in the set of differentially expressed genes are also dramatically affected by the choice of the threshold.

21.4 Unusual ratio

21.4.1 Description

The second widely used selection method involves selecting the genes for which the ratio of the experiment and control values is a certain distance from the mean experiment/control ratio [415, 374, 375]. Typically, this distance is taken to be $\pm 2\sigma$, where σ is the standard deviation of the ratio distribution. In other words, the genes selected as being differentially regulated will be those genes having an experiment/control ratio at least 2σ away from the mean experiment/control ratio. In practice, this can be achieved very simply by applying a z-transform to the log ratio values. The z-transform essentially subtracts the mean and divides by the standard deviation. In consequence, a histogram of the transformed values will still be centered around 0 (most genes will have a ratio close to the mean ratio), but the units on the horizontal axis will represent standard deviation (see Fig. 21.3). Thus, setting thresholds at $+/-2$ will correspond to selecting those genes which have an unusual ratio, situated at least 2 standard deviations away from the mean ratio.

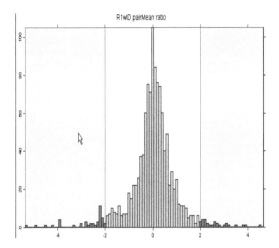

FIGURE 21.3: Selecting by the unusual ratio criterion. The frequencies (vertical axis) of the z-scores of the genes (horizontal axis) are plotted in a histogram. The horizontal axis is now graded in standard deviations. Setting thresholds at the $+/-2$ marks corresponds to selecting genes with a ratio more than 2 standard deviations away from the mean log ratio.

21.4.2 Characteristics

This method is superior to the fold-change method while still simple and intuitive. The advantage of the unusual ratio method is that it will automatically adjust the cut-off threshold even if the number of genes regulated and the amount of regulation vary considerably. Thus, the unusual ratio method uses thresholds on how different the experiment/control ratio of a gene is with respect to the mean of all such ratios instead of thresholds on the values of the ratios themselves. No matter how many genes are regulated and no matter by how much, this method will always pick the genes that are affected most. In particular, as we have seen, if the ratio distribution is close to a normal distribution and the thresholds are set at $+/-2$ standard deviations, this method will select the 5% most regulated genes. This is because the probability of having a Z lower than -2 is $P(Z < -2) = 0.0228$, whereas the probability of having a Z value higher than 2 is $P(Z > 2) = 1 - P(Z < 2) = 1 - 0.9772 = 0.0228$. Joining the two cases, the probability of having a Z value more extreme than ± 2 is $0.0228 + 0.0228 = 0.0456$.

However, the unusual ratio method still has important intrinsic drawbacks. Thus, the method will report 5% of the genes as differentially regulated *even if there are no differentially regulated genes*. This happens because in all microarray experiments there is a certain amount of variability due to noise. Thus, if the same experiment is performed twice, the expression values measured for any particular gene will likely not be exactly the same. If the method is applied to study differential regulated genes in two control experiments, the unusual ratio method will still select about 5% of the genes and report them as "differentially" regulated. This is because different measurements for the same gene will still vary a little bit due to the noise. The method will dutifully calculate the mean and standard deviation of this distribution and will select those genes situated +/-2 standard deviations away from the mean. In this case, the null hypothesis *is true* and the method will still reject (incorrectly) the null hypothesis 5% of the time, i.e. will report 5% of the genes as being differentially regulated.

Furthermore, the method will still select 5% of the genes *even if many more genes are in fact regulated*. Thus, while the fold method uses an arbitrary threshold and can provide too many or too few genes, the unusual ratio method uses a fixed proportion threshold that will always report the same proportion of the genes as being differentially regulated. On a scatter plot (such as the one in Fig. 21.2), the ratio method continues to use cutoff boundaries parallel to the diagonal, which will continue to overestimate the regulation at low intensity and underestimate it at high intensity.

A variation of the unusual ratio method selects those genes for which the absolute difference in the average expression intensities is much larger than the estimated standard error ($\hat{\sigma}$) computed for each gene using array replicates. For duplicate experiments, the absolute difference has to be larger than $4.3\hat{\sigma}$ and $22.3\hat{\sigma}$ for the 5% and 1% significance levels, respectively [99]. For triplicate experiments, the requirements can be relaxed to $2.8\hat{\sigma}$ and $5.2\hat{\sigma}$ for the 5% and 1% significance levels, respectively.

A number of other *ad hoc* thresholding and selection procedures have also been used that are equivalent to a selection based on the unusual ratio method. For instance, [374, 375] only considered genes for which the difference between the duplicate measurements did not exceed half their average. Furthermore, the genes considered as differentially regulated were those genes that exhibited at least a two-fold change in expression. Although this criterion seems to use the fold method, it can be shown [99] that the combination of the duplicate consistency condition and the differentially regulated condition can be expressed in terms of mean and standard deviations, and therefore, it falls under the scope of the unusual ratio method.

21.5 Hypothesis testing, corrections for multiple comparisons, and resampling

21.5.1 Description

Another possible approach to gene selection is to use univariate statistical tests (e.g., t-test) to select differentially expressed genes [27, 99, 138]. This ap-

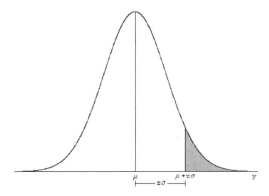

FIGURE 21.4: For a given threshold and a given distribution the *p*-value is the probability of the measured value being in the shaded area by chance. If the method is applied to the distribution of the log ratios, the *p*-value is the probability of making a mistake when calling a gene differentially regulated if its measured log ratio is in the shaded area.

proach essentially uses the classical hypothesis testing methodology discussed in Chapter 11 in conjunction with some correction for multiple comparisons discussed in Chapter 16. This approach will be briefly reviewed here.

Let us consider that the log ratios follow a distribution like the one illustrated in Fig. 21.4. For a given threshold and a given distribution, the confidence level or *p*-value is the probability that the measured value is in the shaded area just by chance. The idea is that a gene whose log ratio falls in the shaded area is far from the mean log ratio and will be called differentially regulated (up-regulated in this case). However, the measured log ratio may be there just due to random factors such as noise. The probability of the measurement being there just by chance is the *p*-value. In this case, calling the gene differentially regulated will be a mistake (Type I error), and the *p*-value is the probability of making this mistake.

Regardless of the particular test used (e.g. *t*-test if normality is assumed), one needs to consider the fact that when many genes are analyzed at one time, some genes will appear as being significantly different just by chance [217, 211, 381, 382, 457]. In the context of the selection of differentially regulated genes, a Type I error (or rejecting a true null hypothesis) is manifested as reporting an unregulated gene as differentially regulated (false positive). In an individual test situation (the analysis of a single gene), the probability of a Type I error is controlled by the significance level chosen by the user. On a high-density array containing many genes, the significance level at gene level does not control the overall probability of making a Type I error anymore. The overall probability of making at least one mistake, or family-wise significance level, can be calculated from the gene-level significance level using several approaches discussed in Chapter 16. Bonferroni [60, 61] and Šidák corrections [439] are quite simple but very conservative. For arrays involving thousands or tens of thousands of genes, these methods are not practical. The Holm step-down group of methods [217, 381, 215, 211], false discovery rate (FDR) [45, 46], permutations [457], and significance analysis of microarray (SAM) [427] are all suitable methods for multiple comparison corrections in the context of microarray data. SAM in particular is more than a method for multiple comparison correction and can probably be considered a gene selection method on its own.

A univariate testing procedure (e.g. *t*-test or Wilcoxon) followed by a Westfall and Young adjustment for multiple testing [457] has been used by Dudoit et al. [138].

21.5.2 Characteristics

A drawback of the methods based on hypothesis testing is that they tend to be a bit conservative. As discussed in Chapter 11, not being able to reject a null hypothesis and call a gene differentially regulated does not necessarily mean that the gene is not so. In many cases, it is just that insufficient data do not provide adequate statistical proof to reject the null hypothesis. However, those genes that are found to be differentially regulated using such methods will most likely be so.

The classical hypothesis testing approach also has the disadvantage of assuming that the genes are independent, which is clearly untrue in the analysis of any real data set. Fortunately, combining a classical hypothesis testing approach with a re-sampling or bootstrapping approach (e.g., in step-down or SAM) tends to lose the conservative tendencies and also takes into consideration dependencies among genes. If the experiment design and the amount of data available allows it, the use of these methods is recommended.

21.6   ANOVA

21.6.1   Description

A particularly interesting approach to microarray data analysis and selecting differentially regulated genes is the ANalysis Of VAriance (ANOVA) [11, 69, 209]. This was discussed in detail in Chapter 13 but will be reviewed briefly here. The idea behind ANOVA is to build an explicit model of all sources of variance that affect the measurements and use the data to estimate the variance of each individual variable in the model.

For instance, Kerr and Churchill [259, 258, 257] proposed the following model to account for the multiple sources of variation in a microarray experiment:

$$\log\left(y_{ijkg}\right) = \mu + A_i + D_j + G_g + (AD)_{ij} + (AG)_{ig} + (VG)_{kg} + (DG)_{jg} + \varepsilon_{ijkg} \quad (21.6)$$

In this model, μ is the overall mean signal of the array, A_i is the effect of the i-th array, D_j represents the effect of the j-th dye, G_g is the variation of the g-th gene, $(AD)_{ij}$ is the effect of the array-dye interaction, $(AG)_{ig}$ is the effect of a particular spot on a given array (array-gene interaction), $(VG)_{kg}$ represents the interaction between the k-th variety and the g-th gene, $(DG)_{jg}$ is the effect of the dye-gene interaction and ε_{ijkg} represents the error term for array i, dye j, variety k, and gene g. In this context, a variety is a condition such as healthy or disease. The error is assumed to be independent and of zero mean. Finally, $\log\left(y_{ijkg}\right)$ is the measured log-ratio for gene g of variety j measured on array i using dye j.

Sums of squares are calculated for each of the factors above. Then mean squares will be obtained by dividing each sum of squares by its degrees of freedom. This will produce a variance-like quantity for which an expected value can be calculated if the null hypothesis is true. These mean squares can be compared the same way variances can: using their ratio and an F distribution. Each ratio will have the MS_E as the denominator. Essentially, each individual test asks the question whether a certain component MS_x has the variance significantly different from the variance of the noise MS_E. The differentially regulated genes will be the genes for which the $(VG)_{kg}$ factor representing the interaction between the variety and gene is significant.

21.6.2 Characteristics

The advantage of ANOVA is that each source of variance is accounted for. Because of this, it is easy to distinguish between interesting variations such as gene regulation and side effects such as differences due to different dyes or arrays. The caveat is that ANOVA requires a very careful experiment design [258, 257] that must ensure a sufficient number of degrees of freedom. Thus, ANOVA cannot be used unless the experiments have been designed and executed in a manner consistent with the ANOVA model used.

21.7 Noise sampling

21.7.1 Description

A full-blown ANOVA requires a design that blocks all variables under control and randomizes the others. In most cases, this requires repeating several microarrays with various mRNA samples and swapping dyes if a multichannel technology is used. A particular variation on the ANOVA idea can be used to identify differentially regulated genes using spot replicates on single chips to estimate the noise and calculate confidence levels for gene regulation. The noise sampling method [123, 131, 132] modifies the Kerr-Churchill model as follows:

$$\log R(gs) = \mu + G(g) + \varepsilon(g,s) \tag{21.7}$$

where $\log R(gs)$ is the measured log ratio for gene g and spot s, μ is the average log ratio over the whole array, $G(g)$ is a term for the differential regulation of gene g, and $\varepsilon(g,s)$ is a zero-mean noise term.

In the model above, one can calculate an estimate $\hat{\mu}$ of the average log ratio μ:

$$\hat{\mu} = \frac{1}{n \cdot m} \sum_{g,s} \log (R(g,s)) \tag{21.8}$$

which is the sum of the log ratios for all genes and all spots divided by the total number of spots (m replicates and n genes). An estimate $\widehat{G(g)}$ of the effect of gene g can also be calculated as:

$$\widehat{G(g)} = \frac{1}{m} \sum_{g} \log (R(g,s)) - \hat{\mu} \tag{21.9}$$

where the first term is the average log ratio over the spots corresponding to the given gene. Using the estimates above, one can now calculate an estimate of the noise as follows:

$$\widehat{\varepsilon(g,s)} = \log (R(g,s)) - \hat{\mu} - \widehat{G(g)} \tag{21.10}$$

For each spot (g,s), this equation calculates the noise as the difference between the actual measured value $\log (R(g,s))$ and the estimated array effect $\hat{\mu}$ plus the gene effect $\widehat{G(g)}$. This will provide a noise sample for each spot. The samples collected from all spots yield an empirical noise distribution.[2] A given confidence level can be associated with a deviation from the mean of this distribution. To avoid using any particular model, the distance from the mean can be calculated by numerically integrating the area under the distribution. This distance on the noise distribution can be put into correspondence to a distance y on the measured distribution (Fig. 21.5) by bootstrapping [132, 457]. Furthermore, the dependency between intensity and variance can be taken into account by constructing several such models covering the entire intensity range and constructing nonlinear confidence boundaries similar to those in Fig. 21.6.

---

[2]Note that no particular shape (such as Gaussian) is assumed for the noise distribution or for the distribution of the gene expression values, which makes this approach very general.

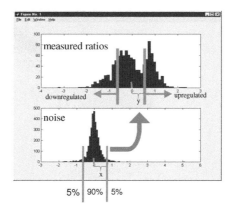

FIGURE 21.5: Using the empirical noise distribution to find differentially expressed genes. The confidence intervals found on the noise distribution (lower panel) can be mapped onto confidence intervals on the distribution of the expression values (upper panel). The confidence interval can be different between the up and down-regulated parts of the distribution if the noise is skewed.

21.7.2   Characteristics

The noise sampling method has the important advantage that its nonlinear selection boundaries adapt automatically both to various amounts of regulation and different amounts of noise for a given confidence level chosen by the user. Fig. 21.6 illustrates the difference between the selection performed by the fold-change method (1) and the selection performed by the noise sampling method (2). Using the nonlinear boundaries (2) takes into consideration the fact that the variance changes with the intensity (lower variance at higher intensities and higher variance at low intensities).

A full-blown ANOVA requires a special experimental design but provide error estimates for all variables considered in the model. The noise sampling method does not require such a special experiment design but only provides estimates for the log ratios of the genes (see Section 21.16.1 in Appendix). It has been shown that the noise sampling method provides a better sensitivity than the unusual ratio method and a much better specificity than the fold change method [132]. The method can be used equally well for Affymetrix data as well as for experiments involving several arrays.

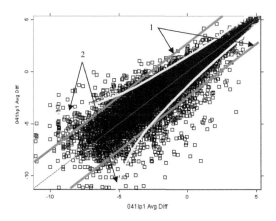

FIGURE 21.6: A scatter plot representing the experiment values plotted against the control values. Unchanged genes will appear on the diagonal as the two values are similar. Selecting genes with a minimum fold change is equivalent to setting linear boundaries parallel to the $y = x$ diagonal at a distance from it equal to the minimum fold change requested (1). The noise sampling method performs a better selection by using nonlinear boundaries (2) that adapt to the increased noise variance at low intensities.

21.8 Model-based maximum likelihood estimation methods

In order to explain this approach, we will consider a simpler problem. Instead of considering two conditions and try to identify genes that are up-regulated, down-regulated, and unchanged in the experiment versus the control, we will consider a single array and will use this approach in order to distinguish the genes that are expressed from the genes that are not expressed [276]. The description of the method will assume the use of cDNA arrays using j replicate spots for each gene. However, this approach is by no means limited to this type of array or this specific problem [89, 172, 368, 276].

21.8.1 Description

We will consider two events. $\mathscr{E}_g$ will represent the event that mRNA for gene g in the array is contained in the target sample tissue, i.e. the gene is expressed. $\overline{\mathscr{E}_g}$ will represent the complement of $\mathscr{E}_g$, i.e. the event in which the gene in unexpressed. Let us denote by p the *a priori* probability of a gene being expressed. This is in fact equal to the fraction of expressed genes present in the mRNA sample. Since there are only two possibilities, either the gene is expressed or not, the probability of the gene being unexpressed is $1 - p$.

The method considers the probability density function (pdf) f_{E_j} associated with the expressed genes and the pdf f_{U_j} associated with the unexpressed genes. The probability density functions will describe the probability of measuring a given intensity value[3] y for a gene. If the gene is expressed, the intensity y will follow f_{E_j} whereas if the gene is not expressed, the intensity y will follow f_{U_j}. This is illustrated in Fig. 21.7. The horizontal axis represents intensity values; the vertical axis represents the probability. The graph to the left represents f_{U_j}. The graph to the right represents f_{E_j}. The placement of the two graphs is in line with our expectations: the maximum probability (the peak of the pdf) for unexpressed genes is at a lower intensity than the intensity corresponding to the maximum probability for the expressed genes. In other words, most of spots corresponding to unexpressed genes will be characterized by an intensity lower than that of most spots corresponding to expressed genes. The intersection of the two graphs corresponds to those occasional spots that belong to expressed genes but have intensities lower than those of some unexpressed genes and, conversely, spots that belong to unexpressed genes that happen to have higher intensities than some of those expressed genes (area A in Fig. 21.7). The two distributions can be combined in the following statistical model:

$$f_j(y) = p \cdot f_{E_j}(y) + (1 - p) \cdot f_{U_j}(y) \tag{21.11}$$

where $f_j(y)$ is the pdf of observed intensity value y. This is **a mixture model** for the distribution of observed log ratios. This is called a mixture model because the observed intensity values are assumed to be distributed as a mixture of two distributions.

Equation 21.11 is straightforward to interpret. The first term in Eq. 21.11 corresponds to the case in which the observed intensity comes from an expressed gene. This will happen with probability p and in this case, the observed intensity will follow the pdf of an expressed gene, which is $f_{E_j}(y)$. The second term in Eq. 21.11 corresponds to the case in which the observed intensity comes from an unexpressed gene. This will happen with probability $1 - p$ and in this case, the observed intensity will follow the pdf of an unexpressed gene, which is $f_{U_j}(y)$. A gene is either expressed or not expressed so the probabilities are added together.

The goal of the approach is to decide whether a given spot j corresponds to an expressed or unexpressed gene. In other words, given that the measured intensity Y_{gj} of the gene g on spot j has the value y, we are to calculate the probability of gene g being expressed at spot j.[4] This is exactly the definition of the conditional probability (see Chapter 8):

---

[3] More precisely, the value is the log ratio of the background corrected mean of the pixels in a given spot. See Chapter 20 for a more complete discussion of the preprocessing and normalization.

[4] It may seem intuitive that if a gene g is expressed this will happen for all its spots.

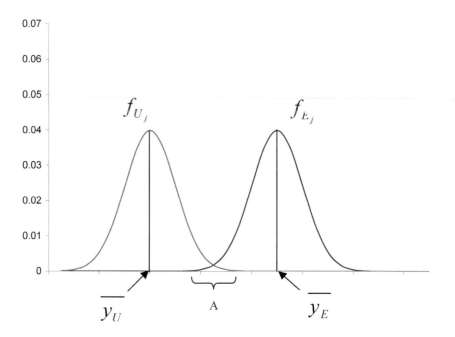

FIGURE 21.7: A mixture model is the result of superposing two distributions. The horizontal axis represents the values of the observed intensity y. The vertical axis represents the probability of observing a given y value. In this case, the distribution to the left corresponds to the unexpressed genes and the distribution to the right corresponds to the expressed genes. Note that most unexpressed genes will have an observed intensity around $\bar{y}_U$ while most expressed genes will have an observed intensity around $\bar{y}_E$. In region A, the two distribution overlap.

$$Pr\{\mathscr{E}_g|Y_{gj}=y\} \tag{21.12}$$

From Bayes theorem (see Chapter 8), this can be expressed as:

$$Pr\{\mathscr{E}_g|Y_{gj}=y\} = \frac{p \cdot f_{E_j}(y)}{f_j(y)} \tag{21.13}$$

Equation 21.13 states that the probability of the gene being expressed given that the observed intensity is y is the probability of a gene being expressed, p, times the probability that an expressed gene has the intensity y, and divided by the probability of observing the intensity y.

If we knew the exact expressions of p, $f_{E_j}(y)$ and $f_{U_j}(y)$ we could use Eq. 21.11 to calculate $f_j(y)$, and then Eq. 21.13 to calculate the probability of a gene being expressed from its observed intensity values.

In order to obtain $f_{E_j}(y)$ and $f_{U_j}(y)$, we can just assume that they are normal distributions. In doing so, the model becomes **a mixed normal probability density function**. In this case, each distribution is completely described by only two values: a mean μ and a variance σ^2 (see Chapter 9 and in particular Eq. 9.36 for the analytical expression of the normal distribution). The values p, μ_{E_j}, $\sigma^2_{E_j}$, μ_{U_j}, and $\sigma^2_{U_j}$ can be estimated numerically, using **a maximum likelihood** approach. The maximum likelihood method searches various combinations of the parameters until the obtained equation fits the data as best as possible. Once obtained, the maximum likelihood estimates are used in conjunction with Eq. 21.11 and Eq. 21.13 to calculate the probability of a gene being expressed from its observed intensity values.

This approach has been used in conjunction with a special set of 288 genes, of which 32 genes were expected to appear highly expressed, in order to study the importance of replication in cDNA microarray gene expression studies [276]. Notably, this work proved several important facts:

1. Any individual spot can provide erroneous results. However, the probability of three or more spots being consistently wrong is negligible.

2. The intensities do seem to be normally distributed. This indicates that a mixed normal probability model may be useful in the analysis of other experiments.

3. The probability of a false negative (an expressed gene that either fails to be represented as probe or, if it is represented as probe, fails to be hybridized to the cDNA that are deposited on the slide) is as high as 5% for any single replicate.

4. The probability of a false positive (an unexpressed gene for which the fluorescence intensity suggests that the gene is expressed) is as high as 10% for any single replicate.

However, as we will see, there is a non-zero probability that an expressed gene will not appear as such when examining a single spot.

21.8.2 Characteristics

This approach is very general and very powerful. Advantages include the fact that the maximum likelihood estimators (MLE) become unbiased minimum variance estimators as the sample size increases and that the likelihood functions can be used to test hypotheses about models and parameters.

There are also several disadvantages. The first one is that they require solving complex nonlinear equations. This is very computationally intensive and requires specialized software. Furthermore, a more realistic situation involving two or more experiments in which the genes have to be divided into up-regulated, unchanged and down-regulated requires a more complex mixture model:

$$f_j(y) = p_1 \cdot f_{UP_j}(y) + p_2 \cdot f_{DOWN_j}(y) + p_3 \cdot f_{UNCHANGED_j}(y) \qquad (21.14)$$

involving three a priori probabilities p_1, p_2, and p_3 corresponding to the up-regulated genes, down-regulated genes, and unchanged genes, respectively. Also, the computation of the MLE becomes more complex due to the increased number of parameters.

Finally, another disadvantage of the maximum likelihood estimate approach is that the results quickly become unreliable as the sample size decreases. Indeed, for small samples, the estimates may be very much different from the real underlying distribution. Furthermore, MLE estimates can become unreliable when data deviates considerably from normality.

21.9 Affymetrix comparison calls

The comparison calls calculated by the Affymetrix software are designed to answer the question whether a particular transcript on one array is changed significantly with respect to another array. Usually, the two arrays are called the experiment array and the baseline array. The calls are categorical classifications into one of the following classes: "Decreased," "Marginally Decreased," "No Change," "Marginally Increased," and "Increased." As in hypothesis testing, "No Change" does not actually mean that the gene has not changed but rather that the amount of change in the context of the measured variance does not allow a rejection of the null hypothesis that the gene is unchanged. In other words, the data are not sufficient to prove that the gene has changed, which may not necessarily mean that the gene has not changed.

If all probe pairs of a probe set are saturated (PM or MM ≥ 46000), the software will report that no comparative calls can be made. If any one of the four cells, PM and MM in baseline and PM and MM in experiment, are saturated the corresponding probe pair is not used in further computations.

The remaining probe pairs are used to form two vectors. The first vector will contain the differences between PM and MM:

$$q = (PM_1 - MM_1, PM_2 - MM_2, \cdots, PM_n - MM_n) \qquad (21.15)$$

The second vector will contain the differences between the PM and the background level computed as described in Sec. 20.4.1:

$$z = (PM_1 - b_1, PM_2 - b_2, \cdots, PM_n - b_n) \qquad (21.16)$$

However, the distribution of q and z values over all probe pairs on a given array are different from each other. Two balancing factors are used to match these two distributions [10]. Finally, a Wilcoxon signed rank test is used to test a two-tailed set of hypotheses on the median of the adjusted differences between the vectors q and z in the baseline and experiment, respectively:

$$f_1 \cdot q_E[i] - q_B[i] \qquad (21.17)$$

and

$$C \cdot (f_2 z_E[i] - z_B[i]) \qquad (21.18)$$

where f_1 and f_2 are empirically determined balancing factors and C is a scaling constant ($C = 0.2$ by default).

The hypotheses tested are:

$$H_0: \quad median(v) = 0$$
$$H_a: \quad median(v) > 0$$

The cutoff values γ_1 and γ_2 for the p-values are adjusted for multiple comparisons using a Bonferroni-like dividing factor only that the exact value of the factor is determined empirically. It is important to note that the p-values produced by this approach are actually overestimates of the probability of a Type I error because the two values used $PM - MM$ and $PM_i - b_i$ are not independent. However, this is acceptable because we only seek an ordering of the values and the thresholds between increase and marginally increase and decrease and marginally decrease, respectively, are more or less arbitrary.

21.10   Significance Analysis of Microarrays (SAM)

A simple t statistic/t-test is not optimal when the number of samples is small. With only three samples in each group, the estimation of the variance in the calculation of t statistic can be highly unstable. Two popular methods have been proposed to address such issue, namely, *SAM* and a *moderated t-statistic* using empirical Bayes.

As explained in more detail in Section 16.4.6, SAM uses a statistic called "relative difference" d_i for gene i. This is very similar to a t statistic with equal variance, except that the "gene-specific scatter" (the standard deviation of the difference) s_i in the denominator is offset by a "fudge factor" s_0. This s_0 is an "exchangeability" factor chosen to minimize the coefficient of variation of d_i, and is computed with a sliding window approach across the data (quantiles of s_i). This small positive exchangeability factor stabilizes the d_i values for genes with low expression levels, genes that can have a denominator close to zero in Equation 16.5. A permutation test is then used to assess significance of d_i, as well as estimate the FDR. More details can be found in the SAM manual available at `http://www-stat.stanford.edu/~tibs/SAM/`.

21.11   A moderated t-statistic

Although offsetting the standard deviation in the denominator of t statistics in SAM makes it more robust to noisy genes with low expression level, this is a nonparametric method. Furthermore, the offset in the denominator is universal to all genes rather than gene specific. Smyth et.al [396, 397] took a different approach and modified the ordinary t statistic by replacing the usual standard deviation with a posterior residual standard deviation, thus creating a "**moderated t statistic**." The model was set up in the context of the general linear models with an arbitrary number of treatment conditions and RNA samples. The Bayes meta-parameters in the model were estimated from data, which justified the name of "empirical Bayes."

Let us assume we have a set of n microarrays with an expression vector (log-intensities for single-channel data; log-ratios for two channels) $y_g^T = (y_{g1}, \cdots, y_{gn})$. The general linear model is assumed as: $E(y_g) = X\alpha_g$, with $var(y_g) = W_g\sigma_g^2$, where X is a design matrix, α_g is a coefficient vector, and W_g is a known weight matrix. Arbitrary contrasts of biological interest β_g can be extracted from the coefficient vector with a contrast matrix C, i.e. $\beta_g = C^T\alpha_g$.

Fitting the linear model to data will generate estimators $\hat{\alpha}_g$ of α_g, s_g^2 of σ_g^2, and $var(\hat{\alpha}_g)$. The estimators of contrast $\hat{\beta}_g$ and its variance estimators $var(\hat{\beta}_g)$ can be derived from above linear model estimators using $\beta_g = C^T\alpha_g$. Two assumptions about the underlying distributions are made here: i) that the contrast estimators $\hat{\beta}_g$ are normally distributed; and ii) that the residual variances s_g^2 follow a scaled chi-square distribution.

At this point, we can derive the ordinary t statistic for the contrast of interest β_{gj} (that is the j-th contrast for g-th gene) through the contrast estimators $\hat{\beta}_g$ and its variance estimators. The null hypotheses $H_0 : \beta_{gj} = 0$ can be tested in the usual context of hypothesis testing.

The above process of linear model estimation is in essence a gene-wise model fitting ignoring the parallel structure of dependent gene expression. A hierarchical Bayes' model is therefore set up to take advantage of such information in the assessment of differential expression. This can be done through specifying prior distributions for the unknown parameters β_{gj} and σ_g^2 in the above linear model. The meta-parameters introduced in the prior distributions can be estimated from data. The posterior residual standard deviation $\hat{s}_g^2$ can be derived from the above models, and the moderated t statistic with s_g^2 replaced by $\hat{s}_g^2$ can be shown to follow a t distribution under null hypothesis.

21.12 Other methods

Another maximum likelihood estimation approach for two color arrays is described in [89]. This approach is based on the hypothesis that the level of a transcript depends on the concentration of the factors driving its selection and that the variation for any particular transcript is normally distributed and in a constant proportion relative to most other transcripts. This hypothesis is then exploited by considering a constant coefficient of variation c for the entire gene set and constructing a 3rd degree polynomial approximation of the confidence interval as a function of the coefficient of variation c. This approach is also interesting because it provides the means to deal with signals that are uncalibrated between the two colors by using an iterative algorithm that compensates for the color difference.

Sapir et al. [368] present a robust algorithm for estimating the posterior probability of differential expression based on an orthogonal linear regression of the signals obtained from the two channels. The residuals from the regression are modelled as a mixture of a common component and component due to differential expression. An expectation maximization algorithm is used to deconvolve the mixture and provide estimates of the probability that each gene is differentially regulated as well as estimates of the error variance and proportion of differentially expressed genes.

Two hierarchical models (Gamma-Gamma and Gamma-Gamma-Bernoulli) for the two-channel (color) intensities are proposed in [321]. One advantage of such an approach is that the models constructed take into consideration the variation of the posterior probability of change on the absolute intensity level at which the gene is expressed. This particular dependency is also considered in [355], where the values measured on the two channels are assumed to be normally distributed with a variance depending on the mean. Such intensity dependency reduces to defining some curves in the green-red plane corresponding to the two channels and selecting as differentially regulated the genes that fall outside the equiconfidence curves (see Fig. 21.6).

Another multiple experiment approach is to identify the differentially expressed genes by comparing their behavior in a series of experiments with an expected expression profile [183, 166]. The genes can be ranked according to their degree of similarity to a given expression profile. The number of false positives can be controlled through random permutations that allow the computation of suitable cutoff thresholds for the degree of similarity. Clearly, these approaches can only be used in the context of a large data set, including several microarrays for each condition considered.

Other methods used for the selection of differentially regulated genes include gene shaving [196], assigning gene confidence [297] or significance [427], bootstrap [155, 257] and Bayesian approaches [31, 291, 456].

Finally, more elaborate methods for the data analysis of gene expression data exist. Such methods include singular value decomposition [17], independent component analysis [283], and many others. The goals of such methods go well beyond the selection of differentially regulated genes and, as such, they are outside the scope of this chapter.

21.13   Reproducibility

A natural question is: how reproducible are the lists of DE genes produced by various methods? In other words, if we were to take one sample, divide it into two aliquotes, and send them to two different laboratories to be hybridized with the same protocols and analyzed with the same methods, what sort of overlap will we get? The answer to this exact question was provided by the MicroArray Quality Assessment project [392]. Figure 21.8 shows that the results are perhaps not as good as expected if only a relatively small (<50) number of DE genes are to be reported. In fact, the top 10 DE genes overlapped only up to 75% even when the most lenient methods (fold change) were used. If the number of genes selected as DE is increased to 100, the overlap also increased to about 80–85%. Finally, even if the number of DE genes was increased to 10,000, the overlap never reached 100%. In other words, one should remember that in any list of reasonable length (e.g., fewer than 500), at least 10% of the genes are probably not reproducible.

The fact that the *p*-values alone provide a disappointingly low overlap does not necessarily mean that the genes that do not overlap are false positives. More likely, the problem is due to the ranking of the genes, which is very sensitive to the specific *p*-values calculated, which in turn are very sensitive to noise. One of the ways to address this is to select DE genes using criteria based on both *p*-values and fold changes, as in volcano plots. This dual-fold approach tends to eliminate genes that have artifactually low *p*-values associated with low fold changes, as well as genes that have high fold changes but associated

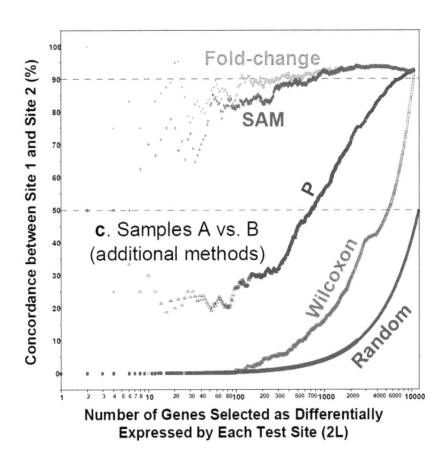

FIGURE 21.8: Some results from the MAQC-I project comparing two samples: A= Universal Human Reference RNA from Stratagene and B= Human Brain Reference RNA from Ambion. Note that if only 50 genes are to be selected as differentially expressed (DE), the overlap between the list of DE genes obtained at two sites ranges from zero (Wilcoxon), to 20% (various p-values, to 70% (SAM), and to about 80% (fold change). Note that these are experiments performed on exactly the same samples.

with high (insignificant) *p*-values. Both will increase the amount of overlap and therefore the reproducibility of the results of this analysis.

21.14   Selecting differentially expressed (DE) genes in R

In this section, we will illustrate how most of the methods discussed above from a theoretical perspective can be actually implemented in R and applied to real data. Many of these methods (e.g., fold change, unusual ratio, etc.) are not good methods and should never be used in practice, for reasons that have been extensively discussed above. Here, will focus on merely illustrating these methods through their R implementation without dwelling again on why most of these methods should not be used in practice.

21.14.1   Data import and preprocessing

As discussed in the chapters on normalization and quality control, the selection of DE genes must be performed on normalized data. In principle, any microarray data set can be stored as a summarized and normalized expression matrix with columns corresponding to arrays and rows corresponding to probes or genes. As we have seen in Chapter20, this data matrix can be stored as objects of class "ExpressionSet" for Affymetrix array or "ExpressionSetIllumina" for illumina array. Here, we will use the data set used to illustrate the normalization procedure for Affymetrix data.

```
> library(affy)
> datDir<-"./Normalization/GSE9412"
> CELfiles<-list.celfiles(path=datDir)
> rawdata<-ReadAffy(filenames=CELfiles, celfile.path=datDir)
> rawdata

AffyBatch object
size of arrays=732x732 features (11 kb)
cdf=HG-U133A_2 (22277 affyids)
number of samples=6
number of genes=22277
annotation=hgu133a2
notes=

> eset<-rma(rawdata)

Background correcting
Normalizing
Calculating Expression
```

```
> eDat<-exprs(eset) #expression matrix
> head(eDat)
```

	GSM239309.CEL	GSM239310.CEL	GSM239311.CEL
1007_s_at	10.398234	10.454271	10.701278
1053_at	8.061821	8.174187	8.047657
117_at	3.969766	4.173768	4.222711
121_at	7.979312	7.955708	8.117934
1255_g_at	2.759197	2.489639	2.702799
1294_at	6.302492	6.407216	6.590398
	GSM239312.CEL	GSM239313.CEL	GSM239314.CEL
1007_s_at	10.234088	10.268217	10.272284
1053_at	7.381790	7.587974	7.561222
117_at	4.098058	4.185551	4.171396
121_at	8.324093	8.198131	8.095500
1255_g_at	2.871059	2.550177	2.415219
1294_at	5.500470	5.533965	5.804020

As seen above, this data set uses the HG-U133Av2 array which includes 22,277 probe identifiers. There are six samples divided into two groups. The first group includes the first three arrays while the second group includes the last three arrays.

21.14.2 Fold change

The idea here is to calculate a mean expression level for each condition, and subsequently select the genes that have a fold change greater than an arbitrarily selected threshold. Since in the above expression matrix, the values are in log base 2 scale, we need to revert to the raw values before calculating the arithmetical means, and then the log fold change corresponding to the log ratio of the mean intensities of two groups.

```
> eDatexp<-2^eDat
> FCM<-log2(rowMeans(eDatexp[,1:3])/rowMeans(eDatexp[,4:6]))
> FCA<-log2(rowMeans(eDatexp))
> mycol<-rep("black", length(FCM))
> mycol[which(FCM > 1)]<-"red"
> mycol[which(FCM < -1)]<-"blue"
> sum(FCM > 1); sum(FCM < -1) #number of genes DE

[1] 479

[1] 519

> plot(FCA, FCM, xlab="log(mean intensity)", ylab="log(fold
+ change)", col=mycol)
> abline(h=c(-1,1), col="purple", lty="dashed")
```

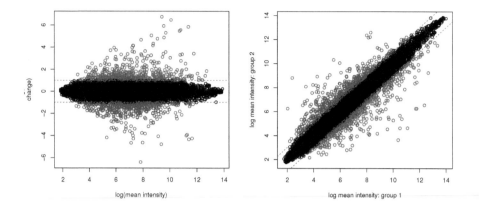

FIGURE 21.9: Scatter plot of log mean intensities and log fold changes. The left panel shows log fold change versus mean intensity. The right panel shows a scatter plot of log mean intensities from the two groups. Probes/genes up-regulated in group 1 (the first three arrays) are red colored, blue for down-regulated. The dashed purple lines show the two-fold threshold applied.

In the code above, we first apply an exponential function to revert the log, then we use `rowMeans` to calculate the means for the two conditions, then we calculate the ratios and their logs. These value are stored in FCM. The overall means of the genes are also calculated and stored in FCA. Colors are used to distinguish between the genes that meet our selection criteria, i.e. have a fold change that is greater that ± 2 (in log2 scale this corresponds to ± 1). The code then plots a graph similar to an M-A plot and draws two lines corresponding to the thresholds used. This graph is shown in the left panel of Fig. 21.9.

We can also construct a scatter plot that shows one group versus the other. This can be done as follows:

```
> meanX<-log2(rowMeans(eDatexp[,1:3]))
> meanY<-log2(rowMeans(eDatexp[,4:6]))
> plot(meanX, meanY, xlab="log mean intensity: group 1",
+ ylab="log mean intensity: group 2", col=mycol)
> abline(-1, 1, col="purple", lty="dashed")
> abline(1, 1, col="purple", lty="dashed")
```

This graph is shown in the right panel of Fig. 21.9. As shown above, the counts of the genes that met each threshold show that using an arbitrary threshold of two-fold, we have 998 differentially expressed genes (479 up-regulated and 519 down-regulated in the first group), out of a total of 22,277 genes.

21.14.3 Unusual ratio

In order to use the unusual ratio criterion, we have to use log transformed fold changes, i.e. log2(mean of experiment / mean of control), instead of raw fold change. This is because the distribution of raw fold change is most often severely skewed to the right (i.e., it has a very long right tail). A log transformation will usually yield a distribution that is more or less symmetrical.

In order to illustrate this, we will plot the distributions of the raw and log data side by side, as follows:

```
> histDens<-function(data,NumBin=200,scale=37,xlab="fold change",
+ ylab="density") {
+ dens<-density(data, n=NumBin)
+ hist(data, xlim=range(dens$x), breaks=dens$x,
+ xlab=xlab, ylab=ylab, freq=F, main=NULL)
+ lines(dens$x, dens$y*dens$bw*scale, col="red", lty=2)
+ }
> par(mfrow=c(1,2))
> histDens(2^FCM, scale=30)
> histDens(FCM, xlab="log2(fold change)")
```

The result produced by the code above is shown in Fig. 21.10. In this figure, the left panel shows the histogram of the data before the log. Note the very log tail that goes beyond 100. The right panel of the same figure shows the distribution of the log data. Note how the distribution has become symmetrical and now has reasonable tails up to ± 6.

If the mean of log ratios is very close to zero (as it would be expected in general), the order of the genes ranked by unusual ratio criterion should be almost identical to that yielded by the fold change criterion. This can be easily checked as follows:

```
> mean(FCM); sd(FCM)

[1] -0.0005980738

[1] 0.4839753

> FCZ<-(FCM-mean(FCM))/sd(FCM)
> head(FCZ[order(abs(FCZ), decreasing=T)])

  202237_at   201641_at   205523_at   202086_at   202411_at
   13.93833    13.33420   -13.26598    12.19747    12.03549
201601_x_at
   11.44285

> head(FCM[order(abs(FCM), decreasing=T)])
```

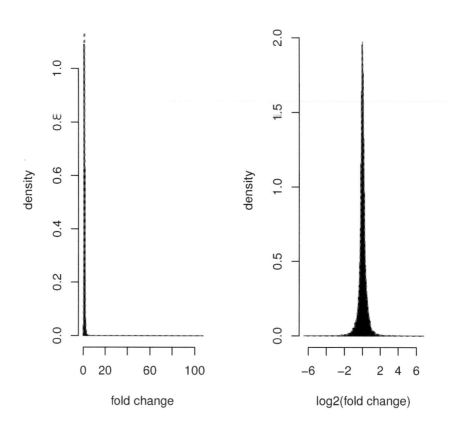

FIGURE 21.10: Log transformation of the distribution of fold changes. Left panel: data before the log. Right panel: data after the log. Note the very long tail to the right in the left panel and the beautiful symmetry achieved after the log transformation.

202237_at	201641_at	205523_at	202086_at	202411_at
6.745207	6.452826	-6.421006	5.902677	5.824282

201601_x_at
5.537457

```
> # ranking with fold change and ranking with unusual ratio are
> # highly correlated
> cor(order(order(abs(FCZ))), order(order(abs(FCM))),
+ method="spearman")
```

```
[1] 0.9999867
```

However, note that the fact that the order is very similar does *not* mean that the two methods will produce similar sets of DE genes. In fact, as discussed in the previous sections, the fold change threshold is arbitrarily selected before hand and can produce any number of DE genes, including zero. In contrast, the unusual ratio will always produce a relatively constant proportion of DE genes. For instance, if the data is normally distributed and a threshold of ± 2 is used on the unusual ratio, about 5% of the genes will be reported as DE, no matter what the samples were. In fact, this will happen even if two arrays are hybridized with two samples from the very same biological specimen.

21.14.4   Hypothesis testing, corrections for multiple comparisons, and resampling

As discussed above, a plethora of methods exist for the selection of DE genes. These can be divided into: i) parametric methods (e.g., t-test and its variations, ANOVA/linear models, empirical Bayes, etc.) and ii) nonparametric methods (e.g., SAM, permutation tests, etc.). Note that nonparametric methods can be constructed by borrowing statistics from parametric methods (e.g., t statistic) but estimating the p-values through permutations or bootstrapping without assuming a parametric distribution for the test statistic. Generally, if the sample size is relatively small, parametric methods are preferred because of their high power. However, small sample sizes constitute exactly the situations in which the underlying assumptions may not hold so extra care is needed from this perspective.

As discussed in Chapter 16, the multiple testing issue is a primary concern for microarray data. In principle, the best way to deal with the multiple comparison testing is to avoid making too many tests. This can be achieved by trying to make sure that we only do the tests that are really needed. Hence, before proceeding to differential expression analysis using hypothesis testing, a gene filtering step can be used in order to remove array spike controls, and probes that are either expressed at a very low level or exhibit little variability among samples. Such a preliminary filtering step that eliminates irrelevant probes will reduce the number of hypothesis to be tested, as well as uninteresting differentially expressed genes. In order to avoid introducing any bias, sample label/group information cannot be used in this filtering step. The package "genefilter" or customized functions can be used for this type of filtering. In the example analysis here, we will keep probes for which i) at least 2 out of 6 samples have an intensity of 100 or above; and ii) ratio of maximal/minimal intensity is at least 1.5. This filtering can be achieved as follows:

```
> library(genefilter)
> f1<-pOverA(1/3, log2(100))
> f2<-function(x) (diff(range(x, na.rm=T))>log2(1.5))
> ff<-filterfun(f1,f2)
> index<-genefilter(eDat, ff)
> sum(index)
```

```
[1] 3145
```

```
> eDatSet<-eDat[index,]  #filtered expression matrix
```

The filtering implemented above reduced the total number of probes from 22,277 to 3,145, while still maintaining the ability to find genes that are DE even though they may not be present at all in one of the two groups.

The first method we will illustrate here is a t statistic followed by a permutation test. In reality, a permutation test is not appropriate for our data because we only have three samples for either of two groups. As discussed in section 16.4.7, here the total number of distinct permutations is not sufficient to calculate meaningful p-values. Nevertheless, we will do it here just to illustrate how something like this could be done in R. The parametric p-values associated with the t statistic in a t-test are also calculated.

```
> perm<-combn(1:6, 3)[,1:10] #generate unique combinations
> tPermFun<-function (x) {
+    tval<-rep(NA, ncol(perm))
+    for (i in 1:ncol(perm)){
+      tmp<-t.test(x[perm[,i]], x[!(1:6) %in% perm[,i]])
+      tval[i]<-tmp$statistic
+    }
+    return(tval)
+ }
> tvalues<-apply(eDatSet, 1, tPermFun)
> #calculate t statistic for each permutation
> pPerm<-rep(NA, ncol(tvalues))
> #store permutation p values
> for (j in 1:ncol(tvalues)) {
+ pPerm[j]<-sum(abs(tvalues[-1,j]) >=
+ abs(tvalues[1,j]))/(nrow(tvalues)-1)
+ }
> summary(pPerm)
```

```
   Min. 1st Qu.  Median    Mean 3rd Qu.     Max.
0.00000 0.00000 0.00000 0.04671 0.00000 1.00000
```

```
> ptTest<-rep(NA, nrow(eDatSet)) #storing p values from t-test
> for (k in 1:nrow(eDatSet)){
+ ptTest[k]<-t.test(eDatSet[k,1:3], eDatSet[k,4:6])$p.value
+ }
> summary(ptTest)
```

```
     Min.    1st Qu.    Median     Mean    3rd Qu.      Max.
0.0000002 0.0028690 0.0143300 0.0654200 0.0562000 0.9996000
```

The "samr" package can be used to perform a SAM analysis. We first create a data list by specifying an expression data matrix x, and a response variable y.[5] As before, genes are in rows, samples are in columns. For a simple case of a two-group unpaired comparison, we only need to call function samr which takes as arguments the data list, the problem type and number of permutations to be performed. For our example, we set the number of permutations to 10, which is the maximum number of distinct permutations for this data. Conveniently, the function samr will automatically execute only the maximum number of possible permutations even if the number provided as an argument exceeds it.

```
> library(samr)
> data<-list(x=eDatSet, y=c(rep(1,3), rep(2,3)),
+ genenames=rownames(eDatSet), logged2=T)
> samr.obj<-samr(data, resp.type="Two class unpaired",nperms=10,
+ random.seed=123)
```

After computing the SAM statistic d_i, we need to choose an appropriate Δ value as the threshold for significance. The Δ value represents the distance from the expected d_i computed from permutations. The package samr includes the function samr.compute.delta.table that can be used for this purpose.

As shown in Fig. 21.8, the lists of DE based solely on p-values are not very reproducible. Furthermore, life scientists tend to be skeptical about a gene that has a very significant p-value but a small fold change that cannot be easily checked with other techniques. Hence, we can require our DE genes to also have a minimum fold change of 1.5, *in addition to* a significant p-value.

```
> delta.table<-samr.compute.delta.table(samr.obj,
+ min.foldchange=1.5)
```

The table returned by the function samr.compute.delta.table provides permutation-based estimates for: the number of median false positive, the 90th

---

[5]Here, the group labels are coded as 1 and 2. The SAM manual provides details regarding various response coding formats.

percentile number of false positive, the number of genes called as significant, the median FDR, the 90th percentile FDR, and the lower and higher cut d_i values.

```
> delta.table[c(1:3, 17:18),]
```

```
          delta # med false pos 90th perc false pos # called
[1,] 0.000000000        9.342448             36.559364      1612
[2,] 0.001503562        9.342448             36.559364      1612
[3,] 0.006014247        9.342448             36.559364      1612
[4,] 0.384911781        0.000000              8.284388       729
[5,] 0.434529316        0.000000              6.764833       594
       median FDR 90th perc FDR      cutlo      cuthi
[1,] 0.005795563    0.02267951 -0.3636249 0.3656959
[2,] 0.005795563    0.02267951 -0.3636249 0.3656959
[3,] 0.005795563    0.02267951 -0.3636249 0.3656959
[4,] 0.000000000    0.01136404 -0.6402669 0.6386214
[5,] 0.000000000    0.01138861 -0.7078573 0.7098941
```

We can actually query the same function for a specific value of Δ that may not be in the table. For instance, interrogating with a Δ value of 0.4 will result in 688 significant genes with a 90th percentile FDR around 1%.

```
> tmp<-samr.compute.delta.table(samr.obj, min.foldchange=1.5,
+ dels=0.4)
```

```
> tmp
```

```
      delta # med false pos 90th perc false pos # called
[1,]    0.4                0             7.822893       688
      median FDR 90th perc FDR      cutlo      cuthi
[1,]          0    0.01137048 -0.6613962 0.6573724
```

The function samr.plot visualizes the effect of the chosen Δ threshold on the selection of significant DE genes.

```
> samr.plot(samr.obj, del=0.4, min.foldchange=1.5)
```

The plot produced by this function is shown in Fig. 21.11.

Finally, we can now extract the list of significant DE genes by calling the function samr.compute.siggenes.table:

```
> siggenes.table<-samr.compute.siggenes.table(samr.obj,del=0.4,
+ data=data, delta.table=delta.table, min.foldchange=1.5)
> dim(siggenes.table$genes.up)
```

```
[1] 376    8
```

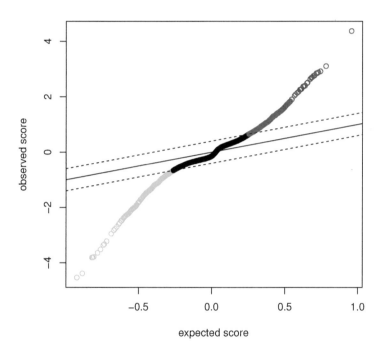

FIGURE 21.11: SAM plot of significance thresholding with Δ. Colored genes have a difference greater than the chosen Δ (in this case 0.4), between observed d_i value and expected d_i value. Green and red denotes down-regulation and up-regulation, respectively.

```
> dim(siggenes.table$genes.lo)

[1] 312    8

> siggenes.table$genes.lo[1:2,]

      Row    Gene ID     Gene Name  Score(d)
[1,] "568" "202237_at" "g567"      "-4.53428948885633"
[2,] "419" "201641_at" "g418"      "-4.38537058185797"
     Numerator(r)            Denominator(s+s0)
[1,] "-6.75053912540836" "1.4887755054014"
[2,] "-6.45344498056177" "1.47158486611355"
     Fold Change              q-value(%)
[1,] "0.00928720936989834" "0"
[2,] "0.0114108891238965"  "0"


> SAMupGene<-siggenes.table$genes.up[,"Gene ID"]
> SAMdownGene<-siggenes.table$genes.lo[,"Gene ID"]
> eDatSet[SAMupGene[1:2],]


             GSM239309.CEL GSM239310.CEL GSM239311.CEL
205523_at         2.265831      2.327470      2.556503
209699_x_at       8.061523      7.972841      7.984117
             GSM239312.CEL GSM239313.CEL GSM239314.CEL
205523_at         8.732534      8.79279       8.899062
209699_x_at      12.366302     12.33964      12.357379

> eDatSet[SAMdownGene[1:2],]


             GSM239309.CEL GSM239310.CEL GSM239311.CEL
202237_at        10.33777      10.33529      10.50503
201641_at        11.12204      10.95320      11.18777
             GSM239312.CEL GSM239313.CEL GSM239314.CEL
202237_at         3.431190      3.758905      3.736374
201641_at         4.553666      4.564932      4.784074
```

The linear model using the empirical Bayes' approach described in Section 21.11 is implemented in the package "limma." The design matrix can be manually constructed or created from function model.matrix. Typically, there are two types of parametrizations: "treatment-contrasts" parametrization, where some coefficients estimate the contrast of interest, and "group-means" parametrization, where each coefficient estimates the mean gene expression within a particular group. After the desired parametrization is specified, the function lmFit is called to estimate the parameters of the linear model. If we use the latter type of parametrization, we need to further extract the contrast of interest β_{gj} from the coefficient vector α_g. This can be manually done in simple cases, but the function makeContrasts is more convenient in more complex situations. After extracting the contrast(s) of interest, the function `contrast.fit` can be used to compute the estimated contrast effects $\hat{\beta}_{gj}$ and standard errors. The function `eBayes` is subsequently applied to build the hierarchical Bayes' model and estimate the moderated statistics and associated p-values as evidence of differential expression. Finally, the function `topTable` extracts the list of genes with their estimated statistics.

```
> library(limma)
> TS <- gl(2,3, labels=c("control", "treatment"))
> design1 <- model.matrix(~TS)
> #"treatment-contrasts" parametrization
> colnames(design1) <- c("control", "contrast")
> design1

  control contrast
1     1        0
2     1        0
3     1        0
4     1        1
5     1        1
6     1        1
attr(,"assign")
[1] 0 1
attr(,"contrasts")
attr(,"contrasts")$TS
[1] "contr.treatment"

> fit1 <- lmFit(eDatSet, design1)
> fit1 <- eBayes(fit1)
> topall1<-topTable(fit1, coef="contrast", number=nrow(eDatSet),
+ adjust="fdr", p.value=1, lfc=log2(1))
> head(topall1)

              ID     logFC  AveExpr          t       P.Value
567     202237_at -6.750539 7.017426 -49.70034 7.647502e-20
418     201641_at -6.453445 7.860947 -48.71487 1.072365e-19
1151    205523_at  6.424861 5.595699  48.60217 1.115116e-19
408   201601_x_at -5.532987 9.021857 -42.87678 9.229616e-19
1187    205869_at -5.290836 8.026878 -40.91998 2.026629e-18
614     202411_at -5.842713 9.444865 -39.32542 3.955468e-18
          adj.P.Val        B
567   1.169014e-16 35.34147
418   1.169014e-16 35.04198
1151  1.169014e-16 35.00722
408   7.256785e-16 33.09236
1187  1.274749e-15 32.36318
614   1.898500e-15 31.73688
```

```
> design2 <- model.matrix(~0+TS)  #"group-means" parametrization
> colnames(design2) <- levels(TS)
> design2

  control treatment
1    1        0
2    1        0
3    1        0
4    0        1
5    0        1
6    0        1
attr(,"assign")
[1] 1 1
attr(,"contrasts")
attr(,"contrasts")$TS
[1] "contr.treatment"

> fit2 <- lmFit(eDatSet, design2)
> cont.matrix <- makeContrasts(contrast=treatment-control,
+ levels=design2)
> #or manually create with "cont.matrix <- c(-1,1)"
> cont.matrix

          Contrasts
Levels       contrast
  control       -1
  treatment      1

> fit2 <- contrasts.fit(fit2, cont.matrix)
> fit2 <- eBayes(fit2)
> topall2<-topTable(fit2, coef="contrast", number=nrow(eDatSet),
+ adjust="fdr", p.value=1, lfc=log2(1))
> head(topall2)

            ID     logFC  AveExpr          t      P.Value
567    202237_at -6.750539 7.017426 -49.70034 7.647502e-20
418    201641_at -6.453445 7.860947 -48.71487 1.072365e-19
1151   205523_at  6.424861 5.595699  48.60217 1.115116e-19
408  201601_x_at -5.532987 9.021857 -42.87678 9.229616e-19
1187   205869_at -5.290836 8.026878 -40.91998 2.026629e-18
614    202411_at -5.842713 9.444865 -39.32542 3.955468e-18
        adj.P.Val       B
567   1.169014e-16 35.34147
418   1.169014e-16 35.04198
1151  1.169014e-16 35.00722
408   7.256785e-16 33.09236
```

```
1187 1.274749e-15 32.36318
614  1.898500e-15 31.73688
```

Using a threshold of 0.05 for FDR adjusted *p*-value and minimal fold change of 2, this limma analysis produces a total number of 674 DE probes (359 up and 315 down), 662 of which are also present in the DE list from SAM analysis.

```
> limmaDEgene<-topTable(fit2, coef="contrast",
+ number=nrow(eDatSet), adjust="fdr", p.value=0.05, lfc=log2(2))
> dim(limmaDEgene)

[1] 674   7

> sum(limmaDEgene$ID %in% c(SAMupGene, SAMdownGene))

[1] 662
```

An interesting question to ask is how the sets of genes identified as DE by various methods compare with one another. Let us compare first the sets of genes found by the unusual ratio, SAM, and the moderated *t*-test.

```
> fcDEid<-names(FCM)[abs(FCM)>1]        # fold change DE genes
> ratioDEid<-names(FCZ)[(abs(FCZ)>2)]   # unusual ratio DE genes
> samDEid<-c(SAMupGene, SAMdownGene)    # SAM DE genes
> limmaDEid<-limmaDEgene$ID             # moderated t DE genes
> allDEid<-union(ratioDEid, union(samDEid, limmaDEid))
> tmp1<-allDEid %in% ratioDEid
> tmp2<-allDEid %in% samDEid
> tmp3<-allDEid %in% limmaDEid
> vennDat<-cbind(tmp1, tmp2, tmp3)
> vennCt<-vennCounts(vennDat)
> vennDiagram(vennCt, names=c("usual ratio", "SAM",
+ "moderated t"), cex = 1, counts.col = "red")
```

A Venn diagram showing the overlap between these three sets is shown in Fig. 21.12. There are 662 genes that have been found by all three methods. These are likely to be truly DE genes. There are a few genes found by two of the three methods but the numbers are small, ranging from 0 between SAM and moderated *t*-test, to 20 between SAM and unusual ratio. The striking number in this comparison is the number of genes reported as DE solely by the unusual ratio method. There are no fewer than 374 such genes. This illustrates with a practical example on a real data set the theoretical weaknesses of this method discussed in Section 21.4.

```
> tmp1<-allDEid %in% fcDEid
> tmp2<-allDEid %in% samDEid
```

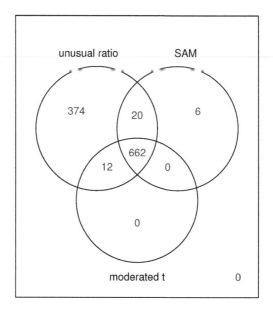

FIGURE 21.12: A Venn diagram showing a comparison between unusual ratio, SAM, and moderated t-test. There are 662 genes identified by all 3 methods and very likely to be truly DE. The genes identified by a single method are more likely to be false positives.

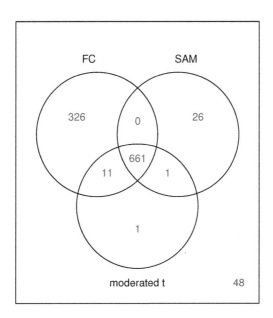

FIGURE 21.13: A Venn diagram showing a comparison between fold change, SAM, and moderated *t*-test. There are 661 genes identified by all three methods and very likely to be truly DE. The genes identified by a single method are more likely to be false positives.

```
> tmp3<-allDEid %in% limmaDEid
> vennDat<-cbind(tmp1, tmp2, tmp3)
> vennCt<-vennCounts(vennDat)
> vennDiagram(vennCt, names=c("FC","SAM","moderated t"),
+ cex = 1, counts.col = "red")
```

Let us now compare the results of the fold change, SAM, and moderated *t*-test. This comparison is shown in Fig. 21.13. The results are very similar to the ones above. The number of gene found by all three methods is almost identical to the one before: 661 compared to 662. Once again, the fold change report a very large number of genes that are not identified by any of the other two methods. Finally, let us compare all four methods at the same time. Fig. 21.14 shows a Venn diagram with all four methods. There are 661 genes on which all methods agreed. The various other groups contain genes that are

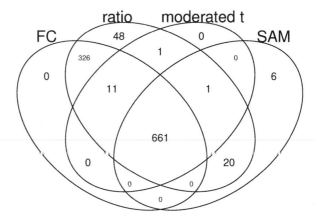

FIGURE 21.14: A Venn diagram showing a comparison between fold change, unusual ratio, SAM and moderated *t*-test. There are 661 genes identified by all three methods and very likely to be truly DE.

less and less likely to be truly differentially express as there are fewer methods that agree on them.

```
> vennDat<-list("FC"=fcDEid, "ratio"=ratioDEid,
+ "moderated t"=limmaDEid,"SAM"=samDEid)
> library(gplots)
> venn(vennDat)
```

21.15   Summary

A plethora of refined methods is available to select differentially expressed genes. Although still in widespread use, the early methods of selection by fold change and unusual ratio are clearly inadequate. Using a fold change without a clear biological justification is just a blind guess. The unusual ratio method will always report some genes as regulated even if two identical tissues are studied (false positives). These two methods suffer from severe drawbacks and their use should be discontinued as methods for selecting differentially regulated genes. However, studying the fold change of genes of known function is and will continue to remain important. In order words, computing statistics as required by biological reasons is fully justified (e.g., how do apoptosis-related genes change in immortalization?). However, drawing biological conclusions based on an arbitrary choice of fold change is not (for instance concluding that gene X is relevant to immortalization because it has a fold change of 2).

Selecting the differentially regulated genes can be done using a classical hypothesis testing approach. When using univariate statistical tests for hundreds or thousands of genes (e.g., with data coming from most commercial chips) the correction for multiple comparisons is absolutely crucial. A number of methods use this approach.

For arrays using oligos, like Affymetrix or Illumina, an alternative approach is to perform the selection at the probe level and map to genes only at the very end. This avoids some of the issues related to the mapping but has more stringent requirements. This is because the number of probes is usually at least an order of magnitude higher than the number of genes, which means a more stringent threshold for the raw p-values. Furthermore, performing the analysis at the probe level and then extrapolating the results to the gene the probe maps to may provide incorrect results if that probe is not unique to the given gene (e.g. it hybridizes with mRNA from other genes), and/or if the probe is present in different alternatively spliced isoforms.

ANOVA is a general data analysis approach that can also provide information able to discern the differentially regulated genes. Kerr and Churchill have proposed several ANOVA models and experiment designs (see also Chapters 13 and 15).

The noise sampling method yields only $2n$ residuals (as opposed to $4n$ residuals when ANOVA is applied on individual channels) and no estimates for the other factors (e.g. array effect, gene effect, etc.). However, the approach does provide the differential expression terms for all genes, as well as noise estimates for calculating confidence intervals. Furthermore, the noise sampling method is suitable for use on individual arrays, which has two important consequences. First, the application of this technique does not require repeating experiments[6] and thus is advantageous when either the number of arrays and/or the amount of mRNA are limited. Second, because the noise estimates come from a single array, they can be used to construct an overall noise distribution which will characterize the experimental quality of the array and thus provide the experimental biologist with a very useful feedback regarding the quality of the laboratory process.

Current statistical methods offer a great deal of control and the possibility of selecting genes within a given confidence interval. However, all such methods rely essentially on a careful experiment design and the presence of replicate measurements. A good way to obtain reliable results is, arguably, some version of the ANOVA method. However, in most cases, this will probably mean involving a statistician from the very beginning and designing the experiment in such a way that enough degrees of freedom are available in order to answer the relevant biological questions.

Model-based maximum likelihood methods use a statistical model involving one or several known distributions. A maximum likelihood approach is used to estimate the parameters of the model in order to fit it best to the given data. Advantages include flexibility and large applicability while disadvantages include the amount of computation and uncertain results for small sample sizes.

Methods such as SAM and moderated *t*-test try to stabilize the variance used in the *t*-test, thus providing more reliable results.

The chapter also included a short description of the approach used to calculate the fold changes for Affymetrix arrays. The methods used include a number of empirical estimates and adjustments that have been shown to work well with real data. The decisions are ultimately based on a nonparametric Wilcoxon signed-rank test for paired data where the pairing is done at probe level between the baseline and experiment arrays.

A final section illustrates the implementation of all the methods discussed in this chapter using R.

21.16   Appendix

21.16.1   A comparison of the noise sampling method with the full-blown ANOVA approach

The noise sampling method presented here is a particular application of the analysis of variance (ANOVA) method. Another ANOVA variation was proposed by Kerr-Churchill [258, 259]. It is informative to compare the two approaches. Let us consider the following model proposed by Kerr and Churchill:

---

[6]However, repeating experiments is very strongly recommended.

$$\log\left(y_{ijkg}\right) = \mu + A_i + D_j + V_k + G_g + (AG)_{ig} + (VG)_{kg} + \varepsilon_{ijkg} \qquad (21.19)$$

In this equation, i indexes the array, j indexes the dye, k indexes the mRNA, and g indexes the genes. Let us assume that there are two arrays in a flip-dye experiment. A flip-dye experiment involves two arrays. In the first array, the experimental mRNA is labeled with one dye, typically cy3, and the control mRNA is labeled with a second dye, typically cy5. In the second array, the colors are reversed: cy5 for experiment and cy3 for control. In this case, we have $(i = 1, 2, \, j = 1, 2, \, k = 1, 2, \, g = 1..n)$. The flip-dye experiment will provide the following $4n$ values: y_{111g}, y_{122g}, y_{212g}, and y_{221g}. According to the Kerr-Churchill approach, the values for μ, A_i, D_j, V_k, G_g, VG_{kg}, and AG_{ig} are obtained using a least-square fit. In consequence, there will be $4n$ residuals ε_{ijkg}. Subsequently, the mean differential expression can be calculated as:

$$\overline{dE_g} = \frac{(y_{111g} - y_{122g}) + (y_{221g} - y_{212g})}{2} - bias \qquad (21.20)$$

and the $4n$ residuals can be used to add error bars. In contrast, the approach proposed here goes straight to ratios, which correspond to difference of log values, as follows:

$$y_{111g} = \mu + A_1 + D_1 + V_1 + G_g + VG_{1g} + AG_{1g} + \varepsilon_{111g} \qquad (21.21)$$

$$y_{122g} = \mu + A_1 + D_2 + V_2 + G_g + VG_{2g} + AG_{1g} + \varepsilon_{122g} \qquad (21.22)$$

which can be subtracted to yield:

$$y_{111g} - y_{122g} = (D_1 - D_2) + (V_1 - V_2) + (VG_{1g} - VG_{2g}) + (\varepsilon_{122g} - \varepsilon_{122g}) \quad (21.23)$$

Taking $(D_1 - D_2) + (V_1 - V_2) = \mu$, $(VG_{1g} - VG_{2g}) = G(g)$ and $(\varepsilon_{122g} - \varepsilon_{122g}) = \varepsilon(g, s)$, we obtain Eq. 21.7. In conclusion, the approach presented here yields only $2n$ residuals (as opposed to $4n$ residuals when ANOVA is applied on individual channels) and no estimates for μ, A_i, G_g, and AG_{ig}. However, the approach does provide the differential expression terms for all genes $(VG_{1g} - VG_{2g})$ as well as noise estimates for calculating confidence intervals. Furthermore, the approach presented here is suitable for use on individual arrays, which has two important consequences. First, the application of this technique does not require repeating experiments[7] and thus is advantageous when either the number of arrays and/or the amount of mRNA are limited. Second, because the noise estimates come from a single array, they can be used to construct an overall noise distribution which will characterize the experimental quality of the array and thus provide the experimental biologist with a very useful feedback.

---

[7]However, repeating experiments is very strongly recommended.

Chapter 22

The Gene Ontology (GO)

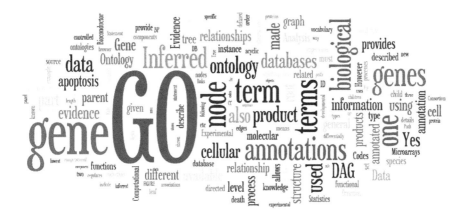

The whole is simpler than the sum of its parts.

—*Willard Gibbs*

22.1 Introduction

This text has focused so far on the numerical analysis of the microarray data and the nitty-gritty of the number crunching. However, the ultimate purpose of the gene expression experiments is to produce biological knowledge, not numbers.[1] Independently of the methods used, the result of a microarray experiment is, in most cases, a set of genes found to be differentially expressed between two or more conditions under study. The challenge faced by the researcher is to translate this list of differentially regulated genes into a better understanding of the biological phenomena that generated such changes. Although techniques aimed at this goal have started to appear (e.g., inferring

---

[1]Some of the material in this chapter is reprinted from *Genomics*, Vol. 81, No. 2, Draghici et al., "Global functional profiling of gene expression," p. 98–104, Copyright (2003), with permission from Elsevier.

gene networks [113, 115, 355, 419], function prediction[160, 167, 270, 466], etc.) such approaches are very difficult and challenged by the amount of noise in the data. However, a good first step in this direction is the translation of the list of differentially expressed genes into a functional profile able to offer insight into the cellular mechanisms acting in the given condition. Even if our information about the genes were complete and accurate, the mapping of lists of tens or hundreds of differentially regulated genes to biological functions, molecular functions, and cellular components is not a trivial matter.

In this and the following chapters, we will shift the focus from the exclusive numerical analysis performed so far to a more integrated bioinformatics approach that will try to accomplish the goal stated above. While the means will continue to consist of statistical and data analysis tools similar to the ones discussed so far, the emphasis will shift towards refining the data analysis results into biological knowledge. We will discuss a number of problems related to this higher level analysis as well as some tools designed to address such problems.

22.2   The need for an ontology

Many biologists now believe that probably there is a limited set of genes and proteins, many of which are conserved in most or all living cells. Recognizing this unification of biology and the information about genes and proteins shared by different organisms led to the development of a comparative genomics approach. Knowledge of a biological role of such a shared gene or protein in one organism may be used to infer its role in other organisms. The ability to infer biological roles provides an opportunity that annotation from one organism can be transferred to another organism in an automated fashion. However, the problem with any automated processing is that many of the existing databases use different formats and, most importantly, different vocabularies in order to describe functional annotation [22].

One way of meeting these challenges is to develop an ontology for gene annotations. The immediate advantage of such an ontology is the ability to explore functional annotations of genomes of different organisms in an automatic way. An ontology is also helpful in describing attributes of genes. For example, genes may have more than one product. Gene products possess one or more biochemical, physiological, or structural functions. Often gene products are located in specific cellular compartments. The structured, controlled vocabulary of an ontology provides a picture that is bigger than just sequences of nucleotides or amino acids [21]. Such an ontology has been developed over the past few years by the Gene Ontology (GO) Consortium [22, 23, 179].

22.3   What is the Gene Ontology (GO)?

The stated goal of the GO Consortium was to produce a dynamic, structured, precisely defined, common, controlled vocabulary useful to describe the roles of genes and gene products in any organism even as knowledge of gene and protein roles in cells is accumulating and changing [22, 23, 179].

There are two major components to GO: 1) the **ontologies** themselves, used to define the terms and describe the structure between them, and 2) the **associations between gene products and the GO terms**, which serve to annotate the genes based on existing biological knowledge. The term GO is commonly used to refer to both, which is sometimes a source of potential confusion. Here, we will use the term "GO ontology" to describe the set of terms and their hierarchical structure and "GO annotations"[2] to describe the set of associations between genes and GO terms. The GO consortium provides both ontologies and annotations for three distinct biological domains, the molecular function (MF), biological process (BP), and cellular location or cellular component (CC).

The main features of the GO are as follows:

1. It has a controlled vocabulary and therefore it is machine readable.

2. It is multidimensional, with more than one axis of classification.

3. It can be used across species.

4. It allows multiple classes of relationships, e.g., "is a" and "part of" relationships.

5. It allows expression of regulatory and reaction relationships.

6. It allows representation of incomplete knowledge.

Our knowledge of what genes and their proteins do is very incomplete and is changing rapidly everyday. To keep up with the new information being available everyday, the GO consortium provides links for each node in the GO ontologies, to related databases, including GenBank, EMBL, DDBJ, PIR, MIPS, YPD, WormPD, SWISS-PROT, ENZYME, and other databases [104]. However, providing links to the external databases is not sufficient. The ontologies themselves must be updated continuously as more information becomes available.

---

[2]Although somewhat redundant, the term "GO ontology" is used in order to eliminate potential confusions between the ontology provided by the GO consortium, which is the focus of this chapter, and other ontologies from other sources.

22.4   What does GO contain?

The GO ontology includes three independent branches: **biological process (BP)**, **molecular function (MF)** and **cellular component (CC)**. They are all attributes of a gene, a gene product, or a gene product group. A gene product can have one or more molecular functions, be used in one or more biological processes and may be associated with one or more cellular components. The biological process is defined as a biological objective to which the gene or gene product contributes. A process is the result of one or more ordered assemblies of molecular functions. The molecular function is defined as the biochemical activity of a gene product. The molecular function only describes what is done without specifying where or when the actual activity takes place. The cellular component refers to the place in the cell where the gene product is active. It must be noted that not all terms are applicable to all organisms [22].

The GO ontology is a database independent of any other. The GO ontology itself is not populated with gene products of any organism, but rather GO terms are used as attributes of genes and gene products by related databases. These databases use GO terms to annotate objects such as genes or gene products stored in their repositories and to provide references and the kind of evidence that is available to support the annotations. They also provide tables of cross-links between GO terms and their database objects to GO, which are then made available publicly [224]. The list of all crossed-referenced databases along with their abbreviated names in GO is available at `http://www.geneontology.org/doc/GO.xrf\_abbs`.

As of February 2011, GO release 1.1778 included 20,299 biological processes, 2,806 cellular components, and 8,992 molecular functions for a total of 33,587 terms.

22.4.1   GO structure and data representation

In general, an ontology such as GO consists of a number of explicitly defined *terms* that are names for biological objects or events. These terms are depicted as *nodes* (also called vertices) in a *directed acyclic graph* (DAG) connected by *edges* to define the relationships between the nodes. In the context of GO, there are several important features of DAGs. First, the edges are *directed* i.e. there is a source and a destination for each edge. In GO, the source is referred to as the *parent* term and the destination is referred to as the *child* term.

Second, unlike a general graph, a DAG does not have cycles, which is to say that one cannot complete a loop by following the directed edges. Among other things, this restriction means that two terms cannot be both parents and children of each other (otherwise, they would form a loop between themselves), and that there must be at least one node that has no children, ie. a root. A

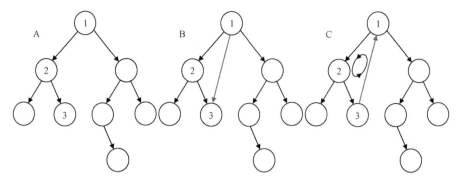

FIGURE 22.1: An example of a tree (A) versus a DAG (B) versus a general graph (C). The main difference between a tree and a DAG is that in a DAG a node is allowed to have more than one parent (see the edge between nodes 1 and 3 in B). However, unlike a general graph, a DAG cannot have cycles such as the one involving nodes 1, 2, and 3 in C.

DAG is similar to a tree with the difference that in a tree each node can have only one parent, while in a DAG a node can have multiple parents. Fig. 22.1 illustrates the differences between a tree, a DAG, and a general graph. A DAG will always have: 1) at least one **root** (also called a source) that is a node with no incoming edges i.e. a top-level node with no parent, and 2) at least one **leaf** (also called a sink), or a node with no outgoing edges i.e. a terminal node with no child. In the tree shown in Fig. 22.1 A, node 1 is the root, node 3 is a leaf, node 2 is the parent of node 3, 1 is the parent of 2, etc. The **depth** of a node is the length of the longest path from the root to that node, while the **height** is the length of the longest path from that node to a leaf. In the tree shown in the panel A of Fig. 22.1, the depth of node 3 is 2 since the length of the (only) path from the root 1 to the leaf 3 is 2. In the DAG shown in panel B, the depth of node 3 remains 2, as the length of the longest path from the root to that node. In a general graph like the one shown in panel C, there is no root so the notion of depth cannot be defined. In this panel, node 1 may look like the root but note that the tree can be redrawn by placing node 3 above node 1 which will make *it* look like the root.

   Initially, the relationship between a parent and a child in GO could be either "**instance of**" or "**part of**." The "instance of" relationship is also referred to as an "**is a**" relationship. If a child has more than one parent, it may have different relationships with its different parents. GO uses "%" to denote the "instance of" relationship and "<" to denote the "part of" relationship. The GO Consortium describes the "is a" relationship as a subclass-class relationship. For instance, the "mitochondrial chromosome" is a "chromosome." On the other hand, the "part of" relationship is different. C "part of" D means that whenever C is present it will also be D, however, C does not always need to be present. For instance, the nucleus is "part of" the cell. This means that every time there is a nucleus, this will be part of the cell. However, not all cells have nuclei.

In order to increase the ontology's ability to better described biological processes, in March 2008 the GO Consortium introduced three new relationship types: "**regulates**," "**negatively regulates**," and "**positively regulates**" into the Biological Process ontology. These substituted some of the relationships formerly described by "part of." Currently, GO uses three qualifiers that further refine annotations, although this number is likely to increase in the future. These are: "**contributes to**," "**colocalizes with**," and "**NOT**"[3].

Each GO term is given a unique identifier of the form "*GO:nnnnnnn*," where *nnnnnnn* is a zero-padded integer of seven digits. The unique identifier is used for cross-referencing in the databases using GO.

Gene Ontology is dynamic and changes as more information is made available. If a term is retired, it is removed from the DAG and made a child of the meta term "obsolete." When two terms, for example term A and term B are merged, for example as term A, the GO ID of term B is made a secondary GO ID of term A. Similarly, if a term is split each new term is assigned a new GO ID and the original GO ID is made a secondary GO ID of the new terms.

Each GO term is also cross-referenced with a number of databases. The links to such external databases are provided as "database-abbreviation:id," where id is an identifier in the given database [228].

One of the goals of GO is to provide an ontology that is applicable across species. However, some of the functions, processes and components are specific to species. In those cases when a text string has a different meaning in different species, GO differentiates them lexically [224]. Definitions are provided in a single text file [226].

22.4.2 Levels of abstraction, traversing the DAG, the "True Path Rule"

The GO is organized in a hierarchical structure that uses the types of relationship described above: "is a," "part of," and "regulates." For instance, "induction of apoptosis by extracellular signals" *is a* type of "induction of apoptosis," which in turn *is a* "positive regulation of apoptosis" which in turn *is a* kind of "regulation of apoptosis," etc. Generally speaking, traversing the DAG following "is a" relationships as above can be seen as moving across levels of abstraction. The root, BP (or "All"), would correspond to the highest level of abstraction, or to the lowest level of details. In contrast, leaf nodes such as "induction of apoptosis by hormones" would correspond to the lowest level of abstraction, with the most details. Similarly, the "cell outer membrane" is part of the "cell envelope," etc. Traversing "part of" relationships could be interpreted as changing scales. In general, terms closer to the roots (BP, CC,

---

[3]Details at: `http://www.geneontology.org/GO.annotation.conventions.shtml`

MF) are more general, while the ones closer to the leaves are more specific. In GO Consortium's terminology, the children are *more specialized* than the parents.[4]

When annotating genes with GO terms, efforts are made to annotate the genes with the highest level of details possible. In a general way, this corresponds to the lowest level of abstraction. For example, if a gene is known to induce apoptosis in response to hormones, it will be annotated with the term "induction of apoptosis by hormones" and not merely with one of the higher-level terms such as "induction of apoptosis" or "apoptosis." When a gene is annotated with a term, all inferences that can be inferred from the structure of the GO must also hold true. In other words, if the child term describes a gene product, then all its parent terms must also apply to that gene product. For instance, if a gene is annotated as having a role in "apoptosis" then it will necessarily be involved in "developmental process" because "apoptosis" is a type of "cell death," which is a type of "death," which in turn is a "developmental process" (see Fig. 22.2). This property is known as the "**True Path Rule**." If an annotation is warranted by experimental evidence and yet one of the relationships that can be inferred by propagating the gene through the hierarchy does not hold, this is an indication that the hierarchy of concepts described by GO is flawed and the structure of the GO itself has to be changed in order to maintain the validity of the True Path Rule.

22.4.3 Evidence codes

A GO annotation is a statement about a gene or gene product associating it with one or more GO terms. Such annotations can be made by curators or generated automatically using computational means. This metadata[5] provides information about the basis on which a specific gene was associated with a given GO term. In essence, this information tries to describe the type of evidence that was used when the decision to make a given association was made. is described using *evidence codes*. The GO Consortium divides the types of evidence in the following categories [225, 227]:

1. Curator-assigned Evidence Codes

    (a) Experimental Evidence Codes

        i. IDA: Inferred from Direct Assay
        ii. IPI: Inferred from Physical Interaction
        iii. IMP: Inferred from Mutant Phenotype
        iv. IGI: Inferred from Genetic Interaction
        v. IEP: Inferred from Expression Pattern

---

[4]http://www.geneontology.org/GO.doc.shtml.

[5]Metadata are data describing the data itself. Such data could include information about how the data were collected, when they were last updated, by whom, etc.

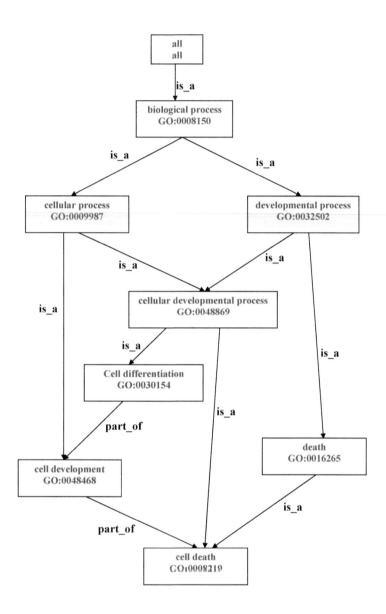

FIGURE 22.2: The placement of "cell death" in the GO hierarchy. The DAG structure of GO is clearly apparent here. This structure shows all the ancestors of cell death and their relationships. Here, the term "all" is the root of the entire GO having BP, CC, and MP as its children.

(b) Computational Analysis Evidence Codes

    i. ISS: Inferred from Sequence or Structural Similarity

    ii. IGC: Inferred from Genomic Context

    iii. RCA: inferred from Reviewed Computational Analysis

(c) Author Statement Evidence Codes

    i. TAS: Traceable Author Statement

    ii. NAS: Non-traceable Author Statement

(d) Curator Statement Evidence Codes

    i. IC: Inferred by Curator

    ii. ND: No biological Data available

2. Automatically-assigned Evidence Codes

(a) IEA: Inferred from Electronic Annotation

3. Obsolete Evidence Codes

(a) NR: Not Recorded

Table 22.1 shows more details about each type of evidence above, as well as the number of annotation from each type.[6]

Among these evidence codes, IEA (Inferred from Electronic Annotation) is of the lowest quality simply because it was generated automatically, without any assessment from an expert curator.

External databases that choose to cross-reference GO by providing links to the genes and gene products in their repository must also provide information about the type of evidence supporting the functional assignments used.

22.4.4 GO coverage

The number of annotations available varies drastically from one organism to another. Table 22.2 shows the distribution of GO annotations for species with more than 5,000 annotations as of January 2008. Of course, these numbers change with each new release of GO.

22.5 Access to GO

GO Consortium provides the AmiGO browser that allows the user to browse the GO database in tree or directed acyclic graph view. AmiGO allows searches

---

[6] As of January 2008.

Code	Evidence code description	Source of evidence	Manually checked	Current number of annotations
IDA	Inferred from direct assay	Experimental	Yes	71,050
IEP	Inferred from expression pattern	Experimental	Yes	4,598
IGI	Inferred from genetic interaction	Experimental	Yes	8,311
IMP	Inferred from mutant phenotype	Experimental	Yes	61,549
IPI	Inferred from physical interaction	Experimental	Yes	17,043
ISS	Inferred from sequence or structural similarity	Computational	Yes	190,043
RCA	Inferred from reviewed computational analysis	Computational	Yes	103,792
IGC	Inferred from genomic context	Computational	Yes	4
IEA	Inferred from electronic annotation	Computational	No	15,687,382
IC	Inferred by curator	Indirectly derived from experimental or computational evidence made by a curator	Yes	5,167
TAS	Traceable author statement	Indirectly derived from experimental or computational evidence made by the author of the published article	Yes	44,564
NAS	Non-traceable author statement	No "source of evidence" statement given	Yes	25,656
ND	No biological data available	No information available	Yes	132,192
NR	Not recorded	Unknown	Yes	1,185

TABLE 22.1: Annotation types in GO. The annotations marked as inferred from electronic annotation (IEA) constitute the vast majority of annotations in GO. Note that this type of annotations are not checked by human experts. Data from [353].

Species (NCBI taxon ID)	Genes with exp annotations	Total annotated genes	genes with exp annotations	Total Genes	% annotations	% known in genome
human (9606)	4780	17021	28.1%	20887	81.5%	22.9%
mouse (10090)	10621	18386	57.8%	27289	67.4%	38.9%
rat(10116)	3566	17243	20.70%	17993	95.80%	19.8%
fission yeast (4896)	4482	4930	90.9%	4930	100%	90.9%
baker's yeast (4932)	4947	5794	85.4%	5794	100%	85.4%
worm (6239)	4614	14154	32.6%	20163	70.2%	22.9%
Arabidopsis (3702)	5530	26637	20.8%	27029	98.5%	20.5%
fruit fly (7227)	2790	9563	29.2%	14141	67.6%	19.7%
Candida albicans (5476)	806	3756	21.4%	6166	60.9%	13.0%
Pseudomonas aeruginosa (208964)	491	2506	19.6%	5568	45.0%	8.82%
slime mold (44689)	797	6892	11.6%	13625	50.6%	5.9%
Trypanosoma brucei (5691)	449	3914	11.5%	9154	42.8%	4.92%
zebrafish (7955)	1235	13574	5.8%	21322	63.7%	3.7%
Plasmodium falciparum (5833)	188	3243	5.8%	5420	59.8%	3.47%
rice (39947)	654	29877	2.2%	41908	71.3%	1.57%
cow (9913)	96	8536	1.1%	21756	39.2%	0.4%
chicken (9031)	75	6063	1.2%	16737	36.2%	0.4%

TABLE 22.2: Distribution of GO annotations for species with more than 5,000 annotations (as of January 2008). The genes include only those that encode proteins. The experimental annotations include those only with IDA, IEP, IGI, IMP, and IPI. The percentage of annotated gene is determined by dividing the number of genes annotated by total gene products. Data from [353].

using a GO term or a gene product. The results of the query performed with a GO term will contain all gene products annotated with the term used for the query. If a query is made for a gene product, the results will provide the gene and all its associations [229]. Some of the related databases also provide browsers to search for GO terms in their own database. Mouse Genome Informatics provides a GO browser [230] that allows search on a GO term and displays all mouse genes annotated to the term. European BioInformatics Institute (EBI) provides the QuickGO browser [222] integrated into InterPro [223] and the EP:GO browser [221] integrated into the Expression Profiler [220].

22.6   Other related resources

A tremendous amount of genetic data is available online from several public databases (DBs). NCBI provides sequence, protein, structure and genome DBs, as well as a taxonomy and a literature DB. Of particular interest are UniGene (non-redundant set of gene-oriented clusters) and LocusLink (genetic loci). SWISS-PROT is a curated protein sequence DB that provides high-level annotation and a minimal level of redundancy [28, 295]. Kyoto Encyclopedia of Genes and Genomes (KEGG) contains a gene catalogue (annotated sequences), a pathway DB containing a graphical representation of cellular processes and a LIGAND DB [246, 247, 327]. GenMAPP is an application that allows the user to create and store pathways in a graphic format, includes a multiple species gene database, and allows a mapping of a user's expression data on existing pathways [108]. Other related databases and online tools include: PathDB (metabolic networks [451]), GeneX (NCGR) (source independent microarray data DB [298]), Arrayexpress [143], SAGEmap [273], μArray [143], ArrayDB [1], ExpressDB [5], and Stanford Microarray Database [389, 2]. Two meta-sites containing information about various genomic and microarray online DBs are [390] and [102].

Data format standardization is necessary in order to automate data processing [67]. The Microarray Gene Expression Data Group (MGED) created a standard for the Minimum Information About a Microarray Experiment (MIAME), the format (MAGE) and ontologies and normalization procedures related to microarray data [68, 144].

22.7 Summary

This chapter briefly described the Gene Ontology (GO) and the main concepts associated with it. GO is used to refer to two different concepts: i) a controlled vocabulary (organized in a directed acyclic graphs structure) that defines the terms used to described the function of genes, their cellular localization, and the way they interact; ii) a set of annotations made using the above controlled vocabulary. These annotations describe what is known about each known gene and its product(s). A direct acyclic graph is a tree-like structure in which a node is allowed to have more than one parent. The annotations in GO respect the True Path Rule, which says that if the association between a gene or gene product and a given annotation X holds true, then also the associations between the given gene or gene product and any other annotation situated on a node that is an ancestor of X in the GO directed acyclic graph must also hold true. In essence, a gene can be seen as being annotated with all terms from each of its explicitly associated GO terms to the root.

GO is an essential tool in the interpretation of the results of high-throughput genomic experiments. The following chapters will explain in details various uses of GO, ranging from the functional profiling used to aid the interpretation of microarray data, to choosing the appropriate custom array to be used in a focused experiment, and to the design of custom microarrays.

Chapter 23

Functional analysis and biological interpretation of microarray data

There are no facts, only interpretations.

—*Friedrich Wilhelm Nietzsche*

23.1 Over-representation analysis (ORA)

The availability of GO and many cross-referencing databases seems to solve the problem of interpreting the results of a microarray experiment from a biological point of view. Most databases provide efficient search mechanisms that return quickly all annotation information associated to any specific gene or gene product of interest. However, the problem is that most relevant databases are oriented towards a manual, gene by gene querying. If the processing of the list of differentially regulated genes were to be done manually, one would take each accession number corresponding to a regulated gene, search various public databases and compile a list with, for instance, the biological processes that the gene is involved in. The same type of analysis could be carried out for other functional categories, such as biochemical function, cellular role, etc. This task can be performed repeatedly, for each gene, in order to construct a master list of all biological processes in which at least one gene was involved. Further processing of this list can provide a list of those biological processes that are common between several of the regulated genes. It is intuitive to expect that those biological processes that occur more frequently in this list would be more relevant to the condition studied. For instance, if 200 genes have been found to be differentially regulated and 160 of them are known to be involved in, let us say, mitosis, it is intuitive to conclude that mitosis is a biological process important in the given condition. As we shall see in the following example, this intuitive reasoning is incorrect and a more careful analysis must be done in order to identify the truly relevant biological processes.

EXAMPLE 23.1

Let us consider that we are using an array containing 2,000 genes to investigate the effect of ingesting a certain substance X. Using some of the classical statistical and data analysis methods discussed in Chapter 21, we conclude that 200 of these genes are differentially regulated by substance X. For each of these 200 genes, one can query the available public databases containing information about the biochemical function, biological process, cellular role, cellular component, molecular function and chromosome location. Let us focus on the biological process for instance, and assume that the results for the 200 differentially regulated genes are as follows: 160 of the 200 genes are involved in mitosis, 80 in oncogenesis, 60 in the positive control of cell proliferation, and 40 in glucose transport.

If we now look at the functional profile described above, we might conclude that substance X may be related to cancer since mitosis, oncogenesis and cell proliferation would all make sense in that context. However, a reasonable question is: what would happen if all the genes on the array used were part of the mitotic pathway? Would mitosis continue to be significant? Clearly, the answer is no. Therefore, in order to draw correct conclusions, it is necessary to compare the actual number of occurrences with the expected number of occurrences for each individual category.

This comparison is shown in Table 23.1 for the example considered. Now, the functional profile appears to be completely different. There are indeed 160 mitotic genes but, in spite of this being the largest number, we actually expected to observe 160 such genes so this is not better than chance alone. The same is true for oncogenesis. The positive control of cell proliferation starts to be interesting because we expected 20 and observed 60. This is three times more than expected. However, the most interesting is the glucose transport. We expected to observe only 10 such genes and we observed 40, which is four times more than expected. Taking into consideration the expected numbers of genes radically changed the interpretation of the data. In light of these data, we may want to consider the correlation of X with diabetes instead of cancer.

This example illustrates that the simple frequency of occurrence of a par-

biological process	genes found	genes expected	
mitosis	160	160	not better than chance
oncogenesis	80	80	not better than chance
positive control of cell proliferation	60	20	better than chance
glucose transport	40	10	much better than chance

TABLE 23.1: The statistical significance of the data mining results. The number of genes that are involved in a given biological process can be misleading. Mitosis may appear to be the most important process affected since 160 of the 200 differentially regulated genes are involved in mitosis. In fact, this is no better than chance alone. In comparison, there are only 40 genes involved in glucose transport but this is four times more than expected by chance alone.

ticular functional category among the genes found to be regulated can be misleading. In order to draw correct conclusions, one must analyze the observed frequencies in the context of the expected frequencies. The problem is that an event such as observing 40 genes when we expect 10 can still occur just by chance. This is unlikely, but it can happen. The next section explains how the significance of these categories can be calculated based on their frequency of occurrence in the initial set of genes M, the total number of genes N, the frequency of occurrence in the list of differentially regulated genes x, and the number of such differentially regulated genes K. The statistical confidence thus calculated will allow us to distinguish between significant events and possibly random events.

23.1.1   Statistical approaches

Several different statistical approaches can be used to calculate a p-value for each functional category F. Let us consider there are N genes on the array used. Any given gene is either in category F or not. In other words, the N genes are of two categories: F and non-F (NF). This is similar to having an urn filled with N balls of two colors such as red (F) and green (not in F). M of these balls are red and N-M are green. The researcher uses their choice of data analysis methods to select which genes are regulated in their experiments. Let us assume that they picked a subset of K genes. We find that x of these K genes are red, and we want to determine the probability of this happening by chance. So, our problem is: given N balls (genes) of which M are red and N-M are green, we pick randomly K balls, and we ask what is the probability of having picked exactly x red balls. This is sampling without replacement because once we pick a gene from the array, we cannot pick it again.

The probability that a category occurs exactly x times just by chance in the list of differentially regulated genes is appropriately modeled by a hypergeometric distribution with parameters (N, M, K) [82]:

$$P(X = x | N, M, K) = \frac{\binom{M}{x}\binom{N-M}{K-x}}{\binom{N}{K}} \tag{23.1}$$

Based on this, the probability of having x *or fewer* genes in F can be calculated by summing the probabilities of picking 1 or 2 or ... or $x - 1$ or x genes of category F [419]:

$$p_u(x) = P(X = 1) + P(X = 2) + \cdots + P(X = x) = \sum_{i=0}^{x} \frac{\binom{M}{i}\binom{N-M}{K-i}}{\binom{N}{K}} \tag{23.2}$$

This corresponds to a one-sided test in which small p-values correspond to under-represented categories. The p-value for over-represented categories can be calculated as $p_o(x) = 1 - p_u(x - 1)$:[1]

$$p_o(x) = 1 - \sum_{i=0}^{x-1} \frac{\binom{M}{i}\binom{N-M}{K-i}}{\binom{N}{K}} \tag{23.3}$$

The hypergeometric distribution may be difficult to calculate when the number of genes is large (e.g., arrays such as Affymetrix HGU133A contain 22,283 genes). However, when N is large, the hypergeometric distribution tends to the binomial distribution [82]. If a binomial distribution is used, the probability of having x genes in F in a set of K randomly picked genes is given by the classical formula of the binomial probability in which the probability of extracting a gene from F is estimated by the ratio of genes in F present on the array M/N and the corresponding p-value can be, respectively, calculated as:

$$P(X = x | K, M/N) = \binom{K}{x}\left(\frac{M}{N}\right)^x \left(1 - \frac{M}{N}\right)^{K-x} \tag{23.4}$$

and

$$p_u(x) = \sum_{i=0}^{x} \binom{K}{i}\left(\frac{M}{N}\right)^i \left(1 - \frac{M}{N}\right)^{K-i} \tag{23.5}$$

Alternative approaches include a χ^2 (chi-square) test for equality of proportions [157] and Fisher's exact test [296]. For the purpose of applying these tests, the data can be organized as shown in Fig. 23.1. The dot notation for an index is used to represent the summation on that index (see Sec. 13.1.2). In

---

[1] The $x - 1$ comes from the fact that we are dealing with a discrete distribution in the right tail. For instance, $P(X \geq 8) = 1 - P(X < 8) = 1 - P(X \leq 7)$.

	Genes on array	Diff. regulated genes	
having function F	n_{11}	n_{12}	$N_{1.} = \sum_{j=1}^{2} n_{1j}$
not having F	n_{21}	n_{22}	$N_{2.} = \sum_{j=1}^{2} n_{2j}$
	$N_{.1} = \sum_{i=1}^{2} n_{i1}$	$N_{.2} = \sum_{i=1}^{2} n_{i2}$	$N_{..} = \sum_{i,j} n_{ij}$

FIGURE 23.1: The significance of a particular functional category F can be calculated using a 2×2 contingency table and a chi-square or Fisher's exact test for equality of proportions. The N genes on an array can be divided into genes that are involved in the functional category of interest F ($n_{11} = M$) and genes that are not involved in F (n_{21}). The K genes found to be differentially regulated can also be divided into genes involved ($n_{21} = x$) and not involved (n_{22}) in F.

this notation, the number of genes on the microarray is $N = N_{.1}$, the number of genes in functional category F is $M = n_{11}$, the number of genes selected as differentially regulated is $K = N_{.2}$, and the number of differentially regulated genes in F is $x = n_{12}$. Using this notation, the chi-square test involves calculating the value of the χ^2 statistic as follows:

$$\chi^2 = \frac{N_{..} \left(|n_{11}n_{22} - n_{12}n_{21}| - \frac{N_{..}}{2} \right)^2}{N_{1.}N_{2.}N_{.1}N_{.2}} \tag{23.6}$$

where $\frac{N_{..}}{2}$ in the numerator is a continuity correction term that can be omitted for large samples [178]. The value thus calculated can be compared with critical values obtained from a χ^2 distribution with $df = (2-1) \cdot (2-1) = 1$ degree of freedom.

However, the χ^2 test for equality of proportion cannot be used for small samples. The rule of thumb is that all expected frequencies: $E_{ij} = \frac{N_{i.} \cdot N_{.j}}{N_{..}}$ should be greater than or equal to 5 for the test to provide valid conclusions. If this is not the case, Fisher's exact test can be used instead [157, 253, 403]. Fisher's exact test considers the row and column totals $N_{1.}$, $N_{2.}$, $N_{.1}$, $N_{.2}$ fixed and uses the hypergeometric distribution to calculate the probability of observing each individual table combination as follows:

$$P = \frac{N_{1.}! \cdot N_{2.}! \cdot N_{.1}! \cdot N_{.2}!}{N_{..}! \cdot n_{11}! \cdot n_{12}! \cdot n_{21}! \cdot n_{22}!} \tag{23.7}$$

Using this formula, one can calculate a table containing all the possible combinations of $n_{11}n_{12}n_{21}n_{22}$. The p-value corresponding to a particular occurrence is calculated as the sum of all probabilities in this table lower than the observed probability corresponding to the observed combination [296].

Finally, Audic and Claverie have used a Poisson distribution and a Bayesian approach [26] to calculate the probability of observing a given number of tags in SAGE data. As noted by Man et al. [296], this approach can be used directly to calculate the probability of observing n_{12} genes of a certain functional category F in the selected subset given that there are n_{11} such genes on the microarray:

$$P(n_{12}|n_{11}) = \left(\frac{N_{.2}}{N_{.1}}\right)^{n_{12}} \cdot \frac{(n_{11} + n_{12})!}{n_{11}! \cdot n_{12}! \cdot \left(1 + \frac{N_{.2}}{N_{.1}}\right)^{n_{11}+n_{12}+1}} \qquad (23.8)$$

The p-values are calculated as a cumulative probability distribution function (cdf) as follows [26, 296]:

$$p = \min \left\{ \sum_{k=0}^{k \leq n_{12}} P(k|n_{11}), \sum_{k=n_{12}}^{\infty} P(k|n_{11}) \right\} \qquad (23.9)$$

Extensive simulations performed by Man et al. compared the chi-square test for equality of proportions with Fisher's exact test and Audic and Claverie's test and showed that the chi-square test has the best power and robustness [296].

23.2 Onto-Express

Onto-Express (OE) is a database and a collection of tools designed to facilitate the analysis described above [129, 264]. This is accomplished by mining known data and compiling a functional profile of the experiment under study. OE constructs a functional profile for each of the Gene Ontology (GO) categories [22]: cellular component, biological process, and molecular function as well as biochemical function and cellular role, as defined by Proteome [343]. The precise definitions for these categories and the other terms used in OE's output have been discussed in Chapter 22. More details can be found in GO [22]. As biological processes can be regulated within a local chromosomal region (e.g., imprinting), an additional profile is constructed for the chromosome location. OE uses a database with a proprietary schema implemented and maintained in our laboratory [126]. Onto-Express uses data from GenBank, UniGene, LocusLink, PubMed, and Proteome.

23.2.1 Implementation

Onto-Express provides implementations of the χ^2 test, Fisher's exact test as well as the binomial test discussed in Section 23.1.1. Fisher's exact test is required when the sample size is small and the chi-square test cannot be used. For a typical microarray experiment with $N \simeq 10,000$ genes on the array and $K \simeq 100 = 1\%N$ selected genes, the binomial approximates very well the hypergeometric and is used instead. For small, custom microarrays (fewer than 200 genes), the χ^2 is used. The program calculates automatically the expected

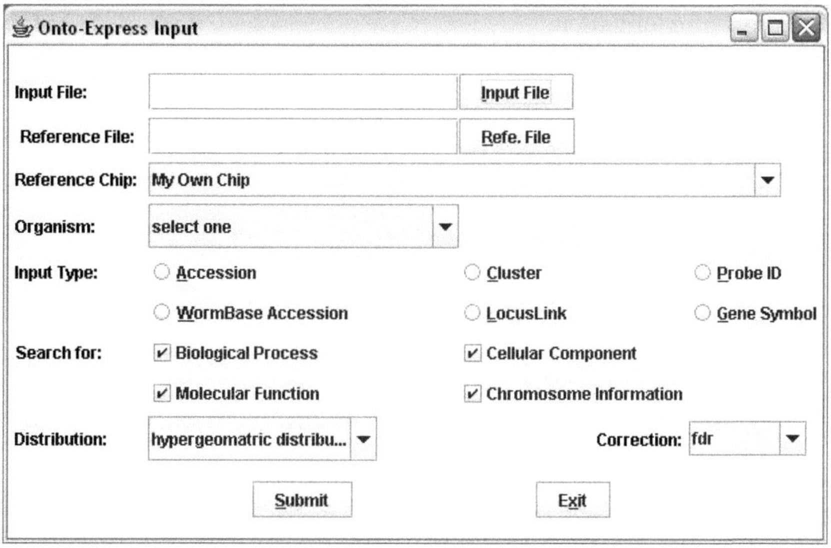

FIGURE 23.2: The input necessary for the Onto-Express analysis.

values and uses Fisher's exact test when χ^2 becomes unreliable (expected values less than 5). Thus, the choice between the three different models is automatic, requiring no statistical knowledge from the end-user. We did not implement Audic and Claverie's test because: i) it has been shown that χ^2 is at least as good [296] and ii) while very appropriate for the original problem involving ESTs, the use of a Poisson distribution may be questionable for our problem.

23.2.2 Graphical input interface description

The input of Onto-Express is a list of genes specified by either accession number, Affymetrix probe IDs or UniGene cluster IDs. In fact, the utility of this approach goes well beyond DNA microarrays since such a list can be also constructed using any alternative technology such as protein arrays, SAGE, Westerns blots (e.g., high-throughput PowerBlots [3]), Northerns blots, etc. Clearly, the utility of this analysis depends considerably on the amount of annotation data available for a given organism. The more data are available, the more useful this analysis will be. The content of the Onto-Express database is updated monthly or whenever one of the primary databases (GenBank, Uni-Gene, GO, etc.) are updated. Thus, the results of the analysis will constantly improve, as more annotation data becomes available.

The input interface is shown in Fig. 23.2. The analysis requires the following pieces of information:

1. The list of differentially regulated genes identified in the given condition.

2. The organism studied.

3. The list of genes on the array used. The database contains information about most mainstream commercial arrays. If such an array was used, the user can simply pick the appropriate array from a drop-down menu.

4. The method to be used for the correction for multiple comparisons (Bonferroni, Šidák, Holm, FDR).

5. The type of input (GenBank accession, Affymetrix probe IDs, UniGene clusters, Entrez Gene IDs, gene symbols, any of the organism specific IDs used in Gene Ontology).

6. A specification of the output required by the user. Currently, the user can select any (or all) of: cellular component, biological process, molecular function, and chromosome location.

The results are provided in graphical form as bar graphs, as well as pie charts and can be saved locally as a file. By default, the functional categories are sorted in alphabetical order. The functional categories can also be sorted by total number of genes, p-values, and corrected p-values with the exception of the results for chromosomes, where the chromosomes are always displayed in their natural order. Fig. 23.3 illustrates the various types of output that can be produced by Onto-Express and its main features.

In the flat view and flat pie chart view, there is one graph for each of the biological process, cellular component, and molecular function categories. A specific graph can be requested by choosing the desired category from the pull-down menu. Specific items within one functional category can be selected by clicking on the corresponding bar graph. A composite graph, including items from various functional categories (e.g. biological process and cellular component and chromosome location) can be obtained by clicking the button "Draw selected." This way, the user can create their own view of the results, view which can later be saved as an image file using the "save as GIF image" feature (see below).

As discussed above, GO is organized in a hierarchical structure that uses the types of relationship described above: "is a," "part of," and "regulates". For instance, "induction of apoptosis by hormones" is a type of "induction of apoptosis," which in turn is a subcategory of "apoptosis." Similarly, the "cell membrane" is part of the "cell." "Apoptosis" and "cell" represent a higher level of abstraction, more general, whereas "induction of apoptosis by hormones" and "cell membrane" represent a lower level of abstraction, more specific. When annotating genes with GO terms, efforts are made to annotate the genes with the highest level of details (lowest level of abstraction) possible. For example,

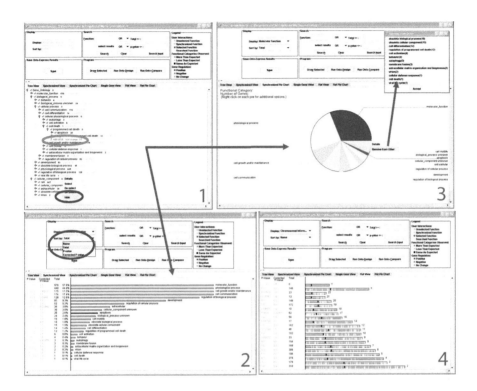

FIGURE 23.3: The main features of the Onto-Express output: tree view (panel 1), synchronized bar-chart view (panel 2), synchronized pie-chart view (panel 3), and chromosome view (panel 4). The user can select different levels of abstraction for different branches of the GO hierarchy in the tree view by expanding or collapsing the desired nodes (panel 1). Switching to the synchronized bar-chart view performs the analysis on at the custom level of abstraction defined in the tree view (panel 2). The tree view also displays the fold change value for the input genes, as provided by the user (highlighted in panel 1). In the synchronized view, results can be sorted by name, by total number of genes for each term, by *p*-value, or by corrected *p*-value (highlighted in panel 2). Categories in the pie chart can be added or removed by the user.

if a gene is known to induce apoptosis in response to hormones, it will be annotated with the term induction of apoptosis by hormones and not merely with one of the higher-level terms such as induction of apoptosis or apoptosis. In the flat view, OE results consider only the lowest level of abstraction with the highest level of detail. As a result, hundreds of GO terms are often included in the result, which makes the interpretation of results rather difficult. For instance, if one tries to ask the question whether apoptosis, as a global process, is significantly impacted, one is forced to scan the entire list of results looking for anything involved in apoptosis. Furthermore, since in many cases these extremely specific categories are only represented by only one or two genes, often they do not appear as statistically significant. In order to answer the question posed, one needs consider a higher level of abstraction (e.g., apoptosis) and calculate the p-value for this level. The integration of a GO browser in OE allows the users to customize the level of abstraction for the given biological hypothesis. Certain branches can now be expanded to provide maximum details and distinguish between specific sub-categories, while others may be kept more general. OE dynamically calculates the new p-values for each term as the user chooses the desired level of abstraction for each category. The GO browser is displayed as "Tree view" in the OE graphical user interface (GUI).

However, the tree view by itself is insufficient. In this view, it is not possible to sort the results by GO terms, total number of genes, or the p-value. This view is also inconvenient since in order to inspect all results, the user has to go through the entire tree manually looking for the significant biological processes, molecular functions, and cellular components. In order to address this, another view, called synchronized view, is added to OE. This view is synchronized with the tree view and displays the categories at the levels of abstraction chosen in the tree view as a bar graph, much like the flat view does. However, in this view the user can sort the results by GO terms, p-value, the total number of genes, or corrected p-value, etc. When a category is collapsed, the number of genes for the category indicates the number of unique genes from the input list for the category itself as well as for its subcategories. In other words, when a subcategory is not visible, the genes annotated with the subcategory are considered as annotated with the current category. When a category is expanded, its subcategories are visible and corresponding bars are added in the synchronized view. When a category is expanded, if there were no genes from the input list annotated with the corresponding term, the number next to it becomes zero. In this situation, the bar corresponding to the term in the synchronized view is made invisible. Also, when a category is collapsed, all bars corresponding to its subcategories are made invisible in the synchronized view. At any moment in time, the user has two types of results available. The flat view contains the results of the analysis (including significance values) at the lowest level of abstraction (most specific annotations). The synchronized view contains the results of the analysis specifically requested by the user in the tree view. Both use the same statistical model (hypergeometric, binomial, chi-square, or Fisher's exact test), as initially selected by the user.

The user can optionally specify the expression values for each gene obtained from their microarray experiment along with the list of genes. The gene ID and its expression value must be separated by a tab character. The expression values as specified by the user are displayed in the tree view, which greatly enhances the interpretation capabilities of the OE results. For example, instead of merely identifying apoptosis as a process that is significantly impacted, the user can quickly understand the type of changes: if the apoptotic genes are mostly up-regulated, the apoptosis is stimulated whereas if they are mostly down-regulated the apoptosis is inhibited.

Onto-Express (OE) now supports functional profiling for 24 organisms as follows: *H. sapiens, Saccharomyces cerevisiae, Drosophila melanogaster, Mus musculus, Arabidopsis thaliana, Caenorhabditis elegans, Rattus norvegicus, Oryza sativa, Danio rerio, Dictyostelium discoideum, Candida albicans, Bacillus anthracis Ames, Coxiella burnetii RSA 493, Geobacter sulfurreducens PCA, Listeria monocssytogenes 4b F2365, Methylococcus capsulatus Bath, Pseudomonas syringae DC 3000, Shewanella oneidensis MR-1, Vibrio cholerae, Leishmania major, Plasmodium falciparum, Schizosaccharomyces pombe, Trypanosoma brucei,* and *Glossina morsitans.*

In addition to broadening its scope by adding new organisms, OE is now able to support 11 more types of input data. Previously, OE allowed the users to submit a list of GenBank accession numbers, UniGene cluster IDs, Affymetrix probe IDs, Entrez Gene IDs, WormBase accession IDs and gene symbols. Now OE allows the users to submit any of the database IDs used in the GO annotations. These ID types include Saccharomyces Genome Database (SGD) IDs, FlyBase IDs, Mouse Genome Informatics (MGI) IDs, The Arabidopsis Information Resource (TAIR) IDs, The Institute for Genomic Research (TIGR) IDs, Rat Genome Database (RGD) IDs, Gramene IDs,, Zebrafish Information Network (ZFIN) IDs, DictyBase IDs, Candida Genome Database (CGD) IDs, and Sanger GeneDB IDs.

In order to facilitate understanding how the expression change of one gene influences various biological pathways and how a given gene interacts with other genes in specific pathways, the gene names in the OE results are now hyperlinked to the KEGG pathway database. Another navigational features include the ability to click on a single gene in the tree view and look up the other GO terms that the given gene is annotated with. This is done using the "Single Gene View." The user can also click on a GO term and look up a list of all genes from the input list that are annotated with the specific GO term in a separate window. In this "Details" window, the UniGene cluster IDs and Entrez Gene IDs of the individual genes are hyper-linked to the NCBI's UniGene and Entrez Gene databases, respectively. The details window also provide literature citations for each gene which includes a brief summary. These are hyperlinked to the PubMed database, as well.

The results of the analysis can be saved by clicking the "Save" button. The output files contain the GO ID of each term, the number of unique UniGene IDs for each GO term on the reference array, the Entrez Gene ID of each gene, the number of unique GenBank accession numbers for each GO term on the reference array and the official gene symbol. The users can also specify the type of information they want to store in the output files. In addition to the text files containing these results, the user can also save the results as images. Each graph can be individually saved or the user can select a set of terms and then save them as a GIF image.

23.2.3 Some real data analyses

Onto-Express has been applied to a number of publicly available data sets [129, 264, 330]. Two examples from [129] will be presented here.

EXAMPLE 23.2

A microarray was recently used to identify 231 genes (from an initial set of 25,000) that can be used as a predictor of clinical outcome for breast cancer [433]. Using a classical approach based on putative gene functions and known pathways, van't Veer et al. identified several key mechanisms such as cell cycle, cell invasion, metastasis, angiogenesis, and signal transduction as being implicated in cases of breast cancer with poor prognosis. The 231 genes found to be good predictors of poor prognosis were submitted to OE using the initial pool of 24,481 genes as the reference set. We concentrated on those functional categories significant at 5% ($p < 0.05$) and represented by two or more genes. These functional categories are presented in Fig. 23.4. It is interesting to note that Onto-Express' results included most of the biological processes postulated to be associated with cancer including the positive control of cell proliferation and anti-apoptosis. The spectacular aspect is related to the fact that these results have been obtained in a matter of minutes while the original analysis required many months.

Interestingly, oncogenesis, cell cycle control, and cell growth and maintenance are not significant at 5% but do become significant if the significance threshold is lowered to 10%. Fig. 23.5 shows these and a few other interesting functional categories. Note that the apoptosis, cell growth and maintenance, etc., do not appear to be significant although they are expected to be related to the condition studied. Two observations are required here. First, the results of these categories should be interpreted cautiously because they are represented by a single gene. Second, even if they were represented by several genes the lack of significance cannot be interpreted as a proof of insignificance (see Sec. 11.4).

Onto-Express also identified a host of novel mechanisms. Protein phosphorylation was one of these additional categories significantly correlated with poor prognostic outcome. Apart from its involvement in a number of mitogenic response pathways, protein phosphorylation is a common regulatory tactic employed in cell cycle progression. PCTK1 [309] and STK6 [266] are among the

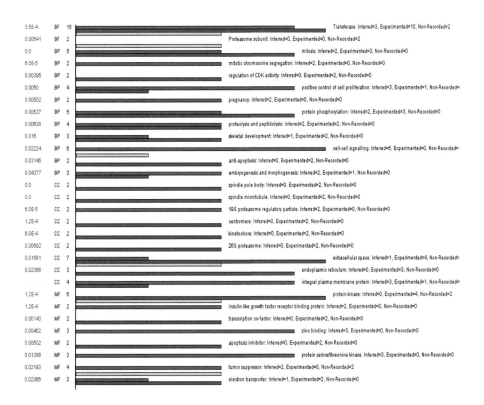

FIGURE 23.4: Significant correlations were observed between the expression level and poor breast cancer outcome for 231 genes [433]. This subset of genes was processed by Onto-Express to categorize the genes into functional groups as follows: BF= biochemical function, BP – biological process, CC – cellular component, MF – molecular function. The figure shows the 30 different functional groups associated with poor disease outcome in a significant way ($p < 0.05$ in left column). Red bar graphs represent genes for which the function was inferred, blue graphs represent genes for which the function was proved experimentally and green graphs represent genes for which this type of information was not recorded in the source database.

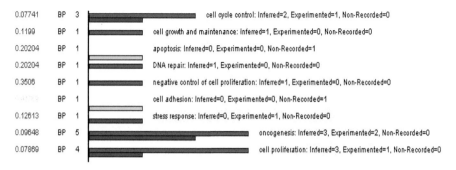

FIGURE 23.5: Some interesting biological processes that are not significant at the 5% significance level. Note that processes commonly associated with cancer such as: cell proliferation, cell cycle control, and oncogenesis *are* significant at 10% significance level. Furthermore, the statistical analysis for apoptosis, cell growth and maintenance, etc., should be interpreted cautiously since they are represented by a single gene. Red bar graphs represent genes for which the function was inferred, blue graphs represent genes for which the function was proved experimentally, and green graphs represent genes for which this type of information was not recorded in the source database.

cell cycle regulatory kinases identified as corollaries to prognostic outcome. Similarly, anti-apoptotic factors, surivin [176, 279], and BNIP3 [64], were identified. Both mechanisms are believed to be intimately linked and active in regulating cell homeostasis and cell cycle progression.

EXAMPLE 23.3

The second data set used to validate our methods was focused around the link between BRCA1 mutations and tumor suppression in breast cancer. The expression of 373 genes was found to be significantly and consistently altered by BRCA1 induction [453].

We submitted this set to OE using the genes represented on the HuGeneFL microarray (aka HU6800; Affymetrix, Santa Clara, CA) as the reference set. This array contains approximately 6800 human ESTs. We divided the genes into up-regulated and down-regulated. The functional categories significantly represented in the set of up-regulated genes are stimulated by BRCA1 over-expression (Fig. 23.6). Functional categories significantly represented in the set of down-regulated genes are inhibited by BRCA1 over-expression (Fig. 23.7). Once again, our approach was validated by the fact that the biological processes found to be significantly affected included several processes known to be associated with cancer: mitosis, cell cycle control, and the control of the apoptotic programme.

This analysis also showed that BRCA1 had somewhat of a homeostatic

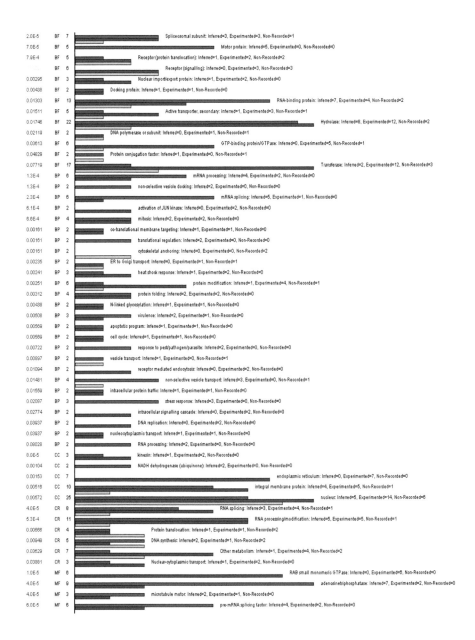

FIGURE 23.6: Functional categories significantly ($p < 0.05$) stimulated by BRCA1 overexpression in breast cancer [453]: BF – biochemical function, BP – biological process, CC – cellular component, CR – cellular role, MF – molecular function. Red bar graphs represent genes for which the function was inferred, blue graphs represent genes for which the function was proved experimentally, and green graphs represent genes for which this type of information was not recorded in the source database.

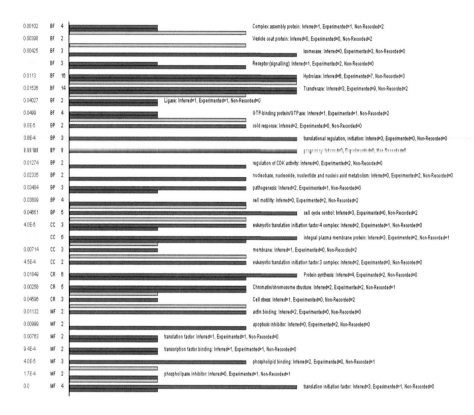

FIGURE 23.7: Functional categories significantly ($p < 0.05$) inhibited by BRCA1 over-expression in breast cancer [453]: BF – biochemical function, BP – biological process, CC – cellular component, CR – cellular role, MF – molecular function. Red bar graphs represent genes for which the function was inferred, blue graphs represent genes for which the function was proved experimentally, and green graphs represent genes for which this type of information was not recorded in the source database.

effect on the cells, promoting many cell survival and maintenance pathways (e.g., mRNA processing, splicing, protein modification, and folding). BRCA1 is known to be involved in the cell cycle checkpoint control (i.e., acting as a tumor suppressor [436]) and significantly down-regulates several genes that normally promote transition through the cell cycle, including CDC2, CDC25B and the c-Ha-ras1 proto-oncogene.

23.2.4　Interpretation of the functional analysis results

Several other factors need to be considered when interpreting the result of this type of functional analysis. First, the data analysis methods used to obtain the list of differentially regulated genes have different error rates. Thus, Onto-Express' *input* can contain false positives (genes reported as being differentially regulated when they are not). Since the presence and number of such false positives can influence the results, it is important to take this into consideration when interpreting the results. If the selection of differentially regulated genes has been done with methods that offer an explicit control of the false positive rates (confidence levels or probability of Type I error), one could repeat the analysis selecting genes at various confidence levels and compare the results.

Second, if a custom array is purposefully enriched with a certain type of genes, the significance of those specific genes will be artificially lowered. This biological bias has to be also taken into consideration when interpreting OE's results. For instance, an apoptosis array will contain many genes related to apoptosis. Because of this, any such genes found to be regulated may have large p-values and may not appear as significant. In interpreting this, it is useful to remember that a high p-value means that *the data are such that the null hypothesis cannot be rejected*, which is different from saying that the null hypothesis is true. We refer the reader to the extensive discussion on this topic in Section 11.4. In other words, having a low p-value is a sufficient, but not necessary, condition for a gene to be interesting. It is very possible for a biological meaningful gene to have a large p-value in certain circumstances (e.g. when using focused arrays).

The exact biological meaning of the calculated p-values depends on the list of genes submitted as input. For example, if the list contains genes that are up-regulated and mitosis appears more often than expected, the conclusion may be that the condition under study stimulates mitosis (or more generally, cell proliferation) in a statistically significant way. If the list contains genes that are down-regulated and mitosis appears more often than expected (exactly as before), then the conclusion may be that the condition significantly inhibits mitosis.

Finally, microarray data are typically obtained from several repeated experiments. If a certain biological process is found to be affected in repeated, independent experiments, it is likely that the process is indeed so, independently of the number of genes representing that process on the array.

23.3   Functional class scoring

More than 20 tools that use the over-representation approach (ORA) above have been developed. In spite of its wide adoption, this approach has a number of limitations related to the type, quality, and structure of the annotations available, as we shall see in Chapter 25.

An alternative approach considers the distribution of the genes associated with a given GO term in the entire list of genes and performs a functional class scoring (FCS) which also allows adjustments for gene correlations [180, 337]. One of the most salient differences between ORA and FCS approaches is that ORA relies on a previous selection of a subset of differentially expressed genes which FCS considers the entire list of genes. Arguably the state-of-the-art in the FCS category, the Gene Set Enrichment Analysis (GSEA)

23.4   The Gene Set Enrichment Analysis (GSEA)

The **Gene Set Enrichment Analysis** (GSEA) [316, 408, 422], is a method for assessing whether a predefined set of genes shows statistically significant differences between two conditions. In principle, GSEA analyzes a genome-wide expression profile from samples belonging to two different phenotypes, for example, normal and disease. The input here is a set of genes S, and a set of gene expression values measured in the two phenotypes. The goal is to establish whether the set S is enriched in genes that are different between the two phenotypes. In the context of GO profiling, the gene set considered S, will include the genes associated with a given GO term. When used in this manner, GSEA can be used to determined the significance of the given GO term.

As a first step, the algorithm ranks the genes with respect to the given profile, given the association of each gene with the phenotype. This association is established using an arbitrary test, for example a t-test. Once the ranked list of genes L is produced, an enrichment score (ES) is computed for each set in the gene set list. The list L is walked from the top to the bottom, and a statistic is increased every time a gene belonging to the set is encountered, and decreased otherwise. The value of the increment (or decrement) depends on the ranking of the gene. At the end of the list, the enrichment score is the maximum distance from zero encountered during the walk. Fig. 23.8 shows how the gene list is walked and the correspondent value of the enrichment score. In this figure, the upper graph shows the enrichment values during the walk through the gene list. The vertical lines represents the genes belonging to the set S at the positions they appear in the ranked list. The lower graph shows the degree to which each gene is correlated with the phenotype.

In principle, higher enrichment scores are yielded when the graph departs considerably from zero. However, the enrichment score by itself cannot be used to assess significance. The significance of the enrichment score is computed with a bootstrap method. There are two possible permutation criteria: permutation of the phenotype samples or permutation of the gene labels. In general, the label permutation method is preferred as it preserves gene-gene correlations. This step of the algorithm produces a null distribution, which allows the computation of an empirical p-value. The empirical p-value is calculated as the number of random bootstrap runs which resulted in an enrichment score equal to or larger than the one observed for the correct labels.

Next and last step, the significance levels are adjusted for multiple hypotheses testing. Each Enrichment Score is normalized with the size of the set obtaining a Normalized Enrichment Score (NES), and then the false discovery rate (FDR) of each NES is computed.

23.5 Summary

The ultimate goal of gene expression experiments is to gain biological knowledge. Many databases exist that contain a wealth of information about various genes, proteins and their interactions. It is useful to exploit the availability of such data to interpret the results of the microarray data analysis. In contrast to the approach of looking for key genes of known specific pathways or mechanisms, global functional profiling is a high-throughput approach that can reveal the biological mechanisms involved in a given condition. Onto-Express is a tool that translates *gene expression profiles* showing how various genes are changed in specific conditions into *functional profiles* showing how various functional categories (e.g., cellular functions) are changed in the given conditions. Such profiles are constructed based on public data and Gene Ontology categories and terms. Furthermore, Onto-Express provides information about the statistical significance of each of the pathways and categories used in the profiles allowing the user to distinguish between cellular mechanisms significantly affected and those that could be involved by chance alone. The functional class scoring uses the entire set of genes rather than a list of DE genes. From this category, the Gene Set Enrichment Analysis (GSEA), calculates an enrichment score that reflects the degree to which a given group of genes is correlated with the phenotype. A bootstrapping approach is used to calculate the significance of the enrichment score thus obtained. GSEA can be used to analyze the relevance of a given GO term by considering the set of genes annotated with that term.

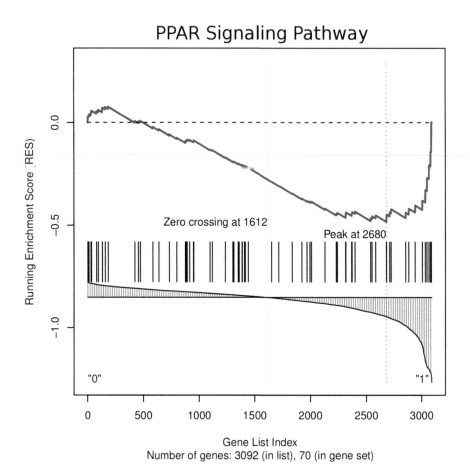

FIGURE 23.8: The Gene Set Enrichment Analysis (GSEA). The upper graph shows the enrichment value during the walk through the gene list. The vertical lines represents the genes belonging to the set S at the positions they appear in the ranked list. The lower graph shows the degree to which each gene is correlated with the phenotype.

Chapter 24

Uses, misuses, and abuses in GO profiling

You must learn from the mistakes of others. You can't possibly live long enough to make them all yourself.

—Sam Levenson (1911 – 1980)

I used to think that the brain was the most wonderful organ in my body. Then I realized who was telling me this.

—*Emo Phillips*, Neuropsychology: Clinical and Experimental Foundations

The human brain is like a railroad freight car – guaranteed to have a certain capacity but often running empty.

—Unknown

24.1   Introduction

The Gene Ontology (GO) filled a crucial gap in the life sciences by providing a controlled and structured vocabulary that allowed computers to start processing genomics-related data in an automated way. Once this gap was filled, a swath of software tools have sprouted almost overnight, performing all sorts of analyses involving GO (see Chapter 25 for a review of such tools). Also, GO annotations have become a great reference for researchers across the entire spectrum of life sciences. But, as in most cases, lots of uses come with a few abuses, as well. The much sought after ability to automate the data analysis involved in this step led to the implementation of tens of tools able to perform a GO analysis at the click of a button. Unfortunately, by automating the process and thus removing the human labor from this step these tools also remove the human intelligence from the loop. And this, as always, is much needed here as well.

This chapter focuses on discussing a number of uses, misuses, and abuses related to the use of GO. In most cases, these issues stem from a lack of understanding of GO's structure and properties. By discussing some of these issues, the chapter will hopefully make the user more GO-savvy and allow them exploit GO at its full potential as well as avoid some of these common mistakes.

24.2   "Known unknowns"

Donald Rumsfeld[1] once famously said: "Reports that say that something hasn't happened are always interesting to me, because as we know, there are known knowns; there are things we know we know. We also know there are known unknowns; that is to say we know there are some things we do not know. But there are also unknown unknowns – the ones we don't know we don't know."[2] According to BBC, this comment about the situation in Iraq won the 2003 Foot in Mouth Award from the Plain English Campaign, which annually hands out the prize for the most nonsensical remark made by a public figure.[3] What Rumsfeld was painfully trying to convey in the quote above is

---

[1] Donald Rumsfeld was the Secretary of Defense in the George W. Bush administration during the period immediately preceding the Iraq War. The context of the quote is the controversy on whether Saddam Hussein had weapons of mass destruction, which was the pretext invoked by the US in order to justify the Iraq invasion.

[2] Feb. 12, 2002, Department of Defense news briefing (http://www.defenselink.mil/transcripts/transcript.aspx?transcriptid=2636).

[3] http://news.bbc.co.uk/2/hi/americas/3254852.stm

that there is a big difference between simply not knowing, and knowing that we do not know. In the first case, the lack of knowledge may be due either to a mere lack of data, or to the fact that we did not bother to – or we could not – find such data. The latter implies an epistemological awareness about the lack of knowledge. This in itself conveys information since in many cases, this awareness comes from a conscious effort to acquire such knowledge that failed.

One of the evidence codes used in GO is ND, which stands for "no biological data available." This indicates that a curator could find no data for the gene in question. These "ND" annotations distinguish genes that are not biologically characterized from those that are not annotated yet. For instance, if a gene is found to have no annotations whatsoever in GO, a researcher may continue to search the literature and indeed retrieve annotations not yet incorporated in GO. In contrast, if the gene is found in GO with an ND annotation, no further searches are needed since ND means that exhaustive attempts to annotate the gene have been made and failed, at least up to the date of the annotation. A common mistake is to fail to correctly interpret the ND as the negative results of a thorough literature search and waste time in further searches. In Donald Rumsfeld's words, these are the "known unknowns," not to be confused with the "unknown unknowns" for which further research might produce results.

24.3   Which way is up?

In Chapter 22, a DAG was defined as a graph containing nodes and directed edges that does not have any loops. This means that a DAG will have at least one leaf, or a node with no outgoing edges, and at least one root, or a node with no incoming edges. Intuitively, the information in a DAG flows in only one way, from the root(s) to the leaf(ves). The problem stems from the fact that if one takes a DAG, D, and reverses the directions of all its edges, the new structure, D', will still be a DAG. This is easy to prove. Let us assume that there is a loop L_k in D' that would prevent it from being a DAG. Without loss of generality, we can rename those nodes starting from an arbitrary one and following the directed edges as: $n_1, n_2, \ldots, n_k$. The means that there is an edge e_i directed from node i to node $i+1$, where $i = 1, n$. Then the same nodes $n_1, n_2, \ldots, n_k$ will also be involved in a cycle in D, since the only difference between the two is the direction of the edges, it follows that there will be an edge e'_i between the same nodes i and $i+1$ in D, but this time, the edge will be oriented from $i+1$ to i. However, this will constitute a loop in D, which contradicts our hypothesis that D is a DAG. Therefore, we can conclude that D' cannot have any loops and is also a DAG. However, when the direction of the edges are reversed, a leaf node in D will become a root node in the D', and a root node in D will become a leaf in D'.

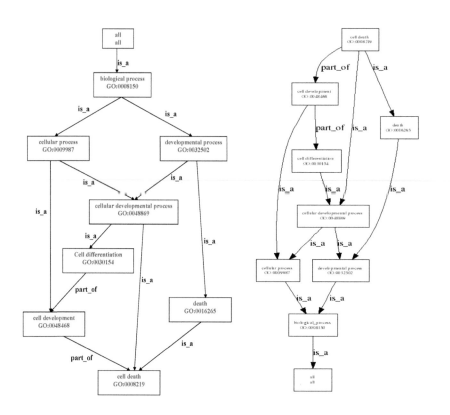

FIGURE 24.1: The placement of "cell death" in the GO hierarchy. The left panel shows the DAG structure involving this term. This structure shows all the ancestors of cell death and their relationships. Here, the term "all" is the root of the entire GO having BP, CC, and MP as its children. This diagram is consistent with intuitive arrangement placing more general terms at the top as ancestors of more specialized terms. However, in this diagram, the direction of the arrows is counterintuitive (e.g. by following the arrows we read that "developmental process" is a 'death" rather than the other way around). The right panel shows the relationship between cell death and its related terms, as shown by the AmiGO browser. In this diagram, the directions of the arrows are reversed to make it more intuitive, but as a DAG this structure would have only one leaf ("all") and the most specific terms would be roots.

From the above, it follows that given the connectivity between the nodes (e.g., the undirected adjacency matrix) the direction of the edges of a DAG is somewhat arbitrary and should be chosen in such a way that it best suits the problem domain. According to the GO Consortium, a child of a term is more specialized than its parent(s).[4] This means that for instance, "cell death" should be a child of "death", which should be a child of "developmental process," which should be a child of "biological process." If we draw the edges according to this criterion, the DAG describing GO will have a unique root, "all," which has three children: BP, CC, and MF, etc.

At the other extreme, there are leaves that correspond to the most specialized terms such as "induction of apoptosis by hormones," etc. The left panel of Figure 24.1 shows these terms in their GO context. This structure shows all the ancestors of cell death and their relationships. As expected, here the term "all" is the root of the entire GO having BP, CC and MP as its children. This diagram is consistent with the intuitive arrangement that places more general terms at the top as ancestors of more specialized terms. This diagram has only one root, "all," and many leaves, corresponding to the most specialized terms, which keeps the structure similar to a tree, which in turn, helps with the interpretation. The DAG is a tree-like structure that grows from the root towards the leaves, as most trees like to do.

All would indeed be good were it not for the fact that in this diagram, the direction of the arrows is extremely counterintuitive. For instance, the edge between cell death and cell development in the left panel of Figure 24.1 goes from biological process towards developmental process and is labeled "is a." This invites the reader to interpret the diagram as "biological process is a developmental process" rather than the other way around. This is particularly confusing when looking solely at two nodes buried somewhere within the GO DAG and carrying unfamiliar terms. Trying to address the problem and eliminate such potential confusions, the AmiGO browser for instance, shows the edges reversed, as in the right panel of Figure 24.1. From here, it follows that any GO diagram that does not include nodes such as "all," "BP," "MF," or "CC" are intrinsically ambiguous, unless one is very familiar with the GO terms involved. Hence, care should be exercised in both the use and the interpretation of such diagrams. If such a diagram is to be included in a paper, it should always be accompanied by a clear explanation of the convention used for the direction of the arrows. Conversely, if one reads a paper that uses such a diagram but does not include an explicit statement about the meaning of the arrows, one has to clarify the convention used in that paper before going any further. Furthermore, one should be aware that the use of terms such as parent, child, descendant, etc., makes an implicit use of the convention used for the edges in the DAG and can be therefore ambiguous.

---

[4]http://www.geneontology.org/GO.doc.shtml

24.4 Negative annotations

One of the qualifiers recently added by GO is NOT. The meaning of this qualifier is the one expected: if gene X is annotated with function F and NOT, it means that the gene X does not have function F. Apparently, nothing could be simpler. However, the problem is that in principle, such an annotation would be superfluous since there is no point in enumerating all the functions that a gene does *not* have. Furthermore, if this were to be done for *all* genes and *all* functions that those genes *do not* have, the GO would be quickly swamped by hundreds of millions of annotations carrying very little information. Hence, the GO curators, the domain specialists that read the literature, interpret it, and add such annotations to individual genes will use a NOT annotation *only* in those situations when a gene is not associated with a given GO term *although it would be expected to*. For instance, if a gene product has a sequence that is similar to proteins having a certain function, such as kinases or proteases, but the given gene product is known not to have the ability to perform the function expected from its sequence, then a NOT annotation is warranted and will be used.

The problem is that most tools performing the GO profiling illustrated in Chapter 23 only do a search for the GO terms themselves and disregard the NOT qualifiers. If this happens, a differentially expressed gene NOT having function F will in fact be counted as evidence that function F is associated with the given phenotype. This is clearly wrong and should be avoided at all costs. One possibility is to use a tool that treats properly the NOT qualifiers. Another alternative is to scan the results of the GO profiling for any NOT annotations, and recalculate the p-values for those terms that appeared with this qualifier. This is a feasible option only because the number of negative annotations is very small at this time.

24.5 Common mistakes in functional profiling

24.5.1 Enrichment versus p-values

In many cases, the result of a high-throughput experiment is a set of genes that are differentially expressed (DE) between different conditions (e.g., cancer versus healthy). The goal of functional profiling is to determine what exactly might be different or interesting about particular sets of genes, and this is often done by determining which GO terms are significantly over- or under-represented within the gene set. For simplicity's sake, we will focus on GO terms that are significantly over-represented in the set of DE genes. The simplest approach is to calculate an "enrichment" value for each GO term – that is, the proportion of DE genes in the set of genes associated with that particular GO term. As thoroughly discussed in Chapter 23, the main problem with this approach is that any enrichment value can occur just by chance; thus, the observation that a GO category is simply enriched or depleted should not be interpreted as unequivocal evidence implicating the GO category in the phenomenon studied.

Hence, the information used in interpreting the results of an experiment should never be the enrichment itself, no matter how this enrichment is calculated. No matter what formula is used, the enrichment is just a statistic. This statistic has to be mapped to the probability of observing a value of this statistic more extreme than a given threshold when there is no real difference between the phenotypes. In other words, the value of the statistics are not meaningful in themselves and they have to be mapped on p-values, which reflect the probability of the null hypothesis being true for a value of the statistic higher than or equal to the observed value. Such p-values could be calculated using a number of different approaches, including a parametric approach (e.g., hypergeometric, χ^2, Fisher's exact test), a bootstrapping approach, or a functional class scoring approach (e.g., GSEA). Independently of the approach used to calculate p-values, it is the p-values, not enrichment or depletions scores of any kind that should be used to assess the significance of GO terms in a given phenotype comparison.

24.5.2 One-sided versus two-sided testing

In addition, many current tools only measure enrichment and ignore depleted GO terms, which could lead to partial answers to the biological problem at hand. For instance, in a cancer data set, the apoptosis pathway will probably have many differentially expressed genes, more so than expected by chance. This is because cancer usually disables the normal apoptotic program of a cell, and the cells proliferate in an uncontrolled manner. This is the justification behind the hunt for enrichment. However, a GO term that is represented significantly less than expected by chance can also have an important biological meaning. A strong under-representation of a biological process in a condition could mean that the normal functioning of that process may be necessary for that particular phenotype. The same reasoning is used when judging the importance of a piece of DNA that is conserved across many species. In essence, that DNA is exhibiting much fewer mutations than expected by chance and therefore, the classical argument goes, it must be important in some way. Many genes for instance, have sequences that are well conserved across related species. A similar argument can be made here. This is not merely an academic argument. If we knew that the homeostasis of a given pathway is crucial for tumor development in a particular type of cancer, and if we could temporarily disturb that pathway without killing the patient, we could, in principle, destroy the tumor. Hence, in principle, the functional profiling should be done for both over- and under-representation. By performing only enrichment, one gives up the ability to ever discover if any GO terms (e.g., biological processes) are essential to the phenotype.

24.5.3 Reference set

All of the statistical models mentioned above calculate the probability of having the observed number of genes annotated to a given GO term when a random draw is performed from the same reference set. Therefore, it is critical that an appropriate reference set is used to calculate enrichment or depletion of GO terms in a gene set. The reference set should *only include the genes that were monitored in the experiment.* This set is often different from the background (total) set of genes in a genome, yet many of the currently available functional profiling tools use an incorrect reference set and hence produce incorrect results.

Although this problem is very simple and easy to understand, using the incorrect reference set is an extremely common mistake in GO analysis. Most software tools performing such an analysis have a number of choices in their input interface, and it is very easy to neglect the lonely text box, menu item, or radio button that asks the user to specify the reference set. Even more dangerously, some tools have a default value for this reference set such as the set of all genes in the genome of interest, or the set of all genes in some reference database such as Entrez Gene, etc. The presence of a default value makes the choice of the reference set appear as optional, and many users fall prey to the temptation to just use their list of differentially expressed (DE) genes and leave everything else with the default choices. In reality, there should be no default choice for the reference set, much as there is no default choice when dining out in a restaurant. Imagine a dialogue with your waiter in which you are asked to specify the silverware you would like to use, the color of the napkins, the size and shape of the plates, height and shape of the glasses, etc., but, unless you remember to ask, the food itself would be a "default" choice. In both cases, this makes no sense. Much the same way it is impossible for the chef to guess what an individual guest might want to eat, it is also impossible for the designers of a piece of software to guess what microarray an individual scientist might use. Various choices for a "default" can be defended but still they do not make sense. For instance, it is possible to use the most common dish ordered in that particular restaurant as the default choice. However, even under these circumstance, that choice will be wrong for a large proportion of customers. In essence, the reference set must be specified *exactly*, either as a list of genes, or as a specific array, and should never be left to any default value (unless of course, that default value happens to be exactly what was used in the current experiment, which is usually unlikely).

24.5.4 Correction for multiple comparisons

Two types of questions can be addressed when performing functional profiling: i) a hypothesis-generating query, such as "what are the GO terms that are significant in this set of genes?" or ii) a hypothesis-driven query, such as "is *apoptosis* significantly enriched or depleted in this experiment?" In the latter case, one can include all of the genes annotated both directly to "apoptosis" and indirectly to all of its children in a single set of genes, and calculate the p-value that corresponds to the observed number of DE genes in this set. This maximizes the statistical power because no correction for multiple comparisons is required and the raw p-value yielded by the computation above can be compared directly with one of the usual significance thresholds (e.g., 1% or 5%).

The hypothesis-generating approach can also be very valuable. This usually implies calculating a p-value for each GO term and selecting those that are significant at the chosen significance level. However, now we are performing a large number of tests in parallel and the raw p-values obtained for each GO term must be corrected, as discussed in Chapter 16. The simplest correction, Bonferroni, would multiply the p-values of all of the terms with the number of tests performed in parallel. If genes are propagated all the way up, the number of tests is equal to the number of terms in the GO hierarchy, currently 33,587 (as of February 2011). In practice, a term would need to have a raw p-value less than $4 \cdot 10^{-6}$ for it to be significant at the 1% significance level. Other corrections such as Holm's [211, 215, 217, 381] and FDR [45, 46] are less conservative but loss of power cannot be completely avoided. Reporting GO terms as significant based on the raw p-values is probably the most common mistake in the current scientific literature.

24.6 Using a custom level of abstraction through the GO hierarchy

The best way to deal with the problem of multiple comparisons is eliminate the multiple tests by only asking the right question, as in the hypothesis-driven example above. If that is not possible, the second best is to ask as few questions as possible. Let us recall that in the hierarchical structure of the GO the genes are annotated at various level of abstraction (see Fig. 24.2). For instance, "induction of apoptosis by hormones" is a type of "induction of apoptosis," which in turn is a part of "apoptosis." Apoptosis represents a higher level of abstraction, more general, whereas induction of apoptosis by hormones represents a lower level of abstraction, more specific. When annotating the genes with the GO terms, efforts are made to annotate the genes with the highest level of details possible which corresponds to the lowest level of abstraction. For example, if a gene is known to induce apoptosis in response to hormones, it will be annotated with the term "induction of apoptosis by hormones" and not merely with one of the higher-level terms such as "induction of apoptosis" or "apoptosis."

The higher the level used to interrogate GO, the fewer multiple comparisons are performed and hence, the higher the statistical power to detect truly significant GO terms. For instance, if an analysis were to be done at the level of apoptosis (shown by the blue dashed line in Fig. 24.2) there will be only seven GO terms to test in parallel.[5] If the analysis were to be done five levels further down, at the level of induction of apoptosis by ionic changes (shown by the purple dash-and-dot line), there will be $6 \cdot 2^5 = 196$ terms.[6] This represents a loss of power of more than 2 orders of magnitude. However, as discussed in Chapter22, the higher levels are far more abstract than the lower ones, and therefore, an analysis at a high level may have a limited benefit in terms of understanding the underlying phenomena.

A better approach is to actually use the biological knowledge and the motivation that stood behind the design of the experiment to also interpret it. In practice, one should formulate a custom cut through GO for each experiment. This custom slicing could use a high level of abstraction for most GO terms that are *not* expected to be involved in the given condition, and a low level of abstraction in those areas of GO in which a detailed understanding is necessary.

Let us consider for instance, an experiment aiming at investigating the mechanism through which apoptosis is induced in a given phenotype. Such an experiment could use the inverted custom cut shown in Fig. 24.3. Here, the GO terms that are relevant to the specific biological question are expanded up to a level of detail that allows the results to provide the answer: a significant enrichment of one (or more) of the lower terms will indicate the mechanism(s) through which apoptosis is induced here. The other GO terms are collapsed to a much higher level at which they will still provide information about the other branches of GO, just in case. Note that this custom cut through GO is independent of the actual experiment results and should be chosen before the experiment is performed, much like in hypothesis testing the hypothesis to be tested should be chosen before any statistic is calculated (see Chapter 11).

This approach would group all genes associated with the descendants of the terms not expected to be related to the condition, to the terms themselves, greatly improving the statistical power by: i) grouping the evidence in the most effective way and ii) reducing the number of parallel tests. For the specific terms that are important for condition studied, the level of abstraction should be as low as necessary to answer the scientific question posed but not lower.

---

[5] The numbers are referring to Fig. 24.2, which does not include all terms at that level of GO.

[6] This is a conservative estimation by taking only the six other nodes shown in the figure on the same level with apoptosis and very conservatively assuming that each node has only two children on each level.

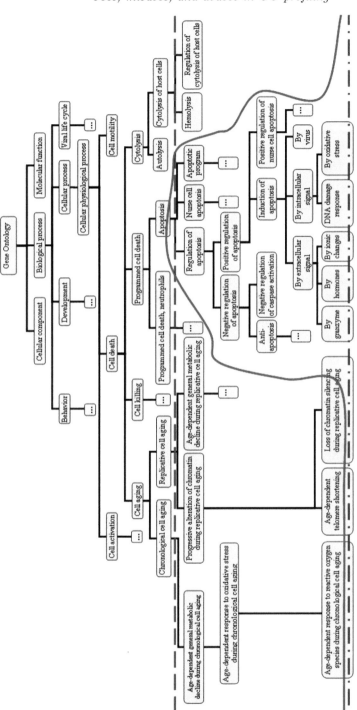

FIGURE 24.2: Different levels of abstraction in GO. The analysis can be performed at a lowest level of abstraction (purple dash-and-dot line), at a fixed level of abstraction chosen by the user (blue dashed line), or at a custom level of abstraction that can go to different depths in various subtrees of the GO (red continuous line). In this case, the custom cut represented by the red line will test the specific biological question "is apoptosis (as a whole) significantly involved in this condition"? Note that this custom cut is not optimal from the point of view of multiple comparisons (see the discussion in the text).

This practice maintains the ability to screen the entire genome and the entire GO DAG for potential new phenomena (not expected when the experiment was designed), while also maximizing the statistical power available to detect significant terms.

24.7 Correlation between GO terms

As explained in Chapter 22, the True Path Rule implies that if the association between a gene g_i and a term t_j is true, then all the associations between g_i and any term that is an ancestor of t_j, all the way to the root, must also hold. Hence, when performing a GO profiling, one could use one of the following choices:

1. A GO analysis **with direct annotations only**: for each term, T_i, consider only the genes directly associated to that particular term.

2. A GO analysis **with complete propagation**: for each term, T_i, consider the genes directly associated to that particular term as well as all genes associated with any of the descendants of T_i.

The first choice above, the GO analysis with direct annotations only, is proper and provides solid results. This approach is shown in Fig. 24.4. Those GO terms that are significant after this type of analysis followed by the correction for multiple comparison are perfectly reliable. The criticism here is that this procedure may be conservative. For instance, it is possible that none of the GO terms in a given sub-graph (e.g., that for biological processes) is significant after this analysis. This means that none of the GO nodes had enough DE genes to make it significant. However, this may be – the argument goes – because various DE genes have been spread by the annotation process in various low-level, high-specificity nodes. For instance, there could be some DE genes associated with induction of apoptosis by intracellular signals, some others associated with induction of apoptosis by extracellular signals, etc. While none of these terms have enough DE genes to be significant, all these genes are involved in apoptosis. If the analysis were to be done at that level, considering all genes associated with any of its descendants, apoptosis would be significant. This argument leads to the second choice described above.

This second choice is to perform the analysis after propagating all genes all the way throughout the entire DAG. Essentially, in this case, we either do not know a priori where to do this analysis for the results to meaningful (or we do not care to make a custom cut as suggested in the previous section). This situation is illustrated in Fig. 24.5. In this case, all genes are propagated up throughout the entire GO hierarchy. The argument for doing this is the True Path Rule. Since, all the relationships between a g_i and all ancestors of

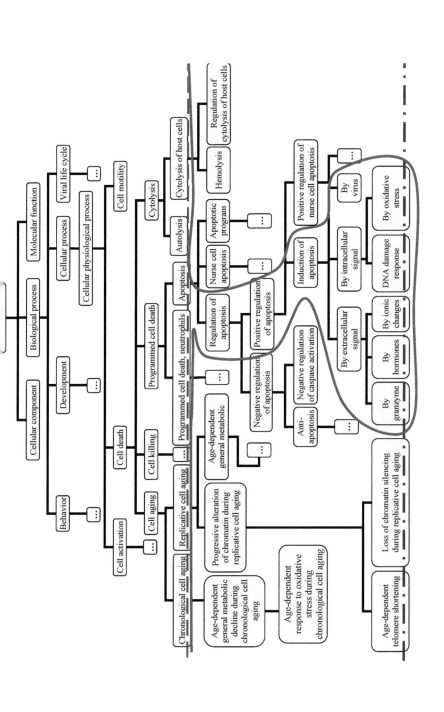

FIGURE 24.3: An inverted cut through GO. In this example, the experiment aims to clarify the mechanism of apoptosis induction. This type of custom cut uses the lowest levels of abstraction in those branches of GO that can provide the answer to the specific biological question posed. In the other directions, the level is kept relatively high in order to maximize the statistical power by minimizing the number of parallel tests.

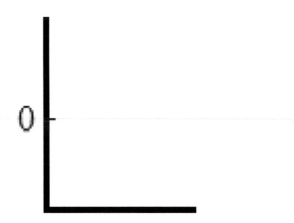

FIGURE 24.4: GO analysis with direct annotations only. Genes g1 through g7 have been found to be differentially expressed (DE) in this experiment and constitute the input of this analysis. Each gene is shown next to the node corresponding to the GO node with which the gene has been annotated. These are the most specific terms that are true for the given genes. For instance, g4 is known to be involved in t2 but not in t4, nor t5. A valid choice is to do the GO analysis considering only the set of genes annotated with each specific GO term: t1 will be analyzed for enrichment considering only g5, t2 will be analyzed considering only g4, t4 will be analyzed considering g1 and g2, etc.

t_j are true, g_i are justified to appear in all these ancestors. Subsequently, the entire set of genes collected in each term after the propagation of all genes all the way up to the root is analyzed for over- or under-representation of DE genes. One issue related to this practice of propagating genes up through the hierarchy is that the tests performed for nodes on any given path will be clearly correlated because the same genes appear in many of these nodes. For instance, in Fig. 24.5 g6 is used as evidence for t8, t6, t3, t1, and all the other nodes from t1 to the root. It is entirely possible that certain terms higher in the GO hierarchy (such as t1, t2, and t3 in Fig. 24.5) appear to be significant only due to the genes propagated to them from below. This is acceptable as long as the terms below them were not significant due to the very same genes. However, if multiple related terms such as t6 and t8 are all significant due to the same genes (g6 in this case), there seems to be little benefit from reporting t6 as significant, once t8 was reported as such.

Furthermore, as discussed above, the number of parallel tests has now increased substantially since basically all 33,587 GO terms[7] are now indiscriminately tested at the same time. Unfortunately, not only that these GO terms are all tested in parallel but the assumption of independence does not hold anymore because now we actually know for a fact that evidence is shared between many terms. Not all multiple comparisons correction methods perform well under such circumstances.

Nonetheless, GO's structure is important because the lack of independence comes from a very clear inheritance phenomenon that could be used to decorrelate the analysis of various terms. Various methods have been proposed for the decorrelation of GO terms [13, 188].

A good unbiased search for significant GO associations is the **elim** algorithm proposed by Alexa and Grossman [13]. This uses a bottom-up approach as follows: for every leaf term, calculate p-values with the genes directly associated to it. If any term is significant, do not propagate its genes above. This would provide the most specific node that is significant in that particular branch. If a term is *not* significant, propagate its genes to its parent(s). The genes will propagate upwards until a significant node is found or until the root is reached. Since this process is applied bottom up, all children of a node are processed before the node itself. A careful analysis is still necessary to properly correct for multiple comparisons.

An alternative approach proposed by the same group is the **weight** algorithm. In this variation, the aim is to establish what are the GO terms that best represents a set of genes. In this context, a GO term represents better a given set of genes if it is more enriched in genes from this set than its neighbors. In order to decide whether a GO term t better represents the given set of genes, the enrichment score of the term t is compared with the enrichment score of the children. The children with a better score than t represent the

---

[7]In the GO version 1.1778, there are 20,299 biological processes, 2,806 cellular components, and 8,992 molecular functions.

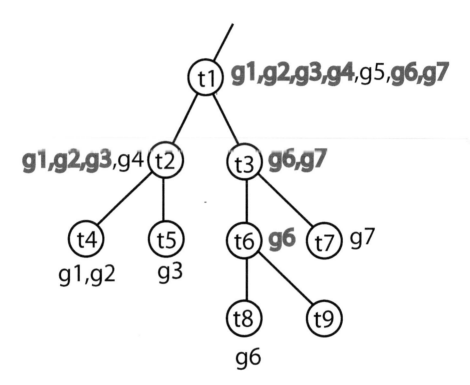

FIGURE 24.5: GO analysis with complete propagation. In this approach, all genes are propagated throughout the GO DAG, all the way to the root. In this example, the p-value of t1 will be calculated with g1, g2, g3, g4, g5, g6, and g7, the p-value of t2 with g1, g2, g3, and g4, etc. The red genes are the genes that were propagated up through the graph. Note how g6 is used as evidence for t8, t6, t3, t1, and all the other ancestors up to the root. The consequence of this is that the GO terms are now correlated to various extents, and their p-values are not independent anymore.

interesting genes better. Hence, their significance will be increased while the significance of the children with a lower score than t will be reduced. Some of the issues related to this approach are that: i) the choice of the weights is more or less arbitrary and ii) although the redundant evidence is weighted down, the redundancy is still present and rather difficult to quantify or to control. As always, a correction for multiple comparisons is still required.

A third alternative proposed to deal with the problem of correlation between GO terms is the **parent-child analysis** proposed by Grossmann et al. [188]. This approach takes into consideration the structure of the GO by conditioning the probability of observing the given number of DE genes in a term t, on the event that a random set of the same size would overlap exactly as observed with the set of genes annotated with the parent(s) of the term t. A complication is given by the fact that in a DAG a node can have more than one parent. To deal with this, Grossmann et al. considered two variations, one using the set of genes in the intersection of the parents (i.e., genes annotated with all the parents), and the other one using the union of the genes associated with the parents (i.e., genes annotated with any of the parents). Their experiments suggest that the parent-child-intersection variant performs better than the parent-child-union.

24.8 GO slims and subsets

GO slims are smaller versions of GO that contain only a subset of terms from the entire GO. These terms are custom sets of terms that reflect the interest of its creator(s). By using a SLIM, one improves the statistical power by reducing the number of parallel tests and also focuses the attention of the user to only what is relevant in the given context. This is done by eliminating those terms that are not relevant to the organism, experiment, or context. Fig. 24.1 shows the GO slims available as of February 2011 and their creators.

A GO subset is a similar concept with a different implementation. The idea is similar: to use only those terms that are relevant for a given category of organisms. For instance, there is a GO subset that was created for prokaryotes. This slim will not contain terms such as nucleus, mitochondrion, etc. Unlike GO slims, which are generated as separate files, GO subsets are stored as categories in the GO ontology file.

Organism or Usage	Developer
Generic GO slim	GO Consortium
UniProtKB-GOA and whole proteome analysis	N. Mulder, M. Pruess
Plant slim	The Arabidopsis Information Resource
Candida albicans	Candida Genome Database
Protein Information Resource slim	D. Natale of PIR
Schizosaccharomyces pombe slim	V. Wood
Yeast slim	Saccharomyces Genome Database

TABLE 24.1: GO slims available as of February 2011.

24.9 Summary

This chapter discussed a number of issues and common mistakes related to the use of GO for the analysis and interpretation of high throughput data. Some of the important issues that tend to be neglected by many users of GO are related to: the meaning of the ND annotation, the NOT annotation and the direction in the GO DAG. The ND (no biological data available) annotation means that considerable efforts have been made to search the literature for that particular gene or gene product and nothing has been found. This is very different from a gene that merely does not have annotations yet, which may be due simply to the fact that this gene has not been looked at by curators. The NOT annotation is used very sparingly and indicates a function or property that *was expected to be associated* with the given gene or gene product, but was found not to be true. Most GO profiling tools disregard this type of annotation and use incorrectly the information associated to the NOT qualifier. The direction of the edges in any subgraph of the GO varies from one source to another. Directing the edges as outgoing from the less specific terms to the more specific terms is more consistent with the usual notation in trees and DAGS (it yields one root and many leaves) but is counterintuitive from a biological point of view. Directing the edges in the opposite way fits the biological interpretation but yields a DAG with many roots and only one leaf.

Common mistakes in GO profiling include: i) using some arbitrary "enrichment" measure rather than a statistically sound *p*-value, ii) testing only for enrichment, iii) using a default or an incorrect reference set, iv) failing to correct for the multiple tests performed in parallel. This chapter discussed these issues in details.

Arguably the best way to perform the GO profiling is to formulate specific biological questions before the experiment. A custom cut through GO that is tailored to the given questions asked will provide the maximum statistical power by reducing the number of parallel tests.

Finally, the GO profiling can be done in two fundamentally different ways: i) using for each GO term only the genes specifically annotated with those term (profiling with direct annotations), or ii) using for each GO term all genes annotated with either that term directly or to any of its descendants (profiling with complete propagation). The first choice above is sound but may miss higher-level terms for which the evidence is distributed among its descendants. The second choice above is justified by the true path rule but introduces serious correlations between the GO terms since all of the evidence used for a term will be also used redundantly for all its ancestors. Three methods have been proposed to deal with this. A first algorithm, elim, does the analysis bottom-up and only propagates up the DAG the genes of those terms that are not significant. A second algorithm, weight, uses a weighting mechanism to reduce the correlation between the terms. A third approach, parent-child, conditions the probability of observing a given number of DE genes just by chance to the event of observing the same overlap among the parents. No matter what type of profiling is performed (with direct annotations only or with complete propagation, with or without correction for correlations, etc.) the results of the GO profiling must be corrected for multiple testing.

Chapter 25

A comparison of several tools for ontological analysis

25.1 Introduction

Independently of the platform and the analysis methods used, the result of a microarray experiment is, in most cases, a list of genes found to be differentially expressed.[1] The common challenge faced by the researchers is to translate such lists of differentially regulated genes into a better understanding

---

[1]Some of the material in this chapter is reprinted from Khatri P., Draghici S., "Ontological analysis of gene expression data: current tools, limitations, and open problems," Bioinformatics, 2005, Epub ahead of print, with permission from Oxford University Press, and Khatri P., Draghici S., "A comparison of existing tools for ontological analysis of gene

of the underlying biological phenomena. This challenge led to the development of the automatic ontological analysis approach discussed in Chapter 23. Currently, this over-representation (ORA) approach is the *de facto* standard for the secondary analysis of high throughput experiments and a large number of tools have been developed for this purpose. Since 2001 when the first such tool appeared, over a dozen other tools have been proposed for this type of analysis and more tools continue to appear every day (see Fig. 25.1). Although these tools use the same general approach, they differ greatly in many respects that influence in an essential way the results of the analysis. In most cases, researchers using such tools are either unaware of, or confused about certain crucial features.

This chapter presents a comparison of 15 tools currently available in this area. A detailed analysis of the capabilities of these tools, of the statistical models deployed as well as of their back-end annotation databases (if applicable), is included here in order to help researchers choose the most appropriate tool for a given type of analysis.

More importantly, we will also discuss some of the issues associated with the current ontological analysis approach. Since all existing tools implement the same approach, these drawbacks are also associated with all tools discussed and represent conceptual limitations of the current state-of-the-art in ontological analysis.

Due to the amazing speed with which things evolve in this area, the specific details of any particular tool are likely to become obsolete rather quickly after the publication of this book. However, a discussion of this type is still very useful in order to illustrate the criteria that can be applied to assess various tools, as well as the main issues that are relevant when choosing one particular tool. These criteria and issues are likely to retain their importance over time while the specific capabilities and features of each tool described here are likely to change from one version to another.

25.2 Existing tools for ontological analysis

The ontological analysis approach described in Chapter 23 was first introduced in 2001 by Onto-Express [264]. A discussion of the various statistical approaches that can be used used to calculate the significance of various categories followed one year later [129]. During the following two years, from 2003 to 2005, 15 other tools have been proposed for this type of analysis. By and large, all these tools use the same approach. The tools discussed here are: FatiGO [12], GOstat [40], eGOn [151], DAVID [110], GeneMerge [83], FuncAs-

expression data," published in "Encyclopedia of Genetics, Genomics, Proteomics, and Bioinformatics" copyright John Wiley & Sons, 2005, with permission.

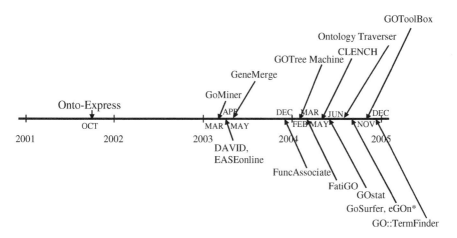

FIGURE 25.1: The evolution of GO-based functional analysis software. At the time of publication, eGOn was not published in a peer-reviewed journal yet.

sociate [49], CLENCH [383], GOToolBox [300], GOSurfer [487], EASE [218], OntologyTraverser [478], GoMiner [482], GO::TermFinder [65], and GOTree Machine [483]. Fig. 25.1 provides a time-line of the development of these tools. The comparison described in this chapter is based on the review of the published manuscripts, user manuals, and our experience with each tool. In the following, we first describe each tool very briefly and then compare their capabilities and features.

Onto-Express (OE) was the first tool to propose the automated functional profiling through the integration of dbEST [58, 57], UniGene [379], and Entrez Gene [294] databases in the Onto-Tools database. The current version of OE supports KEGG, Entrez Gene, PubMed, RefSeq, GenBank, and Gene Ontology databases. OE's back-end database is currently being expanded to integrate GenPept, Protein Information Resource (PIR), Protein Data Bank (PDB), Swiss-Prot, TrEMBL, Online Mandelian Inheritance in Mammals (OMIM), HomoloGene, and Eukaryotic Promoter Database (EPD) databases.

At present, OE can process lists of genes specified as GenBank accession IDs, UniGene cluster IDs, Entrez Gene IDs, Affymetrix probe IDs, gene symbols or the organism-specific database IDs used in GO. OE's output is organized in various views that include: i) a flat view, which is a bar-graph view that only considers the annotations at the most specific level, ii) a tree view, which represents the results in a custom cut through the GO hierarchy as discussed in Chapter 23 and iii) a synchronized view which is a custom bar-graph synchronized with the tree view. The combination of the tree and synchronized views allows the user to select a custom level of abstraction for each branch of the gene ontology. This allows the user to ask specific questions about the phenomenon studied by implementing an arbitrary user-designed cut through the GO hierarchy. For instance, one could ask whether the apoptotic pathway as a whole is affected without making the distinction between various sub-processes such as positive or negative regulation of apoptosis, induction of apoptosis by various causes (e.g., extracellular signals, intracellular signals, hormones), etc. For this purpose, the user can use OE's tree view to collapse the categories that are unnecessarily specific into more abstract categories. For instance, the approximatively 40 subcategories of apoptosis can be collapsed into a single *apoptosis* node. Subsequently, the analysis will be done at this level of abstraction directly providing the answer to the desired question. A more detailed discussion of this issue and its consequences are included in Section 25.3.

The Onto-Express database currently includes 349 arrays from 8 manufacturers that can be selected as the reference. Any other array can be uploaded by the user as a custom reference array. A distinguishing feature of OE is the ability to sort the entire GO tree by the total number of genes, p-values, or terms. Another unique feature is a chromosomal location profile that shows the location of the various differentially regulated genes on various chromosomes.[2] Onto-Express also features an application programming interface (API) that allows any other bioinformatic tool to place queries and display the results of the analysis. This capability is currently used by two commercial companies (Insightful and SAS), which integrated OE with their respective software packaged for microarray data analysis: S+ ArrayAnalyzer and SAS's Microarray Solution.

GoMiner relies on the Gene Ontology database and can only process gene symbols as input. The same group provides a separate tool, MatchMiner, which converts other types of IDs into gene IDs that are appropriate for use with the GoMiner. This does allow the user to perform the analysis using other types of IDs but adds a separate step to the analysis pipeline since the user has to manually take the results from MatchMaker and submit them to GoMiner.

By default, GoMiner's analysis is not organism specific (Fig. 25.3). This means that if a gene is annotated with function A in one organism and function B in another organism, GoMiner will treat it as if it were annotated with A *and* B, independent of the organism that is actually studied. The user has to choose the desired organism and/or the appropriate data source from the menu to restrict the results to a specific organism. GoMiner is able to represent the results graphically as a directed acyclic graph (DAG). This is achieved using an Adobe scalable vector graphics (SVG) web browser plug-in, which is only compatible with MS Windows. In consequence, GoMiner is platform

---

[2]However, this feature is currently available only for human genes.

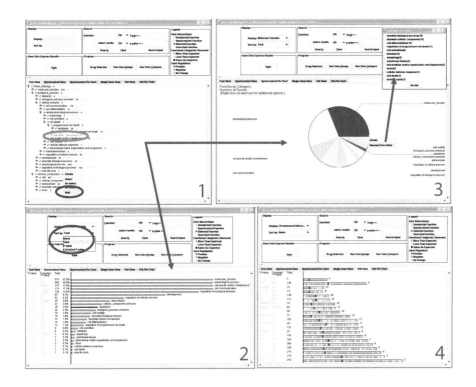

FIGURE 25.2: Onto-Express output interface: tree view (panel 1), synchronized bar chart view (panel 2), synchronized pie chart view (panel 3), and chromosome view (panel 4). The user can select different levels of abstraction for different branches of the GO hierarchy in the tree view by expanding or collapsing the desired nodes (panel 1). Switching to the synchronized bar chart view performs the analysis on the GO terms selected in the tree view (panel 2). The tree view also displays the fold change value for an input gene as provided by the user (highlighted in panel 1). In the synchronized view results can be sorted by name, by total number of genes for each term, by *p*-value or by corrected *p*-value (highlighted in panel 2). Categories in the pie chart can be added or removed by the user.

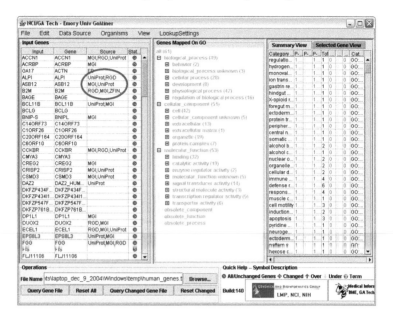

FIGURE 25.3: GoMiner output. By default, GoMiner results lack organism specificity. In response to a query with human genes, GoMiner can return annotations from either one of MGD (mouse), RGD (rat), ZFIN (*D. rerio*), and FlyBase (*D. melanogaster*).

dependent (Windows only) although it is written in Java and could in principle be used on other platforms, as well.

DAVID and EASEonline. DAVID (Database for Annotation, Visualization and Integrated Discovery) provides an HTML-based interface that allows the user to query using a set of genes. The results are presented as a bar chart, indicating the percentage of the input genes for each GO term. The results are very simple, do not consider the GO hierarchical structure and do not calculate any p-value to characterize the statistical significance of the observed results. On the plus side, DAVID does allow the user to specify a depth of analysis, which can be used to perform a horizontal cut through the GO hierarchy.

EASE (Expression Analysis Systematic Explorer) is a companion to DAVID. EASE allows the user to submit Affymetrix IDs, UniGene cluster ID, Entrez Gene ID or GenBank accession IDs and calculates the p-values of various categories using Fisher's exact test or an EASE score. As shown in Fig. 25.4, EASE does not consider or show the ontological categories in the context of the hierarchical structure of GO. EASE currently supports 84 Affymetrix arrays as possible reference arrays.

GeneMerge only accepts organism specific gene IDs used by the various

FIGURE 25.4: EASE input (left panel) and output (right panel) interface. EASE supports GenBank accession IDs, Affymetrix probe IDs, UniGene cluster IDs, and Entrez Gene IDs as input. The user can either submit a file containing the input IDs or paste the input IDs in the text area (left panel). EASE does not consider the hierarchical structure of the GO in the analysis (right panel).

organism-specific annotation groups to deposit functional annotations in the GO Consortium database. Although in principle, it supports more types of IDs as input, its usefulness is somewhat limited since it does not support commonly used IDs such as GenBank accession numbers, RefSeq mRNA, protein IDs, etc. In order to address this, GeneMerge's authors have recently provided a gene name converter that translates a larger set of IDs into the required format accepted by GeneMerge. As a limitation, GeneMerge only allows to query one of the three principal GO categories at a time. Outstanding features for GeneMerge include the ability to query a few organism-specific features such as KEGG metabolic and signalling pathways for yeast and fruit fly, and deletion viability data for yeast, etc. (Fig. 25.5).

FuncAssociate uses Fisher's exact test to calculate the *p*-values and a Monte Carlo simulation to correct for multiple hypothesis testing (Fig. 25.6). FuncAssociate allows the user to submit a rank-ordered list of genes as input, such as a list of genes ranked by fold changes. Given this, FuncAssociate ranks the GO terms in the result according to which initial segment of the input list gives the most significant degree of over- or under-representation, where the initial segment for each gene *g* in the original ordered set *Q*, is the subset of *Q* consisting of *g* plus all the genes that precede *g* in the given order [49].

GOTree Machine (GOTM) has the ability to visualize the results both as a hierarchical tree as well as a bar chart (Fig. 25.7). However, GOTM requires the user to select a specific depth in the GO hierarchy, before displaying a bar chart. GOTM supports 37 Affymetrix arrays as reference arrays. A key aspect

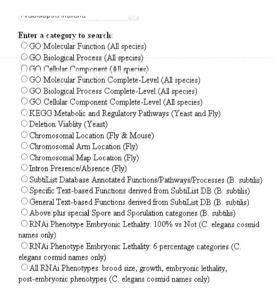

FIGURE 25.5: GeneMerge input interface. In addition to functional profiling, GeneMerge also allows to query for KEGG pathways and chromosomal location for yeast and fly, RNAi phenotypes for *C. elegans*, and functions, processes and pathways for *B. subtilis*.

OVERREPRESENTED ATTRIBUTES

Rank	N	X	LOD	P	P-adj	GO Attribute
1	111	4678	0.703	6.6e-29	<0.001	0007154: cell communication
2	95	3847	0.670	6.3e-25	<0.001	0007165: signal transduction
3	55	1359	0.809	5.1e-23	<0.001	0006793: phosphorus metabolism
4	55	1359	0.809	5.1e-23	<0.001	0006796: phosphate metabolism
5	46	995	0.851	1.8e-21	<0.001	0006468: protein amino acid phosph
6	44	954	0.846	1.7e-20	<0.001	0004672: protein kinase activity
7	141	9175	0.563	1.7e-19	<0.001	0009987: cellular process
8	46	1132	0.790	3.1e-19	<0.001	0016310: phosphorylation
9	45	1093	0.794	4.9e-19	<0.001	0016773: phosphotransferase activ
10	63	2246	0.650	3.1e-18	<0.001	0006464: protein modification
11	46	1299	0.725	6.3e-17	<0.001	0016301: kinase activity/phosphok
12	47	1367	0.713	8.5e-17	<0.001	0016772: transferase activity, tr
13	79	3728	0.546	4.8e-16	<0.001	0004871: signal transducer activi
14	11	31	1.841	8.6e-16	<0.001	0004907: interleukin receptor act
15	11	32	1.821	1.3e-15	<0.001	0019965: interleukin binding/IL b
16	13	72	1.452	3.4e-14	<0.001	0019955: cytokine binding
17	24	413	0.914	1.3e-13	<0.001	0004713: protein-tyrosine kinase
18	13	81	1.391	1.7e-13	<0.001	0004896: hematopoietin/interferon-
19	13	88	1.348	5.2e-13	<0.001	0004714: transmembrane receptor p
20	82	4623	0.459	2.5e-12	<0.001	0019538: protein metabolism/prote
21	32	887	0.706	4e-12	<0.001	0050794: regulation of cellular p
22	59	2733	0.512	5.2e-12	<0.001	0017076: purine nucleotide bindin
23	52	2219	0.539	7e-12	<0.001	0030554: adenyl nucleotide bindin
24	59	2767	0.506	8.7e-12	<0.001	0000166: nucleotide binding
25	13	110	1.237	9.7e-12	<0.001	0019199: transmembrane receptor p
26	51	2186	0.535	1.4e-11	<0.001	0005524: ATP binding
27	35	1207	0.609	1.5e-10	<0.001	0008283: cell proliferation
28	151	12633	0.412	1.5e-10	<0.001	0005488: binding/ligand
29	14	169	1.066	1.9e-10	<0.001	0007167: enzyme linked receptor p
30	35	1242	0.596	3.2e-10	<0.001	0005887: integral to plasma membr
31	42	1716	0.541	3.2e-10	<0.001	0005886: plasma membrane/bacteria
32	12	120	1.155	4.5e-10	<0.001	0007169: transmembrane receptor p

FIGURE 25.6: FuncAssociate output interface. FuncAssociate allows the user to submit a rank-ordered list of genes as input. The GO terms in the results are then ordered in the increasing order of the p-values. The ranks of the GO terms are displayed in the first column and their p-values are displayed in the fifth column.

of GOTM is the performance. The output interface of GOTM is based on HTML which means that every user interaction, such as a request to expand a GO term or a request to retrieve a list of input gene IDs, is an HTTP request over the Internet. This is reflected in a slower performance compared to GoMiner or to Onto-Express which only query their respective back-end database once, after which all processing is done locally.

FatiGO is also a web-based tool. FatiGO performs the analysis on only one of the three principal GO categories per request and requires the user to select the depth in the GO at the beginning of the analysis. However, the analysis can only be performed up to the sixth level in the GO hierarchy. FatiGO displays the results both as a bar chart as well as a completely expanded GO tree in a single HTML page. If the input only contains a relatively small number of genes (e.g., fewer than 50) and the number of GO terms is limited, this display is informative and convenient since little or no page scrolling is required. However, if the input list contains a few hundreds of genes, which usually yields a functional profile with a few thousand GO terms, the single-page display becomes overwhelming.

CLENCH is a command-line, stand-alone tool specifically designed for *Arabidopsis thaliana*. At the moment, CLENCH uses TAIR as a unique source of annotations. In consequence, *A. thaliana* annotations originated at TIGR may not be considered in the analysis. CLENCH retrieves the annotations for

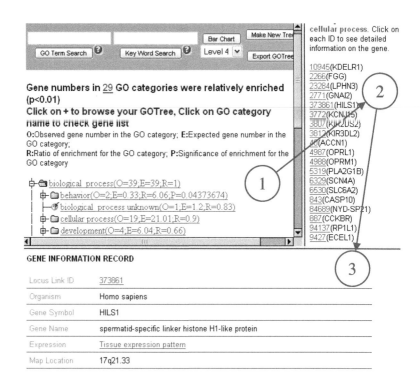

FIGURE 25.7: GOTree Machine (GOTM) output. GOTM results are organized in three frames as numbered in the figure. The first frame displays the results in the GO hierarchy. Clicking a GO category in the first frame displays the input genes annotated with the GO category in the second frame. Clicking on a gene in the second frame presents more details about the gene such as the gene name and symbol, Entrez Gene ID, location on a chromosome, etc., in the third frame.

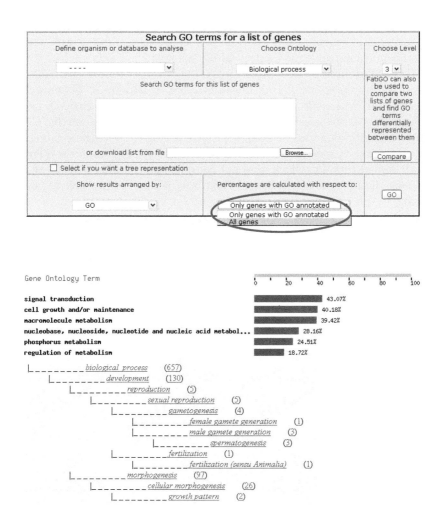

FIGURE 25.8: FatiGO input (upper panel) and output (lower panel) interface. FatiGO only allows to query one of the three principal GO categories at a time. As seen in the highlighted area, FatiGO either uses the entire genome or only the genes annotated with GO terms as the reference. FatiGO presents the results as bar charts as well as a completely expanded GO tree in a single HTML page (lower panel).

A. thaliana at runtime from the TAIR web site. Similar to FatiGO, CLENCH also outputs the results as a single HTML file. Unlike FatiGO, it does not show or consider the hierarchical structure of GO. The local installation of CLENCH requires a previous installation of Perl as well as specific Perl modules. Hence, the initial deployment of CLENCH as well as its command-line interface may be a bit uncomfortable for the typical life scientist.

GOstat is another web-based tool. It supports *Homo sapiens, Mus musculus, Rattus norvegicus, Arabidopsis thaliana, Danio rerio, Caenorhabditis elegans*, and *Drosophila melanogaster*. Similar to EASE and FatiGO, GOstat requires the user to select the level of abstraction in the GO hierarchy. GOstat supports GenBank accession number, UniGene cluster IDs, gene symbols, and various organism specific IDs. GOstat represents the results in an HTML or tab-delimited text file, which does not consider the GO hierarchy. GOstat offers false discovery rate (FDR) and Holm's correction for multiple hypothesis testing.

GOToolBox is also a web-based tool and supports *Arabidopsis thaliana, Drosophila melanogaster, Mus musculus, Homo sapiens, Caenorhabditis elegans*, and *Saccharomyces cerevisiae*. GoToolBox only supports the organism-specific IDs used in the GO database. GOToolBox either requires the user to select a specific depth in the GO hierarchy or considers the lowest level in the GO hierarchy.

GoSurfer is available as a standalone application and supports Affymetrix probe IDs, Entrez Gene IDs, and UniGene cluster IDs as input. It requires the user to download a specifically formatted GO structure file and at least one gene information file, which is also specifically formatted. It displays the results as a DAG on which zoom operations are possible. However, none of the nodes in the DAG are labeled, which makes it very difficult for the user to interpret the results or look for a specific GO term.

OntologyTraverser is described[3] as a web-based tool as well as an R package. Its web-based version allows the user to query one GO category at a time and calculates the *p*-values using a hypergeometric distribution. However, the model seems to be applied rather differently: instead of considering the number of genes for a given GO term, this tool seems to use the number of genes annotated with the terms at the same level in GO [478]. If this description is accurate, this usage of the hypergeometric model may be difficult to justify from a biological point of view.

eGOn. Similar to EASE, GeneMerge, FatiGO and OntologyTraverser, eGOn only allows the user to query one of the three main GO categories at a time. It only supports GenBank accession numbers, UniGene cluster IDs, and clone IDs as input. eGOn only provides the binomial model for calculating *p*-values and does not offer correction for multiple hypothesis. The analysis performed by eGOn is centered around UniGene clusters. The fundamental

---

[3]This is the only tool that we could not use ourselves. In spite of numerous attempts over several weeks we always encountered the following error: "Error unmarshaling return header; nested exception is: java.io.EOFException."

FIGURE 25.9: GOstat input and output interface. GOstat requires that the user must upload a reference gene list. Otherwise, it uses all genes in the GO database as the reference. The gene names in the GOstat output are hyperlinked to GeneCards database.

FIGURE 25.10: GOToolBox input interface. GOToolBox allows to filter the annotations by evidence codes. If the user does not specify the reference list of genes, GOToolBox uses the genome as the reference.

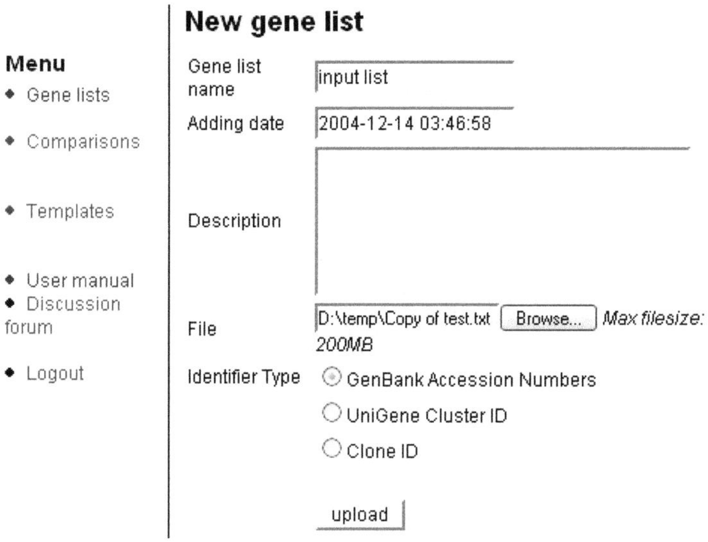

New gene list

Menu
- Gene lists
- Comparisons
- Templates
- User manual
- Discussion forum
- Logout

Gene list name	input list
Adding date	2004-12-14 03:46:58
Description	
File	D:\temp\Copy of test.txt [Browse...] *Max filesize: 200MB*
Identifier Type	⦿ GenBank Accession Numbers ○ UniGene Cluster ID ○ Clone ID

upload

FIGURE 25.11: eGOn input interface. eGOn requires that each gene list and each data analysis is given a unique name. The gene lists and analysis are stored on the server for 30 days.

implicit assumption here is that each UniGene cluster corresponds to distinct genes. However, this assumption may not always be accurate. as it will be discussed in Section 25.3.10.

GO::TermFinder. GO::TermFinder comprises of a set of downloadable Perl modules. A web-based version of GO::TermFinder is also available on Saccharomyces Genome Database (SGD) web-site. The web-based version only allows to query one of the three main GO categories, supports only SGD IDs as input and does not calculate p-values. (Fig. 25.12). However, the stand-alone version of GO::TermFinder supports any organisms in the GO using the organism specific IDs in GO, uses hypergeomatric distribution to calculate p-values, and provides a choice among bootstrapping, Bonferroni, and FDR for multiple hypothesis correction.

25.3 Comparison of existing functional profiling tools

The comparison between the tools currently available for the ontological analysis of high throughput gene expression experiments is summarized in Ta-

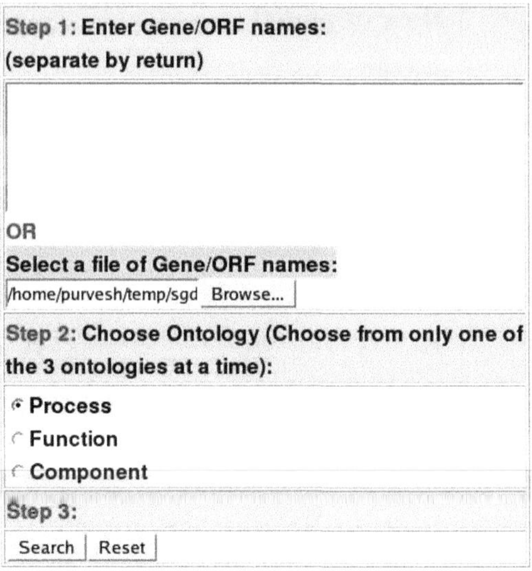

FIGURE 25.12: GO::TermFinder input interface. GO::TermFinder only allows to query one GO category at a time. The web-based version at SGD does not calculate *p*-values.

bles 25.1 and 25.2. The criteria used in these tables are described in details in the following.

25.3.1   The statistical model.

The ontological analysis can be performed with a number of statistical models, including hypergeometric [93], binomial, χ^2 (chi-square) [157], and Fisher's exact test [296]. These tests were discussed in details in Chapter 23 but will be briefly summarized here. The probability that a certain category occurs x times just by chance in the list of differentially regulated genes is appropriately modeled by a hypergeometric distribution. However, the hypergeometric distribution can be more difficult to calculate when large arrays (e.g., Affymetrix HGU133A) are involved. However, the hypergeometric distribution tends to the binomial distribution when the number of genes is large. Therefore, the binomial model is perfectly usable when larger arrays are used. Alternative approaches include a χ^2 test for equality of proportions and Fisher's exact test. In most cases, the difference between the models will not be dramatic.

FatiGO does not use a statistical model as such but does calculate percentages with respect to the genes annotated with GO terms or all known genes in an organism. GoMiner, EASEonline, GeneMerge, FuncAssociate, GOTM, GOSurfer, Ontology Traverser, and eGOn only support one statistical test.

GOstat allows the user to choose between two tests (χ^2, and Fisher's exact test), CLENCH and GOToolBox allow a choice between three tests (χ^2, hypergeometric, and binomial for CLENCH and hypergeometric, binomial, and Fisher's exact test for GOToolBox), while Onto-Express implements all four tests (χ^2, hypergeometric, binomial, and Fisher's exact test).

25.3.2 The set of reference genes

An important consideration when identifying statistically significant GO terms is the choice of the reference list of genes against which the *p*-values for each GO term in the results are calculated. Several tools such as GOToolBox, GOstat, GoMiner, FatiGO, and GOTM[4] use the total set of genes in a genome as the reference [40, 300, 482, 483] or the set of genes with GO annotations [12, 65]. Either of these may be an inappropriate choice when the input list of genes to these tools is a list of differentially expressed genes obtained from a microarray experiment, since the genes that are not present on a microarray do not ever have a chance of being selected as differentially regulated. The fundamental idea is to assign significance to various functional categories by comparing the observed number of genes in a specific category with the number of genes that might appear in the same category if a selection performed *from the same pool* were completely random. If the whole genome is considered as the reference, the pool considered when calculating the random choice includes all genes in the genome. At the same time, the pool available when actually selecting differentially regulated genes includes only the genes represented on the array used, since a gene that is not on the array can never be found to be differentially regulated. This represents a flagrant contradiction of the assumptions of the statistical models used.

25.3.3 Correction for multiple experiments

Another crucial factor in the assessment of a functional category is the correction for multiple experiments, discussed in details in Chapter 16. This type of correction must be performed in all situations in which the functional category is *not* selected *a priori* and many such categories are considered at the same time. The importance of this step cannot be overstated and has been well recognized in the literature [12, 40, 49, 83, 124, 383, 482]. In spite of this, several of the tools reviewed here do not perform such a correction: GoMiner, DAVID, GOTM, CLENCH, and eGOn. GoMiner provides a "relative enrichment" statistic calculated as $R_e = \frac{n_f/n}{N_f/N}$, where n and N are the numbers of genes in the selected and reference sets, respectively, and n_f and N_f are the number of genes in the functional category of interest in the selected and reference sets, respectively [482]. However, this relative enrichment cannot be used

---

[4]GOTM also allows the users to upload their own list of genes or use one of 37 Affymetrix arrays as the set of reference genes.

in any way as a correction for multiple experiments[5] but rather as another indication of the significance of the given category, somewhat redundant to, but less informative than the p-value. This statistic can be misleading because the user will be tempted to assign biological meaning to all those categories that are enriched. In reality, any particular relative enrichment value can actually appear with a nonzero probability just by chance. It is the magnitude of the probability that should be used to decide whether a category is significant or not, rather than the relative enrichment.

All remaining tools deal with the problem of multiple comparisons in some way. EASEonline and GeneMerge support the Bonferroni correction. Bonferroni and Šidák are perfectly suitable in many situations, in particular, when not very many functional categories are involved (e.g., fewer than 50). However, these corrections are known to be overly-conservative if more categories are involved [124]. A family of methods that allow less conservative adjustments of the p values is the Holm step down group of methods [211, 215, 217, 381].

Bonferroni, Šidák, and Holm's step-down adjustment are statistical procedures that assume the variables are independent, which is known to be false for this type of analysis.[6] When it is known that dependencies exist, methods such as false discovery rate (FDR) are more appropriate [45, 46, 124]. Another suitable approach is that of bootstrapping, which actually calculates the null distribution by performing many resamplings from the same data, thus taking into consideration all existing dependencies. Great care should be taken in those situations in which only few categories are involved because the number of distinct resamplings may be insufficient for a reliable conclusion (see Chapter 9 in [124]). In those instances, even Bonferroni or Šidák may be a better choice instead of bootstrapping.

The tools offering more than one correction method effectively allow the researcher to adapt the analysis to the number of categories and degree of known dependencies between them. Bonferroni and Šidák are suitable if few, not directly related categories are involved. If more unrelated categories are involved, Holm's may be a good compromise. If there are several functional categories that are clearly related, FDR is probably the best choice. If the dependencies are very strong (e.g., several sub-processes of the same larger process), a bootstrap or Monte Carlo simulation approach may be better able to capture these dependencies, but only if enough categories are present to make the simulation meaningful.

The one tool standing out regarding this criterion is FuncAssociate which uses a more original Monte Carlo simulation. FatiGO and GOstat implement Holm's and FDR corrections. Onto-Express offers Bonferroni, Šidák, Holm's

---

[5]Note that this statistic does not take into consideration the number of experiments performed in parallel.

[6]The very hierarchy of the GO on which this type of analysis relies, shows that many biological categories are very closely related, sometimes as children of the same node on the next level up.

and FDR, whereas GOToolBox offers FDR, Bonferroni, Holm, Hochberg, and Hommel corrections.

25.3.4 The scope of the analysis

An important factor in assessing the usefulness of a tool is its ability to provide a complete picture of the phenomenon studied. In terms of functional profiling using GO, a complete analysis should include all three primary GO categories: molecular function, biological process, and cellular component as well as other information if available. Among the tools reviewed, eGOn, FatiGO, GeneMerge, and Ontology Traverser only analyze one category at a time. The other tools allow the user to analyze all three categories simultaneously. Extra features are present in GeneMerge and Onto-Express. GeneMerge shows KEGG metabolic and signalling pathways for yeast and fruit fly, and deletion viability data for yeast. Onto-Express also shows KEGG signalling pathway data, as well as a chromosome location of differentially regulated genes (linked to NCBI's Mapviewer for further analysis).

25.3.5 Performance issues

We compared the speed of the tools by submitting four sets of 100, 200, 500, and 1,000 human genes, respectively, to each of the tools (Fig. 25.13). We started with a list of genes containing gene symbols because this type of ID is accepted by most tools in this group (4 tools). Since several tools work only with specific types of IDs, we had to translate these lists of genes into the appropriate type. This was done with Onto-Translate [127, 261]. We translated the lists into Entrez gene IDs for three tools, TrEMBL IDs for three tools and GeneBank accessions for two other tools. The times shown here do not include such translations. We do not report response times for CLENCH, GO:TermFinder and OntologyTraverser. This is because CLENCH only supports *A. thaliana*, GO:TermFinder only supports *Saccharomyces cerevisiae* and OntologyTraverser was unavailable in spite of our numerous attempts over several weeks at the time we gathered these data.

The three fastest tools, GoSurfer, GeneMerge, and Onto-Express, perform the analysis of 200 genes in 2, 6, and 8 seconds, respectively. Interestingly, two of these top three tools (Gene-Merge and Onto-Express) are web-based, which is somewhat counter-intuitive since one would have expected the stand-alone tools to be faster.

25.3.6 Visualization capabilities

The GO is organized as a directed acyclic graph (DAG), which is a hierarchical structure similar to a tree. However, unlike a tree, a DAG allows a node to have several parents. However, the DAG structure may not be the best choice for navigational purposes in GO since it tends to clutter the display [482].

Tool	Scope of the analysis	Level of abstraction	User interface	Application type	Platform	Supported input IDs
Onto-Express	All GO categories	Fully flexible; different levels of abstractions in different GO subtrees	Java GUI	web-based	any	GenBank, UniGene, Entrez Gene, Affymetrix, Gene symbol
GoMiner	All GO categories	static global analysis	Java GUI	stand-alone	Windows only	Organism specific IDs in GO
DAVID	All GO categories	only lowest level of GO	HTML GUI	web-based	any	GenBank, UniGene, Entrez Gene, Affymetrix, RefSeq, UniProt, PIR
EASEonline	All GO categories	user-selected, fixed level	HTML GUI	both	any	Affymetrix, GenBank, UniGene, Entrez Gene
GeneMerge	One category	only lowest level of GO	HTML GUI	both	any	Only supports organism specific IDs used in GO
FuncAssociate	All GO categories	only lowest level of GO	HTML GUI	web-based	any	MODB gene products
GOTM	All GO categories	only lowest level of GO	HTML GUI	web-based	any	Affymetrix, UniGene, ENSEMBL, Swiss-Prot, Entrez Gene
FatiGO	One category	user-selected, fixed level and static global analysis	HTML GUI	web-based	any	Affymetrix, GenBank
CLENCH	All GO categories	only lowest level of GO	command-line input, HTML output	stand-alone	any	A. thaliana MIPS IDs
GOstat	All GO categories	user-selected, fixed level	HTML GUI	web-based	any	GenBank, UniGene, Gene symbol, Organism specific IDs in GO
GOToolBox	All GO categories	user-selected, fixed level	HTML GUI	web-based	any	Only organism specific IDs in GO
GoSurfer	All GO categories	only lowest level of GO	C/C++ GUI	stand-alone	Windows only	Affymetrix, UniGene, Entrez Gene
Ontology Traverser	One category	only lowest level of GO	HTML GUI	web-based	any	Affymetrix
eGOn	One category	only lowest level of GO	HTML GUI	web-based	any	GenBank, UniGene, Clone
GO::TermFinder	One category	static global analysis	HTML GUI	both	any	Only yeast SGD IDs

TABLE 25.1: A comparison of the tools reviewed. Scope of the analysis refers to the number of GO categories that can be analyzed simultaneously. Level of abstraction refers to the depth in GO at which genes are associated with annotations. Note that some tools (e.g., GoMiner) allow the user to expand and collapse nodes in the results, but the analysis is only performed once, without reassigning the genes as nodes are collapsed or expanded by the user. This is described as a static

Tool	Statistical model	Correction for multiple experiments	GO Visualization	Microarrays supported	Time to process 200 genes (sec.)
Onto-Express	χ2, binomial, hypergeometric, Fisher's exact test	Šidák, Holm, Bonferroni, False Discovery Rate (FDR)	Flat, Tree	172 commercial arrays (Affymetrix, SuperArray, Sigma-Genonsys, ClonTech, PerkinElmer, Operon, Takara, NIA); can also upload a user-defined list	7, 8, 16, 28
GoMiner	Fisher's exact test	relative enrichment	Tree, DAG	uploads from user	77, 123, 223, 340
DAVID	none	none	not available	not applicable	15, 17, 27, 54
EASEonline	Fisher's exact test	Bonferroni	not available	27 arrays (Affymetrix only); can also upload a user-defined list	15, 19, 34, 74
GeneMerge	hypergeometric	Bonferroni	Flat, no hierarchical structure	uploads from user	6, 6, 6, 8
FuncAssociate	Fisher's exact test		not available	uploads from user	22, 27, 29, 50
GOTM	hypergeometric	none	Tree	37 arrays (Affymetrix only); uploads from user	59, 60, 157
FatiGO	percentage	step-down minP, FDR [45], FDR [46]	Flat, Tree	uploads from user	15, 49, 69, 105
CLENCH	hypergeometric, χ2, binomial	none	Flat, no hierarchial structure	uploads from user	NA
GOstat	χ2, Fisher's exact test	FDR, Holm	not available	uploads from user	12, 20, 46, 80
GOToolBox	hypergeometric, binomial, Fisher's exact test	Bonferroni, Holm, Hochberg, Hommel, FDR	not available	uploads from user	22, 81, 145, 270
GoSurfer	χ2	q-value	DAG	22 arrays (Affymetrix only); uploads from user	2, 2, 2, 3
Ontology Traverser	hypergeometric	FDR	not available	5 arrays (Affymetrix); uploads from user	NA
eGOn	binomial	none	Tree	uploads from user	20, 45, 80, 95
GO::TermFinder	hypergeometric	Bonferroni, bootstrap, FDR	DAG	uploads from user	NA

TABLE 25.2: More comparison criteria. In the GO visualization column, "flat" indicates that the tool does not represent the hierarchical structure of the GO when displaying the results; "tree" indicates that the tool displays the GO hierarchy as a tree, whereas "DAG" indicates that the tool displays the GO as a directed acyclic graph. The other columns are self-explanatory.

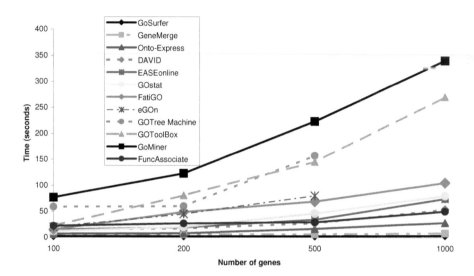

FIGURE 25.13: A speed comparison of the tools reviewed. Four sets of 100, 200, 500, and 1,000 human genes were submitted to each tool. The three fastest tools, GoSurfer, GeneMerge, and Onto-Express, are all able to perform the analysis of up to 200 genes in under 8 seconds. Note that GOTree Machine allows to upload only up to 500 genes.

An alternative way to visualize the DAG structure of the GO is to represent and visualize it as a tree structure in which a node with several parents is represented in the tree multiple times, once under each parent. Any tool using GO for functional profiling of a list of genes should be able to represent the hierarchical relationships between various functional categories. A graphical representation of the analysis results in the hierarchical context of GO allows the user to better understand the phenomenon studied. Furthermore, the functional analysis can be continued and refined by exploring certain interesting subgraphs of the GO hierarchy.

Among the tools reviewed, CLENCH, DAVID, EASEonline, FuncAssociate, GOstat, GOToolBox, and OntologyTraverser do not show the results in the context of the hierarchical structure of GO. Onto-Express, eGOn, FatiGO, GoMiner, and GOTM represent the results in their GO context. Onto-Express, GOMiner, GOTM and eGOn allow the user to manually collapse/expand nodes. Onto-Express also allows sorting and searching operations in the hierarchy, automatically expanding and/or collapsing nodes if necessary.

25.3.7 Custom level of abstraction.

A very valuable capability of a functional profiling tool is to let the user select a custom level of abstraction. From this point of view, the tools reviewed fall into one of the following three categories. The first category includes the tools able to perform the analysis only with the specific terms associated with each gene. This corresponds to an analysis undertaken at the lowest possible level of abstraction or the highest level of specificity (see the dash-and-dot line in Fig. 24.2). This type of analysis is essentially a one-shot look-up into the annotation database used. Each data set can only be analyzed once since any further analysis can only provide the exact same results. The analysis cannot be directed to answer specific biological questions and cannot be refined in any way.

The second category of the tools include those tools that allow the user to select a predetermined depth, or level of abstraction in GO. Once this level is selected, these tools will consider any genes below the chosen level associated with the corresponding category at the chosen level. This is illustrated by the dashed line in Fig. 24.2. Care should be taken here in order to make sure that each category is propagated up through all its parents, by following the DAG structure and not the tree structure that may be used for visualization. The capability of choosing a predetermined depth allows the user to refine this analysis by performing it repeatedly, at various levels of abstraction, thus forcing various very specific terms to be grouped into more general, and perhaps more informative categories. When this is done, several genes that are associated with very specific categories (e.g., induction of apoptosis by X, Y, and Z) are now grouped together under a more general category such as "positive regulation of apoptosis." It is often the case that each specific category does not appear to be significant because there are only few genes associated with it, while the more general category becomes highly significant once all genes associated with specific subcategories are analyzed together as representing the more general category. Tools having this capability allow a more complex and detailed analysis that can be directed to ask specific biological questions.

Finally, the third category includes tools that allow a completely custom cut through the GO, at different levels of abstraction in different directions. If the analysis is performed at a fixed depth of nine for instance, the analysis can distinguish between the various subtypes of apoptosis induction: by hormones, by extracellular signals, by intracellular signals, etc. (see Fig. 24.2). However, for a fixed depth, the same analysis will also be performed on several other thousands of functional categories situated at the same level. If the results are presented in a bar graph, the interesting categories will be cluttered by all the extra categories that just happened to be at the same depth in GO, even though they may not be interesting to the researcher. At the same time, other phenomena may be missed because the chosen level may be too specific for certain GO categories. A tool that allows a full customization is most powerful since it will allow the user to perform the analysis at different depths in various parts of the GO hierarchy, as required by the specific biological hypothesis investigated. This is illustrated by the red continuous line in Fig. 24.2.

Most of the existing tools only perform the analysis at the lowest level in GO, with the specific categories that genes have been annotated with, and do not allow any further refinement. Among the tools reviewed, FatiGO, EASEonline, GOToolBox, and GOstat allow the user to select a specific level of abstraction before submitting the input list of genes. FatiGo and GOMiner also calculate a p-value for all nodes throughout the GO. This corresponds to a global static analysis in which all genes under a certain node are considered to be associated with that node. Onto-Express is, at this time, the only tool that allows a fully customized analysis by allowing any node to be collapsed or expanded in the GO. Collapsing a node is equivalent to reassigning to this node all genes associated with any of its descendants. The p-value calculated for a collapsed node in Onto-Express corresponds to the p-value calculated in the global static analysis performed by GoMiner and FatiGo. Expanding a node will distinguish between genes associated with the node itself and the genes associated to any of it descendants. The p-value of an expanded node will be based only on the genes directly associated with it. This p-value is not provided by any other tool from those reviewed here. A current drawback in Onto-Express is that if a user wishes to perform the analysis at a fixed depth throughout GO, the user is required to manually expand the nodes up to this level.

25.3.8 Prerequisites and installation issues

Another important factor is the amount of effort necessary in order to install and use a tool. The web-based services provide the experimental biologists a convenient solution by avoiding the problems usually associated with a local installation of a program [483]. On the other hand, tools available over the web may be initially be obstructed by security issues. For instance, if the tool uses a specific TCP/IP port and the researcher is behind a firewall, the required port must be open on the firewall before the tool can be used.

Stand-alone tools such as CLENCH, GoMiner, and GoSurfer force the user to understand the complexities of a software installation. For example, prerequisites for CLENCH include the prior installation of perl modules for: i) HTTP request handling, ii) file and console access, iii) common gateway interface (CGI), iv) database access, v) statistical computation, and vi) graphical display (http://www.personal.psu. edu/faculty/n/h/nhs109/Clench/Clench_2.0 /Prerequisites.txt). As another example, GoMiner requires the user to install the Adobe scalable vector graphics (SVG) plug-in in order to view the results as a DAG, and the NCBI Cn3D browser plug-in in order to view molecular structures from the Entrez structure database (http://discover.nci. nih.gov/gominer/requirements.jsp). However, the core GoMiner application works without the plug-ins.

In principle, web-based tools such as Onto-Express, EASEonline, DAVID, GeneMerge, Ontology Traverser, GOTree Machine (GOTM), FuncAssociate, FatiGO only require that the user has a web-browser with an internet connection. In practice though, even the web-based tools suffer from some platform compatibility issues. For instance, the Microsoft Virtual Machine (VM) included in the Internet Explorer browser does not fully implement the Java standard [285]. In consequence, some Java-based tools such as Onto-Express, will require the installation of the Sun Java Runtime Environment (JRE).

Another issue is related to the availability of the tool and the requirement for an internet connection. Web-based tools can be used from any computer, but they cannot be used without an internet connection. Stand-alone tools require local installations on all computers from which they are to be used, but in principle, they can be used without network access. In practice though, among the stand-alone tools in Table 25.1 GoSurfer is the only tool that allows the user to actually analyze data without network access, after the initial download of the application and the required data files. The other stand-alone tools in Table 25.1, either use a local database server (GoMiner, EASE, DAVID) or actually retrieve annotation data at runtime (CLENCH). Unless both the client and the database are on the same computer, these stand-alone tools are essentially the same as web-based tools in as much as the client tool requires some network access to connect to the database server. The specific requirements of an application as well as the user preferences will probably be the determining factors from this perspective.

The most important problem in this category is the version control. From this point of view, the web-based tools are far superior because the researcher can always be assured that they are using the very latest version of the software. The software or database updates are always done on the server, by the team who initially wrote the tool. For stand-alone tools, the burden of version control usually rests on the user who is required to check periodically for new releases and updates. Once such an update becomes available, the burden of software or data update rests again with the user who has to go over the installation process again. In many cases, the updated version works worse than the older version due to issues related to the local environment. In principle, stand-alone tools can try to address this issue by providing automatic software updates. However, this approach means that the updating software must correctly identify, and appropriately deal with, various local software environment issues that tend to be slightly different over the potential hundreds or thousands of different installations.

25.3.9   Data sources

Most of the available tools use annotation data from a single public database. This has the advantage that the data are always as up to date as the database used. The disadvantage is that no single database offers a complete picture. For primary GO annotation data, the GO database is a comprehensive and up-to-date source since the contributing databases commit their data directly there. Other sources such as Entrez Gene derive their data from the GO database, so there is little advantage from the point of view of GO annotations to derive this data from several sources. However, the secondary analysis discussed here is more powerful if more types of data are integrated in a coherent way. A dedicated annotation database that integrates various types of data from various sources (e.g., KEGG pathways) is potentially more useful than any single database. The drawback is that such a database is: i) difficult to design and ii) will need to be updated every time any one of its source databases is. This places a heavy burden on the shoulders of the team maintaining it. Given this, it is understandable that most tools use only one of the available annotation databases and most of them use only GO annotations. EASEonline, GOSurfer, eGOn and GOTM use Entrez Gene, whereas GeneMerge, GoMiner, and GOToolBox use the GO database. Onto-Express uses its own Onto-Tools database, which is currently the only attempt to integrate resources from several annotation databases. Currently, Onto-Tools uses data from, and it is linked to: GenBank, dbEST, UniGene, Entrez Gene, RefSeq, GO, and KEGG. Onto-Tools also uses data from NetAffx and Wormbase without being linked to them.

25.3.10   Supported input IDs

Each probe on a microarray identifies a specific nucleotide sequence, which in turn identifies a specific gene. The annotations databases typically use genes to provide functional annotations. Hence, in order to create a functional profile for a list of differentially expressed genes, one first needs to convert the list of probe IDs into a list of genes. Similar condition also exists when the functional annotations are provided using proteins, where one needs to further map genes to proteins. An ontological analysis tool that supports more than one type of

IDs as input will be more useful since it will relieve the user from translating one type of IDs (e.g., Affymetrix probe IDs) to appropriate IDs (e.g., gene IDs).

Onto-Express, GoMiner and GeneMerge provide separate tools that allow the user to convert from other ID types (e.g., GenBank accession number, RefSeq IDs, etc.) to the type(s) of ID used by the application. These translation tools are Onto-Translate, Match-Miner, and the Gene Name Converter, respectively. Although in principle, these tools support more types of IDs as input, this design adds a separate step to the analysis pipeline since the user has to manually take the results from the conversion tool and submit them to the ontological analysis tool. For the purpose of this comparison, we only considered the capabilities of the ontological analysis tool itself.

GoMiner, GeneMerge, and GOToolBox only allow the user to submit organism-specific IDs used in the GO database as input. FuncAssociate, Ontology Traverser, and CLENCH only allow one type of IDs as input. These tools support MODB gene products, Affymetrix probe IDs, and *A. thaliana* MIPS IDs, respectively. FatiGO supports Affymetrix probe IDs and GenBank accession numbers, and GOToolBox supports GenBank accession numbers and gene symbols. Onto-Express, DAVID, EASEonline and GOTM support the most different types of IDs: Affymetrix probe IDs, GenBank accession numbers, UniGene cluster IDs and Entrez Gene IDs. In addition, Onto-Express and GOTM also support gene symbols, whereas DAVID supports GenPept, PIR and UniProt protein IDs, and RefSeq IDs.

A tool supporting more than one type of IDs must use an appropriate and correct type of identifiers to create functional profiles. For instance, the analysis performed by eGOn is centered around UniGene clusters, which assumes that each UniGene cluster corresponds to distinct genes. However, this assumption may not always be accurate. UniGene has been created by comparing expressed sequence tags (ESTs) in the dbEST database [379]. However, due to the alternative splicing of the mRNA, it is entirely possible that ESTs from the same gene cluster in different groups, which will result in several UniGene clusters being associated with the same gene. As an example, the SET8 gene (*SET8: PR/SET domain containing protein 8*, Entrez Gene ID: 387893) is associated to UniGene clusters Hs.443735 and Hs.536369. If a study includes any such genes, treating each UniGene cluster as a distinct gene may not always be appropriate. In such circumstances, the results can be skewed towards those GO terms that are associated to genes from which more than one UniGene clusters are derived. This may become particularly important if further research will confirm the current estimates that more than half of the human genes may have alternative splice variants.

25.4   Drawbacks and limitations of the current approach

Each of these tools uses one or more annotation databases and creates a list of functional categories known to be associated with the genes from the input list. The functional categories that are overly represented in a statistically significant way in the list of differentially regulated genes are inferred to be meaningful related to the condition under study. However, this approach of translating a list of differentially expressed genes into a list of functional categories using annotation databases suffers from a few important limitations. Since these limitations are related to the approach itself, all current tools exhibit them.

First, the existing annotations databases are **incomplete**. For virtually all sequenced organisms only a subset of known genes are functionally annotated [267]. Furthermore, most annotation databases are built by curators who manually review the existing literature. Although unlikely, it is possible that certain known facts might get temporarily overlooked. For instance, we found references in literature published in early 1990s, for 65 functional annotations that are yet not included in the current functional annotation databases. As an example, the gene HMOX2 was shown to be involved in the process of pigment biosynthesis in 1992 [304] and is still not annotated as such today. More commonly, recent annotations are not in the databases yet because of the time lag necessary for the manual curation process.

Second, certain pieces of information may also be **imprecise or incorrect**. In the GO, out of 19,490 total biological process annotations available for *Homo sapiens*, 11,434 associations are inferred exclusively from electronic annotations (i.e., without any expert human involvement) (http://www.geneontology.org/GO.current.annotations.shtml). The vast majority of such electronic annotations are reasonably accurate [77]. However, many such annotations are often made at very high-level GO terms, which limits their usefulness. Furthermore, some of these inferences are incorrect [267, 444]. Even though in some cases the error is very conspicuous to a human expert, currently, there are no automated techniques that could analyze, discover, and correct such erroneous assignments. At the present time, **none of the tools allows any type of weighting by the type of evidence** which is a limitation since experimentally derived annotations are more trustworthy than electronically inferred ones.

The current approach used for ontological analysis is limited to looking up existing annotations and performing a significance analysis for the categories found. This approach cannot discover **previously unknown functions for known genes** even if there is data justifying such inferences. For example, the gene SLC13A2 (*solute carrier family 13 (sodium-dependent dicarboxylate transporter), member 2 [Homo sapiens]*) encodes the human Na(+)-coupled citrate transporter and is annotated in GO for the molecular function *organic* *anion transporter activity.* However, it is *not* annotated for the corresponding biological process, *organic anion transport.* This is not a problem for the curator, and the human expert querying GO for this specific gene. For them, it is obvious that a gene that has organic anion transporter activity will be involved in the organic anion transport. However, a query that tries to find all genes involved in the process of organic anion transport will fail to retrieve this gene. Similarly, any ontological analysis software trying to find out what underlying processes are represented by a given list of genes containing this gene, will either fail to consider the organic anion transport if no other genes are involved in it, or will calculate its statistical significance incorrectly by ignoring this gene.

Another limitation is related to those genes that are involved in several biological processes. For such genes, all current tools weight all of the biological processes equally. At the moment, it is not possible **to single out the more relevant ones** by using the context of the other genes differentially expressed in the current experiment. BRCA1, for instance, is a well-known tumor suppressor but is also known to be involved in the carbohydrate metabolism. If most other genes found to be changed in the current experiment are involved in processes such as DNA damage response, apoptosis, induction of apoptosis, and signal transduction, it is perhaps more likely that in this experiment BRCA1 is playing its usual tumor suppressor role. However, if most other genes are involved in carbohydrate mediated signaling, carbohydrate transport and metabolism, etc., then it is perhaps more likely that BRCA1's role in the carbohydrate metabolism is more relevant.

The existing GO-based functional profiling approaches are currently **decoupled from the gene expression data** obtained from the microarray experiment in the previous step. In any given biological phenomenon, different genes are regulated to different extents. The data providing information about different amount of regulation for one gene versus another gene can be useful in assigning different weights to the corresponding biological processes they are involved in and hence, can help inferring if one biological process is more relevant than the other(s).

The usefulness of the existing functional profiling approaches is impacted by the **annotation bias** present in the ontological annotation databases. Some biological processes are studied in more detail than the others (e.g., apoptosis), thus generating more data. If more data about a specific biological process is available, more of the genes associate with it will be known, and hence, the process is more likely to appear as significant than the others.

An important issue related to the ontological analysis is the **name-space mapping from one resource to another**. At the moment, the existing knowledge about known genes is spread out over a number of databases and other resources. Different databases are maintained by various independent groups that many times have very different interest and research foci. Each such resource often uses its own type of identifiers, which creates a worldwide Babel's tower. For instance, GenBank uses accession numbers, UniGene uses cluster identifiers (IDs), Entrez Gene uses gene IDs, SWISSPROT uses protein IDs, TrEMBL accession numbers, etc. Furthermore, genes are also referred to using various company-specific gene IDs, which adds to the general confusion. A typical example would be Affymetrix, which uses its own probe IDs to represent various genes. As an illustrative example, the gene beta actin in mouse is referred to as MGI:87904 in Mouse Genome Informatics (MGI), Actb (Gene ID: 11461) in Entrez Gene, Mm.297 in UniGene, ACTB_MOUSE (primary accession number: P60710) in UniProt, and TC1242885 in the TIGR gene index. In addition, the beta actin gene in mouse is referred to by 29 mRNA sequences and 4552 ESTs in dbEST, 5 secondary accession numbers in UniProt, 4 other accession IDs in MGI, and 5 probe IDs on 4 different Affymetrix mouse arrays. The burden of mapping various types of ID on each other is left entirely on the shoulders of the researchers, who often have to revert to cutting-and-pasting lists of IDs from one database to another.

Various resources try to address the problem by maintaining other types of IDs together with their own and by providing ad hoc tools able to map from one type of ID to another. For instance, besides its own gene names, Entrez-Gene database also contains UniGene cluster IDs, and Affymetrix's NetAffyx provides RefSeq and GenBank accession numbers, besides its own array specific probe IDs.

The name-space issue becomes crucial when trying to translate from lists of differentially regulated genes to functional profiles because the mapping from one type of identifier to another is not one-to-one. In consequence, the type of IDs used to specify the list of differentially regulated genes can potentially affect the results of the analysis [124, 261]. While GO represents a viable, long-term solution to the problem of inconsistent vocabulary, the name space problem is yet to be solved.

Novel ideas have started to appear in this area addressing some of the issues above. An SVD approach has been proposed to analyze the semantic content of annotation databases and find incomplete and incorrect annotations [262, 119]. GoToolBox offers a different tool (GO-Proxy) to identify clusters of related terms. MAPPFinder [120], Pathway-Express [265], Cytoscape [385], Pathway Tools [250], and Pathway Processor [189] are only a few of the tools trying to expand the secondary analysis by including metabolic or regulatory pathway information. Other related tools can be found on the tools page of the GO (http://www.geneontology.org/ GO.tools.shtml).

25.5   Summary

This chapter presented a comparison of several ontological analysis tools implementing a very similar functional profiling approach. The comparison used the following criteria: statistical model(s) used, type of correction for multiple comparisons, reference microarrays available, scope of the analysis, visualization capabilities, capabilities for analysis at a custom level of abstraction, prerequisites and installation issues and the sources of annotation data. This comparison emphasized the characteristics of each tool as well as a number of limitations and drawbacks of the approach as a whole.

Chapter 26

Focused microarrays – comparison and selection

If the only tool you have is a hammer, everything starts to look like a nail...

—Unknown

The secret is to know your customer. Segment your target as tightly as possible. Determine exactly who your customers are, both demographically and psychographically. Match your customer with your medium. Choose only those media that reach your potential customers, and no others. Reaching anyone else is waste.

—Robert Grede, Naked Marketing, the Bare Essentials

26.1 Introduction

Microarrays have been introduced as powerful tools able to screen a large number of genes in an efficient manner.[1] The typical result of a microarray

---

[1]Some of the material in this chapter is reprinted with permission from Eaton Publishing from the article: "Assessing the Functional Bias of Commercial Microarrays Using the

experiment is a number of gene expression profiles, which in turn are used to generate hypotheses, and locate effects on many, perhaps apparently unrelated pathways. This is a typical hypothesis generating experiment. For this purpose, it is best to use comprehensive microarrays that represent as many genes of an organism as possible. Currently, such arrays include tens of thousands of genes. For example, the HGU133 (A+B) set from Affymetrix, Inc., contains 44,928 probes that represent 42,676 unique sequences from GenBank database, corresponding to 30,264 UniGene clusters.

Typically, after conducting a microarray experiment, one would select a small number of genes (e.g., 10–50) that are found to be differentially expressed. These genes are analyzed from a functional point of view, either going through online databases manually or by using an automated data mining tool such as Onto-Express [129, 264]. This step identifies the biological processes, molecular functions, biochemical function, and gene regulatory pathways impacted in the condition under study and generates specific hypotheses involving them. In many cases, only a small number of such pathways are identified. The next logical step is to focus on such specific pathways.

In many cases, it is desirable to construct a molecular classifier able to diagnose or classify samples into different categories based on their gene expression profiles [20, 51, 183, 393, 349, 401, 413]. This involves a training process that suffers from a "curse of dimensionality" [42]. In short, the curse of dimensionality refers to the fact that the difficulty of building such a classifier increases exponentially with the dimensionality of the problem, i.e., the number of genes involved. Furthermore, constructing a classifier requires many more training examples (i.e., samples or patients) than variables (genes). Both issues strongly suggest that the number of genes used to build the classifier has to be reduced to a minimum. In other words, it is best if the set of genes is restricted to strictly relevant genes.

Therefore, focusing on a smaller number of genes is both: i) the logical step that follows the initial screening experiment that generated the hypotheses as well as ii) a step required if molecular classifiers are to be constructed. Unlike the first step of exploratory search in which hypotheses are generated, the second steps should be a "hypothesis driven experiment" in which directed experiments are performed in order to test a small number of very specific hypotheses. However, specific hypotheses and a small number of pathways may still involve hundreds of genes. This is still too many for RT-PCRs, Western blotting and other gene-specific techniques and therefore the microarray technology is still the preferred approach.

Many commercial microarray manufacturers have realized the need for such **focused arrays** and have started to offer many such arrays. For instance, ClonTech currently sells focused human microarrays for the investigation of the cardiovascular system, cell cycle, cell interaction, cytokines/receptors,

---

Onto-Compare Database" by Draghici et al. published in the BioTechniques supplement "Microarrays and Cancer: Research and Applications," March 2003.

hematology, neurobiology, oncogenes, stress, toxicology, tumors, etc. Many other companies have picked the same trend and offer focused arrays: Perkin-Elmer, Takara Bio, SuperArray Inc., Sigma Genosys, etc. Literally tens of focused arrays are available on the market with several companies offering customized arrays for the same pathways. Typically, a focused array includes a few hundreds of genes covering the biological mechanism(s) of choice. However, two microarrays produced by different companies are extremely unlikely to use the same set of genes. In consequence, various pathways will be represented to various degrees on different arrays even if the arrays are all designed to investigate the same biological mechanisms. This is an unavoidable **functional bias**. Such a bias will be associated with each and all arrays including less than the full genome of a given organism.

26.2   Criteria for array selection

The general criteria used to select an array usually include several categories of reasons. One such category includes reasons related to the **availability** of a particular array. For instance, one laboratory or core facility may have certain arrays readily available because they have been purchased in a larger lot, or because they remained from previous experiments, etc. Another large category of reasons is related to the **cost**. Even if the cost of the array itself is a relatively small component of the overall cost of the experiment, smaller arrays do tend to cost less and spending more money for a larger array is usually expected to be associated with a benefit. Another set of factors influencing the choice of the array is related to **technological preferences**. For instance, a researcher might prefer cDNA versus oligonucleotide arrays, filters versus glass, etc. Finally, the choice might be influenced by **data analysis issues**. For instance, certain normalization techniques, such as dividing by the mean of all genes on a given array, assume most genes do not change. This is probably true for an array containing thousands of genes but may not be true for smaller arrays. Furthermore, data analysis may be easier and more reliable if fewer genes are present on the array. Considerably fewer genes means fewer difficulties related to correction for multiple experiments in statistical hypothesis testing, much less computation in model fitting (e.g., expectation maximization approaches), less of a curse of dimensionality in the construction of a classifier for diagnosis purposes and in general a better ratio between number of dimensions (number of genes) and number of data points (mRNA samples).

The interplay of these factors eventually decides the choice of the array when, in fact, the choice of the array should be made primarily based on the scientific question at hand. If array A contains 10,000 genes but only 80 are related to a given pathway and array B contains only 400 genes but 200 of them are related to the pathway of interest, the experiment may provide more information if performed with array B instead of A. Furthermore, using a smaller array can also translate into significant cost savings. Since an array can contain thousands of probes and a typical user would compare several arrays at once, manually annotating the probes and comparing the arrays becomes an unfeasible task. Onto-Compare is a tool that allows a researcher to perform this task in a fast and convenient way using terms from the Gene Ontology (GO) [22, 23].

26.3 Onto-Compare

In order to assess the biological bias of various arrays, we designed and implemented a custom database. This Onto-Compare (OC) database is populated with data collected from several online databases, as well as lists of genes (GenBank accession numbers) for each microarray as provided by their manufacturers. From the list of accession numbers, a list of unique UniGene cluster identifiers is prepared for each microarray and then a list of Entrez Gene identifiers is created for each microarray in our database. UniGene is a system for automatically partitioning GenBank mRNA sequences and ESTs into a non-redundant set of gene-oriented clusters with identical 3′ untranslated regions (3′ UTRs) [379]. Entrez Gene [294] is a database of official gene names and other gene identifiers. Each gene in the Entrez Gene database is annotated using ontologies from the Gene Ontology Consortium [104] and ontologies from other researchers and companies. Gene Ontology Consortium provides ontologies for biological process, molecular function, and cellular component. The data from these databases and gene lists are parsed and entered into our Onto-Compare relational database. The OC database is implemented in Oracle using a schema designed to allow for efficient querying. A group of Java programs and Perl routines are used to download, parse, and enter the data into the OC database as well as to update the database on a regular basis[2] with minimal human intervention.

After creating a list of locus identifiers for each array, the list is used to generate the following profiles: biological process, cellular component, and molecular function. The profile of each microarray is stored in the database. The list of genes deposited on a microarray is static, as long as the manufacturer maintains the array in production, but the annotations for those genes keep changing and are updated automatically. A Java program is used to facilitate the update of the database, which re-creates these profiles as more information becomes available.

Since an array can contain thousands of probes and a typical user would

---

[2]Currently this is done monthly.

compare several arrays at once, annotating the probes becomes a daunting computational task. For this reason, we precalculate the functional annotations for each array and store the results in the database. Thus, this very time-consuming computation is only done after database updates. During user interaction, the data are merely queried from the database. However, a researcher might choose to use a set of arrays (e.g., the Affymetrix HGU133 is actually a set of two arrays). In order to accommodate for this, we allow the user to merge arrays and calculate functional profiles of the sets of arrays. Since a user can merge an arbitrary number of arrays, we cannot realistically precalculate the functional annotation for every possible union of arrays. Thus, this computation is done every time a user merges arrays.

Onto-Compare runs as a Java applet in a web browser on a user computer. The input screen is shown in Fig. 26.1. The results are presented as a table, in which the first column displays an ontology term and the rest of the column corresponds to one of the arrays from the set of selected arrays (see Fig. 26.2). For each ontology term, Onto-Compare displays the total number of unique GenBank sequences found on each of the selected arrays. Onto-Compare also displays the total number of unique genes (i.e., Entrez Gene IDs) found on each of the selected arrays for each ontology term in square brackets. Clicking an array name sorts the entire table by the total number of sequences for that array. Clicking on a value for the total number of sequences displays accession numbers of the sequences, their corresponding cluster ID and Entrez Gene ID along with the cluster's official gene symbol. Selecting check boxes and then clicking "Show selected functions" only displays the selected terms. The user can merge two or more arrays, by selecting the check boxes next to the array name and clicking the "Merge selected arrays" button.

Onto-Compare's results are displayed in a GO tree similar to the tree view in Onto-Express (see Fig. 26.2). In the tree view, the genes annotated with a collapsed subcategory are considered to be annotated with it, and the total number of genes for the collapsed category is increased appropriately. The results displayed under the table view consider each gene on the arrays associated only to those GO terms the gene is directly annotated with.

26.4 Some comparisons

In order to illustrate the utility of Onto-Compare, we can consider the example of an anticancer drug candidate that inhibits *bcl2*, which is an antiapoptotic factor. The drug candidate was obtained from a large-scale screening so the exact mechanism through which this drug promotes apoptosis is not yet known. We would like to use a microarray approach to study the interaction between the various genes and their respective proteins on the apoptotic pathway as this pathway is affected by the drug. This is a typical example of a hypothesis

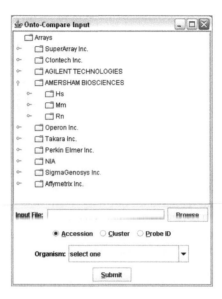

FIGURE 26.1: The input screen of Onto-Compare. The arrays are organized by manufacturer and organism. The user can select any number of arrays to be compared.

FIGURE 26.2: A sample output screen of Onto-Compare. The functional categories analyzed are: biological process, cellular component, molecular function, and chromosomal location. The results are displayed in an integrated GO browser (left panel) as well as a table (right panel). The user can further merge arrays (e.g., for HG133A and B, not shown here), or focus the analysis on a subset of functional categories.

driven research. We have strong reasons to believe that a certain pathway or cellular mechanism is involved, and we would like to focus our experiments on it.

There are several companies, such as ClonTech, Sigma-Genosys, and Perkin-Elmer, that provide arrays designed for research involving apoptosis. The ClonTech human apoptosis array contains probes for 206 UniGene clusters; the Perkin-Elmer apoptosis microarray contains probes for 322 UniGene clusters and the Sigma-Genosys human apoptosis microarray contains probes for 200 UniGene clusters. There are 74 clusters common between all arrays. There are 167 clusters common between ClonTech and Perkin-Elmer array, 92 clusters common between Perkin-Elmer and Sigma-Genosys array, and 82 clusters common between Clontech and Sigma-Genosys array. Comparison of these three apoptosis specific arrays using Onto-Compare is shown in Table 26.1. A biological process such as induction of apoptosis and a molecular function such as caspases are clearly relevant to the apoptosis [205, 477, 87]. Indeed, the table shows that all three arrays have similar numbers of genes representing induction of apoptosis, caspases, and tumor necrosis factor receptors. Various interleukins are also reported to be mechanistically associated with apoptosis at both protein and gene levels [18, 212, 378]. However, neither Clontech nor Perkin-Elmer microarray contains any interleukin related gene. On the other hand, the Sigma-Genosys microarray contains 14 genes related to various interleukins. Clearly, among the three arrays considered, the Sigma-Genosys would be a better choice for testing any hypothesis involving the role of interleukins in apoptosis. Other processes, such as immune response, cell-cell signalling, cell surface receptor linked signal transduction are also better represented on the Sigma-Genosys array. However, it is important to emphasize that the Sigma-Genosys array is not necessarily better than the other two arrays. In fact, since the Sigma-Genosys and the ClonTech arrays have almost the same number of genes, there must exist some functional categories that are represented better on the ClonTech array. Examples include processes such as cell cycle control, oncogenesis, and negative control of cell proliferation.

As another example, we compared commercially available oncogene arrays. Such arrays are available from ClonTech (Atlas Select Human Oncogene 7831-1 and Human Oncogene/Tumor Suppressor 7745-1), Perkin-Elmer and SuperArray. The numbers of distinct sequences available on these arrays are 514, 199, 335, 26, respectively. A comparison between these arrays is shown in Table 26.2. It is interesting to note that the PE array contains 73 sequences (71 UniGene clusters) related to oncogenesis. At the same time, the ClonTech Oncogene array contains only 22 sequences (21 UniGene clusters) representing the same biological process. The results are remarkable in light of the fact that that PE array contains only 335 genes compared to the ClonTech Oncogene array, which contains 514 genes. The same holds true for protein phosphorylation represented by 19 genes on the PE array and 6 genes on the ClonTech Oncogene array and negative control of cell proliferation represented by 15 genes on the PE array and only 6 genes on the ClonTech Oncogene array.

Ontology Term	ClonTech	PerkinElmer	Sigma-Genosys
Total genes on array	214	346	210
induction of apoptosis	17 [17]	28 [27]	24 [24]
antiapoptosis	15 [15]	20[20]	23[23]
immune response	0 [0]	1 [1]	19 [19]
cell-cell signaling	9 [9]	9 [9]	18 [18]
cell surface receptor linked signal transduction	5 [5]	9 [9]	17 [17]
oncogenesis	23 [23]	29 [29]	16 [16]
cell cycle control	30 [30]	31 [31]	12 [12]
positive control of cell proliferation	5 [5]	5 [5]	12 [12]
negative control of cell proliferation	16 [16]	20 [20]	10 [10]
induction of apoptosis by DNA damage	2 [2]	3 [3]	3 [3]
induction of apoptosis by extracellular signals	8 [7]	12 [12]	7 [7]
induction of apoptosis by hormones	1 [1]	1 [1]	1 [1]
induction of apoptosis by intracellular signals	2 [2]	2 [2]	2 [2]
induction of apoptosis by oxidative stress	0 [0]	0 [0]	1 [1]
induction of apoptosis via death domain receptors	4 [4]	5 [5]	7 [7]
caspase	1 [1]	3 [3]	2 [2]
caspase activator	1 [1]	3 [3]	2 [2]
caspase-1	1 [1]	1 [1]	1 [1]
caspase-10	1 [1]	2 [2]	1 [1]
caspase-2	1 [1]	1 [1]	1 [1]
caspase-3	1 [1]	1 [1]	1 [1]
caspase-4	1 [1]	1 [1]	1 [1]
caspase-8	2 [2]	1 [1]	1 [1]
caspase-9	1 [1]	1 [1]	1 [1]
caspase-activated deoxyribonuclease	1 [1]	2 [2]	2 [2]
tumor necrosis factor receptor	2 [2]	2 [2]	2 [2]
tumor necrosis factor receptor ligand	1 [1]	1 [1]	1 [1]
tumor necrosis factor receptor, type I	1 [1]	1 [1]	1 [1]
interleukin receptor	0 [0]	0 [0]	2 [2]
interleukin-1 receptor	0 [0]	0 [0]	2 [2]
interleukin-1, Type I, activating receptor	0 [0]	0 [0]	1 [1]
interleukin-10 receptor	0 [0]	0 [0]	2 [2]
interleukin-12 receptor ligand	0 [0]	0 [0]	2 [2]
interleukin-2 receptor	0 [0]	0 [0]	3 [3]
interleukin-2 receptor ligand	0 [0]	0 [0]	1 [1]
interleukin-4 receptor	0 [0]	0 [0]	2 [2]
interleukin-4 receptor ligand	0 [0]	0 [0]	1 [1]
interleukin-7 receptor	0 [0]	0 [0]	1 [1]
Total distinct genes in categories above	99 [98]	133 [132]	129 [129]

TABLE 26.1: A comparison of three apoptosis specific microarrays: Clon-Tech human apoptosis, Perkin-Elmer apoptosis and Sigma-Genosys human apoptosis. Biological processes such as induction of apoptosis and molecular functions such as caspases and tumor necrosis are almost equally represented on each of the chips, but none of the interleukins are represented on Clon-Tech as well as Perkin-Elmer microarray. Processes such as immune response, cell-cell signalling, cell surface receptor linked signal transduction are better represented on the Sigma-Genosys array. Processes such as cell-cycle control, oncogenesis, and negative control of cell proliferation are better represented on the ClonTech array. The numbers represent sequences present on the arrays; the numbers in brackets represent distinct UniGene clusters.

Biological Process	ClonTech Oncogene	ClonTech Tumor Suppressor	PE Oncogene Tumor Suppressor	SuperArray Oncogene
Total genes on array	514	199	335	26
oncogenesis	22 [21]	40 [40]	73 [71]	8 [8]
signal transduction	46 [46]	39 [39]	54 [54]	6 [6]
cell proliferation	23 [23]	19 [19]	24 [24]	8 [8]
protein phosphorylation	6 [6]	17 [17]	19 [19]	5 [5]
cell cycle control	20 [20]	31 [31]	19 [19]	1 [1]
negative control of cell proliferation	6 [6]	14 [14]	16 [15]	0 [0]
cell-cell signalling	10 [9]	7 [7]	13 [13]	0 [0]
Total genes in categories above	91 [90]	109 [109]	144 [142]	15 [15]

TABLE 26.2: A comparison of the ClonTech (Atlas Select Human Oncogene 7831-1 and Human Oncogene/Tumor Suppressor 7745-1), Perkin-Elmer and SuperArray (Human Cancer/Oncogene) arrays. PE is a better choice for conditions potentially involving oncogenesis, protein phosphorylation, and negative control of cell proliferation. The numbers represent sequences present on the arrays; the numbers in brackets represent distinct UniGene clusters.

Finally, we performed another analysis in which we contemplated the use of both ClonTech Oncogene and ClonTech Tumor suppressor as a two-array solution and compared this ClonTech selection of genes with the genes used on the unique PE Oncogene/Tumor suppressor array. Note that when sets of arrays are used, many genes may be present on more than one array. Thus, the coverage of a given pathway or biological process cannot be inferred by simply summing the number of genes covering the given pathway on each array. Table 26.3 shows the comparison between the set of two ClonTech arrays and the PE array. When the two ClonTech arrays are used together, there is a good representation of general signal transduction, protein phosphorylation and negative control of cell proliferation. Remarkably, oncogenesis is still better represented on the PE array in spite of the fact that the set of two ClonTech arrays deploy 676 genes while the unique PE array uses only 335 genes.

The examples above show that each array or set of arrays has a certain biological bias. These examples should not be interpreted as proving that one particular array is better or worse than other similar arrays. However, these examples do show that in those situations in which a certain hypothesis exist, the choice of the array must be made based on a comprehensive functional analysis of the biological processes, cellular components and chromosomal locations of the genes represented on each of the arrays considered.

Biological Process	ClonTech Oncogene and Tumor Suppressor	PE gene Suppressor	Onco- Tumor	SuperArray Oncogene
Total number of sequences	676	335		26
signal transduction	81 [81]	54 [54]		6 [6]
oncogenesis	55 [54]	73 [71]		8 [8]
cell cycle control	43 [43]	19 [19]		1 [1]
cell proliferation	38 [38]	24 [24]		8 [8]
developmental processes	21 [21]	13 [13]		3 [3]
protein phosphorylation	21 [21]	19 [19]		5 [5]
negative control of cell proliferation	19 [19]	16 [15]		0 [0]
cell-cell signalling	17 [16]	13 [13]		0 [0]
signal transduction	10 [10]	6 [6]		0 [0]

TABLE 26.3: A comparison of the ClonTech set of arrays (Atlas Select Human Oncogene 7831-1 and Human Oncogene/Tumor Suppressor 7745-1) with the Perkin-Elmer and SuperArray (Human Cancer/Oncogene) arrays. When the two ClonTech arrays are used together, there is a good representation of general signal transduction, protein phosphorylation and negative control of cell proliferation. However, oncogenesis is still better represented on the PE array. The numbers represent sequences present on the arrays; the numbers in brackets represent distinct UniGene clusters.

26.5 Summary

The first step of an experiment involving microarrays will probably be of an exploratory nature aimed at generating hypotheses. Comprehensive arrays including as many genes as possible are useful at this stage. In most cases, the hypothesis generating phase will be followed by successive steps of focused research. Such hypothesis-driven research often concentrates on a few biological mechanisms and pathways. However, even a single biological mechanism may still involve hundreds of genes, which may make the microarray approach the preferred tactic. If microarrays are to be involved in any follow-up, focused, hypothesis-driven research, we argue that one should use the array(s) that best represent the corresponding pathways. Functional analysis can suggest the best array or set of arrays to be used to test a given hypothesis. This can be accomplished by analyzing the list of genes on all existing arrays and providing information about the pathways, biological mechanisms, and molecular functions represented by the genes on each array. Onto-Compare is a tool that allows such comparisons. This tool is available at: http://vortex.cs.wayne.edu/Projects.html.

Chapter 27

ID Mapping issues

That is why it was called Babel – because there the Lord confused the language of the whole world.

—Genesis 11:9

27.1   Introduction

Gene annotations databases are widely used as public repositories of biological knowledge.[1] Understanding the results of almost any molecular biology experiment involves consulting such annotation databases. Our current knowledge is spread out over a number of databases (DBs), such as: Entrez Gene [294], UniProt [29], Protein Data Bank [47], RefSeq [344], RGD, SGD, WormBase, and Gene Ontology (GO) [25], to name just a few. Many such databases

---

[1]Some of the material in this chapter is reprinted from Sorin Draghici, Sivakumar Sellamuthu, Purvesh Khatri. Babel's tower revisited: a universal resource for cross-referencing across annotation databases. *Bioinformatics*, 22(23):2934–2939, December 2006, with permission from Oxford University Press.

support multiple organisms but are specialized on a subset of specific biological entities. For instance, UniProt focuses on proteins, Entrez Gene focuses on genes, EPD focuses on eukaryotic promoters, etc. Other databases aim to provide a wider angle but focus on specific organisms. Examples could include RGD for rat, SGD for yeast, WormBase for *C. Elegans*, etc. Obtaining a complete understanding of an experiment, usually requires combining information from several such annotations databases. Unique key identifiers (IDs) in the internal structure of each such database represent biological entities such as genes, proteins, and mRNAs. Design and implementation restrictions specific to each database ensure that, within each database, the data are consistent, coherent, and non-redundant. However, most of these annotation databases have been developed by independent groups, which have used completely different designs and completely different sets of key identifiers for the same biological entities. Because of this, the ensemble of such annotation databases, which is the current repository of all our biological knowledge is inconsistent, incoherent, and highly redundant.

At the same time, the old-fashion, gene-centric approach of research in life sciences has been all but substituted by more high-throughput approaches involving entire sets of genes, sometimes entire genomes. In many current life science experiments, researchers obtain results identifying many genes that are interesting in a given condition. In order to fully interpret such results, researchers must combine annotations from several different databases, which essentially requires mapping tens or hundreds of IDs across all databases involved. If performed manually, this mapping often leads to incomplete and incorrect results, and is time consuming and error prone even for short lists of genes. Even if performed automatically, querying various databases for the same data often yields different results. This represents a very important problem that has not been satisfactorily addressed yet.

27.2   Name space issues in annotation databases

Identifiers used in different databases often represent different types of biological entities (e.g., genes, ESTs, mRNAs, proteins, etc.) Usually, there is a very clear and biologically meaningful mapping from one such entity to another. For instance, in the simplest case, a gene has a unique DNA sequence, which in turn can be mapped to an mRNA sequence, which is translated into a protein sequence, which perhaps has a known protein structure. However, the problem is further complicated by one-to-many mappings at various levels. For instance, several ESTs can represent the same gene, several alternatively spliced mRNAs can be constructed from the same gene DNA sequence, several structures corresponding to alternative folding patterns or different possible ligands can be associated with the same protein, etc. Specific annotations are available

Database	Identifier (ID)
UniGene	Hs.437638
HGNC	12801
Entrez Gene	7494
Swiss-Prot	P17861
ENSEMBL	ENSG00000100219
PharmGKB	PA37400
RefSeq	NM_005080, NP_005071
NetAffx	RC_W90128_s_at (HU35ksubd), 200670_at (HGU 133), 71584_at (HGU95e), M31627_at (HG FL), 39756_g_at (HGU 95av2), 39755_at (HG U95a), g4827057_3p_s_at (U133 X3P)

TABLE 27.1: Human gene **XBP1** is represented by six additional distinct identifiers (IDs) in six different databases, as well as by one nucleotide sequence ID, one protein sequence ID and seven different probe IDs on several different Affymetrix arrays.

at each level (gene, mRNA, protein, structure, etc.). Given for instance, a set of genes found to be differentially expressed in a specific condition of interest, one wishes to quickly find all known annotations about this set of genes, at all levels: the known GO categories associated with each of these genes, their proteins, the annotations associated with these proteins, etc. This information is currently spread out over many different databases, and each such database uses its own type of IDs. For instance, Table 27.1 shows eight different IDs used to refer to the same XBP1 gene in seven different databases, as well as seven different probe IDs used on several Affymetrix arrays. Because the same biological entity is referred to by many different IDs, one needs to first map these IDs from one database to another and then query each database with its own specific IDs. This apparently trivial problem has become a challenge because various databases contain redundant information about the same biological entity. For instance, the GO categories known to be associated to a specific gene are stored in many databases such as UniProt[29],[2] Entrez Gene [294], NetAffx[287], and GO itself [24, 25]. In spite of everybody's best efforts, because these databases are managed separately and they have different release and maintenance cycles, any data stored in more than one database creates very serious consistency and coherency problems.

A brief example will hopefully illustrate the gravity of the issues involved. Let us consider, for instance, the example of a microarray experiment involving Affymetrix's GeneChips. Let us assume that a specific probe ID, 39755_at, corresponding to the human gene XBP1, is found to be differentially

---

[2]Swiss-Prot, TrEMBL, and PIR have been recently merged as a single database in UniProt.

expressed. The researcher may be interested in finding the corresponding Uni-Gene [379] cluster ID for the selected probe ID, 39755_at. This can be achieved by querying NetAffx [287] with the given probe ID, 39755_at, which yields the Hs.437638 cluster ID. Alternatively, one can find the cluster ID by querying NCBI's UniGene database with the gene name, XBP1. In this example, there exist at least two paths that yield the required information and following both paths yields the same final result. However, let us now assume that one is interested in the GO annotations associated with this gene. Querying each of the resources above with the IDs representing the same gene, XBP1, yields very different results. UniProt queried with P17861 provides two unique GO terms: *transcription factor activity* and *immune response*; QuickGO queried with the same P17861 provides eight unique GO terms: *protein dimerization activity, sequence-specific DNA binding, immune response, DNA-dependent regulation of transcription, transcription, DNA binding, transcription factor activity and nucleus*; NCBI's Entrez Gene entry XBP1 provides seven unique GO terms: *immune response, protein dimerization activity, sequence-specific DNA binding, DNA-dependent regulation of transcription, transcription, transcription factor activity and nucleus*; PIR's iProClass [465] entry P17861 provides five unique GO terms: *immune response, nucleus, DNA-dependent regulation of transcription, transcription factor activity, DNA binding*, whereas GO (XBP1_HUMAN) provides only two unique GO terms: *immune response* and *transcription factor activity*. Essentially, querying five different resources can yield anything between two and eight GO terms for the same gene. This situation is nothing short of disastrous. When one retrieves annotations for a set of genes from a particular source, one is always left to wonder whether the results obtained are really the entire picture or just a part of it, and whether one should continue to query other sources or just use the data retrieved so far.

Until the various resources currently available are organized into a real semantic web, free of coherency and consistency problems, arguably the best approach to retrieving annotations for a set of given biological entities is to query the authoritative source of such annotations for the given entity. In turn, in order to do this, one must map various types of IDs onto each other. This is also a tremendous challenge since various IDs can be mapped onto each other by traversing a number of alternative paths from one database to another. Since no unified map of the various databases exists, one is forced to rely on one's inherently limited personal understanding of the relationships between such databases in order to determine such a path on a case by case basis. Unfortunately, due to the lack of global consistency and coherency, the path used to travel from one resource to another often influences dramatically the results obtained.

Another important problem is related to the cross-referencing between various annotation databases. Databases such as Entrez Gene and HGNC provide gene information and are supposed to cross-reference each other. For example, gene SMCR (Smith-Magenis syndrome chromosome region) has the identifier 11113 in HGNC. The same gene is identified by Entrez Gene as gene 6600. Entrez Gene cross-references HGNC i.e. the entry 6600 contains a field with the HGNC ID 11113. However, the reverse is not true. HGNCs entry for this gene does not contain the appropriate Entrez Gene ID. Here the data are mapped only one way, from Entrez Gene to HGNC number. If the user queries HGNC using its IDs, (s)he will not be able to link to NCBI and thus will not have access to all the annotations regarding this gene available in Entrez Gene.

This problem is more widespread than one would like to believe. For instance, both UniGene and Entrez Gene focus on non-redundant genes. However, only 69.53% of the genes in UniGene can be mapped on Entrez Gene entries. Furthermore, only 43.54% of the IDs in Entrez Gene can be mapped back to UniGene. An even more striking example is the mapping between GenBank dbEST and GenPept. GenPept is supposed to contain the protein translations of the sequences in GenBank dbEST, so going back and forth between these resources should be trivially simple. However, this is far from being the case. At the moment, 91.6% of the entries in dbEST can be mapped to GenPept entries, but the reverse mapping is possible only for 1.82% of the entries. Clearly, translations and mappings that are theoretically both meaningful and useful, cannot always be performed just by querying the resources that are supposed to allow them. These examples strongly support the idea that ID mappings cannot be done casually, by ad hoc, need-driven queries, or quick-and-dirty Perl scripts, as most researchers currently do. These quick solutions might satisfy an immediate need for *a translation* but offer no guarantees that the translation performed is the best possible mapping, nor that the results are correct or complete. At this time, the issues of incoherent name spaces between various databases represent a serious impediment to using the existing annotations at their full potential. Navigating between various such name spaces by mapping IDs from one database to another is a very important issue that must be addressed in a thorough and systematic way.

A promising way to address these issues is proposed by the semantic web community. Efforts in this community are focused on making data available in a standard format that is accessible and suitable for automatic queries. The resource description framework (RDF) is the standard model that aims to make the data interchangeable on the web. RDF has features that facilitate data processing and object mapping even if the underlying schemas are different. Furthermore, RDF allows each database developer to make changes in the schema of their own database without requiring the users of their data to change the way they access the data. Uniform resource identifiers (URIs) are strings of characters used to identify a resource on the internet. Such URI can be uniform resource locators (URLs) or uniform resource names (URNs). SPARQL is the query language that is used to query data available in RDF from various URIs. In principle, ID mapping can be done by relatively simple SPARQL queries to appropriate databases that make their data available in RDF. The great advantage of this approach is that it eliminates the consis-

Software name	Types of input IDs	Number of translations
Onto-Translate	22	462
MatchMiner	11	137
SOURCE	6	96
GeneMerge	12	30
RESOURCERER	1	16

TABLE 27.2: A comparison of the scopes of Onto-Translate, SOURCE, MatchMiner, GeneMerge and RESOURCERER: types of input IDs supported and number of possible translation types.

tency and coherency issues related to ID mapping approaches that use intermediate databases. The results of the query are always obtained with the most recent data available from all URIs involved. No maintenance or updating is needed once the queries have been defined, as long as the URIs used remain available. The disadvantage is that any query requires that all URIs be available at the time the query is executed. A single URI that is temporarily unavailable can compromise any query that involves it. Very complex queries involving many URIs have an increased likelihood of failing from time to time. Nevertheless, the semantic web approach is probably the way of the future.

27.3   A comparison of some ID mapping tools

Clearly, the need for a reliable way of mapping IDs from one database to another has been felt in the community. In response to these needs, several approaches have been proposed to deal with this issue although none of them addressed the problem to its full extent. The best known resources currently able to perform a nontrivial mapping of various biological entities are: SOURCE [118] from Stanford University, MatchMiner [75] from NCI, RESOURCERER [425] from TIGR, GeneMerge [83] from Harvard, and Onto-Translate [134] from our laboratory. At the time of writing this material, we were not aware of any widely used ID mapping tool based on RDF and SPARQL.

Table 27.2 shows a comparison of these tools in terms of scope, accuracy of translation, speed, and scaling capabilities [134]. We define scope as the number of different mappings between types of IDs supported by a given resource. Figure 27.1 shows the specific translations that can be performed by each of the resources considered.

Of course, the scope is irrelevant if the accuracy of the mappings performed is inadequate. In order to compare the accuracy of the existing resources, we performed a number of translations using OT, SOURCE, and MatchMiner

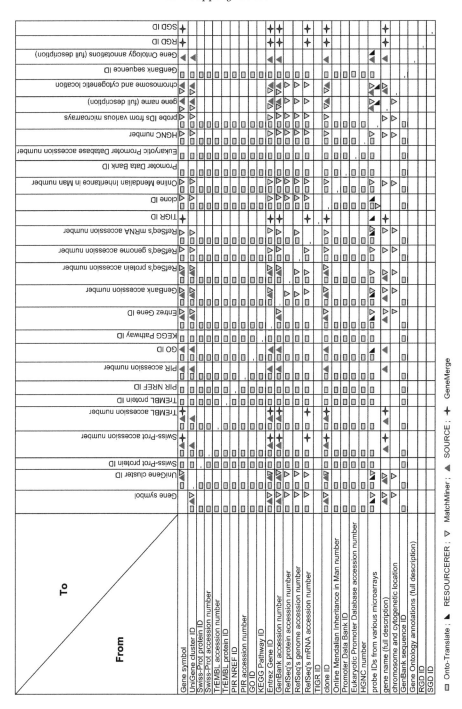

FIGURE 27.1: A comparison of the scopes of Onto-Translate, RE-SOURCERER, MatchMiner, SOURCE, and GeneMerge, in terms of possible mappings between various types of IDs.

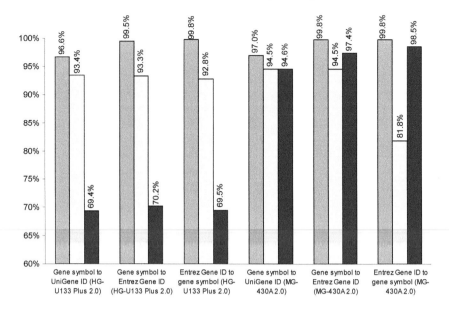

FIGURE 27.2: A comparison of the accuracy of Onto-Translate, Match-Miner and SOURCE. The input file included 19,248 gene symbols (19,562 Entrez Gene IDs) for human, and 12,991 gene symbols (13,023 Entrez Gene IDs) for mouse, from the respective Affymetrix arrays. The graph shows the percentages of the input genes successfully translated in each case.

(top 3 in terms of scope), and compared the number of input IDs correctly mapped by each resource for each data set. The sets of genes to be translated were taken from popular human and mouse Affymetrix arrays. The set of genes contained on the HG-U133 Plus 2.0 array was used to test the translations from gene symbols to UniGene IDs, gene symbols to Entrez Gene IDs, and Entrez Gene IDs to gene symbols. Finally, for the translations involving mouse genes, we used the set of genes contained on the MG-430A 2.0 arrays. These genes were translated from gene symbols to UniGene IDs, gene symbols to Entrez Gene IDs and Entrez Gene IDs to gene symbols. Fig. 27.2 shows a comparison of the accuracy of these translations. OT was the most accurate resource in all cases, with accuracies between 96% and 99%. For human data, SOURCE is second best with an accuracy hovering around 93%. MatchMiner is weaker with an accuracy of around 70%. For mouse data, MatchMiner is better than SOURCE: 94–98% for MM, compared to 81–94% for SOURCE.

Fig. 27.3 shows a comparison of the time (in seconds) necessary to perform a sample translation from gene symbols to gene IDs with Onto-Translate, MatchMiner and SOURCE. The time necessary to translate fewer than 1,000 genes is approximately the same for the three resources. However, when longer lists are involved, OT is approximately two times faster than SOURCE and approximatively 10 times faster than MatchMiner, in all translations performed.

27.4 Summary

This chapter discusses various issues related to name space inconsistencies between existing annotation databases. The distribution of our knowledge over several databases forces researchers to navigate from one such database to another, in order to construct the correct interpretation of any given experiment. Currently, the lack of the ability to map correctly various IDs from one DB to another creates very substantial problems in annotation retrieval. A number of ID mapping tools have been compared in terms of: i) number of translations possible, ii) types of IDs supported, iii) accuracy, and iv) speed. These tools are: SOURCE [118] from Stanford University, MatchMiner [75] from NCI, RESOURCERER [425] from TIGR, GeneMerge [83] from Harvard, and Onto-Translate [134] from our laboratory (http://vortex.cs.wayne.edu/Projects.html).

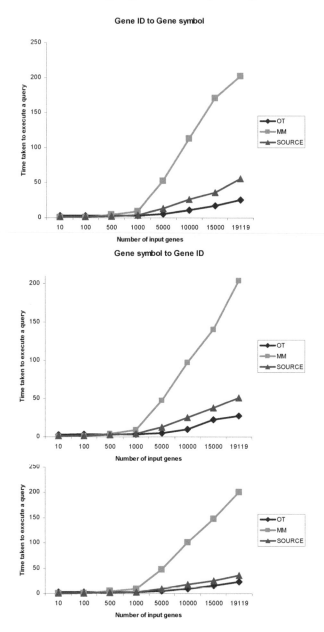

FIGURE 27.3: Scaling properties of Onto-Translate (OT), MatchMiner (MM) and SOURCE. The graph shows the time (in sec) necessary to translate various sets containing between 10 and 19,119 distinct genes from Affymetrix 133 Plus 2.0. At fewer than 1,000 genes, the three resources have very comparable query times of under 10 seconds. When larger sets are involved, there is a substantial performance difference.

Chapter 28

Pathway analysis

That's the holy grail, to understand how everything is interrelated.

—Douglas Emig

28.1   Introduction

Once heralded as the holy grail, the capability of obtaining a comprehensive lists of genes, proteins, and metabolites that are different between the disease and normal phenotypes is routine today. And yet, the holy grail of high throughput has not delivered so far. Even though such high-throughput comparisons are relatively easy to perform, understanding the phenomena that determine the measured changes is as challenging as ever, if not more so. There is a large gap between our ability to collect data and our ability to interpret it. This gap is virtually impossible to close through manual means due to the sheer amount of data involved. Hence, computer analysis holds the only hopes in this direction.

At the same time, living organisms are complex systems whose emerging phenotypes are the results of thousands of complex interactions taking place on various metabolic and signaling pathways. These complex interactions between many actors must be taken into account if we were to fully understand complex diseases. Soon after gene annotations started to be widely used, networks of gene/protein interactions and metabolic reactions started to be assembled in order to describe our understanding of the complex phenomena that govern life. Pathway databases such as KEGG [248], BioCarta [52], and Reactome [243] currently describe metabolic pathways and gene signaling networks offering the potential for a more complex and useful analysis.

As we already pointed out, independent of the platform and the analysis methods used, the result of a high-throughput experiment is, in many cases, a list of differentially expressed (DE) genes. The common challenge faced by all researchers is to translate such lists of DE genes into a better understanding of the underlying biological phenomena and in particular, to put this in the context of the whole organism as a complex system. Being able to correctly infer the perturbed pathways interactions that cause the disease from a list of differentially expressed (DE) genes or proteins may be the key to transforming the now abundant high-throughput expression data into biological knowledge. In this chapter, we will explore techniques and methods that are aimed at analyzing high throughput data with the goal of identifying the pathways that are significantly impacted in a given condition.

28.2 Terms and problem definition

We can define a **pathway** as a subsystem of a larger system that has the following properties:

1. Has some **components** such as genes/gene products/biochemical compounds, etc.

2. These components are linked by some **interactions** such as signals between genes, reactions between components, etc.

3. This **subsystem** interacts with the rest of the system through well-defined **inputs** and outputs.

4. There are more **shared properties** between the components of the sub-system than between these and the rest of the system (all genes in the insulin signaling pathways are related to insulin, all compounds and reaction on the glycolysis pathway will be involved in carbohydrates break down, etc.).

Note that given a specific organism, the definition above does not yield a unique set of pathways. However, this accurately reflects the reality. In fact,

Reactome [243] shows a unique, very large pathway diagram that includes all known facts about the given organism. When the mouse is moved over different areas of this diagram, certain subgraphs are highlighted as specific pathways. This unique diagram emphasizes two important aspects, as follows: i) pathways do not exist in isolation but rather they are highly interconnected, with the outputs of one pathway being the inputs of the next pathway downstream; and ii) the boundaries between the pathways are often arbitrary. In some situations, one could easily move the boundary in such a way that a few genes and edges move from one pathway to another.

In spite of the lack of a precise formulation, the definition above does clarify the concept of pathway, at least intuitively. Such a pathway can be modeled well by a graph in which the components of the pathway are represented by nodes and the interactions between the components are represented by directed edges. The edges and their direction can represent information propagation, causality, temporal succession, etc.

Currently, there are at least two fundamentally different types of pathways used to represent our understanding of how biological systems work: signaling pathways and metabolic pathways.

Signaling pathways use nodes to represent genes or gene products and edges to represent signals that go from one such gene to another. There are various types of edges that represent various types of signals such as activation, inhibition, phosphorylation, etc. Most edges in a signaling pathway will be directed, since usually signals go from one source to one or several destinations. As an example, Fig. 28.12 shows the complement and coagulation pathway as described by KEGG. Note the various types of edges, including the undirected one between MBL and MASP1/2. The various types of edges used by KEGG are shown in Fig. 28.1.

Metabolic pathways use nodes to represent biochemical compounds and edges to represent biochemical reactions that combine, transform, consume or produce such compounds. Such reactions are usually catalyzed or carried out by enzymes. In a typical metabolic pathway, such as those provided by KEGG, the edges are usually annotated with the Enzyme Commission (EC) identifier(s) of the enzyme(s) that carry out the respective reactions. Although the edges on a metabolic pathway are also directed, many reactions are reversible which can be represented either as an edge with arrows at both ends, or as two directed edges going in opposite directions between the same nodes. Note that this representation cannot be interpreted as a loop since usually the reaction happens predominantly in one direction at any given time.

From the description above, it is clear that signaling pathways are fundamentally very different from metabolic pathways, even though they are both represented by directed graphs and they may even look somewhat similar in certain databases, such as KEGG. One fundamental difference is that in signaling pathways genes are associated with the nodes, whereas in a metabolic pathway, genes are associated with edges through the enzymes that carry out the respective reactions.

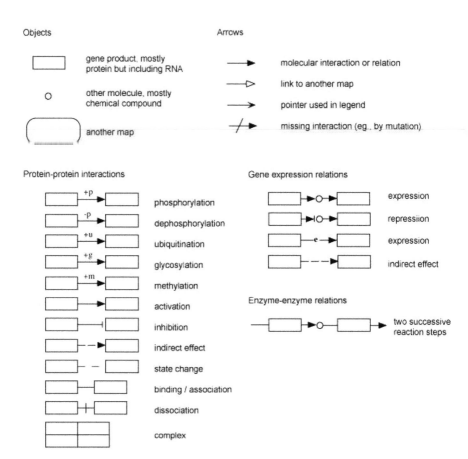

FIGURE 28.1: Various types of edges, most of them directed, used by KEGG (from *http://www.genome.jp/kegg/*).

A third type of pathways can also be found in various pathway databases. These pathways are elaborated diagrams that explain a certain process or cellular component without describing any of the underlying gene signaling or biochemical reactions that might be involved. Two such "pretty picture" pathways are shown below. Fig. 28.2 shows the flagellar assembly in a bacteria. Note the presence of nodes that indicate the various proteins that form various parts of the flagellum but the absence of signals between these. Fig. 28.3 shows a similar diagram describing the lysosome. Again, no gene signals or biochemical reactions are present here.

In the following, we will focus exclusively on gene signaling pathways. The **pathway analysis problem** that we will tackle here focuses on a comparison between two different phenotypes. The archetypal comparison is the disease versus healthy on which we will focus in order to simplify the language. However, the exact same approach and methods can be used in any other comparisons such as: treated versus untreated, drug A versus drug B, before and after treatment, etc. The **goal** of the pathway analysis is to identify the pathways that are significantly impacted in the given condition. The **input** of this problem include:

1. A set of variables for which there are significant measured differences between the given phenotype and control (e.g., mRNA levels, protein abundance, metabolites, etc.)

2. A set of pathways describing sub-systems involving the given variables

As the **output** of the analysis, we would like the analysis to produce:

1. A ranking of the pathways in the decreasing order of the amount of disruption suffered by each of them

2. If possible, a quantification of the probability of observing the given amount of disruption just by chance, i.e., a p-value. This can help us identify those subsystems for which the disruption is "significant," ie. unlikely to be due to noise alone.

In the following, we will discuss various approaches and methods available to address this problem. As it should be very clear from the problem definition above, we focus here on methods and techniques that aim to identify the pathways that are impacted in a given condition. We are not concerned here on how these pathways have been created, and we will not discuss here any methods that aim to build pathways or gene networks from data.

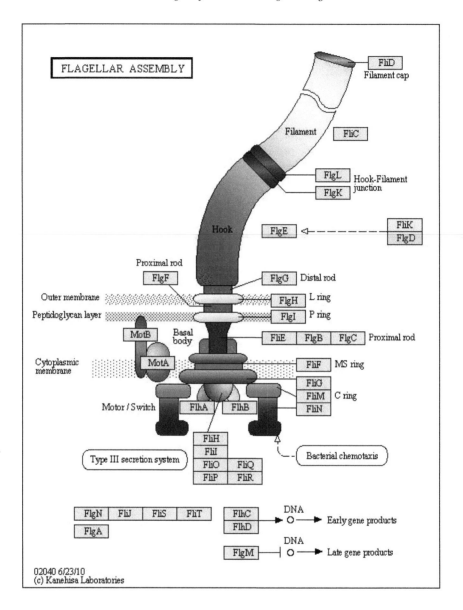

FIGURE 28.2: A "pretty picture" pathway describing the flagellar assembly (from *http://www.genome.jp/kegg/*). The main purpose of such diagrams is to describe the components of a certain part of the organism (in this case, the flagellum), or to describe a given process. Sometimes such pathways use nodes to denote genes or gene products that form various anatomical parts. However, usually very few or no reactions or signals are included in such pathways.

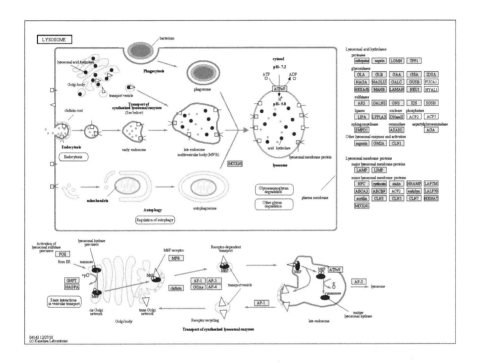

FIGURE 28.3: Another type of "pretty picture" pathway, in this case describing the lysosome (from *http://www.genome.jp/kegg/*). The main purpose of such diagrams is to describe a given process or anatomical part. Again, very few or no reactions or signals are included in such pathways and the genes/gene products involved are only enumerated.

28.3   Over-representation and functional class scoring approaches in pathway analysis

Pathway databases such as KEGG [248], BioCarta [52], and Reactome [243] currently describe metabolic pathways and gene signaling networks offering the potential for an analysis that is both more complex but also more useful than the GO analysis. When such pathway databases started to became available, the methods originally developed for GO analysis (and thoroughly discussed in Chapter 23) were immediately used to analyze pathways. The extrapolation was very simple: consider that a pathway is merely the set of genes that are involved in it (discarding the interactions), and perform exactly the same analysis used for GO annotations.

Let us briefly summarize these approaches here. A first approach takes a list of DE genes and uses a statistical analysis to identify the pathways in which the DE genes are over- or under-represented. This is done by comparing the number of DE genes found in each pathway with the number of genes expected to be found in that pathway just by chance. If these numbers are substantially different, the pathway is reported as significant. A statistical model (e.g., hypergeometric, chi-square, etc.) can be used to calculate the probability of observing the actual number of genes just by chance. Since most of the time the analysis focuses on the pathways in which the DE genes are over-represented, this approach is also known as **over-representation analysis** (ORA).

An alternative approach considers the distribution of the pathway genes in the entire list of genes and performs a **functional class scoring** (FCS), which also allows adjustments for gene correlations. One of the most salient differences between ORA and FCS approaches is that ORA relies on a previous selection of a subset of differentially expressed genes, which FCS considers the entire list of genes. Arguably the state-of-the-art in the FCS category, the Gene Set Enrichment Analysis (GSEA) [316, 408, 422], ranks all genes based on the correlation between their expression and the given phenotype, and calculates a score that reflects the degree to which a given pathway is represented at the extremes of the ranked list. The score is calculated by walking down the list of genes ordered by expression change. The score is increased for every gene that belongs to P and decreased for every gene that does not. Statistical significance is established with respect to a null distribution constructed by permutations.

Considering the pathways as simple sets of genes was easy and provided a quick solution to the pathway analysis problem but, like many quick solutions, this was not that good. Both ORA and FCS approaches have serious limitations when used for pathway analysis. First, both approaches are limited by the fact that each functional category is analyzed independently without a unifying analysis at a pathway or system level [422]. This approach is not well suited for an method that aims to account for system-level dependencies and interactions, as well as identify perturbations and modifications at the pathway or organism level [402].

Unfortunately, most tools currently available for signaling pathways use one of the ORA or FCS approaches above and fail to take advantage of the much richer data contained in the currently available pathways. GenMAPP, MAPPFinder [120, 108] and GeneSifter use a standardized Z-score. PathwayProcessor [189], PathMAPA [332], Cytoscape [385], and PathwayMiner [335] use Fisher's exact test. MetaCore uses a hypergeometric model, while ArrayXPath [97] offers both Fisher's exact test and a false discovery rate (FDR). Finally, VitaPad [214] and Pathway Studio [322] focus on visualization alone and do not offer any analysis. In essence, virtually all these tools available for pathway analysis fail to consider the topology of the pathways, topology which is there to describe our current understanding of the phenomena that take place on these pathways.

28.3.1 Limitations of the ORA and FCS approaches in pathway analysis

The approaches currently available for the analysis of gene signaling networks share a number of important limitations. First, these approaches consider only the set of genes on any given pathway and ignore their position in those pathways. This may be unsatisfactory from a biological point of view. If a pathway is triggered by a single gene product or activated through a single receptor and if that particular protein is not produced, the pathway will be greatly impacted, probably completely shut off. A good example is the insulin pathway shown in Fig 28.4. If the insulin receptor (*INSR*) is not present, the entire pathway is shut off. Conversely, if several genes are involved in a pathway but they only appear somewhere downstream, changes in their expression levels may not affect the given pathway as much.

Second, some genes have multiple functions and are involved in several pathways but with different roles. For instance, the above *INSR* is also involved in the adherens junction pathway as one of the many receptor protein tyrosine kinases (see Fig. 28.5). However, if the expression of *INSR* changes, this pathway is not likely to be heavily perturbed because *INSR* is just one of many receptors on this pathway. Once again, all these aspects are not considered by any of the existing approaches.

Probably the most important challenge today is that the knowledge embedded in these pathways about how various genes interact with each other is not currently exploited. The *very purpose* of these pathway diagrams is to capture some of our knowledge about how genes interact and regulate each other. However, the existing analysis approaches consider only the sets of genes involved on these pathways, without taking into consideration their topology. This is a crucial weakness of the current approach. In fact, our understanding of various pathways is expected to improve as more data is gathered. Path-

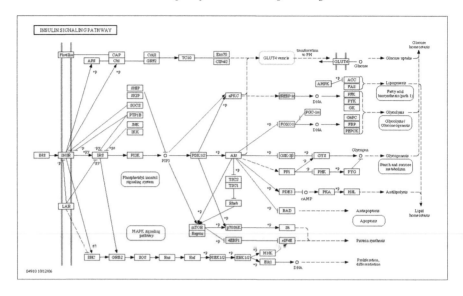

FIGURE 28.4: The insulin pathway. If the insulin receptor INSR is not present or disabled, this pathway will be severely impacted since, INSR is the only entry point in this pathway.

ways will be modified by adding, removing, or redirecting links on the pathway diagrams. Most existing techniques are completely unable to even sense such changes. Thus, these techniques will provide identical results as long as the pathway diagram involves the same genes, even if the interactions between them are completely redefined over time.

Finally, up to now the expression changes measured in these high through-put experiments have been used only to identify differentially expressed genes (ORA approaches) or to rank the genes (FCS methods), but not to estimate the impact of such changes on specific pathways. Thus, ORA techniques will see no difference between a situation in which a subset of genes is differentially expressed just above the detection threshold (e.g., two-fold) and the situation in which the same genes are changing by many orders of magnitude (e.g., 100-fold). Similarly, FCS techniques can provide the same rankings for entire ranges of expression values, if the correlations between the genes and the phenotypes remain similar. Even though analyzing this type of information in a pathway and system context would be extremely meaningful from a biological perspective, currently there is no technique or tool able to do this.

We propose a radically different approach for pathway analysis that attempts to capture all aspects above. An *impact factor* (IF) is calculated for each pathway incorporating parameters such as the normalized fold change of the differentially expressed genes, the statistical significance of the set of pathway genes, and the topology of the signaling pathway. We show on a number

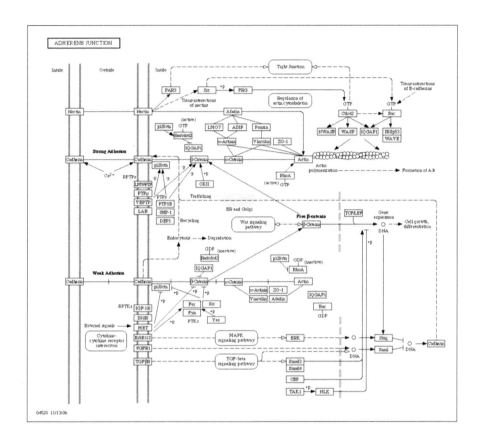

FIGURE 28.5: The adherens junction pathway. Although the same receptor INSR is also present in this pathway in the same position as it was in the insulin pathway, its role here is not as important since it is only one of many tyrosine kinase receptors. A lack of INSR will not completely shut down this pathway, as it would the insulin pathway. None of the existing ORA or FCS approaches is able to take such information into consideration since they all consider the pathways just sets of genes, disregarding their topology.

of real data sets that the intrinsic limitations of the classical analysis produce both false positives and false negatives, while the impact analysis provides biologically meaningful results.

28.4 An approach for the analysis of metabolic pathways

A recent technique, ScorePage, has been developed in an attempt to take advantage of this type of data for the analysis of metabolic pathways [347]. ScorePage calculates a weight for every pair of enzymes, with the maximum weight being given to enzymes acting in consecutive reactions. However, although ScorePage has been shown to be superior to other ORA approaches, its results depend on the choice of a scoring function. In particular, even for two biologically similar data sets, different scoring functions may be necessary to obtain meaningful results [347].

28.5 An impact analysis of signaling pathways

28.5.1 Method description

Our goal is to develop an analysis model that would require both a statistically significant number of differentially expressed genes *and* biologically meaningful changes on a given pathways. In this model, the *impact factor* (IF) of a pathway P_i is calculated as the sum of two terms:

$$IF(P_i) = log\left(\frac{1}{p_i}\right) + \frac{\sum_{g \in P_i} |PF(g)|}{|\Delta E| \cdot N_{de}(P_i)} \qquad (28.1)$$

The first term is a probabilistic term that captures the significance of the given pathway P_i from the perspective of *the set of genes* contained in it. This term captures the information provided by the currently used classical statistical approaches and can be calculated using either an ORA (e.g., z-test [120], contingency tables [332, 335], etc.), a FCS approach (e.g., GSEA [316, 408]) or other more recent approaches. The p_i value corresponds to the probability of obtaining a value of the statistic used at least as extreme as the one observed, when the null hypothesis is true. The results presented here were obtained using the hypergeometric model [419, 129] in which p_i is the probability of obtaining at least the observed number of differentially expressed gene, N_{de}, just by chance.

The second term in Eq. 28.1 is a functional term that depends on *the identity of the specific genes* that are differentially expressed as well as on the interactions described by the pathway (i.e., its topology). In essence, this term sums up the absolute values of the *perturbation factors* (PF) for all genes g on the given pathway P_i. The perturbation factor of a gene g is calculated as follows:

$$PF(g) = \Delta E(g) + \sum_{u \in US_g} \beta_{ug} \cdot \frac{PF(u)}{N_{ds}(u)} \tag{28.2}$$

In this equation, the first term captures the quantitative information measured in the gene expression experiment. The factor $\Delta E(g)$ represents the signed normalized measured expression change of the gene g determined using one of the available methods [98, 123, 345, 473]. The second term is a sum of all perturbation factors of the genes u directly upstream of the target gene g, normalized by the number of downstream genes of each such gene $N_{ds}(u)$, and weighted by a factor β_{ug}, which reflects the type of interaction: $\beta_{ug} = 1$ for induction, $\beta_{ug} = -1$ for repression.[1] US_g is the set of all such genes upstream of g. The second term here is similar to the PageRank index used by Google [73, 198, 331] only that we weight the downstream instead of the upstream connections (a web page is important if other pages point to it whereas a gene is important if it influences other genes).

Under the null hypothesis which assumes that the list of differentially expressed genes only contains random genes, the likelihood that a pathway has a large impact factor is proportional to the number of such "differentially expressed" genes that fall on the pathway, which in turns is proportional to the size of the pathway. Thus, we need to normalize with respect to the size of the pathway by dividing the total perturbation by the number of differentially expressed genes on the given pathway, $N_{de}(P_i)$. Furthermore, various technologies can yield systematically different estimates of the fold changes. For instance, the fold changes reported by microarrays tend to be compressed with respect to those reported by RT-PCR [78, 128]. In order to make the impact factors as independent as possible from the technology, and also comparable between problems, we also divide the second term in Eq. 28.1 by the mean absolute fold change $\overline{|\Delta E|}$, calculated across all differentially expressed genes. Assuming that there are at least some differentially expressed genes anywhere in the data set,[2] both $\overline{|\Delta E|}$ and $N_{de}(P_i)$ are different from zero so the second term is properly defined.

28.5.2 An intuitive perspective on the impact analysis

The differences between the existing techniques and the proposed impact analysis are substantial and worthy of a brief discussion. In order to illustrate these

---

[1] In KEGG, which is the source of the pathways used here, this information about the type of interaction is available for every link between two genes in the description of the pathway topology.

[2] If there are no differentially expressed genes anywhere, the problem of finding the impact on various pathways is meaningless.

differences, consider the task of the life scientist who is given i) a list of differentially regulated genes and ii) an atlas of known pathways describing known connections between genes. The task at hand is to identify the most interesting pathway(s) to study further, perhaps with different types of laboratory assays, most often expensive and time consuming. Let us consider the analogy of a person given a road map atlas containing the usual graph of roads linking localities, a list of known gold ore deposits and the task to find the geographical area(s) with the most promising potential for further mining. Following this analogy, the classical statistical approach (hypergeometric, etc.) is to identify the "interesting" maps by only considering the number of such deposits on any given map, without looking at the amount estimated at each location or anything else. Any map that has more deposits than expected by chance (according to a given model) will appear as interesting. GSEA adds information about the amount of gold at each location i.e. the expression changes for each gene. ScoreMap enhances this further by adding weights corresponding to the distance between deposits. However, all these approaches are still very limited compared to what could be done if the information on the map were to be fully exploited.

In contrast, the impact analysis considers all elements above, but also analyzes the full graph of the maps, actually analyzing all possible routes between known deposits and identifying those parts of each maps that are locally enriched, those that are locally depleted, and those that cannot be reached at all. Interesting aspects that could not be considered previously are now accessible for analysis. Maps containing several deposits in the same restricted geographical area, perhaps signaling a possible common large underground source, appear as more promising. A thorough prospective miner can now assign specific weights for known deposits (alphas in Equation 28.1), as well as costs associated with traveling between them (betas in Equation 28.1) in order to better determine the potential of each area. Such costs (or efficiencies) can be assigned based on the type of the roads connecting them (e.g. freeways allow faster travel than regional roads) or more specific data that might be available for specific pairs of locations (e.g. traffic monitoring data). The impact analysis can be performed on the same maps/pathways at various moments in time: the best route in the summer may not be the best in the winter, and the fastest itinerary at 2am may not be so during the rush hour. If alternative access ways between genes are discovered in the future, i.e. new roads are added to the existing maps, the impact analysis will automatically account for that and adjust the significance of each map accordingly. In contrast, none of the classical approaches can take into consideration either the entire graph of interactions or how these change over time. As long as the locations, amount of gold and shortest distances between locations are the same, all existing techniques will yield exactly the same results. None of the current techniques can include in the analysis specific knowledge about genes or gene-gene interactions, even if such knowledge is available. In essence, none of the classical approaches allows the life scientist to perform even the simplest assessments that we routinely do when we are given similar multi-dimensional data and a similar task in our day-to-day life.

The comparison above hopefully illustrates the nature of the qualitative differences between the existing approaches and the impact analysis. Given the multitude of additional factors that are incorporated in the impact analysis, it is reasonable to expect that this approach will consistently provide better results than any of those obtained by looking only at a limited subset of problem descriptors. What we described here may not be perfect, and certainly is not the ultimate model ever to be developed for this category of problems. However, based on the nature of the differences discussed above, it is reasonable to consider the impact analysis a step forward with respect to other current methods for pathway analysis.

28.5.3 A statistical perspective on the impact analysis

The approach proposed here evaluates the strength of the null hypothesis H_0 (that the pathway is not significant), by combining two types of evidence. In a first analysis, a classical over-representation analysis (ORA) approach provides a p-value defined as the probability that the number of differentially expressed genes, X, is larger than or equal to the observed number of differentially expressed genes, N_{de}, just by chance (when the null hypothesis H_0 is true):

$$p_i = P(X \geq N_{de}|H_0) \tag{28.3}$$

Next, in a separate perturbation analysis, the impact of topology, gene interactions, and gene fold changes come into play and are captured thought the pathway perturbation factor:

$$PF = \frac{\sum_{g \in P_i} |PF(g)|}{|\overline{\Delta E}| \cdot N_{de}(P_i)} \tag{28.4}$$

where $N_{de}(P_i)$ is the number of differentially expressed genes on the given pathway P_i, $PF(g)$ is the perturbation of the gene g:

$$PF(g) = \Delta E(g) + \sum_{u \in US_g} \beta_{ug} \cdot \frac{PF(u)}{N_{ds}(u)} \tag{28.5}$$

and $|\overline{\Delta E}|$ is the mean fold change over the entire set of N differentially expressed genes:

$$|\overline{\Delta E}| = \frac{\sum_{k=1}^{N} |\Delta E|}{N} \tag{28.6}$$

Let PF denote the perturbation factor as a random variable and pf be the observed value for a particular pathway. The score pf is always positive, and the higher its value, the less likely the null hypothesis (that the pathway is not significant). Moreover, this likelihood decays very fast as pf gets away from zero. These features point to the exponential distribution as an appropriate model for the random variable PF:

$$f(x) = \lambda e^{-\lambda x} \tag{28.7}$$

Under the null hypothesis, differentially expressed genes would fall on the pathway randomly, and would not interact with each other in any concerted way. In other words, in the second term in Eq. 28.5 (which captures the influence of the genes upstream) roughly half of those influences will be positives, and half negative, canceling each other out. In such circumstances, the perturbation of each gene would be limited to its own measured fold change (due to random unrelated causes):

$$PF(g) = \Delta E(g) + \sum_{u \in US_g} \beta_{ug} \cdot \frac{PF(u)}{N_{ds}(u)} = \Delta E(g) + 0 = \Delta E(g) \tag{28.8}$$

Consequently, under the same null hypothesis, the expected value for the perturbation of a pathway (from Eq. 28.4) will be:

$$E(PF) = E\left(\frac{\sum_{g \in P_i} |PF(g)|}{|\Delta E| \cdot N_{de}(P_i)} \right) = E\left(\frac{1}{|\Delta E|} \frac{\sum_{k=1}^{N_{de}(P_i)} |\Delta E(g)|}{N_{de}(P_i)} \right) = \tag{28.9}$$

$$= E\left(\frac{\overline{|\Delta E_{P_i}|}}{\overline{|\Delta E|}} \right) = 1 \tag{28.10}$$

The last fraction above is the ratio between the mean fold change on the given pathway, P_i, and the mean fold change in the entire data set. Under the null hypothesis, the genes are distributed randomly across pathways and the two means should be equal. Since this expected value is 1, the distribution of the random variable PF can be modeled by the exponential of mean 1, $exp(1)$.

If we use the PF score as a test statistics and assume its null distribution is exponential with mean 1, then the p-value p_{pf} resulting from the perturbation analysis will have the form:

$$p_{pf} = P(PF \geq pf|H_0) = e^{-pf} \tag{28.11}$$

This is the probability of observing a perturbation factor, PF, greater or equal to the one observed, pf, when the null hypothesis is true.

Let us now consider that for a given pathway we observe a perturbation factor equal to pf and a number of differentially expressed genes equal to N_{de}. A "global" probability p_{global}, of having just by chance both a higher than expected number of differentially expressed genes AND a significant biological perturbation (large PF in the second term), can be defined as the joint probability:

$$p_{global} = P(X \geq N_{de}, PF \geq pf|H_0) \tag{28.12}$$

Since the pathway perturbation factor in Eq. (28.4) is calculated by dividing the total pathway perturbation by the number of differentially expressed genes on the given pathway, the *PF* will be independent of the number of differentially expressed genes X, and the joint probability above becomes a product of two single probabilities:

$$p_{global} = P(X \geq N_{de}|H_0) \cdot P(PF \geq pf|H_0) \tag{28.13}$$

This p_{global} provides a global significance measure that requires both a statistically significant number of differentially expressed genes on the pathway, N_{de}, and at the same time, large perturbations on the same pathway as described by pf. Using equations (28.3) and (28.11), the formula (28.13) becomes:

$$p_{global} = p_i \cdot e^{-pf} \tag{28.14}$$

We take a natural log of both sides and obtain:

$$\log(p_{global}) = \log(p_i) - pf \tag{28.15}$$

which can be rewritten as:

$$-\log(p_{global}) = -\log(p_i) + pf \tag{28.16}$$

in which we can substitute the definition of pf from (28.4) above to yield:

$$-\log(p_{global}) = -\log(p_i) + PF \tag{28.17}$$

The right-hand side of this expression is exactly our definition of the impact factor:

$$IF = -log(p_i) + \frac{\sum_{g \in P_i} |PF(g)|}{|\Delta E| \cdot N_{de}(P_i)} \tag{28.18}$$

This shows that the proposed impact factor, IF, is in fact the negative log of the global probability of having both a statistically significant number of differentially expressed genes and a large perturbation in the given pathway.

Ignoring the discrete character of the hypergeometric distribution, under the null hypothesis $p_i = P(X \geq N_{RP}|H_0)$ has a uniform distribution. By taking negative log, the distribution changes into exponential with parameter 1, similar to the distribution we assumed for PF, the second term in IF formula.

$$-log(p_i) \sim exp(1); \quad PF \sim exp(1); \quad exp(1) = \Gamma(1,1) \tag{28.19}$$

Then, as the sum of two independent exponential random terms, the IF will follow a Gamma distribution $\Gamma(2,1)$ [213]. The pdf of this distribution is:

$$f(x) = xe^{-x}, \quad x \geq 0 \tag{28.20}$$

Finally, the *p*-value corresponding to the observed value if of the statistic *IF* can be easily computed by integrating the density (28.20):

$$p = P(IF \geq if|H_0) = \int_{if}^{\infty} f(x)dx = \int_{if}^{\infty} xe^{-x}dx = (if+1) * e^{-if} \tag{28.21}$$

28.5.4 Calculating the gene perturbations

Note that (28.2) essentially describes the perturbation factor PF for a gene g_i as a linear function of the perturbation factors of all genes in a given pathway. In the stable state of the system, all relationships must hold, so the set of all equations defining the impact factors for all genes form a system of simultaneous equations. Equation 28.2 can be rewritten as:

$$PF(g_i) = \alpha(g_i) \cdot \Delta E(g_i) + \beta_{1i} \cdot \frac{PF(g_1)}{N_{ds}(g_1)} + \beta_{2i} \cdot \frac{PF(g_2)}{N_{ds}(g_2)} + \cdots + \beta_{ni} \cdot \frac{PF(g_n)}{N_{ds}(g_n)} \quad (28.22)$$

Rearranging (28.22) gives

$$PF(g_i) - \beta_{1i} \cdot \frac{PF(g_1)}{N_{ds}(g_1)} - \beta_{2i} \cdot \frac{PF(g_2)}{N_{ds}(g_2)} - \cdots - \beta_{ni} \cdot \frac{PF(g_n)}{N_{ds}(g_n)} = \alpha(g_i) \cdot \Delta E(g_i) \quad (28.23)$$

Using (28.23), a pathway P_i composed of n genes can be described as follows:

$$
\begin{pmatrix}
1 - \frac{\beta_{11}}{N_{ds(g_1)}} & -\frac{\beta_{21}}{N_{ds(g_2)}} & \cdots & -\frac{\beta_{n1}}{N_{ds(g_n)}} \\
-\frac{\beta_{12}}{N_{ds(g_1)}} & 1 - \frac{\beta_{22}}{N_{ds(g_2)}} & \cdots & -\frac{\beta_{n2}}{N_{ds(g_n)}} \\
\cdots & \cdots & \cdots & \cdots \\
-\frac{\beta_{1n}}{N_{ds(g_1)}} & -\frac{\beta_{2n}}{N_{ds(g_2)}} & \cdots & 1 - \frac{\beta_{nn}}{N_{ds(g_n)}}
\end{pmatrix}
\begin{pmatrix}
PF(g_1) \\
PF(g_2) \\
\cdots \\
PF(g_n)
\end{pmatrix}
=
$$
$$
=
\begin{pmatrix}
\alpha(g_1) \cdot \Delta E(g_1) \\
\alpha(g_2) \cdot \Delta E(g_2) \\
\cdots \\
\alpha(g_n) \cdot \Delta E(g_n)
\end{pmatrix}
\quad (28.24)
$$

This can also be written as:

$$(\mathbf{I} - \mathbf{B}) \cdot \mathbf{PF} = \blacksquare\mathbf{E} \quad (28.25)$$

where $\mathbf{B}$ represents the normalized weighted directed adjacency matrix of the graph describing the gene signaling network:

$$
\mathbf{B} =
\begin{pmatrix}
\frac{\beta_{11}}{N_{ds(g_1)}} & \frac{\beta_{12}}{N_{ds(g_2)}} & \cdots & \frac{\beta_{1n}}{N_{ds(g_n)}} \\
\frac{\beta_{21}}{N_{ds(g_1)}} & \frac{\beta_{22}}{N_{ds(g_2)}} & \cdots & \frac{\beta_{2n}}{N_{ds(g_n)}} \\
\cdots & \cdots & \cdots & \cdots \\
\frac{\beta_{n1}}{N_{ds(g_1)}} & \frac{\beta_{n2}}{N_{ds(g_2)}} & \cdots & \frac{\beta_{nn}}{N_{ds(g_n)}}
\end{pmatrix}
\quad (28.26)
$$

$\mathbf{I}$ is the identity matrix, and

$$
\blacksquare\mathbf{E} =
\begin{pmatrix}
\alpha(g_1) \cdot \Delta E(g_1) \\
\alpha(g_2) \cdot \Delta E(g_2) \\
\cdots \\
\alpha(g_n) \cdot \Delta E(g_n)
\end{pmatrix}
\quad (28.27)
$$

is the vector of measured expression changes for genes $g_1, g_2, \ldots, g_n$.

With this notation, the gene perturbation factors can be calculate as:

$$\mathbf{PF} = (\mathbf{I} - \mathbf{B})^{-1} \cdot \blacksquare \mathbf{E} \qquad (28.28)$$

Since the perturbations of the genes are obtained as the solution of a linear system, this approach aims to characterize the steady state of the system rather than rapidly transient states before an equilibrium has been established. Once the perturbation factors of all genes in a given pathway are calculated, Equation 28.1 is used to calculate the impact factor of each pathway.

For some pathways, the matrix describing the interactions between the genes may be singular. In such cases, either the connectivity matrices can be modified slightly to make them non-singular, or the perturbation factors can be calculated by propagating the perturbations as previously described [130].

28.5.5 Impact analysis as a generalization of ORA and FCS

The impact analysis proposed includes and extends the classical approach both with respect to individual genes and with respect to pathways. Interestingly, when the limitations of the existing approaches are forcefully imposed (e.g., ignoring the magnitude of the measured expression changes or ignoring the regulatory interactions between genes), the impact analysis reduces to the classical statistics and yields the same results. We will discuss briefly a few interesting particular cases.

28.5.5.1 Gene perturbations for genes with no upstream activity

In our analysis, the gene perturbation factor for a gene g is defined as:

$$PF(g) = \Delta E(g) + \sum_{u \in US_g} \beta_{ug} \cdot \frac{PF(u)}{N_{ds}(u)} \qquad (28.29)$$

If there are no measured differences in the expression values of any of the genes upstream of g, $PF(u) = 0$ for all genes in US_g, and the second term becomes zero. In this case, the perturbation factor reduces to:

$$PF(g) = \Delta E \qquad (28.30)$$

This is exactly the classical approach, in which the amount of perturbation of an individual gene in a given condition is measured through its expression change ΔE. Examples could include the genes FN1 and CD14 in Fig. 28.6.

28.5.5.2 Pathway impact analysis when the expression changes are ignored

The pathway analysis framework can also be used in the framework in which the ORA approach is usually used. If the expression changes measured for the pathway genes are to be ignored (as they are in the ORA approach), the

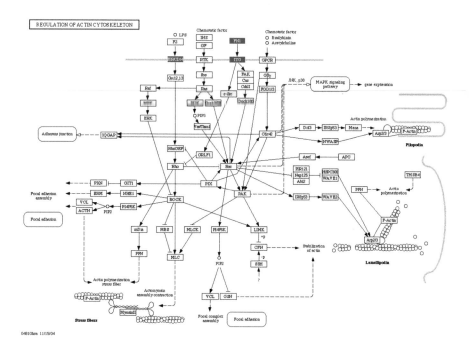

FIGURE 28.6: The hepatic cell line treated with palmitate – differentially expressed genes on the actin cytoskeleton pathway. For genes with no measured changes upstream, such as FN1 and CD14, the gene perturbation will be equal to the measured expression change. The perturbation of genes such as ITG will be higher in absolute value, reflecting both its own measured change as well as the fact that the FN1 gene immediately upstream is also differentially expressed.

pathway impact analysis can still be used to assess the impact of a condition on specific pathways. This is achieved by setting all measured expression changes $\Delta E(g) = 0$ for all genes on the given pathway $g \in P_i$. This will make all gene perturbation factors zero:

$$PF(g) = \Delta E(g) + \sum_{u \in US_g} \beta_{ug} \cdot \frac{PF(u)}{N_{ds}(u)} = 0 \qquad (28.31)$$

Assuming that there are at least some differentially expressed genes somewhere in this data set,[3] (i.e., $\overline{|\Delta E|} \neq 0$), the pathway impact factor in Eq. 28.18 becomes:

$$IF(P_i) = log\left(\frac{1}{p_i}\right) + \frac{\sum_{g \in P_i} |PF(g)|}{\overline{|\Delta E|} \cdot N_{de}(P_i)} = log\left(\frac{1}{p_i}\right) + 0 = -log(p_i) \qquad (28.32)$$

Since the expression now involves a single random variable, the IF values will follow a $\Gamma(1,1) = exp(1)$, rather than a $\Gamma(2,1)$ distribution, and our p-value can be calculated as:

$$p = P(IF \geq -log(p_i)|H_0) = \int_{-log(p_i)}^{\infty} e^{-x} dx = -e^{-x}|_{-log(p_i)}^{\infty} = e^{log(p_i)} = p_i \quad (28.33)$$

This expression shows that, in this particular case, the impact analysis reduces to exactly the classical approach which measures the impact of a pathway by looking exclusively at the probability of the given number of differentially expressed genes occurring just by chance, i.e., the p-value yielded by an analysis in which only the set of genes is considered.

28.5.5.3 Impact analysis involving genes with no measured expression change

Due to the very nature of the technology used, high throughput data is usually noisy. Furthermore, the selection of the differentially expressed genes as well as the computation of the normalized expression value is often done in different ways by different researchers. It is entirely possible that certain genes are in fact changing their expression level but the change is below the sensitivity threshold of the technology, or below the threshold used to select differentially expressed genes. It is also possible that the regulation between genes happens at levels other than that of the mRNA (e.g., phosphorylation, complex formation, etc.). Hence, signals should be allowed to be propagated around the pathway even through those genes for which no expression change has been detected at the mRNA level. The perturbation factor model accounts for these situations. If the measured expression change is zero, the perturbation of the gene becomes:

---

[3] If there are no differentially expressed genes anywhere, the problem of finding the impact on various pathways is meaningless.

$$PF(g) = \sum_{u \in US_g} \beta_{ug} \cdot \frac{PF(u)}{N_{ds}(u)} \qquad (28.34)$$

In this case, the perturbation of a given gene is due to the perturbations of the genes upstream, propagated through the pathway topology.

28.5.5.4 Impact analysis in the absence of perturbation propagation

In certain situations, one might not wish that the analysis propagate the gene perturbations through specific graph edges or types of graph edges (e.g., for edges corresponding to indirect effects or state changes). This can be easily achieved by setting $\beta = 0$ for the desired edges or edge types.

If no perturbation propagations are to be allowed at all, the expression of the gene perturbation in Eq.28.2 reduces to:

$$PF(g) = \Delta E \qquad (28.35)$$

and the impact factor for the pathway becomes:

$$IF(P_i) = \log\left(\frac{1}{p_i}\right) + \frac{\sum_{g \in P_i}|PF(g)|}{|\Delta E| \cdot N_{de}(P_i)} = \log\left(\frac{1}{p_i}\right) + \frac{1}{|\Delta E|}\frac{\sum_{k=1}^{N_{de}(P_i)}|\Delta E(g)|}{N_{de}(P_i)} = \qquad (28.36)$$

$$= \log\left(\frac{1}{p_i}\right) + \frac{\overline{|\Delta E_{P_i}|}}{\overline{|\Delta E|}} \qquad (28.37)$$

In this case, the impact analysis would assess the pathways based not only on the number of differentially expressed genes that fall on each pathway but also based on the ratio between the average expression change on the pathway and the average expression change in the entire set of differentially expressed genes.

In essence, the various particular cases discussed in the previous sections show that the impact analysis is a natural generalization of the classical approaches. As mentioned before, the p-values incorporated in the first term of the impact factor can be obtained with any ORA or FCS approach, from hypergeometric to GSEA.

28.5.5.5 Adding a new dimension to the classical approaches

A two-dimensional plot illustrating the relationship between the two types of evidence considered in the impact analysis is shown in Fig. 28.7. In this figure, the horizontal axis is used to plot the $-\log$ of the p-value yielded by the ORA or FCS evidence (the first term in the definition of the impact factor, Equation 28.1). Small p-values will map onto large $-\log$ values and will be situated towards $+\infty$. For instance, an usual significance threshold such as 0.01 would map onto the value $-\log_{10}(0.01) = 2$.

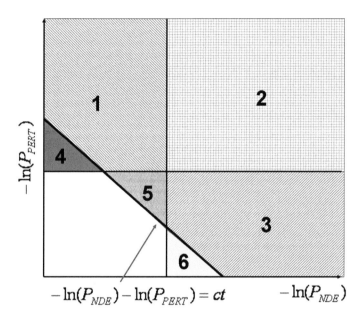

FIGURE 28.7: A two-dimensional plot illustrating the relationship between the two types of evidence considered by SPIA. The x-axis shows the over-representation evidence while the y-axis shows the perturbation evidence. In the top-left plot, areas 2, 3, and 6 together will include pathways that meet the over-representation criterion ($P_{NDE} < \alpha$). Areas 1, 2, and 4 together will include pathways that meet the perturbation criterion ($P_{PERT} < \alpha$). Areas 1, 2, 3, and 5 will include the pathways that meet the combined SPIA criteria ($P_G < \alpha$). Note how SPIA results are different from a mere logical operation between the two criteria (OR would be areas 1, 2, 3, 4, and 6; AND would be area 2). Interestingly, SPIA removes those pathways that are supported by evidence of any one single type that is just above their corresponding thresholds but not supported by the other type of evidence (areas 4 and 6), but adds pathways that are just under the individual significance thresholds but supported by both types of evidence (area 5).

The vertical axis will similarly show the perturbation evidence (second term in the impact factor). Areas 2, 3, and 6 together will include pathways that meet the over-representation criterion ($P_{NDE} < \alpha$), i.e. are larger than the threshold chosen for this type of evidence. Areas 1, 2, and 4 together will include pathways that meet the perturbation criterion ($P_{PERT} < \alpha$), i.e. have a -log of the perturbation p-value that is beyond the threshold for this type of evidence. Areas 1, 2, 3, and 5 will include the pathways that meet the combined SPIA criteria ($P_G < \alpha$).

Let us note that the impact analysis results are different from a mere logical operation between the two criteria: a logical OR would yield areas 1, 2, 3, 4, and 6 while a logical AND would yield area 2. Let us consider for a minute what is happening here. In essence, the impact analysis removes those pathways that are supported by evidence of any one single type (either over-representation or perturbation) that is just above their corresponding thresholds, but not supported by the other type of evidence (areas 4 and 6). However, the impact analysis adds pathways that are just under the individual significance thresholds but supported by both types of evidence (area 5). This is very nice because at the end of the day, any threshold is more or less arbitrary. If a pathway has *both* a relatively large number of DE genes, *as well as* a sufficient amount of perturbation, it will be significant even if neither the number of DE genes, not the amount of perturbation would make it significant by itself. Conversely, if a pathway has only evidence of one single type, either number of DE genes or amount of perturbation, but this amount of evidence is barely above its respective threshold, the impact analysis will not report this pathway as significant.

28.5.6   Some results on real data sets

We have used this pathway analysis approach to analyze several data sets. A first such set includes genes associated with better survival in lung adenocarcinoma [39]. These genes have the potential to represent an important tool for the therapeutical decision and, if the correct regulatory mechanisms are identified, they could also be potential drug targets. The expression values of the 97 genes associated with better survival identified by Beer et al. were compared between the cancer and healthy groups. These data were then analyzed using a classical ORA approach (hypergeometric model), a classical FCS approach (GSEA), and our impact analysis. Fig. 28.8 shows a comparison between the results obtained with the three approaches.

From a statistical perspective, the power of both classical techniques appears to be very limited. The corrected p-values do not yield any pathways at the usual 0.01 or 0.05 significance levels, independently of the type of correction. If the significance levels were to be ignored and the techniques used only to rank the pathways, the results would continue to be unsatisfactory. According to the classical ORA analysis, the most significantly affected pathways in this data set are *prion disease, focal adhesion*, and *Parkinson's disease*. In re-

A

Pathway name	ORA (hypergeometric)		
	p-value	FDR	Bonferroni
Prion disease	0.149649	0.627132	1
Focal adhesion	0.155424	0.627132	1
Parkinson's disease	0.164842	0.627132	1
Dentatorubropallidoluysian atrophy	0.179767	0.627132	1
Calcium signaling pathway	0.262884	0.627132	1
Alzheimer's disease	0.277100	0.627132	1
Apoptosis	0.283744	0.627132	1
TGF-beta signaling pathway	0.303663	0.627132	1
Huntington's disease	0.327491	0.627132	1
Toll-like receptor signaling pathway	0.330069	0.627132	1
Wnt signaling pathway	0.369145	0.637613	1
Regulation of actin cytoskeleton	0.439390	0.695701	1
MAPK signaling pathway	0.560814	0.762988	1
Phosphatidylinositol signaling system	0.572396	0.762988	1
Adherens junction	0.602359	0.762988	1
Complement and coagulation cascades	0.680333	0.766820	1
Cell cycle	0.686102	0.766820	1
Cytokine-cytokine receptor interaction	0.820650	0.866242	1
Neuroactive ligand-receptor interaction	0.972996	0.972996	1

B

Enriched in cancer

Pathway Name	NOM p-val	FDR q-val	FWER p-val
Cell cycle	0.038	0.118	0.140
Huntington's disease	0.074	0.217	0.546
Dentatorubropallidoluysian atrophy (DRPLA)	0.149	0.291	0.751
Alzheimer's disease	0.189	0.344	0.877
Parkinson's disease	0.373	0.485	0.984
Adherens junction	0.583	0.651	0.998
Wnt signaling pathway	0.861	0.785	1

Enriched in normal

Pathway Name	NOM p-val	FDR q-val	FWER p-val
MAPK signaling pathway	0.007	0.170	0.361
Apoptosis	0.019	0.175	0.304
Complement and coagulation cascades	0.037	0.255	0.298
Phosphatidilinositol signaling system	0.189	0.343	0.823
Regulation of actin cytoskeleton	0.010	0.356	0.223
Focal adhesion	0.160	0.384	0.817
Cytokine-cytokine receptor interaction	0.241	0.420	0.910
Toll-like receptor signaling pathway	0.330	0.451	0.963
Calcium signaling pathway	0.308	0.489	0.960
Prion disease	0.474	0.563	0.986
TGF-beta signaling pathway	0.631	0.699	0.998
Neuroactive ligand-receptor interaction	0.947	0.957	1

C

Pathway name	Impact Factor			
	IF	p-value	FDR	Bonferroni
Cell cycle	19.26	8.76E-08	1.66E-06	1.66E-006
Focal adhesion	7.414	0.005072	0.048180	0.0956831
Wnt signaling pathway	6.780	0.008840	0.055988	0.1679642
Dentatorubropallidoluysian atrophy	5.535	0.025788	0.122495	0.4899810
Huntington's disease	4.543	0.058985	0.203925	1
Apoptosis	4.407	0.065921	0.203925	1
Regulation of actin cytoskeleton	4.246	0.075130	0.203925	1
TGF-beta signaling pathway	3.511	0.134730	0.319984	1
Complement and coagulation cascades	3.161	0.176357	0.354145	1
Adherens junction	2.953	0.206279	0.354145	1
Alzheimer's disease	2.752	0.239378	0.354145	1
Parkinson's disease	2.631	0.261455	0.354145	1
Toll-like receptor signaling pathway	2.576	0.272054	0.354145	1
Prion disease	2.572	0.272839	0.354145	1
Calcium signaling pathway	2.538	0.279588	0.354145	1
Cytokine-cytokine receptor interaction	2.353	0.318815	0.366952	1
Phosphatidylinositol signaling system	2.311	0.328326	0.366952	1
MAPK signaling pathway	2.205	0.353353	0.372984	1
Neuroactive ligand-receptor interaction	0.576	0.885936	0.885936	1

FIGURE 28.8: A comparison between the results of the classical probabilistic approaches (A – hypergeometric, B – GSEA) and the results of the impact analysis (C) for a set of genes differentially expressed in lung adenocarcinoma. The pathways in green are considered most likely to be linked to this condition. The ones in red are unlikely to be related.

ality, both prion and Parkinson's diseases are pathways specifically associated to diseases of the central nervous system and are unlikely to be related to lung adenocarcinomas. In this particular case, *prion disease* ranks at the top only due to the the differential expression of *LAMB1*. Since this pathway is rather small (14 genes), every time any one gene is differentially expressed, the hypergeometric analysis will rank it highly. A similar phenomenon happens with *Parkinson's disease*, indicating that this is a problem associated with the method rather than with a specific pathway. At the same time, pathways highly relevant to cancer such as *cell cycle* and *Wnt signaling* are ranked in the lower half of the pathway list. The most significant pathways reported as enriched in cancer by GSEA [408] are: *cell cycle, Huntington's disease, DRPLA, Alzheimer's* and *Parkinson's* (see Fig. 28.8). Among these, only *cell cycle* is relevant, while *Huntington's, Alzheimer's*, and *Parkinson's* are clearly incorrect. However, although ranked first, *cell cycle* is not significant in GSEA, even at the most lenient 10% significance and with the least conservative correction.

In contrast, the impact analysis reports *cell cycle* as the most perturbed pathway in this condition and also as highly significant from a statistical perspective ($p = 1.6 \cdot 10^{-6}$). Since early papers on the molecular mechanisms perturbed in lung cancers [395, 320], until the most recent papers on this topic [334, 103], there is a consensus that the cell cycle is highly deranged in lung cancers. Moreover, cell cycle genes have started to be considered both as potential prognostic factors and therapeutic targets [438]. The second most significant pathway as reported by the impact analysis is *focal adhesion*. An inspection of this pathway (shown in Fig. 28.9) shows that in these data, both *ITG* and *RTK* receptors are perturbed, as well as the *VEGF* ligand. Because these three genes appear at the very beginning and affect both entry points controlling this pathway, their perturbations are widely propagated throughout the pathway. Furthermore, the *CRK* oncogene was also found to be up-regulated. Increased levels of *CRK* proteins have been observed in several human cancers and over-expression of CRK in epithelial cell cultures promotes enhanced cell dispersal and invasion [356]. For this pathway, the impact analysis yields a raw *p*-value of 0.005, which remains significant even after the FDR correction ($p = 0.048$), at the 5% level. In contrast, the ORA analysis using the hypergeometric model yields a raw *p*-value of 0.155 (FDR corrected to 0.627), while the GSEA analysis yields a raw *p*-value of 0.16 (FDR corrected to 0.384). For both techniques, not even the raw *p*-values are significant at the usual levels of 5% or 10%. This is not a mere accident but an illustration of the intrinsic limitations of the classical approaches. These approaches completely ignore the position of the genes on the given pathways and therefore, they are not able to identify this pathway as being highly impacted in this condition. Note that any ORA approach will yield the same results for this pathway for any set of four differentially expressed genes from the set of genes on this pathways. Similarly, GSEA will yield the same results for any other set of four genes with similar expression values (yielding similar correlations with the phenotype). Both techniques are unable to distinguish between a situation in which these genes are upstream, potentially commandeering the entire pathway as in this example, or randomly distributed throughout the pathway.

The third pathway as ranked by the impact analysis is *Wnt signaling* (FDR corrected p=0.055, significant at 10%). The importance of this pathway is well supported by independent research. At least three mechanisms for the activation of Wnt signaling pathway in lung cancers have been recently identified: i) overexpression of Wnt effectors such as Dvl, ii) activation of a non-canonical pathway involving *JNK*, and iii) repression of Wnt antagonists such as *WIF-1* [303]. Mazieres et al. also argue that the blockade of Wnt pathway may lead to new treatment strategies in lung cancer.

In the same data set, *Huntington's disease, Parkinson's disease, prion disease*, and *Alzheimer's disease* have low impact factors (corrected *p*-values of above 0.20), correctly indicating that they are unlikely to be relevant in lung adenocarcinomas.

A second data set includes genes identified as being associated with poor prognosis in breast cancer [433]. Fig. 28.10 shows a comparison between the classical hypergeometric approach, GSEA, and the pathway impact analysis. On this data, GSEA finds no significantly impacted pathways at any of the usual 5% or 10% levels. In fact, the only FDR-corrected value below 0.25, in the entire data set is 0.11, corresponding to the *ubiquitin mediated proteolysis*. Furthermore, GSEA's ranking does not appear to be useful for this data, with none of the cancer-related pathways being ranked towards the top. The most significant signalling pathway according to the hypergeometric analysis, *cell cycle* is also the most significant in the impact analysis. However, the agreement between the two approaches stops here. In terms of statistical power, according to the classical hypergeometric model, there are no other significant pathways at either 5% or 10% significance on the corrected *p*-values. If we were to ignore the usual significance thresholds and only consider the ranking, the third highest pathway according to the hypergeometric model is *Parkinson's disease*. In fact, based on current knowledge, Parkinson's disease is unlikely to be related to rapid metastasis in breast cancer. At the same time, the impact analysis finds several other pathways as significant. For instance, *focal adhesion* is significant with an FDR-corrected *p*-value of 0.03. In fact, a link between focal adhesion and breast cancer has been previously established [184, 432]. In particular, *FAK*, a central gene on the focal adhesion pathway, has been found to contribute to cellular adhesion and survival pathways in breast cancer cells which are not required for survival in nonmalignant breast epithelial cell [50]. Recently, it has also been shown that Doxorubicin, an anticancer drug, caused the formation of well defined focal adhesions and stress fibers in mammary adenocarcinoma MTLn3 cells early after treatment [431]. Consequently, the *FAK/PI-3 kinase/PKB* signaling route within the focal adhesion pathway has been recently proposed as the mechanism through which Doxorubicin triggers the onset of apoptosis [431].

TGF-beta signaling (p=0.032) and *MAPK* (p=0.064) are also significant.

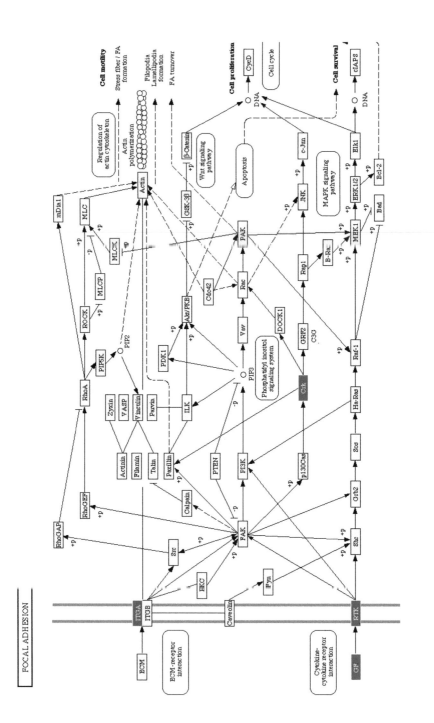

FIGURE 28.9: The focal adhesion pathway as impacted in lung adenocarcinoma versus normal. In this condition, both *ITG* and *RTK* receptors are perturbed, as well as the *VEGF* ligand. Because these three genes appear at the very beginning and affect both entry points controlling this pathway, their perturbations are widely propagated throughout the pathway and this pathway appears as highly impacted. All classical approaches

Both fit well with previous research results. *TGF-beta1*, the main ligand for the *TGF-beta signaling* pathway, is known as a marker of invasiveness and metastatic capacity of breast cancer cells [424]. In fact, it has been suggested as the missing link in the interplay between estrogen receptors and *HER-2* (human epidermal growth factor receptor 2) [424]. Furthermore, plasma levels of *TGF-beta1* have been used to identify low-risk postmenopausal metastatic breast cancer patients [323]. Finally, *MAPK* has been shown to be connected not only to cancer in general, but to this particular type of cancer. For instance, the proliferative response to progestin and estrogen was shown to be inhibited in mammary cells microinjected with inhibitors of MAP kinase pathway [90]. Also, it is worth noting the gap between the *p*-values for *regulation of actin cytoskeleton* ($p=0.111$), which may be relevant in cancer, and the next pathway, *Parkinson's disease* ($p=0.239$), which is irrelevant in this condition.

A third data set involves a set of differentially expressed genes obtained by studying the response of a hepatic cell line when treated with palmitate [412]. Fig. 28.11 shows the comparison between the classical statistical analysis (ORA) and the pathway impact analysis.[4] The classical statistical analysis yields three pathways significant at the 5% level: *complement and coagulation cascades*, *focal adhesion* and *MAPK*. The impact analysis agrees on all these, but also identifies several additional pathways. The top four pathways identified by the impact analysis are well supported by the existing literature. There are several studies that support the existence of a relationship between different coagulation factors, present in the *complement and coagulation cascades* pathway, and palmitate. Sanders et al., for instance, demonstrated that a high palmitate intake affects factor VII coagulant (*FVIIc*) activity [366]. Interestingly, Fig. 28.12 shows not only that this pathway has a higher than expected proportion of differentially expressed genes, but also that 6 out of 7 such genes are involved in the same region of the pathway, suggesting a coherently propagated perturbation. The focal adhesion and tight junction pathways involve cytoskeletal genes. Swagell et al. [412] considered the presence of the cytoskeletal genes among the differentially expressed genes as very interesting and hypothesized that the down-regulation of these cytoskeletal genes indicates that palmitate decreases cell growth. Finally, the link between *MAPK* and the palmitate was established by Susztak et al. who showed that *p38* MAP kinase is a key player in the palmitate-induced apoptosis [411].

---

[4]The GSEA analysis requires expression values for all genes. Since this experiment was performed with a custom array and not all values are publicly available, GSEA could not be applied here.

A

Pathway name	ORA (hypergeometric)		
	p-value	FDR	bonferroni
Cell cycle	3.1E-07	2.8E-06	2.765E-06
MAPK signaling pathway	0.02513	0.11309	0.2261834
Parkinson's disease	0.10752	0.32255	0.9676532
Cytokine-cytokine receptor interaction	0.24736	0.47992	1
Focal adhesion	0.29628	0.47992	1
Calcium signaling pathway	0.37158	0.47992	1
Regulation of actin cytoskeleton	0.40691	0.47992	1
TGF-beta signaling pathway	0.42660	0.47992	1
Neuroactive ligand-receptor interaction	0.58749	0.58749	1

B

Enriched in poor prognosis

Pathway Name	NOM p-val	FDR q-val	FWER p-val
Ubiquitin mediated proteolysis	0.031	0.113	0.111
Prion disease	0.352	0.570	0.802
Alzheimer's disease	0.279	0.625	0.683
Tight junction	0.848	0.749	0.974
Parkinson's disease	0.638	0.795	0.958

Enriched in good prognosis

Pathway Name	NOM p-val	FDR q-val	FWER p-val
Notch signaling pathway	0.082	0.277	0.636
Neuroactive ligand-receptor interaction	0.050	0.280	0.542
Adherens junction	0.136	0.400	0.829
Wnt signaling pathway	0.058	0.410	0.534
Circadian rhythm	0.078	0.582	0.960
Complement and coagulation cascades	0.232	0.638	0.997
Apoptosis	0.212	0.691	0.996
MAPK signaling pathway	0.046	0.693	0.479
Amyotrophic lateral sclerosis	0.244	0.738	0.993
Jak-STAT signaling pathway	0.913	0.952	1
Dentatorubropallidoluysian atrophy	0.792	0.987	1
Cytokine-cytokine receptor interaction	0.913	0.987	1
Calcium signaling pathway	0.522	1	1
Focal adhesion	0.556	1	1
Regulation of actin cytoskeleton	0.575	1	1
Phosphatidylinositol signaling system	0.735	1	1
TGF-beta signaling pathway	0.815	1	1
Cell cycle	0.859	1	1
Huntington's disease	0.885	1	1

C

Pathway name	Impact Factor			
	IF	p-value	FDR	Bonferroni
Cell cycle	10.0	1.0E-07	1.2E-06	1.10E-006
Focal adhesion	7.06	0.00692	0.03112	0.0622412
TGF-beta signaling pathway	6.56	0.01075	0.03225	0.0967557
MAPK signaling pathway	5.40	0.02886	0.06493	0.2597164
Regulation of actin cytoskeleton	4.49	0.06180	0.11125	0.5562285
Parkinson's disease	3.12	0.18207	0.23946	1
Cytokine-cytokine receptor interaction	3.09	0.18624	0.23946	1
Neuroactive ligand-receptor interaction	2.87	0.21942	0.24685	1
Calcium signaling pathway	2.44	0.30047	0.30047	1

FIGURE 28.10: ORA (A) versus GSEA (B) versus impact analysis (C) in breast cancer. GSEA fails to identify any pathway as significant. The hypergeometric model pinpoints *cell cycle* as the only significant pathway. The impact analysis also identifies three other relevant pathways at 5% significance.

Pathway name	ORA (hypergeometric)		
	p-value	FDR	Bonferroni
Complement and coagulation cascades	1.26958E-07	2.28525E-06	2.28525E-06
Focal adhesion	4.03691E-05	0.000363322	0.000726643
MAPK signaling pathway	0.000523961	0.003143765	0.009431295
TGF-beta signaling pathway	0.011698758	0.052644412	0.210577648
Toll-like receptor signaling pathway	0.018714569	0.067372448	0.336862241
Calcium signaling pathway	0.024575814	0.072600598	0.442364654
Tight junction	0.028233566	0.072600598	0.508204185
Wnt signaling pathway	0.050174237	0.100857467	0.903136270
Phosphatidylinositol signaling system	0.058285692	0.100857467	1
Prion disease	0.060516063	0.100857467	1
Jak-STAT signaling pathway	0.061635119	0.100857467	1
Apoptosis	0.106427143	0.146873866	1
Cell cycle	0.106427143	0.146873866	1
Regulation of actin cytoskeleton	0.115415266	0.146873866	1
Alzheimer's disease	0.122394888	0.146873866	1
Huntington's disease	0.146968097	0.165339109	1
Neuroactive ligand-receptor interaction	0.233787848	0.247540075	1
Cytokine-cytokine receptor interaction	0.429908167	0.429908167	1

A

Pathway name	Impact Factor			
	IF	p-value	FDR	Bonferroni
Complement and coagulation cascades	19.374	7.85335E-08	1.41360E-06	1.44761E-06
Focal adhesion	13.791	1.51580E-05	1.36422E-04	3.01180E-04
MAPK signaling pathway	9.475	8.03922E-04	0.004823531	0.014470593
Tight junction	7.128	0.006521277	0.029345745	0.117382981
TGF-beta signaling pathway	6.868	0.008187095	0.029473543	0.147367717
Toll-like receptor signaling pathway	6.391	0.012391594	0.037174781	0.223048688
Calcium signaling pathway	5.774	0.021048873	0.052496861	0.378879719
Apoptosis	5.653	0.023331938	0.052496861	0.419974887
Regulation of actin cytoskeleton	5.225	0.033492741	0.066985482	0.602869334
Jak-STAT signaling pathway	4.983	0.041004319	0.073807774	0.738077735
Wnt signaling pathway	4.313	0.071158653	0.116441431	1
Phosphatidylinositol signaling system	3.975	0.093427025	0.133344438	1
Prion disease	3.937	0.096304316	0.133344438	1
Huntington's disease	3.839	0.104111596	0.133857767	1
Alzheimer's disease	3.387	0.148324272	0.171694058	1
Cell cycle	3.350	0.152616940	0.171694058	1
Neuroactive ligand-receptor interaction	2.414	0.305405348	0.323370368	1
Cytokine-cytokine receptor interaction	2.208	0.352624224	0.352624224	1

B

FIGURE 28.11: A comparison between the results of the classical probabilistic approach (A) and the results of the impact analysis (B) for a set of genes found to be differentially expressed in a hepatic cell line treated with palmitate. Green pathways are well supported by literature evidence while red pathways are unlikely to be relevant. The classical statistical analysis yields three pathways significant at the 5% level: *complement and coagulation cascades*, *focal adhesion* and *MAPK*. The impact analysis agrees on these three pathways but also identifies several additional pathways. Among these, *tight junction* is well supported by the literature.

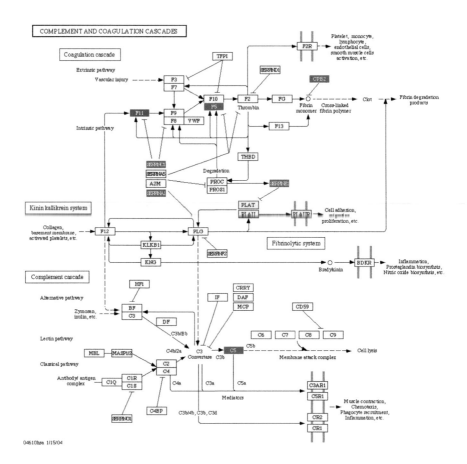

FIGURE 28.12: The complement and coagulation cascade as affected by treatment with palmitate in a hepatic cell line. There are 7 differentially expressed genes (up-regulated in red, down-regulated in blue) out of 69 total genes. All classical ORA models would give any other pathway with the same proportion of genes a similar p-value, disregarding the fact that 6 out of these 7 genes are involved in the same region of the pathway, closely interacting with each other. Both ORA and GSEA would yield exactly the same significance value to this pathway even if the diagram were to be completely redesigned by future discoveries. In contrast, the impact factor can distinguish between this pathway and any other pathway with the same proportion of differentially expressed gene, as well as take into account any future changes to the topology of the pathway.

28.6 Variations on the impact analysis theme

28.6.1 Using the perturbation accumulation

A number of improvements can be brought to the impact analysis as described above. One such improvement uses the **perturbation accumulation** at the level of each gene, instead of the gene perturbation, as the statistic that captures the perturbation of the pathway [417]. This net perturbation accumulation Acc_g, is the difference between the perturbation factor PF of a gene and its observed log fold change:

$$Acc(g_i) = PF(g_i) - \Delta E(g_i) \tag{28.38}$$

With the notation from Equation 28.28, the accumulations can be calculated as:

$$Acc = B \cdot (I - B)^{-1} \cdot \Delta E \tag{28.39}$$

The motivation for doing this is that the measured expression changes are already taken into account by the first term in equation 28.1. By considering only the accumulation at the gene level as defined above, one further decouples the two p-values, strengthening the support for the assumption of independence used in Equations 28.12 and 28.13.

28.6.2 Calculating the perturbation p-value using bootstrapping

As another variation one can use a **bootstrapping** approach in order to calculate the perturbation p-value. By not relying on the exponential distribution, one obtains a more robust method, able to provide sound results even when the data does not follow a particular distribution. Furthermore, the bootstrapping is able to deal better with dependencies and interactions. In fact, our first implementation of this approach in the Pathway-Express tool [265], used this bootstrapping approach. The web implementation was switched to the parametric model using the exponential distribution to ensure a rapid analysis for the online version. Eventually, a version using the total accumulation as the statistic used to assessed the perturbation, and bootstrapping as the method to calculate the corresponding p-value was implemented as the Signaling Pathways Impact Analysis (SPIA) package in Bioconductor.

The computation of P_{PERT} used in SPIA is based on a bootstrap procedure in which we want to test if the observed global activation or inhibition of the pathway computed with the real data, t_A is unusual compared to a multitude of random scenarios. The step by step procedure used is:

1. An iteration counter k is initialized ($k = 1$).

2. A set of $N_{de}(P_i)$ gene IDs is selected at random from the pathway P_i where the $N_{de}(P_i)$ is the number of DE genes observed on the pathway with the real data. The log fold changes for these random gene IDs are assigned by drawing a random sample with replacement from the distribution of all DE genes to be analyzed. item Equation 28.39 is used to compute the perturbation accumulations Acc, for each gene in P_i. The net total accumulation is computed as the sum of all perturbation accumulations across each pathway: $T_A(k) = \sum_i Acc(g_{ik})$.

3. Steps 2 and 3 above are repeated a large number of times ($N_{ite} = 2000$).

4. The median of T_A is computed and subtracted from $T_A(k)$ values centering their distribution around 0. The resulting corrected values are denoted with $T_{A,c}(k)$. The observed net total accumulation is also corrected for the shift in the null distribution median to give, $t_{A,c}$.

5. If $t_{A,c}$ is positive then we conclude that the pathway is activated (or positively perturbed). If $t_{A,c}$ is negative then we assume that the pathway is inhibited (or negatively perturbed).

6. The probability to observe such total net inhibition or activation just by chance, P_{PERT}, is computed as:

$$P_{PERT} = \begin{cases} 2 \cdot \frac{\sum_k I(T_{A,c}(k) \geq t_{A,c})}{N_{ite}} & \text{if } t_{A,c} \geq 0 \\ 2 \cdot \frac{\sum_k I(T_{A,c}(k) \leq t_{A,c})}{N_{ite}} & \text{otherwise} \end{cases}$$

where the identity function $I(x)$ returns 1 if x is true and 0 otherwise. The multiplication by 2 accounts for a two-tailed test, since we do not have a particular expectation regarding the pathway status (inhibited or activated).

28.6.3 Combining the two types of evidence with a normal inversion approach

After computing a p-value for both types of evidence, P_{NDE} and P_{PERT}, we need to combine these two probabilities into one global probability value, P_G, that will be used to rank the pathways and test the research hypothesis, that the pathway is significantly impacted in the condition studied. The probability that a pair of p-values, (P_{NDE}, P_{PERT}), is observed when the null hypothesis is true, can be computed based on the fact that, under the null hypothesis, a p-value is a uniformly distributed random variable on the interval $(0, 1)$. The surface of all theoretically possible values that the variables P_{NDE} and P_{PERT} can take is a square with unity area. The two probability values obtained for a given pathway P_i can be represented as a point within this square $(P_{NDE}(i), P_{PERT}(i))$, as shown in Fig. 28.13.

Since under the null hypothesis $P_{NDE}(i)$ and $P_{PERT}(i)$ are independent probabilities, they can be multiplied to give the joint probability of obtaining the

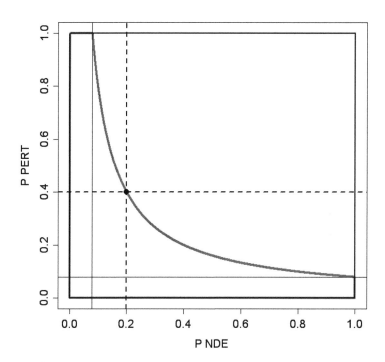

FIGURE 28.13: Combining P_{NDE} and P_{PERT} into a single probability value, P_G. The black rectangle $[0,1]\mathrm{x}[0,1]$ contains all possible values that P_{NDE} and P_{PERT} can take. The curve shown is the locus of all combinations of 2 p-values that have the same product $P_{NDE} \cdot P_{PERT} = c$ (which for this example is: $c = 0.2 \cdot 0.4 = 0.08$). The points under and to the left of this curve represent all combinations that would yield a product less than 0.08. The red contour designates the surface whose area is P_G for the chosen example of the pair ($P_{NDE} = 0.2$ and $P_{PERT} = 0.4$) (black dot), under the null hypothesis. The P_G is the probability to have such a combination which can be quantified as the ratio of the area under the curve divided by the entire area of the square (which is 1). In this case, $P_G = 0.282$.

observed number of DE genes *and* the observed perturbation at the same time. The geometrical locus of the points with the same joint probability is the hyperbola $P_{NDE}(i) \cdot P_{PERT}(i) = c$. The probability to obtain a set of p-values as extreme or more extreme than $(P_{NDE}(i), P_{PERT}(i))$, is the area under and to the left of this hyperbola. The sought global probability P_G is the probability to have such a combination with a product less than or equal to that observed. Hence, P_G can be quantified as the ratio of the area under the curve divided by the entire area of the square (which is 1):

$$P_G = \int_0^c 1 \cdot dx + \int_c^1 \frac{1}{x} \cdot dx = c + c \cdot \ln x \big|_c^1 = c - c \cdot \ln c \qquad (28.40)$$

In the example shown in Fig. 28.13, $P_{NDE}(i) = 0.2$ and $P_{PERT}(i) = 0.4$, which yields $P_G(i) = 0.282$. Eq. 28.40 can be used to calculate the constant c for any desired significance threshold α. For instance, for the the customary $\alpha = 0.05$, the product of the two individual probabilities can be calculated as $c - 0.0087$, a value that has been independently obtained by others [292].

An alternative way to combine these two p-values uses a normal inversion approach. In this approach, the two independent p-values, p_{PERT} and p_{NDE}, are mapped onto Z-scores using the inverse of a random normal cumulative distribution function (cdf). These Z scores are then combined as follows:

$$Z = \frac{Z_{PERT} + Z_{NDE}}{\sqrt{2}} \qquad (28.41)$$

The method is illustrated in Fig 28.14. The differences between the two methods of combining the p-values are shown in Fig. 28.15. In the upper panel, the two p-values are combined with our original method (Equation 28.40), which yields the same results as Fisher's method. In the lower panel, the two p-values are combined with the normal inversion method described above (Equation 28.41). Note how in the panel above one very significant value will make the combination significant independently of what the other p-value is. This does not happen anymore in the lower panel, when a minimum of evidence of the other type is necessary in order to make a pathway significant. In this particular example, Parkinson's disease and Alzheimer's disease are false positive that have very low p-values because that they are small pathways with a few random DE genes. By using the normal inversion method, the two false positives are eliminated, while a new true positive, focal adhesion, is gained.

28.6.4 Correcting for multiple comparisons

Since several pathways are tested simultaneously, we also need to consider adjusting the nominal $P_G(i)$ values for multiple comparisons. For the convenience of the user, both the package implementing SPIA, as well as our web tool, Pathway-Express, provide both Bonferroni- and FDR-corrected p-values.

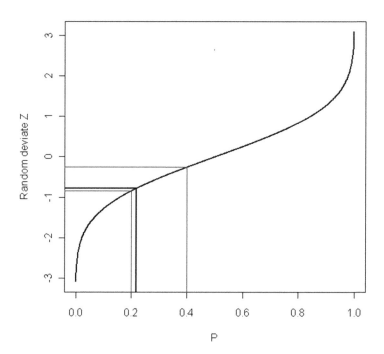

FIGURE 28.14: The normal inversion method maps the independent p-values onto Z-scores using the inverse of a random normal cdf (blue and red lines). Then, these Z-scores are combined and the result mapped back to a p-value using the random normal cdf (black line).

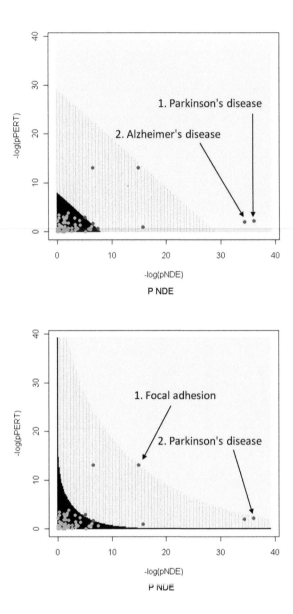

FIGURE 28.15: Combining p-values for the two types of evidence. In the upper panel, the two p-values are combined with our original method which yields the same results as Fisher's method. In the lower panel, the two p-values are combined with the normal inversion method. Note how in the panel above one very significant p-value will make the combination significant. For this data set, this behavior yields two false positives. This does not happen anymore in the lower panel, where the two false positives are eliminated and one true positive (focal adhesion) is gained.

28.6.5 Other extensions

The impact analysis method described in this chapter can be extended to use the entire set of gene expression measurements, thus eliminating the need for the preliminary selection of DE genes. This increases the robustness of the results since they will not depend anymore on the methods and thresholds used to select the genes.

The impact analysis can also take into consideration information regarding the level of significance of each individual gene measurement. To date, no other method is able to include this type of information in the analysis of a pathway.

28.7 Pathway Guide

Our research group is currently funded by both NSF and NIH to extend the impact analysis by adding the capability to automatically identify signaling cascades that are coherent with the measured gene expression changes and could be targeted for potential drug interventions. A software package that is able to identify such signaling cascades in a given condition, Pathway Guide, is currently available (see http://www.advaitacorporation.com/). The output of this package incorporates a lot of data and features, including:

1. A list of all signaling pathways that involve any genes from the input list together with their number of DE genes, their hypergeometric, perturbation and impact p-values.

2. A list of significantly impacted pathways that can be sorted by any of the calculated p-values.

3. The distribution of the perturbations throughout each pathway, showing which areas of the pathways are up-regulated or down-regulated.

4. All coherent perturbation propagation chains that are compatible with all measured gene expression changes, which can be seen as candidate mechanistic explanations for the observed phenomenon and starting points for the search for drug targets.

5. The ability to browse each pathway and see the placement of and relationships between various DE genes.

6. The ability to instantly retrieve annotations from UniProt, NCBI, KEGG, Ensemble, and GeneCards.

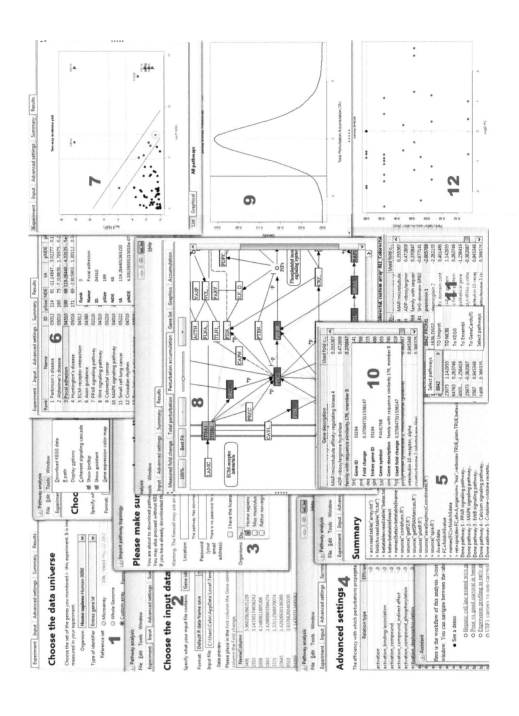

28.7.1 Data visualization capabilities

Pathway-Guide provides an enhanced data visualization module, allowing the users to efficiently identify the components of a pathway and to locate a specific gene. Different pathway annotations allow the user to visualize the conditions observed through the experiment, the computed perturbation induced by the experiment and the total perturbation, computed as a sum of the previous two. A planar chart is provided for each pathway, each gene being represented as a point in a plane where the x dimension is the log of the fold change and the y axis shows the perturbation accumulation. The significance degree of the perturbation value is computed with a bootstrapping technique, that determines how extreme is the obtained value comparing to random permutations of the genes belonging to the analyzed pathway. The result is presented as a probability density function together with the current statistic value. Fig. 28.16 shows the main stages of the analysis performed by Pathway-Guide, as well as some of its results. The first stage consists of: (1) selecting the universe of the experiment (organism, reference list, type of IDs used); (2) providing the list of DE genes; (3) loading pathway data (if necessary); and (4) selecting additional settings such as the efficiency of the propagation for various types of edges, or the significance threshold. The *impact analysis* stage is performed by an underlining *R* process. The log of this process can be reviewed in the Summary section (5). Pathway Guide makes use of both tables and graphics to present the *results* of the analysis. Table (6) provides individual pathway statistics including the number of DE genes, total number of genes, hypergeometric p-value, total accumulation, perturbation p-value, and global p-value. The table can be sorted by any of these. Graphical representations include a two-dimensional evidence plot (7) showing the hypergeometric p-value versus the perturbation p-value for all pathways. All tables and graphics are sensitive to user interaction. If one of the pathways is selected, the list of genes contained in that respective pathway (10) is updated. The gene list contains the gene ID, description, fold change and total perturbation. Additional information related to each individual gene can be obtained using the external links to online data sources including Uniprot, NCBI, KEGG, Ensembl, and GeneCards (11). A pathway diagram with all genes and their connections (8) is available for each pathway. This is also sensitive to user interaction and is updated whenever the user selects a different pathway. The pathway topology diagram uses color to represent fold change, total perturbation, or perturbation accumulation for each pathway gene (8). The relation between the total perturbation and the fold change on a pathway can also be viewed as a two-way plot (12). To better understand the perturbation p-value of the selected pathway, its actual total accumulation is presented in relation to the distribution of the total accumulations under the null hypothesis (9).

28.7.2   Portability

Pathway Guide is highly portable. The user interface uses Java Swing while the computational and graphical parts are based on R. The software requires Java 1.6 or newer and R version 2.9 or newer. Both are widely available on most modern operating systems (OS). Pathway Guide was tested the software on Windows (XP and Vista), Mac OS X and Linux (Fedora and Ubuntu). The integration between Java and R is done using both OS streams and temporary files. There are no Java native libraries involved since such an approach would have coerced both Java and R versions to be part of a predefined subset.

28.7.3   Export capabilities

Pathway Guide has full export capabilities. The software allows the user to cut and paste the selected content of any displayed table into any document processing software such as OpenOffice and Microsoft Office (both text and spreadsheet editors). Furthermore, the software allows the user to save the content of any displayed table as HTML, CSV, or TeX. Any image displayed by the software can be saved in any of the common formats used in publishing, such as BMP, JPEG, PNG, and GIF. Additionally, some synthetic images can be saved in the vectorial EPS format. The layout of these graphs is such that the positions of the genes/proteins is as close as possible to the position of their corresponding elements in the original pathway such as the user can follow the topology (see Fig. 28.17). In addition, any of the gene networks can be saved in either GraphML or GML formats, which allows other network analysis and/or visualization algorithms to run on the same network. The software also converts the list of pathways in a data structure loadable in R, in the form of adjacency matrices and node lists. This allows the user to further extend the analysis capabilities of the software by writing their own custom R scripts.

28.7.4   Custom pathways support

Pathway Guide is able to read custom pathways described in the KGML format (XML with a schema provided by KEGG). A custom pathway is specified as a tabular description of the pathway (describing the genes and their interactions) as well as a graphical image of the pathway, if desired. The format allows a rich description of the interactions between genes, and is able to model gene/protein complexes. To support the enhanced graphical display, the file describing the custom pathway can contain the x and y coordinates location of the entities within the associated image file.

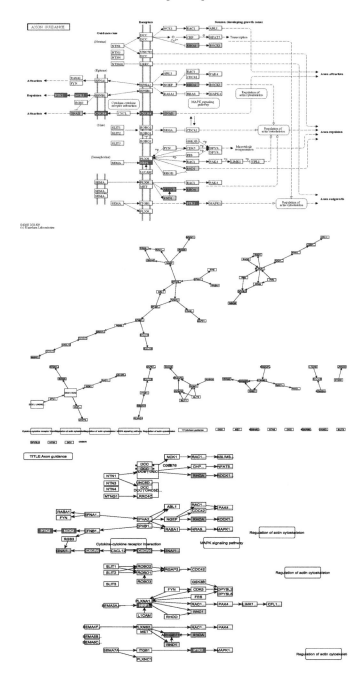

FIGURE 28.17: Pathway layouts. Top: as provided by KEGG, middle: default GraphML layout, bottom: with reformatted layout allowing the user to follow the original topology.

28.7.5 Reliability benchmarks: speed, number of FPs, distribution of *p*-values under the null distribution

Pathway Guide was benchmarked and performed very well in terms of both speed and accuracy. In terms of speed, a code optimization provided a 10-fold increase in the processing speed with respect to SPIA, the previously released Bioconductor package implementing the impact analysis [417].

In order to assess the reliability of this analysis software, we ran 36,000 random data sets generated according to the three scenarios shown in Fig. 28.18: fold changes generated from a normal distribution of mean m=0 and standard deviation sd=1; fold changes from a normal distribution with m=3 and sd=0.5; fold changes from the tails of the normal distribution of m=0 and sd=1. The first scenario represents purely random genes with no bias, the second one represents random genes with a strong positive bias, while the third scenario represents a typical selection of DE genes due only to noise. The three scenarios were used to ensure that Pathway Guide is not susceptible to bias introduced by the data. For each experiment, we picked a random set of 100, 300, 500, 1,000, 2,000 and 5,000 "differentially expressed" (DE) genes, from a reference set of 20,000 genes. The reference set was the subset of human genome containing all the genes present in at least one KEGG pathway. For each size of the DE set, the analysis was performed 2,000 times. The goal here was to make sure that under the null hypothesis the software does not yield any false positives (FP) beyond the expected significance level.

In each case, the number of pathways with an impact analysis *p*-value lower than the chosen significance threshold is counter. Since the input fold changes were random, these pathways are FPs. In all experiments, their number was very close to what is theoretically expected for the chosen significance level (Fig. 28.19). In all cases, a Kolmogorov-Smirnoff as well as a χ^2-test, failed to find any significant differences between the empirical distribution of the *p*-values and the uniform distribution (for the data shown $p=0.52$ and $p=0.43$, respectively).

28.8 Kinetic models versus impact analysis

Kinetic models based on the molecular mechanisms of interaction have been used for more than 25 years to simulate biochemical phenomena. These models use differential equations to simulate the evolution of specific reactions or ensembles of reactions over time. Such models are indispensable in order to make any quantitative predictions, but they are drastically limited by the need to known the precise initial concentration for most reactants, exact reactions constants for all reactions, as well as the appropriate time scale for the studied phenomenon. The goal of such kinetic models is to fully describe the biochem-

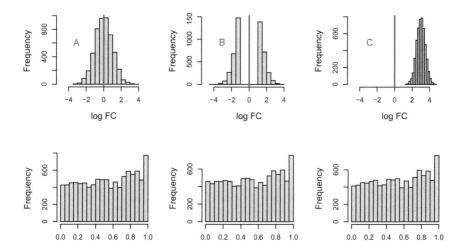

FIGURE 28.18: Reliability testing results for Pathway Guide. On top: the three scenarios used to test the behavior under the null distribution. For each of these distributions 100, 300, 500, 1,000, and 5,000 random input genes were picked. On the bottom: the distribution of the p-values under the null hypothesis for 1,000 input genes. The higher column near 1 is due to pathways that did not have any input genes at all (the probability of having 0 or more input genes on a pathway is 1). Excluding the pathways with zero genes, the distributions are uniform, as they should (testing for differences with respect to the uniform distribution: Kolmogorov-Smirnoff $p=0.52$, χ^2 $p=0.43$).

Experiment	Input genes	Percent of FPs 1%	Percent of FPs 5%
1 A	100	1	5.11
2 B	100	0.97	5.03
3 C	100	0.8	3.97
4 A	300	0.92	4.79
5 B	300	1.09	5.11
6 C	300	0.76	3.99
7 A	500	0.97	4.96
8 B	500	1.01	4.98
9 C	500	0.84	4.13
10 A	1000	0.92	4.98
11 B	1000	0.94	4.99
12 C	1000	0.85	4.33
13 A	2000	0.95	5.03
14 B	2000	0.98	5.02
15 C	2000	0.89	4.44
16 A	5000	1.04	5.06
17 B	5000	1.01	5.06
18 C	5000	0.89	4.72

FIGURE 28.19: Reliability test results for Pathway Guide. For each of the 18 experiments shown, 2,000 trials were performed. In each trial, the number of input genes shown (100, 300, 500, 1,000, 2,000, or 5,000) was chosen as "differentially expressed" from one of the three scenarios A, B, and C, shown in Fig 28.18. In all cases, the percentages of observed false positives are very close to the significance threshold, as they should be.

ical phenomena involved and to make quantitative predictions about some of the reaction products involved. In contrast, our goal here is to be able to identify the most significantly impacted pathways from a large collection of heterogenous pathways, based on incomplete information (relative fold changes for a few genes rather than exact concentrations for most reactants). Even if detailed kinetic models were available for every existing pathway, such models would not be able to tell us which pathways are significant in a given condition. Furthermore, kinetic models work for biochemical pathways describing reactions of the same type (biochemical) with known reaction constants. The pathways we are working with are gene signaling pathways containing heterogeneous "signals" (inhibition, activation, phosphorylation, methylation, etc.) happening at many levels (transcription, translation, post-translational, etc.) between heterogeneous components (mRNA, DNA, protein, metabolites, etc.). For these reasons, kinetic models cannot be used to identify the significant signaling pathways in a given condition, which is what we are trying to do here. Our goal is to depart from the kinetic models and their limitations. The approach we are proposing is a rigorous statistical framework that can work with heterogeneous nodes and incomplete information as they are found in signaling pathways and real experiments. One of the main advantages of the impact analysis approach is that it eliminates the need for the exact concentrations, reaction constants, etc. The limitation is that unlike a kinetic model, our approach cannot make any quantitative predictions about any of the nodes involved. However, this is not part of our goal. Our goal is to identify the pathways that are significant in a given condition, not to predict anything specific on any such pathway.

28.9   Conclusions

A statistical approach using various models is commonly used in order to identify the most relevant pathways in a given experiment. This approach is based on the set of genes involved in each pathway. Here we discussed a number of additional factors that are important in the description and analysis of a given biological pathway. We also described an impact analysis method that uses a systems biology approach in order to identify pathways that are significantly impacted in any condition monitored through a high throughput gene expression technique. The impact analysis incorporates the classical probabilistic component but also includes important biological factors that are not captured by the existing techniques: the magnitude of the expression changes of each gene, the position of the differentially expressed genes on the given pathways, the topology of the pathway which describes how these genes interact, and the type of signaling interactions between them. The results obtained on several independent data sets show that the classical approaches can provide both false positives and false negatives, while the impact analysis is able to provide biologically meaningful results. This analysis method has been implemented as a web-based tool, Pathway-Express, freely available as part of the Onto-Tools (http://vortex.cs.wayne.edu).

28.10   Data sets and software availability

The data sets used in this chapter are available for download from the following sources:

1. Breast cancer data set - van't Veer *et al.* Nature, 415(6871):530-536, January 2002 - associated data files.

2. Lung cancer data set - Beer *et al.* Nature Medicine, 8(8):816-824, Jul 2002 - associated data files.

3. Hepatic cell line data set - Swagell *et al.* Biochemical and Biophysical Research Communications, 328: 432-441, 2005. The list of differentially expressed genes and their corresponding fold changes were obtained from table 2 in the manuscript.

The input files used in Pathway-Express for each of these data sets are also available at **http://vortex.cs.wayne.edu**. Pathway-Express is the free version of the software implementing the impact analysis approach. Pathway-Express is available at **http://vortex.cs.wayne.edu**. Pathway-Guide is the commercial version that includes the ability to find coherent signaling cascade and optimal points of therapeutic intervention (e.g., drug targets). A fully featured time-limited demo version of Pathway-Guide is available from **www.advaitacorporation.org**.

Chapter 29

Machine learning techniques

I hear, I forget
I see, I remember
I do, I understand

—*Chinese proverb*

29.1 Introduction

The term[1] **machine learning** refers to a set of topics dealing with the creation and evaluation of algorithms that facilitate pattern recognition, classification, and prediction, based on models derived from existing data. Two facets of

---

[1] Parts of this chapter are adapted from the review article "Machine Learning and its Applications to Biology" by Adi L. Tarca, Vincent J. Carey, Xue-wen Chen, Roberto Romero, Sorin Draghici, *PLoS Computational Biology* 3(6): 953-963, June 2007. The Public Library of Science (PLoS) applies the Creative Commons Attribution License (CCAL) to all works published. Under the CCAL, authors retain ownership of the copyright for their article, but authors allow anyone to download, reuse, reprint, modify, distribute, and/or copy articles in PLoS journals, so long as the original authors and source are cited.

mechanization should be acknowledged when considering machine learning in broad terms. First, it is intended that the classification and prediction tasks can be accomplished by a suitably programmed computing machine. That is, the product of machine learning is a classifier that can be feasibly used on available hardware. Second, it is intended that the creation of the classifier should itself be highly mechanized, and should not involve too much human input. This second facet is inevitably vague, but the basic objective is that the use of automatic algorithm construction methods should minimize the possibility that human biases could affect the selection and performance of the algorithm. Both the creation of the algorithm and its operation to classify objects or predict events are to be based on concrete, observable data.

The history of the relationship between life sciences and the field of machine learning is long and complex. An early technique [358] for machine learning called *perceptron* constituted an attempt to model actual neuronal behavior. The entire field of artificial neural network (ANN) emerged from this attempt. Early work on the analysis of translation initiation sequences [405] employed the perceptron to define criteria for start sites in *E. coli*. Further ANN architectures such as the adaptive resonance theory (ART) [80] and neocognitron [164] were inspired from the organization of the visual nervous system. In the intervening years, the flexibility of machine learning techniques has grown along with mathematical frameworks for measuring their reliability. It was natural to hope that machine learning methods will improve the efficiency of discovery and understanding in the mounting volume and complexity of biological data and indeed, machine learning techniques have proved invaluable in a number of life science applications. Indeed, life science applications of unsupervised and/or supervised machine learning techniques abound in the literature. For instance, gene expression data was successfully used to classify patients in different clinical groups or identify new disease groups [14, 338, 16, 360], while genetic code allowed prediction of the protein secondary structure [362]. Continuous variable prediction with machine learning algorithms was used to estimate bias in cDNA microarray data [416].

This chapter is divided in four main parts. First, a brief section reviews definitions and mathematical prerequisites. Second, the area of supervised learning is briefly described. Third, methods of unsupervised learning are reviewed. Finally, the last section reviews methods and examples as implemented in R.

29.2   Main concepts and definitions

Two main paradigms exist in the field of machine learning: **supervised** and **unsupervised** learning. Both are relevant and have potential applications related to high throughput data.

In **supervised learning**, objects in a given collection are classified using a set of attributes, or features. The classes are known in advance and the goal is to build a model that can assign an object to a class based solely on the values of its features. In a biological context, examples of *object*-to-*class* mappings are tissue gene expression profiles to disease group, and protein sequences to their secondary structures. The features in these examples are the expression levels of individual genes measured in the tissue samples and the presence/absence of a given amino acid symbol at a given position in the protein sequence, respectively. The goal in supervised learning is to design a system able to accurately predict the class membership of new objects based on the available features. Besides predicting a categorical characteristic such as class label, (similar to classical *discriminant analysis*), supervised techniques can be applied as well to predict a continuous characteristic of the the objects (similar to *regression analysis*).

In any application of supervised learning, it would be useful for the classification algorithm to return a value of "doubt" (indicating that it is not clear which one of several possible classes the object should be assigned to) or "outlier" (indicating that the object is so unlike any previously observed object that the suitability of any decision on class membership is questionable).

In contrast to the supervised framework, in **unsupervised learning**, the classes are not known in advance. In this case, the goal is to explore the data and discover similarities between objects. Such similarities are used to define groups of objects, referred to as *clusters*. In other words, unsupervised learning is intended to unveil natural groupings in the data. Thus, the two paradigms may informally be contrasted as follows: in supervised learning, the data come with class labels, and we learn how to associate unlabeled data with classes; in unsupervised learning all the data are unlabeled, and the learning procedure consists of both defining the labels and associating objects with them.

In some applications, such as protein structure classification, only a few labeled samples (protein sequences with known structure class) are available, while many other samples (sequences) with unknown class are available as well. In such cases, **semi-supervised learning** techniques can be applied to obtain a better classifier than could be obtained if only the labeled samples were used [458]. This is possible, for instance, by making the "cluster assumption," i.e., that class labels can be reliably transferred from labeled to unlabeled objects that are "nearby" in feature space.

In general, problems can be divided into three categories: i) **class prediction** problems, ii) **class comparison** problems, and iii) **class discovery** problems. In a **class prediction** problem, the classes are defined in advance, and we are given a number of examples from each class. These examples are usually vectors of values for a number of features. In the class prediction paradigm, each example belongs to a known class (i.e., has a known class label). The goal is to build a classifier, which is a device that is able to take a previously unseen input vector and correctly assign it to its class.

In a **class comparison** problem, the classes are still defined in advance, but the goal is to find the main differences between the classes. In other words, the goal is to find those features that distinguish the classes.

In a **class discovery** problem, the classes are not known in advance. In this situation, the input vectors come without any labels. The goal here is to exploit the redundancy in the data and identify those input vectors that share certain features. As mentioned above, these vectors will form clusters.

A natural impulse is to try to compare the difficulty of the three problems above. Unfortunately, such a comparison does not have a simple answer. Building a classifier as required in a class prediction problem requires a **feature selection** step. In this step, one evaluates the various features of the input vectors and select those that are useful in order to build a classifier. Since this feature selection step is also required in a class comparison, one might argue that class prediction is more difficult than class comparison because it involves the extra step of actually building the classifier. However, we must point out that building a classifier does not necessarily require the identification of *all* features that are different between the classes. In principle, if only one feature allows the construction of a perfect classifier, there is no need to use any other features. For instance, if were to build a classifier to distinguish between Kubota and John Deere tractors, the color will be the only feature needed since Kubotas are all orange while John Deeres are all green. In contrast, the class comparison problem seeks a thorough comparison of the two classes, much like a potential buyer who would be interested to know *all* differences between two comparable models, not the bare minimum that is needed in order to distinguish them. From this perspective, the class comparison problem may be considered more difficult since failing to find even one significant difference would be a serious drawback in this context.

Finally, **class discovery** is a task that is very different from both class prediction and class comparison. In class discovery, the classes are not know in advance and, in most cases, (slightly) different classes or clusters can be obtained from the same data by applying different methods. Note that redundancy is absolutely needed for class discovery. If only one representative from each cluster is present in the data available for analysis, the correct clusters can never be identified.

It is crucial that the type of problem confronted is identified first since the three classes of problems should be tackled with different sets of tools. Class comparison problems should be tackled with techniques developed for feature selection, including statistical methods such as those discussed in Chapter 12 and Chapter 21. Class prediction problems should be tackled with machine learning techniques able to build a classifier, as those described in this chapter (although a feature selection step may be necessary before actually building a classifier). Class discovery problems should be tackled with unsupervised machine learning techniques, including clustering techniques, as those described in Chapter 18. Using a class of methods that is inappropriate for the given problem cannot possibly produce anything useful. For instance, as it was already explained in Chapter 18, performing a clustering on labeled data after feature selection is useless since the features have been selected based on their ability to differentiate the classes. Hence, such a clustering can only verify that the feature selection was performed correctly. In contrast, performing a clustering of the data *using all available features* (but ignoring the known classes) can be informative. If the results of such clustering match the known classes, this is an indication that the classes are relatively easily distinguishable. If the results of such a clustering are very different from the known classes, it may indicate that either the data are very noisy and many features need to be eliminated before a class separation is possible, or that the classes are not distinguishable.

Since the clustering methods have already been discussed in Chapter 18, the rest of this chapter will focus on supervised machine learning methods.

29.3 Supervised learning

29.3.1 General concepts

Let us consider the general case in which we want to classify a collection of objects $i = 1, \ldots, n$ into K predefined classes. Without loss of generality, data on features can be organized in a $n \times p$ matrix $X = (x_{ij})$, where x_{ij} represents the measured value of the variable (feature) j in the object (sample) i. Every row of the matrix X is therefore a vector $\mathbf{x}_i$ with p features to which a class label y_i is associated, $y = 1, 2, \ldots, c, \ldots, K$. For instance, if one wanted to distinguish between different types of tumors based on gene expression values, then K would represent the number of known existing tumor types, n would be the number of tumor samples available, and p would be the number of genes measured.

In such multi-class classification problems, a classifier $\mathscr{C}(\mathbf{x})$ may be viewed as a collection of K discriminant functions $g_c(\mathbf{x})$ such that the object with feature vector $\mathbf{x}$ will be assigned to the class c for which $g_c(\mathbf{x})$ is maximized over the class labels $c \in \{1, \ldots, K\}$. The feature space $\mathscr{X}$ is thus partitioned by the classifier $\mathscr{C}(\mathbf{x})$ into K disjoint subsets.

There are two main approaches to the identification of the discriminant functions $g_c(\mathbf{x})$ [452]. The first assumes knowledge of the underlying class conditional probability density functions (the probability density function of $\mathbf{x}$ for a given class) and assigns $g_c(\mathbf{x}) = f(p(\mathbf{x}|y = c))$, where f is a monotonically increasing function, for example the logarithmic function. Intuitively, the resulting classifier will classify an object $\mathbf{x}$ in the class in which it has the highest membership probability. In practice $p(\mathbf{x}|y = c)$ is unknown, and therefore needs to be estimated from a set of correctly classified samples that form the **training set**. Both parametric and nonparametric methods for density estimations can be used for this purpose. From the parametric category, we will discuss **linear** and **quadratic discriminants** while from the nonparametric one, we will describe the **k-nearest neighbor** technique.

The second approach is to use data to estimate the class boundaries directly, without explicitly calculating the probability density functions. Examples of algorithms in this category include **decision trees** and **support vector machines**.

29.3.2 Error estimation and validation

Suppose the classifier $\mathscr{C}(\mathbf{x})$ was trained to classify input vectors $\mathbf{x}$ into two distinct classes, 1 and 2. The results of the classification of a collection of input objects $\mathbf{x}_i$, $i = 1, \ldots, n$ can be summarized in a **confusion matrix** that contrasts the predicted class labels of the objects $\hat{y}_i$ with the true class labels y_i. The confusion matrix, as well as the performance measured used to assess a two-class classifier were discussed in details in Chapter 8.

The goal behind developing classification models is to use them to predict the class membership of *new samples*. If the data used to build the classifier is also used to compute the error rate, then the resulting error estimate, called the **resubstitution** estimate, will be optimistically biased [145]. In fact, this type of error estimate assesses only the ability of the model built to *memorize* the data samples used during the training. However, if memorization of a number of data points were the only goal, we would use a database to just store those data. The reason for building a model in the first place was related to the ability of a model to provide outputs, or *predictions*, for data not previously seen during the model building. This ability is called **generalization**. From this perspective, it is intuitive that if we assess the error of the model on the very data used to built it, we will obtain an overoptimistic estimation of the performance of the model. In fact, these assessment will contain very little if any information about the prediction capabilities of the model.

A better way to assess the error is the **hold-out** procedure in which one splits the data into two disjunct parts. The first part is used to train the classifier and is referred to as the **training set**. The remaining part is used to asses the error and is referred to as the **test set** or **validation set**. The size of the training set is crucial because there must be enough information in the training set to allow the construction of a good model. Unfortunately, no specific size of the training set can be calculated because the number of the data points needed depends on the complexity of the problem including the number of dimensions of the data, the number of variables in the model, etc. A common mistake is to try to build a model with a large number of parameters using a training set with an insufficient number of data points. In principle, there should be much more data points than the number of dimensions of the input space, which in turn has to be smaller than the number of parameters of the model built. Furthermore, all aspects of the phenomenon must be represented in the training set for a good model to be even possible. In concrete terms, a typical microarray experiment will have tens of thousands of genes and tens or, in the best case hundreds of samples. Under these circumstances, it is hopeless to build a model using the entire set of genes. In this particular context, hopeless actually means *too easy*, rather than too difficult. In fact, it will always be possible to build a model that has very low error on the training set. This is similar to finding a plane that goes through one or two points in a 3D space. The problem resides in the fact that the model is under-constrained: there are in fact an infinite number of planes that can be fit to pass through any two points in a 3D space. If we were to report the error on the training set, this error will be zero or very small, misleading us to believe that we have a good model. However, as soon as somebody provides more points that should have been on our plane, we will discover that in all likelihood, our chosen plane that provided zero error on the training set will produce very large errors on this additional points. This, in fact, is the role of the validation set: these additional points not used in any way during the training will provide information regarding the ability of our model to provide the correct results for new, previously unseen data, as expected in the real application. In contrast, if we had 10 points in our training set, it is likely that there will be only one plane that will provide the minimum error with respect to these 10 points. If these 10 points had been chosen to correctly and fully represent the problem (e.g., if they had been chosen randomly to cover the entire region of interest of the 3D space), the plane fitted to these 10 points is likely to provide a low error even if more points are provided from the same phenomenon. This is why it is important to have more (and preferably much more) data points than dimensions. Going back to our typical microarray experiment, this means that a **feature selection** step has to be performed before a classifier is built in order to reduce the number of dimensions of the input space. In practice, as a rule of thumb, if the number of samples in the training set is about 100, it is best to keep the number of genes to 10 or fewer.

If there is enough data to allow for a hold-out validation procedure that still has enough data points in the training set, this is the most reliable way of assessing the performance of the classifier in the real world. In essence, since the validation set does not include any of the data used to build the model, it constitutes a realistic proving ground for the newly built classifier. However, many times the number of data points available (e.g., samples) is not sufficient. For instance, if we only had 100 samples and we wanted to use 40 as the validation set, we would only be left with 60 as the training set. This in turn will force us to further reduce the number of genes we consider which may be undesirable. In this case, an alternative is the **n-fold cross validation** method. In this method, the data are divided into n **folds**. The training and testing will be performed n times. Each time, one of these folds will be used as the validation set, while the other $n-1$ folds will form the training set in that particular training run. The performance measures are calculated each time and averaged over the n training runs. These average values (e.g., average accuracy, average specificity, etc.) will provide a good estimate of the performance of the that type of classifier in the real world.

As an example, a 10-fold cross validation will train 10 times, each time using 90% of the data as the training set and 10% of the data as the validation set. A 4-fold cross-validation will train 4 ties, each time using 75% of the data as the training set and 25% of the data as validation set.

However, in some situations, the data may be too scarce even for n-fold cross-validation. If we want to make sure that we use all available data for training and still assess the performance on data that was previously unused, there is one other method: the **leave-one-out cross-validation** (LOO) procedure. The leave-one-out cross-validation method is nothing but an n-fold cross-validation where the number of folds is equal to the number of data points available. This procedure trains a number of times equal to the number of data points available, in such a way that in each training run, a different data point is left out of the training set and used as a minimal validation set. Although the estimate of the error obtained with the LOO procedure gives low bias, it may show high variance [197].

If the number of samples are very different between the two classes, the data set is said to be **unbalanced**. Some models are biased towards the richer class and have a tendency to favor it. In turn, this can produce suboptimal classifiers. Techniques exist to weight or sample the classes differently in order to avoid this. If desired, attention can be paid during the cross-validation to make sure that the ratio between the number of data point from each class remains as desired. In this case, it is said that the training and validations sets are **stratified**.

Note that unlike the hold-out procedure, any cross-validation procedure (e.g., any variation of n-fold, or the leave-one-out validation) actually builds several classifiers. Hence, a natural question is to ask which of these is to be used in practice, in the real world, after this training and validation procedure has been executed. The answer is: none of them. The goal of this procedure was to assess the performance of this type of classifier on previously unseen data. This is accomplished in each training run. However, the danger here is that due to the limited number of training data points in each such run, certain features of the phenomenon were not captured. This is why the procedure is repeated n times, rolling through the entire set of available data. Nevertheless, none of these classifiers were built using all the data available and therefore, they are all suboptimal. A new classifier will be built using the same type of model but this time using the entire set of data available. This will be the model to be used in the real-world application. Note, however, that the error obtained with this model would be a over-optimistic assessment of the performance because the error is measured on the same data the model was trained on. Hence, the performance of this classifier is *not* to be used as an indication of how the classifier will perform in the real world.

29.3.3   Some types of classifiers

29.3.3.1   Quadratic and linear discriminants

A standard classification approach, applicable when the features are continuous variables (e.g., gene expression data), assumes that for each class c, $\mathbf{x}$ follows a multivariate normal distribution $N(\mu_c, \blacksquare_c)$ having the mean μ_c and covariance matrix $\blacksquare_c$.

Using the multivariate-normal probability density function and replacing the true class mean and covariance matrices with sample-derived estimates ($\mathbf{m}_c$ and $\hat{\blacksquare}_c$ respectively), the discriminant function for each class can be computed as:

$$g_c(\mathbf{x}) = -(\mathbf{x} - \mathbf{m}_c)\hat{\blacksquare}_c^{-1}(\mathbf{x} - \mathbf{m}_c)^T - \log(|\hat{\blacksquare}_c|) \tag{29.1}$$

where

$$\mathbf{m}_c = \frac{1}{n_c}\Sigma_{i=1}^{n_c}\mathbf{x}_i \tag{29.2}$$

and

$$\hat{\blacksquare}_c = \frac{1}{n_c}\Sigma_{i=1}^{n_c}(\mathbf{x}_i - \mathbf{m}_c)(\mathbf{x}_i - \mathbf{m}_c)^T \tag{29.3}$$

The discriminant functions are monotonically related to the densities $p(\mathbf{x}|y = c)$, yielding higher values for larger densities. The values of the discriminant functions will differ from one class to another only on the basis of the estimates of the class mean and covariance matrix. A new object $\mathbf{z}$ will be classified in the class for which the discriminant function is the largest. This classification approach produces nonlinear (quadratic) class boundaries, giving the name of the classifier as *quadratic discriminant* rule or *Gaussian classifier*.

An alternative to this quadratic classifier is to assume that the class covariance matrices $\blacksquare_c$, $c = 1, \ldots, K$ are all the same. In this case, instead of using a different covariance matrix estimate for each class, a single, pooled covariance matrix is used. This can be especially useful when the number of samples per class is low. In this case, calculating a covariance matrix from only a few samples may produce very unreliable estimates. Better results may be obtained by assuming a common variance and using all samples to estimate a single covariance matrix. The resulting classifier uses hyperplanes as class boundaries, hence the name *normal-based linear discriminant*.

To cope with situations when the number of features is comparable with the number of samples, a further simplification can be done to the normal-based linear discriminant, by setting all off-diagonal elements in the covariance matrix to 0. This implies that between-features co-variation is disregarded. Such a *diagonal linear discriminant* was found to outperform other types of classifiers on a variety of microarray analyses [137].

The above presented classifiers work optimally when their underlying assumptions are met, such as the normality assumption. In many cases, some of the assumptions may not be met. However, what matters in the end for a practical application is how close the estimated class boundaries are to the true class boundaries. This can be assessed through a cross-validation process.

More recently, Guo and colleagues have presented a regularized linear discriminant analysis procedure useful when the number of features far exceeds the number of samples [190].

29.3.3.2 k-Nearest neighbor classifier

The nearest neighbor (NN) classifier can be seen as a nonparametric method of density estimation [452] and uses no assumption on the data distribution, except for the continuity of the feature variables. The k-nearest neighbor classifier does not require model fitting but simply stores the training data set with all available vector prototypes of each class. When a new object z needs to be classified, the first step in the algorithm is to compute the distance between z and all the available objects in the training set, x_i, $i = 1, \ldots, n$. A popular choice of distance metric is the Euclidean distance.[2] The samples are then ordered in the decreasing order of the distance to the new object z and the top k training samples (closest to the new object to be predicted) are retained. Let us denote with n_c the number of objects in the top k that belong to the class c. The k-nearest neighbor classification rule assigns the new object z to the class with the largest n_c, i.e. the class that is most common among those k neighbors. The k-nearest neighbor discriminant function can be written as $g_c(\mathbf{x}) = n_c$. When two or more classes are equally represented in the vicinity of the point z, the class whose prototypes have the smallest average distance to z may be chosen.

Note how the k-nearest neighbors technique shifts the computational burden from the training stage to the utilization stage. In fact, this technique does not really build any model, it simply stores the training data points. This requires no computation whatsoever but is very inefficient in terms of space when the training set is large. Not only that all computation is done when a new sample is provided, but this computation needs to be repeated every time such a new sample is to be classified. In contrast, most other supervised techniques actually build a model. This is done with a lot of computation during the training phase. On the upside, the resulting model is usually very compact and requires very little space to store. Furthermore, when a new sample is to be classified, the computation involved in applying the model to the new sample is usually very minimal.

---

[2] A thorough discussion of distance functions with application to microarray analysis was provided in Chapter 18.

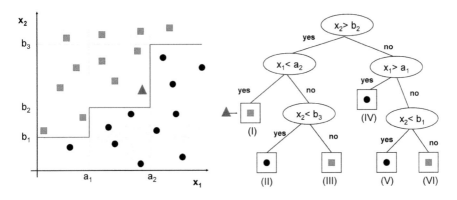

FIGURE 29.1: A binary decision tree. The left panel shows the data for a two-class decision problem, with dimensionality $p = 2$. The points known to belong to classes 1 and 2 are displayed with filled circles and squares, respectively. The decision boundary is shown as the blue thick line in the left panel. The triangle designates a new point, $\mathbf{z}$ to be classified. The right panel shows the decision tree derived for this data set whereas the new point $\mathbf{z}$ is classified in class 2 (squares). The regions in the input space covered by nodes I and IV in the tree are shown as colored areas in the left panel.

29.3.3.3 Decision trees

A special type of classifier is the decision tree [71]. A decision tree is constructed by an iterative selection of individual features that are the most salient at each node in the tree. At each level, the input space $\mathscr{X}$ is repeatedly split into descendant subsets, starting with $\mathscr{X}$ itself. There are several heuristic methods for constructing decision-tree classifiers. Usually, such trees are usually constructed top-down, beginning at the root node and successively partitioning the feature space. The construction involves three main steps:

- Selecting a splitting rule for each internal node. This involves determining the feature together with a threshold on its values that will be used to partition the data set at that particular node.

- Determining which nodes are terminal nodes. This means that for each node we must decide whether to continue splitting or to make the node terminal and assign it a class label.

- Assigning class labels to terminal nodes by minimizing the estimated error rate.

The most commonly used decision tree classifiers are binary. They use a single feature at each node, resulting in decision boundaries that are parallel

to the feature axes (see Figure 29.1). Although they are intrinsically suboptimal, they provide an easy way to interpret set of rules in terms of individual
features.

Building a decision tree can be a very lengthy process because it involves
testing many candidate questions for each node in the tree. For instance, Classification and Regression Trees (CART) [71] uses a standard set of candidate
questions with one candidate test value between each pair of data points. As
seen in Fig. 29.1, the candidate questions are of the form $\{\text{Is } x_m < c \}$, where x_m
is a variable and c is the test value for that particular variable. At each node,
CART searches through all the variables x_m, finding the best split c for each.
Then the best of the best is found [71]. For a problem in a high-dimensionality
space and many input patterns, this can be a very time-consuming process.
Usually, in the process of building the tree, some measures are taken to ensure that the splits optimize some factors such as the information gain. For
an excellent review of decision tree techniques, see [318].

29.3.3.4 Neural Networks

The most common neural network architecture used in classification problems
is a fully connected, three-layered structure of nodes in which the signals
are propagated from the input to the output layer via a hidden layer (see
Fig. 29.2). The hidden layer is called hidden because it can no connections to
the outside. The input layer only feeds the values of the feature vector $\mathbf{x}$ to
the hidden layer. Each hidden unit weights differently all outputs of the input
layer, adds a bias term, and transforms the result using a nonlinear function,
usually the logistic sigmoid:

$$\sigma(z) = \frac{1}{1 + exp(z)} \tag{29.4}$$

Similarly to the hidden layer, the output layer processes the output of the
hidden layer. A simple architecture uses one output unit for each class. The
discriminant function implemented by the k-th output unit of such a neural
network can be written as:

$$g_k(\mathbf{x}) = \sigma[\sum_{j=1}^{J} \alpha_{j,k} \sigma(\sum_{i=1}^{p} x_i w_{i,j} + b_j^h) + b_k^o] \tag{29.5}$$

In this equation, $w_{i,j}$ is the weight from the i-th input unit to the j-th
hidden node, $\alpha_{j,k}$ is the weight from the j-th hidden unit to the k-th output
node, b_j^h is the bias term of the j-th hidden unit, b_k^o is the bias term of the
k-th output unit. They all represent adjustable parameters and are estimated
(learned) during the training process that minimizes a loss function. A commonly used loss function is the sum of squared errors between the predicted
and expected signal at the output nodes, given a training data set.

Consider that N_T training samples are available to train a neural network

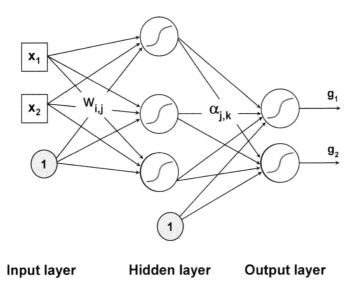

Input layer   **Hidden layer**   **Output layer**

FIGURE 29.2: A schematic representation of a neural network learning to classify two-dimensional data points (p=2) into C=2 known classes. The sigmoid hidden and output units are shown as white circles containing an S-like red curve.

with C output units. The error of the neural network on the training set can be computed as:

$$E(\omega) = \sum_{s=1}^{N_T} e_s(\omega) \tag{29.6}$$

where ω represents all the adjustable parameters of the neural network (weights and biases), which are initialized with small random values, and e_s is the error obtained when the s_{th} training sample is used as input into the network. The error e_s is defined as proportional to the sum of squared differences between the expected outputs of the network and the actual outputs, given the current values of the weights, i.e.

$$e_s = \frac{1}{2} \sum_{i=1}^{C} (d_{s,k} - g_{s,k})^2 \tag{29.7}$$

Here $g_{s,k}$ represents the actual output of the unit k for the sample s, while $g_{s,k}$ is the desired (target) output value for the same sample. When a sample belongs to the class k, it is desired that the output unit k fires a value of 1, while all the other output units fire 0. The learning process is done by updating the parameters ω such that global error decreases in an iterative process. A popular update rule is the back-propagation rule [364], in which the adjustable parameters ω are changed (increased or decreased) towards the direction in which the training error $E(\omega)$ decreases the most.

The equation (6) above can be modified in such a way that the training process not only minimizes the sum of squared errors on the training set, but also the sum of squared weights of the network. This *weights regularization* enhances the generalization capability of the model by preventing small variations in the inputs to have an excessive impact on the output. The underlying assumption of the weights regularization is that the boundaries between the classes are not sharp.

More details on theory and practical use of neural networks, can be found in Duda, Hart and Stork [136], Hagan [192], Venables and Ripley [435], etc.

29.3.3.5 Support vector machines

Consider a two-class classification problem, as shown in the left panel of Fig. 29.3. Such a problem is called **linearly separable** because the two classes can be separated with a straight line.[3] While many decision boundaries exist that are capable of separating all the *training* samples into two classes correctly, a natural question to ask is: are all the decision boundaries equally good? Here the goodness of decision boundaries is to be evaluated as described previously by cross-validation. Among these decision boundaries, **support vector machines** (**SVMs**) find the one that achieves maximum margin between the two classes. From statistical learning theory, the decision functions

---

[3] In the general case of more than two dimensions, a problem is linearly separable if the two classes can be separated by a hyperplane.

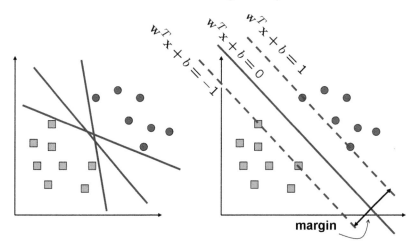

FIGURE 29.3: Maximum-margin decision boundary in support vector machines. The left panel shows several possible decision boundaries between two classes of two-dimensional vectors (denoted by dots and squares). The right panel shows the maximum-margin decision boundary implemented by the support vector machines (SVM). The distance from the decision boundary to the nearest pattern is γ. Samples along the dash lines are called support vectors.

derived by maximizing the margin[4] minimize a theoretical upper bound on the expected risk and are thus expected to generalize well [434].

When building an SVM classifier, the input (x_i, y_i) is a set of labeled points in n-dimensional input space such that $x_i \in R^n$. In case of a binary classification, the label $y_i \in \{-1, 1\}$, and for an m-class classification $y_i \in \{1, 2, \ldots, m\}$. Without loss of generality the following discussion will refer to a binary classification. In case of m-class classification the same procedure can be repeated for each pair of the m classes.

Using the labeled input points, SVM finds a function $f : R^n \rightarrow R$ such that for a given input x, $f(x) \geq 0$, if x belongs to the class denoted by 1, otherwise $f(x) < 0$. The equation $f(x) = 0$ defines a hyperplane that is used for classification of unknown samples. When the input consists of linearly separable classes, it is easy to find such a hyperplane using Equation 29.8:

$$f(x) = \langle w \cdot x \rangle + b = \sum_{k=1}^{n} w_k x_k + b \qquad (29.8)$$

where, w is the normal vector of the hyperplane defined by $f(x) = 0$ and b is the offset from the origin. If $b = 0$, the hyperplane passes through the origin.

---

[4]The margin is defined as the distance between a planar decision surface that separates two classes and the closest training samples to the decision surface (e.g., Figure 29.3, right panel).

However, finding such a function is more complex in case of nonlinearly separable data set. The complexity of the target function $f(x)$ depends on the way input data are represented. A common preprocessing practice in machine learning is to represent the input data by mapping it to another space as:

$$x = (x_1, \ldots, x_n) \longmapsto \phi(x) = (\phi_1(x), \ldots, \phi_N(x)) \qquad (29.9)$$

The new space is often referred to as a *feature space* in the literature. In the new feature space, $f(x)$ is represented as:

$$f(x) = \langle w \cdot \phi(x) \rangle + b = \sum_{q=1}^{N} w_q \phi_q(x) + b \qquad (29.10)$$

It can be shown that Eq. 29.10 can be represented as a linear combination of the input data set as follows [106, 387]:

$$f(x) = \sum_{i=1}^{l} \alpha_i y_i \langle \phi(x_i) \cdot \phi(x) \rangle + b \qquad (29.11)$$

where l is the number of samples in the training set. The parameter α_i is referred to as the *embedding strength* of the pattern x_i, which is proportional to the number of times the pattern x_i is misclassified during the training.

However, mapping the input space into a new feature space can be a time-consuming operation since the feature space is very likely to have more dimensions than the input space. In addition, due to higher dimensionality, it is more difficult to find the hyperplane for classification. Therefore, SVMs employ a *kernel* function that allows us to compute the inner product in Equation 29.11 without explicitly mapping the input points into the feature space. A kernel function is defined as:

$$K(x, z) = \langle \phi(x) \cdot \phi(z) \rangle \qquad (29.12)$$

Using Equation 29.12, we can rewrite Eq. 29.11 as:

$$f(x) = \sum_{i=1}^{l} \alpha_i y_i K(x_i, x) + b \qquad (29.13)$$

The definition of $f(x)$ determines the type of the classifier. The simplest classifier is a *maximal margin* classifier, which is only suitable for linear separable input in the feature space. The geometric margin γ_i of a pattern x_i is defined as its Euclidean distance from a hyperplane. The margin of a training set S with respect to a given hyperplane is the minimum geometric margin of all patterns in the training set S. The margin γ of a training set S is the maximum geometric margin over all hyperplanes. The maximal margin classifier finds the hyperplane that realizes the margin of the training set. It can be shown that finding the maximal margin hyperplane is equivalent to the following optimization problem [106]:

$$minimize_{w,b} \langle w \cdot w \rangle$$
$$subject \ to \ y_i(\langle w \cdot x_i \rangle + b) \geq 1$$
$$where \ i = 1, \ldots, l$$

The hyperplane obtained by solving this optimization problem realizes the margin

$$\gamma = \frac{1}{\|w\|_2} \tag{29.14}$$

It is often easier to solve an optimization problem in its dual form than its primal form because of the inequality constraints. The optimization problem above can be expressed in its dual form as:

$$maximize_{\alpha} \ \sum_{i=1}^{l} \alpha_i - \frac{1}{2} \sum_{i,j=1}^{l} y_i y_j \alpha_i \alpha_j \langle x_i \cdot x_j \rangle$$
$$subject \ to \ \sum_{i=1}^{l} y_i \alpha_i = 0, \alpha_i \geq 0, i = 1, \ldots, l$$

Because the dual only consists of an inner product of the input data, we can use the kernel as defined in Eq. 29.12 to find the optimal hyperplane in the feature space. Hence, the optimization problem can be rewritten as [106]:

$$maximize_{\alpha} \ \sum_{i=1}^{l} \alpha_i - \frac{1}{2} \sum_{i,j=1}^{l} y_i y_j \alpha_i \alpha_j K(x_i, x_j)$$
$$subject \ to \ \sum_{i=1}^{l} y_i \alpha_i = 0, \alpha_i \geq 0, i = 1, \ldots, l$$

The optimal hyperplane obtained using the dual optimization problem realizes the margin

$$\gamma = \frac{1}{\|w\|_2} = \left(\sum_{i \in SV} \alpha_i^* \right)^{-\frac{1}{2}} \tag{29.15}$$

where α^* is the solution of the dual optimization problem and SV represents the set of support vectors (i.e., subset of the input data for which $\alpha_i^* \neq 0$).

The maximal margin classifiers have limited applications since they can only be applied to data sets that are linearly separable in the feature space. This limitation can be avoided by using a *soft margin optimization* technique. The classifier that uses the soft margin optimization allows misclassification of some of the samples during training. This training error is controlled by slack variables. In other words, we need to find a hyperplane such that the constraint of the primal optimization problem above is modified as:

$$\text{subject to } y_i(\langle w \cdot x_i \rangle + b) \geq 1 - \xi_i$$
$$\xi_i \geq 0, i = 1, \ldots, l$$

The generalization error is bounded by either the 2-norm or the 1-norm of the slack vector [106]. In this chapter, we used 1-norm [85] soft optimization. The 1-norm soft margin optimization problem is described as:

$$\text{minimize}_{w,b} \langle w \cdot w \rangle + C \sum_{i=1}^{l} \xi_i$$

$$\text{subject to } y_i(\langle w \cdot x_i \rangle + b) \geq 1 - \xi_i, \ \xi_i \geq 0$$

and its dual as.

$$\text{maximize}_\alpha \sum_{i=1}^{l} \alpha_i - \frac{1}{2} \sum_{i,j=1}^{l} y_i y_j \alpha_i \alpha_j K(x_i, x_j)$$

$$\text{subject to } \sum_{i=1}^{l} y_i \alpha_i = 0, C \geq \alpha_i \geq 0, i = 1, \ldots, l \qquad (29.16)$$

The hyperplane obtained by solving this optimization problem realizes the margin:

$$\gamma = \left(\sum_{i,j \in SV} y_i y_j \alpha_i^* \alpha_j^* K(x_i, x_j) \right)^{-\frac{1}{2}} \qquad (29.17)$$

Using the γ obtained from Eq. 29.17, we will consider the subspace of the feature space with the property:

$$-\gamma \leq f(x) \leq \gamma \qquad (29.18)$$

This subspace will include the set of points that are situated at a distance from the class boundary, which is less than or equal to the distance from the boundary to the closest point in the training set. Clearly, the training set contains no evidence that the points in this area belong to a class rather than the other one. This area is defined as the uncertainty region U:

$$U \equiv \{x \in R^n \text{ such that } -\gamma \leq f(x) \leq \gamma\} \qquad (29.19)$$

and can be graphically illustrated as in Fig.29.3.

29.3.4   Feature selection

As we have already discussed, in many situations the dimensionality p of the input space is too high to allow a reliable estimation of the classifier's internal parameters with a limited number of samples ($p >> n$). In such situations, a dimensionality reduction step is necessary. There are two main categories of approaches to dimensionality reduction. The first one is to obtain a reduced number of new features by combining the existing ones, e.g., by computing a linear combination. Principal component analysis (PCA), discussed in Chapter 17, is one particular method in this branch, in which new variables (principal directions) are identified and may be used instead of the original features. The second type of dimensionality reduction involves a **feature selection** process that aims at finding a subset of the original variables that are adequately predictive.

A serious difficulty arising when $p >> n$ is **overfitting**. Most of the procedures examined in this chapter include a set of tunable parameters. The size of this set increases with p. When more tunable parameters are present, very complex relationships present in the sample can often be fit very well, particularly if n is small. Generalization error rates in such settings typically far exceed training set error rates. Reduction of the dimensionality of the feature space can help to reduce risks of overfitting. However, automated methods of dimension reduction must be employed with caution. The utility of a feature in a prediction problem may depend on its relationships with several other features, and simple reduction methods that consider features in isolation may lead to loss of important information.

The statistical pattern recognition literature classifies the approaches to feature selection into **filter methods** and **wrapper methods**. In the former category a statistical measure of the marginal relevance of the features (e.g., a t-test) is used to filter out the features that appear irrelevant using an arbitrary threshold. For instance, marker genes for cancer prediction were chosen based on their correlation with the class distinction and then used as inputs in a classifier [183].

Although fast and easy to implement, such filter methods cannot take into account the joint contribution of the features. Wrapper methods use the accuracy of the resulting classifier to evaluate either each feature independently or multiple features at the same time. For instance, the accuracy of a KNN classifier has been used to guide a genetic algorithm that searched a optimal subset of genes in a high combinatorial space [240]. The main disadvantage of such methods trying to find optimal subsets of features is that they may be computationally demanding. Main advantages of wrapper methods include the ability to: a) identify the most suited features for the classifier that will be used in the end to make the decision, and b) detect eventual synergistic feature effects (joint relevance). More details on feature selection methods and classification can be found in the literature [137, 357, 418].

29.4   Practicalities using R

For a comprehensive list of machine learning learning methods implemented in R, the reader is referred to the CRAN Task View on machine learning (`http://cran.r-project.org/src/contrib/Views/MachineLearning.html`).

The Bioconductor project includes a software package called `MLInterfaces`, which aims to simplify the application of machine learning methods to high-throughput biological data such as gene expression microarrays. In this section, we will review some examples that can be carried out by the reader who has an installation of R 2.4.0 or later. First, the CRAN package `ctv` is installed and loaded. A rich collection of machine learning tools is obtained by executing `install.views("MachineLearning")`. The `biocLite` function is then made available (e.g., through `source(http://www.bioconductor.org/biocLite.R")`), followed by `biocLite("MLInterfaces")` which installs a brokering interface to a substantial collection of machine learning functions, tailored to analysis of expression microarray datasets.

29.4.1   A leukemia dataset

After obtaining `biocLite` as described above, the command `biocLite("ALL")` installs a data structure representing samples on 128 individuals with acute lymphocytic leukemia (ALL) [92]. The following dialogue with R will generate a subset that can be analyzed to understand the transcriptional distinction between B-cell ALL cases in which the BCR and ABL genes have fused, and B-cell ALL cases in which no such fusion is present:

```
> #source("http://www.bioconductor.org/biocLite.R")
> #biocLite("ALL")
> #biocLite("MLInterfaces")
> library(ALL)
> library(MLInterfaces)
> data(ALL)
> # restrict to BCR/ABL or NEG
> bio = which( ALL$mol.biol %in% c("BCR/ABL", "NEG"))
> # restrict to B cell
> isb = grep("^B", as.character(ALL$BT))
> bfus = ALL[, intersect(bio,isb)]
> bfus

ExpressionSet (storageMode: lockedEnvironment)
assayData: 12625 features, 79 samples
  element names: exprs
protocolData: none
phenoData
```

```
   sampleNames: 01005 01010 ... 84004 (79 total)
   varLabels: cod diagnosis ... date last seen (21 total)
   varMetadata: labelDescription
featureData: none
experimentData: use 'experimentData(object)'
   pubMedIds: 14684422 16243790
Annotation: hgu95av2
```

There are 79 samples present, 37 of which present BCR/ABL fusion.

29.4.2 Supervised methods

Supervised methods of learning, such as trees, neural networks, and support vector machines, will be illustrated in this section.

The following example uses 50 random samples from **bfust** data to train a neural network model, which is used to predict the class on the remaining 29 samples from **bfust**. The confusion matrix is computed to assess the classification accuracy. Indices of the training sample are supplied to the **trainInd** parameter of the **nnetB** interface of the **MLInterfaces** package.

```
set.seed(1234) # repeatable random sample/nnet initialization
smp = sample(1:79, size = 50)
nn1 = nnetB(bfus, "mol.biol", trainInd=smp, size = 5,
maxit = 1000, decay = 0.01)
confuMat(nn1)
          predicted
given      BCR/ABL NEG
   BCR/ABL      4   7
   NEG          2  16
```

The **size** parameter in the function **nnetB** above specifies the number of units in the hidden layer of the neural network, and larger values of the **decay** parameter impose stronger regularization of the weights. The **maxit** parameter should be set to a relatively high number to increase the chance that the optimization algorithm converges to a solution. The confusion matrix is computed using the **confuMat** method on the 29 samples forming the complement of the training set specified by **smp**. This shows a misclassification rate of 31% = 9/29.

A tree-structured classifier derived from the 50-gene extract from the ALL data is shown in Figure 29.4. The procedure defines a single split on a single gene (Kruppel-like factor 9), which does a reasonable job of separating the fusion cases – the estimated misclassification rate seems to be about 30%.

Figure 29.5 depicts the decision regions after learning was carried out with training sets based on two randomly selected genes from ALL data. Qualitative aspects of the decision regions are clear: the tree-structured classifier delivers rectangular decision regions; the neural network fit leads to a smooth, curved

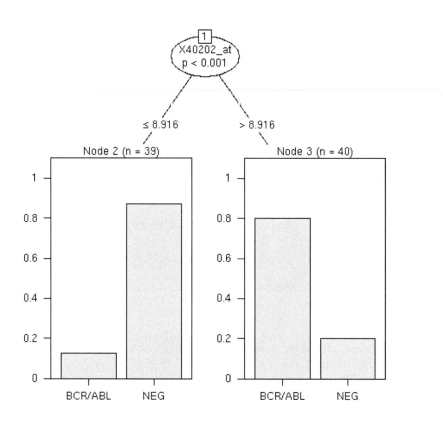

FIGURE 29.4: Rendering of a conditional tree obtained with the **ctree** function of the **party** package.

decision boundary; the 3-NN fit is very jagged; and the SVM fit is similar to but more compact than the neural net. Of note: considerable interpolation and extrapolation is performed to generate the full decision region representation, and decisions are rendered for feature values for which data are very sparse. Boundaries are sharp, and there is no provision for declaring doubt (although one could be introduced with modest programming for those procedures that do return information on posterior probabilities of class membership). Finally, the fine structure of the regions provided by CART and 3-NN are probably artifacts of overfitting, as opposed to substantively interesting indications of gene interaction.

29.4.3 Variable importance displays

Several machine learning procedures include facilities for measuring relative contribution of features in successful classification events. The random forest [70] and boosting [162] methods involve iteration through random samples of variables and cases, and if accuracy degrades when a certain variable is excluded at random from classifier construction, the variable's importance measure is incremented. Code illustrating an application follows, and Figure 29.6 shows the resulting importance measures.

```
ggg = gbmB(bfust, "mol.biol", 1:50)
confuMat(ggg)
         predicted
given     BCR/ABL NEG
  BCR/ABL      11   1
  NEG           6  11
library(hgu95av2)
par(las=2, mar=c(6,9,5,5))
plot(getVarImp(ggg), resolveenv=hgu95av2SYMBOL )
```

29.4.4 Summary

This chapter reviews concepts and tools related to machine learning (including computational pattern recognition, classification, and prediction). The key objectives of this chapter include: a) to acquaint the reader with general concepts related to machine learning; b) to provide a formalism that allows comparison of general approaches to machine learning; c) to review the main classes of machine learning methodology, including linear discriminant analysis, nearest-neighbor methods, neural networks, support vector machines; d) to illustrate directly the execution of several machine learning methods using open source methods from Bioconductor.

It is argued [194] that the success or failure of machine learning approaches on a given problem is sometimes a matter of the quality indices used to evaluate the results, and these may vary strongly with the expertise of the user. Of

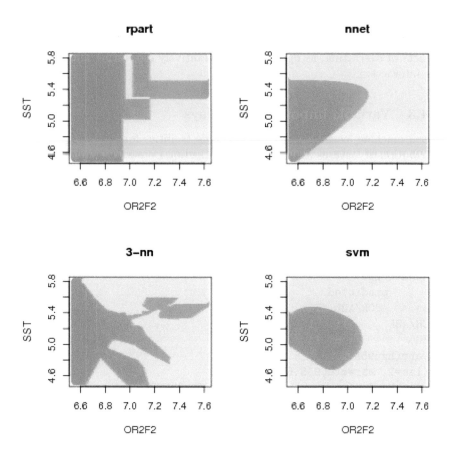

FIGURE 29.5: Displays of four two-gene classifiers. Top left: CART with minsplit tuning parameter set to 4; top right: a single layer feed forward neural network with 8 units; bottom left, $k = 3$ nearest neighbors; bottom right, the default SVM from the e1071 package. The `planarPlot` function of the `MLInterfaces` package can be used to construct such displays.

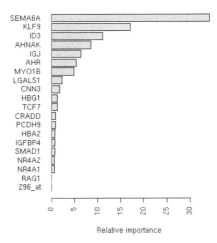

FIGURE 29.6: Display of relative variable importance as computed in a gradient boosting machine run.

special concern with supervised applications is that all steps involved in the classifier design (selection of input variables, model training, etc.) should be validated in order to obtain unbiased estimates of the classifier performance. For instance, selecting the features using all available data and subsequently cross-validating the classifier training will produce an optimistically biased error estimate. Because of inadequate validation schemes many studies published in the literature as successful have been shown to be overoptimistic [312]. It should be clear from the examples used in this chapter that the choice, the tuning, and the diagnosis of machine learning applications require a high level of attention and some understanding of the techniques used.

The R system includes a large number of machine learning methods in easily installed and well-documented packages; the Bioconductor MLInterfaces brokering package simplifies application of these methods to microarray datasets. We have illustrated a number of methods with a demonstration dataset that was obtained by selecting a reduced number of features out of a few tens of thousands that are available in the ALL data set. The features selected were those varying the most among the samples regardless of their class membership. While convenient for the purpose of demonstrating the techniques here, the filtering is not theoretically required by any of the unsupervised methods. For the purpose of developing supervised classification models, in addition to practical limitations related to the amount of memory available, there may not be enough degrees of freedom to estimate the parameters of the models. In such supervised applications filtering should be used as described in Section 29.3.4. More details on machine learning applications with R can be found in the literature [79].

Chapter 30

The road ahead

It is difficult to say what is impossible, for the dream of yesterday is the hope of today and the reality of tomorrow.

—*Robert Goddard*

30.1   What next?

This book barely scratched the surface of the things that can be achieved using microarrays. This is a very new field even by the standards of the 21st century. Ten short years ago, there were a handful of laboratories using this technology. Today, most life scientists are familiar with microarrays and thousands of laboratories are using them routinely. Because of the breathtaking speed with which this technology has been adopted, and the large number of researchers just entering this field, this book has focused on the most basic questions that can be answered using high-throughput gene expression techniques. However, there are several important issues that seem to be the next logical steps in this field. Some results in these areas have already been obtained, while others seem just around the corner.

Arguably, one of the most important directions of future research related to microarrays is molecular diagnosis. The main idea behind this direction of research is that the gene expression profile constructed using high-throughput techniques, such as microarrays, can be used to identify sets of genes that are intimately related to the development and onset of various ailments. A natural target for this approach is the range of illnesses that are both life threatening and difficult to diagnose. Many cancers, for instance, fall into this category. A stronger motivation is related to the fact that most cancers are easier to treat if they are discovered early. Since cancer development is a complex process that usually involves several disruptions in the normal biological functioning of the organism, gene and/or protein microarrays can be potentially used to identify markers for early cancer detection. Furthermore, genomics and proteomics screening can also be used to differentiate between different types of cancer or for treatment selection and/or prognostic purposes.

Another technology that is expected to revolutionize – yet again – the way research is done in life sciences is the family of next generation sequencing methods. These methods are characterized by the ability of sequencing a large amount of DNA in a very short time, at an affordable price. However, the other side of the coin is that the sequencing results are obtained as an extremely large number of very short reads. While the previous sequencing methods were able to produce reads of DNA sequences of a few hundreds and up to a couple of thousands of nucleotides, the next-generation methods use reads of only about 30 nucleotides. This has advantages and disadvantages. The main advantage is the low cost and the high throughput. The disadvantages come from the fact that the algorithms used to assemble genomes using long reads cannot be used for short reads. Essentially, the computing paradigm associated with genome assembly needs to be reinvented.

Methods such as Illumina's RNA-Seq allow researchers to obtain a transcriptomic profile for any organism. This approach works by simply counting the abundance of a given mRNA sequence in the given sample. Thus, there is no need to have a hybridization with prefabricated arrays containing specific sequences, etc. Another advantage is the customizable coverage depth. A deeper coverage can simply be achieved by sequencing more of the sample. Thus, the same technology can be used for expression profiling with low read counts or rare transcript of variant discovery with deep coverage.

The wealth of available software and resources is a clear indication of the amount of effort currently directed at developing new tools for gene and protein expression data analysis. These technologies have already proven invaluable tools able to shed new light on subtle and complex phenomena that happen at the genetic level. However, in order to realize their full potential, the laboratory work using such technologies needs to be complemented by a careful and detailed analysis of the results. The early years of these technologies were characterized by a wealth of spectacular results obtained with very simple data analysis tools. Those were some of the low-hanging fruits of the field. A careful analysis performed with suitable tools may reveal that those low-hanging fruits were only a very small fraction of the potential crop.

Furthermore, it is likely that the most spectacular results, deeply affecting the way we currently understand the genetic mechanism, are farther up in the knowledge tree and their gathering will require a close symbiosis between biologists, computer scientists, and statisticians.

Bibliography

I not only use all the brains I have, but all that I can borrow.

—Woodrow Wilson

[1] National Human Genome Research Institute (NHGRI). ArrayDB. Technical report, National Human Genome Research Institute, 2001. http://genome.nhgri.nih.gov/arraydb/schema.html.

[2] SMD – stanford microarray database. Technical report, Stanford University, 2001. http://genome-www4.Stanford.EDU/MicroArray/SMD/.

[3] BD powerblot western array screening service. Technical report, BD Biosciences, 2002. http://www.bdbiosciences.com.

[4] Yeast sporulation expression data set. Technical report, Harvard-Lipper Center for Computational Genetics, 2002. http://arep.med.harvard.edu/cgi-bin/ExpressDByeast/EXDDisplayEDS?EDSNo=0.

[5] J. Aach, W. Rindone, and G. M. Church. Systematic management and analysis of yeast gene expression data. *Genome Research*, 10:431–445, 2000. http://arep.med.harvard.edu/ExpressDB.

[6] M. D. Adams, S. E. Celniker, R. A. Holt, C. A. Evans, J. D. Gocayne, P. G. Amanatides, S. E. Scherer, P. W. Li, R. A. Hoskins, R. F. Galle, R. A. George, S. E. Lewis, S. Richards, M. Ashburner, S. N. Henderson, G. G. Sutton, J. R. Wortman, M. D. Yandell, Q. Zhang, L. X. Chen, R. C. Brandon, Y. H. Rogers, R. G. Blazej, M. Champe, B. D. Pfeiffer, K. H. Wan, C. Doyle, E. G. Baxter, G. Helt, C. R. Nelson, G. L. Gabor, J. F. Abril, A. Agbayani, H. J. An, C. Andrews-Pfannkoch, D. Baldwin, R. M. Ballew, A. Basu, J. Baxendale, L. Bayraktaroglu, E. M. Beasley, K. Y. Beeson, P. V. Benos, B. P. Berman, D. Bhandari, S. Bolshakov, D. Borkova, M. R. Botchan, J. Bouck, P. Brokstein, P. Brottier, K. C. Burtis, D. A. Busam, H. Butler, E. Cadieu, A. Center, I. Chandra, J. M. Cherry, S. Cawley, C. Dahlke, L. B. Davenport, P. Davies, B. de Pablos, A. Delcher, Z. Deng, A. D. Mays, I. Dew, S. M. Dietz, K. Dodson, L. E. Doup, M. Downes, S. Dugan-Rocha, B. C. Dunkov, P. Dunn, K. J. Durbin, C. C. Evangelista, C. Ferraz, S. Ferriera, W. Fleischmann, C. Fosler, A. E. Gabrielian, N. S. Garg,

W. M. Gelbart, K. Glasser, A. Glodek, F. Gong, J. H. Gorrell, Z. Gu, P. Guan, M. Harris, N. L. Harris, D. Harvey, T. J. Heiman, J. R. Hernandez, J. Houck, D. Hostin, K. A. Houston, T. J. Howland, M. H. Wei, C. Ibegwam, M. Jalali, F. Kalush, G. H. Karpen, Z. Ke, J. A. Kennison, K. A. Ketchum, B. E. Kimmel, C. D. Kodira, C. Kraft, S. Kravitz, D. Kulp, Z. Lai, P. Lasko, Y. Lei, A. A. Levitsky, J. Li, Z. Li, Y. Liang, X. Lin, X. Liu, B. Mattei, T. C. McIntosh, M. P. McLeod, D. McPherson, G. Merkulov, N. V. Milshina, C. Mobarry, J. Morris, A. Moshrefi, S. M. Mount, M. Moy, B. Murphy, L. Murphy, D. M. Muzny, D. L. Nelson, D. R. Nelson, K. A. Nelson, K. Nixon, D. R. Nusskern, J. M. Pacleb, M. Palazzolo, G. S. Pittman, S. Pan, J. Pollard, V. Puri, M. G. Reese, K. Reinert, K. Remington, R. D. Saunders, F. Scheeler, H. Shen, B. C. Shue, I. Sidén-Kiamos, M. Simpson, M. P. Skupski, T. Smith, E. Spier, A. C. Spradling, M. Stapleton, R. Strong, E. Sun, R. Svirskas, C. Tector, R. Turner, E. Vontor, A. H. Wang, X. Wang, Z. Y. Wang, D. A. Wassarman, G. M. Weinstock, J. Weissenbach, S. M. Williams, WoodageT, K. C. Worley, D. Wu, S. Yang, Q. A. Yao, J. Ye, R. F. Yeh, J. S. Zaveri, M. Zhan, G. Zhang, Q. Zhao, L. Zheng, X. H. Zheng, F. N. Zhong, W. Zhong, X. Zhou, S. Zhu, X. Zhu, H. O. Smith, R. A. Gibbs, E. W. Myers, G. M. Rubin, and J. C. Venter. The genome sequence of *Drosophila Melanogaster*. *Science*, 287(5461):2185–95, Mar. 2000.

[7] Affymetrix. Genechip analysis suite. User guide, version 3.3, Affymetrix, 1999.

[8] Affymetrix. Expression analysis technical manual. Technical report, Affymetrix, 2000.

[9] Affymetrix. Genechip expression analysis. Technical manual, Affymetrix, 2000.

[10] Affymetrix. Statistical algorithms description document. Technical report, Affymetrix Inc., 2002.

[11] A. Aharoni, L. C. P. Keizer, H. J. Bouwneester, Z. Sun, et al. Identification of the SAAT gene involved in strawberry flavor biogenesis by use of DNA microarrays. *The Plant Cell*, 12:647–661, May 1975.

[12] F. Al-Shahrour, R. Diaz-Uriarte, and J. Dopazo. FatiGO: a web tool for finding significant associations of Gene Ontology terms with groups of genes. *Bioinformatics*, 20(4):578–580, 2004.

[13] A. Alexa, J. Rahnenfuhrer, and T. Lengauer. Improved scoring of functional groups from gene expression data by decorrelating GO graph structure. *Bioinformatics*, 22(13):1600–7, 2006.

[14] A. Alizadeh, M. Eisen, R. Davis, C. Ma, I. Lossos, A. Rosenwald, J. Boldrick, H. Sabet, T. Tran, X. Yu, J. Powell, L. Yang, G. Marti,

T. Moore, J. Hudson, L. Lu, D. Lewis, R. Tibshirani, G. Sherlock, W. Chan, T. Greiner, D. Weisenburger, J. Armitage, R. Warnke, L. Staudt, et al. Distinct types of diffuse large b-cell lymphoma identified by gene expression profiling. *Nature*, 6769(403):503–511, 2000.

[15] U. Alon. *An introduction to systems biology: Design principles of biological circuits.* CRC Press, 2007.

[16] U. Alon, N. Barkai, D. A. Notterman, K. Gish, S. Ybarra, D. Mack, and A. J. Levine. Broad patterns of gene expression revealed by clustering of tumor and normal colon tissues probed by nucleotide arrays. *Proc. Natl. Acad. Sci.*, 96:6745–6750, 1999.

[17] O. Alter, P. Brown, and D. Botstein. Singular value decomposition for genome-wide expression data processing and modeling. *Proc. Natl. Acad. Sci.*, 97(18):10101–10106, 2000.

[18] R. Aqeilan, R. Kedar, A. Ben-Yehudah, and H. Lorberboum-Galski. Mechanism of action of interleukin-2 (IL-2)-Bax, an apoptosis-inducing chimaeric protein targeted against cells expressing the IL-2 receptor. *Biochemical Journal*, 370(Pt 1):129–40, October 2003.

[19] Arabidopsis. Normalization method comparison. Technical report, Stanford University, 2001. http://afgc.stanford.edu/ finkel/talk.htm.

[20] S. Armstrong, J. Staunton, L. Silverman, R. Pieters, M. den Boer, M. Minden, S. Sallan, E. Lander, T. Golub, and S. Korsmeyer. MLL translocations specify a distinct gene expression profile that distinguishes a unique leukemia. *Nature Genetics*, 30(1):41–47, January 2002.

[21] M. Ashburner. On the representation of gene function in genetic databases. In *Proc. of ISMB, Montreal*, 1998.

[22] M. Ashburner, C. A. Ball, J. A. Blake, D. Botstein, H. Butler, J. M. Cherry, A. P. Davis, K. Dolinski, S. S. Dwight, J. T. Eppig, M. A. Harris, D. P. Hill, L. Issel-Tarver, A. Kasarskis, S. Lewis, J. C. Matese, J. E. Richardson, M. Ringwald, G. M. Rubin, and G. Sherlock. Gene ontology: tool for the unification of biology. *Nature Genetics*, 25:25–29, 2000.

[23] M. Ashburner, C. A. Ball, J. A. Blake, H. Butler, J. M. Cherry, J. Corradi, K. Dolinski, J. T. Eppig, M. Harris, D. P. Hill, S. Lewis, B. Marshall, C. Mungall, L. Reiser, S. Rhee, J. E. Richardson, J. Richter, M. Ringwald, G. M. Rubin, G. Sherlock, and J. Yoon. Creating the Gene Ontology Resource: Design and Implementation. *Genome Research*, 11(8):1425–1433, August 2001.

[24] M. Ashburner et al. Gene Ontology: Tool for the unification of biology. *Nature Genetics*, 25:25–29, May 2000.

[25] M. Ashburner et al. Creating the Gene Ontology Resource: Design and Implementation. *Genome Research*, 11:1425–1433, 2001.

[26] S. Audic and J.-M. Claverie. The significance of digital gene expression profiles. *Genome Research*, 10(7):986–995, Oct. 1997.

[27] S. Audic and J.-M. Claverie. Visualizing the competitive recognition of TATA-boxes in vertebrate promoters. *Trends in Genetics*, 14:10–11, 1998.

[28] A. Bairoch and R. Apweiler. The SWISS-PROT protein sequence database and its supplement TrEMBL in 2000. *Nucleic Acids Research*, 28(1):45–48, January 2000.

[29] A. Bairoch, R. Apweiler, C. H. Wu, W. C. Barker, S. F. Brigitte Boeckmann, E. Gasteiger, H. Huang, R. Lopez, M. Magrane, M. J. Martin, D. A. Natale, C. O'Donovan, N. Redaschi, and L.-S. L. Yeh. The universal protein resource (uniprot). *Nucleic Acids Research*, 33:D154–D159, 2005.

[30] M. Bakay, Y.-W. Chen, R. Borup, P. Zhao, K. Nagaraju, and E. Hoffman. Sources of variability and effect of experimental approach on expression profiling data interpretation. *BMC Bioinformatics*, 3:4, January 2002.

[31] P. Baldi and A. D. Long. A Bayesian framework for the analysis of microarray expression data: regularized t-test and statistical inferences of gene changes. *Bioinformatics*, 17(6):509–519, 2001. Accompanying web page at 128.200.5.223/CyberT/.

[32] T. Bammler, R. Beyer, S. Bhattacharya, G. Boorman, A. Boyles, B. Bradford, R. Bumgarner, P. Bushel, K. Chaturvedi, D. Choi, et al. Standardizing global gene expression analysis between laboratories and across platforms. *Nature Methods*, 2:351–6, 2005.

[33] A. Banerjee, I. Dhillon, J. Ghosh, S. Merugu, and D. S. Modha. A generalized maximum entropy approach to bregman co-clustering and matrix approximation. *J. Mach. Learn. Res.*, 8:1919–1986, 2007.

[34] S. Barkow, S. Bleuler, A. Prelic, P. Zimmermann, and E. Zitzler. BicAT: a biclustering analysis toolbox. *Bioinformatics*, 22(10):1282–1283, 2006.

[35] J. C. Barrett and E. S. Kawasaki. Microarrays: the use of oligonucleotides and cDNA for the analysis of gene expression. *Drug Discovery Today*, 8(3):143–141, 2003.

[36] L. R. Baugh, A. A. Hill, E. L. Brown, and H. C. P. Quantitative analysis of mRNA amplification by in vitro transcription. *Nucleic Acids Research*, 29(5):E29, Mar 2001.

[37] T. Bayes. An essay towards solving a problem in the doctrine of chances. *Philosophical Transactions of the Royal Society*, 53:370–418, 1763.

[38] T. Bayes. An essay towards solving a problem in the doctrine of chances – reprint of the 1763 article. *Biometrica*, 45:293–298, 1958.

[39] D. G. Beer, S. L. Kardia, C.-C. Huang, T. J. Giordano, A. M. Levin, D. E. Misek, L. Lin, G. Chen, T. G. Gharib, D. G. Thomas, M. L. Lizyness, R. Kuick, S. Hayasaka, J. M. Taylor, M. D. Iannettoni, M. B. Orringer, and S. Hanash. Gene-expression profiles predict survival of patients with lung adenocarcinoma. *Nature Medicine*, 8(8):816–824, Jul 2002.

[40] T. Beissbarth and T. Speed. GOStat: find statistically overrepresented gene ontologies within a group of genes. *Bioinformatics.*, 20:1464–1465, June 2004.

[41] A. J. Bell and T. H. Sejnowski. An information-maximisation approach to blind separation and blind deconvolution. *Neural Computation*, 7(6):1004–1034, 1995.

[42] R. Bellman. *Adaptive Control Processes: A Guided Tour.* Princeton University Press, 1961.

[43] A. Ben-Dor, B. Chor, R. Karp, and Z. Yakhini. Discovering local structure in gene expression data: the order-preserving submatrix problem. *Journal of computational biology*, 10(3–4):373–384, 2003.

[44] A. Ben-Dor, R. Shamir, and Z. Yakhini. Clustering gene expression patterns. *J. of Computational Biology*, 6(3/4):281–297, 1999.

[45] Y. Benjamini and Y. Hochberg. Controlling the false discovery rate: A practical and powerful approach to multiple testing. *Journal of the Royal Statistical Society B*, 57(1):289–300, 1995.

[46] Y. Benjamini and D. Yekutieli. The control of the false discovery rate in multiple testing under dependency. *Annals of Statistics*, 29(4):1165–1188, August 2001.

[47] H. M. Berman, J. Westbrook, Z. Feng, G. Gilliland, T. N. Bhat, H. Weissig, I. N. Shindyalov, and P. E. Bourne. The protein data bank. *Nucleic Acids Research*, 28(1):235–242, 2000.

[48] C. Bernard. *An Introduction to the Study of Experimental Medicine.* Dover Publications, New York, 1957.

[49] G. F. Berriz, O. D. King, B. Bryant, C. Sander, and F. P. Roth. Characterizing gene sets with FuncAssociate. *Bioinformatics*, 19(18):2502–2504, 2003.

[50] L. Beviglia, V. Golubovskaya, L. Xu, X. Yang, R. J. Craven, and W. G. Cance. Focal adhesion kinase n-terminus in breast carcinoma cells induces rounding, detachment and apoptosis. *Biochem J*, 373(Pt 1):201–10, 2003.

[51] A. Bhattacharjee, W. Richards, J. Staunton, C. Li, S. Monti, P. Vasa, C. Ladd, J. Beheshti, R. Bueno, M. Gillette, M. Loda, G. Weber, E. Mark, E. Lander, W. Wong, B. Johnson, T. Golub, D. Sugarbaker, and M. Meyerson. Classification of human lung carcinomas by mrna expression profiling reveals distinct adenocarcinoma subclasses. *Proc. Natl. Acad. Sci. USA*, 98(24):13790–5, Nov. 2001.

[52] BioCarta.   BioCarta – Charting Pathways of Life. http://www.biocarta.com.

[53] BioDiscovery. Imagene – User's manual. Technical report, BioDiscovery Inc., 2001.

[54] M. Bittner et al. *Genomics and Proteomics*, chapter Obtaining and evaluating gene expression profiles with cDNA microarrays, pages 5–25. Kluwer, 2000.

[55] F. R. Blattner, G. Plunkett, C. A. Bloch, N. T. Perna, V. Burland, M. Riley, J. Collado-Vides, J. D. Glasner, C. K. Rode, G. F. Mayhew, J. Gregor, N. W. Davis, H. A. Kirkpatrick, M. A. Goeden, D. J. Rose, B. Mau, and Y. Shao. The complete genome sequence of *Escherichia coli K-12*. *Science*, 277(5331):1453–74, Sept. 1997.

[56] T. Blundell and S. Fortier. Data mining in crystallography – report on the 29th Crystallographic Course and Summer School. Technical report, E. Majorana Center, 1999. http://www.geomin.unibo.it/orgv/erice/DataMini.htm.

[57] M. S. Boguski, T. M. J. Lowe, and C. M. Tolstoshev. dbEST – database for expressed sequence tags. *Nature Genetics*, 4:332–333, August 1993.

[58] M. S. Boguski and F. McCormick. Proteins regulating Ras and its relatives. *Nature*, 366(6456):643–54, 1993.

[59] B. Bolstad, R. Irizarry, M. Astrand, and T. Speed. A comparison of normalization methods for high density oligonucleotide array data based on variance and bias. *Bioinformatics*, 19(2):185, 2003.

[60] C. E. Bonferroni. *Il calcolo delle assicurazioni su gruppi di teste*, chapter "Studi in Onore del Professore Salvatore Ortu Carboni", pages 13–60. Rome, 1935.

[61] C. E. Bonferroni. Teoria statistica delle classi e calcolo delle probabilità. *Pubblicazioni del Istituto Superiore di Scienze Economiche e Commerciali di Firenze*, 8:3–62, 1936.

[62] C. Bouton, G. W. Henry, and J. Pevsner. Database referencing of array genes online – DRAGON. Technical report, Kennedy Krieger Institute, 2001. http://pevsnerlab.kennedykrieger.org/dragon.htm.

[63] C. M. Bouton and J. Pevsner. DRAGON View: information visualization for annotated microarray data. *Bioinformatics*, 18:323–324, 2002.

[64] J. Boyd. Adenovirus E1B 19 kDa and Bcl-2 proteins interact with a common set of cellular proteins. *Cell*, 79(2):341–351, Dec. 1994.

[65] E. I. Boyle, S. A. Weng, J. Gollub, H. Jin, D. Botstein, J. M. Cherry, and G. Sherlock. Go::termfinder – open source software for accessing gene ontology information and finding significantly enriched gene ontology terms associated with a list of genes. *Bioinformatics*, 20(18):3710–3715, 2004.

[66] A. Brazma. Mining the yeast genome expression and sequence data. *The BioInformer*, (4), 1998. http://bioinformer.ebi.ac.uk/newsletter/archives /4/lead_article.html.

[67] A. Brazma. On the importance of standardisation in life sciences. *Bioinformatics*, 17(2):113–114, 2001.

[68] A. Brazma, P. Hingamp, J. Quackenbush, G. Sherlock, P. Spellman, C. Stoeckert, J. Aach, W. Ansorge, C. A. Ball, H. C. Causton, T. Gaasterland, P. Glenisson, F. C. P. Holstege, I. F. Kim, V. Markowitz, J. C. Matese, H. Parkinson, A. Robinson, U. Sarkans, S. Schulze-Kremer, J. Stewart, R. Taylor, J. Vilo, and M. Vingron. Minimum information about a microarray experiment (MIAME)-towards standards for microarray data. *Nature Genetics*, 29(4):365–371, December 2001.

[69] A. Brazma, A. Robinson, G. Cameron, and M. Ashburner. One-stop shop for microarray data. *Nature*, 403(6771):699–700, 2000. 0028-0836.

[70] L. Breiman. Random forests. *Machine Learning*, 45(1):5–32, 2001.

[71] L. Breiman, J. H. Friedman, R. A. Olsen, and C. J. Stone. *Classification and regression trees.* Wadsworth and Brooks, New York, 1984.

[72] H. Brightman and H. Schneider. *Statistics for Business Problem Solving.* Southwestern Publishing Co., 1994.

[73] S. Brin and L. Page. The anatomy of a large-scale hypertextual Web search engine. *Computer Networks and ISDN Systems*, 30(1–7):107–117, Apr. 1998.

[74] C. C. Brown and T. R. Fears. Exact significance levels for multiple binomial testing with application to carcinogenicity screens. *Biometrics*, 37:763 –774, 1981.

[75] K. J. Bussey, D. Kane, M. Sunshine, S. Narasimhan, S. Nishizuka, W. C. Reinhold, B. Zeeberg, Ajay, and J. N. Weinstein. Matchminer: a tool for batch navigation among gene and gene product identifiers. *Genome Biology*, 4(4):R27, March 2003.

[76] S. Busygin, O. Prokopyev, and P. M. Pardalos. Biclustering in data mining. *Comput. Oper. Res.*, 35(9):2964–2987, 2008.

[77] E. B. Camon, D. G. Barrell, E. C. Dimmer, V. Lee, M. Magrane, J. Maslen, D. Binns, and R. Apweiler. An evaluation of GO annotation retrieval for BioCreAtIvE and GOA. *BMC Bioinformatics*, 6(Suppl 1):S17, 2005.

[78] R. D. Canales, Y. Luo, J. C. Willey, B. Austermiller, C. C. Barbacioru, C. Boysen, K. Hunkapiller, R. V. Jensen, C. R. Knight, K. Y. Lee, Y. Ma, B. Maqsodi, A. Papallo, E. H. Peters, K. Poulter, P. L. Ruppel, R. R. Samaha, L. Shi, W. Yang, L. Zhang, and F. M. Goodsaid. Evaluation of DNA microarray results with quantitative gene expression platforms. *Nature Biotechnology*, 24(9):1115–22, September 2006. 1087-0156.

[79] V. Carey. *Bioinformatics and Computational Biology Solutions Using R and Bioconductor*, volume 16, chapter Machine Learning Concepts and Tools for Statistical Genomics, pages 273–292. Springer, 2005.

[80] G. A. Carpenter and S. Grossberg. The ART of adaptive pattern recognition by a self-organizing neural network. *Computer*, 21(3):77–88, 1988.

[81] S. Carter, A. Eklund, B. Mecham, I. Kohane, and Z. Szallasi. Redefinition of affymetrix probe sets by sequence overlap with cDNA microarray probes reduces cross-platform inconsistencies in cancer-associated gene expression measurements. *BMC Bioinformatics*, 6(107), 2005.

[82] G. Casella. *Statistical Inference*. Duxbury, Pacific Grove, 2002.

[83] C. I. Castillo-Davis and D. L. Hartl. GeneMerge-post-genomic analysis, data mining, and hypothesis testing. *Bioinformatics*, 19(7):891–892, May 2003.

[84] J. E. Celis, M. Kruhoffer, I. Gromova, C. Frederiksen, M. Ostergaard, T. Thykjaer, et al. Gene expression profiling: monitoring transcription and translation products using DNA microarrays and proteomics. *Federation of European Biochemical Societies Letters*, (23892):1–15, 2000.

[85] C. Chang and C. Lin. LIBSVM: a library for support vector machines. 2001.

[86] M. Chatterjee, S. Drăghici, and M. A. Tainsky. Immunotheranostics: breaking tolerance in immunotherapy using tumor autoantigens identified on protein microarrays. *Curr Opin Drug Discov Devel*, 9(3):380–5, 2006.

[87] F. Chen, R. Chang, M. Trivedi, Y. Capetanaki, and V. Cryns. Caspase proteolysis of desmin produces a dominant negative inhibitor of intermediate filaments and promotes apoptosis. *J. Biol. Chem.*, Dec. 2002. e-pub ahead of print.

[88] J. Chen, H. Hsueh, R. Delongchamp, C. Lin, and C. Tsai. Reproducibility of microarray data: a further analysis of microarray quality control(MAQC) data. *BMC Bioinformatics*, 8(1):412, 2007.

[89] L. Chen, S. Zheng, and M. C. Willingham. Mechanism of taxol-induced apoptosis in human breast cancer cells. *Zhonghua Zhong Liu Za Zhi*, 19:103–6, 1997.

[90] Z. Chen, T. B. Gibson, F. Robinson, L. Silvestro, G. Pearson, B. Xu, A. Wright, C. Vanderbilt, and M. H. Cobb. MAP kinases. *Chemical Reviews*, 101(8):2449–2476, Aug. 2001.

[91] Y. Cheng and G. M. Church. Biclustering of expression data. In *Proceedings of the Eighth International Conference on Intelligent Systems for Molecular Biology*, pages 93–103. AAAI Press, 2000.

[92] S. Chiaretti, X. Li, R. Gentleman, A. Vitale, M. Vignetti, F. Mandelli, J. Ritz, and R. Foa. Gene expression profile of adult t-cell acute lymphocytic leukemia identifies distinct subsets of patients with different response to therapy and survival. *Blood*, 103.

[93] R. J. Cho, M. Huang, M. J. Campbell, H. Dong, L. Steinmetz, L. Sapinoso, G. Hampton, S. J. Elledge, R. W. Davis, and D. J. Lockhart. Transcriptional regulation and function during the human cell cycle. *Nat Genet*, 27(1):48–54, 2001.

[94] S. Choe, M. Boutros, A. Michelson, G. Church, and M. Halfon. Preferred analysis methods for affymetrix genechips revealed by a wholly defined control dataset. *Genome Biol*, 6(R16), 2005.

[95] S. Chu, J. DeRisi, M. Eisen, J. Mulholland, D. Botstein, P. Brown, and I. Herskowitz. The transcriptional program of sporulation in budding yeast. *Science*, 282(5393):699–705, 1998.

[96] E. Chudin, R. Walker, A. Kosaka, S. Wu, D. Rabert, T. Chang, and D. Kreder. Assessment of the relationship between signal intensities and transcript concentration for affymetrix genechip arrays. *Genome Biology*, 3(RESEARCH0005), 2002.

[97] H.-J. Chung, M. Kim, C. H. Park, J. Kim, and J. H. Kim. ArrayXPath: mapping and visualizing microarray gene-expression data with integrated biological pathway resources using scalable vector graphics. *Nucleic Acids Research*, 32(Web Server issue):W460–W464, Jul 2004.

[98] G. A. Churchill. Fundamentals of experimental design for cDNA microarrays. *Nat. Genet.*, 32(Suppl. S):490–495, Dec. 2002.

[99] J.-M. Claverie. Computational methods for the identification of differential and coordinated gene expression. *Human Molecular Genetics*, 8(10):1821–1832, 1999. Available online at http://biosun01.biostat.jhsph.edu/ gparmigi/ 688/claverie1999.pdf.

[100] W. Cleveland. Robust locally weighted regression and smoothing scatterplots. *Journal of the American Statistical Association*, 74:829–836, 1979.

[101] W. Cleveland and S. Devlin. Locally weighted regression: An approach to regression analysis by local fitting. *Journal of the American Statistical Association*, 83:596–610, 1983.

[102] CNRS. Microarray databases. Technical report, Centre National de la Recherche Scietifique, 2001. http://www.biologie.ens.fr/en/genetiqu/puces/bddeng.html.

[103] B. P. Coe, W. W. Lockwood, L. Girard, R. Chari, C. Macaulay, S. Lam, A. F. Gazdar, J. D. Minna, and W. L. Lam. Differential disruption of cell cycle pathways in small cell and non-small cell lung cancer. *Br J Cancer*, 94(12):1927–35, 2006. 0007-0920.

[104] G. O. Consortium. http://www.geneontology.org.

[105] F. Crick, L. Barnett, S. Brenner, and R. Watts-Tobin. General nature of the genetic code for proteins. *Nature*, 192:1227–32, Dec 1961.

[106] N. Cristianini and J. Shawe-Taylor. Support vector machines, 2000.

[107] T. Czechowski, R. P. Bari, M. Stitt, W.-R. Scheible, and M. K. Udvardi. Real-time rt-pcr profiling of over 1400 Arabidopsis transcription factors: unprecedented sensitivity reveals novel root- and shoot-specific genes. *The Plant Journal*, 38(2):366–379, 2004.

[108] K. Dahlquist, N. Salomonis, K. Vranizan, S. Lawlor, and B. Conklin. GenMAPP, a new tool for viewing and analyzing microarray data on biological pathways. *Nature Genetics*, 31(1):19 20, May 2002.

[109] J. Delehanty and F. S. Ligler. Method for printing functional protein microarrays. *Biotechniques*, 34(2):380–385, Feb 2003.

[110] G. Dennis Jr., B. T. Sherman, D. A. Hosack, J. Yang, W. Gao, H. C. Lane, and R. A. Lempicki. DAVID: Database for annotation, visualization, and integrated discovery. *Genome Biology*, 4:P3, 2003.

[111] M. Deodhar, G. Gupta, J. Ghosh, H. Cho, and I. Dhillon. A scalable framework for discovering coherent co-clusters in noisy data. In *ICML '09: Proceedings of the 26th Annual International Conference on Machine Learning*, pages 241–248, New York, NY, USA, 2009. ACM.

[112] S. Der, B. Williams, and R. Silverman. Identification of genes differentially regulated by interferon alpha, beta, or gamma using oligonucleotide arrays. *Proc. Natl. Acad. of Sci.*, 26(95):15623–15628, 1998.

[113] J. L. DeRisi, V. R. Iyer, and P. O. Brown. Exploring the metabolic and genetic control of gene expression on a genomic scale. *Science*, 278:680–686, 1997.

[114] J. L. DeRisi, L. Penland, P. O. Brown, M. L. Bittner, et al. Use of a cDNA microarray to analyse gene expression patterns in human cancer. *Nature Genetics*, 14(4):457–460, 1996.

[115] P. D'haeseleer, S. Liang, and R. Somogyi. Genetic network inference: From co-expression clustering to reverse engineering. *Bioinformatics*, 16(8):707–726, 2000.

[116] P. D'haeseller. *Genetic Network Inference: From Co-Expression Clustering to Reverse Engineering*. PhD Thesis, University of New Mexico, 2000.

[117] P. D'haeseller, S. Liang, and R. Somogyi. Genetic network inference: From co-expression clustering to reverse engineering. *Bioinformatics*, 8(16):707–726, 2000.

[118] M. Diehn, G. Sherlock, G. Binkley, H. Jin, J. C. Matese, T. Hernandez-Boussard, C. A. Rees, J. M. Cherry, D. Botstein, P. O. Brown, and A. A. Alizadeh. SOURCE: a unified genomic resource of functional annotations, ontologies, and gene expression data. *Nucleic Acids Research*, 31(1):219–223, July 2003.

[119] B. Done, P. Khatri, A. Done, and S. Drăghici. Predicting novel human Gene Ontology annotations using semantic analysis. *IEEE/ACM Transactions on Computational Biology and Bioinformatics*, 7(1):91–99, 2010.

[120] S. W. Doniger, N. Salomonis, K. D. Dahlquist, K. Vranizan, S. C. Lawlor, and B. R. Conklin. MAPPFinder: using gene ontology and genmapp to create a global gene expression profile from microarray data. *Genome Biology*, 4(1):R7, 2003.

[121] K. Drlica. *Understanding DNA and Gene Cloning*. John Wiley and Sons, 1996.

[122] S. Drăghici. Piecewise linearization method for the normalization of cDNA and protein microarrays in multi-channel experiments. Technical report, Biodiscovery Inc., 2001. Patent Application.

[123] S. Drăghici. Statistical intelligence: effective analysis of high-density microarray data. *Drug Discovery Today*, 7(11):S55–S63, 2002.

[124] S. Drăghici. *Data Analysis Tools for DNA Microarrays*. Chapman and Hall/CRC Press, 2003 (first print), 2006 (second print).

[125] S. Drăghici, M. Chatterjee, and M. A. Tainsky. Epitomics: serum screening for the early detection of cancer on microarrays using complex panels of tumor antigens. *Expert Review of Molecular Diagnostics*, 5(5):735–743, September 2005.

[126] S. Drăghici and P. Khatri. Onto-Express web site. Technical report, Wayne State University, 2002. http://vortex.cs.wayne.edu.

[127] S. Drăghici, P. Khatri, P. Bhavsar, A. Shah, S. A. Krawetz, and M. A. Tainsky. Onto-tools, the toolkit of the modern biologist: Onto-express, onto-compare, onto-design and onto-translate. *Nucleic Acids Research*, 31(13):3775–81, July 2003.

[128] S. Drăghici, P. Khatri, C. A. Eklund, and Z. Szallasi. Reliability and reproducibility issues in DNA microarray measurements. *Trends Genet*, 22(2):101–9, 2006.

[129] S. Drăghici, P. Khatri, R. P. Martins, G. C. Ostermeier, and S. A. Krawetz. Global functional profiling of gene expression. *Genomics*, 81(2):98–104, February 2003.

[130] S. Drăghici, P. Khatri, A. L. Tarca, K. Amin, A. Done, C. Voichita, C. Georgescu, and R. Romero. A systems biology approach for pathway level analysis. *Genome Research*, 17(10):1537–1545, 2007.

[131] S. Drăghici, A. Kuklin, B. Hoff, and S. Shams. Experimental design, analysis of variance and slide quality assessment in gene expression arrays. *Current Opinion in Drug Discovery and Development*, 4(3):332–337, 2001.

[132] S. Drăghici, O. Kulaeva, A. Petrov, B. Hoff, A. Kuklin, et al. Noise sampling method: an ANOVA approach allowing robust selection of differentially regulated genes measured by DNA microarrays. *Bioinformatics*, 11(19):1348–1359, 2003.

[133] S. Drăghici and B. Potter. Predicting HIV drug resistance with neural networks. *Bioinformatics*, 19(1):98–107, January 2003.

[134] S. Drăghici, S. Sellamuthu, and P. Khatri. Babel's tower revisited: a universal resource for cross-referencing across annotation databases. *Bioinformatics*, 22(23):2934–2939, 2006.

[135] P. Du, W. Kibbe, and S. Lin. lumi: a pipeline for processing Illumina microarray. *Bioinformatics*, 24(13):1547, 2008.

[136] R. Duda, P. Hart, and D. Stork. *Pattern classification*. John Wiley and Sons, New York, 2000.

[137] S. Dudoit, J. Fridlyand, and T. Speed. Comparison of discrimination methods for the classification of tumors using gene expression data. *Journal of the American Statistical Association*, 97(457):77–87, 2002.

[138] S. Dudoit, Y. H. Yang, M. Callow, and T. Speed. Statistical models for identifying differentially expressed genes in replicated cDNA microarray experiments. Technical Report 578, University of California, Berkeley, 2000. Available at www.stat.berkeley.edu/tech-reports/index.html.

[139] D. Duggan, M. Bittner, Y. Chen, P. Meltzer, and J. Trent. Expression profiling using cDNA microarrays. *Nature Genet.*, 21(1 Suppl):10–14, 1999.

[140] M. Dunning, N. Barbosa-Morais, A. Lynch, S. Tavaré, and M. Ritchie. Statistical issues in the analysis of Illumina data. *BMC Bioinformatics*, 9(1):85, 2008.

[141] M. Dunning, M. Smith, M. Ritchie, and S. Tavare. beadarray: R classes and methods for Illumina bead-based data. *Bioinformatics*, 23(16):2183, 2007.

[142] D. Durand. A dictionary for statimagicians. *The American Statistician*, 24(3):21, 1970.

[143] EBI. Arrayexpress. Technical report, European Bioinformatics Institute, 2001. http://www.ebi.ac.uk/arrayexpress/index.html.

[144] EBI. Microarray gene expression database group. Technical report, European Bioinformatics Institute, 2001. http://www.mged.org/.

[145] B. Efron. Estimating the error rate of a prediction rule: improvement on cross-validation. *J. Am. Stat. Assoc.*, 78:316–331, 1983.

[146] B. Efron and R. J. Tibshirani. *An Introduction to Bootstrap*. Chapman and Hall, London, UK, 1993.

[147] M. Eisen, P. Spellman, P. O. Brown, and D. Botstein. Cluster analysis and display of genome-wide expression patterns. In *Proc. Natl. Acad. Sci. USA*, volume 95, pages 14863–14868, December 8 1998. Available online at http://biosun01.biostat.jhsph.edu/ gparmigi/688/eisen1998.pdf.

[148] D. Eisenberg, E. M. Marcotte, I. Xenarios, and T. O. Yeates. Protein function in the post-genomic era. *Nature*, 405:823–826, 2000.

[149] S. M. Elbashir, J. Harborth, W. Lendeckel, A. Yalcin, K. Weber, and T. Tuschl. Duplexes of 21-nucleotide RNAs mediate RNA interference in cultured mammalian cells. *Nature*, 411(6836):494–8, 2001.

[150] R. M. Ewing, A. B. Kahla, O. Poirot, F. Lopez, S. Audic, and J.-M. Claverie. Large-scale statistical analyses of rice ESTs reveal correlated patterns of gene expression. *Genome Research*, 9:950–959, 1999.

[151] Explore Gene Ontology (eGOn). http://nova2.idi.ntnu.no/egon/.

[152] J.-B. Fan, X. Chen, M. Halushka, A. Berno, X. Huang, T. Ryder, R. Lip-shutz, D. Lockhart, and A. Chakravarti. Parallel genotyping of human SNPs using generic high-density oligonucleotide tag arrays. *Genome Research*, (10):853–860, 2000.

[153] J.-B. Fan, J. M. Yeakley, M. Bibikova, E. Chudin, E. Wickham, J. Chen, D. Doucet, P. Rigault, B. Zhang, R. Shen, C. McBride, H.-R. Li, X.-D. Fu, A. Oliphant, D. L. Barker, and M. S. Chee. A versatile assay for high-throughput gene expression profiling on universal array matrices. *Genome Res*, 14(5):878–85, 2004.

[154] G. Fang, R. Kuang, G. Pandey, M. Steinbach, C. L. Myers, and V. Ku-mar. Subspace differential coexpression analysis: problem definition and a general approach. *Pacific Symposium on Biocomputing. Pacific Symposium on Biocomputing*, pages 145–156, 2010.

[155] J. Felsenstein. Confidence limits on phylogenies: an approach using the bootstrap. *Evolution*, 39:783–791, 1985.

[156] D. B. Finkelstein, R. Ewing, J. Gollub, F. Sterky, S. Somerville, and J. M. Cherry. Iterative linear regression by sector. In S. M. Lin and K. F. Johnson, editors, *Methods of Microarray Data Analysis*, pages 57–68. Kluwer Academic, 2002.

[157] L. D. Fisher and G. van Belle. *Biostatistics: A Methodology for Health Sciences*. John Wiley and Sons, New York, 1993.

[158] R. A. Fisher. Frequency distribution of the values of the correlation coefficient in samples from an indefinitely large population. *Biometrika*, 10(4):507–521, May 1915.

[159] R. A. Fisher. *The Design of Experiments*. Oliver and Boyd, 1942.

[160] W. Fleischmann, S. Moller, A. Gateau, and R. Apweiler. A novel method for automatic functional annotation of proteins. *Bioinformatics*, 15(3):228–233, March 1999.

[161] J. E. Forman, I. D. Walton, D. Stern, R. P. Rava, and M. O. Trulson. Thermodynamics of duplex formation and mismatch discrimination on photolithographically synthesized oligonucleotide arrays. In N. B. Leontis and J. J. SantaLucia, editors, *Molecular Modeling of Nucleic Acids*, Washington, D.C., 1998.

[162] Y. Freung. A decision-theoretic generalization of on-line learning and an application to boosting. *J. Computer and System Science*, 55:119–139, 1997.

[163] S. H. Friend. How DNA microarrays and expression profiling will affect clinical practice. *British Medical Journal*, 319:1–2, 1999.

[164] K. Fukushima. Neocognitron: a self-organizing neural network model for a mechanism of pattern recognition unaffected by shift in position. *Biological Cybernetics*, 36(4):193–202, 1980.

[165] C. C. Gaither and A. E. Cavazos-Gaither. *Statistically Speaking*. Institute of Physics, 1996.

[166] T. Galitski, A. J. Saldanha, C. A. Styles, E. S. Lander, and G. R. Fink. Ploidy regulation of gene expression. *Science*, 285:251–254, 1999.

[167] A. Gavin, M. Bosche, K. R., P. Grandi, M. Marzioch, A. Bauer, J. Schultz, J. Rick, A. Michon, C. Cruciat, M. Remor, C. Hofert, M. Schelder, M. Brajenovic, H. Ruffner, A. Merino, K. Klein, M. Hudak, D. Dickson, T. Rudi, V. Gnau, A. Bauch, S. Bastuck, B. Huhse, C. Leutwein, M. Heurtier, R. Copley, A. Edelmann, E. Querfurth, V. Rybin, G. Drewes, M. Raida, T. Bouwmeester, P. Bork, B. Seraphin, B. Kuster, G. Neubauer, and G. Superti-Furga. Functional organization of the yeast proteome by systematic analysis of protein complexes. *Nature*, 415(6868):141–147, 2002.

[168] R. Gentleman. *Bioinformatics and computational biology solutions using R and Bioconductor*. Springer Verlag, 2005.

[169] R. C. Gentleman, V. J. Carey, D. M. Bates, B. Bolstad, M. Dettling, S. Dudoit, B. Ellis, L. Gautier, Y. Ge, J. Gentry, K. Hornik, T. Hothorn, W. Huber, S. Iacus, R. Irizarry, F. Leisch, C. Li, M. Maechler, A. J. Rossini, G. Sawitzki, C. Smith, G. Smyth, L. Tierney, J. Y. Yang, and J. Zhang. Bioconductor: open software development for computational biology and bioinformatics. *Genome Biology*, 5(10):R80, 2004.

[170] T. George and S. Merugu. A scalable collaborative filtering framework based on co-clustering. In *Proceedings of the Fifth IEEE International Conference on Data Mining*, ICDM '05, pages 625–628, Washington, DC, USA, 2005.

[171] G. Getz, E. Levine, and E. Domany. Coupled two-way clustering analysis of gene microarray data. *Proc. Natl. Acad. Sci.*, 97(22):12079–12084, 2000.

[172] D. Ghosh and A. M. Chinnaiyan. Mixture modelling of gene expression data from microarray experiments. *Bioinformatics*, 18:275–286, 2002.

[173] R. A. Gibbs, G. M. Weinstock, M. L. Metzker, D. M. Muzny, E. J. Sodergren, S. Scherer, G. Scott, D. Steffen, K. C. Worley, P. E. Burch, G. Okwuonu, S. Hines, L. Lewis, C. DeRamo, O. Delgado, S. Dugan-Rocha, G. Miner, M. Morgan, A. Hawes, R. Gill, R. A. Holt, M. D. Adams, P. G. Amanatides, H. Baden-Tillson, M. Barnstead, S. Chin, C. A. Evans, S. Ferriera, C. Fosler, A. Glodek, Z. Gu, D. Jennings, C. L. Kraft, T. Nguyen, C. M. Pfannkoch, C. Sitter, G. G. Sutton, J. C. Venter, T. Woodage, D. Smith, H.-M. Lee, E. Gustafson, P. Cahill, A. Kana, L. Doucette-Stamm, K. Weinstock, K. Fechtel, R. B. Weiss, D. M. Dunn, E. D. Green, R. W. Blakesley, G. G. Bouffard, P. J. D. Jong, K. Osoegawa, B. Zhu, M. Marra, J. Schein, I. Bosdet, C. Fjell, S. Jones, M. Krzywinski, C. Mathewson, A. Siddiqui, N. Wye, J. McPherson, S. Zhao, C. M. Fraser, J. Shetty, S. Shatsman, K. Geer, Y. Chen, S. Abramzon, W. C. Nierman, P. H. Havlak, R. Chen, K. J. Durbin, A. Egan, Y. Ren, X.-Z. Song, B. Li, Y. Liu, X. Qin, S. Cawley, K. C. Worley, A. J. Cooney, L. M. D'Souza, K. Martin, J. Q. Wu, M. L. Gonzalez-Garay, A. R. Jackson, K. J. Kalafus, M. P. McLeod, A. Milosavljevic, D. Virk, A. Volkov, D. A. Wheeler, Z. Zhang, J. A. Bailey, E. E. Eichler, E. Tuzun, E. Birney, E. Mongin, A. Ureta-Vidal, C. Woodwark, E. Zdobnov, P. Bork, M. Suyama, D. Torrents, M. Alexandersson, B. J. Trask, J. M. Young, H. Huang, H. Wang, H. Xing, S. Daniels, D. Gietzen, J. Schmidt, K. Stevens, U. Vitt, J. Wingrove, F. Camara, M. M. Alba, J. F. Abril, R. Guigo, A. Smit, I. Dubchak, E. M. Rubin, O. Couronne, A. Poliakov, N. Hubner, D. Ganten, C. Goesele, O. Hummel, T. Kreitler, Y.-A. Lee, J. Monti, H. Schulz, H. Zimdahl, H. Himmelbauer, H. Lehrach, H. J. Jacob, S. Bromberg, J. Gullings-Handley, M. I. Jensen-Seaman, A. E. Kwitek, J. Lazar, D. Pasko, P. J. Tonellato, S. Twigger, C. P. Ponting, J. M. Duarte, S. Rice, L. Goodstadt, S. A. Beatson, R. D. Emes, E. E. Winter, C. Webber, P. Brandt, G. Nyakatura, M. Adetobi, F. Chiaromonte, L. Elnitski, P. Eswara, R. C. Hardison, M. Hou, D. Kolbe, K. Makova, W. Miller, A. Nekrutenko, C. Riemer, S. Schwartz, J. Taylor, S. Yang, Y. Zhang, K. Lindpaintner, T. D. Andrews, M. Caccamo, M. Clamp, L. Clarke, V. Curwen, R. Durbin, E. Eyras, S. M. Searle, G. M. Cooper, S. Batzoglou, M. Brudno, A. Sidow, E. A. Stone, J. C. Venter, B. A. Payseur, G. Bourque, C. Lopez-Otin, X. S. Puente, K. Chakrabarti, S. Chatterji, C. Dewey, L. Pachter, N. Bray, V. B. Yap, A. Caspi, G. Tesler, P. A. Pevzner, D. Haussler, K. M. Roskin, R. Baertsch, H. Clawson, T. S. Furey, A. S. Hinrichs, D. Karolchik, W. J. Kent, K. R. Rosenbloom, H. Trumbower, M. Weirauch, D. N. Cooper,

P. D. Stenson, B. Ma, M. Brent, M. Arumugam, D. Shteynberg, R. R. Copley, M. S. Taylor, H. Riethman, U. Mudunuri, J. Peterson, M. Guyer, A. Felsenfeld, S. Old, S. Mockrin, and F. Collins. Genome sequence of the Brown Norway rat yields insights into mammalian evolution. *Nature*, 428(6982):493–521, Apr. 2004.

[174] G. Gigerenzer and U. Hoffrage. How to improve bayesian reasoning without instruction: frequency formats. *Psychological Review*, 102(4):684–704, 1995.

[175] R. Gill, S. Datta, and S. Datta. A statistical framework for differential network analysis from microarray data. *BMC Bioinformatics*, 11(1):95, 2010.

[176] A. Giodini, M. Kallio, N. Wall, G. Gorbsky, S. Tognin, P. Marchisio, M. Symons, and D. Altieri. Regulation of microtubule stability and mitotic progression by survivin. *Cancer Research*, 62(9):2462–2467, May 2002.

[177] M. D. Giulio. The origin of the genetic code: theories and their relationships, a review. *Biosystems*, 80(2):175–84, 2005.

[178] T. Glover and K. Mitchell. *An Introduction to Biostatistics*. McGraw-Hill, New York, 2002.

[179] GO. Gene ontology. Technical report, Gene Ontology Consortium, 2001. http://www.geneontology.org/.

[180] J. J. Goeman, S. A. van de Geer, F. de Kort, and H. C. van Houwelingen. A global test for groups of genes: testing association with a clinical outcome. *Bioinformatics*, 20(1):93–99, 2004.

[181] A. Goffeau, B. G. Barrell, H. Bussey, R. W. Davis, B. Dujon, H. Feldmann, F. Galibert, J. D. Hoheisel, C. Jacq, M. Johnston, E. J. Louis, H. W. Mewes, Y. Murakami, P. Philippsen, H. Tettelin, and S. G. Oliver. Life with 6000 genes. *Science*, 274(5287):546, 563–7, Oct. 1996.

[182] D. Gold, K. Coombes, D. Medhane, A. Ramaswamy, Z. Ju, L. Strong, J. Koo, and M. Kapoor. A comparative analysis of data generated using two different target preparation methods for hybridization to high-density oligonucleotide microarrays. *BMC Genomics*, 5(2), 2004.

[183] T. R. Golub, D. K. Slonim, P. Tamayo, C. Huard, M. Gaasenbeek, J. P. Mesirov, H. Coller, M. L. Loh, J. R. Downing, M. A. Caligiuri, C. D. Bloomfield, and E. S. Lander. Molecular classification of cancer: class discovery and class prediction by gene expression monitoring. *Science*, 286(5439):531–537, 1999.

[184] V. Golubovskaya, L. Beviglia, L. H. Xu, H. S. Earp, R. Craven, and W. Cance. Dual inhibition of focal adhesion kinase and epidermal growth factor receptor pathways cooperatively induces death receptor-mediated apoptosis in human breast cancer cells. *J. Biol. Chem.*, 277(41):38978–87, 2002.

[185] I. J. Good. What are degrees of freedom? *The American Statistician*, 27(5):227–228, Dec 1973.

[186] W. Gosset. The probable error of a mean. *Biometrika*, 6:1–25, 1908.

[187] T. Grim. A possible role of social activity to explain differences in publication output among ecologists. *Oikos*, doi:10.1111/j.2008.0030-1299.16551.x, 2008.

[188] S. Grossmann, S. Bauer, P. N. Robinson, and M. Vingron. Improved detection of overrepresentation of Gene-Ontology annotations with parent child analysis. *Bioinformatics*, 23(22):3024–31, 2007.

[189] P. Grosu, J. P. Townsend, D. L. Hartl, and D. Cavalieri. Pathway processor: a tool for integrating whole-genome expression results into metabolic networks. *Genome Research*, 12:1121–1126, 2002.

[190] Y. Guo, T. Hastie, and R. Tibshirani. Regularized linear discriminant analysis and its application in microarrays. *Biostatistics*, 8(1):9–31, 2007.

[191] D. Gusfield. Bioinformatics FTE proposal. Technical report, University of California, Davis, 1999. http://genomics.ucdavis.edu/~gusfield/bioinfomaster.

[192] M. T. Hagan, H. B. Demuth, and M. H. Beale. *Neural Network Design*. Brooks Cole, Boston, 1995.

[193] R. G. Halgren, M. R. Fielden, C. J. Fong, and T. R. Zacharewski. Assessment of clone identity and sequence fidelity for 1189 IMAGE cDNA clones. *Nucleic Acids Research*, 29(2):582–588, 2001.

[194] D. J. Hand. Classifier technology and the illusion of progress. *Statistical Science*, 21(1):1–14, 2006.

[195] J. Harbig, R. Sprinkle, and S. A. Enkemann. A sequence-based identification of the genes detected by probesets on the affymetrix u133 plus 2.0 array. *Nucleic Acids Research*, 33:e31, 2005.

[196] T. Hastie, R. Tibshirani, M. B. Eisen, A. Alizadeh, R. Levy, L. Staudt, W. Chan, D. Botstein, and P. Brown. "Gene shaving" as a method for indentifying distinct sets of genes with similar expression patterns. *Genome Biology*, 1(2):1–21, 2000.

[197] T. Hastie, R. Tibshirani, and J. Friedman. *The Elements of Statistical Learning.* Springer, 2001.

[198] T. Haveliwala. Efficient computation of PageRank. Technical Report 1999-31, Database Group, Computer Science Department, Stanford University, Feb. 1999.

[199] P. Hedge, R. Qi, K. Abernathy, C. Gay, S. Dharap, R. Gaspard, J. Earle-Hughes, E. Snasrud, N. Lee, and J. Quackenbush. A concise guide to cDNA microarray analysis. *Biotechniques*, 29(3):548–562, 2000.

[200] D. Hekstra, A. Taussig, M. Magnasco, and F. Naef. Absolute mrna concentrations from sequence-specific calibration of oligonucleotide arrays. *Nucleic Acids Research*, 31:1962–8, 2003.

[201] B. Henderson. *The Gospel of the Flying Spaghetti Monster.* Villard, 2006.

[202] J. Herrero, A. Valencia, and J. Dopazo. A hierarchical unsupervised growing neural network for clustering gene expression patterns. *Bioinformatics*, 17(2):126–136, Feb 2001.

[203] J. Hertz, A. Krogh, and R. G. Palmer. *Introduction to the Theory of Neural Computation.* Perseus Books, 1991.

[204] R. Herwig, A. Poustka, C. Muller, C. Bull, H. Lehrach, and J. O'Brien. Large-scale clustering of cDNA-fingerprinting data. *Genome Research*, 9(11):1093–1105, 1999.

[205] C. Hetz, M. Hunn, P. Rojas, V. Torres, L. Leyton, and A. Quest. Caspase-dependent initiation of apoptosis and necrosis by the fas receptor in lymphoid cells: onset of necrosis is associated with delayed ceramide increase. *J. Cell. Sci.*, 115:4671–4683, Dec. 2002.

[206] L. J. Heyer, S. Kruglyak, and S. Yooseph. Exploring expression data: Identification and analysis of coexpressed genes. *Genome Research*, 9:1106–1115, 1999.

[207] T. Heyerdahl. *Kon-Tiki: Across the Pacific in a Raft.* Pocket, 1990.

[208] J. Heyse and D. Rom. Adjusting for multiplicity of statistical tests in the analysis of carcinogenicity studies. *Biometrical Journal*, 30:883–896, 1988.

[209] A. A. Hill, C. P. Hunter, B. T. Tsung, G. Tucker-Kellogg, and E. L. Brown. Genomic analysis of gene expression in *C. elegans. Science*, 290:809–812, 2000.

[210] S. Hilsenbeck, W. Friedrichs, R. Schiff, P. O'Connell, R. Hansen, C. Osborne, and S. W. Fuqua. Statistical analysis of array expression data as applied to the problem of Tamoxifen resistance. *Journal of the National Cancer Institute*, 91(5):453–459, 1999.

[211] Y. Hochberg and A. C. Tamhane. *Multiple Comparison Procedures*. John Wiley and Sons, Inc., New York, 1987.

[212] S. Hodge, G. Hodge, R. Flower, P. Reynolds, R. Scicchitano, et al. Upregulation of production of tgf-β and il-4 and down-regulation of il-6 by apoptotic human bronchial epithelial cells. *Immunology and Cell Biology*, 80(6):537–543, December 2002.

[213] R. V. Hogg and A. T. Craig. *Introduction to Mathematical Statistics*. New York: Macmillan, 1978.

[214] M. Holford, N. Li, P. Nadkarni, and H. Zhao. VitaPad: visualization tools for the analysis of pathway data. *Bioinformatics*, 21(8):1596–1602, Apr 2004.

[215] B. Holland and M. D. Copenhaver. An improved sequentially rejective Bonferroni test procedure. *Biometrica*, 43:417–423, 1987.

[216] M. J. Holland. Transcript abundance in yeast varies over six orders of magnitude. *The Journal of Biological Chemistry*, 277(17):14363–14366, 2002.

[217] S. Holm. A simple sequentially rejective multiple test procedure. *Scandinavian Journal of Statistics*, 6:65–70, 1979.

[218] D. A. Hosack, G. Dennis Jr., B. T. Sherman, H. C. Lane, and R. A. Lempicki. Identifying biological themes within lists of genes with EASE. *Genome Biology*, 4(6):P4, 2003.

[219] T. M. Houts. Improved 2-color exponential normalization for microarray analyses employing cyanine dyes. In S. Lin, editor, *Procedings of CAMDA 2000, "Critical Assesment of Techniques for Microarray Data Mining," December 18–19*, Durham, NC, 2000. Duke University Medical Center.

[220] http://ep.ebi.ac.uk/EP. Expression profiler – analysis and clustering of gene expression and sequence data.

[221] http://ep.ebi.ac.uk/EP/GO. EP:GO – Browser and analysis for Gene Ontology.

[222] http://www.ebi.ac.uk/ego. GO at EBI – QuickGO.

[223] http://www.ebi.ac.uk/interpro. Ebi interpro database.

[224] http://www.geneontology.org /doc/GO.doc.html. GO General Documentation.

[225] http://www.geneontology.org/doc/GO.annotation.html.

[226] http://www.geneontology.org/doc/GO.defs. GO Term Definitions.

[227] http://www.geneontology.org/doc/GO.evidence.html. GO Evidence Codes.

[228] http://www.geneontology.org/doc/GO.xrf_abbs. GO Collaborative Database Abbreviations.

[229] http://www.godatabase.org/dev. Gene Ontology Tools.

[230] http://www.informatics.jax.org/searches/GO_form.shtml. GO Browser at Mouse Genome Informatics.

[231] E. Hubbell, W. Liu, and R. Mei. Robust estimators for expression analysis. *Bioinformatics*, 18:1585–92, 2002.

[232] J. Ihmels, S. Bergmann, and N. Barkai. Defining transcription modules using large-scale gene expression data. *Bioinformatics*, 20(13):1993–2003, 2004.

[233] A. Inc. *GeneChip Expression Analysis Algorithm Tutorial*. Affymetrix Inc., Santa Clara, CA, 1999.

[234] R. Irizarry, B. Hobbs, F. Collin, Y. Beazer-Barclay, K. Antonellis, U. Scherf, and T. Speed. Exploration, normalization, and summaries of high density oligonucleotide array probe level data. *Biostatistics*, 4(2):249, 2003.

[235] R. A. Irizarry, B. Hobbs, F. Collin, Y. D. Beazer-Barclay, K. J. Antonellis, U. Scherf, and T. P. Speed. Exploration, normalization, and summaries of high density oligonucleotide array probe level data. *Biostatistics*, to appear, 2003.

[236] A.-K. Jarvinen, S. Hautaniemi, H. Edgren, P. Auvinen, J. Saarela, O.-P. Kallioniemi, and O. Monni. Are data from different gene expression microarray platforms comparable? *Genomics*, 83(6):1164–1168, 2004.

[237] T.-K. Jenssen, M. Langaas, W. P. Kuo, B. Smith-Sorensen, O. Myklebost, and E. Hovig. Analysis of repeatability in spotted cdna microarrays. *Nucleic Acids Research*, 30(14):3235–3244, 2002.

[238] L. Ji and K.-L. Tan. Mining gene expression data for positive and negative co-regulated gene clusters. *Bioinformatics*, 20(16):2711–2718, 2004.

[239] C. H. Jiang, J. Tsien, P. Schultz, and Y. Hu. The effects of aging on gene expression in the hypothalamus and cortex of mice. *PNAS*, 98(4):1930–1934, 2001.

[240] T. Jirapech-Umpai and S. Aitken. Feature selection and classification for microarray data analysis: evolutionary methods for identifying predictive genes. *BMC Bioinformatics*, 6:148, 2005.

[241] J. Johnson, S. Edwards, D. Shoemaker, and E. Schadt. Dark matter in the genome: evidence of widespread transcription detected by microarray tiling experiments. *Trends Genet*, 21:93–102, 2005.

[242] R. A. Johnson and D. W. Wichern. *Applied Multivariate Statistical Analysis*. Prentice-Hall, 1998.

[243] G. Joshi-Tope, M. Gillespie, I. Vasrik, P. D'Eustachio, E. Schmidt, B. de Bone, B. Jassal, G. R. Gopinath, G. R. Wu, L. Matthews, S. Lewis, E. Birney, and L. Stein. Reactome: a knowledgebase of biological pathways. *Nucleic Acids Research*, 33(Database issue):D428–432, 2005.

[244] G. Kamberova and S. Shah, editors. *DNA Array Image Analysis – NutsandBolts*. DNA Press, Eagleville, PA, 2002.

[245] M. D. Kane, T. A. Jatkoe, C. R. Stumpf, J. Lu, J. D. Thomas, and S. J. Madore. Assessment of the sensitivity and specificity of oligonucleotide (50mer) microarrays. *Nucleic Acids Research*, 28(22):4552–4557, 2000.

[246] M. Kanehisa and S. Goto. KEGG: Kyoto encyclopedia of genes and genomes. *Nucleic Acids Research*, 28(1):27–30, January 2000.

[247] M. Kanehisa, S. Goto, S. Kawashima, and A. Nakaya. The KEGG databases at GenomeNet. *Nucleic Acids Research*, 30(1):42–46, January 2002.

[248] M. Kanehisa, S. Goto, S. Kawashima, Y. Okunom, and M. Hattori. The KEGG resource for deciphering the genome. *Nucleic Acids Research*, 32(Database isuue):277–280, Jan 2004.

[249] L. Kari. DNA computing: arrival of biological mathematics. *The Mathematical Intelligencer*, 19(2):9–22, 1997.

[250] P. D. Karp, S. Paley, and P. Romero. The pathway tools software. *Bioinformatics*, 18:S225–32, 2002.

[251] L. Kaufman and P. Rousseeuw. *Finding Groups in Data: An Introduction to Cluster Analysis*. John Wiley and Sons, New York, 1990.

[252] G. C. Kennedy, H. Matsuzaki, S. Dong, W. M. Liu, J. Huang, G. Liu, X. Su, M. Cao, W. Chen, J. Zhang, W. Liu, G. Yang, X. Di, T. Ryder, Z. He, U. Surti, M. S. Phillips, B. M. T. Jacino, S. P. Fodor, and K. W.

Jones. Large-scale genotyping of complex DNA. *Nature Biotechnology*, 21(10):1233–7, 2003.

[253] J. W. Kennedy, G. W. Kaiser, L. D. Fisher, J. K. Fritz, W. Myers, J. Mudd, and T. Ryan. Clinical and angiographic predictors of operative mortality from the collaborative study in coronary artery surgery (CASS). *Circulation*, 63(4):793–802, 1981.

[254] T. Kepler, L. Crosby, and K. Morgan. Normalization and analysis of DNA microarray data by self-consistency and local regression. *Submitted to Nucleic Acids Research*, 2001.

[255] K. Kerr, E. Leiter, and G. Churchill. Analysis of a designed microarray experiment. In *Proceedings of the IEEE–Eurasip Nonlinear Signal and Image Processing Workshop*, June 3–6 2001.

[256] M. K. Kerr, C. A. Afshari, L. Bennett, P. Bushel, J. Martinez, N. J. Walker, and G. A. Churchill. Statistical analysis of a gene expression microarray experiment with replication. *Statistica Sinica*, 12(1):203–218. www.jax.org/research/churchill/pubs/index.html.

[257] M. K. Kerr and G. A. Churchill. Experimental design for gene expression analysis. *Biostatistics*, (2):183–201, 2001. www.jax.org/research/churchill/pubs/index.html.

[258] M. K. Kerr and G. A. Churchill. Statistical design and the analysis of gene expression microarray data. *Genetical Research*, 77(2):123–128, Apr 2001. www.jax.org/research/churchill/pubs/index.html.

[259] M. K. Kerr, M. Martin, and G. A. Churchill. Analysis of variance for gene expression microarray data. *Journal of Computational Biology*, 7:819–837, 2000.

[260] J. Khan, L. H. Saal, M. L. Bittner, Y. Chen, J. M. Trent, and P. S. Meltzer. Expression profiling in cancer using cDNA microarrays. *Electrophoresis*, 20(2), 1999.

[261] P. Khatri, P. Bhavsar, G. Bawa, and S. Drǎghici. Onto-tools: an ensemble of web-accessible, ontology-based tools for the functional design and interpretation of high-throughput gene expression experiments. *Nucleic Acids Research*, 32:W449–56, Jul 2004.

[262] P. Khatri, B. Done, A. Rao, A. Done, and S. Drǎghici. A semantic analysis of the annotations of the human genome. *Bioinformatics*, 21(16):3416–3421, 2005.

[263] P. Khatri and S. Drǎghici. Ontological analysis of gene expression data: current tools, limitations, and open problems. *Bioinformatics*, 21(18):3587–3595, 2005.

[264] P. Khatri, S. Drăghici, G. C. Ostermeier, and S. A. Krawetz. Profiling gene expression using Onto-Express. *Genomics*, 79(2):266–270, February 2002.

[265] P. Khatri, S. Sellamuthu, P. Malhotra, K. Amin, A. Done, and S. Drăghici. Recent additions and improvements to the Onto-Tools. *Nucleic Acids Research*, 33(Web server issue), Jul 2005.

[266] M. Kimura, S. Kotani, T. Hattori, N. Sumi, T. Yoshioka, K. Todokoro, and Y. Okano. Cell cycle-dependent expression and spindle pole localization of a novel human protein kinase. *Journal of Biological Chemistry*, 272(21):13766–13771, May 1997.

[267] O. D. King, R. E. Foulger, S. S. Dwight, J. V. White, and F. P. Roth. Predicting gene function from patterns of annotation. *Genome Research*, 13.890–904, 2003.

[268] T. Kohonen. Learning vector quantization. *Neural Networks*, 1(suppl. 1):303, 1988.

[269] T. Kohonen. *Self-Organizing Maps*. Springer, Berlin, 1995.

[270] E. Kretschmann, W. Fleischmann, and A. R. Automatic rule generation for protein annotation with the C4.5 data mining algorithm applied on SWISS-PROT. *Bioinformatics*, 17(10):920–926, Oct. 2001.

[271] W. P. Kuo, T.-K. Jenssen, A. J. Butte, L. Ohno-Machado, and I. S. Kohane. Analysis of matched mRNA measurements from two different microarray technologies. *Bioinformatics*, 18(3):405–412, 2002.

[272] J. Larkin, B. Frank, H. Gavras, R. Sultana, and J. Quackenbush. Independence and reproducibility across microarray platforms. *Nature Methods*, 2:337–44, 2005.

[273] A. E. Lash, C. M. Tolstoshev, L. Wagner, G. D. Shuler, R. L. Strausberg, G. J. Riggins, and S. F. Altschul. SAGEmap: A public gene expression resource. *Genome Research*, 10:1051–1060, 2000.

[274] J. K. Lee and M. O'Connell. *The Analysis of Gene Expression Data: Methods and Software*, chapter An S-PLUS Library for the Analysis and Visualization of Differential Expression. Springer, N.Y., 2003.

[275] M. L. Lee, W. Lu, G. A. Whitmore, and D. Beier. Models for microarray gene expression data. *Journal of Biopharm Stat.*, 12(2):1–19, 2002.

[276] M.-L. T. Lee, F. C. Kuo, G. A. Whitmore, and J. Sklar. Importance of replication in microarray gene expression studies: Statistical methods and evidence from repetitive cDNA hybridizations. *Proc. Natl. Acad. Sci.*, 97(18):9834–9839, 2000.

[277] C. Li and W. H. Wong. Model-based analysis of oligonucleotide arrays: Expression index computation and outlier detection. *Proc. Natl. Acad. Sci.*, 98(1):31–36, January 2001. Software available at http://www.dchip.org.

[278] C. Li and W. H. Wong. Model-based analysis of oligonucleotide arrays: Model validation, design issues and standard error application. *Genome Biology*, 2(8):1–11, 2001. Software available at http://www.dchip.org.

[279] F. Li, G. Ambrosini, E. Chu, J. Plescia, S. Tognin, P. Marchisio, and D. Altieri. Control of apoptosis and mitotic spindle checkpoint by survivin. *Nature*, 396(6711):580–584, Dec. 1998.

[280] J. Li, K. Sim, G. Liu, and L. Wong. Maximal quasi-bicliques with balanced noise tolerance: Concepts and co-clustering applications. In *Proc. SIAM Int. Conf. on Data Mining SDM'08*, pages 72–83, Apr. 2008.

[281] Y. W. Li and F. H. Sarkar. Gene expression profiles of genistein-treated pc3 prostate cancer cells. *Journal of Nutrition*, 132(12):3623–3631, 2002.

[282] R. Liang, W. Li, Y. Li, C. Tan, J. Li, Y. Jin, and K. Ruan. An oligonucleotide microarray for microrna expression analysis based on labeling rna with quantum dot and nanogold probe. *Nucleic Acids Research*, 33:e17, 2005.

[283] W. Liebermeister. Independent component analysis of gene expression data. In *Proc. of German Conference on Bioinformatics GCB'01*, 2001. http://www.bioinfo.de/isb/gcb01/poster/index.html.

[284] S. Lin, P. Du, W. Huber, and W. Kibbe. Model-based variance-stabilizing transformation for Illumina microarray data. *Nucleic acids research*, 36(2):e11, 2008.

[285] T. Lindholm and F. Yellin. *The Java(TM) Virtual Machine Specification*. Addison-Wesley Professional, 2nd edition edition, 1999.

[286] R. Lipshutz, S. Fodor, T. Gingeras, and D. Lockhart. High density synthetic oligonucleotide arrays. *Nature Genetics*, 21(1):20–24, January 1999.

[287] G. Liu, A. E. Loraine, R. Shigeta, M. Cline, J. Cheng, V. Valmeekam, S. Sun, D. Kulp, and M. A. Siani-Rose. NetAffx: Affymetrix probesets and annotations. *Nucleic Acids Research*, 31(1):82–86, January 2003.

[288] J. Liu, Z. Li, X. Hu, and Y. Chen. Biclustering of microarray data with mospo based on crowding distance. *BMC Bioinformatics*, 10(Suppl 4):S9, 2009.

[289] D. J. Lockhart, H. Dong, M. C. Byrne, M. T. Follettie, M. V. Gallo, M. S. Chee, M. Mittmann, C. Wang, M. Kobayashi, H. Norton, and E. L. Brown. DNA expression monitoring by hybridization of high density oligonucleotide arrays. *Nature Biotechnology*, 14(13):1675–1680, 1996.

[290] D. J. Lockhart and E. A. Winzeler. Genomics, gene expression and DNA arrays. *Nature*, 405:827–836, 2000.

[291] A. Long, H. Mangalam, B. Chan, L. Tolleri, G. W. Hatfield, and P. Baldi. Improved statistical inference from DNA microarray data using analysis of variance and a Bayesian statistical framework. *J. Biol. Chem.*, 276(23):19937–19944, 2001.

[292] T. Loughin. A systematic comparison of methods for combining p-values from independent tests. *Computational Statistics and Data Analysis*, 47(3):467–485, 2004.

[293] S. C. Madeira and A. L. Oliveira. Biclustering algorithms for biological data analysis: a survey. *IEEE/ACM Transactions on Computational Biology and Bioinformatics*, 1(1):24–45, 2004.

[294] D. Maglott, J. Ostell, K. D. Pruitt, and T. Tatusova. Entrez Gene: gene-oriented information at NCBI. *Nucleic Acids Research*, 33:D54–D58, 2005.

[295] M. Magrane and R. Apweiler. Organisation and standardisation of information in SWISS-PROT and TrEMBL. *Data Science Journal*, 1(1):13–18, 2002.

[296] M. Z. Man, Z. Wang, and Y. Wang. POWER_SAGE: comparing statistical tests for SAGE experiments. *Bioinformatics*, 16(11):953–959, 2000.

[297] E. Manduchi, G. R. Grant, S. E. McKenzie, G. C. Overton, S. Surrey, and C. J. Stoeckert. Generation of patterns from gene expression data by assigning confidence to differentially expressed genes. *Bioinformatics*, 16(8):685–698, 2000.

[298] H. Mangalam, J. Stewart, J. Zhou, K. Schlauch, M. Waugh, G. Chen, et al. GeneX: An open source gene expression database and integrated tool set. *IBM Systems Journal*, 40(2):552–569, 2001. http://www.ncgr.org/genex/.

[299] L. Margulis. *Origin of Eukaryotic Cells*. Yale University Press, 1971.

[300] D. Martin, C. Brun, E. Remy, P. Mouren, D. Thieffry, and B. Jacq. GOToolBox: functional analysis of gene datasets based on gene ontology. *Genome Biology*, 5:R101, 2004.

[301] D. Martinvalet, D. M. Dykxhoorn, R. Ferrini, and J. Lieberman. Granzyme A cleaves a mitochondrial complex I protein to initiate caspase-independent cell death. *Cell*, 133(4):681–92, 2008.

[302] M. J. Marton, J. L. DeRisi, H. A. Bennett, V. R. Iyer, M. R. Meyer, C. J. Roberts, R. Stoughton, J. Buchard, D. Slade, H. Dia, D. Bassett, Jr., L. H. Hartwell, P. O. Brown, and S. H. Friend. Drug target validation and identification of secondary drug effects using DNA microarrays. *Nature Medicine*, 4:1293–1302, 1998.

[303] J. Mazieres, B. He, L. You, Z. Xu, and D. M. Jablons. Wnt signaling in lung cancer. *Cancer Lett*, 222(1):1–10, 2005. 0304-3835.

[304] W. J. McCoubrey, J. Ewing, and M. Maines. Human heme oxygenase-2: characterization and expression of a full-length cDNA and evidence suggesting that the two ho-2 transcripts may differ by choice of polyadenylation signal. *Archives of Biochemistry and Biophysics*, 295(1):13–20, 1992.

[305] B. H. Mecham, G. T. Klus, J. Strovel, M. Augustus, D. Byrne, P. Bozso, D. Z. Wetmore, T. J. Mariani, I. S. Kohane, and Z. Szallasi. Sequence-matched probes produce increased cross-platform consistency and more reproducible biological results in microarray-based gene expression measurements. *Nucleic Acids Research*, 32(9):e74, 2004.

[306] B. H. Mecham, D. Z. Wetmore, Z. Szallasi, Y. Sadovsky, I. Kohane, and T. J. Mariani. Increased measurement accuracy for sequence-verified microarray probes. *Physiol Genomics*, 18:308–315, 2004.

[307] R. Mei et al. Probe selection for high-density oligonucleotide arrays. *Proc Natl Acad Sci U S A*, 100:11237–42, 2003.

[308] Merriam-Webster, editor. *Merriam-Webster's Collegiate Dictionary*. Merriam-Webster, Inc., 1998.

[309] M. Meyerson, G. H. Enders, C. L. Wu, L. K. Su, C. Gorka, C. Nelson, E. Harlow, and L. H. Tsai. A family of human cdc2-related protein kinases. *Embo Journal*, 11(8):2909–2917, August 1992.

[310] G. S. Michaels, D. B. Carr, M. Askenazi, S. Fuhrman, X. Wen, and R. Somogyi. Cluster analysis and data visualization of large-scale gene expression data. In *Pacific Symposium on Biocomputing*, volume 3, pages 42–53, 1998. Available online at http://psb.stanford.edu.

[311] R. Michelmore, K. Burtis, and D. Gusfield. The UC Davis genomics initiative. Technical report, University of California, Davis, 2000. http://genomics.ucdavis.edu/what.html.

[312] S. Michiels, S. Koscielny, and C. Hill. Prediction of cancer outcome with microarrays: a multiple random validation strategy. *Lancet*, 365(9458):488–92, 2005.

[313] K. Mir and E. Southern. Determining the influence of structure on hybridization using oligonucleotide arrays. *Nature Biotechnol*, 17:788–92, 1999.

[314] B. Modrek, A. Resch, C. Grasso, and C. Lee. Genome-wide detection of alternative splicing in expressed sequences of human genes. *Nucleic Acids Res*, 29:2850–9, 2001.

[315] D. C. Montgomery. *Design and Analysis of Experiments*. John Wiley and Sons, New York, 2001.

[316] V. K. Mootha, C. M. Lindgren, K.-F. Eriksson, A. Subramanian, S. Sihag, J. Lehar, P. Puigserver, E. Carlsson, M. Ridderstråle, E. Laurila, N. Houstis, M. J. Daly, N. Patterson, J. P. Mesirov, T. R. Golub, P. Tamayo, B. Spiegelman, E. S. Lander, J. N. Hirschhorn, D. Altshuler, and L. C. Groop. PGC-1 alpha-responsive genes involved in oxidative phosphorylation are coordinately downregulated in human diabetes. *Nature Genetics*, 34(3):267–273, Jul 2003.

[317] T. M. Murali and S. Kasif. Extracting conserved gene expression motifs from gene expression data. In *Pacific Symposium on Biocomputing*, pages 77–88, 2003.

[318] K. V. S. Murthy. *On Growing Better Decision Trees from Data*. PhD thesis, Johns Hopkins University, 1995.

[319] F. Naef and M. Magnasco. Solving the riddle of the bright mismatches: labeling and effective binding in oligonucleotide arrays. *Phys Rev E Stat Nonlin Soft Matter Phys*, 68(011906), 2003.

[320] M. M. Nau, B. J. Brooks, J. Battey, E. Sausville, A. F. Gazdar, I. R. Kirsch, O. W. McBride, V. Bertness, G. F. Hollis, and J. D. Minna. L-myc, a new myc-related gene amplified and expressed in human small cell lung cancer. *Nature*, 318(6041):69–73, 1985. 0028-0836.

[321] M. Newton, C. Kendziorski, C. Richmond, F. R. Blattner, and K. Tsui. On differential variability of expression ratios: Improving statistical inference about gene expresison changes from microarray data. *Journal of Computational Biology*, 8:37–52, 2001.

[322] A. Nikitin, S. Egorov, N. Daraselia, and I. Mazo. Pathway Studio – the analysis and navigation of molecular networks. *Bioinformatics*, 19(16):2155–2157, 2003.

[323] D. Nikolic-Vukosavljevic, N. Todorovic-Rakovic, M. Demajo, V. Ivanovic, B. Neskovic, M. Markicevic, and Z. Neskovic-Konstantinovic. Plasma tgf-beta1-related survival of postmenopausal metastatic breast cancer patients. *Clin Exp Metastasis*, 21(7):581–5, 2004.

[324] A. Nimgaonkar, D. Sanoudou, A. Butte, J. Haslett, L. Kunkel, A. Beggs, and I. Kohane. Reproducibility of gene expression across generations of affymetrix microarrays. *BMC Bioinformatics*, 4(27), 2003.

[325] O. Odibat and C. K. Reddy. A generalized framework for mining arbitrarily positioned overlapping co-clusters. In *Proceedings of the SIAM International Conference on Data Mining (SDM)*, 2011.

[326] O. Odibat, C. K. Reddy, and C. N. Giroux. Differential biclustering for gene expression analysis. In *Proceedings of the ACM Conference on Bioinformatics and Computational Biology (BCB)*, pages 275–284, 2010.

[327] H. Ogata, S. Goto, K. Sato, W. Fujibuchi, H. Bono, and M. Kanehisa. KEGG: Kyoto encyclopedia of genes and genomes. *Nucleic Acids Research*, 27(1):29–34, 1999.

[328] C. L. Ogden, C. D. Fryar, M. D. Carroll, and K. M. Flegal. Mean body weight, height, and body mass index, United States 1960–2002. *Adv Data*, NIL(347):1–17, 2004.

[329] Y. Okada and T. Inoue. Identification of differentially expressed gene modules between two-class DNA microarray data . *Bioinformation*, 4(4):134–137, 2009.

[330] G. Ostermeier, D. Dix, D. Miller, P. Khatri, and S. Krawetz. Spermatozoal RNA profiles of normal fertile men. *The Lancet*, 360(9335):773–777, Sept 2002.

[331] L. Page, S. Brin, R. Motwani, and T. Winograd. The PageRank citation ranking: Bringing order to the web. Technical report, 1998.

[332] D. Pan, N. Sun, K.-H. Cheung, Z. Guan, L. Ma, M. Holford, X. Deng, and H. Zhao. PathMAPA: a tool for displaying gene expression and performing statistical tests on metabolic pathways at multiple levels for Arbidopsis. *BMC Bioinformatics*, 4(1):56, Nov 2003.

[333] K.-H. Pan, C.-J. Lih, and S. N. Cohen. Effects of threshold choice on biological conclusions reached during analysis of gene expression by DNA microarrays. *Proc Natl Acad Sci USA*, 102(25):8961–5, 2005.

[334] A. D. Panani and C. Roussos. Cytogenetic and molecular aspects of lung cancer. *Cancer Lett*, 239(1):1–9, 2006. 0304-3835.

[335] R. Pandey, R. K. Guru, and D. W. Mount. Pathway Miner: extracting gene association networks from molecular pathways for predicting the biological significance of gene expression microarray data. *Bioinformatics*, 20(13):2156–2158, Sep 2004.

[336] T. A. Patterson, E. K. Lobenhofer, S. B. Fulmer-Smentek, P. J. Collins, T. M. Chu, W. Bao, H. Fang, E. S. Kawasaki, J. Hager, I. R. Tikhonova, S. J. Walker, L. Zhang, P. Hurban, F. de Longueville, J. C. Fuscoe, W. Tong, L. Shi, and R. D. Wolfinger. Performance comparison of one-color and two-color platforms within the Microarray Quality Control (MAQC) project. *Nat Biotechnol*, 24(9):1140–50, September 2006. 1087-0156.

[337] P. Pavlidis, J. Qin, V. Arango, J. J. Mann, and E. Sibille. Using the gene ontology for microarray data mining: A comparison of methods and application to age effects in human prefrontal cortex. *Neurochemical Research*, 29(6):1213 –1222, June 2004.

[338] C. Perou, S. Jeffrey, M. van der Rijni, C. Rees, M. Eisen, D. Ross, A. Pergamenschikov, C. Williams, S. Zhu, et al. Distinctive gene expression patterns in human mammary epithelial cells and breast cancers. *Proc. Natl. Acad. Sci., USA*, 96(16):9212–9217, 1999.

[339] C. Perou, T. Sørlie, M. Eisen, M. van de Rijn, S. Jeffrey, C. Rees, J. Pollack, D. Ross, H. Johnsen, et al. Molecular portraits of human breast tumours. *Nature*, 406(6797):747–752, 2000.

[340] M. Pertea and S. Salzberg. Between a chicken and a grape: estimating the number of human genes. *Genome biology*, 11(5):206, 2010.

[341] L. E. Peterson. CLUSFAVOR 5.0: hierarchical cluster and principal-component analysis of microarray-based transcriptional profiles. *Genome Biology*, 3(7):software0002.1 – 0002.8, 2002.

[342] G. Pietu, R. Mariage-Samson, N.-A. Fayein, C. Matingou, E. Eveno, et al. The genexpress IMAGE knowledge base of the human brain transcriptome: A prototype integrated resource for functional and computational genomics. *Genome Research*, 9:195–209, 1999.

[343] Proteome. Proteome BioKnowledge library. Technical report, Incyte Genomics, 2002. http://www.incyte.com/sequence/proteome.

[344] K. D. Pruitt and D. R. Maglott. RefSeq and LocusLink: NCBI gene-centered resources. *Nucleic Acids Research*, 30(1):137–140, January 2001.

[345] J. Quackenbush. Computational analysis of microarray data. *Nature Reviews Genetics*, 2:418–427, 2001.

[346] G. E. Quinn, C. H. Shin, M. G. Maguire, and R. A. Stone. Myopia and ambient lighting at night. *Nature*, 399(6732):113–4, 1999.

[347] J. Rahnenführer, F. S. Domingues, J. Maydt, and T. Lengauer. Calculating the statistical significance of changes in pathway activity from gene expression data. *Statistical Applications in Genetics and Molecular Biology*, 3(1), 2004.

[348] R. Ramakrishnan et al. An assessment of motorola codelink microarray performance for gene expression profiling applicatoins. *Nucleic Acids Research*, 30:e30, 2002.

[349] S. Ramaswamy, P. Tamayo, R. Rifkin, S. Mukherjee, C. Yeang, M. Angelo, C. Ladd, M. Reich, E. Latulippe, J. Mesirov, T. Poggio, W. Gerald, M. Loda, E. Lander, and T. Golub. Multiclass cancer diagnosis using tumor gene expression signatures. *Proc. Natl. Acad. Sci. USA*, 98(26):15149–15154, Dec. 2001.

[350] S. Raychaudhuri, J. M. Stuart, and R. Altman. Principal components analysis to summarize microarray experiments: Application to sporulation time series. In *Proceedings of the Pacific Symposium on Biocomputing*, volume 5, pages 452–463, 2000. Available online at http://psb.stanford.edu.

[351] T. D. Read, S. N. Peterson, N. Tourasse, L. W. Baillie, I. T. Paulsen, K. E. Nelson, H. Tettelin, D. E. Fouts, J. A. Eisen, S. R. Gill, E. K. Holtzapple, O. A. Okstad, E. Helgason, J. Rilstone, M. Wu, J. F. Kolonay, M. J. Beanan, R. J. Dodson, L. M. Brinkac, M. Gwinn, R. T. DeBoy, R. Madpu, S. C. Daugherty, A. S. Durkin, D. H. Haft, W. C. Nelson, J. D. Peterson, M. Pop, H. M. Khouri, D. Radune, J. L. Benton, Y. Mahamoud, L. Jiang, I. R. Hance, J. F. Weidman, K. J. Berry, R. D. Plaut, A. M. Wolf, K. L. Watkins, W. C. Nierman, A. Hazen, R. Cline, C. Redmond, J. E. Thwaite, O. White, S. L. Salzberg, B. Thomason, A. M. Friedlander, T. M. Koehler, P. C. Hanna, A.-B. Kolstø, and C. M. Fraser. The genome sequence of Bacillus Anthracis Ames and comparison to closely related bacteria. *Nature*, 423(6935):81–6, May 2003.

[352] A. Relogio, C. Schwager, A. Richter, W. Ansorge, and J. Valcarcel. Optimization of oligonucleotide-based DNA microarrays. *Nucleic Acids Research*, 30(11):e51, 2002.

[353] Y. S. Rhee, V. Wood, K. Dolinski, and S. Drăghici. Use and misuse of the Gene Ontology annotations. *Nature Reviews Genetics*, 9(7):509–515, July 2008.

[354] C. S. Richmond, J. D. Glasner, R. Mau, H. Jin, and F. R. Blattner. Genome-wide expression profiling in Escherichia coli K-12. *Nucleic Acids Research*, 27(19):3821–3835, 1999.

[355] C. J. Roberts, B. Nelson, M. J. Marton, R. Stoughton, M. R. Meyer, H. A. Bennett, Y. D. He, H. Dia, W. L. Walker, T. R. Hughes, M. Tyers, C. Boone, and S. H. Friend. Signaling and circuitry of multiple MAPK pathways revealed by a matrix of global gene expression profiles. *Science*, 287(5454):873–880, Feb. 2000.

[356] S. Rodrigues, K. Fathers, G. Chan, D. Zuo, F. Halwani, S. Meterissian, and P. M. CrkI and CrkII function as key signaling integrators for migration and invasion of cancer cells. *Molecular Cancer Research*, 3(4):183–194, Apr 2005.

[357] S. Rogers, R. Williams, and C. Campbell. *Bioinformatics Using Computational Intelligence Paradigms*, chapter Class Prediction with Microarray Datasets, pages 119–142. Springer, 2005.

[358] F. Rosenblatt. The perceptron: a probabilistic model for information storage and organization in the brain. *Psychological Review*, 65:386–408, 1958.

[359] A. D. Roses. Pharmacogenetics and the practice of medicine. *Nature*, 405:857–865, 2000.

[360] D. T. Ross, U. Scherf, M. B. Eisen, C. M. Perou, C. Rees, P. Spellman, V. Iyer, S. S. Jeffrey, M. V. de Rijn, M. Waltham, A. Pergamenschikov, J. C. Lee, D. Lashkari, D. Shalon, T. G. Myers, J. N. Weinstein, D. Botstein, and P. O. Brown. Systematic variation in gene expression patterns in human cancer cell lines. *Nature Genetics*, 24(3):227–235, 2000.

[361] P. E. Ross. The making of a 24 billion gene machine. *Forbes*, February 21:98–104, 2000.

[362] B. Rost and C. Sander. Combining evolutionary information and neural networks to predict protein secondary structure. *Proteins*, 19:55–72, 1994.

[363] C. Rubie, K. Kempf, J. Hans, T. Su, B. Tilton, T. Georg, B. Brittner, B. Ludwig, and M. Schilling. Housekeeping gene variability in normal and cancerous colorectal, pancreatic, esophageal, gastric and hepatic tissues. *Molecular and Cellular Probes*, 19(2):101–109, 2005.

[364] D. E. Rumelhart, G. E. Hinton, and R. J. Williams. Learning internal representations by error backpropagation. In D. E. Rumelhart, J. L. McClelland, and the PDP Research Group, editors, *Parallel Distributed Processing: Explorations in the Microstructures of Cognition*, volume 1. MIT Press/Bradford Books, 1986.

[365] G. Ruvkun. Molecular biology. Glimpses of a tiny RNA world. *Science*, 294(5543):797–9, 2001.

[366] T. A. Sanders, T. de Grassi, G. J. Miller, and S. E. Humphries. Dietary oleic and palmitic acids and postprandial factor VII in middle-aged men heterozygous and homozygous for factor VII R353Q polymorphism. *The American Journal of Clinical Nutrition*, 69(2):220–225, Feb 1999.

[367] J. SantaLucia, J., H. Allawi, and P. Seneviratne. Improved nearest-neighbor parameters for predicting dna duplex stability. *Biochemistry*, 35:3555–62, 1996.

[368] M. Sapir and G. A. Churchill. Estimating the posterior probability of differential gene expression from microarray data. Technical Report http://www.jax.org/research/churchill/pubs/, Jackson Labs, Bar Harbor, ME, 2000.

[369] R. Sasik, T. Hwa, N. Iranfar, and W. F. Loomis. Percolation clustering: A noval algorithm applied to the clustering of gene expression patterns in dictyostelium development. In *Pacific Symposium on Biocomputing*, volume 6, pages 335–347, 2001. Available online at http://psb.stanford.edu.

[370] S. Sato and S. Tabata. [The complete genome sequence of Arabidopsis thaliana]. *Tanpakushitsu Kakusan Koso*, 46(1):61–5, Jan 2001.

[371] E. E. Schadt, L. Cheng, C. Su, and W. H. Wong. Analyzing high-density oligonucleotide gene expression array data. *Journal of Cellular Biochemistry*, 80(2):192–202, Oct. 2000.

[372] M. Schena. *DNA Microarrays: A Practical Approach*. Practical Approach Series, 205. Oxford University Press, Oxford, UK, 1999.

[373] M. Schena. *Microarray Biochip Technology*. Eaton Publishing, Sunnyvale, CA, 2000.

[374] M. Schena, D. Shalon, R. W. Davis, and P. O. Brown. Quantitative monitoring of gene expression patterns with a complementary DNA microarray. *Science*, 270(5235):467–470, 1995.

[375] M. Schena, D. Shalon, R. Heller, A. Chai, P. Brown, and R. Davis. Parallel human genome analysis: microarray-based expression monitoring of 1000 genes. *Proc. Natl. Acad. of Sci. USA*, 93:10614–10519, 1996.

[376] U. Scherf, D. Ross, M. Waltham, L. Smith, J. Lee, L. Tanabe, K. Kohn, W. Reinhold, T. Meyers, D. T. Andrews, D. A. Scudiero, M. Eisen, E. Sausville, Y. Pommier, D. Botstein, P. Brown, and J. Weinstein. A gene expression database for the molecular pharmacology of cancer. *Nature Genetics*, 24(3):236–244, 2000.

[377] J. Schuchhardt, D. Beule, E. Wolski, and H. Eickhoff. Normalization strategies for cDNA microarrays. *Nucleic Acids Research*, 28(10):e47i–e47v, May 2000.

[378] S. Schuhknecht, S. Duensing, I. Dallmann, J. Grosse, and M. Reitz. Interleukin-12 inhibits apoptosis in chronic lymphatic leukemia (CLL) B cells. *Cancer Biotherapy and Radiopharmaceuticals*, 17(5):495–499, October 2002.

[379] G. D. Schuler. Pieces of puzzle: Expressed sequence tags and the catalog of human genes. *Journal of Molecular Medicine*, 75(10):694–698, Oct. 1997.

[380] S. Selvey, E. Thompson, K. Matthaei, R. Lea, M. Irving, and L. Griffiths. [beta]-Actin–an unsuitable internal control for RT-PCR. *Molecular and cellular Probes*, 15(5):307–311, 2001.

[381] J. P. Shaffer. Modified sequentially rejective multiple test procedures. *Journal of American Statistical Association*, 81:826–831, 1986.

[382] J. P. Shaffer. Multiple hypothesis testing. *Annual Reviews in Psychology*, 46:561–584, 1995.

[383] N. Shah and N. Fedoroff. CLENCH: a program for calculating cluster enrichment using the gene ontology. *Bioinformatics*, 20(7):1196–1197, May 2004.

[384] D. Shalon, S. J. Smith, and P. O. Brown. A DNA microarray system for analyzing complex DNA samples using two-color fluorescent probe hybridization. *Genome Research*, 6:639–645, 1996.

[385] P. Shannon, A. Markiel, O. Ozier, N. S. Baliga, J. T. Wang, D. Ramage, N. Amin, B. Schwikowski, and T. Ideker. Cytoscape: A software environment for integrated models of biomolecular interaction networks. *Genome Research*, 13:2498–504, 2003.

[386] A. J. Sharp, H. C. Mefford, K. Li, C. Baker, C. Skinner, R. E. Stevenson, R. J. Schroer, F. Novara, M. D. Gregori, R. Ciccone, A. Broomer, I. Casuga, Y. Wang, C. Xiao, C. Barbacioru, G. Gimelli, B. D. Bernardina, C. Torniero, R. Giorda, R. Regan, V. Murday, S. Mansour, M. Fichera, L. Castiglia, P. Failla, M. Ventura, Z. Jiang, G. M. Cooper, S. J. L. Knight, C. Romano, O. Zuffardi, C. Chen, C. E. Schwartz, and E. E. Eichler. A recurrent 15q13.3 microdeletion syndrome associated with mental retardation and seizures. *Nat Genet*, 40(3):322–8, 2008.

[387] J. Shawe-Taylor and N. Cristianini. *Kernel Methods for Pattern Analysis*. Cambridge Univ Press, 2004.

[388] X. She, Z. Cheng, S. Zollner, D. M. Church, and E. E. Eichler. Mouse segmental duplication and copy number variation. *Nat Genet*, 40(7):909–14, 2008.

[389] G. Sherlock, T. Hernandez-Boussard, A. Kasarskis, G. Binkley, et al. The Stanford Microarray Database. *Nucleic Acid Research*, 29(1):152–155, January 2001.

[390] L. Shi. DNA microarray – monitoring the genome on a chip. Technical report, 2001. http://www.gene-chips.com/.

[391] L. Shi, W. D. Jones, R. V. Jensen, S. C. Harris, R. G. Perkins, F. M. Goodsaid, L. Guo, L. J. Croner, C. Boysen, H. Fang, F. Qian, S. Amur, W. Bao, C. C. Barbacioru, V. Bertholet, X. M. Cao, T.-M. Chu, P. J. Collins, X.-H. Fan, F. W. Frueh, J. C. Fuscoe, X. Guo, J. Han, D. Herman, H. Hong, E. S. Kawasaki, Q.-Z. Li, Y. Luo, Y. Ma, N. Mei, R. L. Peterson, R. K. Puri, R. Shippy, Z. Su, Y. A. Sun, H. Sun, B. Thorn, Y. Turpaz, C. Wang, S. J. Wang, J. A. Warrington, J. C. Willey, J. Wu, Q. Xie, L. Zhang, L. Zhang, S. Zhong, R. D. Wolfinger, and W. Tong. The balance of reproducibility, sensitivity, and specificity of lists of differentially expressed genes in microarray studies. *BMC Bioinformatics*, 9 Suppl 9(NIL):S10, 2008.

[392] L. Shi, L. Reid, W. Jones, R. Shippy, J. Warrington, S. Baker, P. Collins, F. De Longueville, E. Kawasaki, K. Lee, et al. The MicroArray Quality Control (MAQC) project shows inter-and intraplatform reproducibility of gene expression measurements. *Nature Biotechnology*, 24(9):1151–1161, 2006.

[393] M. A. Shipp, K. N. Ross, P. Tamayo, A. P. Weng, J. L. Kutok, R. C. T. Aguiar, M. Gaasenbeek, M. Angelo, M. Reich, G. S. Pinkus, T. S. Ray, M. A. Koval, K. W. Last, A. Norton, T. A. Lister, J. Mesirov, D. S. Neuberg, E. S. Lander, J. C. Astera, and T. R. Golub. Diffuse large B-cell lymphoma outcome prediction by gene-expression profiling and supervised machine learning. *Nature Medicine*, 8(1):68–74, January 2002. Supplemental Information at http://www-genome.wi.mit.edu/mpr/lymphoma/.

[394] R. Shippy, T. Sendera, R. Lockner, C. Palaniappan, T. Kaysser-Kranich, G. Watts, and J. Alsobrook. Performance evaluation of commercial short-oligonucleotide microarrays and the impact of noise in making cross-platform correlations. *BMC Genomics*, 5(61), 2004.

[395] R. J. Slebos and S. Rodenhuis. The molecular genetics of human lung cancer. *Eur Respir J*, 2(5):461–9, 1989. 0903-1936.

[396] G. Smyth et al. Linear models and empirical Bayes methods for assessing differential expression in microarray experiments. *Statistical Applications in Genetics and Molecular Biology*, 3(1):1027, 2004.

[397] G. K. Smyth. *Limma: Linear Models for Microarray Data*, pages 397–420. Springer, New York, 2005.

[398] D. L. Souvaine and J. M. Steele. Efficient time and space algorithms for least median of squares regression. *J. Amer. Stat. Assn.*, 82:794–801, 1987.

[399] T. P. Speed. Hints and prejudices – always log spot intensities and ratios. Technical report, University of California, Berkeley, 2000. http://www.stat.berkeley.edu/users/terry/zarray/Html/log.html.

[400] P. T. Spellman, G. Sherlock, M. Q. Zhang, W. Iyer, K. Anders, M. B. Eisen, P. O. Brown, D. Bostein, and B. Futcher. Comprehensive identification of cell cycle-regulated genes of the yeast Saccharomyces cerevisiae by microarray hybridization. *Mol. Biol. Cell*, 9(12):3273–3297, 1998.

[401] J. Staunton, D. Slonim, H. Coller, P. Tamayo, M. Angelo, J. Park, U. Scherf, J. Lee, W. Reinhold, J. Weinstein, J. Mesirov, E. Lander, and T. Golub. Chemosensitivity prediction by transcriptional profiling. *Proc. Natl. Acad. Sci. USA*, 98(19):10787–92, Sept. 2001.

[402] J. Stelling. Mathematical models in microbial systems biology. *Current Opinion in Microbiology*, 7(5):513–8, 2004.

[403] M. E. Stokes, C. S. Davis, and G. G. Koch. *Categorical Data Analysis Using the SAS System*. SAS Institute, Cary, NC, 1995.

[404] R. A. Stone, T. Lin, D. Desai, and C. Capehart. Photoperiod, early postnatal eye growth, and visual deprivation. *Vision Res*, 35(9):1195–202, 1995.

[405] G. Stormo, T. Schneider, L. Gold, and A. Ehrenfeuch. Use of the perceptron algorithm to distinguish translation initiation sites in e. coli. *Nucleic Acids Research*, 10, 1982.

[406] T. Strachan and A. P. Read. *Human Molecular Genetics*. Wiley-LISS, New York, NY, 2nd edition, 1999.

[407] A. Su, M. Cooke, K. Ching, Y. Hakak, J. Walker, T. Wiltshire, A. Orth, R. Vega, L. Sapinoso, A. Moqrich, A. Patapoutian, G. Hampton, P. Schultz, and H. JB. Large-scale analysis of the human and mouse transcriptomes. *Proc. of the National Academy of Science*, 99(7):4465–4470, Mar 2002.

[408] A. Subramanian, P. Tamayo, V. K. Mootha, S. Mukherjee, B. L. Ebert, M. A. Gillette, A. Paulovich, S. L. Pomeroy, T. R. Golub, E. S. Lander, and J. P. Mesirov. Gene set enrichment analysis: A knowledge-based approach for interpreting genome-wide expression profiles. *Proceeding of the National Academy of Sciences of the USA*, 102(43):15545–15550, 2005.

[409] P. Sudarsanam, V. R. Iyer, P. O. Brown, and F. Winston. Whole-genome expression analysis of snf/swi mutants of Saccharomyces cerevisiae. *Proc. Natl. Acad. Sci.*, 97(7):3364–3369, 2000.

[410] N. Sugimoto, M. Nakano, and S. Nakano. Thermodynamics-structure relationship of single mismatches in rna/dna duplexes. *Biochemistry*, 39:11270–81, 2000.

[411] K. Susztak, E. Ciccone, P. McCue, K. Sharma, and E. P. Böttinger. Multiple metabolic hits converge on CD36 as novel mediator of tubular epithelial apoptosis in diabetic nephropathy. *PLoS Medicine*, 2(2):e45, Feb 2005.

[412] C. Swagell, D. Henly, and C. P. Morris. Expression analysis of a human hepatic cell line in response to palmitate. *Biochemical and Biophysical Research Communications*, 328(2):432–441, 2005.

[413] P. Tamayo, D. Slonim, J. Mesirov, Q. Zhu, S. Kitareewan, E. Dmitrovsky, E. S. Lander, and T. R. Golub. Interpreting patterns of gene expression with self-organizing maps: Methods and application to hematopoietic differentiation. *Proceedings of the National Academy of Sciences of the United States of America*, 96(6):2907–2912, 1999.

[414] P. K. Tan, T. J. Downey, E. L. S. Jr., P. Xu, D. Fu, D. S. Dimitrov, R. A. Lempicki, B. M. Raaka, and M. C. Cam. Evalutation of gene expression measurements from commercial microarray platforms. *Nucleic Acids Research*, 31(19):5676–5684, 2003.

[415] H. Tao, C. Bausch, C. Richmond, F. R. Blattner, and T. Conway. Functional genomics: Expression analysis of *Escherichia coli* growing on minimal and rich media. *Journal of Bacteriology*, 181(20):6425–6440, 1999.

[416] A. Tarca, J. Cooke, and J. Mackay. A robust neural networks approach for spatial and intensity-dependent normalization of cdna microarray data. *Bioinformatics*, 21:2674–2683, 2005.

[417] A. L. Tarca, S. Drăghici, P. Khatri, S. S. Hassan, P. Mittal, J. sun Kim, C. J. Kim, J. P. Kusanovic, and R. Romero. A novel signaling pathway impact analysis (SPIA). *Bioinformatics*, 25(1):75–82, 2009.

[418] A. L. Tarca, B. P. A. Grandjean, and F. Larachi. Feature selection methods for multiphase reactors data classification. *Ind. Eng. Chem. Res.*, 44(4):1073 –1084, 2005.

[419] S. Tavazoie, J. D. Hughes, M. J. Campbell, R. J. Cho, and G. M. Church. Systematic determination of genetic network architecture. *Nature Genetics*, 22:281–285, 1999.

[420] E. Taylor, D. Cogdell, K. Coombes, L. Hu, L. Ramdas, A. Tabor, S. Hamilton, and W. Zhang. Sequence verification as quality control step for production of cDNA microarrays. *Biotechniques*, 31(1):62–65, 2001.

[421] J. J. M. ter Linde, H. Liang, R. W. Davis, H. Y. Steensma, J. P. V. Dijken, and J. T. Pronk. Genome-wide transcriptional analysis of aerobic and anaerobic chemostat cultures of *Saccharomyces cerevisiae*. *Journal of Bacteriology*, 181(24):7409–7413, 1999.

[422] L. Tian, S. A. Greenberg, S. W. Kong, J. Altschuler, I. S. Kohane, and P. J. Park. Discovering statistically significant pathways in expression profiling studies. *Proceeding of the National Academy of Sciences of the USA*, 102(38):13544–13549, 2005.

[423] R. Tibshirani, G. Walther, D. Botstein, and P. Brown. Cluster validation by prediction strength. Technical report, Statistics Department, Stanford University, 2000. Manuscript available at http://www-stat.stanford.edu/ tibs/research.html.

[424] N. Todorovic-Rakovic. Tgf-beta 1 could be a missing link in the interplay between er and her-2 in breast cancer. *Med Hypotheses*, 65(3):546–51, 2005.

[425] J. Tsai, R. Sultana, Y. Lee, G. Pertea, S. Karamycheva, V. Antonescu, J. Cho, B. Parvizi, F. Cheung, and J. Quackenbush. Resourcerer: a database for annotating and linking microarray resources within and across species. *Genome Biology*, 2(11):software0002.1–0002.4, October 2001.

[426] S. Tsoka and C. A. Ouzounis. Recent developments and future directions in computational genomics. *Federation of European Biochemical Societies Letters*, (23897):1–7, 2000.

[427] V. G. Tusher, R. Tibshirani, and G. Chu. Significance analysis of microarrays applied to the ionizing radiation response. *Proc. Natl. Acad. Sci.*, 98(9):5116–5121, Apr 2001. Software available for download at http://www-stat.standford.edu/ tibs/SAM/index.html.

[428] M. J. van de Vijver et al. A gene-expression signature as a predictor of survival in breast cancer. *New England Journal of Medicine*, 347:1999–2009, 2002.

[429] M. J. van der Laan, K. S. Pollard, and J. Bryan. A new partitioning around medoids algorithm. *U.C. Berkeley Division of Biostatistics Working Paper Series., http://www.bepress.com/ucbbiostat/paper105*, 2002.

[430] J. van Helden, A. F. Rios, and J. Collado-Vides. Discovering regulatory elements in non-coding sequences by analysis of spaced dyads. *Nucleic Acids Research*, 28(8):1808–1818, 2000.

[431] M. J. van Nimwegen, M. Huigsloot, A. Camier, I. B. Tijdens, and B. van de Water. Focal adhesion kinase and protein kinase b cooperate to suppress doxorubicin-induced apoptosis of breast tumor cells. *Mol Pharmacol*, 70(4):1330–9, 2006.

[432] M. J. van Nimwegen and B. van de Water. Focal adhesion kinase: A potential target in cancer therapy. *Biochem Pharmacol*, 2006.

[433] L. J. van't Veer, H. Dai, M. J. van de Vijver, Y. D. He, A. Hart, M. Mao, H. L. Peterse, K. van der Kooy, M. J. Marton, A. T. Witteveenothers, G. J. Schreiber, R. M. Kerkhoven, C. Roberts, P. S. Linsley, R. Bernards, and S. H. Friend. Gene expression profiling predicts clinical outcome of breast cancer. *Nature*, 415(6871):530–536, January 2002.

[434] V. N. Vapnik. *Statistical Learning Theory*. Wiley, New York, 1998.

[435] B. Venables and B. Ripley. *Modern Applied Statistics with S*. Springer, 2002.

[436] A. R. Venkitaraman. Cancer susceptibility and the functions of BRCA1 and BRCA2. *Cell*, 108(2):171–182, January 2002.

[437] J. C. Venter, M. D. Adams, E. W. Myers, P. W. Li, R. J. Mural, G. G. Sutton, H. O. Smith, M. Yandell, C. A. Evans, R. A. Holt, J. D. Gocayne, P. Amanatides, R. M. Ballew, D. H. Huson, J. R. Wortman, Q. Zhang, C. D. Kodira, X. H. Zheng, L. Chen, M. Skupski, G. Subramanian, P. D. Thomas, J. Zhang, G. L. G. Miklos, C. Nelson, S. Broder, A. G. Clark, J. Nadeau, V. A. McKusick, N. Zinder, A. J. Levine, R. J. Roberts, M. Simon, C. Slayman, M. Hunkapiller, R. Bolanos, A. Delcher, I. Dew, D. Fasulo, M. Flanigan, L. Florea, A. Halpern, S. Hannenhalli, S. Kravitz, S. Levy, C. Mobarry, K. Reinert, K. Remington, J. Abu-Threideh, E. Beasley, K. Biddick, V. Bonazzi, R. Brandon, M. Cargill, I. Chandramouliswaran, R. Charlab, K. Chaturvedi, Z. Deng, V. Di Francesco, P. Dunn, K. Eilbeck, C. Evangelista, A. E. Gabrielian, W. Gan, W. Ge, F. Gong, Z. Gu, P. Guan, T. J. Heiman, M. E. Higgins, R.-R. Ji, Z. Ke, K. A. Ketchum, Z. Lai, Y. Lei, Z. Li, J. Li, Y. Liang, X. Lin, F. Lu, G. V. Merkulov, N. Milshina, H. M. Moore, A. K. Naik, V. A. Narayan, B. Neelam, D. Nusskern, D. B. Rusch, S. Salzberg, W. Shao, B. Shue, J. Sun, Z. Y. Wang, A. Wang, X. Wang, J. Wang, M.-H. Wei, R. Wides, C. Xiao, C. Yan, A. Yao, J. Ye, M. Zhan, W. Zhang, H. Zhang, Q. Zhao, L. Zheng, F. Zhong, W. Zhong, S. C. Zhu, S. Zhao, D. Gilbert, S. Baumhueter, G. Spier, C. Carter, A. Cravchik, T. Woodage, F. Ali, H. An, A. Awe, D. Baldwin, H. Baden, M. Barnstead, I. Barrow, K. Beeson, D. Busam, A. Carver, A. Center, M. L. Cheng, L. Curry,

S. Danaher, L. Davenport, R. Desilets, S. Dietz, K. Dodson, L. Doup, S. Ferriera, N. Garg, A. Gluecksmann, B. Hart, J. Haynes, C. Haynes, C. Heiner, S. Hladun, D. Hostin, J. Houck, T. Howland, C. Ibegwam, J. Johnson, F. Kalush, L. Kline, S. Koduru, A. Love, F. Mann, D. May, S. McCawley, T. McIntosh, I. McMullen, M. Moy, L. Moy, B. Murphy, K. Nelson, C. Pfannkoch, E. Pratts, V. Puri, H. Qureshi, M. Reardon, R. Rodriguez, Y.-H. Rogers, D. Romblad, B. Ruhfel, R. Scott, C. Sitter, M. Smallwood, E. Stewart, R. Strong, E. Suh, R. Thomas, N. N. Tint, S. Tse, C. Vech, G. Wang, J. Wetter, S. Williams, M. Williams, S. Windsor, E. Winn-Deen, K. Wolfe, J. Zaveri, K. Zaveri, J. F. Abril, R. G. M. J. Campbell, K. V. Sjolander, B. Karlak, A. Kejariwal, H. Mi, B. Lazareva, T. Hatton, A. Narechania, K. Diemer, A. Muruganujan, N. Guo, S. Sato, V. Bafna, S. Istrail, R. Lippert, R. Schwartz, B. Walenz, S. Yooseph, D. Allen, A. Basu, J. Baxendale, L. Blick, M. Caminha, J. Carnes Stine, P. Caulk, Y. H. Chiang, M. Coyne, C. Dahlke, A. D. Mays, M. Dombroski, M. Donnelly, D. Ely, S. Esparham, C. Fosler, H. Gire, S. Glanowski, K. Glasser, A. Glodek, M. Gorokhov, K. Graham, B. Gropman, M. Harris, J. Heil, S. Henderson, J. Hoover, D. Jennings, C. Jordan, J. Jordan, J. Kasha, L. Kagan, C. Kraft, A. Levitsky, M. Lewis, X. Liu, J. Lopez, D. Ma, W. Majoros, J. McDaniel, S. Murphy, M. Newman, T. Nguyen, N. Nguyen, M. Nodell, S. Pan, J. Peck, M. Peterson, W. Rowe, R. Sanders, J. Scott, M. Simpson, T. Smith, A. Sprague, T. Stockwell, R. Turner, E. Venter, M. Wang, M. Wen, D. Wu, M. Wu, A. Xia, A. Zandieh, and X. Zhu. The sequence of the human genome. *Science*, 291(5507):1304–1351, February 2001.

[438] B. Vincenzi, G. Schiavon, M. Silletta, D. Santini, G. Perrone, M. Di Marino, S. Angeletti, A. Baldi, and G. Tonini. Cell cycle alterations and lung cancer. *Histol Histopathol*, 21(4):423–35, 2006. 1699-5848 Review.

[439] Z. Šidák. Rectangular confidence regions for the means of multivariate normal distributions. *Journal of the American Statistical Association*, 62:626–633, 1967.

[440] O. G. Vukmirovic and S. M. Tilghman. Exploring genome space. *Nature*, 405:820–822, 2000.

[441] H. Walker. Degrees of freedom. *Journal of Educational Psychology*, 31(4):253–269, 1940.

[442] T. Walsh, J. M. McClellan, S. E. McCarthy, A. M. Addington, S. B. Pierce, G. M. Cooper, A. S. Nord, M. Kusenda, D. Malhotra, A. Bhandari, S. M. Stray, C. F. Rippey, P. Roccanova, V. Makarov, B. Lakshmi, R. L. Findling, L. Sikich, T. Stromberg, B. Merriman, N. Gogtay, P. Butler, K. Eckstrand, L. Noory, P. Gochman, R. Long, Z. Chen, S. Davis, C. Baker, E. E. Eichler, P. S. Meltzer, S. F. Nelson, A. B. Singleton,

M. K. Lee, J. L. Rapoport, M.-C. King, and J. Sebat. Rare structural variants disrupt multiple genes in neurodevelopmental pathways in schizophrenia. *Science*, 320(5875):539–43, 2008.

[443] D. G. Wang, J. B. Fan, C. J. Siao, A. Berno, et al. Large-scale identification, mapping, and genotyping of single-nucleotide polymorphisms in the human genome. *Science*, 280(5366):1077–1082, 1998.

[444] H. Wang, F. Azuaje, O. Bodenreider, and J. Dopazo. Gene expression correlation and gene ontology-based similarity: An assessment of quantitative relationships. In *Proceedings of the 2004 IEEE Symposium on Computational Intelligence in Bioinformatics and Computational Biology*, pages 25–31, 2004.

[445] M. L. Wang, S. Belmonte, U. Kim, M. Dolan, J. W. Morris, and H. M. Goodman. A cluster of ABA-regulated genes on *Arabidopsis thaliana* BAC T07M07. *Genome Research*, 9:325–333, 1999.

[446] W. Wang, J. Lu, R. Lee, Z. Gu, and R. Clarke. Iterative normalization of cDNA microarray data. *IEEE Transactions on Information Technology in Biomedicine*, 6(1):29–37, March 2002.

[447] W. Q. Wang, Y. H. Zhou, and R. Bi. Correlating genes and functions to human disease by systematic differential analysis of expression profiles. In *Advances in Intelligent Computing, Pt 2, Proceedings*, volume 3645 of *Lecture Notes in Computer Science*, pages 11–20. 2005.

[448] R. Waterston. Genome sequence of the nematode C. Elegans: a platform for investigating biology. *Science*, 282(5396):2012–8, Dec. 1998.

[449] D. E. Watkins-Chow and W. J. Pavan. Genomic copy number and expression variation within the C57BL/6J inbred mouse strain. *Genome Res*, 18(1):60–6, 2008.

[450] A. Watson, A. Mazumder, M. Stewart, and S. Balasubramanian. Technology for microarray analysis of gene expression. *Current opinions in Biotechnology*, 9:609–614, 1998.

[451] M. E. Waugh, D. L. Bulmore, A. D. Farmer, P. A. Steadman, S. T. Wlodekand, et al. PathDB: A metabolic database with sophisticated search and visualization tools. In *Proc. of Plant and Animal Genome VIII Conference*, San Diego, CA, January 9–12, 2000.

[452] A. Webb. *Statistical Pattern Recognition*. John Wiley and Sons Ltd., England, 2002.

[453] P. L. Welcsh, M. K. Lee, R. M. Gonzalez-Hernandez, D. J. Black, M. Mahadevappa, E. M. Swisher, J. A. Warrington, and M.-C. King. BRCA1 transcriptionally regulates genes involved in breast tumorigenesis. *Proc. Natl. Acad. Sci. USA*, 99(11):7560–7565, 2002.

[454] S. Welle, A. I. Brooks, and C. A. Thornton. Computational method for reducing variance with Affymetrix microarrays. *BMC Bioinformatics*, 3:23, 2002.

[455] A. Wellmann, C. Thieblemont, S. Pittaluga, A. Sakai, et al. Detection of differentially expressed genes in lymphomas using cDNA arrays: identification of *clusterin* as a new diagnostic marker for anaplastic large-cell lymphomas. *Blood*, 96(2):398–404, 2000.

[456] M. West, J. Nevins, J. Marks, R. Spang, C. Blanchette, and H. Zuzan. Bayesian regression analysis in the "large p, small n" paradigm with application in DNA microarray studies. Technical report, Duke University, 2000.

[457] P. H. Westfall and S. S. Young. *Resampling-based Multiple Testing: Examples and Methods for p-value Adjustment*. Wiley, New York, 1993.

[458] J. Weston, C. Leslie, E. Ie, D. Zhou, A. Elisseeff, and W. S. Noble. Semi-supervised protein classification using cluster kernels. *Bioinformatics*, 21(15):3241–7, 2005.

[459] K. P. White, S. A. Rifkin, P. Hurban, and D. S. Hogness. Microarray analysis of Drosophila development during metamorphosis. *Science*, 286:2179–2184, 1999.

[460] J. Whitfield. DNA-chip firm backs down over upgrade. *Nature*, 424:119, 2003.

[461] S. E. Wildsmith, G. E. Archer, A. J. Winkley, P. W. Lane, and P. J. Bugelski. Maximizing of signal derived from cDNA microarrays. *BioTechniques*, 30:202–208, 2000.

[462] C. Wilson, S. D. Pepper, and C. J. Miller. QC and Affymetrix data. Technical report, Paterson Institute for Cancer Research, Christie Hospital NHS Trust, 2008.

[463] L. D. Wood, D. W. Parsons, S. Jones, J. Lin, T. Sjoblom, R. J. Leary, D. Shen, S. M. Boca, T. Barber, J. Ptak, N. Silliman, S. Szabo, Z. Dezso, V. Ustyanksky, T. Nikolskaya, Y. Nikolsky, R. Karchin, P. A. Wilson, J. S. Kaminker, Z. Zhang, R. Croshaw, J. Willis, D. Dawson, M. Shipitsin, J. K. V. Willson, S. Sukumar, K. Polyak, B. H. Park, C. L. Pethiyagoda, P. V. K. Pant, D. G. Ballinger, A. B. Sparks, J. Hartigan, D. R. Smith, E. Suh, N. Papadopoulos, P. Buckhaults, S. D. Markowitz, G. Parmigiani, K. W. Kinzler, V. E. Velculescu, and B. Vogelstein. The genomic landscapes of human breast and colorectal cancers. *Science*, 318(5853):1108–13, 2007.

[464] C. Wu, R. Carta, and L. Zhang. Sequence dependence of cross-hybridization on short oligo microarrays. *Nucleic Acids Res*, 33:e84, 2005.

[465] C. H. Wu, L.-S. L. Yeh, H. Huang, L. Arminski, J. Castro-Alvear, Y. Chen, Z. Hu, P. Kourtesis, R. S. Ledley, B. E. Suzek, C. Vinayaka, J. Zhang, , and W. C. Barker. The protein information resource. *Nucleic Acids Research*, 31(1):345–347, 2003.

[466] L. Wu, T. Hughes, A. Davierwala, M. Robinson, R. Stoughton, and S. Altschuler. Large-scale prediction of saccharomyces cerevisiae gene function using overlapping transcriptional clusters. *Nature Genetics*, 31(3):255–265, July 2002.

[467] E. P. Xing and R. M. Karp. Clustering of high-dimensional microarray data via iterative feature filtering using normalized cuts. *Bioinformatics*, 17(Suppl. 1):S306–S315, 2001.

[468] J. Yang, H. Wang, W. Wang, P. Yu, U. Ibm, U. Chapel, H. Ibm, T. J. Watson, and T. J. Watson. Enhanced biclustering on expression data. In *Proc. of 3rd IEEE Symposium on BioInformatics and BioEngineering*, pages 321–327, 2003.

[469] W.-H. Yang, D.-Q. Dai, and H. Yan. Finding Correlated Biclusters from Gene Expression Data. *IEEE Transactions on Knowledge and Data Engineering*, 23(4):568–584, Apr. 2011.

[470] Y. Yang, M. J. Buckley, S. Dudoit, and T.P.Speed. Comparison of methods for image analysis on cDNA. Technical report, University of California, Berkeley, 2000. http://www.stat.berkeley.edu/users/terry/zarray/Html/log.html.

[471] Y. Yang, S. Dudoit, P. Luu, and T.P.Speed. Normalization for cDNA microarray data. In *Proc. of SPIE BiOS*, volume 4266, page 31, San Jose, CA, 2001.

[472] Y. H. Yang, M. J. Buckley, S. Dudoit, and T. P. Speed. Comparison of methods for image analysis on cDNA microarray data. *Journal of Computational and Graphic Statistics*, 11:108–136, 2002.

[473] Y. H. Yang and T. Speed. Design issues for cDNA microarray experiments. *Nature Reviews Genetics*, 3(8):579–588, 2002.

[474] C. L. Yauk, M. L. Berndt, A. Williams, and G. R. Douglas. Comprehensive comparison of six microarray technologies. *Nucleic Acids Research*, 32(15):e124, 2004.

[475] K. Y. Yeung, C. Fraley, A. Murua, A. E. Raftery, and W. L. Ruzzo. Model-based clustering and data transformation for gene expression data. *Bioinformatics*, 17(10):977–987, Oct. 2001.

[476] K. Y. Yeung, D. R. Haynor, and W. L. Ruzzo. Validating clustering for gene expression data. *Bioinformatics*, 17(4):309–318, April 2001.

[477] T. Yoshida, N. Koide, T. Sugiyama, I. Mori, and T. Yokochi. A novel caspase dependent pathway is involved in apoptosis of human endothelial cells by shiga toxins. *Microbiol. Immunol.*, 46(10):697–700, 2002.

[478] A. Young, N. Whitehouse, J. Cho, and C. Shaw. Ontologytraverser: an R package for GO analysis. *Bioinformatics*, 21(2):275–6, 2005.

[479] H. Yue, P. Eastman, B. Wang, J. Minor, M. Doctolero, R. L. Nuttall, R. Stack, J. W. Becker, J. R. Montgomery, M. Vainer, and R. Johnston. An evaluation of the performance of cDNA microarrays for detecting changes in global mRNA expression. *Nucleic Acids Research*, 29(8):e41, 2001.

[480] T. Yuen, E. Wurmbach, R. Pfeffer, B. Ebersole, and S. Sealfon. Accuracy and calibration of commercial oligonucleotide and custom cDNA microarrays. *Nucleic Acids Research*, 30(10):e48, May 2002. Available at http://linkage.rockefeller.edu/wli/microarray/yuen02.pdf.

[481] K. Zadnik, L. A. Jones, B. C. Irvin, R. N. Kleinstein, R. E. Manny, J. A. Shin, and D. O. Mutti. Myopia and ambient night-time lighting. CLEERE study group. collaborative longitudinal evaluation of ethnicity and refractive error. *Nature*, 404(6774):143–4, 2000.

[482] B. R. Zeeberg, W. M. Feng, G. Wang, M. D. Wang, A. T. Fojo, M. Sunshine, S. Narasimhan, D. W. Kane, W. C. Reinhold, S. Lababidi, K. J. Bussey, J. Riss, J. C. Barrett, and J. N. Weinstein. Gominer: a resource for biological interpretation of genomic and proteomic data. *Genome Biology*, 4(4):R28, 2003.

[483] B. Zhang, D. Schmoyer, S. Kirvo, and J. Snoddy. GOTree machine (GOTM): a web-based platform for interpreting sets of interesting genes using gene ontology hierarchies. *BMC Bioinformatics*, 5(16), February 2004.

[484] J. Zhang, R. Finney, R. Clifford, L. Derr, and K. Buetow. Detecting false expression signals in high-density oligonucleotide arrays by an in silico approach. *Genomics*, 85:297–308, 2005.

[485] L. Zhang, M. Miles, and K. Aldape. A model of molecular interactions on short oligonucleotide microarrays. *Nature Biotechnol*, 21:818–21, 2003.

[486] M. Q. Zhang. Large-scaled gene expression data analysis: A new challenge to computational biologists. *Genome Research*, 9:681–688, 1999.

[487] S. Zhong, L. Tian, C. Li, K.-F. Storch, and W. H. Wong. Comparative analysis of gene sets in the gene ontology space under the multiple hypothesis testing framework. *Proc IEEE Comp Systems Bioinformatics*, pages 425–435, 2004.

[488] H. Zhu, J. Cong, G. Mamtora, T. Gingeras, and T. Shenk. Cellular gene expression altered by human cytomegalovirus: global monitoring with oligonucleotide arrays. *Proc. Natl. Acad. Sci.*, 24(95):14470–14475, 1998.

[489] J. Zhu and M. Zhang. Cluster, function and promoter: Analysis of yeast expression array. In *Pacific Symposium on Biocomputing*, volume 5, pages 476–487, 2000. Available online at http://psb.stanford.edu.

Index

For Product Safety Concerns and Information please contact our EU representative GPSR@taylorandfrancis.com Taylor & Francis Verlag GmbH, Kaufingerstraße 24, 80331 München, Germany

T - #0175 - 160425 - C1090 - 234/156/47 [49] - CB - 9781439809754 - Matt Lamination